Duck Production and Management Strategies

A. Jalaludeen • R. Richard Churchil •
Elisabeth Baéza
Editors

Duck Production and Management Strategies

Springer

Editors
A. Jalaludeen
Faculty of Poultry Science
Kerala Veterinary and Animal Sciences University
Wayanad, India

R. Richard Churchil
Faculty of Veterinary and Animal Sciences
Tamil Nadu Veterinary and Animal Sciences University
Chennai, Tamil Nadu, India

Elisabeth Baéza
Animal Physiology and Rearing Systems
The French National Research Institute for Agriculture, Food and the Environment
Nouzilly, France

ISBN 978-981-16-6099-3   ISBN 978-981-16-6100-6 (eBook)
https://doi.org/10.1007/978-981-16-6100-6

This Springer imprint is published by the registered company Springer Nature Singapore Pte Ltd.
The registered company address is: 152 Beach Road, #21-01/04 Gateway East, Singapore 189721, Singapore

Dr. Bhagabat Panda
(1929 to 2016)

Dr. Bhagabat Panda obtained his Bachelor of Veterinary Science degree from Bombay Veterinary College (1953) and MS (1960) and PhD (1963) degrees from the University of Maryland College Park. Following a brief assignment in Venezuela, he joined as Scientific Officer at CFTRI, Mysore, India (1963), and later became the head of the Division of Poultry Research of IVRI (1969). He transformed the Division into a 'Center of Excellence' and later as Central Avian Research Institute and served as Founder Director (1979–1990). He was the first Project Coordinator of All India Coordinated

Research Projects on poultry breeding in India. He founded the Indian Poultry Science Association (IPSA) and led it as President (1981–1997). He was the recipient of UNDP fellowship (1972), National Productivity Award from the President of India (1988), IPSA lifetime achievement award (1990), International Poultry Hall of Fame and Lifetime Achievement Award from WPSA (2001), and honorary degree of Doctor from Orissa University of Agriculture and Technology (2009). He has visited poultry science research and educational institutions in France, USSR, UK, USA, Canada, Japan, Scotland and Australia. He published 320 research papers and six textbooks. He served as the editor-in-chief of the Indian Journal of Poultry Science *and a member of the Editorial Board of the* World's Poultry Science Journal. *Dr. Panda is popularly known as the 'Father of Poultry Education' in India.*

Dr. A. Ramakrishnan
(1936 to 2006)

Dr. A. Ramakrishnan was a graduate of Madras Veterinary College. After a brief stint

as Veterinary Surgeon, he did postgraduation and doctorate from agricultural universities of Gujarat and Andhra Pradesh states respectively and postgraduate diploma in Poultry Nutrition from the University of Calcutta. He was one among the early faculty to join the College of Veterinary and Animal Sciences (CoVAS), Thrissur, Kerala, where he made poultry science education to receive adequate focus in the veterinary curriculum. He assumed the head of Department of Poultry Science in 1972. He has done pioneering research in pastoral duck farming system of Kerala. Under his stewardship, the Department of Poultry Science started research as early from 1974 on different aspects of Kuttanad duck production like incubation and embryonic development, management and housing systems, evaluation of production performances and quality assessment of duck products and economics of duck farming. An AICRP on Poultry Breeding Centre was established at Mannuthy in 1976. He was instrumental in the establishment of Centre for Advanced Studies in Poultry Science in 1985, first of its kind in India. He was the founder Director of the Centre and continued till his retirement from service. Dr. A. Ramakrishnan memorial award has been instituted by the Kerala Chapter of Indian Poultry Science Association to recognize eminent scientists for their contribution in the field of poultry production.

Foreword

Feeding the world with animal protein and sustainability of production are the key issues for poultry production. The efficient production of poultry over the last 70 years has made poultry meat and eggs the most affordable high-quality animal protein. All achieved through the application of research-derived scientific principles of breeding, genetics, nutrition and management. In the same period many demographic changes have taken place and these changes continue. At this moment the world is inhabited by almost 7 billion people (9 billion by 2050), and this population will need animal protein for daily use. The changes include increases in absolute wealth of people in countries in Latin America, the Middle East and South-East Asia and are largely the result of economic factors. On a global scale this already has resulted in more people worldwide being raised from poverty than ever before.

The increase in animal protein consumption has its price, as sustainability of poultry production is affected by aspects of diseases, food safety, animal welfare, environmental impact, use of natural resources and loss of biodiversity. Worldwide, large areas of land and large volumes of water have to be used for necessary crop production and for keeping poultry and livestock. A balance has to be found as natural reserve is crucial to mankind. Global meat production therefore will grow at a moderate pace, due to these constraints resulting in higher input costs and competing demand for land and water from alternative crops. It is expected that poultry meat production will increase and that waterfowl, specially duck, production will take a positive role in relation to food security in several parts of the world.

Duck farming has a long history, but the modern duck industry is relative small in most parts of the world. However the duck industry is very dynamic and over the last couple of decades has been through a period of rapid expansion. In some parts of the world, duck production has started to challenge the consumption of other types of poultry. China produces more than 2/3 of all ducks produced worldwide and so it is inevitable that what happens or does not happen in China will have a profound impact on the world of duck production. The role of the duck industry on the development of rural areas is also important, not only in economic terms. For the further development of duck production, the presence of a strong feed industry and a well-working veterinary service is essential.

The World's Poultry Science Association (WPSA) acknowledges the importance of waterfowl production and the science beyond, not only by placing a related picture on the cover of each June issue of the *World's Poultry Science Journal*, but also in supporting the working group Waterfowl, which operates under the umbrella of the Asian Pacific Poultry Federation of WPSA. As knowledge dissemination is our mission, successful World Waterfowl Conferences have been organized over the past years in China, India, Vietnam and Taiwan. The next conference is planned for Indonesia. During our World's Poultry Congresses there also are sessions specially for waterfowl production.

We are very thankful to the editors who initiated the production of this book and to the contributing authors. This book describing all aspects of duck production will be a standard of scientific knowledge for the years to come.

Roel Mulder
Dr R.W.A.W. Mulder
Secretary General
World's Poultry Science Association (WPSA)

Preface

Duck is one of the earliest domesticated species of poultry, which happened around 2000 years ago. Duck is the second important poultry species, next only to chicken in terms of eggs and meat production. Despite the global presence, economic importance and livelihood contribution, ducks failed to attract the attention of scientific book writers so far. Books on general subjects like 'Poultry Production' or 'Livestock Production' devote only a little portion to ducks.

Majority of ducks are reared in Asia (89.7%), followed by Europe (6.5%), Americas (2.3%), Africa (1.4%) and Oceania (0.1%). China alone had a duck population of 712.4 million, *i.e.*, nearly two-third of world's duck population. The scientific research on ducks is concentrated in mainland and provinces (like Taiwan) of China. Considerable research findings are being published in Mandarin (Chinese language) in China and Taiwan. The mule duck production is the major activity of Taiwanese duck industry, but its breeding and rearing methods and nutritive value of meat are not much known outside China/Taiwan. The information about classic Chinese and Taiwanese duck breeds, value addition especially down and feather, is not available in the public domain and is confined only in the domain of China and its provinces due to language barrier.

The core idea of writing a book on ducks was seeded by Dr. Bhagabat Panda. Dr. Bhagabat Panda desired to do this on his own, but could not accomplish due to his old age. He found a right choice in Dr. A. Jalaludeen, the senior editor of this book to do this. Dr. Panda expressed his desire in 2006 to Dr. A. Jalaludeen, who, at that time, was Director of the Centre for Advanced Studies in Poultry Science of College of Veterinary and Animal Sciences, Thrissur, Kerala. This Centre has a rich tradition and vast experience in duck production research since the 1970s; the efforts initiated by Dr. A. Ramakrishnan, the first Director of this Centre. This Centre has to its credit conducted the fourth World Waterfowl Conference in Thrissur in November 2009 in a grand manner. Although Dr. Jalaludeen narrated this to Dr. R. R. Churchil, his junior colleague in 2009, the thought process regained momentum only in October 2017. The visit of Dr. A. Jalaludeen to Taiwan to attend the 6th World Waterfowl conference in October 2017 had yet another agenda of identifying some potential chapter authors from Taiwan. The contribution of Dr. Jeng-Fang Huang, Director, Ilan Branch, Livestock Research Institute, Council of Agriculture, Taïwan, was immense in identifying a team of outstanding resource persons in Taiwan for this purpose and is highly appreciated. Dr. Elisabeth Baéza,

The French National Research Institute for Agriculture, Food and Environment, Nouzilly, France, who attended the conference also accepted to write a chapter, later joined as an editor of the book.

The book *Duck Production and Management Strategies* provides comprehensive insights into the field of Duck Production. This book presents a complete overview of different aspects of global duck production with emphasis on prospects, constraints and recommendations. It represents the complete integration of current knowledge on anatomy, physiology, genetics and breeding, nutrition, incubation and hatching practices of ducks. The information about classic Chinese and Taiwanese duck breeds, value addition especially down and feather and mule duck production confined only within the domain of China and its provinces have been brought out. The duck breeds have been entirely freshly illustrated. Exhaustive details on transhumant duck farming which is successfully practiced in different countries like India, Bangladesh, Viet Nam and Indonesia along with flock movement have been mapped. The advancements in different rearing and housing systems have been compiled. Our recent understanding of the nutritive value, processing and further processing of duck egg and meat has been scripted in different chapters. One chapter covers the important area of disease management including avian influenza. As there is tremendous increase in knowledge on biotechnology, a chapter has been devoted to provide insights on emerging trends in the subject pertaining to ducks.

The authors invited to contribute to the book have all been carefully chosen from the world community of scientists and scholars who are well qualified to do so. It is strongly believed that this book will serve as an encyclopaedia for academicians who teach life sciences, veterinary and animal sciences and allied subjects, a reference book for researchers and a textbook for students.

Finally, the efforts of publishers for their guidance and support in seamless production process of the book and the care they have taken in printing and reaching the end users are highly appreciated.

Wayanad, Kerala, India — A. Jalaludeen
Chennai, Tamil Nadu, India — R. Richard Churchil
Nouzilly, France — Elisabeth Baéza
15 June 2021

Contents

About the Editors

A. Jalaludeen has served as Director of Academics and Research of Kerala Veterinary and Animal Sciences University, and Special Officer to the College of Avian Sciences and Management. He is known for his pioneering work in characterizing Kuttanad ducks. He was the plenary speaker of the 6th World Waterfowl Conference held at Taiwan in 2017. He has developed linkages with the State Institute of Rural Development, Assam, India, in the refinement of backyard duck farming, which resulted in socio-economic empowerment of rural women and was appreciated by Dr. A. P. J. Abdul Kalam, then President of India. He was invited by the University of Brawijaya, Malang, Indonesia, in 2019 as Visiting Professor. As a teacher with 36 years of experience to his credit, he has published many peer-reviewed research articles and has provided guidance to more than 45 postgraduate and doctoral students in poultry science and allied subjects. He has also authored three books and four book chapters. He has organized many seminars/symposia, including International Conference on Waterfowls in India.

R. Richard Churchil is working as Professor at the Directorate of Research, Tamil Nadu Veterinary and Animal Sciences University, Chennai. He was graduated from Madras Veterinary College and pursued his doctoral degree at Indian Veterinary Research Institute, Izatnagar. He received gold medals in postgraduation and the Jawaharlal Nehru Award of ICAR for his doctoral research on development of transgenic spermatozoa. He worked as poultry breeding scientist in Kerala Agricultural University, Thrissur, for a decade before joining Tamil Nadu Veterinary and Animal Sciences University, Chennai. He has published 77 peer-reviewed research articles in highly reputed journals and has written seven book chapters. He was trained on 'Transgenic animals' at the University of Nantes, France. He teaches Poultry Science to undergraduate students and Poultry Breeding and Biotechnology to postgraduate/doctoral students.

Elisabeth Baéza works as Professor at INRAE, Université de Tours. She has obtained a diploma of Engineer in Agronomy (ENSFA, Rennes, France), specialized in Animal Productions, and a Ph.D. in Animal Sciences from the University of Montpellier (France). She was recruited at INRAE Nouzilly Centre (France) in 1993 to work on waterfowl production and poultry meat quality. She has identified significant determinants of duck meat quality and the main factors influencing these parameters. She is a member of the World Poultry Science Association, and she has organized many congresses related to Poultry Meat Quality and Waterfowl Production. Dr. Baéza has more than 120 peer-reviewed research publications in national and international journals, and she has contributed to three book chapters. She was a co-author of a French book on duck production published in 2012. She teaches Poultry Meat Quality to undergraduate students. She is an associate editor for two scientific journals.

Duck Production: An Overview

1

A. Jalaludeen and R. Richard Churchil

Contents

A. Jalaludeen
Faculty of Poultry Science, Kerala Veterinary and Animal Sciences University, Wayanad, Kerala, India

R. R. Churchil (✉)
Faculty of Veterinary and Animal Sciences, Tamil Nadu Veterinary and Animal Sciences University, Chennai, Tamil Nadu, India
e-mail: richardchurchil.r@tanuvas.ac.in

A. Jalaludeen et al. (eds.), *Duck Production and Management Strategies*,
https://doi.org/10.1007/978-981-16-6100-6_1

Abstract

Asia contributes significantly to world's duck population. Asian countries like China, Vietnam, Indonesia, Malaysia, and Bangladesh have a high number of ducks. Cambodia is the duck predominant country with a share of 40.5% ducks in total poultry, while Bangladesh is the most duck dense country in the world with 438.8 ducks per square kilometer area. The world duck population increased by sixfold from 193.4 million heads in 1961 to 1177.4 million heads in 2019. The growth curve has three phases: a slow and steady early growth from 1961 to 1985 with annual growth rate (AGR) of 5.0%, a fast growth phase from 1985 to 2010 (AGR = 7.3%), and a stationary phase from 2010 onwards with AGR of −0.17%. The first human death due to avian influenza (AI) in Hong Kong in 1997, reemergence of highly pathogenic avian influenza (HPAI) in 2003 and spreading to new territories in 2007 caused significant slumps in duck production. France ranks second in duck meat production after China, although it holds seventh position in duck population. Nearly 80% of the down and feathers are produced in China. Out of the total 195.6 million kilos of duck meat exported in 2019, 71% emanated from Europe. The per capita availability of duck meat was high in European countries like Hungary (9570 g) and France (3460 g). Apart, Europe is the major exporter of live birds, with France alone contributing more than half. During the recent 5 years, the reduction of human cases of AI has signaled the revival of duck farming. Location-specific technological interventions are to be carried out to refine the existing practices and to sustain duck farming.

Keywords

Duck population · Growth · Density · Meat and egg · Per capita availability · Export and import

List of Symbols/Abbreviations

AGR	Annual growth rate
AI	Avian influenza
FCR	Feed conversion ratio
g	Gram
h	Hour
HPAI	Highly pathogenic avian influenza
kg	Kilogram
LPAI	Low pathogenic avian influenza
RFID	Radio-frequency identification
R^2	R-squared (coefficient of determination)

1.1 Introduction

Waterfowls are birds that like the water, living by rivers, lakes, and other bodies of freshwater. They have short, webbed feet and a broad, flat bill. Many are game birds, meaning people hunt them for sport. They belong to the order Anseriformes, especially members of the family Anatidae, which includes ducks, geese, and swans. Waterfowls have historically been an important food source and continue to be hunted as game, or raised as poultry for meat and eggs or for ornamental purpose.

Ducks are important poultry species in many developing countries by virtue of their hardiness to diseases, long production year, easy to train, large sized eggs, early morning egg laying character, and natural tendency of foraging on aquatic weeds, algae, green legumes, fungi, earthworms, maggots, snails, various types of insects, etc., which directly reduce the feeding cost. They can also be used to control snails, slugs, and some other harmful insects from garden. Ducks are free from cannibalism and other objectionable tendencies. They are suitable for integrated farming systems such as duck-cum-paddy, duck-cum-fish, duck-cum-pig-cum-fish, and duck-cum-kitchen garden farming. Ducks flourish well in marshy and wetland where most other domestic animals cannot survive. They are good predators of insect pests, therefore are considered as biological pesticides promoting environment friendly agriculture. Flooding rice fields during winter, to make them available to waterfowl, could thus be a way of combining conservation of biodiversity and provision of agronomic benefits to farmers and should be promoted (Brogi et al. 2015). Waterfowls have webbed feet, which help them paddle the water while swimming; an activity adds oxygen to the water in a symbiotic life with aquatic animals.

Ducks, along with other scavenging poultry species play a critical role in meeting daily protein needs and providing household income of farm families in the mixed farming systems of many developing countries. It is considered the women's enterprise in smallholder farming system because woman heads in rural areas hold the sole responsibility of rearing ducks. Despite potential advantages, duck farming is decreasing day by day which could be due to scarce in scavenging areas and natural feed resources, drying up of natural water bodies, excessive use of pesticides in crop fields, etc. Transhumant and nomadic duck husbandry is widely practiced in South and Southeast Asia. Nomadic duck production is still on a traditional system; location-specific technological interventions are to be carried out to refine the existing practices (Rahman et al. 2017). The duck flocks often migrate to nearby territory/districts in search of fresh forage and water resources and the only job of the nomads is to forage the ducks and collect the eggs. Duck and other waterfowl are of great importance for food security of mankind in many parts of the world as elaborated by Pingel (2009).

The duck egg and meat are the excellent sources of essential amino acids micronutrients like vitamins, carotenoid pigments, and trace elements (Baéza and Huang this volume). Duck egg and meat are important in Asian cuisine also. Many special duck products are consumed in Asian countries. Products like processed duck meat, such as roasting Pekin duck, pressed salt duck, ginger-root duck, smoked duck steak and duck roll, tea-smoked duck, herb duck, crispy skin duck, etc., and processed duck eggs such as salted duck eggs, thousand year duck eggs and balut, which is popular in Philippines and Vietnam (Jalaludeen and Churchil 2006; Jalaludeen et al. 2009). Apart from this, parts of the viscera are also processed as a famous appetizer, such as salted duck gizzards and marinated duck tongues.

In this chapter, an overview of structure and status of duck production in global context is discussed.

1.2 Role of Ducks in Food Security

In the developing countries extensive production in small-scale or family farms is dominating. Traditional system of keeping in free range with access to water for bathing is popular in rural areas. The ducks are reared under free-range systems and survive mainly by scavenging, but most of the farmers give extra feed, mainly grain from the own farm. Bleich et al. (2005) concluded from the analysis of monthly income in households in the country Chad, an African country that the ducks and Guinea fowl performed better than indigenous chicken because of higher value of ducks and guinea fowl when sold at specific festive periods, fewer losses associated with ducks and guinea fowl due to their higher disease resistance, especially to Newcastle Disease.

Due to good foraging and incubation behavior, Muscovy ducks are especially suitable for scavenging systems and they have a better adaptability to hot climate than chickens. The Muscovy duck would be suitable for small-scale rural farmers in Africa and Latin America and could contribute to food security. In countries with such climatic conditions more support should be given for waterfowl production in family farms to ensure an increase in productivity as an important contribution to food security (Pingel 2009). Gueye (2009) has described "Family Poultry" as an appropriate system for supplying the fast growing human population with high quality protein and providing additional income to resource poor small farmers, especially women. However, small-scale producers are often constrained by poor access to appropriate technologies and information, as well as markets and support services, which could otherwise substantially improve productivity and income generation. Along with these basic problems, outbreak of diseases like HPAI made the life of the rural duck farmers more miserable (FAO 2009b).

Ducks and fish are widely used in flooded rice production systems in Asia (Cheng-Fang et al. 2008), but less frequently applied jointly. Ducks and fish can feed on azolla, weeds, and pest organisms, thus improving nutrient cycling within the system and suppressing weed and pest populations (Jana and Jana 2003). Khumairoh et al. (2012) reported that enhancing the complexity of a rice production system by adding combinations of compost, azolla, ducks, and fish can contribute to food security in a changing climate. They found that complex system yielded 114% higher net revenue than rice cultivation with only compost as fertilizer.

1.3 Evolution and Domestication of Duck Breeds

Although not firmly established, the evolution of Anseriformes (waterfowl) which includes true ducks could have happened during the Late Cretaceous Period in Antarctica, some 68–66 million years ago from a common ancestor *Vegavis iaai*, a genus of extinct bird. All goose and duck species belong to the order Anseriformes and family Anatidae. Almost all breeds of domestic duck except the Muscovy (*Cairina moschata*) have been derived from the mallard (*Anas platyrhynchos*), which occur over entire Europe, North Africa, and North America, and Muscovy duck (*Cairina moschata*) is found in South America (Crawford 1990). Mallard ducks are the most common and recognizable wild ducks in the Northern Hemisphere. They are found near ponds, marshes, streams, and lakes, where they feed on plants, invertebrates, fish, and insects. Mallards are dabbling or surface-feeding ducks because they eat by tipping underwater for food with head down, feet, and tail in the air. Mallards also forage and graze for food on land.

The first evidence of domestication is available from the excavation of Eligang Culture located in Zhengzhou, Henan. This supports the theory of domestication of ducks dated back to the start of Common Era (first century AD). Domestication has greatly altered their characteristics. Domestic ducks are mostly polygamous, where wild Mallards are monogamous. Domestic ducks have lost the Mallards territorial behavior and are less aggressive than Mallards. Despite these differences, domestic ducks frequently mate with wild Mallard, producing fully fertile hybrid offspring. The wild Muscovy, native to Central and South America was the ancestor of all present day domestic Muscovies. The domestication of Muscovy ducks had happened before the Spanish invaded South Africa in the sixteenth century.

1.4 Duck Genetic Resources

Almost all modern day duck breeds originate from the green-headed Mallard *(Anas platyrhynchos)*. The finest egg-producing breeds in the world are Brown Tsaiya of Taiwan Province of China, the Indian Runner of Malaysia, and the Khaki Campbell of England. The Indian Runner, Khaki Campbell, Pekin, and Muscovy are the most important breeds in rural poultry (Sonaiya and Swan 2004). The popular meat breeds include Pekin, Muscovy, Rouen, and Aylesbury. The detailed description of duck breeds (Su this volume) and evolution, domestication, classification, distribution, and risk status of duck genetic resources (Liu and Churchil this volume) are given in later chapters of this book.

The indigenous duck breeds of Asian countries play a major role in food security. Brown Tsaiya (Tai et al. 1989) of Taiwan, Itik Java (Widiyaningrum et al. 2016) of Malaysia, Chara and Chemballi varieties of Kuttanad breed of India (Jalaludeen 2005) perform excellently well under scavenging conditions. Nageswari, an egg type indigenous duck breed, is distributed in the Barak valley of Assam bordering Meghalaya, Tripura, and Mizoram of India and adjacent regions of Bangladesh (Islam et al. 2002). Other major indigenous layer ducks of Asian countries are layer type Co duck and Bau Ben breeds and meat type KyLua breed of Vietnam (Hoan 2014).

China is the homeland of diverse duck breeds such as Shaoxing, Jingdin, Shan Ma, Liancheng, Bai, and Gaoyou duck (Ma and Zhao 1998). Hybrids of Shaoxing and other local ducks are able to lay 327 eggs in 72 weeks of age (Hou 2016). Another Chinese breed Jinding duck with high cold resistance is larger and suitable for northern China. This breed has been reported to perform well under farm conditions in other Asian countries also (Islam 2015). Shan Ma ducks mainly being raised in Fujian and Zhejiang provinces are the smallest laying duck breed in China shown to produce good number of eggs (Lin et al. 2016). The adult female body weight of Chinese layer duck breeds is around 1.35 kg with egg production during 500 days of age which is more than 300 eggs, and feed efficiency per kilogram egg mass is around 2.80:1. At present, duck egg consumers generally prefer green-shell eggs; fortunately, Shaoxing, Jinding, and Shan Ma ducks primarily lay green-shell eggs. China also has local breeds for meat as Huabian duck, Linwu duck, and Jianhong Sheldrake, whose meat qualities are higher but growth periods are longer. These breeds are suitable for salted duck, halogen duck, sauce duck, and dried salted duck preparations, which are well received by consumers. The number of Muscovy duck is increasing in many provinces of China due to its preferable meat quality. Muscovy drakes are also used as the sire line in the production of mule ducklings.

1.5 Systems of Management

Carlen and Lansfors (2002) noticed the following three main systems of duck production in the Mekong Delta considerably vary depending on their location (rural, urban, and peri-urban). This pattern of farming systems has its presence in many parts of the world.

- *Scavenging systems*: scavenging, free ranging within the farm and the village (garden, home, or village pond); daily herding in rice fields, dikes, rivers, canals, and tidal areas (beyond the farm); seasonal transhumant supervised ranging (beyond the locality).
- *Integrated duck production systems*: rice–fish–duck; fish–duck–pigs (or other domestic species).
- *Confined systems*: semi-commercial and commercial farms (meat and egg production); duck–fish combination in enclosures or floating cage in ponds/canals/ rivers.

1.5.1 Scavenging System

The nomadic system, a subtype of extensive management of ducks is a special way to raise ducks in Asia (Gajendran and Karthickeyan 2009; Rahman et al. 2017). It is considered as a traditional method of keeping ducks. Ducks are allowed to herd on the rice fields after harvesting or allowed to scavenge in the paddy field during certain period of rice growing cycle. Either mobile or home base type of herding system has been practiced in most Asian countries for centuries. Although semi-intensive and intensive duck raising systems have been developing, most Asian countries still keep this traditional method to raise ducks. In some areas with prosperous industry, traditional duck production has gradually vanished. The emphasis on sustainable agriculture advocates the integrated duck–rice system by the scavenging of ducks on rice field to produce organic rice. The nomadic (transhumant) duck rearing practiced in different countries is described in detail by Sankaralingam and Mahanta (this volume) and the constraints faced by integrated duck farming are detailed by Churchil and Jalaludeen (this volume) in later chapters of this book.

1.5.2 Integrated Farming System

Integrated duck production is still widespread in Asia due to the integration of rice-cum-duck or duck-cum-fish production. The duck–rice production systems are common in many Asian countries when the herded ducks forage around the seedlings, controlling weeds, adding manure, and effectively controlling insects. After the rice is harvested, they feed on the fallen grains. The ducks are usually egg layers that have been selected with special foraging characteristics but meat-type ducks are also sometimes raised under these conditions. The ducks are confined overnight; during this time, they lay eggs. The different duck integrated farming systems are described in detail by Sapcota and Begum (this volume) in Chap. 6 of this book.

1.5.3 Confined Systems

Traditional raising models have been transformed rapidly into indoor raising models far away from waters during the recent decades. The housing requirements, equipment, and management in intensive system are described in detail by Gajendran and Veeramani (this volume) in a later chapter of this book. Intensive system of management, although in vogue in many European countries, is becoming more and more popular nowadays in China, Taiwan, Vietnam, Cambodia, and other Asian countries especially in meat-type duck production. Indoor mesh bed and thick cushion grass models have been well received by the farmers due to rapid development of industrialization and increasing awareness of environmental protection. Duck mesh bed with fermentation bed and duck cage-raising models are undergoing rapid developments, the purpose of which is to transform wastes into organic fertilizers in the aerobic fermentation conditions and therefore to reduce environmental pollutions (Hou 2016).

Ducks are sometimes reared in fully enclosed houses wherein the microenvironment is controlled. Jones and Dawkins (2010) concluded that the control of the environment in which ducks are reared, particularly temperature, humidity, litter moisture, and ammonia, is critical for duck welfare. Good ventilation and high straw quality are keys to maintain duck body and plumage conditions and to reduce mortality. In intensive system of management, indoor or outdoor access to swimming is sometimes provided to the ducks. Access to open water in some form was seen to benefit eye and feather cleanliness and foot pad condition. Farghly et al. (2017) concluded that raising Muscovy ducks during hot conditions in an open-sided house with access to a swimming pond at 12:00 to 14:00 h is highly recommended to get high body weight, better immunity, decreased mortality rate, and low body temperature which positively reflected in the health condition of ducks.

The ducks can be raised on slatted, mesh wire, or synthetic weave floors located over pits, which can be drained. A mixture of slatted (30%) and straw (70%) from 14 days to slaughter is also common. The mesh wire size changes with body size. Foot lesions can occur on mesh wire floors. Good ventilation in duck houses is important to keep ammonia concentration below 15 ppm. Comfort zone of mature ducks is about 8–26 °C. In commercial practice after 21 days of age, 21 °C is recommended. Fan ventilation may be necessary to reduce shed temperature and ammonia levels in enclosed housing systems. Fraley et al. (2013) analyzed the predominant flooring systems in the USA, namely wood shavings and raised plastic flooring and the results did not reveal any advantage of one system over the other. The litter grown ducks scored 0 or 1 higher score than their counterparts reared on slatted floor, when the feather cleanliness, feather and foot pad quality, and gait of ducks were compared (Karcher et al. 2013).

1.6 Breeding

Although not extensive as in chicken, selective breeding for genetic improvement of various characters in ducks has yielded desired results. Liu et al. (2015) selected ducks for number of fertile eggs 14 days post-insemination to develop a strain that needs once in a week insemination rather than conventional twice in a week insemination. Selective breeding for genetic improvement has been attempted in many local ducks. In an attempt to increase fertility period after insemination, Shaoxing ducks have many obvious features after 20-year systematic breeding, such as high yield, early mature, and green shell. In multi-line hybrid breeding system of Shaoxing ducks, the age at first egg was 108 days of age, egg production of 72 weeks was 327 eggs, average egg weight was 69.8 g, feed-egg ratio was 2.65: 1, green shell percentage was 98%, and annual duck stock was over 80 million (Hou 2016). Pingel (2011) observed that the improvement of FCR due to genetic selection results in lower emission of N and P via manure. That means selection for feed efficiency is not only an important economical parameter but also an important ecological factor. In order to increase the meat weight in the carcass, meat ducks are selected for body weight, breastbone length, and breast muscle thickness (Mazanowski et al. 2003). They observed that the chest depth is positively and significantly correlated with the weight and content of breast and leg muscles of carcass and negatively with the weight and content of subcutaneous fat with skin, which can be used in selection for increased muscling and decreased carcass fatness.

Industrial breeding of duck had giant leaps in the past to achieve remarkable improvements in the genetic potential of commercial stocks. As the body weight and reproductive characters are negatively correlated in ducks (Cheng et al. 1995), selection for body weight results in decline in egg production. Therefore separate sire and dam lines with more emphases for growth parameters in sire line and reproductive characters in dam line are necessary. Sire lines are primarily selected for growth rate, feed efficiency, carcass fat percentage, breast muscle, and feather yields and if warranted for egg number also (Cherry and Morris 2008). Although there are many breeding companies, the two companies dominate the global duck-breeding market today are Cherry Valley (CV) Farms Ltd. in the UK and Grimaud Freres (GF) in France. Both commercial companies select from strains of white Pekin ducks. Over the past few decades, tremendous improvements have been made in Pekin duck breeding through modern genetic selection, which has been mainly driven by the demand for faster growth rate, higher breast meat yield, and better conversion ratio. Currently, a typical Pekin duck can reach market weight at about 3.1 kg at 5 weeks, with a feed conversion ratio of approximately 1.7:1. Breast meat yield, which is considered the most important selection trait in certain markets, averages 20% of carcass weight. From 1980 to 2009, the live weight of ducks of an American breeding company, the largest producer of White Pekin ducks in North

America, has improved from 2.8 to 3.0 kg, with an increase in carcass yield from 71% to 74% and a decrease in feed conversion ratio from 3.1 to 1.7. Additionally, boneless skinless breast yield almost doubled from 11.5% to 19.3% and carcass skin and fat has reduced from 33% to 20% during the same period, while market age reduced from 50 to 37 days.

1.6.1 Modern Tools in Breeding Data Collection

Feed efficiency has become more and more important for all kinds of animal production, in response to rising feed cost and awareness of limited resources and environmental issues. Selection for feed efficiency in sire lines results in leaner meat also. The carcass fat percentage is adversely affected by selection for feed efficiency, since low carcass fat can affect the eating quality. In the past, individual feed intake of meat type poultry has been recorded in single cage systems, (Thiele 1995), but this system is unfavorable for the ducks and may not reflect feeding behavior under commercial floor conditions. The modern RFID technology enables the breeding companies nowadays to record even small meals of individuals under nearly practical housing conditions. In these testing units, not only the amount of feed consumed can be recorded, it opens the possibility to observe also the feeding behavior of the ducks, with the frequency of meals and the size of a single meal (Bley 2005; Howie et al. 2009). The availability and utilization of these "feeding stations" accelerate the genetic progress in feed efficiency. Cherry Valley invested in this technology several years ago, gained a huge advantage in developing their pure line breeding stock. Records of daily feed intake can be used to improve feed efficiency in combination with weight gain during the testing period as the commonly known FCR or, independent from metabolic body weight and weight gain, as residual feed intake (RFI) (Herd and Arthur 2009; Aggrey et al. 2010).

The laying performance is usually tested over a period of 35 production weeks. Traditionally laying birds are kept in single cage systems. ORVIA is developing a fully automatic nest system to replace single cages in future. With the new nests, the laying performance can be recorded in floor pens. During the testing period, time-wise information about egg weight, shell strength, fertility, and hatchability can be recorded. Sperm quality is analyzed before males are used for pedigree reproduction. For testing the fertility and hatchability of females, the use of pooled sperm is preferred (Thiele 2016).

In commercial breeding programs, all the birds are additionally graded to judge the carcass quality via conformation. The depth of the breast muscle can be precisely detected by using ultrasonic device. Cup-up tests where full sibs or half sibs are slaughtered to record the total body composition would bring additional information but are always combined with a complete loss of valuable pure line birds. Furthermore, the data recorded by the selection crews using ultrasonic device are incorporated directly into the databases. Since this is done very precisely with modern recording devices, the data are immediately available for parameter and breeding value estimation (Thiele 2016).

1.7 Duck Population

1.7.1 Country Scenario of Duck Population

The data on duck populations in top 35 duck producing countries in 2019 is given in Table 1.1. China is the leading duck producing country in the world with more than 60% (712.4 million heads) of the world's duck population reared in this country in 2019. The data reveals that China is the leader in duck production since long time. The data at decade interval reveals that China held first place with 50% or more of total world's duck population all the time: 50% in 1961 and 1971, 52% in 1981, 57% in 1991, 64% in 2001, and 60% in 2011. The duck industries in China are mostly concentrated in East, South, and parts of the Southwest areas. The duck production in provinces such as Shandong, Sichuan, Jiangsu, Anhui, Zhejiang, Hunan, Hubei, Jiangxi, Guangxi, Fujian, Henan, Guangdong, Henan, and Chongqing account for over 90% of Chinese duck production.

Vietnam held a distant second position all through the years; 8% of world's duck population in 1961, 7% in 1971, 6% in 1981 and 1991, 7% in 2001, 8% in 2011, and 7% in 2019. France held third position with 5% of world's stock in 1961 and fifth place with 2% in 2011. Bangladesh held fourth place in 1961; thereafter, it shared third and fourth positions with Indonesia. India ranked fifth in all the years except 2011.

The duck production scenario in many countries changed after the first human infection of HPAI reported in Hon Kong in the year 1997. After this event, many countries experienced recession in duck population or a turbulent growth. Therefore the annual growth rates (AGR) of pre-human-AI (1961–1997) and post-human-AI (1997–2019) eras are examined separately and presented. In pre-human-AI era, countries like Nepal (from 0.01 to 0.42 million heads; AGR = 112.8%), Republic of Korea (from 0.14 to 2.70 million heads; AGR = 50.2%), and Pakistan (from 0.3 to 3.5 million heads; AGR = 29.6%) had huge jump in their duck population. The other countries that had impressive growth of duck population in terms of AGR (%) were Mexico (19.4), Cambodia (18.4), Egypt (17.8), China mainland (15.6), Lao People's Democratic Republic (15.0), Indonesia (13.1), and Philippines (12.8). The countries registered notable negative growths in terms of AGR (%) during this period were Denmark (−2.0), Netherlands (−1.2), Poland (−0.5), and Malaysia (−0.4).

In post-human-AI era, Sierra Leone, the country located at the southwest coast of West Africa improved its duck population many folds to reach 1.05 million heads in 2019 from its 1997 figure of 0.07 million heads registering AGR of 68.9%. The other countries showcased impressive growth during this period were Russian Federation (from 1.8 to 21.1 million heads; AGR = 48.9%), Democratic People's Republic of Korea (from 0.8 to 7.3 million heads; AGR = 35.6%), Myanmar (from 5.6 to 30.7 million heads; AGR = 20.2%), Bulgaria (from 0.4 to 1.4 million heads; AGR = 13.7%), Mozambique (from 0.7 to 2.6 million heads; AGR = 13.3%), and Lao People's Democratic Republic (from 1.0 to 3.6 million heads; AGR = 12.5%). The other countries registered significant growth during this period in terms of AGR (%) were Malaysia (11.4), Australia (10.7), Republic of Korea

Table 1.1 Duck population (million heads) in top 35 countries in the world over the years (FAOSTAT 2021)

Continent	Country	Year								World rank
		2000	2005	2010	2015	2016	2017	2018	2019	(2019)
Africa	Egypt	9.1	5.6	8.7	11.6	10.6	7.7	6.5	5.6	19
	Madagascar	3.9	3.8	4.2	4.3	4.5	4.5	4.5	4.6	21
	Mozambique	1.5	1.5	1.9	2.5	2.5	2.6	2.5	2.6	26
	United Republic of Tanzania	1.3	1.3	1.3	1.4	1.4	1.4	1.4	1.4	33
	Sierra Leone	0.1	0.3	0.8	1.1	1.1	1.0	1.0	1.1	35
Americas	Mexico	8.0	8.1	8.3	8.4	8.5	8.5	8.5	8.5	15
	United States of America	6.6	6.9	6.9	7.6	7.3	7.3	7.3	7.4	17
	Brazil	3.4	3.6	3.7	3.7	3.4	3.5	3.5	3.4	24
	Argentina	2.3	2.4	2.5	2.6	2.6	2.6	2.6	2.6	25
	Canada	1.2	1.2	1.3	1.5	1.5	1.5	1.5	1.5	30
	Paraguay	0.7	0.7	1.1	1.3	1.4	1.4	1.4	1.5	31
Asia	China, Mainland + Hong Kong SAR + Macao SAR	600.0	720.0	787.0	675.4	726.1	712.1	712.2	712.4	1
	Vietnam	58.8	66.0	68.6	69.5	71.3	74.9	76.9	82.5	2
	Indonesia	29.0	32.4	44.3	45.3	47.4	57.6	59.6	61.3	3
	Bangladesh	33.0	37.3	42.7	50.5	52.2	54.0	55.9	57.1	4
	India	30.4	28.8	25.3	24.7	26.6	28.2	29.8	33.5	5
	Myanmar	6.6	9.1	14.1	23.7	23.6	25.4	28.4	30.7	6
	Thailand	27.9	21.5	16.4	15.3	14.1	12.9	12.3	11.8	9
	Philippines	10.0	10.4	10.3	10.1	10.5	10.8	11.2	11.6	11
	Malaysia	5.3	8.1	8.5	9.9	9.6	9.3	9.7	10.5	12
	Cambodia	5.5	7.5	7.5	8.3	8.3	8.8	8.9	9.0	13
	Republic of Korea	5.1	8.4	14.4	9.8	8.1	7.5	9.0	8.6	14
	Taiwan Province of China	11.5	9.8	9.3	8.7	8.1	7.4	7.9	8.0	16
	Democratic People's Republic of Korea	2.1	5.3	5.9	6.4	6.8	6.9	6.8	7.3	18

	Pakistan	3.5	3.5	3.7	3.7	3.8	3.8	3.8	3.9	22
	Lao People's Democratic Republic	1.7	3.2	3.2	3.5	3.5	3.6	3.6	3.6	23
	Islamic Republic of Iran	1.6	1.6	1.6	1.6	1.6	1.6	1.6	1.6	29
Europe	France	24.4	22.4	26.1	27.4	24.4	23.5	24.0	25.0	7
	Russian Federation	4.2	14.8	24.5	29.6	21.3	21.5	21.9	21.1	8
	Ukraine	8.3	8.4	9.8	11.3	10.2	10.9	11.0	11.7	10
	Poland	3.6	4.6	5.9	2.8	2.9	3.2	4.6	5.6	20
	Belarus	0.0	1.0	1.1	1.8	1.8	1.8	1.8	2.0	27
	United Kingdom of Great Britain and Northern Ireland	2.9	2.6	2.2	2.3	2.0	2.3	2.1	1.9	28
	Bulgaria	0.9	1.5	1.0	1.5	1.2	1.1	1.7	1.4	32
Oceania	Australia	0.5	0.9	1.2	1.1	1.3	1.3	1.3	1.3	34

(9.9), Bolivia (9.1), Czechia (7.4), Poland (5.3), Paraguay (4.9), Cambodia (4.8), and Indonesia (4.6). Conversely, the countries registered negative growth during this period in terms of AGR (%) were Netherlands (−4.5), Denmark (−3.8), Turkey (−3.3), Egypt (−3.2), Thailand (−2.1), and United Kingdom of Great Britain and Northern Ireland (−1.8).

1.7.2 Continent and Global Scenarios of Duck Population

The data on duck population of different continents over the years is given in Table 1.2. According to 2019 data, major portion of world's duck population was in Asia (89.60%), followed by Europe (6.54%), Americas (2.33%), Africa (1.38%), and Oceania (0.14%) (Fig. 1.1). The domination of Asia in duck production is mainly due to the food habits of the people. Duck is an important item in their meal for communities of China, Hong Kong, Japan, Korea, and Southeast Asia. Among the major duck production countries in Asia, Taiwan is the only region, which has exhibited a decrease in duck population after 1990. The domestic consumption in Taiwan has not been reducing, but the exports are at disadvantage due to lower wage and cheaper land resource in other exporting countries. Duck egg is also an essential protein source in some Asian countries. In India, Indonesia, and

Table 1.2 Trend in duck population (million heads) over the years in different continents (FAOSTAT 2021)

Year	Africa	Americas	Asia	Europe	Oceania	World
1961	6.2	9.6	151.4	25.8	0.4	193.4
1965	6.7	11.3	178.9	25.4	0.4	222.7
1970	7.9	12.1	208.4	27.6	0.3	256.3
1975	8.2	13.6	245.3	27.7	0.3	295.1
1980	7.8	17.5	301.4	24.8	0.5	352.0
1985	11.5	22.3	361.2	29.0	0.4	424.4
1990	12.9	24.1	487.6	36.3	0.4	561.3
1995	24.7	25.6	700.3	51.3	0.6	802.5
2000	17.1	23.7	834.3	55.9	0.8	931.8
2005	13.9	24.5	974.8	68.5	1.2	1082.9
2010	17.9	25.7	1064.4	85.7	1.5	1195.2
2011	16.3	26.0	972.8	87.3	1.3	1103.7
2012	19.0	26.6	995.5	85.3	1.4	1127.8
2013	21.1	26.8	973.5	86.3	1.4	1109.1
2014	22.3	27.6	936.2	87.6	1.4	1075.1
2015	21.9	27.1	968.0	90.6	1.5	1109.1
2016	21.1	26.9	1023.5	77.0	1.6	1150.1
2017	18.2	27.0	1026.6	77.7	1.6	1151.1
2018	17.1	27.1	1039.3	75.7	1.6	1160.8
2019	16.3	27.4	1055.0	77.0	1.7	1177.4

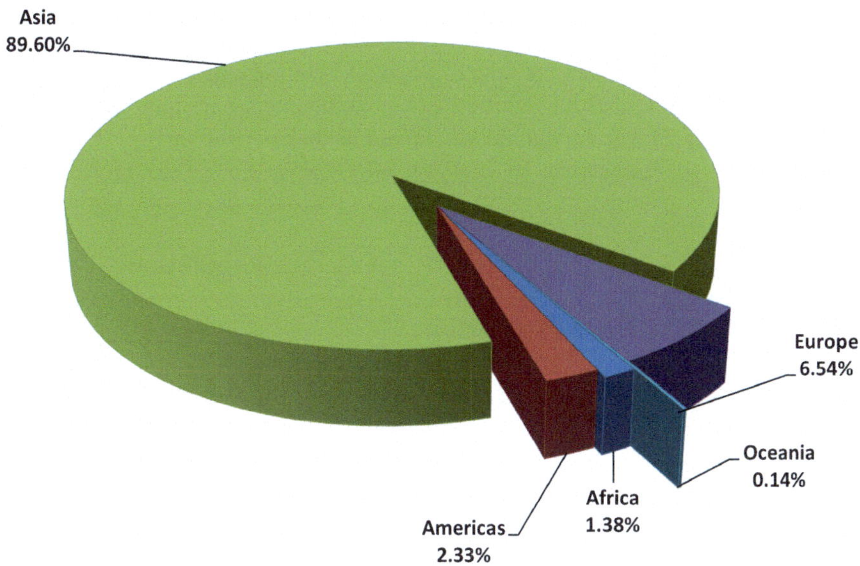

Fig. 1.1 Regional distribution of ducks in the world in 2019

Philippines, ducks are mainly kept for egg production and not for meat supply. France, an important country in Europe has a sizeable number of ducks since long time. Ducks occupy an important place in French cuisine as the speciality products like *foie gras* and fattened breast muscle are a delicacy.

The current world duck population of 1177.4 million ducks was achieved from the stock position of 193.4 million heads in 1961 with the impressive AGR of 8.8%. The AGR for different continents during this period (1961–2019) was 10.3% in Asia, 5.3% in Oceania, 3.4% in Europe, 3.2% in Americas, and 2.8% in Africa. The overall world AGR during pre-human-AI era (1961–1997) was much higher (10.8%) than that of post-human-AI era (1.1%). The AGR of duck population in Asia during pre-human-AI era was much higher (12.7%) compared to that of post-human-AI era (1.1%). The human cases of AI were reported from 17 countries so far, of which 12 countries (WHO 2020) lie in Asia. The list includes major duck producing countries in Asia like China, Vietnam, Bangladesh, Cambodia, Thailand, Indonesia, and Myanmar. All these countries registered turbulent or slowed down growth after human infections reported in the countries, contributing to a slowdown in growth rate of duck population in Asia after 1997. Africa registered negative growth rate (−1.6%) in post-human-AI era compared to its 8.6% AGR during pre-human-AI era. This decline was mainly due to the devastating human-AI outbreaks in Egypt. Out of 861 human cases and 455 deaths happened so far in the world, 359 cases and 120 deaths occurred are confined to Egypt alone. The AGR declined drastically in Americas (from 4.4% to 0.5%) and marginally in Europe (from 2.6% to 2.5%) after 1997. The only continent registered impressive growth during post-human-AI era was Oceania (from 1.7% to 6.8%). None of its countries reported human case of AI so far.

The time trend analysis of duck population during the current decade (2010–2019) also revealed no significant ($P > 0.05$) change in duck populations of Asia and Africa. However, significant increase in duck population over the period has been observed in Americas ($P < 0.05$) and Oceania ($P < 0.01$). Conversely, Europe recorded a significant ($P < 0.05$) decrease in duck population during this decade.

1.8 Growth of Duck Population

The world duck population in 1961 (Table 1.2) was only 193.4 million numbers with Asia contributed a giant share of 78.3%. The world duck population increased to 1177.4 million heads in 2019; sixfold higher from its 1961 status. The analysis of growth pattern of duck population in the world divulges three clear phases, namely phase of slow growth spanned from 1961 to 1985, phase of fast growth spanned from 1985 to 2010, and stationary phase prevailed after 2010 (Fig. 1.2).

1.8.1 Phase of Slow Growth

The period between 1961 and 1985 (24 years) witnessed slow growth in global duck population. During this period the average annual increase was only 9.63 million heads with AGR of 5.0%. It took 24 years for the duck population to rise to 424.4 million heads from the base value of 193.4 million heads (Fig. 1.2). The growth during first phase was steady and stable without much fluctuation as indicated by a very high R^2 value (0.98) in time trend analysis. The analysis also revealed that the growth was significant ($P < 0.01$) but only moderate (coefficient or slope = 9.1). The duck population in Asia, Americas, and Africa grew almost double during this period with AGR of 5.7%, 5.4%, and 3.5%, respectively. However, Europe and Oceania registered only a marginal increase with AGR of 0.5% and 0.2%, respectively. The world human population increased from 3.1 to 4.9 billion heads during this period and it is reasonable to attribute the corresponding growth in duck population as animal protein provider teaming human population.

1.8.2 Phase of Fast Growth

The slow phase of growth was followed by a fast phase of growth from 1985 to 2010 (25 years). This phase witnessed an annual increase of 30.84 million heads (AGR = 7.3%) taking the word duck population to an all-time high of 1195 million heads in 2010. The duck populations in Asia (AGR = 7.8%), Europe (AGR = 7.8%), and Oceania (AGR = 10.1%) also grew substantially, while Africa (AGR = 2.2%) and Americas (AGR = 0.6%) had flat growth rates. This was the period that also

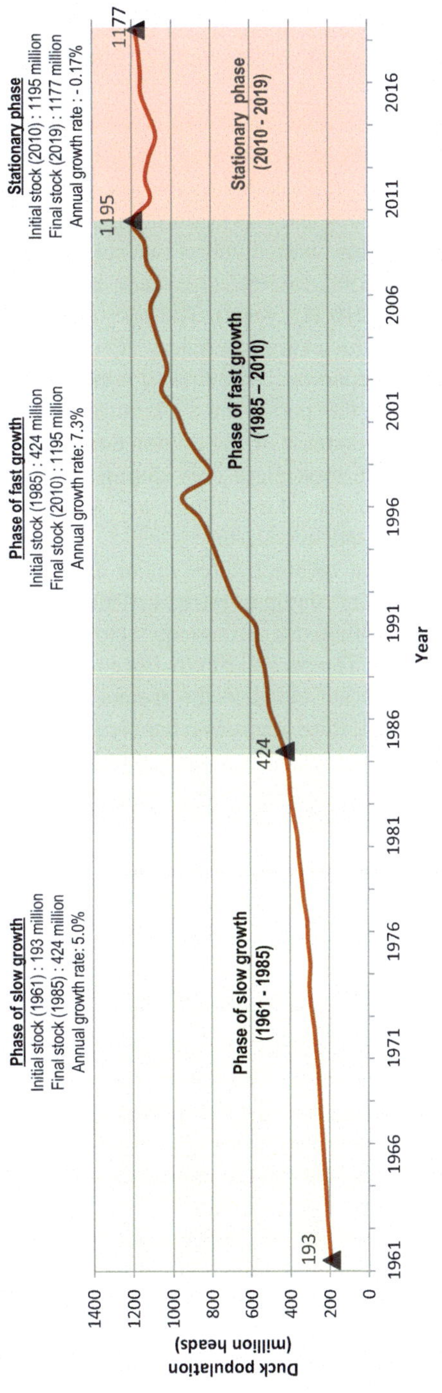

Fig. 1.2 Phases of growth of world duck population over the years

witnessed a huge jump in human population in the world from 4.9 to 7.0 billion human heads. The duck production in the world increased in tandem with human population during this period, mainly to cater the demand for animal protein like egg and meat.

The regression analysis also revealed a significant ($P < 0.01$) growth in this phase. Although the magnitude of growth was much higher (coefficient = 31.0) during this phase, the stability ($R^2 = 0.96$) was poor compared to first phase. The stability was marred by many ups and downs during this phase (Fig. 1.2). A closer look at the growth pattern of this phase divulges two sub-phases: a momentous and steady early sub-phase from 1985 to 1997 (12 years) followed by a sluggish and turbulent one from 1997 to 2010 (13 years). The rate of growth was high (coefficient = 41.5 vs. 23.9) but the fluctuations were less ($R^2 = 0.96$ vs. 0.83) during first sub-phase compared to the second one. The first sub-phase spanning from 1985 to 1997 is the golden era; during this period the duck farming flourished well in most parts of the world. Out of total increase of 984 million ducks between 1961 and 2019 (58 years), 53% of the growth took place only during this 12 years period. This golden era existed until human case of avian influenza was first reported in 1997 in Hong Kong and subsequent pandemic alarm.

The global duck population recorded three major slumps in turbulent second sub-phase (1997–2010). The first slump occurred in 1998; the annual duck population dipped to 800 from 947 million ducks, consequent to the first human causality to HPAI in Hong Kong in 1997. The second dip that occurred in 2003 (from 1049 to 1020 million heads) could be attributed to the reemergence of HPAI with more number of human cases. The third recession happened in 2007 (from 1102 to 1064 million heads) was due to the spread of HPAI to the new territories in Cambodia, Indonesia, Thailand, Vietnam, Egypt, and Turkey during 2004–2006. However, apart from these slumps, in general, the flattening of growth curve was visible throughout this phase.

1.8.3 Stationary Phase

This phase covers the period from 2010 to 2019 (Fig. 1.2). The focused graphic representation of this phase in world and different continents is given in Fig. 1.3. Starting with an initial population size of 1195 million ducks and passing through a wax and wane course, this phase ended with 1177 million ducks in 2019 with the net depletion of 18 million heads and annual depletion of 2 million and AGR of −0.17%. The major duck producing continents like Asia and Europe registered negative AGR of −0.10% and −1.1%, respectively, during this phase. Other continents like Oceania (1.02%), Americas (−0.70%), and Africa (−1.0%) were also registered negative or negligible positive growth. The time trend analysis by regression showed no significant ($P > 0.05$) change in direction in this phase or can be called as "stationary." This phase was unsteady throughout ($R^2 = 0.06$) and the change was trivial (coefficient = 3.08).

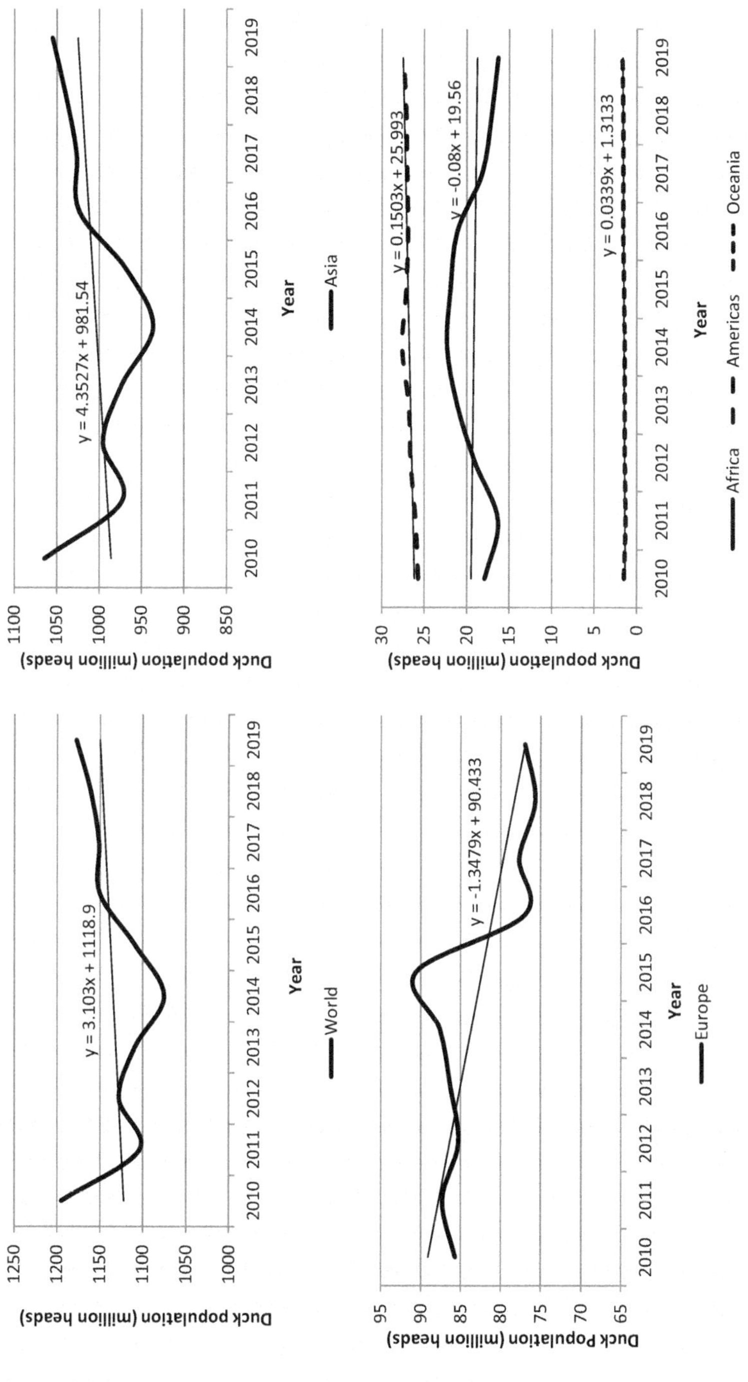

Fig. 1.3 Trend in duck population in the world and continents during the stationary phase (2010–2019)

This first part of the phase, spanning from 2010 to 2014 had a negative annual growth except during 2012, reaching the lowest value of 1075 million ducks in 2014. Although not significant ($P > 0.05$), this period registered an unsteady ($R^2 = 0.68$) but intensive negative growth (coefficient = −23.5). During this period, HPAI continued to cause human infections and casualties with more intensity in major duck producing countries like China, Indonesia, Vietnam, Cambodia, Bangladesh, and Egypt (WHO 2020).

The time trend of duck population in second part of the phase (2014–2019) gives a ray of hope of revival of the global duck production. This 5-year period registered positive growth during all the years (coefficient = 19.1) and the growth is rather steady ($R^2 = 0.89$). The stamping out policy and other avian influenza control strategies (FAO 2009a) adopted by the countries appear to yield the results, as the human HPAI cases showed declining trend during this period (WHO 2020).

1.9 Effect of Avian Influenza on Duck Population

There are four types of influenza viruses, namely A, B, C, and D; of which, type A influenza virus infects birds. Influenza A viruses are divided into subtypes on the basis of hemagglutinin (H) and neuraminidase (N) proteins on its surface. There are 18 known H subtypes and 11 known N subtypes and many combinations of H and N protein subtypes are available. Most of the subtypes are "low pathogenic," infecting only birds. However, some "highly pathogenic" subtypes cause more severe symptoms and high mortality in all avian species, be it domestic or wild. Low pathogenic avian influenza (LPAI) can mutate into highly pathogenic ones (HPAI) (CDC 2017). Avian influenza is primarily a bird disease; however, rarely humans and other animals can also contract this disease on close contact with infected birds. Re-assortment events with other avian viruses have led to the emergence of new HPAI virus subtypes (H5Nx), including H5N2, H5N3, H5N5, H5N6, H5N8, and H5N9 (Pulit-Penaloza et al. 2020). Since 1997, there were 861 laboratory-confirmed human infections in 17 countries; of which, 53% were fatal (WHO 2020).

Avian influenza outbreaks, especially when human infections due to H5N1 virus and other HPAI strains occur, have immediate ramifications on duck production. The AI outbreaks cause loss to the duck farmers directly by

1. causing huge mortalities (CDC 2017),
2. loss of birds in stamping out of all avian species in the infected area, within one to five kilometer radius from epicenter of infection (FAO 2009a),
3. ban on restocking during rest period: the premises must be left without poultry for at least 21 days after stamping out and disinfection (FAO 2009a).

Avian influenza outbreaks cause many indirect effects on the duck production including

1. temporary restriction of movement of poultry and poultry products from and to the restricted area (nearly 10 kilometer radius from the epicenter) for atleast 3 weeks (quarantine period) or till the zone is declared AI free by subsequent surveillance (FAO 2009a),
2. permanent ban or restriction on movement of nomadic duck flocks enforced by the governments in AI endemic areas (FAO 2010),
3. restriction of cross border movement of poultry and poultry products between provinces/states during outbreaks (FAO 2010),
4. ban of export of poultry and its products from the country/zone during outbreaks at least for 21 days after disinfection and subsequent OIE notification based on surveillance results (OIE 2019),
5. above all, the hype among the people on public health and subsequent drop in consumption of poultry and poultry products (FAO 2009b).

The direct effects incur economic losses to the farmers with the loss of their birds. The indirect losses are due to movement restrictions and export ban and slump in demand due to public health concern, thereby accumulation of poultry and poultry products in pockets, ending up in price drops and economic loss to the farmers. Altogether, every outbreak of avian influenza has immediate and cascading effects on duck farming which further leads to slowdown or drop in the growth curve.

1.9.1 During the Phase of Fast Growth (1984–2010)

In 1996, an HPAI virus (H5N1) was found in commercial geese in the Guangdong Province of China, which was thought to originate from H5 viruses in wild migratory birds (Duan et al. 2007). These goose lineage strains of Guangdong gave rise to outbreaks of HPAI (H5N1) in chicken farms in Hong Kong in 1997 that further led to fatal human infections. In 1997, influenza A virus subtype H5N1, which previously had been confined to avian species, was isolated for the first time from a human casualty in May 1997 (Ku and Chan 1999). After 6 months of first human report, the epidemic of influenza A (H5N1) virus was confirmed in 18 individuals in Hong Kong cutting across the age of 1–60 years; six of whom died (Chan 2002). The Hong Kong administration made a decision, on 28 December 1997, to slaughter all the poultry in farms and markets in Hong Kong. The operation started on 29 December 1997 and resulted in the removal of 1.5 million poultry from the territory before it ended on 31 December. The endemic nature of HPAI H5N1 viruses in poultry in China and its subsequent spread to other parts of the world continued to result in outbreaks in poultry and occasional human infections that escalated into a new pandemic threat. In 1998, the world duck population reduced to 800.2 million heads from 946.9 million heads in the previous year (−146.7 million heads). Although this event had no or little impact in terms of change in duck population in Americas (−0.9 million heads), Europe (+1.2 million heads), and Oceania (+0.1 million heads), major impact was perceptible in Asia (−137.4 million heads) and Africa (−9.7 million heads). The population drop in countries like China

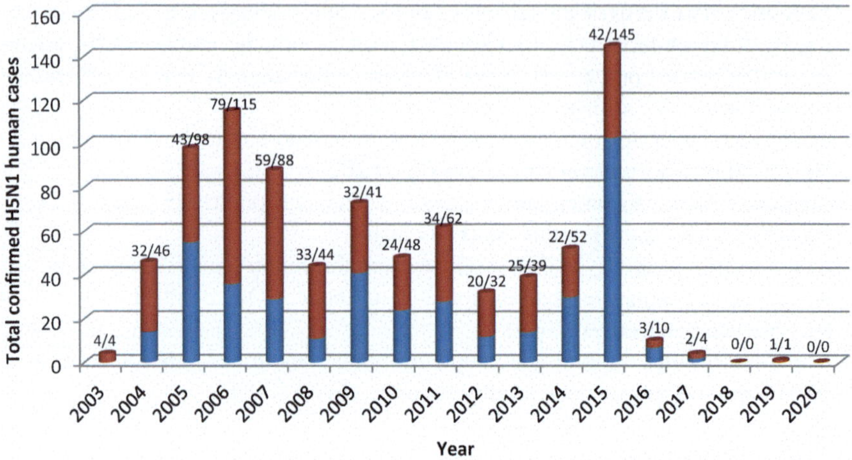

Fig. 1.4 Yearly data on human (death/total confirmed) avian influenza A (H5N1) cases (WHO 2020)

(AGR = −21.4%) and Indonesia (AGR = −14.4%) contributed at large, to the reduction in Asia's duck population and Egypt (AGR = −52.1%) to Africa's decline during this period.

The duck production appeared to be stabilized after fall in 1998 before its lip off the track again in 2003. The widespread reemergence of avian influenza occurred in 2003 with rare but sporadic human infections and death due to H5N1 viruses in Asia and later in Africa, Europe, and the Middle East. The yearly data on total confirmed avian influenza A human cases and death is depicted in Fig. 1.4. The reemergence of H5N1 outbreaks occurred in mainland China, Hong Kong SAR of China, Vietnam, Cambodia, and Thailand during 2003 (Lai et al. 2016) dented or slowed down the growth rate of duck production in these countries. H5N1 infections in poultry flocks were also confirmed in South Korea in December 2003, Japan, Hong Kong, and Laos in January 2004, and Indonesia in February 2004 (MMWR 2004).

The resurgent avian influenza A in 2003 impacted the duck population in China. The numbers plunged to 650 million heads from 676 million heads in the previous year. China had another outbreak in 2005 and its effects were visible in the duck population during 2007 (from 730 to 680 million heads). Thailand and Indonesia reportedly had widespread H5N1 infections in 2003, which the countries attempted to hide (Fidler 2004). In Thailand, the 2003 H5N1 outbreaks caused decline in duck population to 23.8 million heads in 2003, which further deteriorated to 15.6 million heads in 2004 compared to its 2002 status of 25.0 million heads. The duck farming in Thailand continued to suffer as there were H5N1 outbreaks causing high mortality in birds during July and August 2006. The duck production could recover only in 2007 to regain its 2002 status. In Indonesia too, a heavy blow on the duck farming was witnessed in 2003 and the duck population reduced by 24% to reach 34.9 million heads from 46.0 million heads in 1 year. The country suffered recurrent outbreaks of

HPAI from 2005 to 2010 with 162 human cases and 134 casualties (WHO 2020). This impacted duck farming badly for a long period and the duck population could regain its 2002 status only in 2012 after a decade of devastating. The slowdown of growth was also evident in Hong Kong and Cambodia in 2004 due to the reemergence of HPAI. In Vietnam, three human cases of H5N1 and death of all the three patients in 2003 and 29 cases with 20 deaths in 2004 (WHO 2020) made the duck population to shrink to 65 million heads in 2004 from 76 million heads in the previous year. The high number of H5N1 positivity in humans in 2005 (WHO 2020) was reflected in the duck population in the year 2006. The ups and downs in Vietnam duck population were also due to human reports until 2010 (WHO 2020).

The National Influenza Centre (NIC) of Egypt and the WHO Influenza Collaborating Centre in the UK reported the isolation of avian influenza A (H10N7) from two human specimens in May 2004 (PAHO 2004). This influenced the duck farming in the country heavily with the depletion of duck population to almost half in that year (from 9.6 to 4.9 million heads). The duck farming slowly recovered from this blow to reach 5.6 million heads in 2005 and then 7.4 million heads in 2006. Egypt became one of the endemic countries for H5N1 human infection since 2006. The country encountered 18 cases and 10 deaths in 2006, 25 cases and 9 deaths in 2007, 8 cases and 4 deaths in 2008, 39 cases and 4 deaths in 2009, and 29 cases and 13 deaths in 2010 (WHO 2020). Recurrent human HPAI infections heavily impacted duck farming again and the stock position further depleted to 4.0 million heads in 2007 and a long lasting effect subsequently before it could regain its 2003 status only in 2013. Malaysia has experienced waves of H5N1 high pathogenicity avian influenza (HPAI) outbreaks during 2004, 2006, and 2007 leading to slight dip in duck population in 2004 and a significant fall in 2008 (−1.3 million heads; AGR = −16%).

1.9.2 During Stationary Phase (2010–2019)

The avian influenza human cases started to decline after the 2005–2007 peak (Fig. 1.4). The incidences of human cases of AI were reported from many countries only up to 2014; five countries in 2011, six countries in 2012, seven countries in 2013, and five countries in 2014. This widespread infection across the globe has affected the duck farming. In 2015, although the total cases were unprecedentedly high with 145 cases, 42 deaths, the countries contributed to this number were only four with Egypt alone reporting 139 cases and 39 deaths. Post-2015, the human infectivity rate came down drastically but sparsely from Egypt (13 cases, 4 deaths), Indonesia (1 case, 1 death), and Nepal (1 case, 1 death) until 2020. The duck population in Egypt dropped to 7.7 million heads in 2017 from 10.6 in the previous year. The stock position further deteriorated to 6.5 million heads in 2018 and 5.6 million heads in 2019. The human infections in Indonesia caused depletion of duck population to 45.3 million heads in 2014 from 50.9 million heads in the previous year. The condition persisted until it made a comeback in 2017. The HPAI virus (H5N8) spread in the Republic of Korea in January 2014 caused huge mortalities in chicken and turkey farms (FAO 2014) leading to depletion of duck population to 7.5 million heads from 10.9 million heads in the previous year (AGR = −31%).

The disease has been contained in Egypt now and no new case of human infection has been reported in the last 3 years (2018–2020). In fact, no case has been reported in any part of the world during the years 2018 and 2020 (Fig. 1.4). In China also, sporadic cases were reported up to 2015 and no report thereafter. As a result, the duck production in China is steadily improving after 2015. Asian continent as a whole also witnessed steady increase after 2015.

In Europe, highly pathogenic avian influenza A (HPAI) (H5N8) was first reported on 5 November 2014 in a Turkey farm in Mecklenburg-Vorpommern, Germany. Immediately after that, outbreaks were reported from a chicken farm in Hekendorp, northeast of Rotterdam, the Netherlands on 15 November 2014 and in a duck-breeding farm in East Yorkshire, UK on 18 November 2014 (WHO 2014). This impacted duck farming in Germany and its duck population shrinking to 2.4 million heads in 2014 from 2.7 million heads in the previous year. United Kingdom also registered a negative growth of 4 million heads (AGR = −15.1%) in 2014.

Several other countries like Austria, Croatia, Denmark, Germany, Hungary, Israel, the Netherlands, Poland, the Russian Federation, and Switzerland in the European Region reported first infections of HPAI (H5N8) in wild birds and/or domestic poultry in 2014 followed by second outbreak in 2016 (WHO 2016). These outbreaks from 2014 to 2016 in European countries impacted growth of duck population during 2014–2016 in countries like Netherlands (from 1.0 to 0.8 million heads in 2014; AGR = −20%), Austria (from 0.08 to 0.06 million heads in 2014; AGR = −25%), Croatia (from 0.12 to 0.09 million heads in 2014; AGR = −25%), Belgium (from 0.23 to 0.17 million heads in 2015; AGR = −26%), Russian Federation (from 29.6 to 21.3 million heads in 2016; AGR = −28%), Bulgaria (from 1.5 to 1.2 million heads in 2016; AGR = −20%), and Denmark (from 0.25 to 0.18 million heads in 2016; AGR = −208%). In winter 2016–2017, highly pathogenic avian influenza (HPAI) H5N8 virus spread across Europe, causing unprecedented epizootics. France was massively affected, resulting in the culling of over 6 million poultry. A negative growth of 3 million heads of ducks from its previous year status of 27.4 million heads (AGR = −11%) was observed in France in 2016 and the duck population remained low until 2019.

1.10 Density of Duck Population

The overall density of ducks in the world was 8.6 ducks per square kilometer in 2018 (Table 1.3). However, ducks are distributed unequally across continents ranging widely from 23.7 (Asia) to mere 0.2 duck per square kilometer (Oceania). The duck density of Europe was 7.6 ducks per square kilometer matching the world average, while the ducks are sparsely distributed in Africa (0.5 duck per square kilometer) and Americas (0.6 duck per square kilometer). Bangladesh was the most duck dense country in the world with 438.8 ducks per square kilometer area. The other countries of dense duck population per square kilometer in the year 2018 were Vietnam

Table 1.3 Density of ducks in continents and top 25 duck dense countries in the year 2018

Area	Continent/country	Duck population (FAOSTAT 2021)	Area (sq. km)	Density (ducks/ sq. km)	World rank (2018)*
World		**1,177,351,000**	**136,278,000**	**8.6**	
Continents	***Africa***	***16,282,000***	***30,370,000***	***0.5***	***IV***
	Americas	***27,362,000***	***42,549,000***	***0.6***	***III***
	Asia	***1,055,005,000***	***44,579,000***	***23.7***	***I***
	Europe	***77,047,000***	***10,180,000***	***7.6***	***II***
	Oceania	***1655,000***	***8,600,000***	***0.2***	***V***
Countries	Bangladesh	57,119,000	130,168	438.8	1
	Vietnam	82,536,000	310,070	266.2	2
	Taiwan Province of China	8,016,000	32,260	248.5	3
	China, Mainland + Hong Kong SAR + Macao SAR	712,463,000	9,326,410	76.4	4
	Democratic People's Republic of Korea	7,251,000	99,909	72.6	5
	Republic of Korea	8,637,000	120,538	71.7	6
	Cambodia	9,033,000	176,515	51.2	7
	Myanmar	30,681,000	653,508	46.9	8
	Brunei Darussalam	239,000	5265	45.4	9
	France	24,968,000	640,427	39.0	10
	Philippines	11,577,000	298,170	38.8	11
	Indonesia	61,321,000	1,811,569	33.8	12
	Malaysia	10,519,000	329,613	31.9	13
	Thailand	11,765,000	510,890	23.0	14
	Ukraine	11,680,000	579,300	20.2	15
	Poland	5,597,000	311,888	17.9	16
	Lao People's Democratic Republic	3,600,000	230,800	15.6	17
	Sierra Leone	1,050,000	71,620	14.7	18
	Bulgaria	1,408,000	108,612	13.0	19
	India	33,511,000	2,973,190	11.3	20
	Belarus	2,007,000	202,900	9.9	21
	Czechia	768,000	77,247	9.9	22
	Haiti	249,000	27,560	9.0	23
	Israel	180,000	20,330	8.9	24
	Madagascar	4,578,000	581,540	7.9	25

*Roman numerals represent continent rank and Arabian numerals, country rank

(266.2), Taiwan Province of China (248.5), China mainland (76.4), Democratic People's Republic of Korea (72.6), Republic of Korea (71.7), and Cambodia (51.2). All the duck dense countries mentioned above are located in Asia.

1.11 Population Ratio of Duck to Human

The duck-to-human ratio in the world was 152 ducks to 1000 human in the year 2018. Among different continents the ratio was high in Asia followed by Europe, Oceania, Americas, and Africa. The top 20 countries of high duck to human ratio are given in Table 1.4. Among the top 20, 13 countries belong to Asia, five to Europe, and two to Americas. Vietnam of Asia tops the list with 805 ducks per 1000 human being followed by Suriname (695 ducks), one of the smallest countries of South America and Brunei Darussalam (555 ducks), a southeast Asian country.

Table 1.4 Ducks to human ratio in continents and top 20 countries in the year 2018 (FAOSTAT 2021)

Area	Continent/country	Number of ducks per 1000 people	World rank (2018)*
World		**152**	
Continents	***Africa***	***13***	***V***
	Americas	***27***	***IV***
	Asia	***228***	***I***
	Europe	***101***	***II***
	Oceania	***39***	***III***
Countries	Vietnam	805	1
	Suriname	695	2
	Brunei Darussalam	555	3
	Cambodia	548	4
	Myanmar	529	5
	Lao People's Democratic Republic	506	6
	China, mainland + Hong Kong SAR + Macao SAR	496	7
	Hungary	484	8
	France	369	9
	Bangladesh	346	10
	Taiwan Province of China	334	11
	Malaysia	307	12
	Democratic People's Republic of Korea	267	13
	Ukraine	248	14
	Bulgaria	236	15
	Indonesia	223	16
	Paraguay	206	17
	Belarus	189	18
	Thailand	177	19
	Republic of Korea	176	20

*Roman numerals represent continent rank and Arabian numerals, country rank

1.12 Duck Predominant Nations

The countries having high share of ducks in total poultry (chicken, ducks, geese, and Guinea fowl) in the year 2019 are depicted in Fig. 1.5. The overall global picture revealed that the ducks shared 4.2% of total poultry in the world. Among the

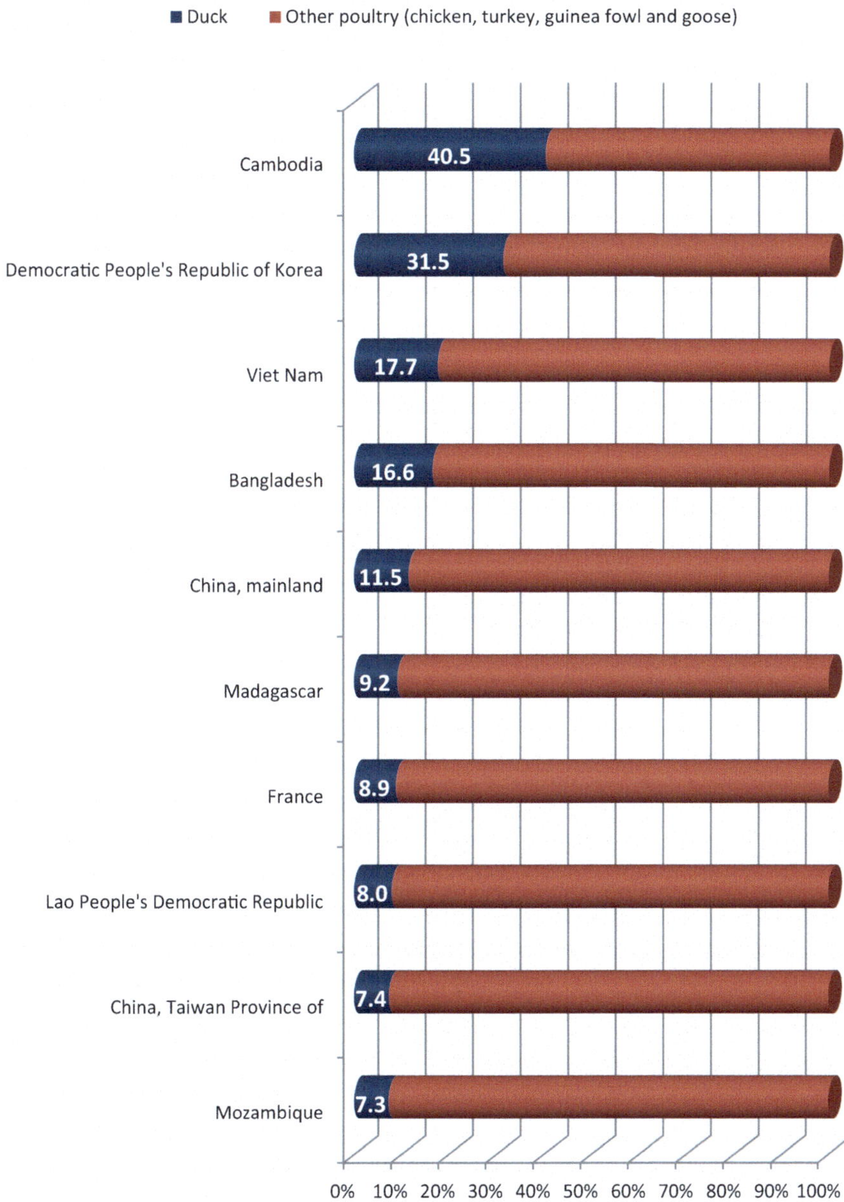

Fig. 1.5 Countries with high share of ducks in total poultry in 2019 (FAOSTAT 2021)

continents, the share of ducks in total poultry was high in Asia (6.1%) followed by Europe (3.5%), Oceania (1.1%), Africa (0.8%), and Americas (0.4%). Among the countries, Cambodia possessed very high proportion of ducks with the ratio of duck to other poultry was around 1:1.5. The ratio was also high in Democratic People's Republic of Korea (1:2.2). Among the top ten nations, seven belong to Asian continent. China although has 712.4 million heads of ducks, this constitutes only 11.5% of its total poultry. The other major duck producing countries having low proportion of ducks to the total poultry were Philippines (5.8%), Ukraine (5.6%), Thailand (4.0%), Russian Federation (4.0%), India (4.0%), Malaysia (3.7%), and Indonesia (1.6%). The ducks in Mexico, Australia, and UK constitute only 1–1.5%of total poultry, whereas Canada, Pakistan, USA, and Brazil have less than 1.0% ducks in their total poultry.

1.13 Duck Meat Production

The total duck meat produced in the world in the year 2019 was 4858.1 million kilograms (Table 1.5). Of this, 84.8% (4120.6 million kilograms) was produced in Asia. With 3299.2 million kilograms, China contributes a giant share of 67.9% duck meat production in the world and contributes 80.1% to the Asia's duck meat production.

The global duck meat production has grown at an annual rate of 2.1% between 2010 and 2019. Among the continents, Asia (2.5%) and Americas (1.7%) recorded high AGR, whereas Oceania (0.3%) and Africa (0.1%) and Europe (−0.1%) recorded stagnated growth between 2010 and 2019. Among the top ten duck meat producing countries (2019), Myanmar (11.2%), Hungary (8.2%), and Indonesia (7.8%) registered impressive growth rate during this period. Other major duck meat producing countries like Republic of Korea (5.2%), Vietnam (3.6%), Taiwan province of China (3.4%), China, mainland (2.7%), and the USA (2.4%) also maintained positive AGR during this period, whereas Malaysia (−6.2%), France (−2.3%), and Egypt (−0.3%) registered negative growth.

The time trend analysis of duck meat production during this decade (2010–2019) revealed that the global duck meat production made a significant ($P < 0.01$) progress although the course of the progress is unsteady ($R^2 = 0.70$). During this decade Asia had high magnitude increase (3354.9–4120.6 million kilograms; AGR = 2.5%), which was significant ($P < 0.05$) but unsteady ($R^2 = 0.74$) growth curve with a dip during 2014–2017. The Asia's performance was contributed by the steady and significant growth of duck meat production in its member countries, China ($R^2 = 0.74$; $P < 0.01$; 2650.5–3299.2 million kilograms; AGR = 2.7%), Myanmar ($R^2 = 0.90$; $P < 0.01$; 96.6–194.1 million kilograms; AGR = 11.2%), and Vietnam ($R^2 = 0.00$; $P > 0.05$; 74.8–98.7 million kilograms; AGR = 3.6%).

Although there are numerous breeds of ducks in China, there are mainly three breeds that are found in commercial meat production: the Pekin, Muscovy (Barbarie), and the mule duck (cross between the Muscovy and common duck like Pekin). The duck farming is concentrated in East, South, and parts of the Southwest

Table 1.5 Duck meat production (million kilograms) in continents and top 30 countries over the years (FAOSTAT 2021)

Area	Country	Year								World rank (2019)*
		2000	2005	2010	2015	2016	2017	2018	2019	
World		**2905.2**	**3363.9**	**4078.5**	**4305.7**	**4402.9**	**4361.4**	**4801.0**	**4858.1**	
Africa		***57.5***	***73.8***	***84.1***	***103.2***	***95.7***	***100.1***	***100.0***	***84.9***	***IV***
Of which	Egypt	39.0	55.0	66.2	83.6	75.7	79.9	79.7	64.5	9
	Madagascar	10.9	10.6	11.8	12.1	12.4	12.5	12.6	12.7	24
Americas		***96.6***	***106.1***	***101.4***	***114.0***	***113.3***	***113.2***	***116.7***	***117.4***	***III***
Of which	United States of America	52.6	61.0	52.6	62.4	61.6	60.9	63.9	64.0	10
	Mexico	20.0	20.3	20.8	20.8	21.0	21.1	21.1	21.2	20
	Argentina	7.3	7.4	9.2	9.3	9.3	9.4	9.6	9.7	28
	Canada	7.4	7.4	7.5	8.8	8.9	9.0	9.1	9.1	29
	Brazil	7.2	7.4	7.7	7.6	7.0	7.3	7.2	7.1	30
Asia		***2318.3***	***2679.6***	***3354.9***	***3527.1***	***3681.2***	***3691.5***	***4030.9***	***4120.6***	***I***
Of which	China, Mainland + Hong Kong SAR + Macao SAR	1803.2	2078.5	2650.5	2760.4	2952.9	2971.1	3244.1	3299.2	1
	Myanmar	28.8	60.1	96.6	135.0	133.5	144.3	181.0	194.1	3
	Vietnam	69.6	72.0	74.8	90.0	83.3	87.2	90.7	98.7	4
	Republic of Korea	44.7	52.0	65.0	80.0	80.0	71.0	91.5	95.2	5
	Taiwan Province of China	64.5	71.5	67.1	80.0	82.8	82.1	84.5	87.5	7
	Malaysia	64.3	104.0	155.5	122.0	79.0	66.6	66.1	68.7	8
	Bangladesh	35.9	38.0	41.5	50.0	52.1	53.6	55.1	56.1	12
	Thailand	102.5	85.0	77.3	66.9	64.3	60.3	57.5	55.1	13
	Indonesia	13.8	21.4	26.0	34.9	41.9	42.3	44.7	44.2	14
	India	40.0	37.7	33.2	32.5	34.9	36.9	39.1	43.9	15
	Philippines	21.6	20.7	28.1	33.9	33.8	32.8	33.0	33.3	17
	Democratic People's Republic of Korea	3.8	10.1	10.8	11.9	12.7	12.9	12.7	13.5	23
	Cambodia	6.6	9.0	8.7	9.4	9.5	9.9	10.0	10.1	27

(continued)

Table 1.5 (continued)

Area	Country	Year								World rank (2019)*
		2000	2005	2010	2015	2016	2017	2018	2019	
Europe		***424.0***	***493.6***	***519.8***	***544.9***	***494.4***	***438.6***	***535.1***	***516.5***	***II***
Of which	France	234.1	260.5	282.3	286.9	239.2	210.8	225.0	225.0	2
	Hungary	43.4	53.1	52.4	78.0	78.8	52.8	92.9	90.9	6
	Poland	11.0	18.2	17.2	23.8	25.2	24.2	61.9	61.9	11
	Germany	31.6	40.1	61.4	47.1	41.2	36.0	39.0	34.0	16
	Ukraine	25.6	24.4	26.5	28.7	26.2	26.3	27.9	28.9	18
	United Kingdom of Great Britain and Northern Ireland	40.5	41.7	29.6	29.2	29.6	30.6	30.4	27.9	19
	Bulgaria	8.0	17.0	19.9	20.3	22.5	19.6	23.1	19.8	21
	Czechia	3.9	8.3	6.9	6.8	7.2	9.1	11.3	11.3	25
	Portugal	0.0	0.0	0.0	0.0	0.0	0.0	10.6	10.5	26
Oceania		***8.8***	***10.8***	***18.3***	***16.5***	***18.3***	***18.0***	***18.4***	***18.8***	***V***
Of which	Australia	8.0	9.9	16.8	15.5	17.2	16.9	17.3	17.6	22

*Roman numerals represent continent rank and Arabian numerals, country rank

China and more than 90% of Chinese duck production comes from the provinces located in these regions. Of total duck meat production in China, meat-type ducks contribute 95% and spent laying ducks 5% in the year 2014 and 92% and 8%, respectively, in 2015. The total values of output of duck industry in 2014 and 2015 were 18.5 billion dollars and 21 billion dollars, respectively (Hou 2016). At present, the major meat-type ducks in China are local Pekin ducks (10–15%), commercial Pekin ducks imported from a UK based Breeding Corporation (more than 65%), and local Sheldrake ducks (20%). Muscovy and mule ducks of Kelimer Group mostly being raised in Fujian province as well as its surrounding areas have a larger market share, nearly 100 million ducks per year. In Taiwan, the inter-generic hybrid of Muscovy drake and common duck, mule has been the major meat duck (75–80%) for many years. Local farmers routinely apply artificial insemination to obtain the hybrids.

The growth curve of Africa's duck meat production was highly unsteady ($R^2 = 0.23$) with three wax and wane cycles ended with very little net gain (from 84.1 to 84.9 million kilograms) during these 9 years (2010–2019). The same pattern of three wax and wane cycles and poor performance ($R^2 = 0.15$; $P > 0.05$; 66.2–-64.5 million kilograms) was seen in the growth curve of Egypt, the major duck meat producer of Africa. The growth in Europe ($R^2 = 0.10$; $P > 0.05$; 519.8–516.5 million kilograms) and Oceania ($R^2 = 0.22$; $P > 0.05$; 18.3–18.8 million kilograms) was also insignificant, inconsistent, and low in magnitude. Although the progress was meager (101.4–117.4 million kilograms; AGR = 1.8%), the growth was stable ($R^2 = 0.93$) and significant ($P > 0.05$) in Americas.

France occupied second place in duck meat production in 2019, although it was in seventh position in terms of number of ducks. Ducks in France and other European countries are primarily reared for meat purpose, therefore are heavier breeds. The average meat yield of ducks in France is 2.5 times heavier (3321 g) compared to 1324 g in China and Asian countries put together (1358 g), where mostly local breeds with low body weight are reared extensively for egg production. The meat yield in most of the leading duck producing Asian countries like China, Vietnam, Indonesia, Bangladesh, and India is less than 1350 g. The average meat yields of ducks in different continents and countries are given in Table 1.6.

1.14 Mule Duck Production

The mule ducks are the sterile progeny from a cross of two species in which common duck females (*Anas platyrynchos*) are inseminated by semen from Muscovy drakes (*Cairina moschata*). Mule ducks are produced because of their large size, quality liver, and reduced meat fat content. Mule duck meat production is developing in Europe, Vietnam, and Southeast China (Cheng et al. 2009). Mule ducks are used in Asia, especially in Taiwan, for meat production (Tai et al. 1999), whereas in Europe, mule duck production is mainly concentrated in France for *foie gras* production. The breeding of mule ducks in Taiwan is in practice for nearly 300 years. In Taiwan the female parent of the mule is the Kaiya, a cross of Pekin drake with the female Tsiaya,

Table 1.6 Average meat yield per duck and the number of ducks slaughtered in top 30 duck producing countries from 2015 to 2019 (5 years) (FAOSTAT 2021)

Area	Continent/country	Average of 5 years (2015–2019)	
		Carcass weight (g)	Birds slaughtered (million no)
World		**1469**	**3094.2**
Continents	***Africa***	***2209 (II)***	***43.8 (IV)***
	Americas	***2173 (III)***	***52.9 (III)***
	Asia	***1358 (V)***	***2804.5 (I)***
	Europe	***2744 (I)***	***184.3 (II)***
	Oceania	***2064 (IV)***	***8.7 (V)***
Countries	Bulgaria	3687 (1)	5.7 (25)
	France	3321 (2)	72.3 (4)
	Republic of Korea	3039 (3)	27.6 (12)
	Hungary	2728 (4)	28.8 (10)
	Argentina	2600 (5)	3.6 (29)
	Malaysia	2536 (6)	32.7 (8)
	Mexico	2500 (7)	8.4 (21)
	Portugal	2456 (8)	4.3 (28)
	Egypt	2439 (9)	31.4 (9)
	Ukraine	2400 (10)	11.5 (19)
	Thailand	2304 (11)	26.5 (14)
	United States of America	2287 (12)	27.4 (13)
	Taiwan Province of China	2262 (13)	36.9 (7)
	Germany	2258 (14)	17.4 (16)
	Poland	2119 (15)	18.8 (15)
	United Kingdom of Great Britain and Northern Ireland	2113 (16)	14.0 (18)
	Australia	2104 (17)	8.0 (22)
	Philippines	2094 (18)	15.9 (17)
	Democratic People's Republic of Korea	2000 (19)	6.4 (23)
	Madagascar	2000 (20)	6.2 (24)
	Canada	1981 (21)	4.5 (27)
	Czechia	1972 (22)	11.1 (20)
	Cambodia	1500 (23)	3.6 (30)
	China, mainland	1324 (24)	2298.2 (1)
	Brazil	1300 (25)	5.6 (26)
	India	1300 (26)	28.8 (11)
	Vietnam	1200 (27)	75.0 (3)
	Myanmar	1174 (28)	133.6 (2)
	Bangladesh	1000 (29)	53.4 (5)
	Indonesia	906 (30)	46.2 (6)

Figures in the parentheses indicate the world rank. Roman numerals represent continent rank and Arabian numerals, country rank

the Taiwanese egg-laying breed. This produces a market duck with extremely lean meat; a fat content of less than 18% as compared to 30% of more fat content of purebred Pekin ducks. In comparison with the parental species, the mule duck is intermediary though it could present a better growth rate (Pingel 1990), a good feed efficiency (Shalev 1995), a lean carcass (Wawro et al. 2004), and nearly no sexual dimorphism (Baéza et al. 2000).

1.15 *Foie gras* Production

About 97% of *foie gras* is produced from ducks and the remaining 3% from goose (Marie-Etancelin et al. 2008). Among ducks, about 95% of duck liver production comes from overfed male mule ducks, the remaining 5% being from male Muscovy ducks. France produces almost 20 million kilos of *foie gras* per year. *Foie gras* production is concentrated in Aquitaine, Midi-Pyrénées, Pays de la Loire, Poitou, and Bretagne regions of France. These regions have perfect weather conditions for the time the animals need to spend outdoors and for local maize production, an essential resource for feeding the ducks and geese. The *foie gras* industry in France supports about 30,000 families and employs 100,000 people. During the year 2013, the volume of overseas trade of *foie gras* at global level was 3.78 million kilograms. Of which, Europe shared an unassailable 98.73%. France alone contributed 62.89% of world's *foie gras* export. The major buyer of *foie gras* at world market was Spain consuming 45.33% of world's imports.

1.16 Duck Down Production

The duck's down is an important by-product of meat-type duck industry. Around 2.2 billion ducks were slaughtered in China in 2018, which was 74.6% of total ducks slaughtered in the world during the same year. Consequently, it is not surprising that China is the world's leading supplier of down and feathers for apparel and bedding. The annual down production in China is around 35,000 tons valued at over 10 billion RMB per year. According to American Down and Feather Council, 80% of the down and feathers used globally are produced in China. The majority of 90% of down feather come from ducks and the remaining 10% from goose. Duck down is also produced in Hungary, Poland, Turkey, the European Union, and the USA.

Despite their enormous popularity, down and feather bedding, apparel and outdoor products have come under increased scrutiny in recent years as more and more consumers seek to familiarize themselves with companies associated with the humane treatment of animals. At a global level, the International Down and Feather Bureau (IDFB) does not accept the illegal and inhumane harvesting of feathers and down from live birds. As such, live plucking is forbidden in China's down industry. All IDFB members, which include country trade associations across Europe, Asia, and America, as well as individual companies from around the globe, follow this same standard regarding the humane treatments of birds in slaughtering and collection of feathers (Schmitz 2016).

1.17 Duck Egg Production

The data on number of eggs produced by poultry other than chicken (other eggs) is given in Table 1.7, as there is no exclusive data available on duck egg production. It is deemed that the duck egg forms the major component of "other eggs" category. World's total production of other eggs in 2019 was 95,943 million numbers, which represents 5.7% of total poultry egg production. Asia contributed 94.1% to the world's other egg production. The share of other eggs in total poultry eggs was also high in Asia (8.2%) followed by Africa (1.3%) and Americas (1.2%), whereas the share of other eggs in total poultry eggs had least significance in Europe (0.4%) and Oceania (0.7%). On individual country basis, China tops the list with 71.3 billion other eggs that was 79.0% of Asia's other egg production and 74.3% of world production. Although at distant low, the countries like Indonesia, Thailand, Brazil, and Bangladesh occupy slots next to China in terms of other eggs production.

The time trend analysis of other egg production from 2010 to 2019 reveals that the growth rate was significant ($P < 0.01$) with poor consistency in world ($R^2 = 0.80$), Asia ($R^2 = 0.84$), and Europe ($R^2 = 0.65$). The growth was instable in all other three continents, although the growth was significant ($P < 0.05$) only in Oceania. Among the top five other eggs producing countries, only Indonesia ($P < 0.01$; $R^2 = 0.99$) and Bangladesh ($P < 0.01$; $R^2 = 0.83$) had significant and steady growth between 2010 and 2019. Although the growth rate was significant in China mainland ($P < 0.01$) and Thailand ($P < 0.05$), both the countries lacked stability of growth. The growth rate in Brazil was neither stable nor significant.

The duck egg production in Asian countries comes from mostly non-intensive systems of management. Most Asian countries have their traditional egg-producing breeds, but the Khaki Campbell and the Pekin have become increasingly popular, although the latter is not well adapted to forage in the rice fields. The layer breeds of ducks in China and Taiwan are capable of laying 320 eggs in 500 days of age with feed-egg ratio of around 2.75:1. In Vietnam and Indonesia, ducks are kept primarily for table eggs. Local breeds of ducks are often herded in flocks of about 100 in the rice fields where they scavenge for their feed and follow the harvest to collect fallen grains and insects, frogs, fish, snakes, and other aquatic animals during rice field cultivation and early plant growth. In India, only indigenous breeds are used for eggs production. Kuttanad breed, native to Kerala state of India has very good egg laying potential (Jalaludeen 2005) and therefore used widely in Kerala and many states outside Kerala including northeast Indian states.

Table 1.7 Egg production (million nos.) from poultry other than chicken (mainly ducks) in continents and top 20 countries over the years (FAOSTAT 2021)

Area	Countries	Year								World ranking (2019)*
		2000	2005	2010	2015	2016	2017	2018	2019	
World		**63,270**	**71,516**	**83,206**	**95,243**	**96,578**	**92,954**	**93,568**	**95,943**	
Africa		***797***	***967***	***888***	***1424***	***887***	***891***	***952***	***952***	***III***
Of which	Namibia	670	840	760	1295	760	760	821	821	7
Americas		***1070***	***1438***	***2817***	***4623***	***3287***	***3674***	***3603***	***3814***	***II***
Of which	Brazil	1046	1412	2789	4595	3259	3646	3576	3787	4
Asia		***60,265***	***68,105***	***78,044***	***87,928***	***91,040***	***87,107***	***88,028***	***90,324***	***I***
Of which	China, mainland	50,350	56,262	63,754	70,323	72,938	67,656	68,213	71,322	1
	Indonesia	2349	3249	4289	5534	5775	6147	6228	6475	2
	Thailand	4200	4370	5560	5650	5560	5560	5700	5700	3
	Bangladesh	1138	1687	1720	3299	3574	4480	4656	3556	5
	Myanmar	183	360	701	1000	1023	1074	1000	1000	6
	Philippines	823	819	564	639	669	688	706	763	8
	Taiwan Province of China	478	488	484	437	448	441	470	437	9
	Republic of Korea	267	350	374	386	386	386	386	386	10
	Pakistan	122	122	139	211	211	211	211	211	13
	Malaysia	142	157	200	200	200	200	200	200	14
	Madagascar	91	91	91	90	90	92	92	92	16
	Singapore	80	80	80	80	80	82	82	82	17
	Cambodia	66	85	82	80	80	80	80	80	18
	Uzbekistan	12	21	31	21	26	33	34	47	19
Europe		***1107***	***975***	***1417***	***1229***	***1324***	***1243***	***943***	***811***	***IV***
Of which	Russian Federation	131	241	207	479	472	539	503	366	11
		196	235	224	237	237	237	237	237	12

(continued)

Table 1.7 (continued)

Area	Countries	Year 2000	2005	2010	2015	2016	2017	2018	2019	World ranking (2019)*
	United Kingdom of Great Britain and Northern Ireland									
	Ukraine	134	91	488	168	302	155	161	167	15
Oceania		***30***	***32***	***40***	***40***	***40***	***40***	***41***	***41***	***V***
Of which	New Zealand	28	29	37	37	37	37	38	38	20

*Roman numerals represent continent rank and Arabian numerals, country rank

1.18 Per Capita Availability of Duck Meat

The perusal of per capita availability of duck meat in different countries (Table 1.8) revealed that European countries like Hungary (9570 g), France (3460 g), Bulgaria (3320 g), and Albania (3280 g) and Asian countries like Taiwan Province of China (3560 g) and Myanmar (3370 g) occupied top six spots with availability of more than 3000 g duck meat per head in the year 2018. The mainland China, the world leader in total duck meat production, however, stands at seventh place with 2260 g. Malaysia, Republic of Korea, Czechia, and Suriname were the other countries with considerable quantity of more than one kilogram per capita duck meat availability. The continents with decreasing order of availability of per capita duck meat availability were Asia followed by Europe, Oceania, Americas, and Africa.

Table 1.8 Per capita duck meat availability in continents and top 20 countries in the year 2018 (FAOSTAT 2021)

Area	Continent/country	Per capita duck meat availability (g)	World rank (2018)*
World		**630**	
Continents	***Africa***	***80***	***V***
	Americas	***120***	***IV***
	Asia	***880***	***I***
	Europe	***720***	***II***
	Oceania	***440***	***III***
Countries	Hungary	9570	1
	Taiwan Province of China	3560	2
	France	3460	3
	Myanmar	3370	4
	Bulgaria	3320	5
	Albania	3280	6
	China, mainland	2260	7
	Malaysia	2100	8
	Republic of Korea	1790	9
	Czechia	1060	10
	Suriname	1040	11
	Vietnam	950	12
	Thailand	830	13
	Egypt	810	14
	Singapore	770	15
	Australia	690	16
	Brunei Darussalam	630	17
	Cambodia	630	18
	Lao People's Democratic Republic	620	19
	Ukraine	560	20

*Roman numerals represent continent rank and Arabian numerals, country rank

1.19 Per Capita Availability of Other Eggs

The data on per capita availability of eggs of other species of poultry (in which duck eggs share a major portion) in major duck producing countries for the year 2018 was examined. The consumption was high in Asian countries like Thailand (82.1 eggs), China Mainland (47.5 eggs), Bangladesh (28.9 eggs), Indonesia (23.3 eggs), Taiwan Province of China (19.8 eggs), and Myanmar (18.6 eggs). The availability in rest of the countries was only less than 10 per head. Among continents, Asia has highest per capita availability of other eggs followed by Americas, Europe, Oceania, and Africa. The overall world average is 12.3 eggs (Table 1.9).

Table 1.9 Per capita availability of other eggs in continents and top 20 countries in the year 2018 (FAOSTAT 2021)

Area	Continent/country	Per capita availability of other eggs (no.)	World rank*
World		**12.3**	
Continents	***Africa***	***0.7***	***V***
	Americas	***3.6***	***II***
	Asia	***19.3***	***I***
	Europe	***1.3***	***III***
	Oceania	***1.0***	***IV***
Countries	Thailand	82.1	1
	China, mainland	47.5	2
	Bangladesh	28.9	3
	Indonesia	23.3	4
	Taiwan Province of China	19.8	5
	Myanmar	18.6	6
	Republic of Korea	7.5	7
	Philippines	6.6	8
	Malaysia	6.3	9
	French Polynesia	5.8	10
	Cambodia	4.9	11
	Belarus	3.9	12
	Ukraine	3.6	13
	Madagascar	3.5	14
	United Kingdom of Great Britain and Northern Ireland	3.5	15
	Russian Federation	3.5	16
	Paraguay	2.4	17
	Seychelles	1.8	18
	Lao People's Democratic Republic	1.4	19
	Albania	1.1	20

*Roman numerals represent continent rank and Arabian numerals, country rank

1.20 Balut and Other Special Duck Egg Products

Balut (embryonated egg) is a special Filipino delicacy which is actually a fertile egg harvested at 15–18 days of incubation. They are common everyday food in some countries in Southeast Asia, such as in the Philippines, Cambodia, and Vietnam. Eggs with relatively smaller embryos are called *balut mamatong*, while those with relatively bigger embryos are called *balut saputi*. Eggs not suited for balut making are processed into salted century eggs. Salted eggs processed through using either brine or clay-salt solution are consumed as regular viand or used as ingredients in *ensaymada* and *puto pao*. Century eggs are common ingredients in several food preparations. Fresh duck eggs are more popular in several Chinese food preparations. Fresh duck eggs are more popular than chicken eggs in making *leche flan*. Boiled duck eggs are also used as components of popular foods such as *pancit palabok* and *siopao* (Bondoc 2008). In the Philippines, about 90% of total duck egg produced is used for processing. Eighty-seven percent of which is processed into balut and another 7% is processed into salted eggs. The remaining 6% consists of century eggs, penoy, and other unidentified forms. This means that balut accounts for about 80% of total egg production. This emphasizes that balut production is a unique feature of the Philippine duck industry (Chang and Dagaas 2004). The annual balut production in Philippines alone is nearly 30,000 metric tonnes considering the total duck egg production of 42,400 metric tons in 2015 (Anonymous 2016).

1.21 Overseas Trade of Live Ducks

The overseas trade of live ducks involves mainly day-old grandparent, parent, and commercial ducklings rather than adult birds. The data on export and import of live ducks shows high variation from year to year; therefore, the continents and countries are ranked based on the last 5 years data.

1.21.1 Export

The data on latest 5-year average (Table 1.10) reveals that Europe is the major exporter of live birds, which constitutes massive 71% of the world's export in terms of number of birds and 44% in terms of value. Furthermore, six out of ten top live duck exporting countries are located in Europe. France tops the country list, alone contributing more than half (51%) of Europe's export and more than one-third (36%) of world's export. Time trend analysis of export data for the last 10 years (2010–2019) revealed that Europe registered an unsteady but significant ($P < 0.01$) growth of4.9 million ducks per year. France also exhibited unsteady but significant ($P < 0.01$) growth of 2.5 million ducks per year during this period. Asia, the leading continent in duck production stands second in terms of live bird export and its country Malaysia stands third in world ranking in terms of number of

Table 1.10 Export of live ducks in continents and top 15 countries over the years (FAOSTAT 2021)

Area	Continent/country	Number of ducks (1000s) 2015	2016	2017	2018	2019	Average (2015–2019)	Rank (2015–2019)*	Total value from 2015 to 2019 (million US$)#
World		**77,347**	**50,844**	**53,886**	**51,816**	**47,060**	**56,191**		**483.69**
Continents	***Africa***	***426***	***252***	***254***	***411***	***110***	***291***	***IV***	***0.98 (IV)***
	Americas	***2462***	***5646***	***3773***	***3753***	***3348***	***3796***	***III***	***72.68 (III)***
	Asia	***34,857***	***6169***	***6206***	***6614***	***6183***	***12,006***	***II***	***195.72 (II)***
	Europe	***39,599***	***38,776***	***43,653***	***41,036***	***37,418***	***40,096***	***I***	***214.26 (I)***
Countries	France	25,738	18,576	20,740	18,598	18,594	20,449	1	100.41 (2)
	Czechia	8506	11,539	11,452	11,576	9623	10,539	2	40.02 (4)
	Malaysia	5623	5563	5483	5460	5277	5481	3	169.73 (1)
	Netherlands	1560	3385	7782	6867	5823	5083	4	21.02 (6)
	Canada	2010	4888	3397	3130	3004	3286	5	68.40 (3)
	Germany	2061	2434	1471	1625	1399	1798	6	29.84 (5)
	Hungary	1426	1607	1350	1544	1358	1457	7	5.17 (8)
	Thailand	312	510	702	786	781	618	8	3.06 (10)
	United States of America	448	758	376	623	344	510	9	4.26 (9)
	Poland	140	691	318	471	229	370	10	0.54 (12)
	UK of Great Britain and Northern Ireland	158	364	412	296	364	319	11	15.87 (7)
	Vietnam	302	92	0	291	0	137	12	0.90 (11)
	Kenya	312	119	24	110	42	121	13	0.44 (13)
	Ethiopia	51	20	159	249	0	120	14	0.19 (14)
	South Africa	18	95	24	30	14	36	15	0.16 (15)

Figures in the parentheses indicate the world rank

*, # Roman numerals represent continent rank and Arabian numerals, country rank

birds but placed first in terms of value of export. The time trend analysis revealed unsteady growth in export of ducks in all continents between 2010 and 2019, although the growth was significant ($P < 0.05$) in Europe and Americas. The growth of world's export was also significant ($P < 0.05$) but unsteady during this period. The country scenario also revealed significant but unsteady growth from 2010 to 2019 in all major live duck exporting countries like France, Czechia, Malaysia, Netherlands, Canada, and Hungary.

1.21.2 Import

The data on import trade of live ducks around the world is given in Table 1.11. Based on the last 5-year average, Europe accounts for major portion (61%) of the total import market of the world. Poland is the top live duck importing country sharing 32% of worlds total import market. Although Singapore stands second, next to Poland, in terms of number of live ducks imported, the value of ducks imported by Singapore was much higher and tops the list.

1.22 Overseas Trade of Duck Meat

1.22.1 Export

The world's total export of duck meat has improved with an AGR of 3.3%from 2010 to 2019. Out of total 195.6 million kilos of duck meat exported in 2019, 96% originated from Asia (25.0%) and Europe (71.0%) together (Table 1.12). The European countries like Hungary (24%), France (15%), Netherlands (9%), and Poland (5%) were the major contributors to the world duck meat export in the year 2019. China contributed 21% of world's total duck meat export.

The analysis of AGR from 2010 to 2019 revealed that Poland has registered an AGR of 484.7% to increase the export from its initial value of 0.2–8.9 million kilograms. The other European countries that exhibited impressive AGR during this period were France (23.7%) and Hungary (11.3%), while Netherlands's export grew at the rate of 2.6%. However, China registered a slight negative growth (−0.6%) during this period. Among continents, only Europe (8.8%) registered positive AGR, whereas Africa (−9.8%), Oceania (−7.6%), Americas (−5.0%), and Asia (−1.7%) witnessed negative growths. The overall AGR in the world was 3.3%. The time trend analysis of export from 2010 to 2019 revealed that although Hungary and Poland registered significant ($P < 0.01$) growth, the consistency of growth in major duck meat exporting countries like China, France, Hungary, Netherlands, and Poland was very poor. On examination of overall data for world or continents, none had neither significant nor consistent growth during this period.

In terms of value of export, 79% world export revenue was received by Europe in the year 2019 (Table 1.13). Although France ranked third in terms of volume, it was placed at first rank in terms of value of export. France and Hungary together earned

Table 1.11 Import of live ducks in continents and top 15 countries over the years (FAOSTAT 2021)

Area	Continent/ country	Number of ducks (1000s)							Total value from 2015 to 2019 (million US$)#
		2015	2016	2017	2018	2019	Average (2015–2019)	Rank (2015–2019)*	
World		**28,477**	**32,740**	**30,967**	**32,547**	**34,751**	**31,896**		**385.2**
Continents	***Africa***	***1678***	***1839***	***666***	***5839***	***5391***	***3083***	***III***	***22.8 (IV)***
	Americas	***1430***	***4201***	***2786***	***1436***	***1409***	***2252***	***IV***	***22.9 (III)***
	Asia	***6212***	***6292***	***6331***	***7375***	***9611***	***7164***	***II***	***198.1 (I)***
	Europe	***18,981***	***20,398***	***21,180***	***17,892***	***18,339***	***19,358***	***I***	***140.0 (II)***
	Oceania	***176***	***10***	***4***	***5***	***1***	***39***	***V***	***1.3 (V)***
Countries	Poland	9880	12,770	10,014	8099	9650	10,083	1	36.9 (2)
	Singapore	5629	5506	5333	5359	5271	5420	2	174.7 (1)
	Egypt	1100	1300	271	4849	5220	2548	3	14.8 (5)
	Germany	705	2514	3275	2058	1463	2003	4	14.4 (6)
	Netherlands	4109	260	1273	1824	1655	1824	5	25.2 (3)
	United States of America	1295	3763	2222	918	896	1819	6	15.9 (4)
	Myanmar	1	0	406	1233	3678	1064	7	1.7 (15)
	Hungary	471	488	1437	1826	1061	1057	8	9.9 (8)
	Romania	820	998	742	1057	1254	974	9	3.3 (13)
	Spain	1052	647	798	1036	1094	925	10	12.9 (7)
	Bulgaria	150	118	925	825	518	507	11	9.2 (9)
	Portugal	483	492	475	364	363	435	12	2.0 (14)
	France	340	1056	211	128	199	387	13	4.5 (10)
	Italy	125	135	1162	43	38	301	14	4.5 (11)
	Trinidad and Tobago	0	240	411	296	405	270	15	3.7 (12)

Figures in the parentheses indicate the world rank

*, # Roman numerals represent continent rank and Arabian numerals, country rank

Table 1.12 Export of duck meat (million kilograms) from continents and top 20 countries over the years (FAOSTAT 2021)

Area	Continent/country	Year								Rank (2019) *
		2000	2005	2010	2015	2016	2017	2018	2019	
World		**107**	**102.4**	**151.1**	**214**	**262.3**	**351.9**	**255.8**	**195.64**	
Continents	***Africa***	***0.3***	***0.4***	***0.9***	***2.1***	***0.2***	***0.2***	***0.1***	***0.11***	***IV***
	Americas	***6.8***	***16.1***	***14.8***	***12.3***	***14.5***	***11.2***	***7.7***	***8.07***	***III***
	Asia	***34***	***29.2***	***57.8***	***61***	***102.1***	***209.6***	***97.8***	***48.78***	***II***
	Europe	***65.3***	***56.6***	***77.4***	***138.4***	***145.3***	***130.8***	***150***	***138.60***	***I***
	Oceania	***0.6***	***0.2***	***0.2***	***0.1***	***0.1***	***0.1***	***0.2***	***0.07***	***V***
Countries	Hungary	24.7	16.1	23.5	42.9	44.8	28.7	49.2	47.3	1
	China, mainland + Hong Kong SAR	23.9	22.3	45.7	52.5	92.5	201.5	87.0	41.4	2
	France	16.5	13.5	9.5	34.0	33.3	31.7	31.5	29.7	3
	Netherlands	8.9	7.5	14.2	24.1	21.0	22.9	18.3	17.5	4
	Poland	0.0	1.2	0.2	5.5	6.9	7.4	9.0	8.9	5
	Germany	4.5	6.5	18.4	12.6	13.7	12.6	9.4	8.7	6
	Bulgaria	0.0	0.0	0.3	6.8	9.4	9.8	10.6	8.4	7
	United Kingdom of Great Britain and Northern Ireland	8.2	5.5	4.7	5.1	7.1	5.9	7.6	6.0	8
	Thailand	8.8	0.7	5.3	2.6	4.4	4.2	4.5	3.5	9
	United States of America	4.9	12.1	6.7	6.5	8.3	5.7	3.5	3.2	10
	Brazil	0.0	1.2	2.2	2.4	3.0	3.4	3.0	3.1	11
	Belgium	0.1	0.2	0.2	2.7	3.1	3.5	3.9	2.9	12
	Czechia	0.0	0.5	0.6	1.0	1.3	2.6	3.0	2.4	13
	Ireland	0.1	2.6	3.0	0.1	0.1	0.1	1.4	2.2	14
	Portugal	0.0	0.1	0.7	1.1	1.5	2.1	1.7	1.8	15
	Canada	1.8	2.8	5.8	3.4	3.3	2.1	1.2	1.8	16

(continued)

Table 1.12 (continued)

Area	Continent/country	Year								Rank (2019)
		2000	2005	2010	2015	2016	2017	2018	2019	*
	Republic of Korea	0.2	0.0	0.2	3.1	2.0	0.3	2.6	1.6	17
	Denmark	1.7	1.9	0.8	0.9	0.7	1.0	1.6	0.9	18
	Taiwan Province of China	0.7	5.0	1.7	1.0	0.9	1.3	1.4	0.6	19
	Spain	0.0	0.2	0.1	0.3	0.6	0.5	0.7	0.5	20

*Roman numerals represent continent rank and Arabian numerals, country rank

Table 1.13 Export value (million US$) of duck meat in continents and top 15 countries the year 2019 (FAOSTAT 2021)

Area	Continent/country	Total export in 2019		Rate (US $/ton)[#]
		Value (in million US$)	Rank (2019)*	
World		**678.34**		**3467**
Continents	***Africa***	***0.39***	*V*	***3527 (IV)***
	Americas	***33.22***	***III***	***4119 (II)***
	Asia	***110.00***	***II***	***2256 (V)***
	Europe	***534.26***	***I***	***3855 (III)***
	Oceania	***0.44***	*IV*	***5701 (I)***
Countries	France	181.4	1	6104 (2)
	Hungary	142.3	2	3008 (9)
	China, mainland	81.1	3	2073 (15)
	Bulgaria	49.5	4	5899 (3)
	Netherlands	45.8	5	2617 (13)
	Poland	24.0	6	2704 (11)
	Germany	24.0	7	2761 (10)
	United Kingdom of Great Britain and Northern Ireland	15.5	8	2591 (14)
	Belgium	15.5	9	5321 (4)
	Thailand	15.1	10	4346 (5)
	Canada	14.0	11	7983 (1)
	United States of America	11.1	12	3422 (7)
	Ireland	8.7	13	4019 (6)
	Brazil	8.2	14	2653 (12)
	Czechia	8.0	15	3274 (8)

Figures in the parentheses indicate the world rank

*, # Roman numerals represent continent rank and Arabian numerals, country rank

61% of Europe and 48% of world's duck meat export revenue. Further the rates of exported duck meat also vary among the countries/continents. Interestingly, Canada, the country ranked 16th in terms of quantity and 11th in terms of total value, ranked first in terms of rate (price per ton) (Table 1.13).

1.22.2 Import

European countries were the major importers (64%) of the duck meat in the world; of which, the major shareholders were Germany (21%), France (8%), UK (6%), and Czechia (5%) in the year 2019. Asia (33%) stood second among continents with major share of 17% of world's import went to Hong Kong during the year. The status of import of duck meat during selected years is given in Table 1.14.

Table 1.14 Import of duck meat (million kilograms) by continents and top 15 countries in the year 2019 (FAOSTAT 2021)

Area	Continent/country	Year								Rank (2019) *
		2000	2005	2010	2015	2016	2017	2018	2019	
World		**164.9**	**115.2**	**153.2**	**181.2**	**190.4**	**180.3**	**182.1**	**179.0**	
Continents	***Africa***	***4.1***	***0.6***	***3.7***	***4.3***	***2.2***	***1.4***	***2.0***	***1.6***	***IV***
	Americas	***3.5***	***4.5***	***3.2***	***6.6***	***8.8***	***6.3***	***4.8***	***4.3***	***III***
	Asia	***99.8***	***63.4***	***81.4***	***64.5***	***62.7***	***54.9***	***53.0***	***58.4***	***II***
	Europe	***55.3***	***46.0***	***64.2***	***105.1***	***115.6***	***116.2***	***120.6***	***114.3***	***I***
	Oceania	***2.1***	***0.7***	***0.7***	***0.6***	***1.1***	***1.5***	***1.7***	***0.5***	***V***
Countries	Germany	27.1	12.8	17.5	30.6	37.9	38.3	40.9	38.4	1
	Hong Kong SAR of China	55.2	45.5	52.4	32.8	35.8	31.6	31.0	30.9	2
	France	1.3	1.4	1.7	14.1	16.5	14.8	15.9	14.9	3
	United Kingdom of Great Britain and Northern Ireland	5.3	7.6	7.1	11.2	10.5	10.3	10.7	10.9	4
	Czechia	0.0	3.7	5.7	8.0	9.5	6.9	8.9	9.6	5
	Japan	14.9	8.3	3.3	5.2	5.1	4.8	6.6	6.8	6
	Democratic People's Republic of Korea	0.0	0.0	0.0	0.0	0.0	0.0	0.0	6.3	7
	Belgium	0.9	1.5	0.7	5.9	5.9	6.5	7.1	6.0	8
	Spain	3.4	5.1	4.4	5.5	6.7	10.5	6.8	6.0	9
	Denmark	2.7	2.6	4.8	7.0	6.9	6.7	6.4	5.8	10
	Netherlands	1.9	0.8	1.0	1.8	1.5	2.6	3.4	2.9	11
	Austria	2.8	2.0	3.5	4.8	3.5	3.3	2.8	2.9	12
	Romania	0.0	0.0	0.1	1.4	1.7	1.7	2.7	2.8	13
	Tajikistan	0.0	0.0	0.0	3.0	0.7	5.9	2.2	2.7	14
	Lao People's Democratic Republic	0.0	0.0	0.0	1.1	1.1	0.5	1.1	2.3	15

*Roman numerals represent continent rank and Arabian numerals, country rank

The AGR of import of duck meat between 2010 and 2019 in Europe (8.7%) and Americas (3.8%) was higher than the world's average of 1.9%. The negative growth rates were registered in other continents, namely Africa (−6.3%), Oceania (−3.2%), and Asia (−3.1%). Among the major duck meat importing countries, the highest AGR was noticed in France (86.3%) followed by Belgium (84.1%), Germany (13.2%), Japan (11.8%), and Czechia (7.6%). Hong Kong registered a negative annual growth of −4.6%. The time trend analysis of annual import of duck meat from 2010 to 2019 revealed that all the continents registered unsteady change; however, the increase was significant ($P < 0.01$) in Europe. The world's overall duck meat import increased significantly ($P < 0.05$) but with poor consistency. Among the major duck meat importing countries, Germany had somewhat steady ($R^2 = 0.81$) and significant ($P < 0.05$) increase in duck meat import during the period. Although the increase was significant ($P < 0.01$) in France, all major duck importing countries like UK and Czechia including France had unsteady growth during this period. Hong Kong and Japan witnessed unsteady negative growth; of which, the decline in Japan was significant ($P < 0.01$).

1.23 Overseas Trade of Duck (Other) Eggs

1.23.1 Export

The quantity and value of export of 20 top countries in the world and continents are given in Table 1.15. The export of other eggs in the world, in which duck eggs form major component had increased over 5.8 fold from 53.6 million kilograms in the year 2010 to reach a figure of 311.5 million kilograms in 2019 (AGR = 53.5%). The major players in world's total export of other poultry eggs in 2019 were Europe (52%) and Asia (45%). The important nations that shared major portion of world's other egg export in 2019 were Jordan (29%), Belgium (17%), and Netherlands (16%). Out of total 459.3 million US$ involved in export of other eggs in the year 2019, Europe spent 58% and Asia 22%.

1.23.2 Import

The value of import market of other eggs was sized at 263.6 million kilograms worth 372.5 million US dollars. The major players in world's import of other poultry eggs in 2019 were Europe (83.1%) and Asia (14.6%); both continents together shared 97.7% of the total import market of other eggs. Between 2010 and 2019, the other eggs import of the world was grown at an AGR of 55.4%. Among continents, the import of other eggs in Europe (129%), Americas (48.8%), and Oceania (74.0%) increased at very high AGR; among these, the growth rate in Americas and Europe was significant ($P < 0.05$). Of the world's total other eggs import market, Belgium alone imported (153.1 million kilograms; 58.1%) in the year 2019. The other major

Table 1.15 Quantity and value of other eggs export in continents and top 20 countries in the year 2019 (FAOSTAT 2021)

Area	Continent/country	Million kg	Million US$
World		**311.5**	**459.3**
Continents	***Africa***	***5.1 (IV)***	***13.8 (IV)***
	Americas	***6.7 (III)***	***71.2 (III)***
	Asia	***138.8 (II)***	***99.7 (II)***
	Europe	***160.5 (I)***	***268.5 (I)***
	Oceania	***0.4 (V)***	***6.1 (V)***
Countries	Jordan	90.8 (1)	18.8 (6)
	Belgium	52.6 (2)	3.7 (17)
	Netherlands	48.7 (3)	58.0 (2)
	China, mainland	22.9 (4)	56.9 (3)
	France	12.3 (5)	90.7 (1)
	Sweden	10.5 (6)	10.7 (9)
	Portugal	8.8 (7)	12.4 (8)
	Kazakhstan	7.8 (8)	5.9 (13)
	Malaysia	6.1 (9)	2.2 (19)
	Pakistan	6.1 (10)	7.7 (12)
	United States of America	4.7 (11)	19.5 (5)
	Spain	4.5 (12)	8.2 (11)
	Denmark	4.2 (13)	13.1 (7)
	Bulgaria	4.1 (14)	4.6 (15)
	Ukraine	2.9 (15)	2.2 (20)
	Czechia	2.5 (16)	4.0 (16)
	Saudi Arabia	2.4 (17)	3.1 (18)
	Morocco	1.8 (18)	10.1 (10)
	Poland	1.8 (19)	5.7 (14)
	Canada	1.6 (20)	51.5 (4)

Figures in the parentheses indicate the world rank. Roman numerals represent continent rank and Arabian numerals, country rank

buying countries of other eggs in the world market were Russian Federation (4.6%), Spain (4.2%), Saudi Arabia (3.7%), and Netherlands (3.6%).

1.24 Interventions Needed

The nomadic duck farming is at crossroads because of recurrent outbreaks of disease and plight of farmers due to changing lifestyle, shrinking grazing land due to draught and urbanization, and less remuneration. The traditional nomadic duck production system needs location-specific technological interventions to refine the existing practices to check vagaries like avian influenza, duck plague, and duck cholera that need to be worked out to support this farming system. It is imperative to make available the vaccines at village level and awareness creation on method and importance of vaccination among people to whom this system of farming provides livelihood. Government initiatives for organized technological training on recent developments in duck production practices along with the assurance of availability of quality ducklings are the need of the hour.

In traditional farming system, the wisdom of farmers like those in India (Jalaludeen 2005; Gajendran and Karthickeyan 2011) of lean season feeding with coryphapalm pith, application of indigenous technical knowledge in brooding with no artificial heating, beak branding for identification, ethno-veterinary practices in herd health management, and other regions of world need to be documented and preserved.

Countries with limited land area require a system to treat duck wastewater before disposal. It is one of the bottlenecks in the development of intensive duck industry. The increasing density of duck rearing and the awakening of environmental protection hinder the traditional duck raising. The wire-floor or slatted-floor housing of ducks provides a possible way to control the disposal of wastewater. Chinese duck industry is in the stage of transformation and upgrading, traditional raising models in waters are being eliminated gradually, on the contrary, raising models such as indoor mesh bed model and thick cushion grass model, which effectively protect waters and efficiently reduce emission have been developed into main raising models rapidly. Meanwhile, some technical advances for ducks under intensive feeding system such as complete-pellet feeds or the improvement of artificial insemination for cross breeding will certainly raise the efficiency of production.

The adaptability of local breeds to the hardy environment, which includes the ability to use local feeds and disease resistance, is precious not only to Asia but also to the whole world. Duck genetic resources face extinction in many parts of the world due to commercialization of duck production with imports of genetically improved Pekin ducks from commercial breeding companies. The duck germplasm conservation needs to be given priority for those local ducks facing extinction. A balance of exotic and indigenous breeds needs to be maintained in Asian countries. Policies or strategies that suit the local circumstances are essential in order to protect the domestic duck industry.

The skyrocketing of price of feed ingredients is another limiting factor in duck production in many developing countries. Feeding of non-conventional raw materials and by-products, such as rice bran, copra meal, peanut meal, palm kernel, canola meal, animal offal, etc., need to be given adequate consideration wherever these feed resources are in abundance. The duck producers need to be more mindful of the effectiveness of nutritional modification, which can manipulate output within strain and produce a variety of products to satisfy specific market. At the same time, it must be noted that significant differences in response to dietary nutrient profile exist between different ducks, and thus evaluation of specific yield and composition response in each population requires further attention. Value addition of duck meat by feed manipulations is another area for future research in duck nutrition. Despite increasing popularity of duck meat in restaurants and families worldwide, preferences with regard to the duck composition suitable for method of preparation should be considered. Western countries predominantly use the duck breast meat and prefer minimal fat content, the famous Peking roast duck; conversely, others prefer higher subcutaneous fat content to create the crispy fleshy texture. Yet other Chinese cuisines demand otherwise; for instance, the equally popular Nanjing roast salted duck utilizes smaller ducks (approximately 2 kg) with more lean meat. Growth performance and carcass traits can be manipulated through modifying dietary protein and energy in Pekin ducks even within strain and genetic lines as reviewed by Chen and Applegate (2016). Subsequent modeling of amino acid and energy inputs and corresponding outputs of growth and carcass characteristics warrants further attention.

Currently, *foie gras* can only be labeled as such if the liver is at least 300 g for ducks and 400 g for geese, a size that is commonly reached by overfeeding but considered a welfare problem by many. The practice of producing smaller and less fatty livers from ducks and geese without force-feeding is becoming popular. These are offered as a substitute for *foie gras* (such as "*Faux Gras*"), sometimes called "ethical *foie gras*" or "humane *foie gras*." "Self-gorging" *foie gras* produced in some European countries like Spain comes from geese kept in very extensive free-range conditions. The geese have enlarged livers but only about half the minimum size of conventional goose *foie gras*. This is a premium product, costing several times more than conventional *foie gras*.

Comprehensive, integrated national surveillance and diagnostic programs need to be implemented for effective control or eradication of avian influenza from duck populations. Enhanced biosecurity should be practiced at all levels of production and processing by all employees of companies, diagnostic laboratories dealing with this disease. Education of poultry farmers and other workers about AI control and sharing of information on surveillance and control strategies at all levels in the production process should be conducted. Accepted OIE sanitary standards, movement restrictions by importing countries as legitimate measures to prevent the importation of HPAI need to be implemented. However, any restriction imposed by an importing country should be justified by a science-based risk analysis shared between the trading partners.

In conclusion, as Pingel (2009) recommended some measurements are necessary to support waterfowl farms to achieve higher productivity. They are providing ducklings and goslings with improved genotypes from parent-stock farms, offering concentrate as additional feed for better utilization of natural feed resources and to ensure a balanced nutrition to achieve higher productivity, improving management especially for protecting ducklings and goslings in the first weeks of life by providing an additional heat source as well as drinking water and protein rich feed, using of veterinary service, vaccination programs, and disease control and improving education, training, and extension by radio programs and demonstration farms.

References

Aggrey SE, Karnuah AS, Bram S, Anthony NB (2010) Genetic properties of feed efficiency parameters in meat-type chickens. Genet Sel Evol 42:*Article # 25*

Anonymous (2016) Duck industry performance report. ISSN: 2012-0664. Livestock and Poultry Statistics Division, Livestock and Poultry Statistics Division, Quezon City, Philippines

Baéza E, Huang JF (this volume) Nutritive value of duck meat and eggs. In: Jalaludeen A, Churchil RR, Baéza E (eds) Duck production and management strategies. Springer, Cham

Baéza E, Salichon MR, Marché G, Wacrenier N, Dominguez B, Culioli J (2000) Effects of age and sex on the structural, chemical and technological characteristics of mule duck meat. Br Poult Sci 41(3):300–307. https://doi.org/10.1080/713654934

Bleich EG, Rhissa Z, Mack S (2005) The FAO special programme for food security: livestock diversification – a case study in Chad. Worlds Poult Sci J 61:23–29

Bley TAG (2005) Untersuchungenzur Variation und Rhythmik der individuellenFutteraufnahmebeiPekingenten in Gruppenhaltung. Arch Geflügelk 69(1):46–47

Bondoc OL (2008) Animal breeding: principles and practice in the Philippine context. University of the Philippines, Press Diliman, Quezon City, Metro Manila, Philippines. http://www.nast.ph/images/pdf%20files/Publications/Outstanding-Awardees%20BOOKS/2010/Animal%20Breeding%20and%20Practice%20in%20the%20Philippine%20Context.pdf. Accessed 3 Apr 2018

Brogi A, Pernollet CA, Gauthier-Clerc M, Guillemain M (2015) Waterfowl foraging in winter-flooded rice fields: any agronomic benefits for farmers? Ambio 44:793–802. https://doi.org/10.1007/s13280-015-0678-0

Carlen E, Lansfors B (2002) Duck production in the Mekong Delta of Vietnam. A survey with emphasis on the development and the sustainability. Minor field studies no. 209. Swedish University of Agricultural Sciences, International Office, Uppsala

CDC (2017) Influenza type A viruses. Centers for Disease Control and Prevention, National Center for Immunization and Respiratory Diseases (NCIRD), US Department of Health and Human Services. https://www.cdc.gov/flu/avianflu/influenza-a-virus-subtypes.htm. Accessed 20 May 2021

Chan PKS (2002) Outbreak of Avian Influenza A (H5N1) Virus Infection in Hong Kong in 1997. Clin Infect Dis 34(Suppl 2):S58–S64. https://doi.org/10.1086/338820

Chang H, Dagaas CT (2004) Working the Philippine duck industry: issues and research needs. In: Paper series in agricultural and resource economics. ISSN 1442 1909. University of New England Graduate School of Agricultural and Resource Economics & School of Economics

Chen X, Applegate TJ (2016) Meat duck nutrition- formulation considerations across genetic and feedstuff resources. In: Proceedings of XXV world's poultry congress- invited papers, 5--9 September 2016, Beijing, China, pp 213–216

Cheng YS, Rouvier R, Poivey JP, Tai C (1995) Genetic parameters of body weight, egg production and shell quality traits in the laying Brown Tsaiya duck. Genet Sel Evol 27:459–472

Cheng Y, Rouvier R, Liu H, Huang S, Huang Y, Liao C, Tai J, Tai C, Poivey J (2009) Eleven generations of selection for the duration of fertility in the intergeneric crossbreeding of ducks. Genet Sel Evol 41:32. https://doi.org/10.1186/1297-9686-41-32

Cheng-Fang L, Cou-gui C, Jin-ping W, Ming Z, Wei-ling Y, Ahmad S (2008) Nitrogen losses from integrated rice-duck and rice-fish ecosystems in Southern China. Plant Soil 307:207–217

Cherry P, Morris TR (2008) Domestic duck production: science and practice. CAB International, Wallingford

Churchil RR, Jalaludeen A (this volume) Duck farming: opportunities, constraints and policy recommendations. In: Jalaludeen A, Churchil RR, Baéza E (eds) Duck production and management strategies. Springer, Cham

Crawford RD (1990) Origin and history of poultry species. In: Crawford RD (ed) Poultry breeding and genetics. Elsevier Amsterdam, Science Publishers BV, Amsterdam, pp 1–41

Duan L, Campitelli L, Fan XH, Leung YHC, Vijaykrishna D, Zhang JX, Donatelli I, Delogu M, Li KS, Foni E, Chiapponi C, Wu WL, Kai H, Webster RG, Shortridge KF, Peiris JS, Gavin M,

Smith JD, Chen H, Guan Y (2007) Characterization of low-pathogenic H5 subtype influenza viruses from Eurasia: implications for the origin of highly pathogenic H5N1 viruses. J Virol 81(14):7529–7539. https://doi.org/10.1128/JVI.00327-07
FAO (2009a) Preparing for highly pathogenic avian influenza (Martin V, Forman A, Lubroth J (eds)). Food and Agriculture Organization of the United Nations. ISBN 978-92-5-106249-4. http://www.fao.org/3/i0808e/i0808e.pdf. Accessed 15 May 2015
FAO (2009b) Investigation of duck production and hatcheries, and duckling supply in Cambodia. Prepared by Khieu Borin, Pok Samkol and Olaf Thieme. AHBL – promoting strategies for prevention and control of HPAI, Rome
FAO (2010) Duck farming systems and avian influenza in the Mekong delta of Vietnam, by Bui Xuan Men. FAO smallholder poultry production paper no. 1, Rome
FAO (2014) Avian influenza A (H5N8) detected in Europe – a journey to the West? Agriculture and Consumer Protection Department, Animal Production and Health, Food and Agriculture Organization of the United Nations. http://www.fao.org/ag/againfo/home/en/news_archive/2014_A-H5N8_detected_in_Europe.html. Accessed 15 May 2021
FAOSTAT (2021) Food and Agriculture Organization Corporate Statistical Database. Food and Agriculture Organization of the United Nations, Rome, Italy. http://www.fao.org/faostat/en/#data. Accessed 09 May 2021
Farghly MFA, Mahrose KM, Ullah Z, Rehman ZU, Ding C (2017) Influence of swimming time in alleviating the deleterious effects of hot summer on growing Muscovy duck performance. Poult Sci 96:3912–3919
Fidler DP (2004) Global outbreak of avian influenza A (H5N1) and international law. American Society of International Law Insights 8(1). https://www.asil.org/insights/volume/8/issue/1/global-outbreak-avian-influenza-h5n1-and-international-law. Accessed 15 May 2021
Fraley SM, Fraley GS, Karcher DM, Makagon MM, Lilburn MS (2013) Influence of plastic slatted floors compared with pine shaving litter on Pekin duck condition during the summer months. Poult Sci 92:1706–1711
Gajendran K, Karthickeyan SMK (2009) Status of indigenous ducks in Tamil Nadu, India: a traditional perspective. Livest Res Rural Dev 21:Article # 175 .http://www.lrrd.org/lrrd21/10/gaje21175.htm. Accessed 21 June 2017
Gajendran K, Karthickeyan SMK (2011) Indigenous Technical Knowledge in duck production in Tamil Nadu, Indian. J Tradit Knowl 10:307–310
Gajendran K, Veeramani P (this volume) Intensive duck rearing. In: Jalaludeen A, Churchil RR, Baéza E (eds) Duck production and management strategies. Springer, Cham
Gueye EF (2009) The role of networks in information to family poultry farmers. Worlds Poult Sci J 65:115–124
Herd RM, Arthur PF (2009) Physiological basis for residual feed intake. J Anim Sci 87:64–71
Hoan ND (2014) Overall assessment of domestic ducks in Vietnam. In: Proceedings of XIV European poultry conference Stavanger, Norway, 23–27 June 2014. http://www.wpsa.com/index.php/publications/wpsaproceedings/2014/xivth-european-poultry-conference/2023-overall-assessment-of-domestic-ducks-in-vietnam/file. Accessed 23 Mar 2018
Hou S (2016) Duck production in China. In: Proceedings of XXV world's poultry congress- invited papers, 5–9 September 2016, Beijing, China, pp 209–212
Howie JA, Tolkamp BJ, Avendano S, Kyriazakis I (2009) The structure of feeding behavior in commercial broiler lines selected for different growth rates. Poult Sci 88:1143–1150
Islam MS (2015) Productive performance of different duck genotypes reared at Regional Duck Breeding Farm, Sonagazi, Feni. Report presented in partial fulfillment of the requirement for the Degree (DVM) (Doctor of Veterinary Medicine) at Chittagong Veterinary and Animal Sciences University. http://101.2.160.165:8080/bitstream/handle/123456789/694/Production%20Report%20Saiful%20final.pdf?sequence=1. Accessed 24 Mar 2018
Islam R, Mahanta JD, Barua N, Zaman G (2002) Duck farming in North-Eastern India (Assam). Worlds Poult Sci J 8:567–572

Jalaludeen A (2005) Duck rearing – a promising enterprise in the changed global scenario. In: Proceedings of annual conference and national symposium of Indian Poultry Science Association (IPSACON-2005), 2–4 February 2005, Hyderabad, India

Jalaludeen A, Churchil RR (2006) Duck eggs and their nutritive value. Poult Line, October 2006, pp 36–39

Jalaludeen A, Churchil RR, Joseph L, Anitha P (2009) Duck meat egg and their products. In: Souvenir of IV world waterfowl conference, Thrissur, Kerala, India, pp 57–64

Jana BB, Jana S (2003) The potential and sustainability of aquaculture in India. J Appl Aquac 13: 283–316

Jones TA, Dawkins MS (2010) Environment and management factors affecting Pekin duck production and welfare on commercial farms in the UK. Br Poult Sci 51:12–21. https://doi.org/10.1080/00071660903421159

Karcher DM, Makagon MM, Fraley GS, Fraley SM, Lilburn MS (2013) Influence of raised plastic floors compared with pine shaving litter on environment and Pekin duck condition. Poult Sci 92: 583–590

Khumairoh U, Groot JCJ, Lantinga EA (2012) Complex agro-ecosystems for food security in a changing climate. Ecol Evol 2:1696–1704

Ku A, Chan L (1999) The first case of H5N1 avian influenza infection in a human with complications of adult respiratory distress syndrome and Reye's syndrome. J Paediatr Child Health 35(2):207–209. https://doi.org/10.1046/j.1440-1754.1999.t01-1-00329.x

Lai S, Qin Y, Cowling BJ, Ren X, Wardrop NA, Gilbert M, Tsang TK, Wu P, Feng L, Jiang H, Peng Z, Zheng J, Liao Q, Li S, Horby PW, Farrar JJ, Gao GF, Tatem AJ, Yu H (2016) Global epidemiology of avian influenza A H5N1 virus infection in humans, 1997-2015: a systematic review of individual case data. Lancet Infect Dis 16(7):e108–e118. https://doi.org/10.1016/S1473-3099(16)00153-5

Lin RL, Chen HP, Rouvier R, Marie-Etancelin C (2016) Genetic parameters of body weight, egg production, and shell quality traits in the Shan Ma laying duck (*Anas platyrhynchos*). Poult Sci 95:2514–2519

Liu MHC, Churchil RR (this volume) Duck genetics and breeding. In: Jalaludeen A, Churchil RR, Baéza E (eds) Duck production and management strategies. Springer, Cham

Liu HC, Huang JF, Lee SR, Liu HL, Hsieh CH, Huang CW, Huang MC, Tai C, Poivey JP, Rouvier R, Cheng YS (2015) Selection for duration of fertility and mule duck white plumage colour in a synthetic strain of ducks (*Anas platyrhynchos*). Asian Australas J Anim Sci 28(5):605–611

Ma C, Zhao Q (1998) China's waterfowl production takes off. World Poult 14:40–41

Marie-Etancelin C, Chapuis H, Brun J, Larzul C, Mialon-Richard MM, Rouvier R (2008) Genetics and selection of mule ducks in France: a review. Worlds Poult Sci J 64:187–208

Mazanowski A, Kisiel T, Gornowicz E, Mazanowski A, Kisiel T, Gornowicz E (2003) Carcass quality, meat traits and chemical composition of meat in ducks of paternal strains A44 and A55. Anim Sci Paper Rep 4:251–263

MMWR (2004) Outbreaks of avian influenza A (H5N1) in Asia and interim recommendations for evaluation and reporting of suspected cases – United States, 2004. Morb Mortal Wkly Rep 53(05):97–100. https://www.cdc.gov/mmwr/preview/mmwrhtml/mm5305a1.htm. Accessed 15 May 2021

OIE (2019) Chapter 10.4. Infection with avian influenza viruses. Terrestrial Animal Health Code, World Organisation for Animal Health, Paris, France. https://www.oie.int/en/what-we-do/standards/codes-and-manuals/terrestrial-code-online-access/?id=169&L=1&htmfile=chapitre_avian_influenza_viruses.htm. Accessed 20 May 2021

PAHO (2004) Avian influenza virus A (H10N7) circulating among humans in Egypt. EID Weekly Updates 2:2. Pan American Health Organization. https://www.paho.org/en/documents/avian-influenza-virus-h10n7-circulating-among-humans-egypt-vol-2-no-18-7-may-2004. Accessed 20 May 2021

Pingel H (1990) Genetic of the growth and meat production in waterfowl. In: Crawford (ed) Poultry breeding and genetics. Developments in animal and veterinary sciences, 22. Elsevier, pp 691–704

Pingel H (2009) Waterfowl production for food security. In: Proceedings of IV world waterfowl conference, 11–13 November 2009, Thrissur, India, pp 3–15

Pingel H (2011) Results of selection for breast muscle percentage and feed conversion ratio in Pekin ducks. Biotechnol Anim Husb 27:769–776

Pulit-Penaloza JA, Brock N, Pappas C, Sun X, Belser JA, Zeng H, Tumpey TM, Maines TR (2020) Characterization of highly pathogenic avian influenza H5Nx viruses in the ferret model. Sci Rep 10:12700. https://doi.org/10.1038/s41598-020-69535-5

Rahman SM, Fouzder KS, Sarkar NR (2017) Nomadic herded duck production as a livelihood tool in some selected areas of Bangladesh: a case study. IOSR J Agric Vet Sci 10:16–20

Sankaralingam S, Mahanta JD (this volume) Nomadic (transhumant) duck farming practices. In: Jalaludeen A, Churchil RR, Baéza E (eds) Duck production and management strategies. Springer, Cham

Sapcota D, Begum K (this volume) Integrated duck farming. In: Jalaludeen A, Churchil RR, Baéza E (eds) Duck production and management strategies. Springer, Cham

Schmitz H (2016) The sustainable and humane practices of the down and feather industry: a global assessment of industry statistics and practices. The International Down and Feather Bureau. https://downandfeathercouncil.com/wp-content/uploads/2018/06/The-Sustainable-and-Humane-Practices-of-the-Down-and-Feather-Industry.pdf. Accessed 17 May 2021

Shalev BA (1995) Comparative growth and efficiency of various avian species. In: Hunton P (ed) World animal science, production system approach, poultry production (C9). Elsevier Science BV, Amsterdam, pp 53–78

Sonaiya EB, Swan SEJ (2004) Species and breeds. In: Small-scale poultry production. FAO animal production and health 1. Food and Agriculture Organization of the United Nations, Rome

Su CH (this volume) Breeds of domestic ducks. In: Jalaludeen A, Churchil RR, Baéza E (eds) Duck production and management strategies. Springer, Cham

Tai C, Rouvier R, Poivey JP (1989) Genetic parameters of some growth and egg production traits in laying Brown Tsaiya (*Anas platyrynchos*). Genet Sel Evol 21(3):377–384

Tai C, Wang CT, Huang CC (1999) Production systems and economic characters in waterfowl. In: Proceedings of I world waterfowl conference, Taichung, 1–4 December 1999, Taiwan, pp 19–31

Thiele HH (1995) Recent tendencies in duck breeding. In: Proceedings of 10th European symposium on waterfowl, 26–31 March 1995, Halle (Saale), Germany, pp 421–428

Thiele HH (2016) Breeding Pekin ducks for meat production. Lohmann Information, vol 5028–33

Wawro K, Wilkiewicz-Wawro E, Kleczek K, Brzozowski W (2004) Slaughter value and meat quality of Muscovy ducks, Pekin ducks and their crossbreeds, and evaluation of the heterosis effect. Arch Anim Breed 47:287–299. https://doi.org/10.5194/aab-47-287-2004

WHO (2014) First European outbreak of avian influenza A (H5N8) confirmed in poultry. World Health Organization – Regional Office for Europe, Copenhagen, Denmark. https://www.euro.who.int/en/health-topics/communicable-diseases/influenza/news/news/2014/11/first-european-outbreak-of-avian-influenza-ah5n8-confirmed-in-poultry#:~:text=Germany%2C%20the%20Netherlands%20and%20the,avian%20influenza%20A(H5N8). Accessed 16 May 2021

WHO (2016) Influenza A (H5N8) virus detected in birds in several countries in the WHO European region. World Health Organization – Regional Office for Europe, Copenhagen, Denmark. https://www.euro.who.int/en/countries/austria/news/news/2016/12/influenza-ah5n8-virus-detected-in-birds-in-several-countries-in-the-who-european-region. Accessed 16 May 2021

WHO (2020) Cumulative number of confirmed human cases for avian influenza A (H5N1) 2003–2020. World Health Organization Reports. https://www.who.int/influenza/human_animal_interface/2020_DEC_tableH5N1.pdf?ua=1. Accessed 15 May 2021

Widiyaningrum P, Lisdiana L, Utami NR (2016) Egg production and hatchability of local ducks under semi intensive vs extensive managements. J Indones Trop Anim Agric 41:77–82

Dr. A. Jalaludeen, PhD (Poultry Science)

Dr. A. Jalaludeen hails from Thiruvananthapuram District, Kerala, India. He earned his graduation from College of Veterinary and Animal Sciences (COVAS), Kerala Agricultural University (KAU) and post-graduation in the discipline of Poultry Science from the same institute. He pursued his doctoral degree in Poultry Science with specialization in Poultry Nutrition. He served as teacher, researcher, and extension worker in Kerala Agricultural University and Kerala Veterinary and Animal Sciences University (KVASU) for more than 36 years. He functioned as Professor and Head of Dept. of Poultry Science and Director of Centre for Advanced Studies in Poultry Science, Kerala Agricultural University for more than 10 years. He also functioned as Dean, College of Veterinary and Animal Sciences, Pookode and later as Director of Academics and Research of Kerala Veterinary and Animal Sciences University, Wayanad, Kerala, India. He was made Special officer in KVASU and was instrumental in establishing a separate faculty of Poultry Science which is the first of its kind in India. He guided many research students both for post-graduation and doctoral degree programs. He has published many peer-reviewed research articles in journals of high repute.

Dr. R. Richard Churchil, PhD (Poultry Science)

Dr. R. Richard Churchil hails from Kanyakumari district, Tamil Nadu, India. He earned his graduation from Madras Veterinary College, doctoral degree from Indian Veterinary Research Institute, many accolades including gold medals in post-graduation, and Jawaharlal Nehru Award of ICAR for his doctoral research on development of transgenic spermatozoa. He worked as poultry breeding scientist in Kerala state for a decade before joining TANUVAS. He was trained on "transgenic animals" at University of Nantes, France. He has published 89 peer-reviewed research papers and seven book chapters. He teaches Poultry Science to undergraduate and Poultry Breeding and Biotechnology to postgraduate/doctoral students in TANUVAS.

Breeds of Domestic Ducks 2

C. H. Su

Contents

Abstract

The ducks were kept domesticated in southern China as early during Han dynasty. This later gave rise to domesticated varieties in Southeast and East Asia, from where ducks moved to the eastern Mediterranean. Pekin is the finest dual-purpose breed evolved from China which is utilized for egg and meat production all over the world. China is also the homeland for dual-purpose breeds like Gaoyou and Linwu. Further Chinese breeds like Shan Ma, Shaoxing, Jingding, and Liancheng excel in egg production. Brown Tsaiya of Taiwan and Indian Runner of East Indies are other two elite germplasm that are capable of laying as many as 330 eggs or more in a laying year. England to its credit has Aylesbury for meat purpose, Campbells for egg purpose, and Orpington for dual

C. H. Su (✉)
Ilan Branch, Livestock Research Institute, Council of Agriculture, Executive Yuan, Wuchieh Township, Yilan County, Taiwan
e-mail: chsu@mail.tlri.gov.tw

A. Jalaludeen et al. (eds.), *Duck Production and Management Strategies*,
https://doi.org/10.1007/978-981-16-6100-6_2

purpose. Rouen of France and Swedish of Germany have the European origins, whereas Cayuga, Crested duck, and East Indie have American lineage. Mandarin is an ornamental breed of Japan. Apart from this, there are many local ecotypes and lesser-known breeds providing livelihood to the local farmers.

Keywords

Duck · Breeds

List of Symbols/Abbreviations

%	Percent
G	Gram
Kg	Kilogram
USA	The United States of America
Viz	Videlicet

2.1 Introduction

The livestock and poultry have always made a significant contribution to the livelihood of human societies by providing food, fertilizer, and other products and services. They are renewable resources, which utilize another renewable resource like plants to produce these products and services. In addition, the manure produced by the animals helps to improve soil fertility and, thus, aids the plants. The genetic isolation and natural adaptation to the environment over centuries lead to refinement of the germplasm to evolve as a breed that is fit to survive the local conditions. A breed is a specific group of domestic animals, which are phenotypically and behaviorally similar. The human need and environmental conditions have resulted in the evolution of location specific breeds of livestock and poultry.

The domestic ducks evolved from a common ancestor called Mallard (*Anas platyrhynchos*) duck which is found over the Europe, North Africa, and North America. The domestic breeds of ducks, other than the Pato (*Cairina moschata*), which is not a true duck, are lineal descendants of the wild common mallard, and all deviations there from are the results of domestication or natural mutations. Muscovy duck (*Cairina moschata*) which is originated from South America is the ancestor for all Muscovies found all over the world. It is generally agreed that Southern China is the origin for ducks. From China, domesticated varieties spread to the Southeast and East Asia: Thailand and Indochina; Malaya; Celebes and the Philippines; Java and Sumatra. China is the homeland for many modern day elite duck breeds including Pekin. Tsaiya from Taiwan, Runner duck from East Indies, Campbell from America, and Aylesbury and Orpington from England are the other notable breeds. Including this, many other local ecotypes/breeds contribute to the duck egg, meat, and feather production worldwide. It is estimated that 1124 million heads of ducks produce 98.6 billion eggs and 4.5 million tonnes of meat per annum.

Muscovies are used in terminal cross with Tsaiya and Pekin as a three-way crossing to produce mule ducks, which are used mainly for meat production in Taiwan. Muscovy ducks are also used in France for meat production and to produce mule ducks dedicated to the production of fatty liver. Ducks' plumage color and biochemical and morphological traits were influenced during domestication. In the early twentieth century, the breeding of ducks was practiced only for standards of type, size, and plumage color for exhibition. The breeding of ducks for meat and eggs was started towards the end of the twentieth century. The genetic improvement programs has led to the development of meat ducks with the capacity of attaining 3.7 kg in 7 weeks and the layer ducks laying more than 330 eggs in 1 year.

Nowadays, almost all native domestic animal and poultry populations are facing the threat of extinction in many countries. Because producers want to maintain high profitability and opting for few high producers resulting in the irreversible loss of diversity of duck populations. The naturally evolved pure breeds of domesticated birds are probably on their way to genetic monoculture.

2.2 Evolution of Ducks

All goose and duck species belong to the order *Anseriformes* and family *Anatidae*. Morphologically, duck as a class is distinguished from other *Anatidae* by their smaller sizes, shorter necks, flatter bodies, shorter legs, and broader bills, in contrast to geese and swans (Grow 1985). The history of domestication of Mallard ducks by human could be traced to 1500–1000 years BC. Muscovies have been domesticated in Central and South America during the sixteenth century before the Spanish invasion of South America (Huang et al. 2012).

Regardless of all speculative claims to the contrary, the preponderance of reliable evidence indicates that domestic breeds of ducks, other than the Pato (*Cairina moschata*), which is not a true duck, are lineal descendants of the wild common mallard. All deviations then are the results of domestication or natural mutations, in as much as the Mallard is the only breed of feral true ducks in which the males are adorned with the so-called sex feathers at the upper base of the tail. In as much as these curled sex feathers disappear, or at least become vestigial, when the Mallard is crossed with other wild breeds and in view of the further fact that all drakes of domesticated true ducks disclose these unique sex feathers, the conclusion is unavoidable that the common domestic breeds are exclusive descendants of the common Mallard (Grow 1985).

2.2.1 Mallard

Mallard is usually thought to be the progenitor of most modern domestic breeds of ducks with the exception of the Muscovy. This species frequents most countries. They are also of interest to bird watchers and hunters in the wild. The shiny green head of the slate-gray drake is legendary. It has earned this fowl the nickname of greenhead duck (Rick and Gail 1978). Mallard is decidedly dimorphic in plumage patterns, it means the male ones show far greater brilliancy than the female (Grow 1985).

Mallards easily acclimatize to domestication and after about four to five generations of unlimited feeding, they lose their desire and ability to fly owing to the dramatically increased body size (Holderread 1978). Because all breeds of domestic duck except the Muscovy were developed from it and consequently bred for recessive factors. The Mallard when crossed with any breed will produce birds, which approximate the "gray" or Mallard pattern (Sheraw 1975).

Mallards have impressive natural instincts to hatch and breed their offspring if not accosted repeatedly while nesting. Mothers can raise their broods with little supplementary feed when the supply of natural foods is insufficient (Holderread 1978) (Fig. 2.1).

Fig. 2.1 Mallard drake (above) and duck (below)

2.3 Domestication

Man has domesticated lots of animals and plants, though the former greatly outnumber the latter, often have well developed social tendencies, and stand closer to man in an evolutionary sense. This comparative lack of success reflects not so much less conscious effort on the part of man (arguably a relatively late feature of the domestication process), as fundamental differences between animals and plants. Animals are mobile and, more importantly, possess a variety of psychological characteristics. Animal domestication is a more complex and a more obviously bilateral operation, and favorable circumstances occur more rarely. The kinds of relationship that can exist between man and animals are more varied that between man and plants, the latter largely economic (Donkin 1989). Muscovy duck (*Cairina moschata*) and common duck (*Anas platyrhynchos*) are the two genera of ducks, which were domesticated for commercial exploitation (Crawford 1990).

The domestication of ducks has attracted relatively little scholarly attention. Domesticated ducks are only rarely of major economic importance. Their non-utilitarian value, too, is usually overshadowed by that of other animal species. The origin may be approximately the same as those of geese. It is generally agreed that southern China was important; ducks were kept under the Earlier Han dynasty, and the Chinese became "perfect masters in the management of ducks." There is less support for two other possible centers, Lower Mesopotamia and East-central Europe. From China, domesticated varieties spread to much of other parts of Asia. India may have obtained domestic ducks from both the east and the west, but, apart from Kashmir, they appear always to have been more important in the South and around the Bay of Bengal (including Burma), suggesting a principal Chinese influence (Donkin 1989).

The domestication of birds (and of animals generally) may be broadly correlated with regions of early cultivation, from Southeast Asia to the eastern Mediterranean. The sub-tropical and warm-temperature lands of Asia contributed most, Africa and the middle and high latitudes of Eurasia comparatively little. In the New World, too, animal domestication was associated with ancient farming communities, within a nuclear zone stretching from central Mexico to the central Andes (Donkin 1989).

After ducks have been domesticated and became one of the important animal protein source, the pedigree farms were established to practice selection for body weight or egg number improvement based on individual records. Some multinational commercial breeding companies had been established and developed rapidly. Small breeding farms focused on the specific local breeds. Comprehensive reviews of breeding for growth, meat production, egg production, and reproduction have been presented by Pingel (1990a, b), Rouvier (1999), and Huang et al. (2012).

The phenomenal genetic improvement has been achieved in growth performance of ducks. The selection target of rapid growth in Pekin ducks has made their body weight at 49 days of age increasing from about 1.7 to nearly 4.5 kg. In 1928, the mean live weight was only 2.7 kg at the age of 75 days. The same body weight was achieved in 35 days due to genetic improvement in 1993. Over the period from 1928 to 1993, there was a reduction of about 27 g of feed per year to produce 1 kg body weight (Cherry and Morris 2008). In 2012, Pekin ducks were generally slaughtered at 6 to 8 weeks of age when they attain 3.3 to 3.7 kg body weight (Huang et al. 2012).

2.3.1 U.K. Standardized Duck Breeds

2.3.1.1 Aylesbury Duck

This large white duck may weigh as much as 8 or 9 pounds (3.6 to 4.1 kg), and its principal use is as a table fowl. Aylesburies are fairly poor layers, as is often true of the heavy meat breeds, and are also poor setters. Because of their large size they have difficulty breeding and so their fertility rate is often low. The Aylesbury is named for the town in Buckinghamshire, England, where it originated. It is as common a meat duck in England in early 19th century as the Pekin is in the USA. This provides an interesting study in the development of consumer preferences, for the chief difference between the Aylesbury and the Pekin as meat ducks is the color of their respective skins—the Aylesbury dresses out white, while the Pekin is yellow-skinned. The British consumer has been brought up to prefer the former, while the American duck-eater has been conditioned to prefer the latter (Rick and Gail 1978) (Fig. 2.2).

Fig. 2.2 Aylesbury drake (above) and duck (below)

2.3.1.2 Call Duck

Call ducks are generally considered ornamental, being too small to be of economic value. Conversely, their small size makes them desirable to duck lovers with limited facilities. Calls require less space than the larger breeds, and being among the most active foragers, they can be relatively inexpensive to feed. They do, however, tend to get larger through overfeeding as well as careless breeding and may lose their characteristic compactness unless selectively bred for smallness. While Calls are quite prolific, the young are delicate and often require special care in order to get them through the first three or four critical days after they hatch. Calls are considered friendly and docile but are known for their tendencies to fly (Rick and Gail 1978).

Hunters to call in wild game ducks once used these bantam ducks, which usually weigh 2 pounds (900 g) or less. The breed's name reflects this utility. In Britain, they are usually called decoy ducks. They can make noise all out of proportion to their size and appropriately are also sometimes known as quack ducks. Their quacking is persistent and high-pitched, and this must be reckoned with by anyone hoping to own them happily (Rick and Gail 1978).

Calls commonly come in two color varieties, the white and the gray, although others have been developed by Call breeders. The gray was the original and is nearly identical to the Mallard in color pattern (Rick and Gail 1978) (Fig. 2.3).

2.3.1.3 Campbell Duck

Originally bred in England out of a desire for a good laying duck that would also have a full body suitable for table use and light plumage for easy plucking, Campbells are among the best general-purpose ducks. Mrs. Adele Campbell of Ulay, Gloucestershire, developed the breed around 1904 by crossing Mallard, Rouens, and Fawn-and-White Runners (Rick and Gail 1978). Campbells weigh around 4.5 pounds (2 kg) and come in three varieties. The original and still most common variety is the Khaki. The female of this variety is full uniform seal-brown color, while the male has the same general body color with darker bronze in some sections. In fact, the Khaki Campbell drake's plumage pattern closely resembles that of the Mallard drake, but instead of green tone, gray, and rust in sepia tones. Campbells acquired an upright carriage, along with their prolific egg-laying ability, from their Indian Runner ancestry (Rick and Gail 1978). They are fairly active foragers and can withstand cool climates. It is believed that Campbell are highly robust and active, along with seem to have the minimum desire to swim (Holderread 1978). The data from Jansen-farm in the Netherlands with 50,000 Khaki Campbell ducks showed that they can lay 335 to 340 eggs per laying season (Hutt 1952; Huang et al. 2012). (Fig. 2.4)

Fig. 2.3 Call duck drake (above) and duck (below)

2.3.1.4 Cayuga Duck

The Cayuga duck is unquestionably of America origin, and they thought to have been developed in the region of Lake Cayuga in New York State. They are excellent forager and a moderately good egg layer capable of laying early in the spring while the weather is still quite cold (Sheraw 1975).

Cayuga lays its first egg of the season coated in a black or dark gray film, which gradually turns light gray or blue as the season goes on. The ducklings have shiny black feet and beaks, black down, and only a little yellow on the breasts. The Oregon Cayuga is one of the strongest and quietest of all duck species. However, if slaughtered during feather growth, the erupting black pinfeathers give unattractive appearance to the carcasses (Holderread 1978).

The weights of adult Cayuga are 7–8 Ibs, and the flesh has an excellent flavor; however, the plumage color is a disadvantage because the plumage of Cayuga duck is their brilliant green-black color. Like all black plumaged breeds of poultry, they

Fig. 2.4 Khaki Campbell drake (above) and duck (below)

also develop more or less white feathers, especially in their necks and breasts when they become old. Also the black plumage tends to become rusty on the surface when exposed for long periods to sun and water, with the result that originally sound-colored individuals become nearer brown than black during the late summer season (Grow 1985) (Fig. 2.5).

2.3.1.5 Crested Duck

The big, round crest on the head gives this beautiful bird its name. When domestic ducks are raised for many years, specimens with small tufts often appear. These tufted ducks have been selectively bred in many parts of the world (including Europe and North America) to evolve the crested ducks (Holderread 1978).

It is a medium-weight breed and therefore is active and forages well. As a commercial duck, the Crested has fine possibilities, if one is willing to work with a large purebred flock. The Crested duck is a very respectable dual-purpose breed

Fig. 2.5 Cayuga drake (above) and duck (below)

because young grow quickly and are good layers. According to the respected waterfowl authority Oscar Grow, some of today's better-bred strains approach the popular Pekin in both rapidity of growth and productiveness. It is unfortunate, therefore, that the craze for the Pekin in North America, for the Aylesbury in Europe, and for crossbreeds everywhere in Commercial duck production, has resulted in the Crested duck becoming quite scarce in this country and dangerously low in England (Sheraw 1975).

An interest feature of crested ducks is that the size and shape of the crest still vary greatly, and as many as one-third of all ducklings do not have head adornment. This phenomenon is the result of the premature death of an embryo caused by a lethal gene, which carries two crest genes. The only ducklings that can hatch are those with only one crest gene or two normal genes (no crest). Due to this situation, the average hatching rate of crested duck eggs is about 25% lower than that of other medium-sized breeds (Holderread 1978) (Fig. 2.6).

Fig. 2.6 White Crested drake (above) and duck (below)

2.3.1.6 East Indie Duck

It is believed that East Indie ducks originated in the USA. However, British waterfowl breeders should be credited for its refinement. The East Indies are usually a bit larger than Calls and show a racier body structure (Holderread 1978).

The East Indies reproduce very well under natural conditions with very little care. If ducklings are slaughtered when they are 8 to 10 weeks of age, they produce attractive carcass, the size of which is just right for dinner for two (Holderread 1978).

There is probably no other breed of domestic ducks whose origin is so completely shrouded in mystery as is that of the East Indie; for the breed appears to be without a truly black ancestor. The so-called feral black ducks of both Europe and North America are far from black in color; they are merely a darker shade of brown than is characteristic of the Mallard and its closely allied species, the Florida, the Mottled, and the New Mexico ducks. Moreover, its earlier appellations, such as the Labrador, the Buenos Ayres, and the Brazilian duck, have served to complicate the mystery, rather than to resolve it. The last two names would suggest that it might have

Fig. 2.7 Black East Indie drake (above) and duck (below)

originated in the southern hemisphere, where it not for the fact that the East Indie undergoes two molts per year, whereas the feral ducks south of the equator renew their feathers once annually (Grow 1985) (Fig. 2.7).

2.3.1.7 Indian Runner Duck

Indian Runner ducks possess long and slender body and assume vertical posture, therefore, referred as the Penguin duck. They do not waddle like other breeds, but run. The speedy gait and agile mature make Runners the best foragers of all domesticated duck breeds (Holderread 1978).

Indian Runners are a small (4 to 4.5 pounds or 1.8 to 2 kg) breed of duck, developed in Scotland from stock originating in the East Indies. Their unique carriage appeals to those who enjoy the unusual. They look like walking wine bottles, and when they have some place to go they scamper rather than waddle. The varieties most commonly seen are white, penciled, and fawn-and-white,

Fig. 2.8 Indian Runner drake (above) and duck (below)

although there are at least half-a-dozen other color varieties. Runners are well known for their ability to produce large eggs in copious quantities. Because they are lightweight and therefore do not eat much, the breed is a favorite of commercial duck egg producers. Runners are an active breed and tend to be somewhat nervous (Rick and Gail 1978) (Fig. 2.8).

2.3.1.8 Muscovy Duck

The much larger body size of Muscovy ducks makes them best suitable meat type ducks. Muscovies are also sexually dimorphic with adult males reaching 4.6 to 6.8 kg, while females only reach 2.7 to 3.6 kg. They mature at around 28 weeks of age. The egg production of females in two reproductive cycles is 150 to 180 eggs. A female of a Muscovy duck reared up to 65 days weighs approximately 2.5 kg, while a male reared for 80 days weighs approximately 4.7 to 5.1 kg (Huang et al. 2012). The laying period of Muscovy is from March up to September, giving about 80 eggs

Fig. 2.9 White Muscovy drake (above) and duck (below)

per laying duck (Romboli et al. 1983). In France, the farmers use the lighting programs. Under intensive conditions of production, Muscovy ducks lay two periods of 5 months each, separated by a 12 weeks molt and giving in total 172 eggs up to 88 weeks of age (Sauveur and de Carville 1985).

Muscovy duck is native to Neotropica, more particularly to Southern Middle America, Central America, and Northern South America. The wild Muscovy's characteristic environment is low, forested swampland, the landward side of fresh or brackish Ciénagas, shady creeks and river banks, and mangrove-bordered coasts. A tree-dweller at night and for much of the day, it tends to avoid open water (in the domesticated state it is often kept far from rivers or lakes) (Donkin 1989).

Muscovy is often observed singly or in pairs, occasionally in flocks of 10 to 15; in the dry season they may congregate temporarily to number 40 to 50. They are not considered to migrate; on the contrary, they appear to have a strong sense of resident territory. They roost in trees, "resorting to the same place night after night," and nest in a hollow of fork some distance from the ground (Donkin 1989) (Figs. 2.9 and 2.10).

2.3.1.9 Orpington Duck

The breed of Orpington ducks, in its several varieties, is a development of the famous English Orpington Farms. Mr. William Cook of Kent, England, was the originator of the Orpington duck. According to existing information, Orpington duck was introduced in North America during the early 1900s. The five verities of Orpingtons breeds are light yellow, blue, silver, black, and chocolate. Except light

Fig. 2.10 Black Muscovy drake (above) and duck (below)

yellow variety, all the other four varieties have white bibs on their chests (Holderread 1978). The Orpington is a fine dual-purpose duck breed like other poultry breeds of England. They are good growers, active foragers, and fine egg producers. Records of one duck laying 260 eggs in 12 months in a British national test have been documented (Sheraw 1975). The Orpington is a medium-sized duck weighing 7 to 8 pounds (3.2 to 3.6 kg). At present, the Orpington duck does not usually constitute one of the larger classes in the showroom and this is unfortunate, considering the excellent utility and show qualities. The breed is now even less frequently raised in England and Europe so it is clear that this breed not only deserves but needs a boost (Sheraw 1975) (Fig. 2.11).

2.3.1.10 Pekin Duck

The Pekin duck was originated and breed for hundreds of years in China. In 1873, a flock of Pekin was exported to Connecticut, USA, where it quickly acclaimed. Since then, this breed has been widely used for table duck production in America. The

Fig. 2.11 Orpington drake (above) and duck (below)

information on whether Pekins came to the UK via the US or directly from China is lacking (MAFF 1980).

The Pekin duck is the most popular table duck in the world. The efforts of continuing breed selection making Pekin duck greatly improve their growth performances. From two commercial Pekin duck breeding companies' introduction, Pekin duck can grow up to 3.27 to 3.55 kg at 42 days of age with 1.88 to 1.97 feed conversion ratio at this moment (Fig. 2.12).

2.3.1.11 Rouen Duck

The Rouen bears the beautiful plumage pattern of the mallard, from which the French developed the breed long ago. It is named for a city in northern France, but the pronunciation is strictly Anglo-Saxon (ROW-en or sometimes ROO-en). It is a very large duck, usually weighing 9 to 11 pounds (4 to 5 kg) or sometimes more, and looks like a Mallard that has been stuffed with buckshot. Rouens drag their bulging

Fig. 2.12 Pekin drake (above) and duck (below)

bodies over the ground like a dinosaur. Rouens are not very active and are generally quiet and docile. The ducks are poor layers and their eggs are low in fertility. The Rouen is generally considered ornamental, but due to its size, it is also suitable as a meat breed (Rick and Gail 1978) (Fig. 2.13).

2.3.1.12 Swedish Duck

It is believed that Swedish duck was developed in Germany and its name is a misnomer. They are three varieties, namely blue, black, and silver. All varieties have white bibs staring from the bottom of the bill to half of the chest. The color of the blue variety is subtle shade of gray-blue and not a bright blue as seen in certain cage or wild birds. Swedish ducks do not breed true, produce offspring of blue, black, and silver varieties at 2:1:1 ratio. (Holderread 1978). The weights of Swedish adult drake and duck are 8 and 7 pounds (3.6 and 3.2 kg), respectively (Fig. 2.14).

Fig. 2.13 Rouen drake (above) and duck (below)

2.3.2 Breeds of Chinese Origin

2.3.2.1 Brown Tsaiya Duck

It is believed that Brown Tsaiya ducks were imported from south area in China, after generations of selection, now Brown Tsaiya duck is the mainly egg-laying breed in Taiwan. Owing to the annual agricultural statistics, the average rearing brown Tsaiya is about 2 million each year in Taiwan and producing nearly 500 million eggs.

The body size of Brown Tsaiya duck is extremely small. The average body weight of male and female is 1310 and 1390 g at 40 weeks of age. They have brown or light brown feathers, dark green or dark brown over the head and neck. Sex feather appears at around 6 weeks of age and the color turns greenish or brown when the birds mature. They possess excellent egg-laying performance. The orange beaks of the female ducks turn to chestnut color as they start egg laying. They can lay up to 330 eggs up to 500 days of age (Huang et al. 2012) with average egg weight 70.1 g at

Fig. 2.14 Swedish drake (above) and duck (below)

40 weeks of age. The characteristics of Brown Tsaiya duck are the tendency to congregate, jittery, easily frightened (Fig. 2.15).

2.3.2.2 Gaoyou Duck

Gaoyou duck is a dual-purpose breed. Gaoyou duck is originated from Gaoyou city, Jiangsu Province, China. There are plenty stream, river, lake in Gaoyou area and making an excellent natural environment for rearing duck; therefore, the history of rearing duck in this area can trace to Qing dynasty, about 400 years ago. In 1990, the number of Gaoyou duck was about 12 million, nevertheless, the number decreased to less than 20,000 in 1997 due to the importing of foreign laying duck. Until 2006, the Gaoyou duck breeding center in Gaoyou city kept 25,000 grandparents Gaoyou ducks and there were 200,000-parent stock and two million commercial Gaoyou ducks, respectively, in the main Gaoyou area (CNCAGR 2011).

The body size of Gaoyou duck is larger than other egg-laying type ducks. The average body weight for male and female ducks is 2660 and 2790 g, respectively, at

Fig. 2.15 Brown Tsaiya drake (above) and duck (below)

470 days of age. They can lay 190 to 200 eggs at 500 days of age with the age of first egg of about 170 to 190 days. The average egg weight is 84 g at 290 days of age. The Gaoyou drake have light blue and yellow beak, deep green with metal shiny on the head and neck, brown black feather on the chest, gray white on the abdomen, and orange-yellow color feet. For female Gaoyou ducks, blue gray or slightly yellow beak, full body covered with light flaxen feather, and the foot color is blue gray (CNCAGR 2011) (Fig. 2.16).

2.3.2.3 Jingding Duck

Jingding duck is an egg-laying type breed. They originated from Jingding village, Longhai city, Fujian province, China. It is believed that the history of breeding Jingding duck is more than 200 years because there are some references recorded this progress from Qing dynasty. Jingding village is also famous by the name "hometown of duck," most villagers are adept of rearing ducks. In 1982, there were 1.66 million Jingding ducks in the main rearing area; however, owing to the

Fig. 2.16 Gaoyou drake (above) and duck (below)

wetland area rapid decrease in Fujian province since 1990s, only 12,000 Jingding ducks existed from the survey in 2005 (CNCAGR 2011).

Jingding ducks have white skin, orange footpad, and black claw. In Jingding drakes, the beak is yellow to green, deep green with metal shiny on the head and neck, gray white color on abdomen and back, brown to black on primary feather. In females, body is covered totally with light brown feather and some short deep brown or even black stripes. The body weights of male and female Jingding ducks are 1614 and 1796 g at 300 days of age, respectively. They can lay 288 eggs at 500 days of age with average egg weight of 72 g (CNCAGR 2011) (Fig. 2.17).

Fig. 2.17 Jingding drake (above) and duck (below)

2.3.2.4 Liancheng White Duck

Liancheng White duck is an egg-laying breed that originated from Liancheng county, Fujian province in China. The history of Liancheng White duck can trace to some references in Qing dynasty. In 1980s, Liancheng White duck barely became extinct until China started a germ preservation in 1983, selection and industrial promotion proposal. There were about 1.5 million Liancheng white ducks in 2007 (CNCAGR 2011).

The body weights of male and female ducks at 300 days of age are 1165 and 1487 g, respectively. The average egg production at 300 days of age is 122.5 with 63 g egg weight. It is difficult to identify the gender of Liancheng White duck

Fig. 2.18 Liancheng drake (above) and duck (below)

because of their similar appearance. Their body are covered totally with white feather, in the contrast, their beak, eye, foot, and pad are totally dark brown or even black (CNCAGR 2011) (Fig. 2.18).

2.3.2.5 Linwu Duck

Linwu duck is a dual-purpose breed, originated from Linwu County, Hunan Province in China. Owing to the inconvenient traffic condition of Linwu County in the past, local farmers bred duck as an additional income. The main breeding target is laying performance and body size (CNCAGR 2011).

The male and female body weights at 300 days of age are 1943 and 1714 g, respectively. They can lay 246 eggs in a year with the average egg weight of 70 g. Linwu ducks have yellow beak, pad, and white skin. Drake is usually covered brown, gray, and yellow feather and female is totally covered with yellow feathers.

Fig. 2.19 Linwu drake (above) and duck (below)

Most of the Linwu ducks have a "white ring" over the neck in both genders (CNCAGR 2011) (Fig. 2.19).

2.3.2.6 Shan Partridge Duck (Shan Ma Duck)

Shan Partridge duck is an egg-laying type breed and originated from Fujian province in China. From some references in Ming dynasty, it has been a long history for human involvement in breeding of Shan Partridge duck since 400 years ago (CNCAGR 2011).

Shan Partridge ducks have extremely small body size that only weigh 1265 and 1440 g for male and female at 300 days of age, respectively. They can lay average 299 eggs at 500 days of age with average 66 to 68 g egg weight. Shan Partridge drake have a blue yellow beak with a black spot on the end, shiny green feather on head and neck with a white ring. The breast is dark brown and there are some white feathers on the tail. Conversely, female ducks have yellow beak, most of them are

Fig. 2.20 Shan Partridge drake (above) and duck (below)

covered with light brown or white or even spotted feathers (CNCAGR 2011) (Fig. 2.20).

2.3.2.7 Shaoxing Duck

Shaoxing duck is also called Shaoxing partridge duck, is an egg-laying type breed. Shaoxing duck is originated from Shaoxing city, Zhejiang Province, China. The history of people in Shaoxing rearing ducks can be traced back to eleventh century. Shaoxing city is also a famous for rice wine production. It is said "rice for winemaking, wine by-products (distiller's grain) for feeding duck, duck's feces as rice field fertilizer," showing the conventional farming and recycling practices in Shaoxing area (CNCAGR 2011).

Shaoxing duck can lay 307 eggs in a year with 67 g average egg weight, the body weight of drake and female duck at 120 days of age is 1506 and 1547 g, respectively. There were about 2.38 million Shaoxing ducks in the end of 2006. There are three

Fig. 2.21 Shaoxing drake (above) and duck (below)

different strains of Shaoxing duck, the predominant strain is the "white wing and ring" strain; the drake's body is covered by flaxen feather, dark green color on the head and neck, orange to yellow beak and pad and white claw. The main color of female body looks like sparrow with a white ring on the middle of the neck (CNCAGR 2011) (Fig. 2.21).

2.3.2.8 Mule Duck

The mule duck meat production was studied since a long time in Taiwan where the selection of parents was acting to get white plumage mule duck at the beginning (Huang 1985). The hybrid mule duck is early maturing than Muscovy duck, and it give less fat than Pekin duck. It was shown that the hybrid mule duck could cumulate for valuable carcass parts, the earliness of growth and maturity of his Pekin mother and the meatiness of his Muscovy father. The genetic mechanisms are unknown and

Fig. 2.22 Three way-crossbred mule ducks of Taiwan

difficult to understand because mule is a hybrid from parents, which do not have the same chromosome complement (Rouvier 1999).

The meat of Mule ducks is extremely lean. They are the offspring of Kaiya mother and a Muscovy father. Kaiya is a cross of Pekin drake with the female of Taiwanese egg-laying duck breed, namely Tsaiya. The French will tolerate black pinfeathers as evidence of the bird being rustic but the Taiwanese demand a cleanly picked bird. Therefore, the plumage of a duck for Taiwanese market must be white so that the pinfeathers blend with the skin color. The recessive white Tsaiya is therefore used in the breeding programs to get the white plumaged mule ducklings. White Tsaiya female was first crossed with a White Pekin male to increase the size of the crossbreed (Kaiya), the mother of the mule. When Kaiya female is mated to Muscovy males, the resultant mule ducklings are white plumaged. After many generations of selection, mule ducks in Taiwan are now totally white color with only some little spots on the head (Fig. 2.22).

2.3.3 Indigenous Duck Breeds of Other Asian Countries

Apart from the recognized duck breeds described above, there are many lesser-known duck breeds which are being reared in many Asian countries, viz. Indonesia, India, Vietnam, Malaysia, Myanmar, and Philippines. These indigenous breeds are believed to be descendants of Mallard and are reared for both egg and meat. Information on many of the breeds is scanty and is limited only to their countries of origin. Details of some of the Indian breeds are mentioned below.

2.3.3.1 Kuttanad Duck

The native duck breed of Kerala state in India is named as Kuttanad duck. They originated from Kuttanad region of Alappuzha district, which is abundant in water bodies. The topography of Kuttanad is peculiar, which lies 1 m below mean sea level making it highly suitable for duck rearing. Many households are practicing duck rearing in a traditional system, which provides major source of their livelihood. The origin of Kuttanad ducks is considered to be from Mallard as they resemble Mallard in their shape, size, and plumage color (Jalaludeen et al. 2004). The two plumage varieties of Kuttanad ducks are Chara and Chemballi. The major difference between Chara and Chemballi drakes is in the plumage color of head region, which is lustrous greenish black in Chara and dull greenish black in Chemballi. The plumage color of back and tail in Chara female is blackish brown and the breast is primarily brownish black; but light brown and white are also seen. The general plumage color of Chemballi female is brownish black and brownish gray in back, tail and wings; wherein, brown is predominant over black and gray (Mahanta et al. 1999). The average egg number per duck up to 50 weeks of age was 116.09 in Chara and 124.95 in Chemballi (Mahanta et al. 1998) (Figs. 2.23 and 2.24).

2.3.3.2 Arni Duck

Arni ducks are the native ducks of Tamil Nadu, India. There are two varieties in Arni breed, viz. Sanyasi and Keeri. The name was coined based on the color of plumage. The Sanyasi ducks have saffron colored plumage with or without white ring like feathers around the neck and drakes are with dark brown plumage mixed with black (Veeramani et al. 2014). The head and neck are covered with lustrous brown plumage. Males have brown colored drake feather. The bill color of female is orange and for males, it is yellowish orange. The shank color is orange for both males and females. The Keeri females have mixture of black and brown plumage characteristically in striations with or without white ring like feathers around the neck and males have a mixture of dark black and while plumage. The head and neck are covered with lustrous black plumage. The bill and shank color of female is gray/orange.

Fig. 2.23 Chara variety of Kuttanad drake (above) and duck (below)

Keeri male duck has dark yellow colored bill and orange colored shank. The drake feather is black in color (Veeramani et al. 2014) (Fig. 2.25 and 2.26).

2.3.3.3 Pati Duck

Pati ducks are registered as a breed in India. They are reared in backyard production system in rural areas of Assam state of India and are used for meat, egg, and ritual sacrifices. In drakes, the plumage color is dark brown with greenish black head; tail with black and white feathers. Ducks are solid brown. A white ring may or may not be present at neck in both sexes. The bill, shank, and feet are predominantly yellow. The average body weight is 1.58 kg. The annual egg production of Pati duck is 75 to 93 eggs. It is a good sitter, hence used in villages for hatching purpose. There is no discernible clutch size and their age at first egg is 160 to 190 days (Nahardeka et al. 2019) (Fig. 2.27).

Fig. 2.24 Chemballi variety of Kuttanad drake (above) and duck (below)

2.3.3.4 Nageswari Duck

Nageswari ducks are locally called "Nagi," the snake deity, may be due to its head assuming a high snake like posture with a white stripe in the neck extending up to the breast. It is considered as an egg type variety confined to Cachar and Karimganj districts of Barak valley of Assam. Its original homeland is Sylhet district of Assam (now in Bangladesh). The head of drake is dull brownish black, while the neck has white plumage. The breast feathers are white in color and of back and tail are grayish black. The plumage color in ducks is blackish brown in back, tail and wing; wherein, blackish brown is predominant over brown and black (Sharma et al. 2002). The body weights in males and females are 1.42 and 1.26 kg, respectively, and the annual egg production is 100 to 120 eggs (Fig. 2.28).

Fig. 2.25 Sanyasi drake (above) and duck (below). (Photo courtesy: P Veeramani)

Fig. 2.26 Keeri drake (above) and duck (below). (Photo courtesy: P Veeramani)

Fig. 2.27 Pati drake (above) and duck (below)

2.3.3.5 Maithili

Maithili duck is a registered duck breed of India. They are distributed in five to six districts of Bihar state. The ducks have uniform light/dark brown feathers throughout the body. Circular spots on the feathers (Mosaic pattern) are seen. The drakes have dark brown to ash colored plumage. It is a low producer with average annual egg production of around 55 eggs and an average egg weight of around 50 g. Body weight at 6 month of age is around 1.2 kg (Fig. 2.29).

2.3.4 Semi-Wild Duck

Mandarin Duck

The Mandarin drake is accepted by most as the most beautiful duck in existence. It is an arboreal or tree dwelling duck closely related to the Wood Duck. Though it is

Fig. 2.28 Nageswari drake (above) and duck (below)

related to the Wood Duck of North America, the Mandarin will never cross with it because of basic genetic differences. The breed (or more properly, species) lives wild in Japan, Eastern Asia, and in the British Isles where it was transplanted (Sheraw 1975). Generally, the Mandarin drake displays a brightly colored head of lustrous green with purple and copper reflections with a broad cream-colored band and a mane or whiskers of chestnut brown. The breast is reddish-purple, the flanks and sides gray-buff with black penciling, and the triangular sails extending over the back are chestnut bay. The shanks and feet are pale orange to orange-yellow (Sheraw 1975).

Except for the Teals, there are few smaller ducks than the Mandarin is, but, notwithstanding to its size, they are a hardy fowl capable of defending themselves against larger breeds. Yet they soon become reconciled to captive conditions and usually reproduce normally when provided with appropriate rations and suitable environs. There is a tendency, however, for captive Mandarins to lose gradually their

Fig. 2.29 Maithili drake (above) and duck (below)

brilliancy of plumage unless protective measures are exercised through the feeding of rations designed to promote brilliancy of feather and the selection of breeding stock potent in this feature (Grow 1985) (Fig. 2.30).

2.4 Duck Genetic Resource Conservation and Prevention of Genetic Erosion

Most of poultry scientists lack awareness of the genetic diversity in domesticated bird species. Conservation and preservation of duck genetic resources is important for their possible future needs. In Asian countries, some minor native breeds of duck are now diminishing drastically due to the introduction of new types of duck breeds for producing massive quantity of eggs and meats. Because producers want to maintain high profitability, this means the duck population is also being

Fig. 2.30 Mandarin drake (above) and duck (below)

homogenized, diversity is being irreversibly lost, and carefully evolved purebred domesticated birds are likely to be heading towards genetic monoculture (Fujihara and Xi 1999).

In the past century, at least 28% of farm animal breeds became extinct, rare, or endangered (WCMC 1992). Most endangered or extinct breeds have been reported in North America and Europe (Bódi et al. 2017). Furthermore, only a few percent of endangered or critical status breeds are maintained in different conservation programs (FAO 2000a). According to Weigend et al. (2004), poultry genetic resources are one of the most endangered groups. Out of the 938 avian breeds of five species, 460 (49%) breeds have been classified as being at risk of loss (Weigend and Romanov 2002). In 1999, the ratio of breeds in endangered and critical state was 20% in duck and 28% in goose in the world and 43% in duck and 25% in goose in Europe (FAO 2000b).

Today, in many countries, almost all local livestock and poultry populations are threatened with extinction (Romanov et al. 1996). Breaking the principle of animal hybridization, local livestock are effectively replaced by hybrids that are usually not adapted to the original environmental conditions and low-quality feed (Primack 1995). As a result, animal populations have changed. They are not only susceptible to local extreme weather conditions, but also susceptible to a variety of diseases which those local breeds have been adapting to these diseases for centuries (Wezyk et al. 1994). Most governments try to protect the native animal species from destruction, but it survives extinction only due to the conservation by farmers and animal fanciers (Fujihara and Xi 1999). Currently, it is difficult to collect complete and accurate information about the state of native poultry populations in Asian countries. Most recent trends in duck farming in Japan are an effective use of crossbred ducks for improving rice production known as combined farming (Manda 1992, 1993, 1996). In addition to Japan, some other countries such as Vietnam, Indonesia, Korea, and China, including even some of African countries have been trying to develop this type of farming (Fujihara and Xi 1999). This integrated duck farming system has already provided a certain profit for miserable farmers in the developing countries, where the daily income is not so high. Therefore, some small population of native ducks may be kept in this production system. Small population of waterfowl species tends to lose genetic diversity. People will face two dilemmas: how to maintain genetic diversity without letting population lose those characteristics and how to maintain genetic diversity in a small population that has probably already lost many alleles (Dobson 1998).

Most recent trends in the production of duck in Asian countries are becoming monoculture of species, such as Pekin, Khaki Campbell, and Muscovy. These highly productive species yield much more eggs and meat compared with other native ducks. This will lead to genetic erosion in duck genetic resources. Most of the countries of Asia are eagerly following up the poultry production systems of economically developed countries (Fujihara and Xi 1999). Different countries require different attributes in their ducks and the breeding firms work with many pure lines, which are constantly improved by genetic selection. The duck breeding industry is worldwide and fast growing, but in some area in Asia there are still traditional markets, which the local duck breeders cater the requirement. In this connection, a well thought out plan should be evolved with the point of view of environmental protection; otherwise, sustainable waterfowl farming would not be realized in Asian countries. In addition, special attention should also be paid for keeping as many different species of waterfowl as possible in the area suitable for rearing them, even in very small countries (Fujihara and Xi 1999). Preserving different kinds of waterfowl is quite important for future waterfowl production, in particular, for maintaining genetic diversity and protecting genetic erosion of the birds necessary to make crossbred birds.

However, it is worrisome to note that some minor native breeds of duck are now diminishing drastically due to the introduction of improved duck breeds for producing massive quantity of eggs and meats. Conservation of different kinds of waterfowl is the need of the hour for their use in any exigency in future waterfowl production and also for maintaining genetic diversity by preventing genetic erosion.

2.5 Conclusion

The duck breeds are diverse due to the geographic difference all over the world. Some duck breeds like Khaki Campbell and Brown Tsaiya have similar egg-laying potential of commercial laying hens. The Pekin, the most popular meat type duck in the world is an excellent animal protein source like broiler chicken. In addition to the recognized duck breeds, many indigenous breeds are being reared in most of the Asian countries. They are of much economic importance to the rural people. Though, many of them are not efficient producers as other popular breeds, they integrated into the local environment and society since long time. For example, the Mandarin duck becomes a symbol of lovers in most parts of Chinese society; Call ducks are drawing more attention as a pet, representing the close relationship between humans and ducks. Owing to their great tolerance to extreme climate and stable feed supply conditions, the ducks will become more popular and important poultry species in the future; therefore, more attention should be paid on preventing local duck breed genetic erosion.

References

Bódi L, Szalay I, Phuong, TNL (2017) Waterfowl gene conservation-importance possibilities and practice in the Carpathian Basin, In: Proceedings of 6th world waterfowl conference, October 22–25, 2017, Taipei, Taiwan pp. 63–69

Cherry P, Morris TR (2008) Domestic duck production: science and practice. CAB International, Wallingford

CNCAGR (2011) China National Commission of Animal Genetic Resources. Animal genetic resources in China: poultry. China Agriculture Publisher, Beijing

Crawford RD (1990) Origin and history of poultry species. In: Crawford RD (ed) Poultry breeding and genetics. Elsevier Science, Amsterdam, pp 1–41

Dobson AP (1998) Conservation and biodiversity. Scientific American Library, New York

Donkin RA (1989) The Muscovy duck, *Cairina moschata domestica* origins, dispersal, and associated aspects of the geography of domestication. Routledge, A. A. Balkema, Rotterdam

FAO (2000a) One third of farm animal breeds face extinction. http://www.fao.org/News/2000/001201-e.htm. Accessed 18 Dec 2019

FAO (2000b) World watch list for domestic animal diversity 3rd edn. Edited by B. D. Scherf. Food and Agriculture Organization of the United Nations Rome. Available via http://www.fao.org/docrep/009/x8750e/x8750e00.htm. Accessed 18 Dec 2019

Fujihara N, Xi YM (1999) Genetic resource conservation in the waterfowl. In: Proceedings of 1st world waterfowl conference, December 1–4, 1999, Taichung, Taiwan, pp 63–69

Grow O (1985) Modern waterfowl management and breeding guide. American Bantam Association, Augusta

Holderread D (1978) Raising the home duck flock: a complete guide. Garden Way Publishing, Pownal

Huang HH (1985) Selection of white mule duck. J Chin Soc Anim Sci 14:111–112

Huang JF, Pingel H, Guy G, Lukaszewicz E, Baéza E, Wang SD (2012) A century of progress in waterfowl production, and a history of the WPSA Waterfowl Working Group. Worlds Poult Sci J 68:551–563
Hutt FB (1952) The Jansen Khaki Campbell ducks. J Hered 43:277–281
Jalaludeen A, Peethambaran PA, Leo J, Manomohan CB (2004) Duck production in Kerala. NATP on Productivity Enhancement of Ducks, Centre for Advanced Studies in Poultry Science, College of Veterinary and Animal Sciences, Kerala Agricultural University, Thrissur, p 3
MAFF (1980) Ministry of Agriculture, Fisheries and Food. Ducks and geese. Her Majesty's Stationery Office, London
Mahanta JD, Ramakrishnan A, Jalaludeen A (1998) Egg production performance of two indigenous types of ducks of Kerala. J Vet Anim Sci 29:30–35
Mahanta JD, Ramakrishnan A, Jalaludeen A (1999) Biometric characteristics of two indigenous ducks of Kerala. Indian Vet J 76:1091–1094
Manda M (1992) Paddy rice cultivation using crossbred ducks. Farm Jpn 26:35–42
Manda M (1993) Definition and utilization of crossbred duck. Anim Hus 47:167–170
Manda M (1996) Aigamo (corssbred duck)-rice farming in Asia. Farm Jpn 30:39–43
Nahardeka N, Mahanta JD, Zaman G, Das GC, Bhattacharyya BN (2019) Monograph on Pati duck of Assam. Published by Director of Research (Vety), Assam Agricultural University, Guwahati
Pingel H (1990a) Genetics of growth and meat production in waterfowl. In: Crawford RD (ed) Poultry breeding and genetics. Elsevier Science, Amsterdam, pp 691–704
Pingel H (1990b) Genetics of egg production and reproduction in waterfowl. In: Crawford RD (ed) Poultry breeding and genetics. Elsevier Science, Amsterdam, pp 771–780
Primack RB (1995) A primer of conservation biology. Sinauer Association Inc., Sunderland
Rick L, Gail L (1978) Ducks and geese in your backyard: a beginner's guide. Rodale Press, Emmaus
Romanov MN, Wezyk S, Cywa-Benko K, Sakhatsky NI (1996) Poultry genetic resources in the countries of Eastern Europe-history and current state. Poult Avian Biol Rev 7:1–29
Romboli I, Mori B, Bagliacca M, Fedeli Avanzi C (1983) Influenza dei fattori ambientali sulla deposizione dell'anatra muschiata. Avicoltura 52:37–39
Rouvier R (1999) Genetics and physiology of waterfowl. In: Proceedings of 1st world waterfowl conference, December 1–4, 1999, Taichung, Taiwan, pp 1–18
Sauveur B, de Carville H (1985) Recent studies on the management of Muscovy breeding ducks in France. In: Farrell DJ, Stapleton P (eds) Duck production science and world practice. University of New England, Amidale, pp 293–305
Sharma SS, Zaman G, Goswamy RN, Mahanta JD (2002) Biometrical characteristics of Nageswari ducks of Assam. Cheiron 31(5&6):128–130
Sheraw D (1975) Successful duck and goose raising. Stromberg Publishing Company, Pine River
Veeramani P, Prabhakaran R, Selvan ST, Sivaselvam SN, Sivakumar T (2014) Morphology and morphometry of indigenous ducks of Tamil Nadu. Glob J Med Res G Vet Sci Vet Med 14 ((3) Version 1.0):16–20
WCMC (1992) World Conservation Monitoring Centre. Global biodiversity: status of the Earth's living resources. Chapman and Hall, London, p 585
Weigend S, Romanov MN (2002) The world watch list for domestic animal diversity in the context of conservation and utilisation of poultry biodiversity. Worlds Poult Sci J 58:411–430
Weigend S, Romanov MN, Rath D (2004) Methodologies to identify, evaluate and conserve poultry genetic resources. In: Book of abstract of XXII world's poultry congress & exhibition, June 8–13, 2001, Istanbul, Turkey, p 84
Wezyk S, Cywa-Benko K, Romanov MN (1994) Protection against the extinction of the native breeds of poultry in the countries of Eastern Europe. In: Proceedings of the international symposium "conservation measures for rare farm animal breeds," May 17–19, 1994, Balice, Poland, pp 89–101

Chin-Hui Su, Ph.D. candidate (Department of Livestock management, Ilan Branch, Livestock Research Institute, Taiwan)

Mr. Chin-Hui Su is a native of Yilan County, Taiwan. He earned his Master's degree (Animal Science) from National Chaiyi University and started Ph.D. (Biotechnology and Animal Science) degree in National Ilan University. Mr. Su had more than 13 years of experience in duck research after joining the research group of Ilan Branch in 2007. He specializes in nutrition, welfare, egg processing, and management of duck. He was also invited by the committee of sixth World Waterfowl Conference to give a speech on "A Review of the Effects of Light-Emitting Diode on Poultry Welfare and Economic Traits."

Duck Genetics and Breeding

3

Martin H. C. Liu and R. Richard Churchil

Contents

M. H. C. Liu
Ilan Branch, Livestock Research Institute, Council of Agriculture, Ilan, Taiwan
e-mail: scliu@tlri.gov.tw; sc12601@gmail.com

R. R. Churchil (✉)
Faculty of Veterinary and Animal Sciences, Tamil Nadu Veterinary and Animal Sciences University, Chennai, Tamil Nadu, India
e-mail: richardchurchil.r@tanuvas.ac.in

A. Jalaludeen et al. (eds.), *Duck Production and Management Strategies*,
https://doi.org/10.1007/978-981-16-6100-6_3

Abstract

The evolution of ducks took place 66 million years ago in Cretaceous–Paleogene (K–Pg) boundary and domestication took place in China (500 BC) and in Western Europe (800 AD). FAO enlists 400 duck and 63 Muscovy duck genetic groups (breeds, strains and lines) spreading across the world; although the list is expanding. Sixteen duck genetic groups are extinct and 34 ducks and 1 Muscovy are endangered. Many more are critical or vulnerable. The Mendelian traits like colour of plumage, skin, shank, web, feet, bill, bean, caruncle, eye and eggshell, bill shape, crest and body carriage have distinct inheritance in ducks. Growth rate, body weight, feed conversion ratio, meat yield and breast thickness are the traits of importance in meat-type ducks, whereas egg production and related traits, fertility, hatchability and eggshell colour are important in egg-type stocks. The mule ducks are sterile intergeneric hybrids, which are produced especially in Taiwan and in France by crossing Pekin or Kaiya (crossbred Pekin × White Tsaiya) ducks with Muscovy drakes. Selection for white plumage and duration of fertility has yielded good results in mule duck production.

Keywords

Phylogeny of ducks · Genetic resources · Mendelian and polygenic · Genetic parameters · Selective breeding · Marker-assisted selection

List of Symbols/Abbreviations

♀	Female
♂	Male
ABW	Average body weight
AFE	Age at first egg
AFLP	Amplified fragment length polymorphism
AI	Artificial insemination
BB	Breast breadth
BMP	Breast muscle percentage
BMT	Breast muscle thickness
BMW	Breast muscle (meat) weight
BMY	Breast muscle (meat) yield
BW	Body weight
EN	Egg number
ES	Eggshell strength
EW	Egg weight

EYW	Egg yolk weight
FCR	Feed conversation ratio
FI	Feed intake
FL	Feather length
g	Gram
GBW	Breeding value of body weight
GEN	Breeding value of egg number
GES	Breeding value of eggshell strength
GEW	Breeding value of egg weight
h^2	Heritability
kg	kilogram
KL	Keel length
K–Pg	Cretaceous–Paleogene boundary
PCR	Polymerase chain reaction
REW	Relative egg weight
REYW	Relative egg yolk weight
RFI	Residual feed intake
RFID	Radio-frequency identification
r_g	Genetic correlation
SNP	Single nucleotide polymorphism
ST	Shell thickness

3.1 Introduction

The ducks are an economically important poultry species, second only to chicken. *Anas platyrhynchos*, one among more than 30 species of dabbling ducks of genus *Anas* and *Cairina moschata,* the only species of genus *Cairina* are the domesticated duck species of economic utility in family *Anatidae* (waterfowls). The common ducks are the descendants of mallard, while the domestic Muscovies evolved from wild Muscovy. A study on 12S rRNA gene polymorphism revealed that duck is the most genetically divergent species among poultry species like chicken, duck, Japanese quail, turkey and guinea fowl (Gupta et al. 2005); the gene polymorphism studies on BLB2 gene also revealed similar results (Singh et al. 2005). There are 400 diverse genetic groups of domestic common ducks and 63 groups in domestic Muscovy ducks registered in DAD-IS of FAO (DAD-IS 2020). A few of them are extinct or endangered or critical. The ducks have the ability to grow fast in their early life compared to chicken; therefore, ducks make a perfect meat animal. They also have the ability to lay a large number of eggs with higher weight than chicken ending up in higher egg mass output. The selection is being carried out in many countries to develop separate specialized lines for egg and meat. The selection experiments have yielded a very good response, as a result, today we have duck attaining a market weight of about 3.1 kg at 5 weeks of age with a feed conversion ratio of approximately 1.7:1. The genetically selected layer lines are capable of producing 327 eggs in 72 weeks of age with average egg weight of 70 g and feed-egg ratio of 2.65:1. The intergeneric hybrids of common and Muscovy ducks are utilized for meat production in some parts of the world and fatty liver production in other parts. The demand from consumers and industry prompted breeding for blue egg colour in layer ducks, lean meat and higher breast meat in broiler ducks and white plumage with auto-sexing ability in mule ducks. This progress was achieved so far through the conventional methods of genetic selection and breeding. A huge volume of data has been generated on their genetic parameters, which can be utilized in any new population for genetic selection. The latest advances in computers are supporting data processing in huge volume at less time and thereby aiding selection from large population with high selection pressure. The new directions in molecular biology provide marker-assisted selection based on quantitative trait loci. Today, the assessment of variations at nucleotide level enables the evaluation of the genetic merit of individual before the phenotypic traits are recorded. These advances in genetic assessment make the genetic improvement possible in otherwise plateaued duck populations also. However, the conservation efforts need to be intensified world over, to protect the duck genetic resources from extinction and probably to utilize them to play a larger role as livelihood provider in changing scenario of consumer demands and human disease pandemics.

3.2 Duck Genetic Resources: Evolution and Diversity

3.2.1 Evolution of Ducks

The evolution of birds began in the Jurassic Period, with the earliest birds derived from a clade of theropod dinosaurs named *Paraves* (Wilford 2016). Archaeopteryx was described in the 1860s, by Thomas Henry Huxley from its fossil specimen as having feathers but with many reptilian features. He proposed that dinosaurs never really vanished from Earth. Most did go extinct, but their evolutionary legacy lives around us as birds, all 18,000 avian species of them. Late Jurassic and Early Cretaceous birds from Northeastern China, including many complete skeletons of *Confuciusornis*, provide evidence for a fundamental dichotomy in the class Aves that may antedate the temporal occurrence of the Late Jurassic *Archaeopteryx* (Hou et al. 1996).

The Cretaceous–Paleogene (K–Pg) boundary was considered as a key episode in recent vertebrae history, marking the end of Dinosaur Era and the evolution and diversification of mammalian and avian taxa (Feduccia 1995). The K–Pg boundary marks the end of the Cretaceous Period, the last period of the Mesozoic Era, and marks the beginning of the Paleogene Period, the first period of the Cenozoic Era. Its age is usually estimated at around 66 Ma (million years ago) with radiometric dating yielding a more precise age of 66.043 ± 0.011 million years (Renne et al. 2013). However, it was a long-standing controversy on whether living bird lineages emerged after non-avian dinosaur extinction at the K–Pg boundary or they co-existed before K-Pg boundary (Feduccia 2003; van Tuinen et al. 2003). It is put forward that from several phylogenetic analyses supported by independent histological data that *Vegavis iaai*, a genus of extinct bird that lived during the Late Cretaceous Period in Antarctica, some 68 to 66 million years ago was a part of Anseriformes (waterfowl) and is most closely related to Anatidae, which includes true ducks. A minimum of five divergences within Aves before the K-Pg boundary are inferred from the placement of *Vegavis*; at least duck, chicken and ratite bird relatives were coexistent with non-avian dinosaurs (Clake et al. 2005). However, the connection of *Vegavis* and modern day waterfowls is challenged by a recent study based on the analysis of plesiomorphic traits of the pterygoid and the mandible and anseriform or galloanserine affinities of Vegaviidae could not be firmly established (Mayr et al. 2018).

3.2.2 Taxonomy of Ducks

Ducks, swans and geese are the members of waterfowl family or Anatidae. The members of the family Anatidae do not represent a monophyletic group (the group of all descendants of a single common ancestral species). It is divided into several subfamilies. The swans and geese are not considered ducks, but the ducks are mostly aquatic birds, normally smaller in size compared to swans and geese. The taxonomic classification of ducks (Sibley and Ahlquist 1990; Cracraft et al. 2004; Clements et al. 2019) is given below.

3.2.2.1 Kingdom: Animalia (Animals)

All members of Animalia (also called as metazoa) are multicellular and heterotrophs (rely directly or indirectly on other organisms for their nourishment). The animals consume organic material, breathe oxygen, are able to move, can reproduce sexually and grow from a hollow sphere of cells, the blastula, during embryonic development. It is estimated that around 9 or 10 million species of animals inhabit the earth.

3.2.2.2 Phylum: Chordata (Animals Possessing Notochord)

The most distinguishing character of all animals belonging to this phylum is that they develop notochord during some period of their life cycle. Notochord is a longitudinal rod that is made of cartilage and runs between the nerve cord and the digestive tract. Its main function is to support the nerve cord. In vertebrate animals, the vertebral column replaces the notochord.

3.2.2.3 Subphylum: Vertebrata (Vertebrates)

Vertebrata (includes fish, amphibians, reptiles, birds and mammals) is one among the three subphyla of phylum Chordata. The other two are Tunicata or Urochordata (sea squirts, salps) and Cephalochordata (which includes lancelets).

3.2.2.4 Superclass: Gnathostomata (Jawed Vertebrates)

Two well-formed jaws encircle the mouth aperture in all members of this superclass. Along with birds, creatures like fishes, amphibians, reptiles and mammals are also grouped under this superclass.

3.2.2.5 Class: Aves (Birds)

The class Aves comprise birds, which are vertebrates and covered with feathers. These creatures have a four-chambered heart and a strong yet lightweight skeleton, which are modified for flight and for active metabolism. Toothless beaked jaws and hard-shelled eggs also characterize them. There are a few kinds of birds that do not fly, but their ancestors did, and these birds have secondarily lost their ability to fly.

3.2.2.6 Subclass: Neornithes (Modern Birds or True Birds)

Neornithes have well-developed sternum; tail is not long; no teeth; forelimbs are modified as wings; teeth are replaced by horny rhamphotheca over bill. They evolved in the Middle to Late Cretaceous period and diversified dramatically around the time of the Cretaceous–Paleogene extinction event, which took place about 66 million years ago. There are about 10,000 known living modern bird species in the world. This is composed of birds of two clades: (1) Palaeognathae (flightless lineages) are termed 'ratites': kiwis, cassowaries, emus, rheas and ostrich and one flying lineage, Tinamou and (2) Neognathae (all modern fowl other than palaeognathae).

3.2.2.7 Infraclass: Neognathae (Modern Palate Birds)

Neognaths (Neognathae) include virtually all living birds, exceptions being their sister taxon (Palaeognathae). Fowl are birds belonging to one of two biological orders, namely the gamefowl or landfowl (Galliformes) and the waterfowl (Anseriformes). Anatomical and molecular similarities suggest these two groups are close evolutionary relatives; together, they form the fowl clade, which is scientifically known as Galloanserae (fowls).

3.2.2.8 Order: Anseriformes (Waterfowl)

There are three families under this order, namely Anhimidae, Anseranatidae and Anatidae.

1. ***Anhimidae***: They are related to the ducks, geese and swans of family Anatidae but have bills looking more like those of game birds. There are two Genera, namely Anhima and Chauna. The birds in this family are screamers, known for making loud sounds.
2. ***Anseranatidae***: This family contains two genera namely *Eoanseranas* and *Anseranas*, having one species each. Two species of magpie geese (*Eoanseranas handae and Anseranas semipalmata*) are kept under this family.
3. ***Anatidae***: The family members of Anatidae are waterbirds or waterfowl that includes ducks, geese and swans. The family has a worldwide presence, occurring on all the continents. These birds are adapted for swimming, floating on the water surface. The family contains around 146 species in 43 genera.

The anatomic features of the members of family Anatidae and Anseranatidae are webbed feet, soft sensitive bills and relatively long necks with short legs, strong wings and plumage that often includes heavy down feathers. The body parts of ducks are shown in Fig. 3.1, and the caruncle an additional feature of Muscovy ducks is shown in Fig. 3.2.

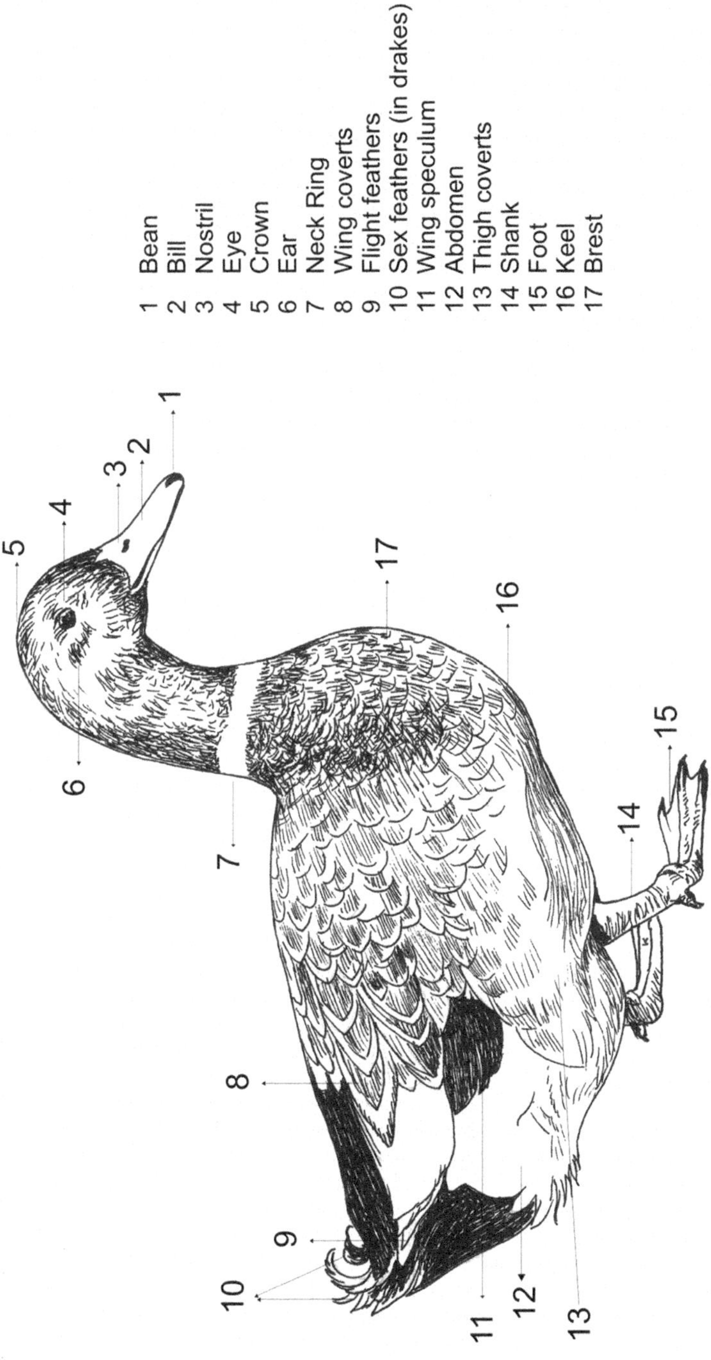

Fig. 3.1 Body parts of duck

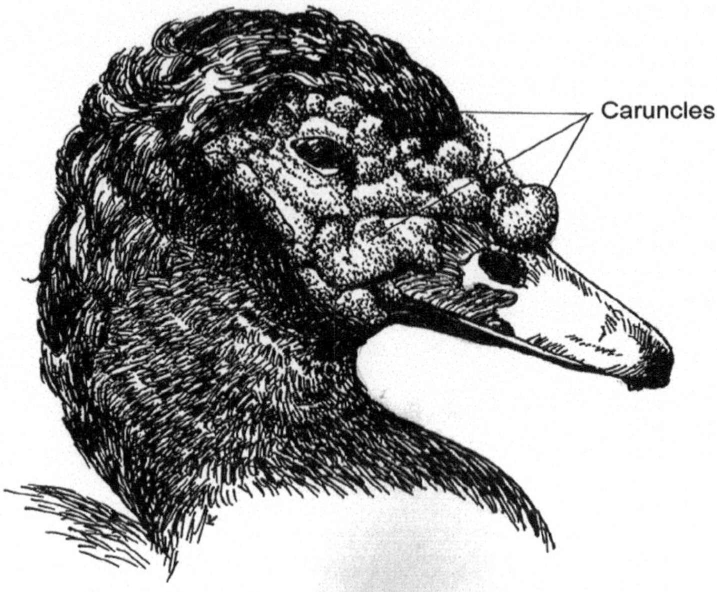

Fig. 3.2 Head of Muscovy duck

The species *A. platyrhynchos* under family *Anas*, commonly called as duck (descendants of Mallard) and *C. Moschata* under family *Cairina* commonly called as Muscovy duck are the domesticated species in the list. The 35 genera classified under the family *Anatidae* and the species under each genus are given below; the list excludes geese and swans. The common names are given in the parentheses.

1. ***Anas (Dabbling ducks)***: *A. sparsa* (African black duck); *A. undulate* (yellow-billed duck); *A. melleri* (Meller's duck); *A. superciliosa* (Pacific black duck); *A. laysanensis* (Laysan duck); *A. wyvilliana* (Hawaiian duck); *A. luzonica* (Philippine duck); *A. poecilorhyncha* (Indian spot-billed duck); *A. zonorhyncha* (Eastern spot-billed duck); *A. platyrhynchos* (mallard); *A. fulvigula* (mottled duck); *A. rubripes* (American black duck); *A. diazi* (Mexican duck); *A. capensis* (Cape teal); *A. bahamensis* (white-cheeked pintail); *A. erythrorhyncha* (red-billed teal); *A. georgic*; (yellow billed pintail); *A. eatoni* (Eaton's pintail); *A. acuta* (Northern pintail); *A. crecca* (Eurasian teal); *A. carolinensis* (green-winged teal); *A. flavirostris* (yellow-billed teal); *A. andium* (Andean teal); *A. gibberifrons* (Sunda teal); *A. albogularis* (Andaman teal); *A. gracilis* (grey teal); *A. castanea* (chestnut teal); *A. bernieri* (Bernier's teal); *A. chlorotis* (brown teal); *A. aucklandica* (Auckland teal); *A. nesiotis* (Campbell teal)
2. ***Aix***: *A. sponsa* (wood duck); *A. galericulata* (Mandarin duck)
3. ***Amazonetta***: *A. brasiliensis* (Brazilian teal)

4. ***Asarcornis***: *A. scutulata* (white-winged duck or white-winged wood duck)
5. ***Aythya (Diving ducks)***: *A. ferina* (common pochard); *A. Americana* (redhead); *A. collaris* (ring-necked duck); *A. australis* (hardhead); *A. baeri (*Baer's pochard); *A. nyroca* (ferruginous duck); *A. innotata* (Madagascar pochard); *A. novaeseelandiae* (New Zealand scaup); *A. fuligula* (tufted duck); *A. marila* (greater scaup); *A. affinis* (lesser scaup)
6. ***Biziura***: *B. lobata* (Musk duck)
7. ***Bucephala (Goldeneye duck)***: *B. clangula* (common goldeneye); *B. islandica* (Barrow's goldeneye); *B. albeola* (bufflehead)
8. ***Cairina:*** *C. moschata* (Muscovy duck)
9. ***Callonetta***: *C. leucophrys* (ringed teal)
10. ***Camptorhynchus***: *C. labradorius* (Labrador duck)
11. ***Clangula***: *C. hiemalis* (long-tailed duck)
12. ***Dendrocygna***: *D. autumnalis* (black-bellied whistling ducks); *D. viduata* (white-faced whistling duck)
13. ***Heteronetta***: *H. atricapilla* (black-headed duck)
14. ***Histrionicus***: *H. histrionicus* (Harlequin duck)
15. ***Hymenolaimus***: *H. malacorhynchos* (blue duck)
16. ***Lophodytes***: *L. cucullatus* (hooded merganser)
17. ***Lophonetta***: *L. specularioides* (crested duck)
18. ***Malacorhynchus***: *M. membranaceus* (pink-eared duck)
19. ***Mareca***: *M. strepera* (gadwall); *M. falcate* (falcated teal); *M. Penelope* (Eurasian wigeon); *M. sibilatrix* (Chiloé wigeon)
20. ***Marmaronetta***: *M. angustirostris* (marbled duck, or marbled teal)
21. ***Melanitta***: *M. boie* (scoters)
22. ***Merganetta***: *M. armata* (torrent duck)
23. ***Mergellus***: *M. albellus* (smew)
24. ***Mergus***: *M. merganser* (common merganser); *M. octosetaceus* (Brazilian merganser); *M. serrator* (red-breasted merganser); *M. squamatus* (scaly sided merganser)
25. ***Netta***: *N. rufina* (red-crested pochard); *N. erythrophthalma* (southern pochard); *N. peposaca* (rosy-billed pochard)
26. ***Nomonyx***: *N. dominicus* (masked duck)
27. ***Oxyura (stiff-tailed ducks):*** *O. australisn* (blue-billed duck); *O. jamaicensis* (ruddy duck); *O. ferruginea* (Andean duck); *O. leucocephala* (white-headed duck); *O. maccoa* (Maccoa duck); *O. vittata* (lake duck)
28. ***Polysticta***: *P. stelleri* (Steller's eider)
29. ***Pteronetta***: *P. hartlaubii* (Hartlaub's duck)
30. ***Salvadorina***: *S. waigiuensis* (Salvadori's teal)
31. ***Sarkidiornis***: *S. sylvicola* (comb duck); *S. melanotos* (knob-billed duck)
32. ***Sibirionetta***: *S. formosa* (Baikal teal)
33. ***Somateria (Eiders)***: *S. mollissima* (common eider); *S. spectabilis* (king eider); *S. fischeri* (spectacled eider)
34. ***Spatula:*** *S. querquedula* (garganey); *S. hottentota* (Hottentot teal); *S. puna* (Puna teal); *S. versicolor* (silver teal); *S. platalea* (red shoveler); *S. cyanoptera*

(cinnamon teal); *S. discors* (blue-winged teal); *S. smithii* (Cape shoveler); *S. rhynchotis* (Australasian shoveler); *S. clypeata* (Northern shoveler)
35. ***Speculanas***: *S. specularis* (bronze-winged duck)
36. ***Tadorna*** **(Shelducks)**: *T. ferruginea* (ruddy shelduck); *T. cana* (South African shelduck); *T. tadornoides* (Australian shelduck); *T. variegata* (paradise shelduck); *T. tadorna* (common shelduck).

3.2.3 Domestication of Ducks

Localized expansion of human populations triggered by climatic changes at the end of the Pleistocene (12,000 to 14,000 BP) led to localization of human population and emergence of crop farming (Larson and Burger 2013). The process of domestication of animals and birds varies from species to species. Larson and Burger (2013) proposed three plausible pathways, namely 'commensal', 'prey' and 'directed'. The first pathway involved animals being attracted to human settlements and then becoming captive as a source of food. The localized expansion of human populations occurred in riverbank area; ducks being a waterfowl, frequent human encounter with ducks could have occurred very early. The avian species possibly connected with human being are chicken, pigeon, duck and turkey along with dog, cat, rat, mouse and guinea pig in this pathway. The second pathway involved the capture of even-toed hoofed animals like cattle, sheep, goats, pigs, camels, etc., for meat purpose and their domestication led to the supply of milk, wool and leather also. The third pathway came into play later in history. This pathway involved deliberate efforts to exploit the specific capabilities of the target species (e.g., their potential as pack, riding or draught animals).

3.2.3.1 Ancestors of Ducks

The two species of domestic duck are the common duck, which was domesticated from the wild mallard (*Anas platyrhynchos*) and the Muscovy duck (*Cairina moschata*). Both are members of the subfamily Anatinae. The same scientific binomial is used for the domestic and the wild forms of common and Muscovy ducks. When the two species are interbred, the resultant offspring are infertile.

Mallard

The wild duck species of mallard was originally described by Carl Linnaeus in the year 1758 in 10th edition of his book Systema Naturae (Linnaeus 1758). He gave the mallard two binomial names: *Anas platyrhynchos* and *Anas boschas* (Gordh and Beardsley 1999). It is a large dabbling duck, very common in all kinds of waters. The mallards have a wide ecological tolerance and distributed across temperate Eurasia and North America.

The male mallards have a green head and neck with a white ring at the bottom of its neck. It has a brown chest, whitish-grey undersides, brown wings and a yellow bill. The females are mottled dull brown in colour, but sport iridescent purple-blue wing feathers that are visible as a patch on their sides. Mallards spend near the surface of the water and dabble to find its feeds like invertebrates, fish, amphibians and a variety of plants. They also graze on land, feeding on grains and plants.

The birds are distributed widely across the Northern and Southern Hemispheres; in North America, across the Palearctic, Siberia, Japan and South Korea. In Southern Hemisphere they are naturally seen in south-east and south-west Australia and New Zealand. Mallards are strongly migratory in the northern parts of its breeding range and to south in winters. The mated pairs of mallards migrate to the northern parts of their range and build nests on the ground or in a protected cavity. The female mallards generally lay around dozen eggs and incubate the eggs, while males abandon the nest in the middle and join a flock of other males. Most mallard species are common and not considered threatened. However, one threat to their populations includes hybridization with other ducks.

Wild Muscovy

The Muscovy duck's ancestry is connected neither with Moscow (Muscovy) nor with Cairo (*Cairina*). A very large forest duck, the Muscovy is taxonomically included with the tropical to subtropical sharp-clawed perching ducks (Tribe: Cairinini). The Cairinini are large anatids that have long tails which facilitate braking for tree landing. They use their sharp claws to roost and nest in hollows from 3 to 20 m above ground (Kaulicke 1991). The name 'Muscovy' probably refers its musky smell (moschata from the Latin moschatus, 'musky'), or Mexico or the muysca Indians of Nicaragua's Mosquito Coast.

The Muscovy duck is notable for its large body size and distinct sexual dimorphism with males weighing more than double of females. Wild Muscovies have greyish black plumage on the ventral side, blackish plumage with green and purple iridescence on the dorsal side and wing coverts have prominent white patches. The face of the drakes is adorned with small fleshy caruncles distributed around from orbital region to the base of the bill, with prominent occipital and nuchal crests.

3.2.3.2 Domestication of Common Ducks

The current state of knowledge on literature, archaeological evidences and historical information reveals two possible locations in which the domestication of ducks took place independently, namely China and Western Europe. The domestication of ducks happened in China as early in around first century BC, while in Western Europe it occurred much later in around 800 AD.

East Asia

China's earliest duck potteries ever found were unearthed from the remains of Yangshao Culture dated about 4000 years ago, located in Shanxian County, central Henan Province. This reveals that ducks were the parts of human life in China as early as 2000 BC, although the duck in portrait could not be clearly identified as domestic one. Clayton (1984) described a report in the Chinese literature by Yeh, who investigated the archaeological evidence and suggested ducks were domesticated in China in around 1000 BC; but clear evidence is lacking.

The first evidence of domestication is available from the excavation of a painted grey duck shaped pottery in 1954 from the remains of Eligang Culture located in Zhengzhou, Henan. This supports the theory of domestication of ducks dated back to the start of Common Era (first century AD). This is the oldest evidence on domestication of ducks. There are first written records of domestic ducks in central China shortly after 500 BC (Kiple 2000). The ducks (*Anas platyrhynchos*) likely reared in large-scale duck farms in the Eastern Zhou dynasty (514–495 BC) in China. Force-feeding techniques for duck production, similar to modern Pekin duck production techniques, are featured in 'Qi Min Yao Shu', a book written by Jia Sixie during the Northern Wei dynasty (386–524 AD), which indicates that Pekin duck breeding was present then (Zhang and Zhu 1986; Chen 1986).

The origin and domestication of Pekin ducks are uncertain but there are two possibilities. The first possibility is a direct domestication from a white plumage mutant of wild mallards. The second possibility is the domestication of Pekin ducks from other indigenous duck breeds of Jiangsu province, after being transferred to Beijing through rivers and canals during the Ming (1368–1644 AD) or Qing dynasties (1644–1911 AD) and then bred into Pekin ducks for the imperial court (Zhang 1989). A recent study with the use of microsatellite and mitochondrial marker analysis supports the theory of Pekin ducks originating directly from wild mallard ducks (Qu et al. 2009).

Western Europe

(i) ***Domestication of common ducks***

The ducks are depicted in paintings and relief carvings as early as 2240 BC in the tombs of Mereruke vizier to pharaoh teti (sixth dynasty) at Saqqara in Egypt. The wildfowl including ducks were trapped and kept in large aviaries and were force-fed before slaughter more than 1000 BC as shown in Karnak Temple in Egypt (Cherry and Morris 2008). Harper (1972) gave the narrative of early history of duck in human life from the classical and medieval literature. He described that the term 'tame duck' has been mentioned in the writings of Theophrastus, a plant biologist who lived from 371 to 287 BC in Eresos of Greece; however, no concrete evidence is available from his writings whether the ducks were truly domesticated and used for economic purpose (Theophrastus 1961). Smith (1969) asserts that the ducks were the common domesticated animals in the ancient Egypt, but offers only the evidence about goose. The mentions of birds like geese, chicken, pigeons and ducks in Egyptian papyri describe that these birds were raised in Egypt from third century BC to seventh century AD. The ducks have been mentioned less frequently in papyri compared to other birds and no clear evidence is available that the ducks in mention were domestic or wild. The suspicion that the Greeks did not possess the domestic duck gains support from the writings of Cato (234–149 BC), who describes fattening hens, geese and pigeons but not ducks. The first evidence of confinement rearing of ducks is available from the writings of Varro (116–27 BC), who mentioned that it is necessary to build a solid enclosure covered with a wide meshed net so that an eagle could not fly in or the duck fly out. Pliny (23–79 AD) recommended hatching of eggs of wild ducks under chickens. Martial (38–104 AD) mentions the presence of birds like geese, peafowl, partridges, guineas, pheasants, chickens, doves and turtle doves, but not ducks. All these suggest that the duck was not a true domesticate, but held captive to be serves as luxury of the Roman table. Capitulare de villis, the text composed in 771–800 AD that guided the governance of the royal estates under Roman Emperor Charles I allows the officials of his empire to keep peafowl, pheasants, ducks, doves, partridges and turtle doves in their villa. This is possibly the oldest evidence that Harper (1972) could show for domestication of ducks and it happened at the end of eighth century AD in Western Europe. A more clear evidence of domestication is available in a document of royal domain in Carolingian time listing peasant holding geese, chickens, peafowl and ducks in the start of ninth century (810 AD). This is the oldest evidence that the domestication of ducks in Western Europe could have taken place in around 800 AD.

The ducks are portrayed in a painting of Dutch painter Jan Steen who lived in Warmond in Western Netherlands when he made this artwork 'The Poultry Yard' in 1660. A child named Jacoba is seen sitting in a poultry yard surrounded by a motley collection of birds with ducks upfront and prominent. This shows that the ducks were a major component of domestic animals in Western Europe in seventeenth century AD. Duck rearing became a major industry in Aylesbury of England in the nineteenth century. The ducks were bred on farms in the surrounding countryside in Aylesbury and transported to the surrounding places from 1839 resulting from the opening of a railway to Aylesbury and later at around 1860s, the duck rearing industry began to move out of Aylesbury to the peripheral towns.

(ii) ***Domestication of Muscovy ducks***

In the wild, Muscovy ducks are distributed Southern North America starting from Mexico to southern South America. Its range continues southward through Central America into forested lowland areas with Peru, Andes, northern Argentina and Uruguay as boundaries. In the Caribbean, there is no evidence of wild populations north of Trinidad. The domestication of Muscovy ducks took place in South American continent earlier than 1500 AD well before the Columbian Era. The Muscovy ducks were domesticated along southern shores of Caribbean (Crawford 1990) and domestication also took place in Paraguay and the Chaco (Donkin 1989). It is also reported that the domesticated Muscovy ducks were found in the northern coast of Columbia and Peru by the Spanish conquerors in the sixteenth century. From these reports, it is clear that the centre of domestication must be around north and central parts of South America. Shortly after Europeans arrived in the New World; domesticated Muscovies were brought back to Europe and subsequently introduced into Africa, Asia and Australia. The diffusion of Muscovy from South America to other parts of world took place very rapidly. Three routes of diffusion from Americas were proposed by Donkin (1989) reaching out to (i) west coast of Africa, (ii) Spain and France and (iii) Southeast Asia.

Table 3.1 The regional distribution of duck genetic resources

Species	Africa	Asia	Europe	Americas	Oceania	World
Common ducks	29	132	181	31	27	400
Muscovy ducks	14	19	12	8	10	63

3.2.4 Duck Genetic Resources

The Domestic Animal Diversity Information System (DAD-IS) is the global databank on Animal Genetic Resources since 1995. The publication of The State of the World's Animal Genetic Resources for Food and Agriculture (first SoW-AnGR) was brought out in 2007 and the Global Plan of Action for Animal Genetic Resources was adopted. The Commission on Genetic Resources for Food and Agriculture, in 2009, prepared a format of DAD-IS for reporting on the status and trends of animal genetic resources (AnGR). There were many modifications made in the reporting format in the due course. DAD-IS maintained and developed by FAO provides access to searchable databases of breed-related information and photos. It allows analysis of the diversity of livestock breeds on national, regional and global levels including the status of breeds regarding their risk of extinction.

The total number of genetic groups of common ducks and Muscovy ducks reported in 2014 was 311 and 21, respectively (FAO 2015). In 2014, 43 countries reported the presence of ducks. The number registered genetic groups (breeds/strain/variety) has risen to 400 and 63, respectively, for common ducks and Muscovy ducks, the region-wise split up is given in Table 3.1.

Among all domestic avian genetic groups registered in DAD-IS, the number of duck genetic groups (400) is second, next only to that of chicken (2566) and higher in number than that of goose (261), turkey (184) and pigeon (82). There are 63 genetic groups in Muscovy ducks, which is higher than that of quail (54), ostrich (22), pheasant (13), partridge (12), emu (5) and cassowary (3) (DAD-IS 2020).

3.2.4.1 Classification and Distribution of Common Ducks

Among the 400 genetic groups of common ducks, 14 are lines, 9 are strains, 4 are varieties, 256 are breeds and 4 are mongrel. The classification of remaining 113 genetic groups is not available in the database. The major duck breeds and their country of origin are given in Table 3.2.

Based on geographical classification, these 400 duck genetic groups are classified as local (272), regional (26) and international (102) (DAD-IS 2020). The domestication status of these genetic groups is noted as domestic (319), feral (3) and wild (2); while for the remaining 76 are not known. Anatra mignon (call duck) of Italy registered in DAD-IS is the smallest duck among all genetic groups weighing 0.8 kg. The breeds reported with higher body weights in DAD-IS are Strkov TTH (Male – 4.0 and female – 3.6 kg), Pekin (Male – 4.1 and female – 3.7 kg), Rouen and Aylesbury (Male – 4.5 and female – 4.0 kg) and Pato of Ecuador (Male – 6.0 and female – 5.0 kg). The characteristics of duck genetic groups with higher annual egg number (>200) are listed in Table 3.3.

Table 3.2 Important duck breeds of different countries of origin

Countries	Common breed name
Antigua and Barbuda	Pekin, Pond Duck, West Indian
Australia	African, Aylesbury, Blue Swedish, Cayuga, Chocolate Bibbed, Elizabeth, Indian Runner, Mallard, Orpington, Pekin, Roueen, Silver Appleyard, Welsh Harlequin, Black East Indian, Call, Crested, Saxony
Bangladesh	Deshi Black, Deshi White, Indian Runner, Jending, Nondescript indigenous type, White Pekin
Belgium	Dendermondse eend, Forest, Huttegem, Huttegemse eend, Laplaigne, Merchtemse eend, Semois
Bolivia	Grimao ermaÃ's, Japanice Criollo, Pekin
Burkina Faso	Burkina Faso
CÃ'te d'Ivoire	Local duck of CÃ'te d'Ivoire
Cambodia	Tea Ankam
Chad	Local duck of Gredaya and Massakory, local duck of Guelateng, local duck of Moulkou and Bongor
Chile	Azul, Mudo, Pekin, Pato Colorado, Pato cuchara, Pato jergÃ³n, Pato juarjual, Pato vapour, Rouan
China	Beijing, Caohu, Dayu, Enshi Partridge, Gaoyou, Guangxi Small Partridge, Hanzhong Partridge, Huainan Partridge, Ji'an Red Duck, Jiangchang, Jianshui brown duck, Jinding, Jingjiang Partridge, Jingxi Large Partridge, Liancheng White, Linwu, Longsheng Cui duck, Mawang duck, Mianyang Partridge, Putian Black, Quemoy, Rongshui Xiang duck, Sansui, Shan Partridge, Shaoxing, Sichuan Partridge, Taiwan Partridge, Weishan Partridge, Wendeng Black, Xingyi, Youxian Partridge, Yunnan Partridge, Zhongshan Partridge
Cuba	Pekin
Cyprus	Barbary, Pekin
Denmark	Sort hvidbrystet dansk
Ecuador	Pato
Egypt	Domiati, Pekin, Sudani
Eswatini	Lidada
Ethiopia	Albet
Fiji	Indigenous duck, Pekin
France	Canard de Bourbourg, Canard de Challans, Canard de Duclair, Canard de Rouen
Gabon	Canard locale
Germany	Deutsche Pekingenten, Pommernenten
Ghana	Local Ghanaian Duck
Guyana	Kunshan, Pekin
Hungary	FehÃ©r magyar kacsa, Pekin, K-94 of Szarvas, Barbarie, Seddin Vital, Tarka magyar kacsa
India	Kuttanadu Chara, Chemballi, Indian Runner, Nageshwari, Sythetemete

(continued)

Table 3.2 (continued)

Countries	Common breed name
Indonesia	Alabio, Bali, Aylesbury, Bayang, Cihateup, Gunsi PKC, Jawa, Kerinci, Kisaran, Magelang, Maros, MojoMaster-1 Agrinak, Mojosari, Pajajaran, Pegagan, Pitalah, Rambon, Talang Benih, Tegal, Tondano, Turi, White Peking, Roueen
Ireland	Aylesbury, Cayuga, Indian Runner, Orpington, Pekin, Roueen, Swedish Blue, Welsh Harlequin, Abacot Ranger, Appleyard
Italy	Anatra mignon, Germanata Veneta
Japan	Osaka duck
Jordan	Jordan Baladi
Kazakhstan	Arman cross, Pekin
Kyrgyzstan	Pekin
Korea, Republic of	Cheongsoo, Pekin
Lao People's Democratic Republic	Pet Kab, Pet Thed
Madagascar	Canard domestique
Malaysia	Nila, Pekin
Netherlands	Krombek, Kuifeend, Kwaker, Noord Hollandse witborsteend, Overbergse eend
Pakistan	Batakh
Peru	Criollo, Pato Criollo FrancÃ©s, Pekin, Roueen
Philippines	Indian Runner, Pekin, Philippine Duck, Tsaiya
Serbia	Domestic Duck
Slovakia	IndickÃ, Ä opasta raca, Kajuga, Orpingtonska raca, PekinPritlikava raca, Raca tekaÄ Ica, Rouanska raca, SaÅ¡ka raca, Smaragdna raca, VisokoleteÄ a raca
Slovenia	Pekin
Solomon Islands	Indian Runner, Pekin
Spain	Annera Mallorquina
Sri Lanka	Petrock, Velovi, Vigro
Sweden	Blekingeanka, Svensk BlÃ¥ Anka, Svensk Gul Anka
Thailand	Paknam Duck (Ped Paknam)
Tonga	Local duck
Turkey	Native, Pekin
Tuvalu	Local duck
Ukraine	Pekin, Ukrainian Black White-Breasted, Ukrainian Clay, Ukrainian Grey
United Kingdom	Abacot Ranger, Appleyard, Aylesbury
Uruguay	Criollo, Marruecos, Pekin
Vanuatu	Vanuatu Duck
Vietnam	Anh Dao, Bau quy, Bauben, Khaki Campbell, Ky Lua, Moc, Nang, Omon, Pekin, Tiep, Vit Bau, Vit Co
Zambia	Aylesbury, Pekin, Madada, Roueen

Source: DAD-IS (2020)

Table 3.3 Description of duck genetic resources with good egg laying potential in descending order for egg number

Common name	Country from reported	EN	ABW (♂)	ABW (♀)	EW	Description
Paknam Duck (Ped Paknam)	Thailand (Eastern, Central, and Western parts)	290	1.7	1.55	NA	Age at first egg: 140–180 days, resistant to diseases ***Males***: dark green head, other parts are similar to that of females, BW at first egg: 1.450–1.500 g, first EW: 62 g ***Females***: black plumage throughout the body, white chest, black mouth, black shank and feet
Alabio	Indonesia (West Java, South Kalimantan, West Nusa Tenggara)	287	1.4	1.3	60	Selected Alabio local duck for egg production at four generations. White skin and yellow shanks, year of origin 2009 ***Males***: brown spotted black body, greyish white brown chest, brown flickering green blue silver wings ***Females***: black with white colour on the head, greyish brown black
Jinding	China (originated in Longhai City, Fujian Province)	280	1.76	1.73	72	Indigenous breed, multi-coloured ***Males***: brown iris, red orange shanks or webs, greenish feathers on the head, red-brown feathers on the front breast, grey brown back, dark brown wing feather, wide breast, long narrow body shape, Kelly beak ***Females***: copper-coloured beak, red orange shanks or webs, black bay

(continued)

Table 3.3 (continued)

Common name	Country from reported	EN	ABW (♂)	ABW (♀)	EW	Description
						feather, long and slim body shape, compact body structure, long and slim neck
Putian black	China (originated in Putian County, Fujian Province)	280	1.34	1.63	70	Indigenous breed, small body size
Brown Tsaiya	China (Taiwan Province)	270	1.4	1.5	NA	Indigenous breed
Itik Pinas Itim	Philippines (Central and South Luzon, Soccsksargen, Cagayan Valley, Western Visayas and Negros Island)	256	1.23	1.22	67	Genetically improved Philippine mallard ducks selected for high egg production. White skin and orange shank, predominantly black with white feathers in neck 'bib', no vaccination is needed, no major disease outbreak reported, more than 90% livability, year of origin – 2008
Khayan Duck	Myanmar (West Java, South Kalimantan, East Java)	250	NA	NA	NA	–
Mojosari	Indonesia (Shaoxing, Xiaoshan and Zhuji Counties, Zhejiang Province; suburbs of Shanghai Municipality; south of Jiangsu Province)	250	1.5	1.4	53	Selected Mojosari egg production duck raised for four generations. White skin and black shank. Age at first egg: 22–24 weeks, year of origin 2009 ***Males***: blackish brown chest and black tail, foot and beak ***Females***: brown chest and tail, and black foot and beak, feather line on the top side of the eyes

(continued)

Table 3.3 (continued)

Common name	Country from reported	EN	ABW (♂)	ABW (♀)	EW	Description
Shaoxing	China (originated in Yilan of Taipei, Dalin of Taizhong, Pingdong of Tainan, Taiwan)	250	1.36	1.26	68	Indigenous breed, male and female have different necks, compact body conformation, long narrow body shape, long beak, slim neck, sparrow feather
Taiwan Partridge	China (Longyanhu township, Fujian Province)	250	1.1	1.4	NA	Indigenous breed
Shan Partridge	China (originated in Wendeng City, Shandong Province; distributed in Rushan and Mouping Counties, Shandong Province)	243	1.43	1.55	55	Indigenous breed, multi-coloured ***Males***: red-brown breast, black tail, white ring on the neck ***Females***: light brown with slight black stripe, orange shanks, black feet, shiny green feather on the head and neck, moderately head size, long and slim neck
Wendeng Black	China	242	1.9	1.8	80	Indigenous breed
Youxian Partridge	China (Beijing, Tianjin and Shanghai municipalities; Guangdong and Liaoning provinces)	225	1.17	1.23	62	Indigenous breed, multi-coloured ***Males***: white ring on the middle of neck, emerald feathers on the head or up-neck, reddish brown on the front breast and lower neck, grey brown wing feather, dark greenish tail ***Females***: brown with oval-shaped black spot, red orange shanks and webs, black feet, small body size, narrow long body shape with boat-shape, compact feather.

(continued)

Table 3.3 (continued)

Common name	Country from reported	EN	ABW (♂)	ABW (♀)	EW	Description
Pekin	China (Hanzhong, Chenggu, Nanzheng, Xixiang, Mianxian and Yangxian counties, Shaanxi province)	220	3.49	3.41	90–95	Indigenous breed, orange beak
Hanzhong Partridge	China	220	1.17	1.16	68	Indigenous breed, multi-coloured: mainly brown, yellow on the back with brown spots: males have several green feathers, yellow beak
White Pekin	Slovakia	220	2.5	2.5	80	Unicoloured – white
Philippine Duck	Philippines (IV – Southern Tagalog)	220	1.7	1.5	NA	The origin of this breed dates back to the Spanish time or fourteenth century, resistant to duck viral enteritis (duck plague) and leg paralysis. Hardy but rather noisy breed, comment: black with white chest and white necklace, which is a dominant phenotypic character
Sansui	China (Sansui County, Guizhou Province)	220	1.69	1.68	65	Indigenous breed, multi-coloured, reddish-brown feathers on the head, white ring on the lower neck, light-brown abdomen ***Males***: long body shape ***Females***: red orange shanks and webs, black feet, long and slim neck, boat-shape body, puce feather

(continued)

Table 3.3 (continued)

Common name	Country from reported	EN	ABW (♂)	ABW (♀)	EW	Description
Jingjiang Partridge	China (originated in Jiangling, Jianli and Mianyang counties, Hubei Province)	214	1.34	1.44	63	Indigenous breed, multi-coloured, male: green feathers on the head, reddish-brown feathers on the front breast and back, slight grey tail, green beak, red orange shanks and webs, white eyebrow, comeliness head
Guangxi Small Partridge	China (Pingnan, Guiping, Tengxian and Cangwu Counties, Guangxi Zhuang Autonomous Region)	210	1.6	1.55	65	Indigenous breed
Vit Co (Co Duck)	Vietnam	210	1.7	1.5	65	Indigenous breed, yellow skin and shank, high resistance, mottled feather pattern, Co duck has quite various plumages, majority in colour of sparrow's wing, part of the population is grey, black and white in colour
Laplaigne	Belgium	200	NA	NA	NA	
Linwu	China (Linwu County, Hunan Province)	200	2.75	2.75	67	Indigenous breed, multi-coloured ***Males***: brown head and neck, few are green, having white loop around neck, belly is brown, and some are white or yellow ***Females***: yellow, large body size
Merchtemse Eend	Belgium (North of Brussels, Central Belgium)	200	3	2.5	NA	Might be related to Aylesbury

(continued)

Table 3.3 (continued)

Common name	Country from reported	EN	ABW (♂)	ABW (♀)	EW	Description
Zhongshan Partridge	China (Guangdong Province; distributed in the delta area of the Pearl River)	200	1.7	1.7	70	Indigenous breed, originated in Zhongshan County, multi-coloured *Males*: light-brown breast feathers, green abdomen *Females*: mostly brown, having a white loop around neck, small or medium body size

Source: DAD-IS (2020)
NA not available

3.2.4.2 Classification and Distribution of Muscovy Ducks

Out of 63 genetic groups of Muscovy ducks registered in DAD-IS, 43 are denoted as breeds, 2 as varieties and the rest of 15 without any detail. Based on domestication status, 59 are reported as domestic, one as feral and rest is unknown. Based on geographical classification, 26 are designated as international and the remaining 37 are local. The Muscovy ducks from Guam and Thailand are reported to produce as high as 180 eggs annually and Philippine Muscovy duck reported to lay only 35 eggs. The lowest body weights of 2.7 and 2.2 kg in males and females, respectively, are reported in Philippine Muscovy duck and highest body weights of 5.5 in males and 3.5 kg in females are recorded in Serati of Malaysia and French Muscovy of Vietnam.

3.2.5 Risk Status of Duck Genetic Resources

The risk status of animal genetic germplasm is classified into following categories (FAO 2015).

1. **Extinct:** A breed in which there are no breeding males or breeding females remaining.
2. **Critical:** A breed in which the total number of breeding females is less than or equal to 100 or the total number of breeding males is less than or equal to five; or the overall population size is less than or equal to 120.
3. **Endangered:** A breed in which the total number of breeding females is greater than 100, and less than or equal to 1000 or the total number of breeding males is less than or equal to 20 and greater than 5; or the overall population size is greater than 80 and less than 100 (with more than 80% females and increasing) or greater

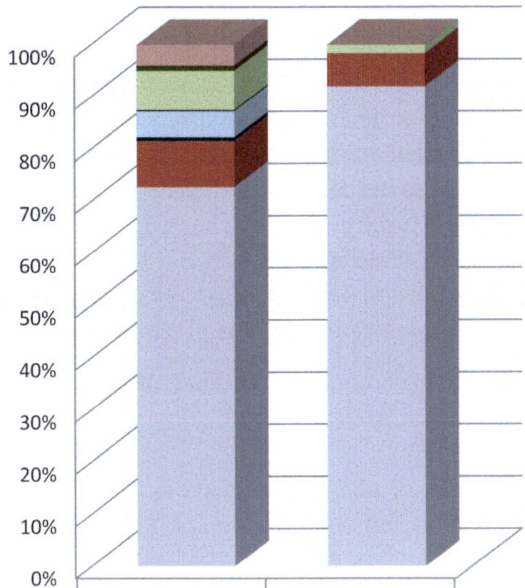

	Common Duck	Muscovy Duck
Extinct	16	0
Endangered maintained	4	0
Endangered	30	1
Critical maintained	1	0
Critical	20	0
Vulnerable	3	0
Not at risk	35	4
Unknown	291	58

Fig. 3.3 Local risk status of common duck and Muscovy duck. (Represented as percentage in graph and as numbers in table)

than 1000 and less than or equal to 1200 (with less than 80% females and decreasing) but not classified as extinct or critical.

4. **At risk:** A breed classified as either critical, critical maintained, endangered or endangered-maintained.

The risk status of duck genetic resources is given in Fig. 3.3.

3.2.6 Genetic Conservation and Improvement

As per the Second Report on the State of World's Animal Genetic Resources for Food and Agriculture (FAO 2015), only 16 (37%) of common duck rearing countries and 33% of Muscovy duck rearing countries have reported at least one genetic selection and improvement programme. Among the countries having conservation programmes for common ducks, only 63% of the countries had in situ conservation, while 50% of the countries reported existence of ex situ in vivo conservation and 13% countries reported ex situ in vitro conservation. The proportion of genetic groups from which material is stored in a gene bank is only 3% for common ducks and 43% for Muscovy ducks. The proportion of genetic groups from which sufficient material is stored in a gene bank to allow the breed to be reconstituted is 3% for common ducks and 43% for Muscovy ducks. The data on use of reproductive and molecular technologies in ducks is given below (FAO 2015).

- Number of countries reporting presence of the species (n) 43
- Artificial insemination (%) 9
- Embryo transfer (%) 2
- Molecular genetic or genomic information (%) 7
- Multiple ovulation and embryo transfer (%) 2
- Semen sexing (%) 2
- In vitro fertilization (%) 2
- Cloning (%) 2
- Genetic modification (%) 5
- Transplantation of gonadal tissue (%) 2

3.2.7 Duck Breed Descriptors

The person preparing data (compiler) should keep in mind the value of information on breed characteristics within defined environments and its future utilization of animal genetic resources in other similar or dissimilar environments. Therefore, the compiler should ensure the collection of valid genetic and associated environmental information and preparation of this information in a form suitable for entry into the Descriptor Lists. The presentation is separately done for physical characteristics (in Master Records) and performance and environmental characteristics (in Slave Records). A detailed breed descriptors to be used in characterizing the duck populations is available from FAO website (http://www.fao.org/3/ah807e/AH807E08.htm; accessed on 30th May 2020).

The Master Record of Descriptor List should contain general information like name of the breed, its synonyms and the strains within the breed. The size of the farm is grouped as those having less than 10, 11–50, 51–100, 101–200 and above 200 birds. The type of stock should be described as indigenous, improved or industrial, whereas the purpose is referred as for eggs, meat, dual or fancy. The colour pattern and other characteristics of feather, colour of skin, skeletal variations like crested, body carriage, heat tolerance, disease tolerance and conservation status are to be recorded.

In Slave Record, the details on terrestrial environment, elevation and topography, climate, temperature, rainfall, relative humidity, etc. are to be included. The details on socio-management system, farm type, mating and incubation methods, flock size and flock composition should also find place. The details on management practices like feeds and feeding, housing, lighting and disease control should also be included. The egg production traits to be studied include age at first egg, 50% and peak egg production, clutch length, hen-day, hen-housed and survivor egg production, size, shape index, shell colour and internal and external egg qualities.

The reproductive characters to be recorded are broodiness, fertility and hatchability of common and Muscovy ducks and their crosses. The growth and conformation traits include body weight, keel length, shank length and breast angle at different growth stage. The deformations in body confirmation to be recorded are breast blisters, crooked keels and other carcass defects. The other production traits to be noted are feed efficiency during different stages of growth, feather and down yields and mortality rate. The data on tolerance to temperature and humidity extremes, industrial floor housing and cage housing can be recorded. The genetic parameters to be calculated are heritability and correlation for various production traits.

For Muscovy ducks, the breed descriptor list is almost similar to that of common ducks. Additionally, the Muscovies are also classified as feral or wild in addition to duck categories. The fatty liver yield should also be recorded.

3.3 Qualitative Genetics

The pictorial guidance for phenotypic characterization of chickens and ducks is provided by Cuesta (2008).

3.3.1 Plumage Colour

In birds, brown, grey and black plumage colours are due to the deposition of eumelanin pigments, whereas reddish-brown feather colours are caused by pheomelanin. Conversely, carotenoid pigments usually generate yellow to red colourations. Roulin and Ducrest (2013) reviewed the genetics of colouration in birds in detail. Both eumelanin and pheomelanin pigments are synthesized in dedicated organelles called melanosomes in melanocytes. When cysteine and cysteinyl-dopa are depleted, eumelanin synthesis starts. In the presence of the dopachrome tautomerase (DCT) or tyrosinase-related enzyme (TRP2), dopachrome produces 5, 6-dihydroxyindole-2-carboxylic acid (DHICA) that is further oxidized by tyrosinase related enzyme 1 (TYRP1) or TYR and polymerized to form eumelanin pigments. Therefore, a mixture of pheomelanin and eumelanin polymers is produced by melanogenesis. The melanocortin-1 receptor (Mc1R) protein mainly regulates the relative amount of eumelanin to pheomelanin pigments.

3.3.1.1 Plumage Colour in Common Ducks

The ducks are characterized by the colour distribution in the head, neck, breast and wings. The term ‘Mallard’ is used to designate the plumage pattern of wild duck or stock reared from it in captivity. There are three alleles responsible for the formation of white spots, namely bib (b), recessive white primaries (w) and white marking (R). The first two loci were found to be simple recessive genes, while R is an incomplete dominant locus. The alleles of the mallard pattern are distinguished in adult ducks by the presence or absence of white in certain regions of the plumage (Jaap 1933a). Another multiple allele locus M^R (restricted Mallard) is dominant to M (Mallard) that in turn is dominant to m^d (dusky Mallard), also involves in controlling the colour pattern (Jaap 1934). In 1984, Campbell and co-workers reported a fourth allele in this locus, M^B, which is dominant to all the other three alleles (Jaap 1934) and responsible for black plumage in the following order of dominance: $M^B > M^R > M > m^d$ (Campbell et al. 1984).

'Bib' is white markings roughly occupy the area of the breast normally covered by claret brown in the Mallard male and extend upwards on to the ventral surface of the neck. The dominant allele of 'dominant bib' (S) locus in Blue Swedish and Blue Orpington breeds is responsible for appearance of bib, while homozygous recessive genotype does not express the bib character (Lancaster 1963). Whereas Jaap (1933a) reported another locus namely 'recessive bib' with its recessive allele (b) causing bib in pure Mallard stock.

The other two loci influencing the feather colour are 'white primaries' (w) and 'runner' (R). The recessive white primaries (w) allele and dominant runner (R) allele are responsible for white primary wing feathers in ducks (Jaap 1933a). The runner gene is also responsible for the pigmented cap and cheek markings, white cheek marking and white neck in Fawn and white varieties of Indian Runner breed.

A summary of the genes, which affect white spotting in the various parts of the body as described by Jaap (1934) is given in Table 3.4.

Jaap (1933b) described an autosomal mutant allele in semi-domesticated Mallards, which he called 'light phase' (li) that lighten the colour of the adult plumage. The dominant allele of this factor, 'dark phase' (Li) allows full expression of any one of the three 'Mallard' alleles described previously.

The M allele allows the expression of the wild plumage colour pattern of the adult Mallard duck. A recessive epistatic locus (cc) prevents the expression of colour. The C locus controls melanin synthesis, the C allele permits melanin synthesis, while cc genotype prevents this.

Table 3.4 The genotypes of mutant alleles causing plumage variations in common ducks

	Genes producing	
Section of the plumage	White	Colour
Anterior breast	b	B
Posterior breast	RR	Rr, rr
Ventral wing surface	M^R, M	m^d
Dorsal wing surface	M^R	M, m^d
Neck ring of male	M^R, M	m^d
Neck	RR	Rr, rr
Primaries	RR, Rr, w	Rr, w
Secondaries	RR	Rr, rr

In addition to the C locus, another autosomal locus (t) carried by the white Liancheng breed of duck also determines white plumage in ducks. The homozygous dominant genotype (CCTT) at both loci can interact with each other to produce the grey phenotype. In F2, the dosage effect of the T allele results in a dark phenotype (Gong et al. 2010).

Apart from the two multiple allele loci already described, an extended black gene and three dilution genes, two of which are sex-linked also control plumage colour. The fully extended black is autosomal dominant allele (E) to non-extended black (e) (Phillips 1915; Jaap and Milby 1944). The dominant allele (E) causes solid black pigment in all areas except those that are controlled by genes for white spotting. The dominant allele of blue dilution (Bl) locus dilutes black colouration of plumage in almost all domesticated poultry (Wright 1902) in an imperfect dominant inheritance and has a greater effect in the homozygous dominant state than in the heterozygous state (Jaap and Milby 1944). The recessive sex-linked allele of brown dilution locus (d) exerts main effect on the black pigment changing it to a dark brown. In day-old harlequin phase ducklings, the homozygous or hemizygous dominant allele (D) causes an apparent smokiness, whereas its recessive allele (d) causes less apparent smokiness due to its dilution action (Lancaster 1990). Conversely, the dominant allele of buff dilution locus (Bu) interacts with the alleles of blue (Bl) and brown (d) dilution loci to cause darker down colour and darker bill, legs and feet in day-old male ducklings than females Lancaster (1963).

A magpie feather pattern is formed when the extended black (E), dominant bib (S) and runner (R) loci are homozygous dominant (E/E S/S R/R). In magpie pattern, most of the body surface is white feathered except a coloured cap, mostly black, and a patch of same colour on the back that runs down to shoulder joint in front and the tail in the back.

Gong et al. (2010) reported a new autosomal locus T that governs melanin transfer to the feathers. Liancheng White breed of Chinese indigenous duck (genotype CCtt) has pure white plumage but black beak, grey black or black shank and webs. Li et al. (2012) demonstrated that *Tyr* (tyrosinase) and *Tyrp1* (tyrosinase-related protein-1) genes are expressed in white feather bulb but not in black feather bulb.

3.3.1.2 Plumage Colour in Muscovy Ducks

The wild type phenotype of Muscovy ducks is similar to their domestic counterparts with black plumage. There are few mutant plumage variants reported. The plumage colour variations are reviewed in detail by Avanzi and Crawford (1990), and the summary of plumage colour variants of Muscovy ducks as described by them is given below.

1. ***Atipico, dusky***: Atipico is a wild type like phenotype in Muscovy ducks but with complete replacement with seal-brown feathers. The alleles for atipico and dusky are homologous and the cross between the two variants produces dusky-atipico. This plumage colour variant is inherited by an autosomal recessive gene (a, A^+).

2. ***Barring***: White transverse bars on a colour background is called barring. This is an autosomal recessive character (b, B^+).
3. ***Brown rippled***: This is the plumage pattern of sepia bars in grey background, caused by an autosomal recessive gene (br, Br^+).
4. ***Chocolate***: This is a brown variant of wild type due to hindered eumelanin production caused by a sex-linked recessive gene (ch, Ch^+).
5. ***Blue dilution***: Blue or slate-grey phenotype is produced by incomplete dominance when back and pearl-grey phenotypes are crossed; inherited by an autosomal incompletely dominant gene (N, n^+).
6. ***Lavender***: An autosomal recessive gene (l, L^+) is responsible for this colour variation. In which, the adults are dull grey in colour without any sheen to the feathers.
7. ***Sepia, Faiogeno***: The plumage of adults is sepia with metallic green sheen, caused by an autosomal recessive gene (f, F^+).
8. ***Self-white***: This plumage pattern results due to autosomal incompletely dominant gene (P, p^+). The homozygous dominant genotype is black, heterozygous individual is black with white speckles and the homozygous recessive genotype is white.
9. ***Duclair piebald***: White plumage with black colouration on the head, back, saddle and tail. This character is inherited by an autosomal recessive gene (d, D^+).
10. ***White head, canizie***: The head and upper third of the neck are white with coloured plumage in the remaining parts; controlled by an autosomal incompletely dominant (C, c^+).

3.3.2 Skin/Shank/Web/Feet/Bill Colour

The skin colour is black or black with yellow tint or yellow in ducks (Morduzzaman et al. 2015) and white or yellow in Muscovy ducks (Oguntunji and Ayorinde 2015). The shank colour may be yellow or slate or black (Cuesta 2008). Other variants of shank colour are green and yellow-green (FAO 2009) and ash colour (Oguntunji and Ayorinde 2015). In general, the web and feet colour shades match with the shank colour. The web colour in ducks could be black or black with yellow taint (Morduzzaman et al. 2015). In addition, ash and grey feet colours were reported in Muscovy ducks (Oguntunji and Ayorinde 2015). The common bill colours in ducks are yellow, pink-white, orange, slate grey and black (Cuesta 2008). Other bill colour variants are green, green-black, pink-black, white, white-black, yellow black and yellow-green (FAO 2009). In addition red, ash, brown and white bill colours are seen in Muscovy ducks (Oguntunji and Ayorinde 2015).

The skin colouration is determined by xanthophyll and melanin; the combined effects of the duo result in varying phenotypes (Mancha et al. 2006). It has been established that yellow skin is influenced by the amount of carotenoids, primarily xanthophylls in the feed (Hutt 1949). The mature drakes have more yellow bill and skin than ducks. Once laying started, the bill colour will bleach due to the depletion

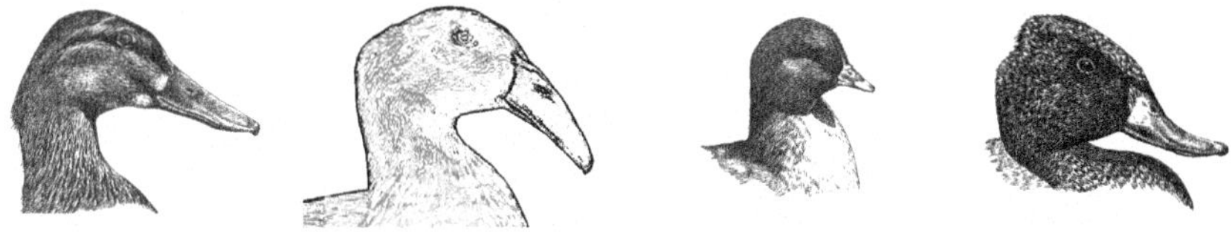

Fig. 3.4 Different shapes of bills in duck

of xanthophyll, the pigment that gives yellow colour to the bill and skin, for yolk formation. Rendel (1940) provided the evidence that yellow bill is inherited by an autosomal recessive gene (yy). The plumage colour genes such as C, R and E are epistatic to Y locus and thus cause different dark skin colours. It is likely that white and ash plumage genes are also responsible for the yellow feet/web (Oguntunji and Ayorinde 2015). Conversely, genes coding for black plumage colour are responsible for black and slate/grey feet colours.

3.3.3 Bill Shape

The bill may be uniform or saddled shaped, or hooked or short (Cuesta 2008; FAO 2009). The saddle bills have upward curve, while the hook bill has downward curve towards the tip. The Hook Bill or Dutch Hook Bill is an ancient breed of domestic duck illustrated by Charles Darwin. The call ducks have short bills. The shape of the bill in ducks is illustrated in Fig. 3.4. The genetic mechanism of control of bill shape in ducks has not been in investigated.

3.3.4 Bean Colour

The small round bump on the end of a duck's upper bill is called as 'bean', also called as 'nail'. The bean can be white, or black or dark brown or white or yellow in colour (Cuesta 2008; FAO 2009), and in addition ash and red colours are reported in Muscovy ducks (Oguntunji and Ayorinde 2015). The colour of the bean may be of matching or contrasting shade to that of beak colour. Sex related variations are seen in bean colour with male's bill in most coloured varieties is paler and sometimes greener than the female's. The beans in some varieties, particularly the Khaki Campbell and Buff Orpington are darker in females. Lancaster (1963) described the influence of buff dilution locus (Bu) on bean colour with dominant allele producing darker shade than the recessive one.

Fig. 3.5 Crested duck

3.3.5 Crest

In some breeds a tuft of feathers grow on the top of the skull is otherwise called 'crest' as shown in Fig. 3.5. Rust (1932, as cited in Lancaster 1963) found that this character is influenced by incompletely dominant autosomal gene (Cr). The heterozygous genotype (Cr/cr) exhibits this character. A double dose of the crested gene (Cr/Cr) will cause the duckling's brain to develop improperly and outside of the skull. This lethal genotype develops fully as embryo but dies inside the shell before hatching. The recessive genotype (cr/cr) results in normal un-crested phenotype.

3.3.6 Eye Colour

The variations of eye colour are yellow and dark (Cuesta 2008), black, blue, brown and grey brown (FAO 2009) and black and ash (Morduzzaman et al. 2015) in ducks. The age and breeding season related changes in eye colour influenced by hormones occur in ducks like that of many other avian species. The pigments like pteridines and purines, carotenoid and melanin are the components of pigment cells in avian iris. Oliphant (1987) demonstrated that pteridines and purines were present in iris pigment cells of most of the birds, but carotenoids could only be extracted from iris of anatidae species. The genetic mechanism that governs the iris colour in ducks is not fully understood so far.

3.3.7 Caruncle Colour in Muscovy Ducks

The caruncle is a small, fleshy excrescence that is present on the head, neck, throat, cheeks or around the eyes of some birds including Muscovy ducks. The colour variations of caruncle are red, black, admixture of black and red and light yellow (Cuesta 2008; Oguntunji and Ayorinde 2015).

3.3.8 Eggshell Colour

The eggshell colour in ducks is generally cream (Kamal et al. 2019). The eggshell colour of Nageswari breed is bluish (Morduzzaman et al. 2015). Chinese breeds like Shaoxing duck, Jinding duck and Shan Ma duck primarily lay green-shell eggs. The first eggs of laying season are black in Cayuga breed; the shell colour may become grey, blue, green or white as season progress. The SLCO1B3 gene codes a membrane transporter, namely OATP1B3 which is considered a liver-specific transporter and is highly expressed in liver where it transports a wide range of substrates including bile salts, which are responsible for the colour of eggshell. Liu et al. (2010) demonstrated that high biliverdin concentration in the uterus fluid but not in shell gland influences the coloured (charcoal, grey, blue and green) or white eggshell in ducks. There could be a mechanism that probably regulates biliverdin transportation from the shell gland into the uterus.

Fig. 3.6 Body carriage in ducks

3.3.9 Body Carriage

The body carriage in ducks can be horizontal, slightly upright or upright (See Fig. 3.6) (Cuesta 2008). The body carriage in Aylesbury and Muscovy ducks is horizontal, while that of Indian Runner and Bali breeds is upright. Most of the other breeds possess slightly upright body carriage.

3.4 Duck Breeding

3.4.1 Phenotypic Values of Important Quantitative Traits in Ducks

The phenotypic values of common quantitative traits are presented in Table 3.5. The body weight of meat-type Pekin ducks ranges between 2290 and 3320 g at marketing age of 6–7 weeks of age, while layer type ducks weigh only around 950 g at 8 weeks of age. During grower stage of 9–20 weeks of age, the local duck breeds in different countries weigh around 1250 g. The body weight reported in birds over 20 weeks in local indigenous breeds ranged around 1250–1600 g. The age at first egg, in common ducks is nearly 120 days and that of Muscovy ducks is 261 days. From the reports available, later ducks have the ability to lay around 140 eggs up to 40 weeks of age and 200 eggs up to 52 weeks of age, while Muscovy ducks can lay only 46 egg in its first year of life. The egg weight of indigenous duck breeds is around 65 g.

3.4.2 Genetic Parameters of Important Economic Traits in Ducks

The genetic parameters of traits include the (i) heritability and (ii) correlation. Heritability is an estimate from a population (represented as h^2) that describes how much of the variation of a trait can be attributed to genetic variation. An estimate of the heritability of a particular trait is specific to one population in one environment. Heritability can change over time due to selection or as any other circumstance changes. Heritability estimates range from zero to one. The heritability values less than 0.2 are described as low, 0.2 to 0.4 as moderate and above 0.4 as high. The genetic correlation (denoted as r_g) is the proportion of variance due to genetic causes that is common to two traits. The linkage disequilibrium (the genes of two traits linked in a same chromosome) and pleiotropy (single gene affects two or more traits) are the two genetic mechanisms that are responsible for correlation. Genetic correlations can be positive or negative, which is indicated by assigning a number in the range from +1 to −1, with zero indicating no genetic correlation. The sign (+ or −) indicates the direction of correlation. The positive values indicate that increase in the value of one trait has positive response, an increase, in the other trait and vice versa. The negative values indicate decrease in the value of one trait will have opposite effect in the other. The numerical value indicates the magnitude of correlation, low, moderate or high.

Table 3.5 Phenotypic values of production, reproduction, egg quality and conformation traits in ducks

Sl. no	Name of the trait	Age/period (trait code)	Breed	Phenotypic value	References
1	Body weight (g)	6 weeks (BW6)	Pekin	2496	Xu et al. (2018b)
			Pekin	3320	Li et al. (2020)
		7 weeks (BW7)	Pekin	2290	Xu et al. (2011)
		8 weeks (BW8)	Brown Tsaiya	997 (♂) and 937 (♀)	Cheng et al. (2002)
		110 days (BW15)	Shan Ma duck	1299	Lin et al. (2016)
		16 weeks (BW16)	Brown Tsaiya	1270 (♂) and 1251 (♀)	Cheng et al. (2002)
		20 weeks (BW20)	Brown Tsaiya	1403 (♂) and1263 (♀)	Cheng et al. (2002)
				1317–1418 (♂) &1248–1329 (♀)	Cheng et al. (1995)
		30 weeks (BW30)	Brown Tsaiya	1397 (♂) and 1315 (♀)	Cheng et al. (2002)
		40 weeks (BW40)	Brown Tsaiya	1405 (♂) and1343 (♀)	Cheng et al. (2002)
				1326–1459 (♂) and 1272–1343 (♀)	Cheng et al. (1995)
		300 days (BW42)	Shan Ma duck	1381	Lin et al. (2016)
2	Age at first egg (d)	(AFE)	Brown Tsaiya	126.0	Cheng et al. (2002)
			Shan Ma duck	109.5	Lin et al. (2016)
			Brown Tsaiya	116.0–125.6	Cheng et al. (1995)
			White Muscovy	261.0	Hu et al. (2004)

(continued)

Table 3.5 (continued)

Sl. no	Name of the trait	Age/period (trait code)	Breed	Phenotypic value	References
3	Egg number	Up to 40 weeks (EN40)	Brown Tsaiya	121.9–147.9	Cheng et al. (1995)
			Shan Ma duck	143.0	Lin et al. (2016)
			White Muscovy	13.0	Hu et al. (2004)
		Up to 52 weeks (EN52)	Brown Tsaiya	197.1–219.2	Cheng et al. (1995)
			White Muscovy	46.0	Hu et al. (2004)
4	Egg weight (g)	30 weeks (EW30)	Brown Tsaiya	62.4–64.9	Cheng et al. (1995)
			Shan Ma duck	65.0	Lin et al. (2016)
		40 weeks (EW40)	Brown Tsaiya	67.4–68.8	Cheng et al. (1995)
			Shan Ma duck	65.9	Lin et al. (2016)
5	Eggshell strength (kg/cm^2)	30 weeks (ES30)	Brown Tsaiya	3.8–3.9	Cheng et al. (1995)
		40 weeks (ES40)	Brown Tsaiya	3.4–3.6	Cheng et al. (1995)
6	Egg yolk weight (g)	40 weeks (EYW40)	Brown Tsaiya	22.9–23.8	Cheng et al. (1995)
7	Relative egg weight % (EW/BW X 100)	40 weeks (REW40)	Brown Tsaiya	4.7–5.1	Cheng et al. (1995)
8	Shell thickness	42 weeks (ST40)	Shan Ma duck	0.31	Lin et al. (2016)
9	Breast muscle thickness (cm)	6 weeks (BMT6)	Pekin	1.75	Xu et al. (2018b)
		7 weeks (BMT7)	Pekin	1.41	Xu et al. (2011)
10	Keel length (cm)	6 weeks (KL6)	Pekin	13.02	Xu et al. (2018b)
		7 weeks (KL7)	Pekin	11.94	Xu et al. (2011)

(continued)

Table 3.5 (continued)

Sl. no	Name of the trait	Age/period (trait code)	Breed	Phenotypic value	References
11	Breast breadth (cm)	6 weeks (BB6)	Pekin	10.71	Xu et al. (2018b)
		7 weeks (BB7)	Pekin	11.30	Xu et al. (2011)
12	Breast muscle weight (g)	6 weeks (BMW6)	Pekin	170.70	Xu et al. (2018b)
13	Breast meat percentage	6 weeks (BMP6)	Pekin	6.69	Xu et al. (2018b)
		7 weeks (BMP7)	Pekin	8.60	Xu et al. (2011)
14	Feather length (cm)	20 weeks (FL30)	Brown Tsaiya	15.3–15.6 (♂) and 15.0–15.5 (♀)	Cheng et al. (1995)

Note: The values reported by Cheng et al. (1995) represent the minimum and maximum values recorded in five generations. The values reported by Hu et al. (2004) represent the pooled mean of seven generations

3.4.2.1 Heritability

The heritability estimates of important traits in ducks are presented in Table 3.6. The data indicate the existence of high degree of additive genetic variance for body weight as exhibited by high heritability values ranging from 0.35 to 0.65. Therefore, this trait can be improved easily by individual selection in ducks.

The other reproductive trait like age at first egg had low heritability values (0.13 to 0.21), while egg number is a low to moderate heritable character (0.12 to 0.44). Conversely, the egg weight had moderate to high heritability estimates ranging from 0.24 to 0.51.

The feeding traits (FI, RFI and FCR) are moderately heritable (0.19–0.41), although higher values for residual feed intake in Muscovy ducks (0.83) and feed intake of Pekin ducks (0.52) are also reported. The egg (0.09–0.33) and carcass quality (0.12–0.28) traits are low to moderately heritable in ducks.

3.4.2.2 Genetic Correlations

The genetic correlations between different trait combinations are given in Table 3.7. The body weights among themselves and egg numbers among themselves have very high positive genetic correlation. In general, the improvement in egg number has negative effect on age at sexual maturity in poultry species. This kind of negative genetic association is evident between these traits in ducks also. The other combination of traits, which have notable correlation between them are body and egg weights as any increase in body weight will bring positive change in egg weight also. Similarly, egg strength had positive genetic association of moderate to high magnitude with egg weight and body weight. The selection for high breast muscle yield (BMY) is possible when selected for breast muscle thickness (BMT), as both the traits are highly correlated with positive sign.

Table 3.6 Heritability estimates of production, reproduction, feeding, egg quality and conformation traits in ducks

Sl. no	Name of the trait	Age/period (trait code)	Breed	Heritability (h^2)	References
1	Body weight	6 weeks (BW6)	Pekin duck	0.48	Xu et al. (2018b)
			Pekin duck	0.48	Li et al. (2020)
			Muscovy duck	0.40 (♂) and 0.51 (♀)	Mignon-Grasteau et al. (1998)
		110 days (BW15)	Shan ma duck	0.35	Lin et al. (2016)
		20 weeks (BW20)	Brown Tsaiya	0.39	Cheng et al. (1995)
		39 weeks (BW39)	Common laying ducks	0.65	Basso et al. (2012)
		40 weeks (BW40)	Brown Tsaiya	0.48	Cheng et al. (1995)
			Brown Tsaiya	0.50	Chen et al. (2003)
		300 days (BW43)	Shan Ma duck	0.54	Lin et al. (2016)
2	Age at first egg	(AFE)	Brown Tsaiya	0.19	Cheng et al. (1995)
			Indonesian native ducks	0.21	Gunawan (1990)
			Shan Ma duck	0.13	Lin et al. (2016)
			White Muscovy	0.20	Hu et al. (2004)
3	Egg number	Up to 30 weeks (EN30)	Shan Ma duck	0.43	Lin et al. (2016)
		Up to 40 weeks (EN40)	Indonesian native ducks	0.39	Gunawan (1990)
			Brown Tsaiya	0.19	Cheng et al. (1995)
			Brown Tsaiya	0.33	Chen et al. (2003)
			Shan Ma duck	0.43	Lin et al. (2016)
			White Muscovy	0.23	Hu et al. (2004)
		Up to 300 days (EN42)	Shan Ma duck	0.38	Lin et al. (2016)
		Up to 52 weeks (EN52)	Brown Tsaiya	0.13	Cheng et al. (1995)
			Brown Tsaiya	0.12	Chen et al. (2003)
			White Muscovy	0.27	Hu et al. (2004)

(continued)

Table 3.6 (continued)

Sl. no	Name of the trait	Age/period (trait code)	Breed	Heritability (h^2)	References
		Up to 72 weeks (EN72)	Indonesian native ducks	0.44	Gunawan (1990)
4	Egg weight	30 weeks (EW30)	Brown Tsaiya	0.35	Cheng et al. (1995)
			Shan Ma duck	0.47	Lin et al. (2016)
		40 weeks (EW40)	Brown Tsaiya	0.30	Cheng et al. (1995)
			Shan Ma duck	0.51	Lin et al. (2016)
		52 weeks (EW52)	Indonesian native ducks	0.24	Gunawan (1990)
		72 weeks (EW72)	Indonesian native ducks	0.42	Gunawan (1990)
5	Feed intake	39 weeks (FI39)	Pekin	0.34	Basso et al. (2012);
		3–6 weeks (FI6)	Pekin	0.33	Zhang et al. (2017)
		Laying stage (FILS)	Shaoxing ducks	0.23	Zeng et al. (2016)
6	Residual feed intake	39 weeks (FI39)	Pekin	0.24	Basso et al. (2012)
		Up to 45 days (RFI6)	Pekin	0.39	Thiele (2016)
		Laying stage (RFILS)	Shaoxing ducks	0.26	Zeng et al. (2016)
		4–7 weeks (RFI7)	Muscovy	0.83	Drouilhet et al. (2016)
		3–6 weeks (RFI6)	Pekin	0.41	Zhang et al. (2017)
		42 46 weeks (RFI46)	Jinyun ducks	0.24	Zeng et al. (2018)
		42 46 weeks (RFI46)	Shaoxing ducks	0.27	Zeng et al. (2018)
7	Feed conversion ratio	3–7 weeks (FCR7)	Pekin	0.52	Pingel (1999)
		4–7 weeks (RFI7)	Muscovy	0.41	Drouilhet et al. (2016)
		Laying stage (FCRLS)	Shaoxing ducks	0.18	Zeng et al. (2016)
		3–6 weeks (FCR6)	Pekin	0.38	Zhang et al. (2017)
		42 46 weeks (RFI46)	Shaoxing ducks	0.19	Zeng et al. (2018)
		42 46 weeks (RFI46)	Jinyun ducks	0.19	Zeng et al. (2018)

(continued)

Table 3.6 (continued)

Sl. no	Name of the trait	Age/period (trait code)	Breed	Heritability (h^2)	References
8	Eggshell strength	40 weeks (ES40)	Brown Tsaiya	0.09	Chen et al. (2003)
		40 weeks of age (ES40)	Brown Tsaiya	0.09	Cheng et al. (1995)
		300 days (ES42)	Shan Ma duck	0.20	Lin et al. (2016)
9	Egg yolk weight	40 weeks (EYW40)	Brown Tsaiya	0.19	Cheng et al. (1995)
10	Relative egg yolk weight %	40 weeks (REW40)	Brown Tsaiya	0.33	Cheng et al. (1995)
11	Eggshell thickness	42 weeks (ST40)	Shan Ma duck	0.28	Lin et al. (2016)
12	Breast muscle thickness	6 weeks (BMT6)	Pekin duck	0.12	Xu et al. (2018b)
13	Keel length	6 weeks (KL6)	Pekin duck	0.26	Xu et al. (2018b)
14	Breast breadth	6 weeks (BB6)	Pekin duck	0.28	Xu et al. (2018b)
15	Breast meat weight	6 weeks (BMW6)	Pekin duck	0.26	Xu et al. (2018b)
16	Breast meat percentage	6 weeks (BMP6)	Pekin duck	0.16	Xu et al. (2018b)

3.4.3 Genetic Selection in Ducks

Pingel et al. (2013) have done comprehensive review on breeding and genetics of growth and meat production, as well as egg production and reproduction in waterfowl, recently. Cheng et al. (2003) also reviewed the genetics and breeding of waterfowl. Marie-Etancelin et al. (2008) have discussed the genetics and selection of ducks in France. The common duck species (*Anas platyrhynchos*) share the wild mallard as a common ancestor, and the second species is the Muscovy duck (*Cairina moschata*) which originated from South America, and the last genotype is mule duck which is sterile hybrid resulting from intercross between a Muscovy drake and a common duck female.

The ducks are domesticated poultry, kept for providing egg, meat and feather for human use. The goal of genetic selection in ducks is to produce desirable characteristics as that of selection program of any other poultry. The dual purpose birds are used primary for egg production and occasionally for meat. More recently, birds began to be selected for either meat production or egg production instead of both. This specialization has allowed for much greater advances in egg or meat production characteristics in separate lines. This allows the use of ducks as an

Table 3.7 Genetic correlation between different traits in ducks

Sl. no.	Trait combinations	Trait codes	Genetic correlation	References
1	Between body weights	BW20 vs BW40	0.99	Cheng et al. (1995)
2	Between egg numbers	EN40 vs EN52	0.95	Cheng et al. (1995)
			0.93	Hu et al. (2004)
3	Body weight vs age at first egg	BW20 vs AFE	0.06	Cheng et al. (1995)
4	Age at first egg vs egg number	AFE vs EN40	−0.76	Cheng et al. (1995)
			−0.98	Hu et al. (2004)
		AFE vs EN52	−0.75	Cheng et al. (1995)
			−0.96	Hu et al. (2004)
5	Body weight vs egg weight	BW20 vs EW30	0.46	Cheng et al. (1995)
		BW20 vs EW40	0.60	Cheng et al. (1995)
		BW40 vs EW40	0.62	Cheng et al. (1995)
6	Body weights vs egg number	BW20 vs EN40	−0.15	Cheng et al. (1995)
		BW40 vs EN52	0.01	Cheng et al. (1995)
7	Egg number vs egg weight	EN40 vs EW30	−0.21	Cheng et al. (1995)
		EN40 vs EW40	−0.35	Cheng et al. (1995)
		EN52 vs EW30	−0.20	Cheng et al. (1995)
		EN52 vs EW40	−0.32	Cheng et al. (1995)
8	Egg number vs egg strength	EN40 vs ES30	−0.11	Cheng et al. (1995)
		EN40 vs ES40	−0.33	Cheng et al. (1995)
		EN52 vs ES30	−0.19	Cheng et al. (1995)
		EN52 vs ES40	−0.20	Cheng et al. (1995)
9	Egg weight vs egg strength	EW30 vs ES30	0.53	Cheng et al. (1995)

(continued)

Table 3.7 (continued)

Sl. no.	Trait combinations	Trait codes	Genetic correlation	References
		EW30vs ES40	0.32	Cheng et al. (1995)
		EW40 vs ES30	0.58	Cheng et al. (1995)
		EW40 vs ED40	0.48	Cheng et al. (1995)
		EW42 vs BS42	0.26	Lin et al. (2016)
10	Body weight vs egg strength	BW20 vs ES30	0.39	Cheng et al. (1995)
		BW20 vs ES40	0.36	Cheng et al. (1995)
11	Breast muscle thickness vs breast muscle percentage	BMT6 vs BMP6	0.63	Xu et al. (2018b)

affordable source of animal protein because of their ability to gain very high body weight with less amount of feed. The high genetic potency also allows the eggs to be relatively inexpensive because of the ability of laying ducks to lay more eggs in a row.

3.4.3.1 Selection to Develop Meat-Type Ducks

The world production of duck meat was 4.4 million tons in 2013. Most of the duck meat produced came from Pekin Ducks, and roughly, 90% of it was based on fattened Pekin ducks. Muscovy ducks produced only 4%, and the remaining 6% was by-product from mule ducks, used for the fatty liver production (Thiele 2016). The traits like body weight, growth rate, skeletal development, feed efficiency, liveability and carcass composition are the important economic characters of meat-type ducks. As feed cost involves nearly 70% of the cost of production, feed efficiency or efficiency of feed conversion by the duck determines profit margin also. A strong skeleton is very important to support the body weight. The bones are synthesized from calcium, so a bird's ability to absorb calcium and convert it into bone is very important. The ability of the bird to withstand disease and other environmental factors is reflected as liveability.

Growth Rate, Body Weight

The waterfowl grow very fast growth during their juvenile life. The sigmoid growth curve of waterfowl is characterized by the initial specific growth followed by asymptotic weight gain with an age at inflection point. In Pekin duck, the inflection point (or maximum growth) occurs early between 24.1 and 26.8 days for females and between 22.5 and 27.6 days for males. The growth rate of mule ducks is intermediate with inflection point occurring between 25 and 30.5 days for males (Marie-Etancelin et al. 2008). In Muscovy duck, this occurs late at 30 days in females and between

34.6 and 37.2 days in males (Marie-Etancelin et al. 2008). In Pekin duck, Pingel (1999) showed that selection to improve food efficiency could be efficient and had a correlated response in decreasing fatness. Heritability estimation of body weight of Pekin duck at slaughter age was high (0.49–0.66) (Thiele 2016; Xu et al. 2011). Heritability of body weight of Muscovy duck at 10 weeks of age was found to be medium (0.24 and 0.31 in males and females, respectively) to high (0.36 and 0.43 for males and females, respectively) for body weight at 18 weeks of age (Hu et al. 1999). The heritability values of the mule duck body weight reported by Larzul (2002) ($h^2 = 0.29$) and Brun and Larzul (2003) ($h^2 = 0.31$) are lower than that of Pekin. Hu et al. (2014) found that after five generations of selection, the cumulated selection responses for tenth week body weight were significant in males (173 g), but not significant in females (60 g). In general, the growth rate has heritability of high magnitude. Therefore, individual selection on this trait is an efficient way of improving duck's growth performance.

Feed Conversion Ratio

The feed cost contributes a major portion to the total cost of duck production. Improvement in feed efficiency would reduce the amount of feed required for growth, the production cost and the amount of nitrogenous waste. The feed efficiency (FE) or feed conversion ratio (FCR) is a trait usually related to productive efficiency, indicating the efficiency of ducks in converting feed into the outputs like meat or egg. FCR of the Pekin duck is moderate ($h^2 = 0.21$ (Powell 1984); 0.38 (Zhang et al. 2017)) to highly heritable ($h^2 = 0.52$, Klemm et al. 1994), individual selection for improving this traits is possible. Retailleau (1999) estimated the FCR at 2.5 and 2.9 for Pekin males and Muscovy males between 0 and 7 weeks of age and between 0 and 12 weeks of age, respectively. For a female mule duck, FCR ranges from 2.12 between 0 and 42 days of age, to 4.43 between 0 and 84 days of age (Guy et al. 1998). Selection on individual feed consumption and constant body weight will result in divergent lines for FCR. As a correlated response, there will be an increase in protein content and decrease in fatty proportions in the carcass of Pekin ducks selected for a low feed conversion ratio. The improvement in feed conversion ratio in Pekin lines was partially transmitted to their mule offspring (3.20 Vs 3.66 at 10 weeks or age for Low FCR and High FCR mule ducks, respectively) (Larzul et al. 2004).

Improvements in feed efficiency can also be achieved by selecting for residual feed intake (RFI). The residual feed intake is defined as the difference between observed feed intake and predicted feed intake calculated based on the estimates of maintenance and production requirements. The residual feed intake is independent from body weight and is a useful parameter to judge the conversion of feed to body mass (Thiele 2016). The heritability values of feed intake and residual feed intake estimated by Basso et al. (2012) were 0.34 and 0.24, respectively. They also found that the residual feed intake was strongly genetically linked to feed intake ($r_g = +0.89$). The heritability estimate of residual feed intake is found to be moderate to high in magnitude ($h^2 = 0.33$) (Thiele 2016); 0.41 (Zhang et al. 2017).

The earlier studies suggested that the animals selected for high feed efficiency, eventually become not only susceptible to stresses and challenges but had negative effect on other productive traits. However, no significant correlation between feed efficiency and feeding behaviour traits was identified. Although the selection for residual feed intake had no effect on mule progeny liver weight and quality, it had slightly deleterious impact on mule progeny meat quality (Drouilhet et al. 2016). Conversely, selection for feed efficiency resulted in significant effect on skin percentage and fat content between high and low feed conversion ratio duck lines after selection for 11 generations. The percentage of skin with subcutaneous fat was markedly reduced. The reduction of carcass fat is an important reason for differences in feed conversion ratio (Pingel 2011).

Meat Yield and Breast Thickness

A growing duckling develops muscles in different body parts in different growth stages. The breast muscle growth happens later after the development of legs, flight muscles and feathers. Selection for breast muscle will indirectly improve feed efficiency. However, the late development of breast muscle is the limiting factor for selection for low marketing age. Further, the meat yield of ducks can be accurately measured only after the slaughter of the bird, that will ends up in loss of pedigreed birds. However, the breast meat yield is significantly correlated with the thickness of breast meat. Thickness of breast meat can be measured in living duck; therefore, selection for higher breast meat yield is possible (Pingel 2011). Furthermore, there exists positive genetic correlation between breast meat thickness and body weight at a fixed age, which means breast meat yield can be improved indirectly when ducks are selected for body weight (Thiele 2016). The heritability values of breast thickness of different lines of Pekin duck were found to be moderate ranging from 0.15 to 0.32 (Duggan et al. 2017; Pingel 1999; Thiele 2016). Breeding for higher breast meat yield is possible, only when it is recorded in the live bird by measuring the thickness of breast meat (Pingel 1999) without slaughter. The combined selection for breast muscle thickness and yield of breast and leg muscle of sibs has improved the proportion of breast and leg muscle from 26.2% to 31.7% after 10 years of selection (Kain 1988, as cited in Pingel 1999). The thickness of the breast muscle had been measured traditionally by a needle probe, which was later replaced by non-invasive ultrasound probing to avoid harming the live ducks. Nevertheless, multi-dimensional measurement in quick time can be done by subjective assessment by experienced technicians; therefore, this method is also in practice today. Xu et al. (2004), however, showed that the breast muscle weight can potentially be improved by employing a selection index combining the measurements of live duck weight, keel weight, breast width and breast meat thickness. The full carcass dissection of smaller number of pedigree birds from sire line is followed by industrial breeding companies to improve the carcass quality (Cherry and Morris 2008). Therefore, the main objective of successful duck breeding is to increase the percentage of breast muscle and to improve feed efficiency by direct selection for feed conversion ratio. As the relationship between breast meat thickness and the percentage of breast muscle to carcass is 0.73, which is high enough for direct individual selection. The realized heritability for both the traits in this series of experiments had been 0.39 and 0.32, respectively, thus confirming the estimated values (Pingel 2011).

3.4.3.2 Selection to Develop Egg-Type Ducks

The FAO statistics indicated that Asia contributes about 95% to the world's duck eggs other than chicken egg production. Most of the Asian countries have their traditional egg-producing breeds. Maximizing duck egg production depends on good genetics and good management. Many studies have aimed to improve important economic traits of egg-type duck, such as medium body size, early maturation, high egg yield, adaptation to high ambient temperature and resistance to disease. The heritability of egg number is low (0.1–0.3), and selection for genetic improvement of egg production is expensive.

The traditional breeding method is still used without pedigree and performance tests in many countries, where people use appearance judgement and random mating. In Taiwan, the farmers select breeders at the end of the production season using their own judgement, the sires and dams then mate randomly (Chou and Huang 1970). Simple selection of females with good egg production by independent culling is one way to improve population's laying performance.

The age at start of lay, body weight at maturity, egg production (egg number), egg weight, egg quality, feed efficiency and liveability are the major traits which a duck breeder likes to improve for increasing duck egg production. When egg production reaches 5%, it is considered as the age at maturity of the flock. The age at sexual maturity of the flock is very important as it decides feed efficiency, egg weight and egg number also. Lower body weight corresponds to poor growth of egg forming female reproductive tract, which in turn will result in poor egg production and low egg weight. Conversely, higher body weight at maturity will lead to higher feed consumption and reduced persistency. Feed efficiency is efficiency of conversion of feed into product, which is calculated in terms of conversion of feed into number of eggs or weight of eggs. The external egg quality traits of the egg will be assessed by egg size, shape, shell colour, volume, specific gravity or surface area. The internal egg quality traits include albumin and yolk indices, Haugh unit, yolk colour, shell thickness and presence of blood and meat spots.

Multi-Trait Selection for Egg Production and Related Traits

In Taiwan, the major laying duck is Brown Tsaiya, and a selection of Brown Tsaiya for the four traits like egg weight at 40 weeks of age (EW40), body weight at 40 weeks of age (BW40), number of eggs laid up to 52 weeks of age (EN52) and eggshell strength at 30 or 40 weeks of age (ES30 or ES40) has been conducted in order to obtain maximum gain for egg laid and keep egg weight and body size constant. Some genetic parameters of the reproduction characters from the Brown Tsaiya duck, for a period of 30 years, were calculated by sire and dam within sire variance components (Tai et al. 1989; Lee et al. 1992).

The genetic parameters were estimated by Cheng et al. (1995) in the base population by REML applied to a multiple trait animal model based on data of the first five generations of selection. The heritability of egg numbers and eggshell strength were low (0.09–0.16), egg weight was moderate (0.33–0.35) and body weight was high (0.42–0.50). The egg number at 52 weeks of age has high positive correlation with egg number at 40 weeks of age ($r_g = 0.95$), uncorrelated with body weight and negatively correlated with egg weight and eggshell strength. Egg weights, body weights and eggshell strength traits had positive genetic correlation among themselves. These results suggested that a linear selection index for EN52 with constraints for EW40, BW40 and ES40 could be an efficient tool for improving the efficiency of egg production. Then, a new linear genetic selection index suggested by Cheng et al. (1996) was used:

$$I = a_0 * \text{GEW40(g)} + a_1 * \text{GBW40(g)} + a_2 * \text{GES40}\left(\text{kg/cm}^2\right) + a_3 * \text{GEN52(eggs)}$$

where GEW40, GBW40, GES40 and GEN52 were the MT-BLUP animal model predictors of the breeding values of EW40, BW40, ES40 and EN52, respectively. The drakes and the ducks were selected according to their own index values, with this new genetic selection index. After four generations of selection, the average predicted genetic responses per generation was 0.05 g for EW40, 0.92 g for BW40, slight increase of 0.035 g/cm^2 for ES40 and 2.13 eggs for EN52. These results indicated that the selection of laying type duck with a restricted genetic selection index could be an efficient way to improve egg production while arresting any change in egg weight and body weight (Chen et al. 2003).

There are a lot of distinctive laying duck breeds in China, such as Shaoxing duck, Jinding duck, Shan Ma duck, Youxian Sheldrake duck and Jingjiang duck, whose egg production of 500 days of age are 280 to 320 eggs and feed-egg ratio is 2.65–2.85. The Shan Ma duck (*Anas platyrhynchos*) is one of the main Chinese indigenous egg laying duck breeds, and heritability estimated using restricted maximum likelihood methodology was high for egg weights (ranging from 0.43 to 0.61), moderate to high for the number of eggs laid (ranging from 0.38 to 0.43) and low for the age at first egg (0.13) (Lin et al. 2016).

Fertility and Hatchability

Fertility and hatchability are usually better in egg-type strains than in meat-type ducks and there may be no need for further selection. If either or both of these traits do need genetic improvement, progeny testing is the only effective method, since both traits have low heritability (0.1–0.2). It should be standard hatchery practice to rejects eggs that are abnormal in appearance, but this culling of extremes rejects only a small proportion of hatching eggs. It therefore exerts only a very small selection pressure and is more effective in preventing a negative drift than in achieving any real improvement in hatchability (Cherry and Morris 2008).

Eggshell Colour Selection

The pigmentation of avian eggshell is a complex trait exhibited by the presence of two pigments, namely biliverdin and protoporphyrines. Biliverdin is a by-product of haemoglobin breakdown pathway and protoporphyrine is the precursor of heme molecule. Biliverdin gives blue or green colour, while protoporphyrines give reddish or brown colour to the eggshells (Kennedy and Vevers 1976; Mikšík et al. 1996). It is often a cultural preference for one colour over another. Some blue-egg lines of ducks have been selected for attractive blue shells for many years, based on subjective evaluation and quantitative measurement of shell colour. The major reason which prompts the breeder to select special egg colour depends on customer's preference.

The study on inheritance of pigment deposition in the Tsaiya duck eggs indicated that the genes controlling blue colour shells have dominant effect over white-shelled eggs (Liu et al. 1998). The heritability of eggshell colour grade at 50 weeks of age and 71 weeks of age was 0.69 and 0.61, respectively (Hu et al. 2002). Liu et al. (2011) implied that expression of heme oxygenase and delta-aminolevulinate synthase mRNA of shell glands in blue-shelled duck and white-shelled duck during the ovulation period did not show significant difference. Results of this study also showed significant differences in the biliverdin concentration in the uterine fluid, but not in the serum between blue-shelled duck and white-shelled duck. The results demonstrated that blue- and white-eggshell colours were influenced by the amount of biliverdin in the uterus fluid and not determined by the amount of biliverdin in the shell gland. This implies the existence of a mechanism that controls the biliverdin transportation from the shell gland into the uterus fluid, thereby playing a key role in regulating duck eggshell colour (Liu et al. 2010). The distinction of DNA fingerprints in Tsaiya ducks with different eggshell colours had been studied by Huang et al. (2006a). The results demonstrated that the DNA fingerprints between blue and white-shelled ducks were different and the blue eggshell ducks also have the highest yield rate of alkalized eggs followed by the high eggshell strength ducks and then the ducks from commercial hatchery.

Although there is a widespread belief that pigmented eggs are stronger than white eggs, conflicting evidence has also been reported. We cannot conclude that eggshell colour alone would be a sufficient estimator of eggshell quality. Although eggshell colour has been studied intensely in chicken species, the origin of eggshell and the molecular mechanisms that control this property are not well understood. Further identifying the molecular basis of duck will advance our understanding on the evolution of the trait in the avian phylogeny and will also aid breeding for the eggshell colour in the ducks (Wang et al. 2017; Xu et al. 2018a).

In Taiwan, a blue-shelled egg line was established from the Brown Tsaiya duck and the colorimeter is employed to define the eggshell colour grade as the criteria for selection. After eight generation of selection, results indicated that selection of laying Brown Tsaiya by 'a' value of eggshell could be an efficient tool for improving the eggshell blue colour while holding egg production performance.

3.4.3.3 Mule Duck Production

The Pekin ducks, with fast growth rate and 30% of more fat content, is the most popular meat duck in the world. Conversely, there is a growing interest for leaner ducks among health conscious people around the world. Until now, the preference of the French for leaner duck meat has been met by the Muscovy which they call

'*Canard de Barbarie*'. The Taiwanese have preferred to produce lean mule ducks to avoid the problems involved in breeding Muscovies (Hoffman 1993). The early 'Mule' ducks were a Tsaiya and Muscovy cross; the Pekin genes were added in later experiments to improve the growth rate. The mating difficulties due to morphological difference were overcome initially by manual assistance in mating the Muscovy drake to the Tsaiya and later the *Kaiya* duck to produce mules or intergeneric hybrid progenies. This obvious inefficient mating procedure was replaced by artificial insemination. This technique is now common practice in Taiwan's breeding farms as a result of agricultural extension work. So it is Taiwan's integrated program of research and extension work that has enabled the 'Mule' duck to become predominant among the island's cumulative meat bird production that accounts for more than 30 million slaughtered per year. There are two advantages of producing lean meat from mule ducks comparing with Muscovy duck: first, because there is sexual dimorphism for growth of Muscovies (female Muscovy ducklings are only 60% as large as males at market age), and second, that duck mother of the mule duck lays many more eggs than the Muscovy females.

In France, 'Mulard' usually implies the cross between a male Muscovy duck and a female Pekin duck. Occasionally, the female may be a Rouen. 'Mulard' ducks are bred especially for '*foie gras*' and for meat production. France is the largest producer and consumer of '*foie gras*', of which, over 95% is being harvested from force-fed mule ducks. Furthermore, '*magret*', which corresponds only to the breast meat from overfed waterfowl, and '*foie gras*' labelling are officially defined and recognized. The resultant females are valued for their large breasts; the males are valued for '*foie gras*' production, because the males are larger and hardier, making them better able to stand up to overfeeding, and the female livers are often veined, causing them to get downgraded.

Selection for White Plumage of Mule Ducks

The plumage colour inheritance in the common duck has been explained in detail by Lancaster (1990). The mule ducks have been produced in Taiwan since the early eighteenth century (Chou and Huang 1970). The feather colour of mule ducks, before 1960s, produced by mating Black Muscovy drake with Brown Tsaiya ducks was black brown (Fig. 3.7). The carcasses of mule ducks with white plumage colour offered a better appearance and popular among consumers. This initiated the selection for parental lines of white mule ducks. For the demand for white plumage mule ducks, one criterion system of six colour levels of mule ducklings' down had been created, and then the improved criterion system of 15 colour levels have been developed since 1984. The realized heritability of white mule duck's plumage colour estimated by Huang (1985) was 0.38. In these studies, the selection of common ducks for 'bleaching' of mule ducks was carried out on progeny by monitoring the extension of the colour of the hatched ducklings. The selection of one of parental lines, White Tsaiya of generation 27, has reached 99.9% of mule duck whose white plumage colour which was accepted by farmers. A percentage of 99.5% better than grade 7 plumage colour was observed in the offspring mules in generation 14 of the other parental Pekin duck line (Lee and Kang 1997). Even plumage colour and

Fig. 3.7 Different plumage colour of mule ducks

patterns are generally characterized by simple inheritance (Lancaster 1990), the mechanism of white mule ducks needs to be established by more studies.

In industrial duck breeding in France, selection for complete bleaching of feather colour of the mule ducks should not be reached in order to perform day-old sexing of duckling, to allow a coloured spot to appear on the head in male ducklings. The chocolate gene, carried by the chromosome Z, whose recessive allele gives a brown colour to the feathers, allows sexing when the chocolate allele is fixed in the Muscovy population. Male mule ducks have a black spot due to double dose of chocolate gene (ch ch) and the female mule ducks have a brown spot due to single dose of this gene (ch_). This gene, already mentioned in Muscovy duck by Hollander (1970), is also present in many avian species (Carefoot 1996, as cited in Marie-Etancelin et al. 2008).

Lin et al. (2014) aimed to investigate the inheritance of plumage colour in three Chinese indigenous egg-type duck breeds, Shan Ma, Putian white, Putian black and some of their crossbreds. Brown Putian-Ma is a crossbreed of three duck breeds. It has been developed with desired plumage colour and it was tested for egg production.

Selection for Duration of Fertility

The mule ducks are sterile intergeneric hybrids which are produced by crossing Pekin or Kaiya (crossbred Pekin × White Tsaiya) female ducks with Muscovy drakes (*Cairina moschata*). The mule duck meat production is gaining momentum in Europe, Vietnam and Southeast China (Cheng et al. 2009). Before artificial

Fig. 3.8 Mule ducks were produced by application of artificial assisted mating between black Muscovy drake and Brown Tsaiya duck

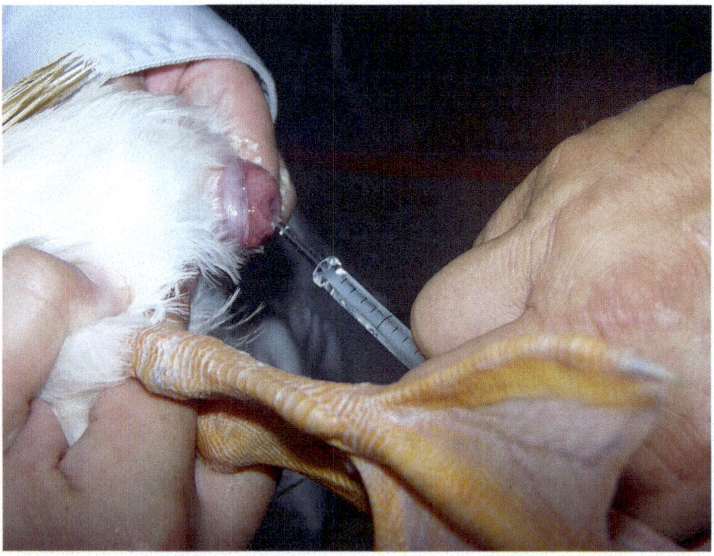

Fig. 3.9 Artificial insemination of duck

insemination (AI) technic was well developed, duck farmers in Taiwan produced mule ducks by artificial assisted mating of Brown Tsaiya ducks with black Muscovy drake and (Fig. 3.8.). This method of mule duck production was labour consuming and the fertility was only about 70%. The reproductive efficiency was dramatically improved over the last 25 years after AI was employed (Fig. 3.9.). This technique is also widely used in France where male mule ducks are overfed to produce fatty liver and the females are used for meat production (Rouvier et al. 1988). Unfortunately, owing to the short duration of fertility in such intergeneric crossbreeding, the ducks have to be inseminated twice a week in order to maintain the fertility rate (Liu et al. 1980; Rouvier et al. 1984). It would be economically beneficial if the duration of the female fertile period could be lengthened and this can enable performing AI once instead of twice a week while maintaining fertility. The time lag between artificial insemination and the last fertile egg laid is termed as the duration of the fertile period (Lake 1975).

A long-term selection program was conducted to improve the number of fertile eggs after single AI in Brown Tsaiya female duck since 1992 (Cheng 1995). The results of this study suggested that the selection for this trait can increase the storage capacity of spermatozoa at the utero vaginal junction. Tai et al. (1994) found that the best selection criterion to improve fertility duration is the number of fertile eggs laid between the 2nd and 15th day after a single AI with pooled Muscovy semen ($h^2 = 0.29$). Poivey et al. (2001) estimated the genetic parameters for the duration of fertility with data been collected from selection of five generations of selection in Tsaiya ducks. Cheng et al. (2005) improved the duration of fertility and hatchability by genetic selection for number of fertile days after AI estimated using logistic models. Cheng et al. (2009) further studied up to 12 generations of selection in selected line and control line, which began in 1992. The mean predicted genetic progress expressed in genetic standard deviation units for number of fertile eggs, maximum duration of fertility and number of hatched mule duckling were0.40, 0.45 and 0.32, respectively. The fertility rate for 2–8 days after single AI was improved to 89.1% in selected line without impairing embryo viability compared to 61.5% of unselected line and not blenching embryo viability by this selection. This selection experiment reveals that selection for the number of fertile eggs after a single AI with pooled Muscovy semen could effectively increase the duration of the fertile period in ducks. The research should now be focused on ways to improve the viability of the hybrid mule duck embryo so as to improve the fertility rate. The parental line of mule duck like White Tsaiya, Pekin and Kaiya ducks was also subjected to genetic selection for improvement of duration of fertilization, and as a result the number of fertile eggs after a single AI with pooled Muscovy semen improved to 5.94, 4.88 and 6.38 eggs, respectively (unpublished data). The results of the study indicated that genetic selection was an effective way to improve duration of fertility in mule ducks (Fig. 3.10) (Liu et al. 2015).

3.4.4 Marker-Assisted Selection (MAS)

Most of the traits economically considered important in duck are quantitative, which means they are affected by many genes with small effect each, thus the benefits of using DNA markers links to genes of interest breeding programs have been looking forward for decades. After the DNA marker technology was well developed, the idea of rapidly uncovering the loci controlling complex, multigenic traits seemed like a

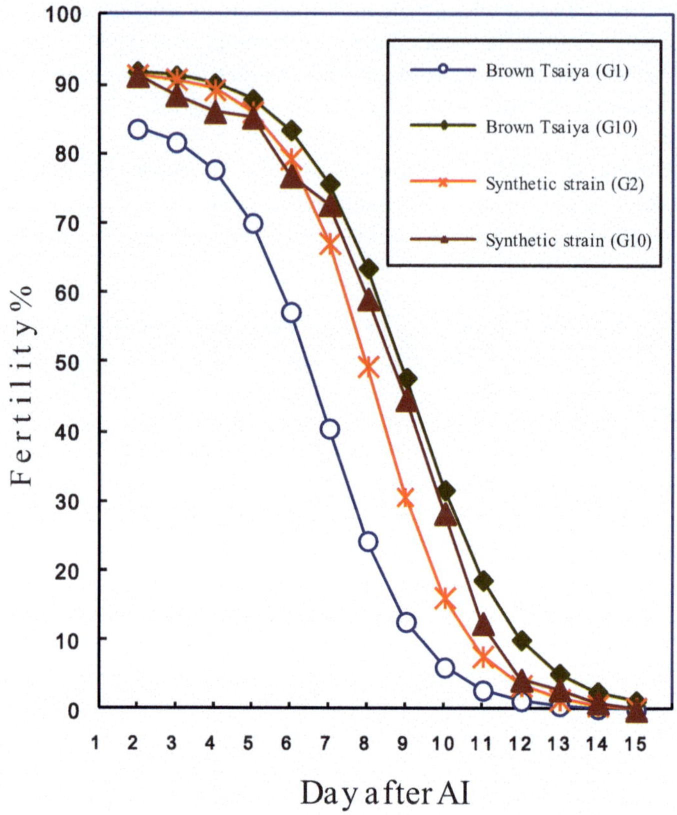

Fig. 3.10 Fertility response for duration of fertility selection in synthetic and Brwon Tsaiya duck

dream come true. An ideal DNA marker should be co-dominated, evenly distributed throughout genome, highly reproducible and having ability to detect higher level of polymorphism. Different kinds of molecular markers were developed, such as restriction fragment length polymorphisms (RFLPs), random amplified polymorphic DNA (RAPDs), amplified fragment length polymorphisms (AFLPs), microsatellites and single nucleotide polymorphisms (SNPs). The information provided to the breeder by the markers varies depending on the type of marker system used.

The use of marker-assisted selection (MAS) is expected to increase the accuracy of breeding value information and is especially useful for traits that have low heritabilities or are difficult to measure. MAS will also allow the improved utilization of available 'selection space'. The approximate time from which the DNA markers are widely used in poultry breeding is the year 2000. The most promising new breeding technologies are DNA related.

Owing to the biological limits which may be reached in the near future, the difficulty and the cost of measuring some new traits, the detection of quantitative trait loci (QTL) by the closely linked DNA markers could improve the efficiency of selective breeding. A few chicken markers could be used as anchor points for a comparative mapping in ducks, developing duck specific microsatellite markers that will be more efficient for the mapping and analysing of the duck genome. The genomic libraries enriched for (GATA)n in Brown Tsaiya duck had been constructed and used in genetic diversity study of duck populations (Hsiao et al. 2008). In addition to microsatellite markers, the amplified fragment length polymorphism (AFLP) markers are studied in Brown Tsaiya duck for duration of fertility. Two hundred and sixty AFLP markers are dispersed in 32 linkage groups. The linkage groups cover 1766 cM, with an average interval distance of 6.79 cM (Huang et al. 2006b). Chang et al. (2000) indicated that the polymorphism of the oestrogen receptor gene in Brown Tsaiya ducks was similar to that observed in pigs. Huang et al. (2011a) indicated that a novel PCR-RFLP polymorphism of ovomucoid gene can be used as a genetic marker for selection to improve hatchability in Tsaiya ducks. The authors performed the first genome-wide expression analysis to compare the differences in gene expression in the female sperm reservoir of the duck reproductive tract between two groups with long and short fertile periods to identify factors that may be associated with the fertile period using an oligonucleotide microarray. The expression of trafficking kinesin-binding protein 1 was increased by 2.33-fold in the long-period group as compared with its counterpart. It has been suggested that this gene product may inhibit the precocious activation of sperm and prolong sperm life in the female sperm reservoir (Huang et al. 2011b).

A novel SNP marker of ovalbumin gene, an important candidate gene associated with duck hatchability can be used for marker-assisted selection to improve hatchability in Tsaiya ducks (Huang et al. 2013). Further, the growth hormone gene polymorphisms are associated with duck fertility-related traits and the SNPs in this gene may be used as potential marker to improve fertility (Chang et al. 2012a). A novel non-synonymous SNP of the COLX gene and its association with duck reproductive traits indicated that selection for the genotype TT and TC ducks might contribute to an improved egg weight in the Tsaiya ducks (Chang et al. 2012b). The whole-genome resequencing was applied to the selected and controlled lines of high feed efficiency Brown Tsaiya duck to screen associated SNPs (single nucleotide polymorphism) over the genome by Chang et al. (unpublished data). After blast to reference sequences of Pekin duck, there were 1.1 million and 1 million SNPs were detected for selected and control lines, respectively. There were 0.3 and 0.2 million private SNPs in the selected and controlled lines, respectively, after deduction of SNPs common in both lines. Out of the SNPs in selected line, 35% belong to the genic region, 4% of those SNPs were located in the coding region and 30% were non-synonymous mutations. A total of 32,628 SNPs were screened out from 335,810 SNPs, of which only 15,455 SNPs were found with final score greater than 0.99. These researchers found several genes located in the above regions may be related with residual feed consumption traits because of the association with growth, reproductive performance and metabolic regulation or located in calcium metabolism or energy utilization pathway. In future, the genomics could well play an important role in supporting breeders in selection programmes. The only relevant qualification of this prediction is on the time scale. Genome-wide association study, which combines SNP chip and performance data, could be applied to investigate the possible regulatory mechanisms of important economic traits. While applying this technique to genome-wide selection, we can improve the accuracy of selection, shorten the generation intervals and accelerate the selection process, furthermore, improve the production benefit.

3.4.5 Modern Devices for Individual Bird's Data Collection

In the last few years tremendous advances were achieved in sensing technology in terms of diversity, accuracy and affordability. Sensors, especially wireless sensors, have a wide range of applications in agriculture. Individual performance recording system is very important for evaluating genetic potential of individuals. In the past, individual productive performance of ducks has been recorded in single cage system, or measured by manpower in specific days. Nowadays, using novel radio-frequency identification (RFID) technology, breeding bodies are able to record even small meals of individuals under nearly practical housing conditions (Fig. 3.11). In these testing units not only the amount of feed consumed can be recorded, it opens the possibility to observe also the feeding behaviour of the ducks, with the frequency of meals and the size of a single meal (Howie et al. 2009). The laying performance is usually tested over a period of production weeks. Traditionally laying birds are kept in single cage systems. Some breeding companies are developing a fully automatic nest system to replace single cages in future. With the new nests the laying performance can be recorded in floor pens. During the testing period, time wise information about egg weight, shell strength, fertility and hatchability are recorded (Thiele 2016). The availability and utilization of these 'new devices' accelerate the genetic progress in duck production traits.

Fig. 3.11 Ducks with RFID tag

Different devices have been developed to measure breast fat cover and muscle depth, and ultrasound probes are widely used. However, these methods are slow and do not simultaneously judge breast shape, keel shape or calcification.

3.4.6 Performance of Commercial Stocks

The performances of commercial stocks of common and Muscovy ducks developed by the commercial breeding firms are given below.

3.4.6.1 Common Ducks (CV 2021)

Parent stock performance of commercial ducks

• Mature live weight of males (g)	4095–4463
• Mature live weight of females (g)	3360
• No. of eggs produced in 50 weeks lay	296
• Settable eggs (%)	97–98
• No. of day-old ducklings/dam	241–244
• Fertility rate (%)	93–95
• Hatchability rate (%)	84–85

Growing performance of commercial ducks

• Live weight at 42 days (g)	3450–3550
• Feed efficiency at 42 days	1.88–1.92
• Liveability at 42 days (%)	98

Carcass quality of commercial ducks

• Breast meat yield at 42 days (%)	17.2–18.3
• Breast fillet yield at 42 days (%)	24.8–25.6
• Skin and fat yield at 42 days (%)	28.8–29.3

3.4.6.2 Muscovy Ducks (GF 2021)

Parent stock performance of Muscovy ducks

• No. of eggs produced (in two laying cycles)	220
• Fertility rate (%)	88–93
• Hatchability rate (%)	90

Growing performance of commercial Muscovy ducks

- Recommended slaughter age (days) 84 (♂); 70 (♀)
- Slaughter weight (g) 4700–5500 (♂); 2550–3000 (♀)
- Feed conversion ratio 2.5–2.8
- Breast fillet yield (%) 16–18

3.5 Conclusion

The evolution of ducks took place in a very ancient period dates back 66 million years in Cretaceous–Paleogene (K–Pg) boundary that marks the end of the Cretaceous Period and the beginning of the Paleogene Period. The current state of knowledge from literature, archaeological evidences and historical information reveals that the domestication of ducks took place independently in China (500 BC) and in Western Europe (800 AD). Conversely, the domestication of Muscovy ducks took place in South American continent at around 1500 AD well before the Columbian Era. As on today, the Domestic Animal Diversity Information System (DAD-IS) of FAO enlists 400 duck and 63 Muscovy duck genetic groups, which includes breeds, strains, lines, etc. Spreading across the world, the list enlarges day by day. The cause for concern today is that as much as 16 duck genetic groups are extinct and 34 ducks and 1 Muscovy genetic groups are endangered. Many more are critical or vulnerable.

The qualitative genetics of ducks are distinct from other birds and the priority to quantitative economic traits for genetic improvement is also different. A great deal of advancement in breeding has been achieved especially in genetic improvement of eggshell colour and elucidation of its mechanism, feather colour and duration of fertility in mule duck production and artificial insemination apart from genetic improvement in conventional egg and meat-type traits. The advances in data collection methods have a great influence in modern breeding research.

It is reasonable to expect that phenotypic selection will continue to be the workhorse of duck breeding, and it seems certain that continued genetic progress in the production potentials of ducks will be made (which means we are still a long way off from the 'ideal' duck production). The real question is the extent to which the continued progress can be achieved with the application of quantitative genetic theory. Certainly computer technology will make it possible to collect and process more data and to employ more sophisticated models, but how far can one go with a technology based on *not knowing* what the genes are actually doing? Meanwhile, molecular genetics (genomics, transcriptomes, chips) technology continues to move forward at speed in the near future. May be new traits could be selected in the future such as the robustness of birds faced to climate changes or instance or the resistance to disease (i.e., avian influenza).

References

Avanzi CF, Crawford RD (1990) Mutations and major variants in Muscovy ducks. In: Crawford RD (ed) Poultry breeding and genetics. Elsevier Science, Amsterdam, pp 389–394

Basso B, Bordas A, Dubos F, Morganx P, Marie-Etancelin C (2012) Feed efficiency in the laying duck: appropriate measurements and genetic parameters. Poult Sci 91:1065–1073

Brun JM, Larzul C (2003) Inheritance of reproductive traits of female common ducks (*Anas platyrhynchos*) in pure breeding and in inter-generic crossbreeding with Muscovy ducks (*Cairina moschata*). Br Poult Sci 44(1):40–45

Campbell RR, Reinhard BS, Jerome FN (1984) A 'new' allele of the mallard plumage pattern in ducks. Poult Sci 63:19–24

Carefoot WC (1996) Chocolate: a sex-linked recessive plumage colour mutant of the domestic fowl. Br Poult Sci 37:867–868

Chang T, Hung TS, Chang LY, Hsieh MH, Chang YS, Tai C (2000) The duration of fertility in ducks: studies on the polymorphism of the oestrogen receptor gene. In: Proceedings of the 9th CAAAAPS and 23rd biennial conference of ASAP, Australia, p 125

Chang MT, Cheng YS, Huang MC (2012a) The SNP genotypes of growth hormone gene associated with reproductive traits in Tsaiya ducks. Reprod Domest Anim 47:568–573

Chang MT, Cheng YS, Huang MC (2012b) A novel non-synonymous SNP of the COLX gene and its association with duck reproductive traits. Mol Cell Probes 26:204–207

Chen MY (1986) Feeding and breeding of Peking duck. Science and Technology Publishing Press, Beijing, pp 1–5

Chen DT, Lee SR, Hu YH, Huang CC, Cheng YS, Tai C, Poivey JP, Rouvier R (2003) Genetic trends for laying traits in the Brown Tsaiya (*Anas platyrhynchos*) selected with restricted genetic selection index. Asia Aust J Anim Sci 16(12):1705–1710

Cheng YS (1995) Sélection de la race Tsaiya Brune sur la ponte et la durée de la fertilité en croisement avec le canard de Barbarie. Thèse. Institut National Polytechnique de Toulouse, France

Cheng YS, Rouvier R, Poivey JP, Tai C (1995) Genetic parameters of body weight, egg production and shell quality traits in the laying Brown Tsaiya duck. Genet Sel Evol 27:459–472

Cheng YS, Rouvier R, Poivey JP, Tai C (1996) Prediction of genetic gains in body weight, egg production and shell quality traits in the laying Brown Tsaiya (*Anas platyrhynchos*). Genet Sel Evol 28:443–455

Cheng YS, Rouvier R, Hu YH, Tai JJL, Tai C (2002) Breeding and genetics of waterfowl. In: Proceedings of 37th world congress on genetics applied to livestock production, August 19–23, Montpellier, France. http://www.wcgalp.org/system/files/proceedings/2002/breeding-and-genetics-waterfowl.pdf. Accessed 25 July 2020

Cheng YS, Rouvier R, Hu YH, Tai JJL, Tai C (2003) Breeding and genetics of waterfowl. Worlds Poult Sci J 59:509–519

Cheng YS, Rouvier R, Poivey JP, Huang SC, Liu HL, Tai C (2005) Selection responses in duration of fertility and its consequences on hatchability in the intergeneric crossbreeding of ducks. Br Poult Sci 46:565–571

Cheng YS, Rouvier R, Liu HL, Huang SC, Huang YC, Liao CW, Tai JJL, Tai C, Poivey JP (2009) Eleven generations of selection for the duration of fertility in the intergeneric crossbreeding of ducks. Genet Sel Evol 41:32

Cherry P, Morris TR (2008) Domestic duck production: science and practice. CAB International, Wallingford, pp 229–234

Chou KY, Huang HH (1970) The duck industry in Taiwan, Animal industry series no 8. Chinese-American Joint Commission on Rural Reconstruction, Taipei

Clake JA, Tambussi CP, Noriega JI, Erickson GM, Ketcham RA (2005) Definitive fossil evidence for the extant avian radiation in the cretaceous. Nat 433:305–308

Clayton GA (1984) Common duck. In: Mason IL (ed) Evolution of domesticated animals. Longman, London, pp 334–339

Clements JF, Schulenberg TS, Iliff MJ, Billerman SM, Fredericks TA, Sullivan BL Wood CL (2019) The eBird/Clements checklist of birds of the world: 2019. https://www.birds.cornell.edu/clementschecklist/download/. Accessed 25 July 2020

Cracraft J, Barker FK, Braun MJ, Harshman J, Dyke G, Feinstein J, Stanley S, Cibois A, Schikler P, Beresford P, García-Moreno J, Sorenson MD, Yuri T, Mindell DP (2004) Phylogenetic relationships among modern birds (neornithes): toward an avian tree of life. In: Cracraft J, Donoghue MJ (eds) Assembling the tree of life. Oxford University Press, New York, pp 468–489

Crawford RD (1990) Chapter 1. Origin and history of poultry species. In: Crawford RD (ed) Poultry breeding and genetics. Elsevier Science, Amsterdam, pp 1–42

Cuesta ML (2008) Pictorial guidance for phenotypic characterization of chickens and ducks. FAO GCP/RAS/228/GER working paper no 15, Rome. http://www.fao.org/3/al702e/al702e00.pdf. Accessed 7 June 2020

CV (2021) Cherry Valley, SM3 product specification. Cherry Valley Farms Ltd, Cherry Valley House. https://bb162a23-e952-48d3-b1f8-8428182b8f6c.filesusr.com/ugd/949670_45e8451249894c97ac1fae0879aaf6dd.pdf. Accessed 4 June 2021

DAD-IS (2020) Domestic animal diversity information system. Food and Agriculture Organization of the United Nations. http://www.fao.org/dad-is/dataexport/en/. Accessed 25 May 2020

Donkin RA (1989) The Muscovy duck, *Cairina moschata domestica*: origin, dispersal and associated aspects of the geography of domestication. A A Balkema, Rotterdam

Drouilhet L, Monteville R, Molette C, Lague M, Cornuez A, Canario L, Gilbert H (2016) Impact of selection for residual feed intake on production traits and behaviour of mule ducks. Poult Sci 95(9):1999–2010

Duggan BM, Rae AM, Clements DN, Hocking PM (2017) Higher heritabilities for gait components than for overall gait scores may improve mobility in ducks. Genet Sel Evol 49:42

FAO (2009) Characterization of domestic duck production systems in Cambodia. Prepared by Dinesh MT, Geerlings E, Sölkner J, Thea S, Thieme O, Wurzinger M, AHBL- Promoting strategies for prevention and control of HPAI. Rome

FAO (2015) The second report on the state of the world's animal genetic resources for food and agriculture. In: Scherf BD, Pilling D (eds). FAO Commission on Genetic Resources for Food and Agriculture Assessments Rome. http://www.fao.org/3/a-i4787e/index.html. Accessed 7 June 2020

Feduccia A (1995) Explosive evolution in tertiary birds and mammals. Science 267:637–638

Feduccia A (2003) 'Big bang' for tertiary birds? Trends Ecol Evol 18:172–176

GF (2021) Products – Muscovy ducks, Grimaud Freres, Sevremoine. https://grimaudfreres.com/wp-content/uploads/2020/11/120720-FICHE_BARBARIE-EN.pdf . Accessed 5 June 2021

Gong Y, Yang Q, Li S, Feng Y, Gao C, Tu G, Peng X (2010) Grey plumage colouration in the duck is genetically determined by the alleles on two different, interacting loci. Anim Genet 41(1):105–108

Gordh G, Beardsley JW (1999) Taxonomy and biological control. In: Bellows TS, Fisher TW (eds) Handbook of biological control: principles and applications of biological control. Academic Press, Amsterdam, pp 45–55. isbn 978-0-12-257305-7

Gunawan B (1990) Endangered breeds of poultry and ducks. In: Wiener G (ed) Animal genetic resources A global programme for sustainable development, Proceedings of an FAO Expert Consultation, Rome, Italy, September 1989. Food and Agriculture Organization of the United Nations, Rome. http://www.fao.org/3/t0284e/T0284E00.htm#TOC. Accessed 25 July 2020

Gupta J, Singh A, Churchil RR, Singh BP, Sharma D (2005) Genetic divergence between red jungle fowl and other domesticated poultry species using 12S rRNA gene polymorphism. Indian J Anim Genet Breed 26(1–2):26–30

Guy G, Rousselot-Pailley D, Rouvier R (1998) Description de la croissance de canettes mulardes. Ann Zootech 45:303–310

Harper J (1972) The tardy domestication of the duck. Agric Hist 46(3):385–389

Hoffman (1993) Mule Duck, Canning, Nova Scotia, Canada. http://www.feathersite.com/Poultry/Ducks/Musc/Mule.html. Accessed 13 Jan 2019
Hollander WF (1970) Sex-linked chocolate coloration in the Muscovy duck. Poult Sci 49:594–596
Hou L, Martin LD, Zhou Z, Hou AF (1996) Early adaptive radiation of birds: evidence from fossils from north-eastern China. Science 274(5290):1164-1167. Gale Academic OneFile. https://go.gale.com/ps/i.do?v=2.1&it=r&sw=w&id=GALE%7CA18928387&prodId=AONE&sid=googleScholarFullText&userGroupName=tel_main&isGeoAuthType=true. Accessed 27 Jan 2021
Howie JA, Tolkamp BJ, Avendano S, Kyriazakis I (2009) The structure of feeding behavior in commercial broiler lines selected for different growth rates. Poult Sci 88:1143–1150
Hsiao MC, Liu HC, Hsu YC, Hu YH, Li SH, Lee SR (2008) Isolation and characterization of microsatellite markers in Tsaiya duck. Asian Aust J Anim Sci 21:624–627
Hu YH, Poivey JP, Rouvier R, Tai C (1999) Heritabilities and genetic correlations of body weights and feather length in growing Muscovy selected in Taiwan. Br Poult Sci 40:605–612
Hu YH, Liu SC, Liao YW, Pau CM (2002) Studies on intra-line and intra-family varieties of green egg color in Brown Tsaiya. Taiwan Lives Res 35(3):231–240
Hu YH, Poivey JP, Rouvier R, Liu SC, Tai C (2004) Heritabilities and genetic correlations of laying performance in Muscovy ducks selected in Taiwan. Br Poult Sci 45:180–185
Hu YH, Liu HC, Rouvier R, Wang CT, Tai C, Poivey JP (2014) Genetic trend and selection response in Muscovy duck over five generations of selection for body weight. J Chin Soc Anim Sci 43(2):13–24
Huang HH (1985) Selection of white mule ducks. J Chin Soc Anim Sci 14(3–4):111–122
Huang CW, Liu SC, Hu YH, Yang KT, Wu CP, Cheng YS, Rouvier R, Huang MC (2006a) The distinction of DNA fingerprints in Tsaiya ducks with different eggshell colours. J Chin Soc Anim Sci 35:153–164
Huang CW, Cheng YS, Rouiver R, Yang KT, Wu CP, Huang MC (2006b) AFLP fingerprinting for paternity testing in ducks. Br Poul Sci 48:323–330
Huang HL, Cheng YS, Huang CW, Huang MC, Hsu WH (2011a) A novel genetic marker of the ovomucoid gene associated with hatchability in Tsaiya ducks (*Anas platyrhynchos*). Anim Genet 42:421–427
Huang HL, Cheng YS, Yang KT, Chen CH, Huang MC, Hsu WH (2011b) Genome-wide transcript expression analysis in the uterovaginal junction in associated with fertile period in Tsaiya ducks. J Reprod Dev 57:731–736
Huang HL, Huang LT, Cheng YS (2013) A novel SNP marker of ovalbumin gene in association with duck hatchability. Theriogenology 79:1218–1223
Hutt FB (1949) Genetics of the fowl. McGraw Hill Book Company Inc, New York
Jaap RG (1933a) Inheritance of white spotting in ducks. Poult Sci 12:233–241
Jaap RG (1933b) Light phase Mallard ducks. J Hered 24:467
Jaap RG (1934) Alleles of the Mallard plumage pattern in ducks. Genet 19:310–322
Jaap RG, Milby TT (1944) Comparative genetics of blue plumage in poultry. Poult Sci 23:3–8
Kain HH (1988) Untersuchungen zur zuchterischen Verbesserung des Fleischansatzes und des Futteraufwands bei Pekinggenten. Diss., Univ. Leipzig
Kamal R, Dey A, Chandran PC, Mohanta RK, Giri SC, Mohanty S, Gupta SK, Kumar S (2019) Phenotypic and morphometric characterization of *desi* duck of Odisha. Indian J Anim Sci 89(3):334–336
Kaulicke P (1991) El peri'odo intermedio temprano en el Alto Piura: avances del proyecto arqueolo'gico 'Alto Piura' (1987–1990). Bulletin de L'Institut Français d'Etudes Andines 20: 381–422
Kennedy GY, Vevers HG (1976) A survey of avian eggshell pigments. Comp Biochem Physiol 55B:117–123
Kiple KF (2000) The Cambridge world history of food. Cambridge University Press, Cambridge
Klemm R, Peiter K, Pingel H (1994) Results and effects of direct selection for feed efficiency in the domestic ducks: 2 report: methodical aspects. Arch Geflügelk 58:176–181

Lake PE (1975) Gamete production and the fertile period with particular reference to domesticated birds. Symp Zool Soc Lond 35:225–244

Lancaster FM (1963) The inheritance of plumage colour in the common duck (*Anas platyrhynchos linne*). Bibliogr Genet 19:317–404

Lancaster FM (1990) Chapter 16. Mutations and major variants in domestic ducks. In: Crawford RD (ed) Poultry breeding and genetics. Elsevier, Amsterdam, pp 381–387

Larson G, Burger J (2013) A population genetics view of animal domestication. Trends Genet 29: 197–205

Larzul C (2002) Genetic parameters for overfed mule duck traits. In: Proceedings of 7th world congress on genetics applied to livestock production, August 19–23, 2002, Montpellier, France

Larzul C, Guy G, Bernadet MD (2004) Feed efficiency growth and carcass traits in female mule ducks. Arch Geflügelk 68(6):265–268

Lee SR, Kang CL (1997) Improvement on percentage of white plumage in mule ducks. J Taiwan Livest Res 30(3):293–299

Lee SR, Lee YP, Chen BJ (1992) Diurnal behavior patterns of cage-reared Brown Tsaiya ducks (*Anas platyrynchos* var *domestica*). Appl Anim Behav Sci 34:255–262

Li S, Wang C, Yu W, Zhao S, Gong Y (2012) Identification of genes related to white and black plumage formation by RNA-Seq from white and black feather bulbs in ducks. PLoS One 7(5): e36592. https://doi.org/10.1371/journal.pone.0036592

Li G, Zhu F, Yang F, Hao J, Hou Z (2020) Selection response and genetic parameter estimation of feeding behavior traits in Pekin ducks. Poult Sci 99(5):2375–2384. https://doi.org/10.1016/j.psj.2020.01.013

Lin RL, Chen HP, Rouvier R, Poivey JP (2014) Selection and crossbreeding in relation to plumage color inheritance in three Chinese egg type duck breeds (*Anas Platyrhynchos*). Asian Aust J Anim Sci 27(8):1069–1074

Lin RL, Chen HP, Rouvier R, Marie-Etancelin C (2016) Genetic parameters of body weight egg production and shell quality traits. Poult Sci 95:2514–2519

Linnaeus C (1758) Systema naturae per regna tria naturae: secundum classes ordines genera species cum characteribus differentiis synonymis locis (in Latin), 10th edn. Laurentius Salvius, Stockholm

Liu JJ, Tai C, Huang HH (1980) Studies on artificial insemination of ducks: II. Effects of sperm number and sperm concentration on fertility. J Chin Soc Anim Sci 9:71–77

Liu SC, Huang JF, Sun TJ, Lee SR, Wang CT (1998) The inheritance of blue eggshell in brown Tsaiya. J Taiwan Livest Res 31(4):373–382

Liu HC, Hsiao MC, Hu YH, Lee SR, Cheng WTK (2010) Eggshell pigmentation study in blue-shelled and white-shelled ducks. Asian Aust J Anim Sci 23:162–168

Liu HCM, Hsiao C, Cheng WTK (2011) Expression of HO and ALAS in blue and white-shelled ducks. J Taiwan Livest Res 44(4):353–362

Liu HC, Huang JF, Lee SR, Liu HL, Hsieh CH, Huang CW, Huang MC, Tai C, Poivey JP, Rouvier R, Cheng YS (2015) Selection for duration of fertility and mule duck white plumage colour in a synthetic strain of ducks (*Anas platyrhynchos*). Asian Aust J Anim Sci 28:605–611

Mancha YP, Mbap ST, Abdul SD (2006) Incidence and frequencies of some qualitative traits in the local chicken population on the Jos Plateau. Trop J Anim Sci 9(1):39–46

Marie-Etancelin C, Chapuis H, Brun JM, Larzul C, Mialon-Richard MM, Rouvier R (2008) Genetic and selection of mule ducks in France: a review. Worlds Poult Sci J 64:187–207

Mayr G, De Pietri VL, Scofield RP, Worthy TH (2018) On the taxonomic composition and phylogenetic affinities of the recently proposed clade Vegaviidae Agnolín et al 2017 – neornithine birds from the Upper Cretaceous of the Southern Hemisphere. Cretaceous Res 86: 178–185

Mignon-Grasteau S, Beaumont C, Poivey JP, Rochambeau H (1998) Estimation of the genetic parameters of sexual dimorphism of body weight in 'label' chickens and Muscovy ducks. Genet Sel Evol 30:481. https://doi.org/10.1186/1297-9686-30-5-481

Mikšík I, Holáň V, Deyl Z (1996) Avian eggshell pigments and their variability. Comp Biochem Physiol 113B:607–612
Morduzzaman M, AKFH B, Rana M, Islam MR, MSA B (2015) Phenotypic characterization and production potentials of Nageswari duck in Bangladesh. Bangladesh J Anim Sci 44(2):92–99
Oguntunji AO, Ayorinde KL (2015) Phenotypic characterization of the Nigerian Muscovy Ducks (Cairina moschata). Anim Genet Resour 56:37–45
Oliphant LW (1987) Pteridines and purines as major pigments of the avian iris. Pigment Cell Res 1: 129–131
Phillips JC (1915) Experimental studies of hybridization among pheasants and ducks. J Exp Zool 18:69–144
Pingel H (1999) Influence of breeding and management on the efficiency of duck production. Lohmann Inf 22:7–13
Pingel H (2011) Results of selection for breast muscle percentage and feed conversion ratio in Pekin ducks. Biotechnol Anim Husb 27(3):769–776
Pingel H, Martin W, Klemm R (2013) Breeding and genetics of meat production and reproduction in waterfowl. In: 'Genetic breeding and reproduction' part of proceeding of the 5th world waterfowl conference, Vietnam Branch WPSA, Vietnam, pp 50–53
Poivey JP, Cheng YS, Rouvier R, Tai C, Wang CT, Liu HL (2001) Genetic parameters of reproductive traits in Brown Tsaiya ducks artificially inseminated with semen from Muscovy drakes. Poult Sci 80:703–709
Powell JC (1984) Selection for improved feed conversion ratio in the domestic ducks. In: Proceeding of the 17th world poultry congress, Helsinki, pp 108–109
Qu L, Liu W, Yang F, Hou Z, Zheng J, Xu G, Yang N (2009) Origin and domestication history of Peking ducks determined through microsatellite and mitochondrial marker analysis. Sci China Ser C Life Sci 52(11):1030–1035. https://doi.org/10.1007/s11427-009-0145-x
Rendel JM (1940) Note on the inheritance of yellow bill colour in ducks. J Genet 40:439–440
Renne PR, Deino AL, Hilgen FJ, Kuiper KF, Mark DF, Mitchell WS III (2013) Time scales of critical events around the cretaceous-Paleogene boundary. Science 339(6120):684–687
Retailleau B (1999) Comparison on the growth and body composition of 3 types of ducks: Pekin Muscovy and mule. In: Proceedings of the 1st world conference of waterfowl, Taichung, Taiwan, pp 597–602
Roulin A, Ducrest A (2013) Genetics of colouration in birds. Semin Cell Dev Biol 24:594–608
Rouvier R, Tai JJL, Tai C (1984) L'insémination artificielle des canes communes pour la production de mulards à Taïwan La situation actuelle InInsémination artificielles et amélioration génétique: bilan et perspectives critiques Toulouse-Auzeville, France, pp 360–367
Rouvier R, Mialon MM, Salzmann F, Poujardieu B (1988) Fertilité et éclosabilité des oeufs d'une souche de cane Pékin (Anas platyrhynchos) en croisement interspécifique avec le Barbarie (Cairina moschata) par insémination artificielle. Ann Zootech 37:73–86
Sibley CG, Ahlquist JE (1990) Phylogeny and classification of birds: a study in molecular evolution. Yale University Press, London, isbn:03-000-40857, p 976
Singh A, Gupta J, Churchil RR, Sharma D, Singh RV (2005) Genetic relatedness between red jungle fowl and other poultry species for BLB2 gene. Indian J Anim Genet Breed 26(1–2):18–20
Smith HS (1969) Animal domestication and animal cult in dynastic Egypt. In: Ucko PI, Dimbleby GW (eds) The domestication and exploitation of plants and animals. Aldine Publishing Company, Chicago, pp 307–316
Tai C, Rouvier R, Poivey JP (1989) Genetic parameters of some growth and egg production traits in laying Brown Tsaiya (*Anas platyrynchos*). Genet Sel Evol 21(3):377–384
Tai C, Poivey JP, Rouiver R (1994) Heritability for duration of fertility traits in Brown Tsaiya female ducks (*Anas platyrnchos*) by artificial insemination with pooled Muscovy (*Cairina moschata*) semen. Br Poult Sci 37:571–577

Theophrastus (1961) Concerning weather signs. In: Enquiry into plants and minor works on odours and weather signs. English translation by Sir Arthur Holt. W Heinemann and Cambridge: Harvard University Press, London
Thiele HH (2016) Breeding Pekin ducks for meat production. Lohmann Inf 50:28–33
van Tuinen M, Paton T, Haddrath O, Baker A (2003) 'Big bang' for tertiary birds? A replay. Trends Ecol Evol 18:442–443
Wang Z, Meng G, Bai Y, Liu R, Du L, Su L (2017) Comparative transcriptome analysis provides clues to molecular mechanisms underlying blue-green eggshell color in the Jinding duck (Anas platyrhynchos). BMC Genomics 18(1):725
Wilford JN (2016) 'Dinosaurs among us' retraces an evolutionary path. The New York Times. https://www.nytimes.com/2016/03/29/science/dinosaurs-birds-evolution-american-museum-of-natural-history.html. Accessed 28 Mar 2020
Wright L (1902) The new book of poultry. Cassell & Co. Ltd, London
Xu TS, Hou SS, Liu XL, Huang W (2004) Study on synthetic selection index of breast meat weight and breast meat percentage with body size and body weight traits in Peking duck. Chine Poult 26:8–10
Xu TS, Liu XL, Huang W, Hou S (2011) Estimates of genetic parameters for body weight and carcass composition in Pekin ducks. J Anim Vet Adv 10(1):23–28
Xu FQ, Li A, Lan JJ, Wang YM, Yan MJ, Lian SY, Wu X (2018a) Study of formation of green eggshell colour in ducks through global gene expression. PLoS One 13(1):e0191564
Xu Y, Hu J, Zhang Y, Gao S, Huang W, Xie M, Liu H, Lei CZ, Hou SS, Liu XL, Zhou ZK (2018b) Selection response and estimation of the genetic parameters for multidimensional measured breast meat yield related traits in a long-term breeding Pekin duck line. Asia Aust J Anim Sci 31(10):1575–1580
Zeng T, Chen L, Du X, Lai SJ, Huang SP, Liu YL, Lu LZ (2016) Association analysis between feed efficiency studies and expression of hypothalamic neuropeptide genes in laying ducks. Anim Genet 47:606–609
Zeng T, Zhang H, Liu J, Chen L, Tian Y, Shen J, Lu L (2018) Genetic parameters of feed efficiency traits and their relationships with egg quality traits in laying period of ducks. Poult Sci 97:758–763
Zhang L (1989) The role of Peking duck in duck production in the world. Chin J Anim Sci 25:56–57
Zhang ZG, Zhu XH (1986) China animal science history collected works. Science and Technology of China Press, Beijing, pp 271–273
Zhang YS, Zhan BG, Xie M, Zhang Z, Hou SS (2017) Genetic parameters for residual feed intake in a random population of Pekin duck. Asian Aust J Anim Sci 30(2):167–170

Dr. Martin Hsiu-Chou Liu, PhD (Ilan Branch, Livestock Research Institute, Taiwan) is a Researcher and Director of Ilan Branch, Livestock Research Institute (Taiwan). He holds a Master's of Agriculture degree in Animal Husbandry from National Taiwan University, and further he gained his PhD degree from the same university, Taipei, Taiwan after working on duck breeding program in Ilan Branch for 16 years. He has worked in the field of duck breeding for over 25 years and nomenclature 7 new species (lines) of duck. He has earned many national awards and has supervised some national students for their graduate and post-graduate dissertation work.

Dr. R. Richard Churchil, PhD (Poultry Science) hails from Kanyakumari district, Tamil Nadu, India. He earned graduation from Madras Veterinary College, doctoral degree from Indian Veterinary Research Institute, many accolades including gold medals in post-graduation and Jawaharlal Nehru Award of ICAR for his doctoral research on development of transgenic spermatozoa. He worked as poultry breeding scientist in Kerala state for a decade before joining TANUVAS. He was trained on 'Transgenic animals' at University of Nantes, France. He has published 89 peer-reviewed research papers and seven book chapters. He teaches Poultry Science to undergraduate and Poultry Breeding and Biotechnology to postgraduate/doctoral students in TANUVAS.

Anatomy and Physiology of Ducks

4

K. M. Lucy and K. Karthiayini

Contents

K. M. Lucy (✉)
Kerala Veterinary and Animal Sciences University, Wayanad, Kerala, India
e-mail: lucy@kvasu.ac.in

K. Karthiayini
Department of Veterinary Physiology, College of Veterinary and Animal Sciences, Pookode, Kerala Veterinary and Animal Sciences University, Wayanad, Kerala, India
e-mail: karthiayini@kvasu.ac.in

A. Jalaludeen et al. (eds.), *Duck Production and Management Strategies*,
https://doi.org/10.1007/978-981-16-6100-6_4

Abstract

The ducks are an important species of waterfowl that spend much of their times on or around the water. Giving their living conditions, ducks have special anatomical features like webbed feet and water resistant feather coat. Further, their skeletal system has many unique features like ossified tracheal rings that enable catching by neck, fused vertebrae called notarium that give rigidity for swimming and so on. Small comb-like structures along the inside of the bill called lamellae act like sieve and look like teeth which are the fascinating adaptation of the ducks. The ducks can separate and expel the non-food items such as mud and water while retaining seeds, bugs and other food items using lamellae when scavenging under water. The copulatory organ of drake is special and different from other avian species to give much needed adaptation to mate on waters. Nevertheless, the anatomical features and functions of digestive, respiratory and lymphatic systems of ducks are different from chicken and other poultry at varying degrees.

Keywords

Ducks · Comparative anatomy · Skeleton · Viscera · Physiology

List of Symbols/Abbreviations

ANP	Atrial natriuretic peptide
Cl^-	Chloride ion
CNS	Central nervous system
FSH	Follicle stimulating hormone
GnRH	Gonadotropin-releasing hormone and LH
H_2O	Water
HCl	Hydrochloric acid
LH	Luteinizing hormone
Na^+	Sodium ion
NaCl	Sodium chloride
PTH	Parathyroid hormone

4.1 Introduction

The ducks are waterfowl and spend a considerable amount of time on water. Their plumage is thick and waterproof. The size and spacing of barbs of feathers in ducks also play a major role in the water repellence property of feathers. The skeletal system shows unique characteristics especially the sternum and hind limbs as an adaptation for swimming. Except for last two thoracic vertebrae, others articulate with each other to give mobility in water. The ducks are diving birds and are filter-feeders with the bills showing much adaptive variation. The ossified trachea permits lifting of the ducks by holding its neck. The copulatory organs, the vagina in females and phallus in males are much larger than those of domestic birds. The vaginal spirals that are coiled check water from inflowing the reproductive tract during copulation. There are many more adaptive variations in ducks compared to chicken. This chapter provides the glimpses of comparative anatomy and physiology of ducks which can be used as a tool to differentiate health and disease conditions and to improve the productivity of this economically important poultry species.

4.2 Skeletal System

The skeletal system of the duck is depicted in Fig. 4.1.

The avian skeleton is characterized by its lightness and strength attained by fusion and/or deletion of certain bones. Fusion is seen mainly in pelvic girdle, cranium, distal parts of wings and legs. Lightness in weight is provided by extension of air sacs to many bones replacing the bone marrow (pneumatic bones).

The cervical part of the vertebral column has 14 vertebrae as in the case of fowl, whereas in goose there are 17 vertebrae. The comb-like dorsal spinous processes of the cranial half of the cervical vertebral column are particularly well developed in ducks and goose compared to that of domestic fowl. They possess nine thoracic vertebrae unlike in fowl and pigeon which have only seven thoracic vertebrae. Last two thoracic vertebrae in ducks and last three in goose form a bony union with synsacrum. The remaining thoracic vertebrae articulate each other by movable saddle joints. In case of domestic fowl and pigeon, the second to fifth vertebrae are fused to form a single bony mass, the *notarium*. There are nine pairs of ribs in ducks. The sternum or keel bone of duck and goose is rectangular in outline and there is a caudolateral process which is separated from the middle piece of sternum by an oval notch (Nickel et al. 1977).

The skeleton of the wing mostly resembles that of domestic fowl. Clavicles of the pectoral girdle show considerable variation among different avian species depending on the function. In ducks, the two branches of the *furcula* or wishbone show a lateral ridge along their entire length. The border of this ridge is curved posteriorly. At the proximal extremity, there is a caudal process which unites with the furcular tuberosity of the coracoid by syndesmosis. The pointed ends of the lateral ridges are connected to the acromion of the scapula by a ligament. Acting like a spring, they maintain the distance between the two shoulder joints during swimming. In the forearm region, shaft of the ulna is distinctly curved in fowl, but it is less so in the duck and goose. The length of *autopodium* (carpals, metacarpals and digits) also shows variation among different birds. In the duck and goose, it exceeds the length of both arm and forearm. In fowl, *autopodium* is shorter than the arm and forearm (Nickel et al. 1977).

The skeleton of the pelvic limb shows the following differences from that of the domestic fowl. The pelvis is narrow and elongated in the duck and goose. The region of ilium cranial to the level of acetabulum is one-third shorter than the post-acetabular part. The ischiatic foramen is elongated oval in the duck, but circular or oval in fowl. The pubis extends considerably beyond the caudal border of ilium and

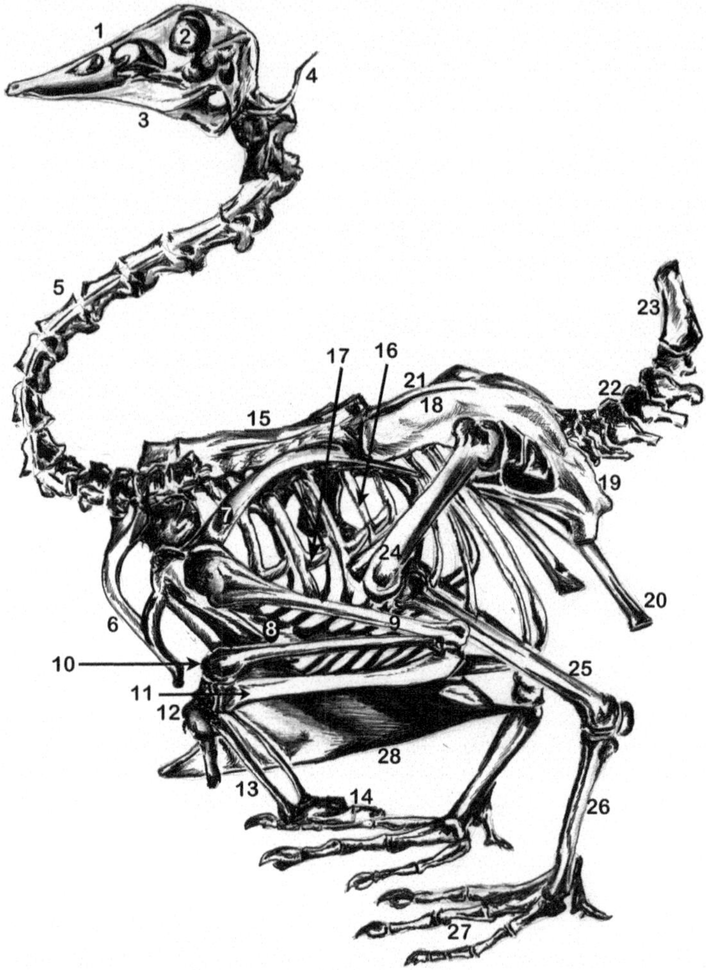

Fig. 4.1 Skeleton of the duck
1. Skull, *2*. Orbit, *3*. Mandible, *4*. Hyoid, *5*. Cervical vertebrae, *6*. Furcula, *7*. Scapula, *8*. Coracoid, *9*. Humerus, *10*. Radius, *11*. Ulna, *12*. Carpals, *13*. Fused metacarpals, *14*. Digits, *15*. Thoracic vertebrae, *16*. Rib, *17*. Uncinate process, *18*. Ilium, *19*. Ischium, *20*. Pubis, *21*. Lumbosacral mass, *22*. Coccygeal vertebrae, *23*. Pygostyle, *24*. Femur, *25*. Tibiotarsus, *26*. Tarsometatarsus, *27*. Digits, *28*. Keel bone

ischium. Femur is remarkably shorter in the duck and goose. This is compensated by the *tibiotarsus*, which is twice longer than the femur. In fowl, the *tibiotarsus* is only one-third longer than that of femur. Length of the tarsometatarsus determines to great extent the 'ground clearance' of the bird in standing position. Among domestic birds, it is shortest in ducks and longest in fowl. Plantar surface of the tarsometatarsus has four longitudinal ridges in the duck and goose, whereas there are only three ridges in fowl. The spur core present in the tarsometatarsus of cockerel is absent in ducks. There are four toes in the leg of domestic birds and the feet are webbed in the duck and goose for swimming. The toes are longer in these species especially the third one (Nickel et al. 1977). The plumage is thick and waterproof. The contour feathers are arranged over distinct feather tracts with underlying coat of down feathers.

The avian skull is one of the most specialized skull among vertebrates. Fusion of several skull bones and pneumatization makes the skull a greatly lightened one. The skull is highly kinetic and the upper jaw is movable with the help of nasal-frontal hinge joint. At its articulation with the brain case, the quadrate bone is also movable and is indirectly connected to upper jaw via palatine and pterygoid bones. All galliform and anseriform birds (including ducks) have well developed kinetic mechanism of the skull. The avian skull is also noted for its very large orbit to accommodate large eye balls (Feduccia 1975).

4.3 Digestive System

An understanding of the avian digestive system is essential for developing an effective and economical feeding program for poultry, for recognizing the digestive problems and to take necessary actions to correct it. The anatomy of digestive system varies among different species of birds based on feeding habits. The digestive system consists of the oral cavity, tongue, pharynx, oesophagus, proventriculus, gizzard, small intestine, large intestine, cloaca, liver and pancreas. The oesophagus is divided into cervical and thoracic parts by the dilated portion, the crop. The caeca are paired. The teeth, lips and soft palate are absent in birds.

4.3.1 Upper Digestive Tract

The upper digestive tract includes the oral cavity, tongue, pharynx, oesophagus, proventriculus and gizzard. Post-hatch development of the upper digestive tract of ducks has been studied in detail by Shyla (1990). Among the different regions, the oesophagus, crop and gizzard attained maximum weight at an early stage (75 days of age), whereas the tongue, pharynx and proventriculus reached maximum weight at 150 days.

The ducks are diving birds and are filter-feeders. The bills are of medium length, broad, soft and often flattened centrally and distally but high at the base and rounded at the tip with a horny nail made up of keratin. Structure of the roof of the oral cavity in ducks is depicted in Fig. 4.2. The bills show much adaptive variation, representing the fundamental apparatus unique to the order Anseriformes. The border of the bill shows a row of lamellae (meaning thin bony plates), which are blade like in duck and thick in goose.

Roof of oral cavity in duck shows a longitudinal mucous ridge which continues caudally as a wide-basal portion with four papillae in it. This ridge lies in the median longitudinal groove of the tongue when the mouth is closed (McLelland 1975).

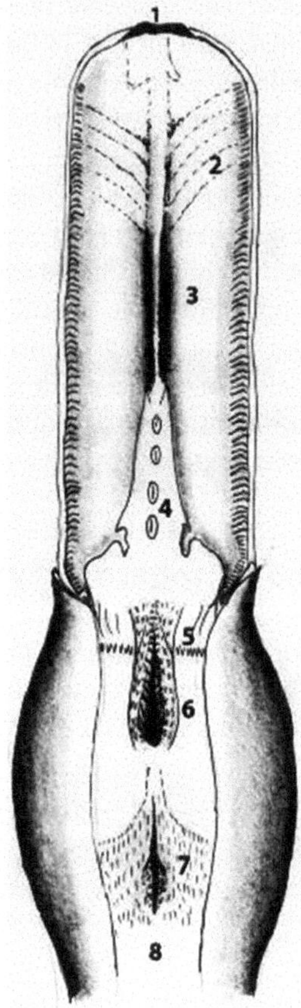

Fig. 4.2 Roof of oral cavity in duck
1. Hard keratin, *2*. Rostro-lateral transverse mucosal ridges, *3*. Median longitudinal ridge, *4*. Wide-basal papillae, *5*. Transverse row of papillae, *6*. Choanal slit, *7*. Infundibulum and pharyngeal papillae, 8. Oesophagus

Towards the rostral part (or the tip portion) of the palate there are several short, broad rostral and rostro-lateral transverse mucosal ridges. The short choanal slit (a slit in the roof of the mouth that connects nasal cavity) has a narrow cranial portion and a wide caudal part and is guarded by thin, pointed caudally directed papillae. A transverse row of papillae extends on either side from the choanal slit at the junction of its narrow and wide portions. The pharynx also shows numerous papillae on its roof.

The anatomy of the floor of the oral cavity in ducks is shown in Fig. 4.3. The tongue is spatula-shaped, thick and fleshy and covered with numerous papillae and horny spines. Rostral part of the dorsum (tip) of tongue presents a median longitudinal groove. A longitudinal row of small papillae is present on either side caudolateral to this median groove. This row of papillae extends caudally to form a wide mucous ridge. These ridges are caudally continuous with the raised root of the tongue. Numerous wide-based papillae are present lateral to this row of papillae and ridge. Lateral edges of dorsum of the tongue have large papillae and many thread-like papillae which interdigitate with lamellae of bill forming a filter. This filter allows

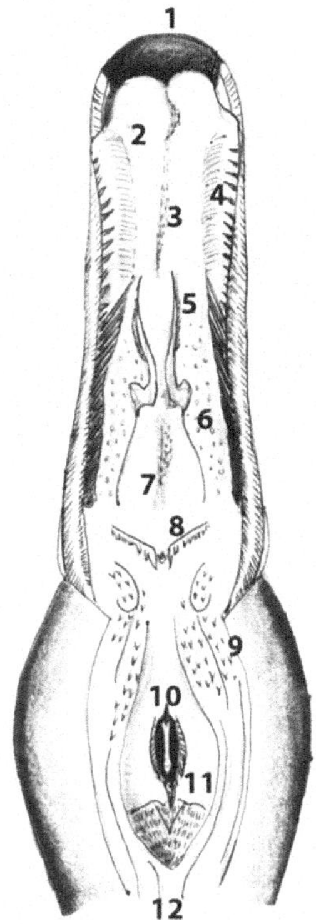

Fig. 4.3 Floor of the oral cavity in the duck
1. Hard keratin, *2*. Tongue, *3*. Median longitudinal groove, *4*. Large papillae, *5*. Row of fine papillae, *6*. Wide-based papillae, *7*. Mucosal ridge, *8*. Transverse rows of papillae, *9*. Pharyngeal papillae, *10*. Laryngeal inlet, *11*. Laryngeal mound, *12*. Oesophagus

water to escape through the sides of bill, retaining the food particles in the mouth. At the base of the tongue, there are thin caudally directed rostral and caudal transverse rows of papillae (McLelland 1975).

The bill acts as a double-action suction-pump where fluid is drawn in at the tip portion and expelled past filter plates at the sides and rear part. The tongue and interior shape of the bill provide an elaborate piston effect. The papillae, common to all members of the order *Anseriformes* (waterfowls like ducks, geese and swans), function as the sieves. These papillae trap the food, which is then brushed free and finally swallowed by the combined actions of tongue and papillae (Marchant and Higgins 1990). In ducks, the bottom of its mouth is not firm and it can flex out the mouth to some extent. This helps the duck to take in food and sift through water to get hold of food particles. Since birds do not have teeth no significant mechanical breakdown of food particles is possible in mouth. But they use their sharp beak to break the food into smaller particles. In waterfowls the flat beak will not contribute much as far as this breaking is concerned.

The salivary glands are well developed in ducks and are of mucous type. The oesophagus shows a spindle-shaped enlargement at the thoracic inlet forming the crop. In ducks, mucous glands are distributed throughout the mucosa of crop, whereas in domestic fowl, these glands are confined to the crop channel. Diffuse and nodular forms of lymphoid tissue can be seen around the glands in the crop of

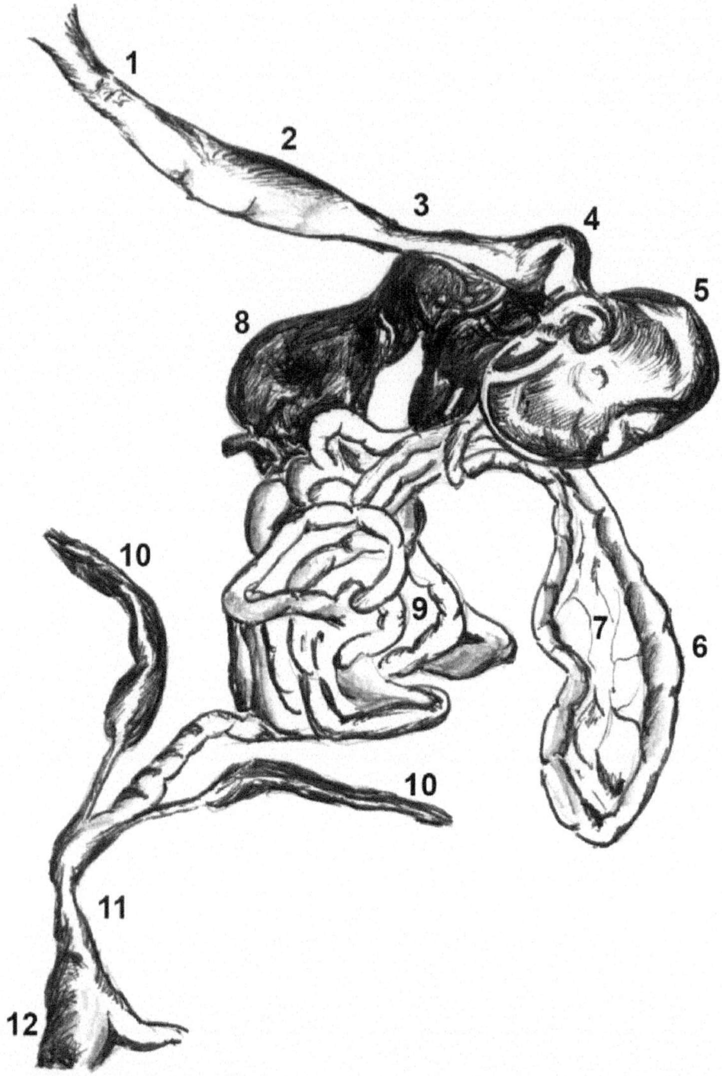

Fig. 4.4 Digestive tract of the duck
1. Cervical part of oesophagus, *2*. Crop, *3*. Thoracic part of oesophagus, *4*. Proventriculus, *5*. Gizzard, *6*. Duodenum, *7*. Pancreas, *8*. Liver, *9*. Small intestine, *10*. Caeca, *11*. Colorectum, *12*. Cloaca

ducks which is absent in the case of fowl. Parts of the digestive tract of the duck are shown in Fig. 4.4. Cervical part of oesophagus has greater diameter than the thoracic part. At the oesophago-proventricular junction, oesophageal tonsil is present in ducks which is an example for gut associated lymphatic tissue.

The proventriculus is elongated with narrow cranial end and a wider caudal end. It has a maximum diameter in the middle region. In adult duck, the average length of proventriculus is 5.72 ± 0.57 cm (Shyla 1990), whereas in adult chicken, the length of proventriculus ranges from 4 to 5 cm (Nickel et al. 1977). The mucous membrane of the proventriculus presents numerous small papillae showing the openings of proventricular glands. In domestic fowl, these papillae are larger and grossly visible.

Proventriculus is the glandular stomach of birds where the enzymatic digestion begins. The oesophagus and proventriculus (considered as the true glandular stomach) belong to the upper digestive tract of waterfowl. In order to study diets of waterfowl, biologists typically obtain food samples from the upper digestive tract area. The food gets softened by enzymes in this area and then enters the gizzard. In proventriculus, hydrochloric acid and digestive enzymes, such as pepsin, are added to the feed. However, no significant digestion takes place in the proventriculus mainly because the food is intact due to lack of grinding of food in the mouth. The cells of proventricular glands are oxynto-peptic cells which secrete both HCl and pepsinogen combining the function of both chief cells and parietal cells of mammalian gastric glands. It resembles amphibian gastric glands in which the glands are not differentiated into acid and enzyme secreting types. Glandular cells in the proventriculus have the terminal bars at a basal position which allow the use of most of the lateral and apical surfaces of the cells for secretion and hence, the surface area is increased by six times. This is analogous to the intracellular canalicular system of mammalian parietal cells which secrete HCl. Canaliculi are deep infoldings of the cell membrane which help to increase the surface area of the cell. Isthmus is the region which divides the proventriculus and gizzard and in isthmus, the proventricular glands are absent. Instead, the lamina propria of isthmus contains glands resembling that of gizzard.

The gizzard attains maximum weight of 58 g by 75 days of age in ducks (Shyla 1990). The gizzard has a greater dorso-ventral diameter than its cranio-caudal diameter except in day-old ducklings, whereas the gizzard of domestic fowl has greater cranio-caudal and lesser dorso-ventral. The musculature of the gizzard is well developed and unlike in the fowl, the mucosa is strongly adherent to the muscularis layer so that it cannot be peeled off easily. The fibrocartilage that is present at the junction of muscle and tendon layer in the case of fowl (at the rim of tendinous centre) is absent in case of ducks. The gizzard has simple tubular glands lined by chief cells with a few basal and intermediate cells. Secretion of the chief cells contains carbohydrates and protein. The secretion enters the gland lumen as fine filament and joins with the secretion from other cells to form compact masses in the glandular lumen which anchors the koilin lining of the mucosa. Koilin is a protein similar to keratin except in the fact that it has less amount of cysteine.

Gizzard, the muscular stomach of birds, performs the function of teeth of mammals. The extremely thick muscles of the gizzard grind large food particles against intentionally swallowed stones (along with food, ducks swallows stones and grit which is taken to gizzard). This grinding aids in the breakdown of cellulose walls in the feed. It also increases the surface area available for digestive juice to act on. Consumed feed along with the digestive juices from the salivary glands and proventriculus pass into the gizzard, where the food is ground, mixed and mashed. The protein digestion begins in the gizzard. The size of gizzard varies with the type of feed it consumes. Feeding hard feedstuffs like molluscs increases the frequency and force of contractions of gizzard and thereby increases size. Thick cuticle, called koilin protects mucosa from acid and proteolytic enzymes from proventriculus. The cuticle is continuously secreted at its base and is continuously worn out on its surface. Waterfowl changes its feeding preferences with change in season. They prefer carbohydrate rich foods during cold season, to meet their high energy requirements. During late winter and spring, ducks choose more protein-rich foods (largely invertebrates) to prepare for the upcoming breeding season.

4.3.2 Intestines

Partially digested food from gizzard passes to small intestine. The initial part of the small intestine is the duodenum followed by lower small intestine. The digestion completes in the duodenum, and the released nutrients are absorbed mainly in the lower small intestine. The pancreas secretes digestive enzymes and bicarbonate to duodenum. Bile from the liver (via the gall bladder) also reaches the duodenum. The digestive juice produced by the pancreas (contains primarily pepsin, trypsin, chymotrypsin and carboxyl peptidase) digests the feed proteins. Bile functions like a detergent and helps in the digestion of lipids and the absorption of fat-soluble vitamins (A, D, E and K). Bicarbonate helps to neutralize the hydrochloric acid from the proventriculus and provides an alkaline pH for the pancreatic enzymes to act. A pH of 6–8 is considered optimal. The lower small intestine includes jejunum and ileum. The Meckel's diverticulum marks the end of the jejunum and the beginning of the ileum.

The ileum has the same length as that of the caeca in the duck, goose and fowl. Average length of jejunum, the longest segment of intestine is 105 cm in both fowl and duck. In the case of goose, it is longer (165 cm). The shape of jejunal loops shows marked species difference. In the case of ducks and goose, the jejunum has six to eight parallel loops of about equal length arranged in long axis of the body. In fowl, the loops are of varying sizes. Meckel's diverticulum, the remnant of yolk sac and stalk, persists for a long period in ducks and goose, whereas it disappears early in life in other avian species. The mucosa of Meckel's diverticulum shows deep folds and contains lympho-reticular tissue. The caeca (singular *caecum*) are two blind pouch like structures located at the junction of small and large intestine. They are 10–20 cm long and show three segments, viz. short narrow neck, long, thin-walled greyish green main part and a short, vesicular, bright red part. The caecal tonsils are located immediately behind the neck region of caeca in fowl, duck and goose and have accumulation of lymph follicles in crypts. Unlike in fowl, the end portion of the caecum in ducks shows only less amount of lympho-reticular tissue. The colon is 7–12 cm long and opens into the cloaca which has three regions namely, an anterior coprodeum, middle urodeum and caudal proctodeum. The proctodeum region of the cloaca has the copulatory organ in drake comparable to that of mammalian penis (Nickel et al. 1977).

Caeca is the site of water absorption. Fermentation (microbial) of undigested coarse materials takes place in caeca. Several fatty acids and B vitamins (thiamine, riboflavin, niacin, pantothenic acid, pyridoxine, biotin, folic acid and vitamin B12) are produced by caecal fermentation. In colon also, the water reabsorption occurs. Colon opens to cloaca where the digestive wastes mix with wastes from the urinary system.

4.3.3 Liver

In ducks, the liver is larger and contributes to 3.1% to 4.1% of its body weight. The corresponding value in domestic fowl is 1.7% to 2.3%, in goose, 1.5% to 3.6% and in pigeon, 2.5% to 2.6%. Absolute weight of the liver ranges from 58 to 113 g in duck, 35 to 51 g in fowl, 85 to 171 g in goose and 8 to 10.5 g in pigeon. The shape of the liver also varies in different species of birds. In ducks, right lobe of the liver is tongue-shaped and extra-long, whereas the left lobe is bean-shaped. The gall bladder is situated in the middle of dorsal surface of the right lobe. It is elongated and tube-like and does not extend to the caudal border as against in fowl (Nickel et al. 1977).

Avian liver is similar to mammalian liver in function. Nearly all the blood circulating through the abdomen flows back through the portal vein to reach liver and then enter the general circulation. The liver functions in carbohydrate metabolism, glycogenesis, glycogenolysis and gluconeogenesis. In birds also, the liver is the major site of synthesis of all plasma proteins (albumins, globulins and fibrinogen). The liver functions in lipogenesis and the synthesis of cholesterol. It also acts as storage site for water soluble vitamins, fat-soluble vitamins, iron, triglyceride and glycogen. The liver is the major site of haemoglobin metabolism. It also converts ammonia to uric acid and allows the management of endogenous waste.

4.4 Respiratory System

4.4.1 Anatomy

Avian respiratory system consists of the nostrils, nasal cavities, pharynx, cranial larynx, trachea, syrinx, bronchi, lungs and air sacs. Pre- and post-hatch development of the respiratory tract has been studied in detail in Kuttanad ducks (Firdous 2014), the indigenous ducks of Kerala. Development of the respiratory tract begins on the third day of incubation in duck embryos. The nasal cavity is formed on the seventh day. Each part of the respiratory system is grossly appreciated by 14th day of embryonic age. The three nasal conchae, larynx with laryngeal openings guarded by the sagittal row of papillae and the firm, wedge-shaped lungs with six indentations of ribs are visible by 21st day of incubation. Parabronchi, atria and infundibula also appear at this age. Most of the development of the respiratory system occurs inside the egg itself and the system starts functioning even before hatch with the beak placed in the air cell of the egg.

In the post-hatch session, chiefly morphometric growth occurs with very little cellular changes. The nostrils open on the day of hatch. The nostrils are situated more caudally in duck and goose compared to that of chicken and are in the form of a large elongated oval opening and there is no operculum in these aquatic species unlike that of the domestic fowl (King 1975). The nasal septum is incomplete at the level of nostrils which allows free communication between both nasal cavities. This open type of nostrils, also called as *nares perviae*, may help in preventing blockage of the nostrils by extraneous material during swimming.

The laryngeal mound is lozenge-shaped and is 2.4 to 2.8 cm long and 1.2 cm wide in adult ducks (King 1975). Numerous papillae are also distributed in this region. The muscles of the larynx are less bulky than that of domestic fowl. The trachea in ducks is long and wide and is made up of 127 hyaline cartilaginous complete rings in the adult birds. The tracheal rings show varying degrees of ossification as age advances. Among the tracheal rings, first three and last six rings are simple, whereas the remaining are signet-ring shaped and overlap with their wider parts forming right and left halves alternately (Firdous 2014).

Ossification of larynx, trachea and syrinx starts at eighth week of age in Kuttanad ducks and at tenth week, all the tracheal rings are ossified on the ventral aspect. In adult birds, trachea is highly ossified except at the dorsal region and this ventral ossification reduces the damage to the trachea and deeper tissues while handling the birds by holding their neck (Firdous 2014).

The syrinx is of tracheo-bronchial type similar to that of other domestic birds. Six caudal rings of trachea and first four to five rings of primary bronchi participate in the formation of syrinx. It consists of cranial tympanum, rigs of the dilated part and the pessulus. *Bulla tympaniformis*, a bony vesicle projecting to the left side of tracheal bifurcation formed by the last six tracheal rings, can be seen in the case of drake which is absent in the female. It is considered as an organ of resonance that increases the volume of voice or is responsible for the production of drake sound (Nickel et al. 1977). Syrinx helps to produce as well as modulate sound. Myers (1917) observed that when a cut is made in the trachea the pitch of the sound increases suggesting that syrinx and trachea can modulate sound.

The lungs in ducks possess more number of secondary bronchi which in turn increase the number of parabronchi and air capillaries that help to increase the surface area for gas exchange. This can be considered as an adaptation for swimming. The air sacs are similar to that of other domestic birds, but they are larger which may assist in holding the breath when dive under water. In addition to the function of ventilation, the air sacs help in thermoregulation and reducing the body weight during swimming (Firdous 2014).

The lung in ducks is trapezoid in shape with a sharp ventral border (Fig. 4.5). In chicken, the shape of the lung is rhomboid. The dorsal border is marked by six deep costal sulci. The medial surface is smooth and flat. There are four categories of secondary bronchi, viz. the medio-ventral, latero-dorsal, latero-ventral and medio-dorsal. The neopulmonic parabronchi form is a felt work on the ventral third of the

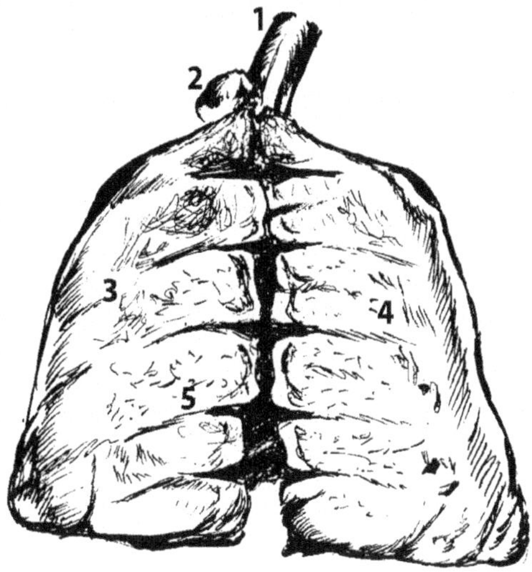

Fig. 4.5 Dorsal view of lungs in the duck
1. Trachea, *2*. Syrinx, *3*. Left lung, *4*. Right lung, *5*. Costal impression

lung. The parenchyma of the lung is organized into cylindrical parabronchi separated by thin septa containing blood vessels. The atria are shallow and show epithelial ridges supported by smooth muscle and give rise to 2 to 6 elongated infundibula. Air capillaries arise directly from the atria or from the infundibula. These are tubular or spherical in shape with thin interconnecting branches. The spatial disposition of the conducting air conduits closely resembles that of the chicken.

In the avian lung, the structural arrangement of gas exchanging tubes facilitates unidirectional airflow. There is remarkable similarity between the three-dimensional arrangement of the secondary bronchi in ducks and chicken and the number of secondary bronchi. The few differences occur in regions supplied by some secondary bronchi and the extent of the neopulmonic region. Minor differences also occur in the smaller conduits. In ducks, atria are generally shallow, infundibula are large and elongate and air capillaries are globular. In chicken and pigeon, the air capillaries are more tubular and their interaction with the blood capillaries is largely orthogonal (Makanya et al. 2014).

As in domestic fowl, the ducks possess nine air sacs, viz. cervical, clavicular (unpaired), cranial thoracic, caudal thoracic and abdominal air sacs (King 1975). Gross anatomical peculiarities of the air sacs were studied in detail in Golden Pekin ducks (*Anas platyrhynchos*) by El-Bably et al. (2014). The cervical air sacs communicate with the lungs through the first medio-ventral bronchi. This sac aerates the first two thoracic vertebrae and all the cervical vertebrae except the first. Last five thoracic vertebrae are aerated straight from the lungs. The clavicular air sac is connected with the lungs through third medio-ventral bronchi. This sac has six diverticula which are arranged intra- and extra-thoracically and aerates the keel bone and the humerus. The cranial thoracic air sacs are smaller than the caudal thoracic sacs and are aerated by the first, second and fourth medio-ventral bronchi and are rectangular in shape. These sacs pneumatise third to seventh sternal ribs by their diverticula. Compared to cranial sacs, caudal thoracic air sacs are twice larger and are with no diverticula. Abdominal air sacs are the largest among all sacs and extend from caudal border of the lungs through cloaca. These sacs are aerated by second, third, fourth and fifth latero-ventral bronchi. The right abdominal air sac is bigger than the left. The left abdominal air sac has two portions, the cranial small and narrow sac and caudal sac, the latter aerating last two ribs, synsacrum and femur. No sex difference could be observed in the morphology and topography of air sacs in Golden Pekin ducks.

The connection between lungs and air sacs differs in ducks compared to that of chicken. (1) The clavicular air sac is directly connected to the root of the first medio-ventral secondary bronchus, whereas in chicken it is to the third medio-ventral secondary bronchus. (2) Cranial thoracic air sac has an additional group of indirect connections that join to form a single tube opening directly to this sac. (3) Direct connection to caudal thoracic air sac may occur from third latero-ventral secondary bronchus (King 1975). The newly described spatial disposition of the conducting air conduits in ducks closely resembles that of the chicken. This remarkable similarity between the categories, numbers and three-dimensional arrangement of the secondary bronchi in the duck and chicken points to a convergence in function-oriented design (Makanya et al. 2014).

Avian respiratory system is much more efficient than mammalian respiratory system of gas exchange. This helps the birds to meet their high oxygen demand especially during flight. In birds pulmonary system consists of two separate and distinct functional components: one for ventilation (includes trachea, bronchi, air sacs, thoracic skeleton and muscles of respiration) and other for gas exchange, the parabronchi.

4.4.2 Functional Aspects

The functional aspects of nasal cavity of birds are based on the reports of Bang (1971) and Pearson (1972). The cilia present in the nasal cavity help to remove the dust particles in the inspired air which will sweep out the dust along with the mucus into the oral cavity through choanal opening and swallowed. Humidification and temperature regulation of inspired air starts in nasal cavity and continued in the trachea and bronchi. The anatomical as well as functional studies of avian nasal cavity showed that they can function in sensing the smell.

Larynx prevents the entry of food into the lungs. Stimulation of interior mucosa of larynx causes reflex closure of larynx (White and Chubb 1967). Larynx also helps in swallowing food. When birds ingest solid food particles, sudden caudal movement of the laryngeal mound happens against the pharyngeal roof. This movement is repeated at a faster rate when the caudally directed papillae of the mount help the mound to drag the food towards oesophagus.

The avian tracheal functions are similar to that of mammals. It conducts air from the external nares and mouth to the bronchi. During this passage, the inspired air is warmed, humidified and screened of particulate matter. However, avian and mammalian trachea differs anatomically. In birds, tracheal rings are complete against the incomplete C-shaped cartilaginous rings of mammals. As discussed earlier, in ducks, trachea is highly ossified except at the dorsal region and this ventral ossification reduces the damage to the trachea and deeper tissues while holding the birds by their neck (Firdous 2014). Unlike mammals, remarkable variations in tracheal anatomy with significant implications for ventilation can be seen in some species (some duck types have an inflatable sac-like diverticulum that opens from the trachea and therefore, the males of many waterfowl species have a bulbous expansion). Bird's trachea is nearly 2.7 times longer and 1.29 times wider than that of a comparably sized mammal. Though increase in length of trachea increases the resistance to airflow, it is compensated by increased radius so that resistance to airflow through the trachea in birds is comparable to that in mammals. The resistance to laminar flow is inversely proportional to fourth power of radius of the trachea. The increased length and radius increase the anatomical dead space which is compensated by slow breathing rate (of one-third of mammals) and much higher tidal volume in birds (possible with the help of air sacs). As a result of these, in spite of four times increase in anatomical dead space in birds, the minute tracheal ventilation (tidal volume x anatomical dead space) is only 1.5 to 1.9 times higher than that of mammal of similar size (John 2015).

The bronchial system in birds consists of three orders of branching before the gas-exchange surface, the air capillaries, is reached. These include a primary bronchus, secondary bronchi and tertiary bronchi, more commonly referred to as parabronchi. The primary bronchus consists of two parts, namely extrapulmonary primary bronchus, that lie outside the lung tissue and intrapulmonary primary bronchus, lying within the lung tissue. In ducks, the secondary bronchi are divided into four groups: medio-ventral, medio-dorsal, latero-ventral and latero-dorsal. The medio-ventral secondary bronchi is the first division of the primary intrapulmonary bronchus originating close to the entry of primary bronchus into the lung, while the medio-dorsal, latero-ventral and latero-dorsal secondary bronchi arise from the caudal curved portion of the primary intrapulmonary bronchus. The three anterior air sacs (cervical, clavicular and cranial thoracic) are connected to the medio-ventral secondary bronchi. The caudal thoracic air sac is connected to the latero-ventral secondary bronchi and the abdominal air sac is connected to the end of the primary bronchus (John 2015).

Tertiary bronchi or parabronchi are long narrow tube-like structures with a high degree of anastomosis. There are two parabronchi in avian lung, the neopulmonic parabronchi that supply and take air from caudal air sacs (abdominal air sacs) and the paleopulmonic parabronchi from where the air moves to air capillaries. Paleopulmonic parabronchi consist of parallel stacks of profusely anastomosing parabronchi, whereas the neopulmonic parabronchi are meshwork of anastomosing parabronchi located in the caudolateral portion of the lung. In paleopulmonic parabronchi, airflow is always unidirectional and in neopulmonic parabronchi, it is bidirectional. During both inspiration and expiration the paleopulmonic parabronchi is supplied with fresh air so that the gas exchange can happen both during inspiration and expiration. A network of smooth muscle controlled by the vagus nerve is present surrounding the entrances to the parabronchi. On electrical stimulation, this smooth muscle contracts and narrows the openings to the parabronchi. At its inner surfaces, parabronchi are pierced by numerous pentagonal or hexagonal openings which leads chambers called atria (John 2015).

The funnel-shaped ducts called infundibula arise from the atrial surface opposite the parabronchial opening and this infundibula lead to air capillaries. The air capillaries form an anastomosing network that is intimately anastomosed with network of blood capillaries. The gas exchange occurs within this mantle of anastomosed air and blood capillaries. Compared to mammalian lung, the surface area available for gas exchange is greater and mean harmonic thickness is less in the avian lung which makes it an exceedingly efficient gas exchanger compared to the mammalian lung. The air capillaries, infundibula and atria are lined by surfactants that lower surface tension and prevent their collapse and prevent transudation of fluid from blood into the air capillaries. The lamellated osmophilic bodies present in the atria of parabronchi secrete the surfactants which spread over the air capillary surface as a trilaminar substance. This trilaminar nature is unique to birds (Pattle 1978).

The air sacs are poorly vascularized and for this reason they do not significantly contribute to gas exchange. Avian lung is relatively rigid structure and the air sacs functions as bellows to the lungs by providing tidal airflow. Since diaphragm is absent, the thoraco-abdominal cavity acts as a single compartment and creation of a subatmospheric pressure in the thoracic cavity is impossible in birds. Based on their connections with bronchial tree, air sacs are grouped into two. The cervical, clavicular and cranial thoracic air sacs form cranial group, while caudal thoracic and abdominal air sacs form the caudal group. The volume of the cranial and caudal air sacs is approximately equal. During ventilation, all the air sacs except cervical air sacs are effectively ventilated, and the ratio of ventilation to volume is similar for all.

Aves depend on cervical, thoracic and abdominal muscles for their respiration. Both inspiration and expiration are active processes as both require muscular activity. During inspiration, the inspiratory muscles contract increasing the internal volume of the thoraco-abdominal cavity as well as air sacs. This reduces the pressure within the air sacs and it becomes negative compared to atmospheric pressure, leading to flow of atmospheric air into the pulmonary system. Upon inspiration about half of tidal volume goes into caudal air sacs via neopulmonic parabronchi, while the other half will move to cranial air sacs via the paleopulmonic parabronchi. The inspired air bypasses the medio-ventral secondary bronchi opening which comes first in the primary bronchus. This peculiar movement of gas is possible by the special aerodynamic valving. The part of the gas that enters the medio-dorsal secondary bronchi directly goes to the paleopulmonic parbronchi in a caudal to cranial direction and enters the cranial air sacs through the cranial secondary bronchi. Whenever tidal volume is large, during the same breath, inspired gas reaches cranial air sac directly from paleopulmonic parabronchi (John 2015).

Upon expiration, when the expiratory muscles contract, the internal volume of the thoraco-abdominal cavity decreases and pressure within the air sacs increases. As a result air from the cranial air sacs leaves out through the primary bronchi and trachea.

Air from caudal air sacs flows in a caudal to cranial direction via neopulmonic parabronchi, medio-dorsal secondary bronchi and paleopulmonic parabronchi. During both inspiration and expiration, in paleopulmonic parabronchi airflow is unidirectional and is from caudal to cranial direction. On expiration little or no airflow happens in the intrapulmonary bronchus (mesobronchus) due to expiratory valving. During the very next inspiration half of the fresh inspired gas again flows to the posterior sacs and next half through medio-dorsal secondary bronchi to paleopulmonic parabronchi and the anterior sacs receive deoxygenated air from the preceding inhalation. In paleopulmonic parabronchi, the airflow is always unidirectional, but in neopulmonic parabronchi it is bidirectional (John 2015).

The primary and secondary bronchi function as mere air conduits and do not participate in gas exchanges (Duncker 1974). Air capillaries are the fundamental units where the gas exchange takes place. Computer assisted avian lung studies show that, as in mammalian lung, a counter-current flow of air in air capillaries and blood in blood capillaries happen within the avian lung. However, its efficiency in gas exchange has not yet fully elucidated. The avian lung is richly innervated with vagal and sympathetic nerves. So it can be presumed that neural pathways can control pulmonary smooth muscle and alter the airflow through the parabronchial lung.

4.5 Urinary System

The urinary system in birds consists of the paired kidneys and ureters which open into the urodeum of cloaca. The urinary bladder is absent. The kidney is relatively longer and narrower in ducks and geese compared to that of domestic fowl. The total length of the kidney in the duck comes around 9 cm and the greatest diameter (across the caudal lobe) is nearly 2.2 cm and the width across the cranial lobe is about 1.2 cm. Each kidney is made up of approximately one million renal corpuscles. The ureter has two regions: the renal part and pelvic part. The renal part emerges from the kidney at a more caudal level compared to that of the chicken (King 1975).

Avian kidney is similar to mammalian kidney in structure and function. However avian kidney consists of two types of nephrons: reptilian type (small-sized, with no loops of Henle), confined to renal cortex and mammalian type (large sized nephrons with long or intermediate length loop of Henle) grouped into a medullary cone. In mallard, numerous mammalian and reptilian types of nephrons are seen. As loop of Henle is absent, the reptilian type nephrons cannot function in concentrating the urine and mammalian type of nephrons alone is associated with concentrating the urine, mainly in response to salt load. The volume of cortex ranges from 50% to 52% and of medulla 48% to 50%. As in mammals, urine is formed by glomerular filtration followed by tubular reabsorption and tubular secretion whereby the filtrate is modified. Also, bird urine can have an osmolality that is above or below that of plasma (similar to mammals). Unlike in mammalian kidney, presence of a renal portal system, formation of uric acid (as the major end product of nitrogen metabolism) instead of urea and post-renal modification of ureteral urine exist in birds. The avian kidney can alternate between the use of reptilian-type and mammalian-type nephrons, depending on the need for water conservation. On salt load a majority (about 80%) of the reptilian-type nephrons shut down to zero filtration and the mammalian type nephrons function to conserve water for diluting the salt (Reece 2015).

The major share (about 70%) of water absorption takes place in proximal tubule by osmosis and this depends on active sodium reabsorption and resultant changes in osmolarity. Regulation of water reabsorption occurs in the collecting ducts where urine, more concentrated than plasma can be formed. It is by the countercurrent mechanism similar to that in mammals. The long loops of Henle of the mammalian-type nephrons function to develop a corticomedullary osmotic gradient in the peritubular fluid of the medullary cone and vasa recta functions to maintain this gradient. The high osmotic gradient developed in the medullary interstitium results in movement of water from collecting ducts to peritubular interstitial fluid and then to peritubular capillaries and thus permits the excretion of urine that has an osmolality greater than that of plasma. The tubular fluid from both mammalian and reptilian type of nephrons is drained to renal pelvis through the same collecting duct, and since the collecting ducts exit through the medullary cone, all tubular fluid, from nephrons of the reptilian or the mammalian type, is exposed to the osmotic gradient. The antidiuretic hormone, arginine vasotocin increases the permeability of the collecting ducts to water (similar to vasopressin of mammals) and helps in conserving water. In birds (as against mammals), urea plays almost no role in the establishment of hypertonicity in the medullary cone interstitial fluid. The hypertonicity is mainly created by NaCl transport from the thick ascending limbs of the loop of Henle. The maximum concentration that urine can achieve is close to the concentration of the interstitial fluids at the tip of the medullary cone. In birds like ducks with free access to water, the urine osmolality is nearly isosmotic with plasma (320 mosmol/kg H_2O). During dehydration, osmotic concentration of urine can increase up to about 600 mosmol/kg H_2O (Reece 2015).

In birds, the nitrogenous waste is excreted as uric acid rather than urea. This helps to conserve water. As uric acid is water insoluble, it will get precipitated and in the precipitated form it will not contribute to effective osmotic pressure and will not cause water flow to tubule by osmosis. Just like urea is formed in the liver of mammals from ammonia, uric acid is formed in the liver of birds from ammonia. Avian kidneys can also form uric acid. Uric acid precipitates in the tubules because of the greater tubular concentration. The extra blood from the renal portal system that perfuses the tubules leads to greater tubular secretion leading to greater tubular concentration of uric acid. When the concentration of uric acid in the tubules exceeds uric acid solubility it precipitates. Uric acid continues in the tubules as urate and appears in the urine as a white coagulum (Reece 2015).

Birds have considerable control over tubular reabsorption of Na^+ and Cl^- and the urinary excretion of NaCl is less than 0.5% to 30% of that filtered in the glomerular capillaries. As in mammals, angiotensin II, aldosterone and atrial natriuretic peptide are the hormones associated with NaCl excretion. As in other animals renin–angiotensin–aldosterone system is well functional in birds. When there is salt and volume depletion, angiotensin II produces a reduction in glomerular filtration rate followed with antidiuresis and antinatriuresis. On salt and volume loading, angiotensin II produces a natriuretic and diuretic effect. Aldosterone is presumed to have an action on the kidney similar to that in mammals, whereby it is associated with sodium reabsorption coupled with potassium secretion in distal convoluted tubule and collecting duct. Aldosterone is likely the principal regulator of plasma potassium concentration. Atrial natriuretic peptide (ANP) from the atrial wall also acts on collecting duct to produce natriuresis depending on salt and water load. Under normal conditions, more than 98% of filtered calcium is reabsorbed by kidney. Reabsorption depends on the amount of parathyroid hormone (PTH) which favours calcium resorption from distal convoluted tubule and is coupled with the phosphate excretion. PTH inhibits phosphate reabsorption and stimulates secretion and assists in the maintenance of an appropriate calcium/phosphorus ratio (Reece 2015).

Unlike in mammals, post-renal modification of ureteral urine is possible in birds. This is possible because after passing through the urethra it reaches directly into the cloaca (urine is not stored as urinary bladder is absent in birds) and from there by reverse peristalsis it moves to colon and caeca. Major post-renal modification of ureteral urine takes place in the colon and the cloaca functions mainly as a storage organ. In the colon Na^+ is reabsorbed actively and to maintain electrical neutrality Cl^- will be taken in passively. As the NaCl increases the cellular fluid osmolarity, water will be moved by osmosis. NaCl and water reabsorption does occur from the caeca too and could involve urine water, provided urine reaches up to caeca on retrograde flow. Uric acid, the nitrogenous component of avian urine is degraded in caeca for utilization of the nitrogen by the microflora. Avian urine is cream in colour and contains thick mucus in it. Mucus secretion probably facilitates transport of the precipitated solutes similar to the mucus in equine urine.

As like other birds, ducks also have glands in the head known as nasal glands, which in many species produces a non-serous, non-mucoidal secretion. The exact function of these glands is not known. However, it is reported that salt-loading provokes the secretion of a hypertonic sodium chloride solution into the nasal cavity. Because of their osmoregulatory function, they are called salt glands. Other than ducks functional salt glands were reported in many birds as ostriches, penguins, pelicans, geese, hawks, eagles and gulls. The salt glands can excrete a salt solution of about twice the concentration of sea water. The salt secreted flows through the salt gland ducts into the nasal cavity and runs out through the nares, to drip from the tip of the beak. The salt glands can secrete only NaCl and function only when there is a salt load (Reece 2015).

4.6 Reproductive System

4.6.1 Male Reproductive System

During reproductive period, testis of drake is roughly cylindrical and measures up to 8.0 cm in length and 4.0–4.5 cm in diameter. In drakes, the copulatory organs are much larger than other domestic birds. In erect state, the copulatory organ measures

Fig. 4.6 Copulatory organ of the drake
1. Phallus, *2*. Muscular base

6.0–8.0 cm. Penis extends at the transition between urodeum and proctodeum on the ventral part of cloaca (King 1975; Nickel et al. 1977).

In waterfowl, the phallus is highly variable in length ranging from 1.25 to even more than 40 cm and in the degree of elaboration varying from smooth or covered with spines and grooves. Across species these differences are positively correlated with the frequency of forced extra-pair copulation. The avian phallus allows male birds to achieve intromission without cooperation of the female bird and help to deposit semen close to the site of sperm storage to increase the rate of fertilization, thus providing male birds with a copulatory advantage over the females (Brennan et al. 2007).

The phallus is spiralled in a counter-clockwise direction when viewed from base of the phallus to the tip (Fig. 4.6) and the vaginal spirals are coiled in the opposite direction when viewed from the cloaca to the uterus or shell gland. The length of the phallus is positively correlated with length of the vagina and the number of vaginal spirals. The vagina shows pouches and spirals and it is presumed that vaginal spirals might check water from inflowing the reproductive tract during copulation if they form a tight seal at the entrance of the uterus or shell gland (Brennan et al. 2007).

Functionally the phallus is similar to that of gallinaceous birds as the semen is carried in an open groove that spirals down the outer surface of the phallus. The phallus has right and left fibro-lymphatic bodies reaching up to the tip of the phallus of which the right is narrower than the left. Between the two bodies is the phallus sulcus on the surface of the phallus in which the semen is flowing to the vagina during copulation. Centre of the phallus contains elastic ligament and a tubular cavity surrounded by glandular tissue (Etches 1996). Erection is brought about by the influx of lymph into the interior of the blind tubular cavity. During copulation, the phallus sulcus is transformed into a tube for the passage of semen. The sperms remain fertile in the female reproductive tract for 8–10 days in duck and goose and 10–21 days in fowl (Nickel et al. 1977).

Spermatogenesis takes place in seminiferous tubules (as in mammals). Within the seminiferous tubules of a reproducing bird, there are distinct compartments with germ cells in different stages of development and maturation proceeds from the periphery to the lumen. The complete series of stages of developing spermatozoa within the seminiferous tubule is called a spermatogenic wave. Although avian testes are internal (unlike the externally located testes of mammals), spermatogenesis is sensitive to temperature. A variety of experiments have demonstrated that spermatogenesis is well adapted for the warm body temperature of birds (~41 °C) and deviation (both increase and decrease) from this temperature affects spermatogenesis. As in mammals, a linear relationship exists in birds between testicular size and sperm production (Patricia 2015).

4.6.2 Female Reproductive System

The post-hatch development of genitalia in ducks has been studied by Vernerova and Burda (1984). The female genitalia shows maximum development between 130 and 160 days of age. As in other birds, only the left ovary and oviduct are functional. The ovary is relatively longer cranio-caudally than that of domestic fowl. The average number of ovarian follicles visible to the unaided eye is reported to be 1279 ± 382 which is considerably less than that of domestic fowl. The oviduct has five regions, namely infundibulum, magnum, isthmus, uterus or shell gland and the vagina (Fig. 4.7). Total length of the oviduct is reported to be 32 to 36 cm. Infundibulum has a fimbriated funnel-shaped part and a narrow neck region. The magnum (longest region) measures 20.0 cm, isthmus, 1.5 cm, uterus, 9.0 cm and vagina, approximately 3.0 cm (King 1975). The vagina is comparatively longer than that of the domestic fowl and has pouches and spirals corresponding to the phallus of the drake. Lymphocytes in the diffuse and nodular form seen in the oviduct of hen are absent in ducks (Das and Biswal 1968). The wall of magnum is markedly thicker than that of infundibulum and showed numerous mucosal folds formed by compact layer of glands secreting albumin located around the central core of smooth muscle, connective tissue and blood vessels. These folds are higher and thicker than those in the other regions increasing the surface area of the mucosa. Utero-vaginal junction has sperm host glands as in other avian species.

The vagina in waterfowl is well developed and shows pouches and spirals. Vaginal spirals are coiled in the opposite direction to that of the phallus and it is presumed that vaginal spirals might check water from inflowing the reproductive tract during copulation if they form a tight seal at the entrance of the uterus or shell gland (Brennan et al. 2007).

Birds are oviparous and the egg should have all nourishment for embryo development. As a result, significant modifications exist in the reproductive anatomy and physiology between birds and mammals. Under standard management conditions,

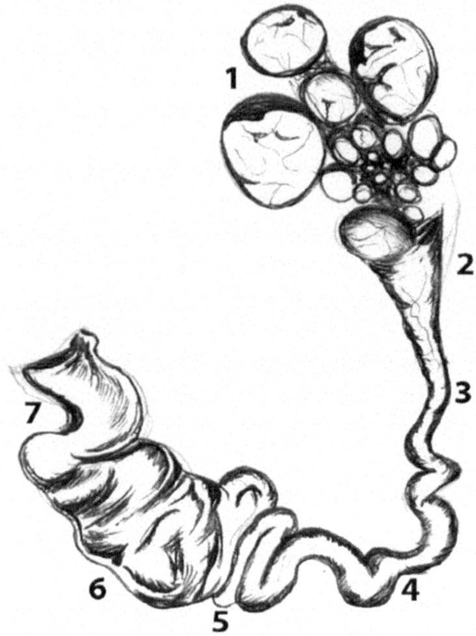

Fig. 4.7 Female reproductive system of the duck
1. Ovary, *2*. Fimbriated end of infundibulum, *3*. Neck of infundibulum, *4*. Magnum, *5*. Isthmus, *6*. Shell gland, *7*. Vagina

light breeds of duck attain sexual maturity early at 17–24 weeks of age, whereas the heavy breeds usually start egg production between 20 and 30 weeks of age.

The bird's reproductive function is highly influenced by duration and intensity of light. Photoperiod is the most important cue that controls the reproductive cycle of birds. They have photoreceptor neurons located in the hypothalamus with photosensitive pigment opsin. Studies show that pineal gland and eyes are not critical for perception of day length, but light must pass through the avian skull and reach these photoreceptors in hypothalamus to produce the effect. Light perception by these photoreceptors results in the release of gonadotropin-releasing hormone (GnRH) and thereby gonadotropins (FSH and LH) which stimulate gonads. Long days can stimulate reproductive development and advance sexual maturity in females. A photoperiod of more than 12 h (though it is not continuous) is perceived as long by almost all domesticated species of birds. However, to produce stimulatory effects the light must fall during the photosensitive phase. The time of dawn sets the photosensitive phase. Usually the photosensitive phase occurs 12 h after the dawn. The response to light is also mainly influenced by the previous photoperiod that the bird experienced. For example, a bird exposed to16L:8D then to 13L:11D will perceive the new light regimen as short day and will develop a photo-induced regression of gonads. As other birds, ducks also exhibit the phenomenon of photo-refractoriness. Prolonged exposure to long day length eventually makes the gonads inactive and leads to regression of gonads. For such birds, exposure to a period of short day length is necessary to break photo-refractoriness and regain photosensitivity (Patricia 2015). However, Jacquet (1997) reported that exposure to short day length is not necessary to regain photosensitivity in Muscovy ducks.

Thyroid gland also plays a role in the photosensitivity of birds. It has been noticed that thyroidectomy in birds prevents development of photo-refractoriness and sustains gonadal stimulation even after continuous exposure to long days for a prolonged period. Now, it is proved that stimulation of hypothalamic photoreceptors causes thyroid-stimulating hormone (TSH) production from adenohypophysis and long days increases concentration of thyroid deiodinase type 2, which increases local production of bioactive thyroid hormone T3 in hypothalamic neurons. The T_3 produces morphological changes in the GnRH nerve terminals and glial cells, facilitating GnRH secretion.

The ovary of a bird consists of follicles in various stages of development. The cell types of avian follicle are similar to those of a mammal. Unlike mammals, oocyte is large and yolk-filled. Under the influence of estradiol, the liver produces the yolk proteins and lipoproteins, taken to the ovary via blood to incorporate into the follicles. The follicles with yolk will be arranged in a hierarchy as and when the bird reaches sexual maturity and start laying. The largest follicle (F1) will ovulate first followed by the second largest (F2) and so on. The factor that regulates this hierarchy is not yet understood. The large hierarchical follicles sometimes undergo atresia or apoptosis, as happens during the transition to broody behaviour. In general, it appears that follicular atresia mainly occurs in small growing follicles, which have not yet been recruited into the hierarchy. Ducks are an indeterminate layer, which means that if eggs are sequentially removed from the nest as they are laid, the bird will continue to ovulate and lay eggs to attain a theoretical clutch (Patricia 2015). Clutch is the sequence of eggs that are laid over consecutive days for a single natural incubation by the bird (around 20 in Muscovy ducks).

Studies by Hall and Goldsmith (1983) showed that plasma prolactin concentration increases significantly during egg laying. The prolactin concentration in plasma remains high from the start through entire incubation, but falls sharply near the time of hatching. It is shown that local anaesthesia of the incubation patches of ducks incubating their eggs for 9 h resulted in significant reduction of prolactin level in plasma. The tactile stimulation of incubation patches due to the contact of the eggs in the nest may be the reason for increased plasma prolactin level during the egg laying period and maintaining its high level during incubation also.

4.7 Circulatory System

The circulatory system consists of the heart, blood vessels and the lymphatic system. Unlike in mammals, the avian heart is larger in proportion to its body weight and is located in the cranial part of thoraco-abdominal cavity. It has three principal surfaces, the sternal, hepatic and pulmonary surfaces. The cranial venae cavae are two in number.

4.7.1 Blood

Blood cell counts in ducks are as follows: Erythrocytes 2.8 million/cubic mm; Leucocytes 23.4 thousands/cubic mm; Lymphocytes 64.25%; Pseudo-eosinophils 21.05%; Eosinophils 5.75%; Basophils 5.40%; Monocytes 3.55% (Nickel et al. 1977).

4.7.2 Lymphatic System

The resistance to diseases mainly depends on the immunity provided by the lymphatic system. Avian lymphatic system has bursa-dependent (responsible for humoral immunity) and thymus-dependent (responsible for cell mediated immunity) components.

4.7.2.1 Lymph Nodes

Unlike in mammals, most birds lack any discrete lymph nodes and possess mural lymph nodes that do not possess filtering capacity. True lymph nodes are present only in water birds. In ducks, there are two pairs of elongated spindle-shaped lymph nodes, namely cervico-thoracic and lumbar lymph nodes. These nodes have similar filtering function as that of mammalian lymph nodes. Hence ducks are the ideal experimental model for studying the phylogenetic immunological relations between birds and mammals.

Post-hatch development of lymph nodes from day-old to 24 weeks of age has been studied in detail in Kuttanad ducks (Arunima 2011). Cervico-thoracic lymph nodes are located close to the jugular vein just caudal to the thyroid and parathyroid glands at the level of confluence of jugular vein with the anterior vena cava. The lumbar lymph nodes are situated ventral to the vertebral column on either side of lumbar aorta near the point of origin of external iliac arteries. In the day-old birds, lumbar lymph nodes were less differentiated than the cervico-thoracic nodes. Weight of lymph nodes increases up to 18 weeks of age, thereafter it decreases.

Histologically, the clear demarcation of cortex and medulla as seen in mammalian lymph node is absent in ducks. However, the spongy peripheral region resembles mammalian medullary region and the compact central part resembles the cortex with more number of primary and secondary lymphatic nodules. In addition to lymphatic nodules, parenchyma consists of lympho-reticular cords (which anastomose with each other and with the nodules) and are separated by lymphatic sinuses. Unlike in mammalian lymph node where the sinuses are seen mostly in the peripheral part, there is a central axial lymphatic sinus and several intermediary and marginal sinuses in the lymph nodes of ducks. Lymph vessels and blood vessels enter the lymph node from different sites and penetrate the capsule. Trabeculae and a distinct hilus are absent in the lymph nodes of ducks (Arunima 2011).

4.7.2.2 Thymus

The structure and development of thymus in Indian Runner duck has been studied by Sreedharanunni (1976). The thymus in duck is confined to the lower half of the neck and has three to seven pairs of white or yellowish white lobes, the caudal most lobe being the largest. These lobes are arranged in chains along the jugular vein up to the level of the thyroid glands. No thoracic part is visible in ducks. Thymic primordium appears in duck embryos by 3rd day of incubation and lobulation pattern is evident by 15th day with lymphocytic accumulation. Each lobule has a cortex and medulla, which appears by 22nd day of incubation, and the medulla shows Hassall's corpuscles on the same day. The thymus contributes to 0.08% of the body weight in day-old ducklings as against 0.312% in day-old chicks. At 180 days of post-hatch life, total weight of thymus is 4.9 g and no regression of thymus is seen up to 180 days in ducks.

4.7.2.3 Bursa of Fabricius

The bursa of Fabricius is characteristic to the avian species and is located on the dorsal aspect of the cloaca. It is also known as cloacal thymus. Post-hatch development of bursa of Fabricius has been studied in the duck by Indu (1999). In day-old ducklings, the bursa can be seen as an elongated yellow blind sac with a tapering apex, whereas, in chicks, it is a rounded or oval blind sac. The bursa communicates with dorsal part of proctodeum by a short stalk. The maximum size and weight of the bursa are recorded at 58 days of post-hatch life and at this age, the organ is cylindrical with a pointed apex. Thereafter, the size and weight reduce gradually. The inner surface of the bursa has numerous plicae. Histologically, the wall has inner mucosa lined by epithelium and the *lamina propria* studded with lymphoid follicles having cortex and medulla. The mucosa is surrounded by the muscular coat and outer serosa. Peak functional activity has been noticed at 2 months of post-hatch stage and thereafter, the organ regresses.

4.7.2.4 Spleen

The spleen of duck is triangular with flattened dorsal and convex ventral surfaces. The weight of the spleen in ducks ranges from 1.5 to 4.5 g which is similar to that of the domestic fowl (Nickel et al. 1977).

4.8 Glycogen Body

The spinal cord in birds including ducks possesses an oval mass of glycogen-filled cells between the dorsal funiculi of spinal cord in the region of lumbosacral plexus which is absent in other vertebrates (Farner et al. 1982). The exact function of this structure is not known. Generally, it is considered to be metabolically inert, but may support the formation of myelin in the central nervous system (CNS) of avian species. The abundant glycogen stores in this tissue, unlike those in the liver or in skeletal muscle, can act as a recyclable substrate for the production of reducing equivalents for the synthesis of myelin lipid cholesterol. Such glycogen may also serve as a source of organic acids which provide alternate substrates to the CNS during metabolic stress (Benzo and Gennaro 1983).

4.9 Preen Gland

The preen gland or uropygeal gland or oil gland is the only true cutaneous gland that is present in birds. This gland is relatively large and well developed in water birds. It is a holocrine gland that structurally and functionally resembles mammalian sebaceous gland. The gland originates from paired invaginations of ectoderm on the dorsal aspect of the tail on the tenth day of incubation in duck embryos (Farner et al. 1982).

The structure and post-hatch development of preen glands have been studied in the duck by Rajathi (2005). The preen gland is a paired organ having a common cylindrical papilla. Lamellar corpuscles and circlet feather follicles are present in the papilla. Each gland is pear shaped and yellowish in colour and together forms a 'V' shaped structure located on the dorsal surface of the pygostyle. The right gland is slightly larger and each gland has separate duct. Histologically, the gland is simple, branched tubular type. The secretory tubules are arranged in a radiating manner from the centre of the gland to the periphery and have an outer zone and an inner zone. The tubule is lined by stratified epithelium consisting of basal, intermediate and transitional layers.

Secretion of the preen gland, the preen gland wax, is a water repellant agent to protect the ducks against wetting. The wax also makes the keratin of the feathers, beak and scales flexible. The secretion is also reported to have fungicidal activity and plays a role in intra-species communication (Jacob et al. 1979).

4.10 Skin and Feathers

Comparative studies on the skin and feathers of broiler Vigova Super-M ducks of 6 to 8 weeks of age and spent Kuttanad ducks above 40 weeks of age were conducted by Alphine (2018). The histomorphology of skin and morphological, physical, chemical and scanning electron microscopic peculiarities of feathers was compared. On histology, skin of duck is found thicker on the ventral part of body and is slightly thicker in male birds. Since the duck being a semi-aquatic bird, greater thickness of the ventral regions of the body is considered as an adaptation to resist heat dissipation during swimming. The epidermis is very thin and formed of only two layers, *stratum germinativum* and *stratum corneum.* The dermis is made up of three layers, viz. *stratum superficiale, stratum profundum* (containing *stratum compactum* and *stratum laxum*) and *lamina elastica. Stratum laxum* is the thickest layer of skin in all groups and all regions under study bear the feather follicles and its associated structures. The presence of lipid in the entire thickness of *stratum corneum* of duck skin is demonstrated using osmium tetroxide. The duck skin has the ability to produce huge amount of keratin and lipids and hence, the whole skin can be considered as a sebaceous secretary organ. Hydroxyproline and collagen contents, which resist the wear and tear, are higher in the skin of aged layer ducks than in that of broiler ducks. The feathers are implanted in the skin in feather follicles and always lay in the epidermis and dermis. Feather performs the requirements of duck such as flight, camouflage, courtship, thermal insulation and water resistance. The multi-scale structure of duck feather owes adequate roughness to the surface for repelling water along with the low energy of the feather surface provided by the preen oil. The average diameter of duck feather barb is 0.06 ± 0.004 mm and this diameter and spacing of barbs have a major role in the water repellence property of feathers. The unequal length of barbs on either sides of the rachis of wing feathers is an adaptation for aerodynamic performance of the duck species. Scanning electron microscopic studies reveal that both rachis and barb of duck feather have an internal structure of a honey comb formed by the hollow cells which is the reason for the low density, light weightiness and thermal resistance of the feather. Duck feather is made up of protein called keratin and the average crude protein content is 87%. In duck, the keratin is made up of higher percentage of hydrophobic amino acids when compared to hydrophilic amino acids and also the presence of cysteine which has the ability to form disulphide bonds provides more hydrophobicity and stable structure to the feather keratin.

4.11 Salient Features of Duck Anatomy and Physiology

In nutshell, certain anatomical features are distinct in ducks compared to other poultry species. The locomotor apparatus and all the organ systems in ducks are developed to carry out necessary functions for the survival and further independent development of this avian species. The second, third and fourth toes of water birds are joined together by a double layer of skin, the webs. The ducks possess nine thoracic vertebrae and the last two vertebrae in ducks and last three in goose form a bony union with the synsacrum. The remaining thoracic vertebrae articulate with each other forming movable joints which provide more mobility in this region. The ducks are diving birds and are filter-feeders with the bills showing much adaptive variation. The crop is spindle-shaped. The proventriculus is longer than that of fowl. The gizzard has a greater dorso-ventral diameter than its cranio-caudal diameter and the mucosa is strongly adherent to the muscularis layer so that it cannot be peeled off easily. In ducks, the liver is larger than in fowl and contributes to 3.1% to 4.1% of its body weight.

The nostrils of waterfowl are of open type which may help in preventing blockage of the nostrils by extraneous material during swimming. In adult birds, trachea is highly ossified except at the dorsal region and this ventral ossification reduces the damage to the trachea and deeper tissues while handling the birds by holding their neck. The *bulla tympaniformis* seen in the case of drake is responsible for the characteristic drake sound. The lungs in ducks possess more number of secondary bronchi which in turn increase the number of parabronchi and air capillaries that help to increase the surface area for gas exchange.

In drakes, the copulatory organs are much larger than other domestic birds. In female birds, the vagina is well developed with vaginal spirals that are coiled in the opposite direction to that of the phallus and it is presumed that vaginal spirals might check water from inflowing the reproductive tract during copulation. Among birds, true lymph nodes are present only in water birds. In ducks, there are two pairs of elongated spindle-shaped lymph nodes, namely cervico-thoracic and lumbar lymph nodes. The spleen of duck is triangular in shape. The 'V'-shaped uropygeal/preen gland is relatively large and well developed in water birds. The preen gland wax is a water repellant agent to protect the ducks against wetting. Greater thickness of the skin of ventral regions of the body is considered as an adaptation in water birds to resist heat dissipation during swimming. The plumage is thick and waterproof. The size and spacing of barbs of feathers in ducks also play a major role in the water repellence property of feathers.

References

Alphine J (2018) Comparative studies on the skin and feathers of broiler and layer ducks. MVSc Thesis submitted to the Kerala Veterinary and Animal Sciences University, Pookot, Wayanad, India, p 99

Arunima SS (2011) Postnatal development of lymph nodes in Kuttanad duck (*Anas platyrhynchos domesticus*). MVSc Thesis submitted to the Kerala Veterinary and Animal Sciences University, Pookot, Wayanad, India, p 66

Bang BG (1971) Functional anatomy of olfactory system in 23 orders of birds. Acta Anat 79(suppl. 58):1–76

Benzo CA, Gennaro DLD (1983) A hypothesis of function for the avian glycogen body: a novel role for glycogen in the central nervous system. Med Hypotheses 10(1):69–76

Brennan PL, Prum RO, McCracken KG, Sorenson MD, Wilson RE, Birkhead TR (2007) Coevolution of male and female genital morphology in waterfowl. PLoS One 2(5):e418. https://doi.org/10.1371/journal.pone.0000418

Das LN, Biswal G (1968) Microanatomy of the reproductive tract of domestic duck (*Anas boscas*). Indian Vet J 45:1003–1009

Duncker HR (1974) Structure of avian respiratory tract. Resp Physiol 22:1–19

El-Bably SH, Rezk HM, Tolba AR (2014) Gross morphological studies on the air sacs (*sacci pneumatici*) of Golden Pekin duck (*Anas platyrhyncha*). Haryana Vet 53(1):13–17

Etches RJ (1996) Reproduction in poultry. CAB International, Wallingford, p 318

Farner DS, King JR, Parkes KC (1982) Avian biology, vol 6. Academic Press, London, p 490

Feduccia A (1975) Aves osteology in 'Sisson's and Grossman's the anatomy of the domestic animals', vol 2. WB Saunder's Company, Philadelphia, pp 1790–1801

Firdous AD (2014) Developmental pattern of the respiratory system in Kuttanad ducks (*Anas platyrhynchos domesticus*). PhD Thesis submitted to the Kerala Veterinary and Animal Sciences University, Pookot, Wayanad, India, p 197

Hall MR, Goldsmith AR (1983) Factors affecting prolactin secretion during breeding and incubation in the domestic duck (*Anas platyrhynchos*). Gen Comp Endocrinol 49(2):270–276

Indu VR (1999) Postnatal development of Bursa of Fabricius in the duck (*Anas platyrhynchos*). MVSc Thesis submitted to the Kerala Agricultural University, Thrissur, Kerala, India, p 78

Jacob J, Balthzart J, Schoffenicls E (1979) Sex differences in the chemical composition of uropygeal gland waxes in domestic duck. Biochem Syst Ecol 7:149–153

Jacquet JM (1997) Photorefractory period of the Muscovy duck (*Cairina moschata*): endocrine and neuroendocrine responses to day length after a full reproductive cycle. Br Poult Sci 38(2):209–216
John WL (2015) Respiration in birds. In: Reece WO (ed) Dukes' physiology of domestic animals, 13th edn. Wiley Blackwell, New York, pp 245–258
King AS (1975) Aves respiratory system and Aves urogenital system. In: Getty R (ed) Sisson's and Grossman's the anatomy of the domestic animals, vol 2. WB Saunder's Company, Philadelphia, pp 1883–1967
Makanya AN, Kavo BM, Djonov V (2014) Three-dimensional structure and disposition of the air conducting and gas exchange conduits of the avian lung: the domestic duck (*Cairina moschata*). https://doi.org/10.1155/2014/621982
Marchant S, Higgins PJ (1990) Handbook of Australian, New Zealand and Antarctic birds. Vol 1, ratites to ducks; part B, Australian pelican to ducks. Oxford University Press, Melbourne, pp 1112–1332
McLelland J (1975) Aves digestive system. In: Sisson's and Grossman's the anatomy of the domestic animals, vol 2. W.B. Saunder's Company, Philadelphia, pp 1857–1882
Myers JA (1917) Studies of the syrinx of *Gallus domestics*. J Morph 29:165–215
Nickel R, Schummer A, Seiferle E (1977) Anatomy of the domestic birds. Verlag Paul Parey, Berlin, p 202p
Patricia AJ (2015) Avian reproduction. In: Reece WO (ed) Dukes' physiology of domestic animals, 17th edn. Wiley Blackwell, New York, pp 715–725
Pattle RE (1978) Lung surfactant and lung lining in birds. In: Pipper J (ed) Respiratory function in birds, adult and embryonic. Springer, New York, pp 22–32
Pearson R (1972) The avian brain. Academic press, London, pp 453–455
Rajathi S (2005) Posthatch development of preen gland in the duck (*Anas platyrhynchos*). MVSc Thesis submitted to the Kerala Agricultural University, Thrissur, Kerala, India, p 71
Reece WO (2015) Kidney function in birds. In: Reece WO (ed) Dukes' physiology of domestic animals, 17th edn. Wiley Blackwell, New York, pp 193–211
Shyla P (1990) Postnatal development of upper digestive tract in the duck. MVSc Thesis submitted to the Kerala Agricultural University, Thrissur, Kerala, India, p 114
Sreedharanunni CK (1976) Studies on the structure and development of thymus in the Indian Runner duck (*Anas platyrhynchos domesticus*). MVSc Thesis submitted to the Kerala Agricultural University, Thrissur, Kerala, India, p 45
Vernerova E, Burda Z (1984) The development of genitalia in ducks. Zivocisna Vyroba 43:281–296
White SS, Chubb JC (1967) The muscles and movements of the larynx of *Gallus domesticus*. J Anat 102:575

Dr. K. M. Lucy, Ph.D. (Veterinary Anatomy, Kerala Veterinary and Animal Sciences University)

Dr. K. M. Lucy hails from Thrissur, Kerala, India. She acquired Bachelor in Veterinary Science and Animal Husbandry in 1992 with first rank, post-graduation in 1995 and Ph.D. in 2005 in Veterinary Anatomy from Kerala Agricultural University. She is currently working as Professor and Head of Department of Veterinary Anatomy, College of Veterinary and Animal Sciences, Mannuthy, Kerala and has 25 years of teaching and research experience including the field of avian anatomy. She has more than 130 peer-reviewed research publications in national and international journals. She has authored textbooks on 'Anatomy of the Rabbit' and 'Comparative Anatomy of Wild Mammals'. She is the recipient of several awards including three best teacher awards.

Dr. K. Karthiayini, Ph.D. (Veterinary Physiology, Kerala Veterinary and Animal Sciences University)

Dr. K. Karthiayini is a native of Kodakara, Thrissur, Kerala, India. She completed her Bachelor of Science in 1986 from Calicut University, Bachelor of Veterinary Science and Animal Husbandry in 1992 from Kerala Agricultural University, post-graduation in Veterinary Physiology from Tamil Nadu Veterinary and Animal Sciences, Chennai, India and Ph.D. from Kerala Agricultural University. She is currently working as Professor in Department of Veterinary Physiology, College of Veterinary and Animal Sciences, Mannuthy, Kerala and has 23 years of teaching and research experience.

Nomadic (Transhumant) Duck Farming Practices

5

S. Sankaralingam and J. D. Mahanta

Contents

S. Sankaralingam (✉)
Department of Poultry Science, College of Veterinary & Animal Sciences, Kerala Veterinary & Animal Sciences University, Thrissur, Kerala, India
e-mail: sankaralingam@kvasu.ac.in

J. D. Mahanta
Department of Poultry Science, College of Veterinary Science, Assam Agricultural University, Guwahati, India
e-mail: jogadev.mahanta@aau.ac.in

A. Jalaludeen et al. (eds.), *Duck Production and Management Strategies*,
https://doi.org/10.1007/978-981-16-6100-6_5

Abstract

The nomadic system of duck rearing is popular in rice growing countries like India, Bangladesh, Vietnam, and Indonesia. The poor, landless farmers with ducks in few hundreds to thousands move to faraway places of nearby states or provinces and very rarely to nearby countries in search of pastures. During lean seasons, the ducks are handfed with locally available feedstuffs. The nomads move back to their native place at the time of harvest or between batches. The nomadic farmers are vulnerable to exploitation as they are financially bonded with hatchery operators, middleman, collector, and itinerant sellers. The lackluster response from the financial institutions in lending loans due to lack of traceability and non-availability of immovable property for security forces them to get microcredit with heavy interest of 50–100% of the loan amount per year. The egg production ranges from 70 to 200 per year due to poor production potential of non-descript ducks, nutrient deficiency, and poor management. The nomadic farming faces problems of diseases like duck plague, duck cholera, and avian influenza, lack of inheritance by new generation people, shrinking of scavenging areas and natural feed resources, drying up of natural water bodies, excessive use of pesticides on paddy fields, cultivation of short duration paddy varieties, shift from paddy cultivation to cash crop cultivation, industrialization, and urbanization.

Keywords

Nomadic · Transhumance · Duck farming · Production and practices · India · Bangladesh · Mekong delta · Indonesia

List of Symbols/Abbreviations

cm	Centimeter
ft	Feet
ha	Hectare
INR	Indian Rupee
kg	Kilogram
km	Kilometer
L	Liter
m	Meter
mg	Milligram
ppm	Parts per million
USD	United States Dollar
@	At the rate of
%	Per cent

<	Less than
>	Greater than
=	Equal to
+	Plus

5.1 Introduction

The present day domestic ducks (*Anas platyrhynchos domesticus*) which originated from Asian wild Mallard ducks (*Anas platyrhynchos*) have been improved and exploited for commercial purpose through scientific breeding programs. The world duck population is estimated to be 1241.388 million (FAOSTAT 2018). Out of this, 1102.66 million ducks (88.82%) are found to be in Asian countries predominantly, China (818.86 million), Vietnam (71.29 million), Bangladesh (52.24 million), and Indonesia (47.36 million). In most of these Southeast Asian countries, indigenous ducks are usually used for producing eggs, meat being only a by-product. Traditional extensive duck rearing systems are still dominant in developing countries although there is a trend towards the more intensive systems which is predominant in the West (Edwards 1986).

Duck farming is a profitable livestock industry in the world because of its egg, meat, and feather. Duck eggs are relatively larger, weighing about 4.5% of duck's body weight, compared to that of chicken, whose egg weight is only about 3.3% of the hen's body weight (Narahari 2009). Moreover, ducks are more prolific than chicken and more adaptable to free-range system of rearing. They also grow faster than chicken in their early stage of life and require no elaborate housing, compared to chicken. Therefore, they are more popular in many Asian and some European countries. Nomadic duck rearing is one of the old traditional farming systems, which contributes significantly to the duck production in India (Tamizhkumaran et al. 2013). Ducks play a critical role in meeting daily protein needs and providing household income of farm families in the mixed farming systems of many developing countries. Duck rearing is considered to be the women's enterprise in smallholder farming system as they hold the sole responsibility of rearing ducks. Despite of all the above mentioned advantages, duck farming is decreasing day-by-day due to reduction in scavenging areas, natural feed resources, drying up of natural water bodies, excessive use of pesticides in the crop fields, shift from paddy cultivation to cash crop cultivation, etc. Transhumant or nomadic duck husbandry is widely practiced in South and South East Asia as a traditional system of duck rearing. Location specific technological interventions in nomadic duck rearing enhance the performance of the birds (Rahman et al. 2017). The duck flocks often migrate to nearby areas through road, waterways, or transportation vehicles in search of fresh forage and water resources. The only job of nomads is to forage the ducks and collect the eggs. Ducks and other waterfowls are of great importance for food security of mankind in many parts of the world as elaborated by Pingel (2009).

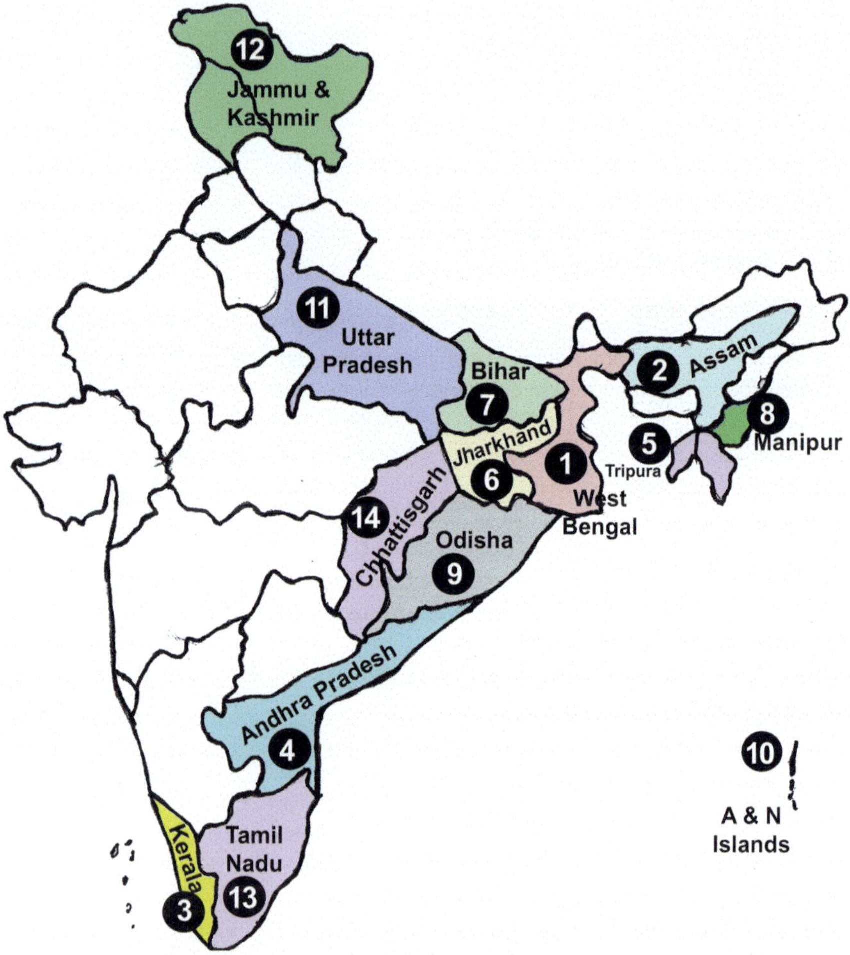

Fig. 5.1 Reference map of India with major duck growing states as per BAHS (2019)

According to provisional data of livestock census (BAHS 2019), the duck population in India was 33.51 million, which is 3.99% of the total poultry population (fowl and duck together). West Bengal and Assam states together contribute 73.82% of the total duck population of India followed by Kerala, Andhra Pradesh, Tripura, Jharkhand, Bihar, and Manipur states (Fig. 5.1). There is 42.35% increase in duck population of India (23.54 million) as compared to livestock census 2012. The credit for this increase goes mainly to the duck farmers of West Bengal and Assam and to some extent to Bihar, Jharkhand, and Manipur. As per FAOSTAT (2018), India ranks 7th in the world in duck population. Duck population of India is in the hands of socially and economically backward section of the people and also mainly depends on backyard, free range, and nomadic system of rearing.

India is an agricultural country with most of the population (65.53%) living in rural areas in which the duck production offers itself as an alternate and additional system of food production. The ducks contribution to India's gross national product (GNP) is about 0.57 million USD per annum (Jalaludeen et al. 2003). In this chapter, an overview of nomadic duck farming practices in global context is discussed.

5.2 Systems of Duck Production

Carlén and Lansfors (2002) recorded the following three main systems of duck production in the Mekong Delta considerably vary depending on their location (rural, urban, and peri-urban). This pattern of farming systems has its presence in many parts of the world.

- ***Scavenging systems***: scavenging, free ranging within the farm and the village (garden, home, or village pond); daily herding in rice fields, dikes, rivers, canals, and tidal areas (beyond the farm); seasonal transhumant supervised ranging (beyond the locality)
- ***Integrated duck production systems***: rice–fish–duck; fish–duck–pigs (or other domestic species)
- ***Confined systems***: semi-commercial and commercial farms (meat and egg production); duck–fish combination in enclosures or floating cage in ponds/canals/rivers

5.3 Nomadic Systems of Duck Production

The nomadic system, a subtype of extensive management of ducks is considered as a traditional method of keeping ducks in Asia (Gajendran and Karthickeyan 2009; Rahman et al. 2017). Ducks are allowed to herd on the rice fields after harvesting or allowed to scavenge in the paddy fields during certain period of rice growing cycle. Either mobile or home based type of herding system has been practiced in most Asian countries for centuries. The area of foraging is gradually vanishing due to industrialization, urbanization, and cultivation of cash crops like sugarcane, tapioca, etc. The emphasis on sustainable agriculture advocates the integrated duck–rice system by the scavenging of ducks on rice field to produce organic rice.

India has 24 breeds of native ducks with 34 varieties within them (Bhat et al. 1980). The nomadic duck farmers move their flocks from one area to another in a cyclical manner depending on the availability of feed from harvested paddy fields, rainfall, and cropping patterns (Nambi 2001). The indigenous practices of management under nomadic duck rearing system in few states of India are discussed below.

5.3.1 Kerala State of India

In Kerala state, Ramachandran and Ramakrishnan (1982) first reported the nomadic or herding type of duck rearing, where flocks of adult ducks are moved from one place to the other in search of post-harvested paddy fields. This type of duck management system in the state of Kerala is unique in the country. Till the 1980s, duck farmers were economically backward and availed loan from the financiers who were mostly traders or commission agents of duck egg market. Most of the time, the duck farmers were unable to repay the loan amount in one laying cycle and forced to continue the duck rearing throughout their life. All the activities of nomadic duck rearing like purchase of ducklings or ready-to-lay pullet ducks, sale of the eggs, drakes and spent ducks, collection and transportation of ducks and eggs, and purchase of feed were done with the help of financiers. The entire members of the joint family were looking after the ducks. At the time of partition of family, the ducks were also divided. The average flock size was 200–400 depending upon the family size. Although some farmers had more number of birds, they were sub-grouped into small flocks of 350–400 ducks which were reared by three persons (Ravindran 1983).

5.3.1.1 Profile of the Duck Farmers

Among the duck producing states of India, Kerala is the only state with a duck population of 1.19 million moving towards the commercialization of the enterprise to some extent. Alappuzha district of Kerala is one of the important places, where state's 49% ducks are available and it lies one meter below mean sea level (MSL). The place is full of brackish water areas including backwaters, canals, ponds, streams, waterlogged areas, and large number of paddy fields. Paddy is the major agricultural crop and duck rearing is closely associated with its cultivation. Jalaludeen et al. (2003) conducted a survey study in Alappuzha district to find out the socio-economic status of farmers involved in nomadic duck rearing (Table 5.1). They found that most of the nomadic farmers were between 35 and 45 years of age (47%). The literacy rate among the farmers was 98%, of which, 60% learned up to upper primary schooling. The nomadic farmers in this area are predominantly Christians (62%) and with regard to land holding, most of them possessed 0.04–0.10 ha of land (62%). Duck farmers of Kerala prefer only Kuttanad duck (88%) for nomadic rearing compared to exotic ducks like Khaki Campbell due to its hardiness and large egg size.

5.3.1.2 Flock Strength

Jalaludeen et al. (2003) reported that majority (66%) of duck farmers were having a flock size below 2000 in Alappuzha area of Kerala. The common flock size was in the range of 1000–2000 birds (40%). Conversely, Ravindran (1983) after conducting a survey on the status of duck farming in Kerala state observed that common flock strength appeared to be 350–400 ducks.

5.3.1.3 Preference of Duck Varieties/Breeds

Under nomadic pattern of duck rearing system, the farmers of Kerala maintain indigenous ducks only. The two major varieties of Kuttanad ducks are referred to as Chara and Chemballi (Jalaludeen 2002). The other minor varieties are Thoovella, Pulli, and Pandi depending on color and spots (KSBB 2019). Majority of farmers (88%) preferred indigenous ducks (Table 5.1) in confirmation of earlier survey

Table 5.1 Socio-economic profile of duck farmers ($N = 200$; Jalaludeen et al. 2003)

Parameters	Classification	Percentage
Age wise distribution	<35 years	17.0
	35–45 years	47.0
	>45 years	36.0
Level of literacy	Illiterate	02.0
	Lower primary	20.0
	Upper primary	60.0
	High school	18.0
Religious status	Hindu	23.0
	Christian	62.0
	Muslim	12.0
	Scheduled caste	03.0
Experience in duck farming	<5 years	12.0
	5–15 years	39.0
	15–30 years	37.0
	>30 years	12.0
Size of land holding	<0.04 ha	21.0
	0.04–0.10 ha	62.0
	0.10–0.20 ha	07.0
	>0.20 ha	10.0
Flock size	<1000	26.0
	1000–2000	40.0
	2000–5000	25.0
	>5000	09.0
Net profit from duck farming per month (USD)	<13.33	25.0
	13.33–40.00	26.0
	40.00–66.67	30.0
	>66.67	19.0
Preference of variety	Kuttanad	88.0
	Others	12.0

conducted by Ravindran (1983). Although the farmers of Kerala were aware of exotic egg type ducks like Khaki Campbell, they prefer indigenous ducks, because of their hardiness, ease of management, and attractive egg size (Jalaludeen et al. 2004).

5.3.1.4 Incubation of Duck Eggs

In traditional system, the duck farmers of India incubated the duck eggs under the broody hens (living incubator) specially reared for this purpose and were able to produce flocks in thousands (Ramachandran and Ramakrishnan 1982; Ravindran et al. 1984; Rithamber et al. 1986; Mahanta et al. 2000; Jalaludeen et al. 2003). Experienced farmers obtained 80–85% hatchability on the total egg set. It was a cottage industry in Kuttanad area practiced by womenfolk, where each woman owned 10–15 broody hens to incubate duck eggs. The same hen was utilized to incubate two sets of duck eggs continuously one after the other. During this period, the hens were fed with cooked rice alone to keep them broody. After the hatch out of second set of ducklings, the hens were released and fed with protein diet too for rejuvenation. After 1–2 months, the hens lay a clutch of eggs before becoming broody again. In each year, the hens were incubating 6 sets of eggs (Ravindran 1983). As the natural incubation had certain limitations, from late 1990s, forced-draft incubators were used by more than half a dozen hatcheries operating in and around Kuttanad region (which covers Alappuzha, Kottayam, and Pathanamthitta districts) of Kerala to hatch about six lakh ducklings. The ducklings were hatched out in three specific months, viz. Vrishchikam (mid-November to mid-December) and Makaram (mid-January to mid-February) to utilize the harvested paddy fields of mundakan (winter) and puncha (summer) crops and Chingam (mid-August to mid-September) to utilize the harvested paddy fields of varsha (rainy season) crops. These three hatch-out seasons coincided with the harvesting of paddy during growing and laying periods.

At present there are around ten hatcheries in Kuttanad, which were located mainly at Haripad, Chennithala, Thiruvalla, and Changanassery areas for hatching the duck eggs. Most of them are working in all seasons except summer (March, April, and May) to avoid hatchability problems and also to avoid rearing grower ducks during rainy season when foraging fields for ducks are not available. The day-old ducklings are sold not only in the state of Kerala but also to neighboring states of Tamil Nadu and Andhra Pradesh. Although the hatcheries produce in all seasons, most of the Kerala nomadic duck farmers purchase the ducklings during Vrishchikam to Makaram hatch outs to utilize the paddy fields of Kuttanad, Thrissur, Malappuram, Kannur, and Palakkad districts for foraging the growing and laying birds. Some farmers of Kuttanad region purchase the ducklings during Chingam month also. Hatch out of all other months is mostly going to Tamil Nadu and Andhra Pradesh states as day-old or one-month-old ducklings.

In the past, the cost of fertile eggs was 8–10% higher than the table eggs (Ramakrishnan 1996) and the same is enhanced now to 10–20%. The number of eggs incubated per batch depends upon the placement of order by the nomadic farmers and not on the capacity of the hatchery. As eggs are incubated at every 3 days interval, there is no need for egg cooler room. Eggs are incubated at the same temperature meant for chicken eggs with little higher humidity which is achieved by sprinkling lukewarm water from second day to 23rd day. The incubators are fumigated with potassium permanganate and formalin before setting the eggs. Although it is very easy to do sexing of day-old ducklings by Japanese method of vent sexing, hatchery operators are not practicing it.

The nomadic farmers attain 75–85% fertility with the sex ratio of 3–5 drakes per 100 ducks (Ramakrishnan 1996) with the hatchability (on total egg set) of 55–70%. Presently, hatching eggs are purchased (1USD = INR 75) by the hatchery owners at the rate of 0.09–0.11 USD (INR 6.5–8.0) per egg and the hatched out ducklings are sold at the rate of 0.24–0.33 USD (INR 18–25) per duckling depending upon the season, demand, and availability of good quality hatching eggs.

There are some custom hatcheries, which incubate the duck eggs at the hatching charge of 4.67–5.33 USD (INR 350–400) per 100 egg set. Here, as the incubation conditions maintained in setter and hatcher are similar to that of chicken eggs, the duck egg hatchability on total egg set is around 50% only due to lower humidity.

The newly hatched ducklings are sold to duck nursery farmers or brooded by the hatchery owners themselves up to 30–45 days and sell it to the needy grower farmers. The transportation charge of the ducklings to the farmers premise is 0.47–0.53 USD (INR 35–40) per kilometer for mini-lorry which can carry around 12,000–15,000 ducklings in plastic boxes. The hatchery owners and the nursery owners sell their ducklings for ready cash or loan. The loan amount is paid back with interest by the grower farmers as 3 months old drakes or eggs after start of lay.

Even now, there are some housewives in Kuttanad area who are doing natural incubation by keeping 15–18 eggs under a broody hen. As the remuneration to them is depending on number of hatch outs, they try to get maximum hatchability by giving utmost care.

5.3.1.5 Brooder Management (Duck Nursery)

The duck nursery is maintained mostly in Kuttanad region by seasonal farmers. They purchase the day-old ducklings at the rate of 0.24–0.33 USD (INR 18–25) per duckling from the hatchery owners and rear them up to 30–45 days of age. Till the 1990s, the ducklings were brooded in the nursery units with an average flock size of 2000–3000 ducklings per brooder shed with the dimension of 2 m × 4 m with day time run space of 150 cm^2 per duckling for feeding and watering (Fig. 5.2). The ridge

Fig. 5.2 Duck nursery. (Photo courtesy: Leo Joseph)

Table 5.2 Feeding schedule of ducklings (Jalaludeen et al. 2004)

First day	No feed
1st week	Cooked rice
2nd week	Cooked rice mixed with chick starter
3rd week	Cooked rice mixed with chick starter and dried unsalted fish
4th week	Poultry feed, dried unsalted fish, and rice bran. Some farmers used to feed cooked wheat and coconut cake also

height of these houses was 2.5 m and eaves' height was 60 cm. The shed was divided into 3–4 partitions lengthwise and or breadth wise to prevent huddling death (Ravindran 1983). In the 2000s, the practice of brooding of 5000–6000 ducklings per shed with the dimension of around 6.6–5.2 m with the ridge height of 2.25 m and eaves' height of 1.25 m was documented by Jalaludeen et al. (2004). The present day duck nursery farmers rear around 5000–15,000 ducklings by increasing the shed size to cope up with the demand for one-month-old ducklings. A 15,000 flock can be reared in a temporary shed with the dimension of 7.5 m × 30 m. The shed is constructed on the elevated dry land near lake, water canal, lagoon, pond, or river. The roof is thatched and/or covered with polythene sheets and the sides are covered with bamboo mat or plaited coconut leaves for 30 cm height and rest of the area is covered with nylon wire net for proper ventilation. The overhang is lengthy enough to cover the open sides. Water for drinking is given in plastic basins or on polythene sheet spread on the shallow pit in the run area. No artificial heat is given to the ducklings during brooding period; instead ducklings are overcrowded to maintain their own body temperature. During nighttime, light sources like candle or kerosene lamps are kept inside the shed for proper vision of all ducklings. After 5–7 days of brooding, the ducklings are introduced to water source for swimming during daytime and the swimming time will be enhanced at the rate of 30 min per day. A flock is maintained by two to four caretakers for feeding, watering, and managing the ducklings during swimming. The mortality percentage during brooding is around 5–10% depending upon management, disease outbreak, and mycotoxin content of the feed.

During brooding, a mixture of cooked rice, chick feed, and dried fish is given as starter feed (Table 5.2). When the ducklings enter the growing stage at 1 month of age, they are taken to paddy field for foraging and the hand feeding is discontinued. The first dose of vaccination against duck plague is done at duck nursery itself.

5.3.1.6 Grower Management

The duck grower farmers purchase 5000–15,000 numbers of 30- to 45-day-old ducklings at the rate of 0.93–1.20 USD (INR 70–90) per duckling (Sankaralingam 2017). Some nursery owners in Kuttanad area are doing both brooding cum growing together and sell them to layer duck farmers. The transhumance of ducks starts at 30–45 days of age itself, the ducklings are reared on the harvested paddy fields,

Fig. 5.3 Caretaker on canoe guiding the ducklings in water canal. (Photo Courtesy: Leo Joseph)

lakes, ponds, water canals, rivers, lagoons, and other water sources up to 4–5 months of age. The herders use a long stick with a piece of red or black cloth tied at one end to guide and control the ducks while foraging. The caretakers use canoe (Kothumbuvallam) for guiding and goading ducks in water (Fig. 5.3). In foraging, ducks eat grasshoppers, snails, slugs, tadpoles, frogs, small fishes, small crabs, earthworms, mosquito larvae, algae, and other small insects. Anitha et al. (2009) found 64.01% paddy grains, 23.15% crab, 4.07% snails, 2.02% worms, and 6.75% weeds in the crop samples of nomadic ducks (Fig. 5.4). At the time of foraging, ducks put back dropping as manure in the paddy field and also they reduce the percolation loss of water from the paddy field by eating crabs and covering crab and rat holes by creating semi-pervious layer of soil with their bill during scavenging. Anitha et al. (2011) reported that the combined residual level of different organochlorine compounds was much below the maximum residue limits (MRL) in crop and body fat (0.0033 and 0.0077 ppm, respectively) of nomadic ducks of Kerala.

The ducklings are bill branded with red hot gunny bag needle or iron ribs of umbrella at or after 2 months of age (Fig. 5.5). Some use thin iron rod with alphabets to brand their birds. This is done to maintain the owner identity of the ducks when more than one flock mix together. Till the 1960s, ducks were foraged on the harvested paddy fields at free of cost. From the 1970s onwards, the duck farmers started paying to the paddy field owners in the form of 20 to 40 eggs or equivalent cash per acre or in the form of drake as foraging charge to maintain good rapport with them (Ravindran 1983). In the areas where the paddy cultivation is done by

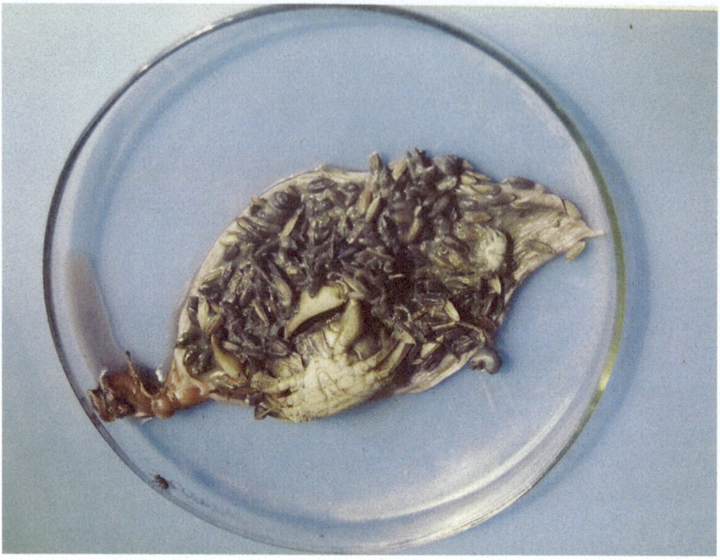

Fig. 5.4 Crop content of a nomadic duck. (Photo Courtesy: Leo Joseph)

Fig. 5.5 Bill branding. (Photo courtesy: Leo Joseph)

cooperatives like that of Kole land paddy cultivation of Thrissur district, the harvested paddy fields are auctioned for duck foraging by the Farmer's Cooperatives (Padasekhara Samithi). The present auction rate is 400–800 USD (INR 30,000–60,000) per 100 acres of foraging paddy field for 3–4 months duration depending upon the demand, number of bidders, availability of feed resources, etc. Each 10,000 ducks need 100 acres of land for foraging in three rotations. The foraging area is divided into three fields and the ducks are foraged rotationally one

Fig. 5.6 Night enclosure. (Photo courtesy: A. Jalaludeen)

after the other. The birds are foraged from 6.30 am to 6.00 pm. During hot sunny days, the ducks are rested under shades from 11.30 am to 3.30 pm to avoid heat stress. After allowing them to take bath in the water source, at around 6.30 pm they are flocked together in an elevated area near paddy fields or water sources by encircling two sets of nylon net with the help of bamboo or wooden poles with sharp bottom (Fig. 5.6). There is one meter gap between two sets of net enclosures to avoid predator attack. The height of enclosure should be three feet (0.9 m). At around 2½–3 months of age, the drakes are identified by their dimorphic color or voice or drake feather and all the excess males are sold for meat purpose at the rate of 2.40–2.67 USD (INR 180–200) per drake. During growing period, one caretaker per 3000–4000 duck is employed and they are paid at the rate of 9.33 USD (INR 700) per day (Sankaralingam 2017). If the flock size is more, one more worker is kept as spare and utilized when any other caretaker becomes sick or goes home for any emergency and also for selling the drakes at 2½–3 months of age. The caretakers dwell inside a small tent made up of polythene sheet with two 4 ft long poles at the ridge and several small sticks at the eaves. During night time one caretaker watches the flock in rotation for predator attack and theft.

During growing period, no hand feeding is done other than foraging but watering is done every night using plastic basins. During lean season like rainy days, birds are hand-fed. In Kuttanad area, Corypha or umbrella palm (*Corypha umbraculifera*) pith which is rich in starch and soluble sugar is also fed along with dried fish or groundnut cake during lean season. During growing period, ducks are reared mainly in Kuttanad area and Thrissur district of Kerala. In normal case, the mortality during growing period (1–5 months of age) is around 10% or more depending on the predator attack, mycotoxin level in the feed, worm infestation, pesticide level in the feed, availability of feed, outbreak of diseases like duck plague, duck cholera, avian influenza, etc.

5.3.1.7 Layer Management

The age at sexual maturity of Kuttanad duck is 126–140 days according to the availability of feed and disease management during growing period. After 4–5 months of age, the grower cum layer duck farmers continue to rear 1000–2000 females along with required males (3–5 drakes per 100 ducks) and sell the excess females to the needy layer duck farmers at the rate of 3.06–3.73 USD (INR 230–280) per pullet duck (Sankaralingam 2017). If the males are more, it will lead to sexual disturbance and higher egg breaking by males. These layer duck flocks are kept at different areas of Kerala mainly in Palakkad district till mid-November and then most of them are sold to another set of farmers (mostly from Tamil Nadu and some from Andhra Pradesh state) at the rate of 2.67–3.06 USD (INR 200–230) per duck. Some of them move back to Kuttanad. The nomadic duck cycle of Kerala starts in the months of Vrishchikam and Makaram and ends in the month of Thulam (mid-November) with a few days gap between two cycles. The poor layer duck farmers who do not have money to purchase ready-to-lay or partly laid pullet ducks borrow the birds from the grower farmers or borrow money from financier cum egg traders to purchase the pullet ducks. These farmers give back the money after few months from the revenue generated by selling eggs; otherwise, the duck egg traders get back their money along with interest by collecting the egg from the farmers' doorstep. Ramakrishnan (1982) reported that 89% of the farmers were borrowing money from the financier cum egg traders and few were only getting loan from co-operative societies and banks. Nowadays, most of the farmers finance from their own source or getting gold loan from bank for rearing ducks. The drakes are kept to control the social order within the flock rather than getting fertile eggs. With this male:female ratio, farmers achieve 75–85% fertility. The layer/breeder management under nomadic system is similar to that of grower management.

In the month of August, all the flocks from Kerala and some flocks from Tamil Nadu, Karnataka, and Andhra Pradesh states move to Palakkad paddy fields. This is the place of annual get-together for the nomadic duck farmers of all four states. Earlier the ducks were herded on foot via road or by swimming through water canal for short distance (Ramakrishnan 1996), but now most of the farmers use specially designed trucks (Fig. 5.7) to transport the ducks especially for long distance (Jalaludeen et al. 2004). The transportation cost is 0.53 USD (INR 40) per kilometer for mini-lorry which can transport around 1400 ducks and 0.73 USD (INR 55) per kilometer for lorry which can transport 2500 ducks. For layer flocks, each 500–700 ducks need one caretaker for foraging and collection of egg and they are paid at the rate of 9.33 USD (INR 700) per day (Sankaralingam 2017). The egg laying starts at 12.00 midnight and ends at 6.00 am. Before 6.00 am, 96–98% of eggs are laid. The egg collection starts at 5.00 am and ends at 6.00 am. After 6.30 am, the ducks are allowed to forage as usual. The duck eggs are packed en masse in the boxes made of low cost wood. The gap between the wooden planks provides aeration and prevents quick deterioration of egg quality (Jalaludeen et al. 2004). Nowadays, the duck

Fig. 5.7 Transportation. (Photo courtesy: Leo Joseph)

farmers started using filler flats of extra-large egg size for collection and transportation of duck eggs too. The spent ducks after 1 or 2 years of lay are sold at the rate of 2 USD (INR 150) per duck.

In some areas of Kerala although there is spilled grains in the paddy field, ducks are not foraged due to the non-availability of water during summer (April and May). Duck needs at least 1–2 inches of water logging on the paddy field to avoid chocking of feed material inside the esophagus. During heavy rain (mid-June to mid-August) also they cannot be foraged due to flooding of the paddy fields. Altogether for 2–4 months duration, duck farmers of Kerala have to hand-feed the birds. The annual egg production of Kuttanad duck under nomadic system of rearing is 200 (Ramachandran and Ramakrishnan 1982).

5.3.1.8 Breeder Management

There is no separate breeder flock of Kuttanad duck. Hatching eggs are collected from the nomadic flocks 6–8 weeks after start of laying. Some nomadic farmers supplement rice, wheat, jowar, groundnut cake, and shell grit additionally while collecting fertile eggs.

The fertile egg collection starts from 3 am and collected frequently to get clean eggs. The inner circle of enclosure is removed at around 11.00 pm to allow the laying ducks to go to the clean outer circle and lay eggs. During rainy days, a temporary shed of 4–5 feet ridge height and 2 feet eave height with approximately 30 × 15 ft size is provided for around 1500 ducks as night shelter especially during laying time to get clean hatching eggs. The sheds are made up of bamboo or wooden poles covered with tarpaulin or thick polythene sheets. During hatching season, hatchery owners purchase the eggs from these nomadic herders and incubate it.

If a farmer suffers genuine loss due to disease outbreak or other calamities, the lenders such as hatchery owners, financier cum traders, and grower farmers waive the loan amount fully or partially and help them to restart the activity; because if a farmer goes out of the cluster, the loss is not only to the farmer but also to all the people involved in the entire chain of activities.

5.3.1.9 Feeding Ducks During Lean Season

In Kerala, Jalaludeen et al. (2003) reported that rearing indigenous ducks under foraging condition is possible only for a period of 7–8 months in 1 year since the availability of post-harvested paddy fields is limited to this period. The remaining period is considered as the lean season for duck farming. Lean season in Kerala includes both monsoon seasons (Southwest and Northeast) and summer months which constitutes about 4–5 months in a year. During monsoon the paddy fields are flooded with water, so that ducks may not be able to get enough feed from the fields. Therefore, duck farmers have to resort to supplementary feeding. Many feed items are tried by farmers depending upon their cost, availability, and regional variations. Some of the common items used by duck farmers of Kerala during lean seasons are palm pith, jowar, wheat, rice (either whole or broken), paddy chaff, unsalted dried fish, ground nut cake, and rice polish (Annual Report 2002). Among the hand-feeding items, the major cheap feed alternative is the chopped pith of umbrella palm (*Corypha umbraculifera*) of the family Palmae (Fig. 5.8). This palm is abundantly distributed in hilly areas of Central Kerala. It grows up to a height of 70 feet and has a massive trunk. It blooms only once in lifetime and after blooming and fruiting, the plant dies. The palm trees intended for feeding ducks are used before the onset of blooming. The trunk is cut into 1.5 m long pieces for sales. On dry matter basis, the central pith of the trunk contains 42.1% moisture and 10.90% crude protein and 65.80% nitrogen free extracts on dry matter basis (Joseph et al. 2004). Exploitation of locally available cheap feed resources helps the farmers to keep the cost of hand-feeding at a minimum level.

5.3.1.10 Flock Movement

The major hatching seasons in Kerala are:

1. Vrishchikam (mid-November to mid-December), the ducklings hatched out during this month spend their brood cum grow period in Kuttanad area and then the ready-to-lay pullet ducks are sold and moved to Thrissur Kole field in the month of April (Fig. 5.9).
2. Makaram (mid-January to mid-February), the ducklings hatched out during this month are brooded for 1–1½ months in Kuttanad area and then most of them are sold and moved to Thrissur for growing and laying or the day-old ducklings are sold and moved to Thrissur Kole field for brooding and growing till June. Some ducklings are maintained in Kuttanad itself for growing and laying.

Fig. 5.8 (**a**) Corypha palm tree, (**b**) Corypha palm trunk pieces, (**c**) Corypha palm pith. (Photo courtesy: Leo Joseph)

In the month of July, the birds forage in the harvested paddy fields of Thrissur district other than Kole land and also hand-fed during rainy days. The entire flock moves to Palakkad district in the month of August and stays there up to November. After November, most of the nomadic farmers of Kerala sell their partly laid pullet ducks to Tamil Nadu farmers at ready cash or for loan. The ducks move to different parts of Tamil Nadu and Mysore district of Karnataka and Nellore district of Andhra Pradesh for the remaining laying period. The eggs and spent ducks after completion of laying move back to Kerala for consumption. Some birds move back from Palakkad to Kuttanad in the month of November for laying.

Some flocks move from Thrissur to Mysore in the months of June and July and come back to Palakkad in the month of August and then go back to Mysore in the month of December. Some flocks move from Kuttanad to Tirur and Manjeri area of Malappuram district, grow there from December to February, and then herded to Thrissur Kole field in the month of March. Some other flocks move from Thrissur to Tirur and Manjeri area during June, July months after start of lay. As the layer flock size is small (1000–2000), it is comfortable to rear them in Malappuram district where the paddy fields are small.

There is another minor route to Wayanad. The duck farmers of Wayanad district move from Thrissur to Wayanad and herd there in the months of May and June, return to Thrissur in July and move to Palakkad in the months of August to October, and finally move back to Wayanad in the month of November.

Fig. 5.9 Foraging of ducks in the Kole field of Thrissur district. (Photo courtesy: Leo Joseph)

Another route is movement of duck flocks from Kuttanad to Pokkali rice fields of Thalassery taluk in Kannur district from October to December and return. However, as these tidal swamps are converted to shrimp farming, the Pokkali paddy cultivation area is reduced and the movement of duck flocks in this route is minimized.

Some flocks of day-old ducklings, grower ducklings, and ready-to-lay pullet ducks move from Kuttanad to Tenkasi and Tirunelveli area of Tamil Nadu via Punalur and some other flocks move to Kanyakumari district of Tamil Nadu via Thiruvananthapuram.

Apart from the above major routes, few flocks move through some minor routes of Kerala and also through Tamil Nadu, Karnataka, and Andhra Pradesh. The flock movement of Kerala is depicted in Fig. 5.10.

5.3.1.11 Healthcare Management

Most common diseases in nomadic duck rearing are duck Plague (duck viral enteritis) and duck cholera (Pasteurellosis). Duck viral hepatitis, duck pox, avian influenza, and some other diseases are reported rarely. At times, duck plague leads to complete devastation of the duck population due to high rate of morbidity and mortality under nomadic system. As duck cholera causes sudden death of active

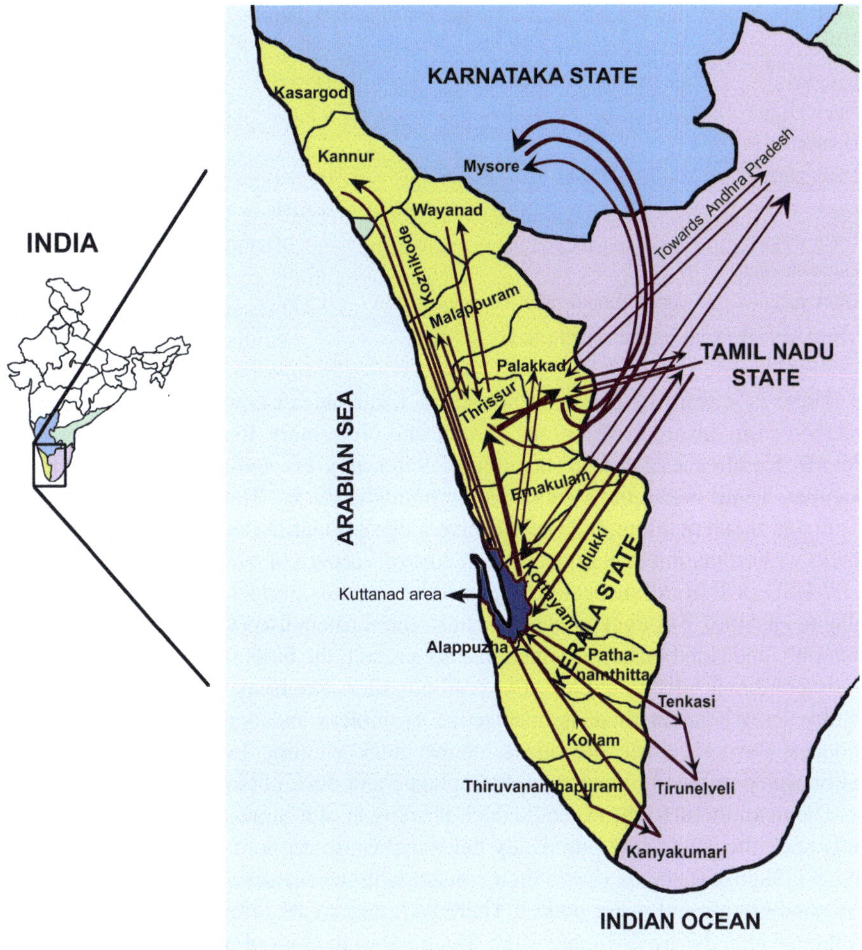

Fig. 5.10 Reference map of movement of duck flocks in and around Kerala state. (Image courtesy: R. Richard Churchil)

flock after fluttering 2–3 times, it is called as "Attack" (similar to heart attack in human being) by nomadic duck farmers of Kerala (Jalaludeen et al. 2004). The mycotoxicosis and worm infestation from the foraging ground increase the severity of the duck plague and duck cholera due to immunosuppression. The present day threat to nomadic duck farming is new duck disease caused by *Riemerella anatipestifer*. It affects young flocks especially after 7 days of age to growing stage. It is seen in nomadic duck population since 2005. The mortality due to new duck disease ranges from 20% to 50% in the affected flocks. Its severity is more when it combines with some other diseases like *E. coli*. The recommended vaccination schedule for nomadic ducks of Kerala is given in Table 5.3.

Table 5.3 Vaccination schedule advised to the nomadic duck farmers of Kerala (Sankaralingam 2017)

Disease	Age (weeks) of vaccination	Dosage[a]	Route of administration
Duck plague	3	0.5 ml	Subcutaneous
Duck cholera	4	0.3 ml	Intramuscular
Duck plague	7	0.5 ml	Subcutaneous
Duck cholera	8	0.3 ml	Intramuscular
Duck plague	15	0.5 ml	Subcutaneous
Duck cholera	16	0.5 ml	Intramuscular
Duck plague	Every 6 months after last vaccination	0.5 ml	Subcutaneous

[a]The dosage depends on the manufacturers' advice

There is a team of people available in Kuttanad area, who gets duck plague vaccine from Institute of Animal Health and Veterinary Biologicals (IAH&VB), Palode, Kerala and also from Institute of Veterinary Preventive Medicine (IVPM), Ranipet, Tamil Nadu and vaccinates the nomadic ducks. They start vaccination at 7 pm and finish at midnight. They receive 5.30–8.00 USD (INR 400–600) per 1000 ducks as vaccination charge apart from cost of vaccine (4.67 USD or INR 350 and 1.07 USD or INR 80, respectively, per 200 dose vials of duck pasteurella and duck plague vaccine). For duck cholera, most of the farmers used to do the first vaccination only and skip booster vaccinations. As a result, the birds succumb to the disease during outbreaks due to unprotective antibody titer. The farmers treat the ducks with antibiotics whenever there is occurrence of cholera and once the mortality level reduces, they vaccinate the ducks against duck cholera. In most of the disease outbreaks, combined infection of duck plague and duck cholera is noticed.

The main threat to the nomadic duck farming is of migratory wild birds; they not only spoil the leased foraging paddy fields and cause economic loss but also lead to avian influenza outbreak due to their contact with the nomadic ducks while foraging the paddy fields and water bodies. There were two avian influenza outbreaks in the Kuttanad area during 2014 and 2016 which led to the loss of number of duck flocks due to death and also massive culling.

The constraints in controlling the duck diseases in Kerala (Jalaludeen et al. 2004) are:

1. Improper storage of vaccine by the flock owners and the nomadic vaccinators.
2. Nonadherence to the vaccination schedule and doing vaccination after the outbreak.
3. Under or overdosing of medicines and vaccines.
4. Mixing two or more types of antibiotics and injecting without knowing the synergistic effect, which causes kidney damage and other consequences.
5. Non-availability of vaccine against new duck disease.

5.3.1.12 Marketing and Financing

Ramachandran and Ramakrishnan (1982) reported that most of the duck farmers of Kerala were indebted to the intermediaries who function as both financers and egg dealers. Only few farmers approach co-operative organizations and similar financing agencies for finance. The marketing of duck eggs was fully under the control of private commission agents, who act as financers to the duck farmers as well (Ravindran 1983). The marketing channels of duck eggs and meat in Kerala are:

1. Producer => Consumer (in case of duck for meat)
2. Producer => Egg Trader => Wholesaler => Retailer => Consumer
3. Producer => Egg Trader => Retailer => Consumer
4. Producer => Retailer => Consumer
5. Producer => Hatchery (in case of hatching egg)

5.3.2 Region I of Tamil Nadu State of India

Tamil Nadu is another southern state of India located east of Kerala. In the 1990s and early 2000s, the herd size in nomadic duck farming of Tamil Nadu was 100–600 ducks and the laborer intensity was one per 200 laying ducks. The duck egg and meat consumption during that period was negligible in Tamil Nadu due to the stronger flavor and greasy taste that was not relishable to the local population and also due to general belief of duck egg causing flatulence and joint pain (Nambi 2001).

There are two regions in Tamil Nadu where two different breeds of ducks are reared. The farmers of western, middle, and southern districts of Tamil Nadu rear "Kuttanad" breed of ducks and all their activities are part of and depend on Kerala. The farmers of northern districts of Tamil Nadu and Puducherry Union Territory rear "Arni" breed of ducks.

The farmers of western middle and southern districts of Tamil Nadu include Erode, Namakkal, Coimbatore, Tiruppur, Trichy, Dindigul, Karur, Madurai, Theni, Virudhunagar, Tenkasi, Tirunelveli, Thoothukudi, and Kanyakumari districts rear Kuttanad breed of ducks. They are the integral part of nomadic duck rearing of Kerala. Till last decade, these farmers used to borrow the ready-to-lay pullet ducks or partly laid ducks on loan from the Kerala farmers or used to borrow money from the financiers cum egg traders of Kerala to purchase pullets and rear them in Tamil Nadu. The grower farmers and egg traders of Kerala got back their money with interest in the form of eggs collected at farmer's doorstep. Some poor farmers are still continuing this lender-debtor relationship.

Nowadays, there are few duck hatcheries in Madurai, Trichy, and Paramathi Velur, which hatch out Kuttanad ducklings. There are some other custom hatcheries which also incubate duck eggs according to the demand of the duck farmers. The duck hatcheries in Tamil Nadu prefer to sell 10- to 15-day-old duckling at little higher rate (0.33–0.40 USD or INR 25–30 per duckling) instead of selling as day-old duckling. Farmers of Tamil Nadu who are specialized in brooding and growing of duckling purchase 3000–15,000 day-old to one-month-old duckling from hatcheries or duck nurseries of Kuttanad area of Kerala or from Tamil Nadu itself by giving ready cash. Most of the farmers make part payment and sell back the 10–12 weeks old drakes to pay the balance amount. At this point of time, the hatchery owners/ duck nursery owners credit the revenue against the pending loan amount and interest till the loan is repaid. Normally, day-old ducklings are purchased in the month of Aadi (mid-July to mid-August) to utilize the harvested paddy fields of Kar and Kuruvai cultivation; Karthigai (mid-November to mid-December) to utilize the Early-Samba and Samba seasons; Thai (mid-January to mid-February) to utilize Late-Samba, Thaladi, and Pishanam seasons; and Panguni (mid-March to mid-April) to utilize the Navarai and Sornavari cultivation during growing period. Out of the above four seasons, one type of farmers purchase ducklings in the months of Aadi, Karthigai, and Panguni and the other type of farmers purchase ducklings in the months of Aadi and Thai. During the first 20 days of age, farmers give broiler starter feed along with cooked rice (the rice from public distribution system is free in Tamil Nadu state) at 50:50 proportion in 3–5 times with 10 min feeding period (Sivakumar et al. 2009). After 20 days, the ducklings are fed with unsalted dry fish along with soaked rice.

The grower management is similar to that of Kerala except family laborers are used in central and western Tamil Nadu. In southern Tamil Nadu, most of the duck farmers employ laborers from both Tamil Nadu and Kerala state for duck herding. The labor charge given is 8.00–10.67 USD (INR 600–800) per person per day, out of this amount 2.67–4.00 USD (INR 200–300) is given daily to meet their day-to-day expenses and balance 5.33–6.67 USD (INR 400–500) is accumulated and handed over to the laborers whenever they go home. In Kerala, a flock of up to 15,000 birds are foraged together in the paddy field; but in Tamil Nadu, the flock is divided into small groups of 4000 ducks each and reared in different regions of the same area to avoid the starvation of the ducklings which graze last in the herd.

The layer duck farmers of Tamil Nadu follow the same nomadic foraging system of Kerala. They purchase 1000–2000 (rarely 3000) ready-to-lay or partly laid ducks from Kerala or Tamil Nadu grower farmers and rear them on the harvested paddy fields throughout Tamil Nadu and nearby states. There are about 70 families of duck herders in a single village called Komaralingam located between Udumalpet and Pazhani in Tamil Nadu who are involved in nomadic duck rearing in the western and central districts of Tamil Nadu. A family of 3–5 persons rear around 1000–2000 layer ducks for a laying year. Their children are taken care of by their grandparents in their native place. Only 2–3 persons take care of ducks while foraging and one, preferably a woman cooks the food and the other one goes to nearby town to purchase essential items. During lean season, they feed the layer ducks with soaked whole grain rice or broken rice, karugai (immature paddy with husk), and rice bran. The egg production is 160–180 which is little less than the production performance of the same breed of duck in Kerala. This may be due to the variation in the quantity and quality of scavengeable feed resources available in foraging. After completion of laying period, the spent ducks are sold and the farmers go back to their native place and stay there for 1 or 2 months before starting the next cycle.

5.3.2.1 Flock Movement

The movement of duck flocks during different seasons in Tamil Nadu is shown in Fig. 5.11.

(a) The nomadic farmers of west and central districts of Tamil Nadu state (Erode, Namakkal, Coimbatore, Tiruppur, Trichy, Dindigul, and Karur) rear the ducks in Cauvery river belts of these areas and also Thanjavur area during the months of mid-June to mid-August. Some flocks are herded in Mysuru district of Karnataka state and some other flocks are herded in Thrissur district of Kerala during this period. The entire flocks move to Palakkad district and stay there from mid-August to mid-November. From Palakkad some flocks move to Mysuru district of Karnataka state and others to nearby places from mid-November to mid-January. In the month of mid-January, the flock herders

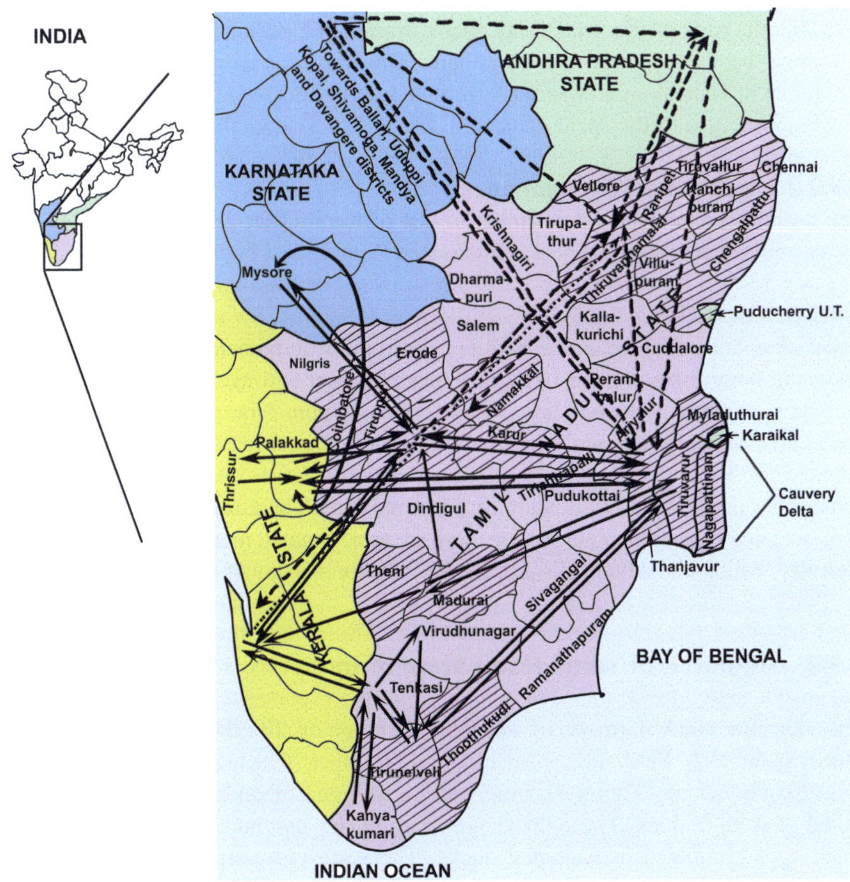

Fig. 5.11 Reference map of movement of duck flocks in and around Tamil Nadu state. (Image courtesy: R. Richard Churchil)

move to their native places of central and western Tamil Nadu and also to Thanjavur area wherever the paddy harvesting is over and stay there up to mid-March and then some flocks move to Thrissur district via Palakkad and stay there from mid-March to mid-June.

(b) The nomads of Madurai and Theni districts forage their ducks in their native districts from mid-September to mid-December, after that they are herded to Cauvery river belt of Thanjavur area and stay there from mid-December to mid-March, then move to Tirunelveli and Thoothukudi districts and forage there in the months of mid-March to mid-May and then move back to Thanjavur and stay there from mid-May to mid-September and finally return to Madurai and Theni districts.

(c) There are around 50 nomadic farmers in Tenkasi district and some from Rajapalayam area, who purchase their day-old or one-month-old ducklings from hatcheries at Kuttanad and rear them at different areas of Tenkasi district from mid-September to mid-December months, then move to Tirunelveli and Thoothukudi districts and stay there from mid-December to mid-May, then move back to Tenkasi and stay there from mid-May to mid-July, and finally move to Kanyakumari district from mid-July to mid-September.

Finally, most of the spent ducks are sold to Kerala state through middlemen.

5.3.2.2 Healthcare Management

The healthcare management is similar to that of Kerala state and most of the flocks are vaccinated by the nomadic vaccinators of Kerala.

5.3.2.3 Marketing

The excess drakes, spent ducks, and duck eggs are sold back to Kerala through the traders or commission agents or to the hatchery owners/grower farmers from whom they have purchased the ducklings. There are some poor farmers, who purchase the ducks after a laying year and keep them for second year of lay. Nowadays, there is demand for 3 months old drakes and spent ducks in Erode, Namakkal, Karur, and some other districts of Tamil Nadu for meat purpose. There is a place in Namakkal district called Paramathi Velur where there are more than 10 motels on the Bangalore highway, which altogether sells around 7000 ducks in a day as several duck recipes.

5.3.3 Region II of Tamil Nadu State of India

The nomadic duck farmers of northern districts of Tamil Nadu viz. Vellore, Thiruvannamalai, Tiruvallur, Kanchipuram, Ranipet, Chengalpattu, Viluppuram, and also Puducherry Union Territory rear Arni breed of duck which is otherwise called as Andhra duck. These ducks are small in size and body length and also lay small eggs compared to Kuttanad ducks. There are two varieties of Arni ducks, namely Keeri and Sanyasi (Veeramani et al. 2016) which are similar in color pattern as that of Chara and Chemballi varieties of Kuttanad duck breed of Kerala. Very few farmers purchase Khaki Campbell ducklings from the Central Poultry Development

Organization (CPDO), Hesaraghatta, Bangalore, and rear them under semi-intensive or intensive system for getting more eggs.

The rich to moderate duck farmers of Northern Tamil Nadu who holds hectares of land of their own move within the district or to nearby districts and does not move to faraway places. At the same time, poor landless farmers with bigger flock size move to faraway places of Tamil Nadu and nearby states.

There are more than 10 hatcheries in the Tiruvallur, Kanchipuram, and Vellore districts which hatch Arni ducklings for 6 months duration from May to January months with 1 or 2 months gap in between. The duck hatchery owners, who are also the financiers cum egg traders supply day-old or 10- to 15-day-old ducklings to the herders on loan and also lend money to the duck herders whenever required for the purchase of feed, medicine, transportation, and also for their personal well-being. The financier cum traders who do not have hatchery purchase the day-old or 10- to 15-day-old ducklings from hatcheries in Northern Tamil Nadu or Nellore and Prakasam (Singarayakonda) districts of Andhra Pradesh at the rate of 0.20 and 0.33 USD (INR 15 and 25) per duckling, respectively, and supply them to different herders for loan. Each hatchery owner/financier has around 10–15 nomadic duck herders under their control. They auction the post-harvested paddy field at the rate of 133.33–400 USD (INR 10,000–30,000) per 500 acres of land for 2–3 months lease from the farmer's associations/village heads/temple authorities of the different localities to rear around 20,000 ducks. The money is utilized for temple maintenance and also for cleaning the water sources like pond, lake, tank, etc. (Palanisami and Balasubramanian 1998). The financiers split the auctioned land into several regions and give them to the different duck herders under their control. The duck farmers in return give them back 3-month-old drakes, spent ducks, and eggs. Whenever the ducks' herds move within the state or to nearby states like Karnataka, Andhra Pradesh, and Puducherry Union Territory for foraging, the hatchery owner/financier arranges vehicles for flock movement and also introduces some egg traders of that locality for selling their eggs. Some financiers go to the area of herding at 3–4 days interval to collect the eggs from the nomads. These eggs are directly transported to Kerala for sales. Out of the sale proceeds, duck farmers are paid for their products after deducting the loan amount with interest.

In Andhra Pradesh, the duck herders lease paddy fields at 6.6 USD (INR 500) per acre to forage ducks for 3–4 months. There are some big lorries designed for carrying around 6000 ducks from Andhra Pradesh to Tamil Nadu and back. The hire charge for such lorry is around 1 USD (INR 75) per kilometer.

In Tamil Nadu and Karnataka states, the paddy field workers hesitate to work in the slushier paddy fields which were foraged by the ducks and also they demand for more wages (Nambi 2001) due to development of itching on the skin. The cause of itching might be due to the ammonia release from the nitrogen rich duck droppings and fungal growth in the field.

The general rearing system of this locality is same as that of nomadic rearing of Kerala, but there are no separate brooding, grower, and layer farmers. Each farmer used to rear around 2000–8000 ducks from day-old to the end of laying period. Out of the above stock, the excess drakes are sold to the hatchery owner or directly in the market at the age of 3 months. As the Arni ducklings are small in size, the brooder management is to be extended to 1½ months instead of 1 month in case of Kuttanad ducklings and the ducklings are allowed for full-time foraging after 1½ months only. The ducklings are fed with broiler starter feed along with cooked whole grain rice or broken rice during brooding period. There is no hand-feeding during growing and laying period. During lean seasons, the ducks are fed with soaked (for 2–3 h) rice, karugai (immature paddy with husk), broken rice, sorghum, bran, etc. The age at sexual maturity of Arni duck is higher (150–180 days) with lower annual egg production (150–170 eggs) compared to Kuttanad ducks due to the breed difference and feed availability (Gopinathan et al. 2015). The age at slaughter is higher (>3 months) and slaughter weight is less in Arni drakes compared to Kuttanad drakes. The sex ratio maintained during laying season is 1:15 to 20 in case of nomadic ducks which are traveling for long distance (Gopinathan et al. 2015), but the same is 1:12 in case of ducks which are maintained locally to produce hatching eggs. The Arni ducks are reared for 1–1½ years after start of laying (Tamizhkumaran et al. 2013).

5.3.3.1 Flock Movement

Most of the flocks start transhumance from northern districts of Tamil Nadu like Thiruvannamalai, Vellore, Tiruvallur, Kanchipuram, Ranipet, Chengalpattu, and Viluppuram (Nambi 2001) and some other flocks start from Tungabhadra river belt of Hospet in Ballari district, Udupi and Koppal districts of Karnataka state during November to January, and move to Nellore, Cuddapah, Kurnool, and Prakasam districts of Andhra Pradesh state and stay there from January to April. Some flocks move interior to Krishna river belt of Andhra Pradesh and herded there from April to July. If there is draught in Tamil Nadu, the flock move even interior to Godavari river belt of Andhra Pradesh. Some other flocks are foraged in different parts of Northern Tamil Nadu from mid-November to mid-April and move directly to Shivamogga, Mandya, and Davangere districts of Karnataka state by skipping Andhra Pradesh. From mid-July/mid-August, the flocks from Andhra Pradesh and Karnataka state move to Cauvery river belt of Tamil Nadu state and Puducherry Union Territory and kept there up to mid-November and then they move back to native districts of Northern Tamil Nadu and Ballari, Udupi, and Koppal districts of Karnataka state (Fig. 5.11).

Although a lot of duck herds are reared in Cauvery river belt of Thanjavur and nearby areas throughout the year, there are no nomadic duck farmers from these areas except very few from Cuddalore and Karaikal. At last, most of the spent ducks are sold to Kerala state through middlemen.

5.3.3.2 Healthcare Management

Vaccine against duck plague is the only vaccination done at the age of 4–6 weeks. Although there is frequent duck pasteurellosis outbreak, Tamil Nadu farmers do not vaccinate the ducks against it; instead, they go for medication after outbreak. Around 38% of the duck population was infested with helminths, namely *Echinostoma, Capillaria, Notocotylus, Coccidia, Raillietina, Choanotaenia, Hymenolepis,* and

Table 5.4 Suggested deworming schedule as per NATP Report (NATP 2004)

Month	Name of dewormer
December to February	Albendazole @ 10 mg/kg body weight
March to May	Oxyclozanide @ 30 mg/kg body weight + Fenbendazole @ 5 mg/kg body weight
June to August	Nil
September to November	Oxyclozanide @ 30 mg/kg body weight + Fenbendazole @ 5 mg/kg body weight

Cotugnia spp. The higher incidence of trematodes and cestodes indicates the presence of intermediate hosts like snail and earthworms in the foraging field. Recommended deworming schedule is given in Table 5.4.

5.3.3.3 Marketing

As most of the duck herders borrow money from the hatchery owners cum egg traders or financier cum traders for purchasing and rearing the ducks and also for their well-being, marketing of ducks and eggs are fully controlled by the traders. The farmers are forced to sell their drakes, duck eggs, and spent ducks to the lenders. The lenders deduct the loan amount with huge interest from the sale proceeds and give them back some amount as final settlement. Most of the time, the net amount to be given to the herders is negative and the herders are not able to come out of the clutch.

The independent farmers are gaining profit by selling their birds and eggs in the local market. Sometimes the excess birds and eggs are sold to Kerala also through the agents. The spent ducks are sold at the rate of 1.60–2.13 USD (INR 120–160) per bird. Some nomadic farmers used to sell both sexes for meat at 2½–3 months of age at the rate of 1.33–1.87 USD (INR 100–140) per bird due to difficulty in rearing up to 5 months. Most of them rear the females up to 5 months and sell the excess females to needy layer farmers at the rate of 3.33 USD (INR 250) after keeping 1000–2000 ducks as their own layer flock.

Some farmers and financiers sell their drakes and spent ducks to the local markets, duck meat shops at Chennai, and also motels at Paramathi Velur in Namakkal district. As the local sale of duck egg is very less, most of the eggs are transported to Kerala and around 5–10% eggs are transported to West Bengal also, whenever the demand is less in Kerala.

5.3.4 Puducherry Union Territory of India

Puducherry Union Territory of India also has nomadic duck herds and is a part of nomadic duck rearing system of both region I (western, central, and southern districts) and region II (northern districts) of Tamil Nadu state, where the flock moves from Puducherry to Tamil Nadu, Kerala, Karnataka, and Andhra Pradesh according to the paddy harvesting season. The family nomadic duck herders get their flock of 500–3000 numbers of 12 weeks old ducks (depending upon the family size) on lease from the contractors/egg traders of Kerala and Northern Tamil Nadu. Tamizhkumaran et al. (2013) reported that the contract duck herders were getting 26.67–40.00 USD (INR 2000–3000) as weekly expense and also 0.4 USD (INR 30) per duck per year as their profit. The sex ratio maintained is 1:30. The eggs are collected by the egg traders of Puducherry Union Territory.

5.3.5 Karnataka State of India

There is no nomadic duck rearing by the Karnataka farmers. However, duck herds of both Tamil Nadu and Kerala states used to go to different parts of Karnataka for foraging in some specific seasons.

5.3.6 Andhra Pradesh State of India

In the 1970s itself, nomadic duck rearing was reported in Coastal Andhra Pradesh by Rithamber (1984) especially near Kolleru lake of Krishna district. He reported that 72% farmers were rearing less than 300 ducks, 24% were rearing less than 1000 ducks, and only 4% were having the herd size of more than 1000 ducks. The sex ratio maintained during laying season varied from 1:7 to 1:57, but majority of the herders were maintaining the sex ratio of 1:10 to 1:30 (Rao et al. 2009).

The duck variety reared in Andhra Pradesh is Arni duck, otherwise called as Andhra duck. The nomadic duck rearing system followed in Andhra Pradesh is similar to that of North Tamil Nadu nomadic duck rearing system, the breed maintained, and its production performance is also similar. The lender–debtor relationship is also similar to that of north Tamil Nadu. Rao (1981) reported the exploitation of nomadic duck farmers by money lenders by the way of huge interest for the loan amount. Out of total nomadic duck population of Andhra Pradesh, 70% are owned by Andhra herders and remaining 30% by herders of Northern Tamil Nadu.

Most of the duck herders are from landless tribal and backward communities. The major areas of duck rearing are:

(i) Krishna and Godavari delta along with Kolleru lake located in between and
(ii) Somasila ayacut area of Nellore and Prakasam districts, Kadapa and Kurnool districts, and Pulicat lagoon located in Nellore district bordering Tamil Nadu.

The ducklings are mainly hatched out in Nellore, Prakasam, Krishna, and West Godavari districts of Andhra Pradesh and to some extent from northern districts of Tamil Nadu and Kerala.

At present, each nomadic duck flock has about 500–3000 layer ducks which are reared by the entire family. Most of the nomadic farmers purchase 10-day-old ducklings from hatcheries (which passed critical period of brooding with artificial heat) at the rate of 0.33 USD (INR 25) per duckling (Pradeep 2017) and remaining farmers purchase ready-to-lay pullet ducks. The ducklings are hand-fed with soaked broken rice along with chicken feed for up to 21 days. The ducklings herded near the Kolleru lake are fed with snails and earthworms collected from the lake shore along with soaked rice. After 21 days, ducklings are allowed to forage on the harvested paddy field and the hand feeding is restricted to one time. After 1½ months, no hand feeding is given. The duck herds move from one place to another using trucks arranged by financier for search of harvested paddy fields. The ducks are reared up to the maximum of two laying cycles before slaughter. The fertility percentage of eggs from nomadic flocks ranges from 65 to 90 and hatchability percentage ranges from 50 to 75. The mortality up to 8 weeks of age is 10–20% and less than 10% thereafter (Rao et al. 2009).

5.3.6.1 Flock Movement

The movement of duck flocks in Andhra Pradesh is shown in Fig. 5.12.

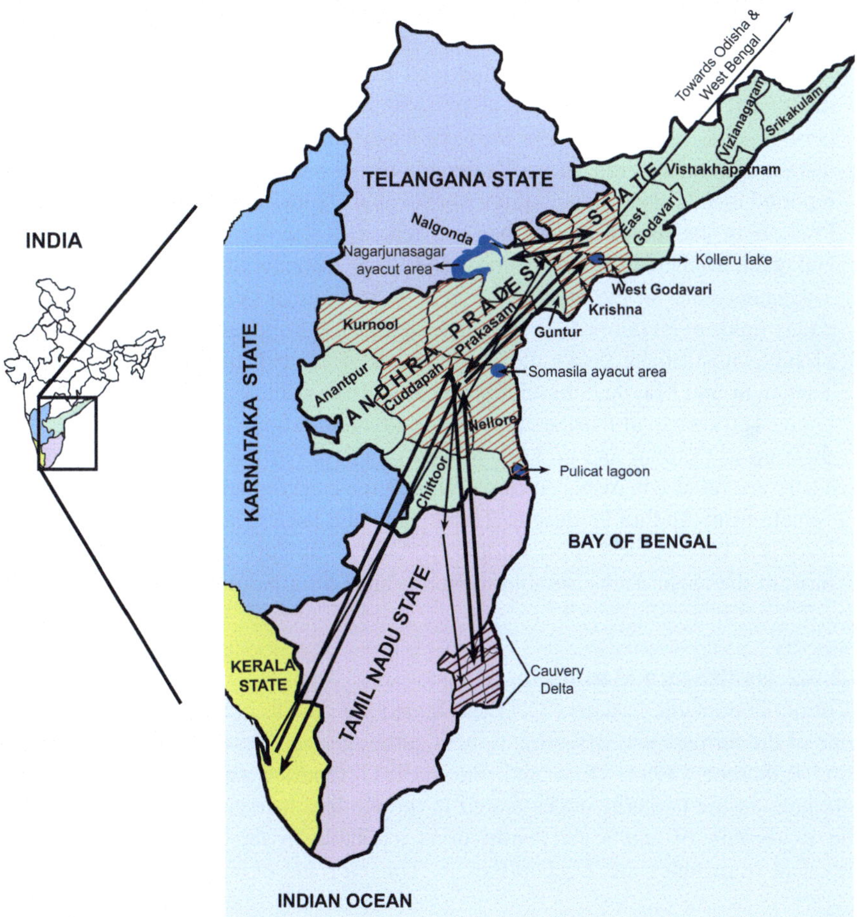

Fig. 5.12 Reference map of movement of duck flocks in and around Andhra Pradesh state. (Image courtesy: R. Richard Churchil)

1. The brooding is done mostly by hatchery owners for 10 days and sold to needy farmers. Some nomads purchase day-old ducklings from Tamil Nadu and Kerala states and brood them. The grower and layer flocks are reared in Krishna and Godavari river belt of Andhra Pradesh from February to June after harvesting of Rabi paddy and then the flocks move to Kolleru lake area and stay there from July to September. In the month of October some flocks move to Nagarjuna Sagar ayacut area of Guntur district of Andhra Pradesh and Nalgonda district of Telangana and stay there up to January and return. Some other flocks move to Somasila ayacut area of Nellore and Prakasam districts and also to Kadapa and Kurnool districts after harvesting of Kharif paddy and remain there up to January and move back to native Krishna and Godavari deltas in the month of February (Rao et al. 2009). In the month of July, some flocks move directly from Krishna and Godavari districts to Cauvery river belt, stay there up to October, then go to Nellore, Prakasam, Kadapa, and Kurnool districts and stay there up to January, and finally go back to Krishna and Godavari districts.
2. The duck herds of Nellore district are foraged in the Somasila ayacut area of Nellore and Prakasam districts and also Kadapa and Kurnool districts in the months of October to January after harvesting of Kharif paddy. In the same period, some landless farmers move to Nagarjuna Sagar ayacut area of Guntur district of Andhra Pradesh and Nalgonda district of Telangana. Pradeep (2017) reported that 25-day-old ducklings were brought from Singarayakonda in Andhra Pradesh in the month of October and reared in the harvested paddy fields of Nakrekal, Kethepally, Nidamanoor, and other mandals of Nalgonda district of Telangana state. In the month of February, some flocks continue to stay on the paddy fields of Nellore and Prakasam districts after Rabi harvesting. In the month of July, most of the flocks move to Cauvery river belt of Tamil Nadu mainly Thanjavur and Mayiladuthurai districts and also Karaikal district of Pondicherry Union Territory to utilize the harvested paddy fields after Kuruvai season and stay there up to October and go back to Nellore district in the month of November. There are hundreds of big lorries designed for carrying around 6000 ducks per vehicle from Andhra Pradesh to Tamil Nadu and back. The hire charge for such lorry is 1 USD (INR 75) per kilometer. After completion of the nomadic cycle, most of the spent ducks are sold to Kerala state through middlemen.

5.3.6.2 Healthcare Management

Most of the nomadic farmers (92%) vaccinated their ducks against duck plague, but none of the farmers was deworming them, although there was heavy worm infestation (Rithamber 1984). Rao and Rao (1981) reported twenty-one species of helminths in the nomadic ducks reared in the Kolleru lake area of Krishna district due to feeding of snails and earthworms available in the lakeshore. The most prevalent trematodes are *Trichobilharzia, Opisthorchis obsequence, Echinostoma revolutum*, and *Prilochammus oxyureous*, most prevalent cestode is *Hymenolepis* and nematodes are *Anidiostomum skyabihi* and *Tetrameres snatis*.

5.3.6.3 Marketing

Most of the duck eggs, spent ducks, and drakes from southern Andhra Pradesh (Nellore, Prakasam, Kadapa, and Kurnool districts) are transported to Kerala through middlemen/commission agents/traders. The herders of Krishna and Godavari delta are selling their egg to West Bengal market. Rao (1981) reported that 90% of the eggs produced in the Krishna and Godavari delta are sold to West Bengal. The farmers of Krishna district sell their eggs to the commission agents of nearby market who are the financiers and middlemen, who in turn sell the eggs to the wholesalers and retailers in Kolkata (Khan et al. 1994). There are three marketing channels:

Producer => Consumer
Producer => Commission agent => Wholesaler => Retailer => Consumer
Producer => Hatchery

5.3.7 Telangana State of India

Earlier, the duck farmers of Telangana region practiced natural incubation of duck eggs under broody hens (Rithamber 1984). Now, there are some duck hatcheries and custom hatcheries, which incubate the duck eggs. Telangana area is having less number of nomadic duck population as compared to coastal Andhra Pradesh. Most of the duck herds are located in the Nalgonda, Warangal, and Khammam districts. The rearing practice is similar to and part of Andhra Pradesh and northern districts of Tamil Nadu. The ducks are brought from faraway places like Tirupati and reared in the harvested paddy fields of Nakrekal, Kethepally, Nidamanoor, and other mandals of Nalgonda district (Pradeep 2017). Srinivas (2018) reported that a nomadic family from Nellore district shifted to Nalgonda district due to the availability of plenty of harvested paddy fields, water source, and paddy farmers' acceptance.

5.3.8 Assam State of India

Islam et al. (2002) reported a duck rearing system similar to that of nomadic in Cachar district of Assam in which some farmers forage the layer ducks (<200 ducks) in the harvested paddy fields with the help of an attendant during day time and keep them inside a night holding pen constructed on an elevated area surrounded by paddy fields. However, they were not moving to faraway places. The farmer or the attendant looks after the flock during night to safeguard them from the predator attack and theft. The farmers used to purchase ready-to-lay ducks (>5 months) and rear them for one laying year. Although Assam is second in duck population in India, entire ducks are reared in the backyard or free-range system with small flock size

(<200 ducks) and ducks scavenge inside the village, paddy field, and water bodies during daytime and return at evening.

5.3.9 West Bengal State of India

Although West Bengal is number one in duck population in India, duck rearing is the subsidiary business after cultivation and they are reared by backyard and scavenging system in the villages. Women are mainly engaged in duck rearing. Around 90% of the duck farmers have less than 20 ducks (most of them have 6–10 ducks) which are hand-fed and also foraged on the harvested paddy field at day time and rested in the night shelter (Halder et al. 2007). There is no transhumance reported in West Bengal. Similarly, all other states of India also follow backyard and scavenging system of duck rearing.

5.3.10 Bangladesh

The duck population in Bangladesh comprised about 16.60% (57.12 million out of 344.02 million total poultry) of the total poultry population (BER 2019). Out of this, a handsome population is reared under scavenging or free-range system which plays critical role in meeting daily protein needs and providing household income of farm families. Pervin et al. (2013) reported that 90% of the scavenging and free-range duck farmers are women and 91.5% rear indigenous ducks. The annual egg production is 51–70 eggs with the maximum production in rainy season. The stationary duck flocks which are reared under scavenging or free-range system had a mean flock size of 6.5 compared to the nomadic flock size of 152 ducks. The flock size of nomadic herds varied from 146 to 687 depending upon the financial status of the farmers; 52.6% farmers had 250 ducks or less, 36.8% had 250–500, and 10.5% farmers had the herd capacity of above 500 ducks (Rahman et al. 2017). However, larger nomadic flock size (105–1100 ducks) has also been reported (Sarkar et al. 2017). The indigenous desi ducks are reared by scavenging or free-range farmers (95.7%); conversely, around 80% of the nomadic farmers rear Khaki Campbell ducks for higher egg production in Jamuna floodplains and Padma basin (Hossain et al. 2019). In Bangladesh, nomadic duck production system has become popular within a decade or two in Sunamganj, Habiganj, Moulvibazar districts of Sylhet division and nearby Netrokona district and also in coastal areas of Chittagong and Noakhali districts of Chittagong division (Sarkar et al. 2017). Recently nomadic duck rearing has been introduced in Jamuna floodplains of Sirajganj and Pabna districts and lower Padma basin of Faridpur and Madaripur districts (Rahman et al. 2017).

5.3.10.1 Incubation of Duck Egg

Although the government duck breeding stations are supplying high yielding Khaki Campbell and Jinding ducklings for egg purpose and Muscovy ducklings for meat purpose, their contribution to the demand for duckling is very meager and the major portion of the demand is met by the local "rice husk hatcheries." This "rice husk hatcheries" alone produce about 1.5 million ducklings per year in Netrokona, Kishoreganj, and Sunamganj districts (Islam et al. 2012). The nomadic farmers with herd size of greater than 500 ducks collect day-old ducklings from the nearby private or government hatcheries, but most of the nomadic farmers with less than 500 herd size and backyard farmers incubate the eggs of their own by using broody hens, Muscovy ducks, and geese. The price of duckling is around 0.3 USD (Rahman et al. 2017) which is comparable to the duckling price of India. However, all the nomadic farmers (whether small or large) of Chittagong and Moulvibazar districts purchase the day-old duckling from the government hatcheries and none is doing natural incubation (Hossain et al. 2019).

Rice husk incubation: The cylindrical egg baskets are set into larger bamboo setter boxes with rice husk filled between the egg baskets and also between the egg baskets and walls of the setter box. The gap between egg baskets should be 8 cm and between egg baskets and the wall of the setter box should be 10 cm. About 50 eggs are packed in black colored cloth, preheated with the help of 100 w incandescent bulb for 1 h, and kept inside the cylindrical egg basket. Likewise 10 egg packages are kept in an egg basket (500 eggs). Paddy husk is preheated and filled between the egg baskets inside the setter box and also between eggs baskets and wall of the setter box. The paddy husk is taken out, reheated, and filled back three times a day (at regular intervals) for the first 3 days of incubation and two times a day for next 3 days of incubation. At the time of reheating the paddy husk, the eggs of one basket are transferred to another basket in such a way that the top layer goes to the bottom of next basket and vice versa.

The eggs are candled on day 5 and 13 to find and remove infertile and dead in shells, respectively, and also to know the stage of development which guides the hatchery owner to adjust the basket temperature. The egg is placed on the upper eyelid and the temperature is assessed. From day 17 onwards, the eggs are transferred and spread as a single layer on the hatching beds till the eggs hatch out. The surface of the hatching bed is covered with thin layer of paddy husk and straw mat on that. The edges are lined with padding to protect the eggs. The eggs are covered with thin or thick cloth according to the room temperature. The eggs are moved from periphery to center and vice versa, sprayed with fine mist of water, and also the thickness of the egg spread is changed to adjust the temperature of the eggs until the hatch out of ducklings (Khalequzzaman et al. 2006).

5.3.10.2 Management of Ducks

The nomadic farmers of Bangladesh are herding their flocks in distant places for 4.62 months, local foraging for 5.20 months, and hand-feeding for 2.19 months in a year (Rahman et al. 2017). The duck herding in each migratory place varies from 1 to 3 months depending upon the availability of feed source. The nomadic ducks lay more eggs during September to February due to paddy harvesting and poor in March–April and July–August due to summer and heavy rain, respectively. The duck flocks are moved to 1–150 kilometers range with an average of seven kilometers from the home village. Most of the herders who move to nearby villages or sub-districts by foot, about 16% of the duck flocks are transported by boat and 5% by the motor vehicle to nearby districts (Sarkar et al. 2017).

The Haor regions of Bangladesh are surrounded by water during rainy and autumn seasons which leads to availability of snails, fishes, pests, and water hyacinth for ducks. The nomadic farmers move towards deep seats of Haors and stay there from January to May and as the Haors become flooded from June to August, the ducks are moved back to the periphery of Hoars. From mid-September to mid-December, natural feed becomes scarce and the nomadic farmers are compelled to hand-feed the birds to maintain the egg production. Poor farmers sell their stock at this time and purchase new batch of ducks at the beginning of next season (Islam et al. 2012).

The hatchery owners and duck farmers need money for purchase of hatching eggs, hatchery operations, purchase of ducklings, pullet ducks, vaccines, etc. As the financial support from scheduled banks is limited, the farmers are more dependent on microcredit providers (Dadon) who levy high interest rate (Islam et al. 2012). The nomadic farmers are forced to sell their products to money lenders for the whole season at lower price till paying their loan amount with interest (1.5–2 times of the loan amount).

The differences between nomadic duck rearing of Bangladesh from other countries are:

- the duck flocks are not moved to faraway places in Bangladesh,
- the flocks are fed with paddy, whole wheat, and maize everyday evening after foraging due to inadequate food material in the paddy field,
- the duck flocks are maintained up to 3 years of age or drastic reduction in egg production whichever is earlier, and,
- the dead ducks are thrown into the adjacent water bodies which causes the spread of diseases like avian influenza.

5.3.10.3 Healthcare Management

Most of the farmers reported that duck plague and duck cholera are the major diseases affecting their flock (Rahman et al. 2017). In addition, respiratory problem, poisoning, and occasionally, bird flu also encountered among nomadic ducks.

Mortality of ducks due to duck cholera was high (48%) in herds of lesser flock size (<500 ducks) compared to larger flocks which were properly vaccinated and dewormed (Hossain et al. 2019). Although farmers are aware of vaccination, insufficient supply of vaccine and improper vaccination results in duck cholera outbreak even after vaccination. Hossain et al. (2019) also reported that sudden death was reported in all nomadic flocks compared to 40% in stationary scavenging or free-range flocks. Since 2011, avian influenza outbreak is common in nomadic flocks due to their contact with the wild waterfowls on the foraging area especially during winter (Sarkar et al. 2017).

The nomadic ducks are affected with trematodes (*Echinostoma revolutum, Notocotylus attenuatus, Hypoderaeum conoideum,* and *Echinoparyphium recurvatum*), nematodes (*Amidostomum anseris, Capillaria contorta)*, cestodes (*Hymenolepis coronula* and *Fimbriaria fasciolaris), and* acanthocephala (*Arythmorhynchus anser* and *Filicollis anatis).* So, treatment with anthelmintics (40%) and vaccination against duck plague (70%) are more common in nomadic duck flocks compared to stationary flocks (8.6% and 25.7%, respectively). Moreover, predators attack is more common in both nomadic and stationary flocks especially of crows and wild cats (Yousuf et al. 2009).

5.3.10.4 Marketing

All the eggs are sold in the village market in the flock owners' sub-districts itself, but most of spent ducks and drakes (92%) are sold through vendors who come to the herding place and purchase them and the remaining ducks (8%) are sold directly to retail customers (Sarkar et al. 2017). Islam et al. (2012) reported that about 70% of the duck eggs are sold to Dhaka and other urban cities and the rest only being consumed by the rural households. The major marketing channel for egg is as follows:

Farmer => -Paikers/local traders/buyers => small trader/buyer => large trader => Wholesalers (Arotdar) => Egg retailers

All the nomadic farmers have their primary income from the duck rearing (from small to large flock size) which accounts for 70% of the total income and the agriculture income comes next. Out of the household income, 36% was saved for next year's investment for nomadic duck rearing.

5.3.11 China

Before Second World War, nomadic duck herders of Hainan Island of China had the practice of foraging their duck flocks on the river banks till reaching the mouth of the river at Haikou city, by the time the ducks became fatty and were sold for meat and the herders go back home (Furuno 2012). Although most of the Chinese duck farmers are doing free-range system of duck rearing, no nomadic duck rearing is reported in China thereafter.

5.3.12 Taiwan

In Taiwan, until the 1970s, duck herding in the harvested paddy field was noticed and a flock size of 500 ducks were herded by two men (Huang 1973). Afterwards, this practice disappeared due to the application of pesticides in the paddy field and also due to the shortage of laborers to look after the nomadic duck flocks (Huang et al. 2008).

5.3.13 Vietnam

Vietnam ranks number two in duck population with 76.91 million ducks in the year 2018 (Doan 2020). In Vietnam, 25% of the households rear waterfowl mainly ducks (Desvaux et al. 2008). In north and central Vietnam, native breeds like Co and Bau are mostly reared by the farmers, but in Mekong delta region Khaki Campbell, Cherry Valley White Pekin and their crosses with native duck breeds are popularly grown. Red river and Mekong river delta are the two regions of highest duck population in Vietnam. Out of the two, Mekong delta region is famous for duck transhumance, where around 50,000–70,000 medium (50–200 ducks) to large (200–4000 ducks) scale nomadic duck flocks are located which accounted for 4 million ducks (75% of the total duck population of Mekong delta region). About 80% of the nomadic flocks of Mekong delta are layers and 20% are meat ducks (Men 2010; Buranakanonda 2011). Henning et al. (2013) reported that about 64% of the nomadic flocks are meant for egg production, 18% are meant for duck meat production, and remaining 18% are meant for hatching egg production.

In Mekong delta, most of the nomadic duck populations are located in Long An, Tay Ninh, Dong Thap, and An Giang provinces. About 68% of the nomads were herding only one flock per year, 18% were rearing two flocks, 9% were rearing three flocks, and 5% were rearing 4 flocks in a year depending upon the breed and purpose of duck rearing (Henning et al. 2013).

The major extensive duck rearing systems followed in Vietnam (Edan 2006) are:

(a) Harvested rice-running duck system: A type of foraging system in the harvested paddy field to pick spilled grains.
(b) Rice–duck pest control system: The ducklings are allowed in the paddy field 20 days after rice transplantation to the start of flowering to catch insects, eat weeding, and muddying the field.
(c) River and coast band collector system: Here the ducks pick tiny fish, shrimps, oysters, etc. from the low tide river and coastal area.
(d) Fish–duck pond system.
(e) Backyard duck raising system.

5.3.13.1 Duck Egg Incubation

In Vietnam, ducklings are incubated and supplied in five ways (Edan 2006):

1. Incubation under broody hen
2. Traditional incubation using sunlight and rice husk
3. State farms owning grandparent and parent stocks under Ministry of Agriculture and Rural Development (MARD)
4. Private hatcheries with small and large incubation capacity
5. Hatched and smuggled from China

Incubation Under Broody Hen

Nomadic farmers with small flock size and scavenging backyard farmers used to incubate duck eggs under broody hens and Muscovy ducks and used to get about 85% hatchability. On day 15, eggs are candled for live embryo. When the ambient temperature is very high, eggs are dipped in water and placed back on the nest in last few days of hatching. After hatch out, the hens are tethered with leg rope for weaning and the hens will start laying after 15 days of weaning. The newly hatched ducklings are brooded artificially using electric bulb (Nind and Tu 1998).

Traditional Incubation

The eggs for incubation are placed on hessian cloths (approximately 50 cm × 50 cm) in front of the hatchery premise under the morning sunlight for 1–3 h and turned 2–3 times. When the egg reaches required temperature (40–42 °C) which is observed by sweating or felt by eyelid or philtrum or cheek, eggs with hessian cloths are placed inside woven bamboo baskets (60 cm diameter × 70–90 cm depth) lined with paper and kept on the paddy husk bed. The procedure is followed up to 15 days of incubation with turning of the egg twice a day (4.00 am to 8.00 am and 4.00 pm to 6.00 pm) by taking out the eggs with hessian cloth and turned by gentle backward and forward movement of the cloth or individual eggs manually. The eggs are set again into the basket by placing the oldest eggs on top, the second oldest on the bottom, and alternative layers of young eggs with older incubated eggs to keep the required temperature. One basket can accommodate 1000–1200 eggs. This basket system continues up to 14–17 days and then the eggs are moved to the next stage of incubation.

In the second stage, about 2000 eggs are kept in each shelf (approximately 1.6 × 4 m) and are located above the baskets. If weather is cold, one to three blankets are placed on the eggs. Eggs on the shelves are turned three times a day (5.00 am, 12.00 noon, and 5.00 pm) until duckling hatches out. When the ambient temperature is high, eggs are spread thinner on the shelves and water is sprayed and the ventilation into the room is increased. The hatched out ducklings are placed in the woven bamboo containers (1 m diameter). In this method, the hatching rate is around 50–65%. The baskets and shelves are cleaned properly between hatches. March to April and November to December are the two main hatching seasons (which is 1 month before transplantation of paddy). A hatchery can produce 5000–7000 ducklings in 2 days interval (Nind and Tu 1998).

The other method of incubation is using forced-draft incubator which is carried out by private farms and state farms. The other source is smuggling of ducklings from China at cheaper rate.

5.3.13.2 Duckling Management

Nomadic duck farmers purchase about 200–8000 ducklings (average 2200) and gradually introduce them to paddy field after 4–10 days of age. Full-time foraging of ducklings starts only after 3 weeks of age. They are taken to paddy field at 6 am and brought back at 6 pm; meantime they are taken to water canals and shades at hot noon hours. During night, they are kept in night holding pens (2.5 to 3.5 × 4.0 to

5.0 m) which usually has a dirty floor and are bounded by nylon netting and wooden support stakes. Some farmers produce ducklings at their home and keep them for the first 1–2 weeks in raised wire and wooden cage (2.0 × 1.0 m) with an electric lamp as a heat source located inside the thatched shed. After 2 weeks, the ducklings were put into pens. Ducklings up to 3 weeks prefer to eat tadpoles, insects, and worms which are palatable to them. During the brooding period of first 2 weeks, the mortality is 0–6% due to diseases and attack of predators like rats and freshwater crabs (Nind and Tu 1998).

5.3.13.3 Grower Management

The ducklings are brought to the rice fields 20 days after rice transplantation to control pests and when the flowering starts, ducklings are driven out of rice fields to canals, ditches, rivers and brought back to the rice fields just after harvesting to forage the weeds, spilled grains, helixes, small frogs, small fishes, insects, freshwater crustaceans, duckweeds, and other water plants. In case of heavy Golden Apple Snail (GAS) infestation, ducks are herded in the paddy field prior to transplantation and also 30–45 days after transplantation to clear the field out of GAS (Edan 2006). The grower ducks are kept inside the open roof enclosures made of fishing nets at night. The number of laborers involved ranges from 1 to 7 (average 4) according to the herd size. The herders live inside the plastic shelters often with their families where they cook and keep the supplementary feed for ducks (Henning et al. 2013). During this period the mortality goes up to 60% due to diseases like duck cholera, duck plague, avian influenza, and some other bacterial diseases and also due to predators such as eels and snakes (Nind and Tu 1998).

The main period of duck production is March to July and September to December. In Red River Delta, ducks are reared mostly for meat; they are left in the paddy field in the morning and brought back to the shed in the evening. There is no herding from one place to another. The ducks are culled at 2½–3 months of age for meat (Desvaux et al. 2008) and the price of meat type duck is 1.7 USD per kilogram (Asian Agribiz 2020).

5.3.13.4 Layer Management

There are two types of farmers (Meyer et al. 2017), the first (70%) rear the ducks from day-old to the end of laying and other type of farmers (30%) purchases the pullet ducks at different ages (75, 150, and 240 days of age). Foraging of layer ducks is similar to that of grower management. The first laying phase ends 2–3 months after the start of lay followed by 1 month molting and so on. Most of the farmers keep their flock for one or two laying years (up to a maximum of 3 years) depending upon the egg prize and availability of foraging land and then cull them for meat (Nind and Tu 1998).

In Mekong delta, ducks are mainly reared for egg and transhumance between districts and provinces is also noticed. The breeds like Tau or Co, Sieu trung, and Khaki Campbell and their crosses at different levels are reared for egg purpose. The native Co duck (grass ducks) reaches 1.3–1.5 kg and drake reaches 1.5–1.8 kg of mature body weight with the age at sexual maturity of 140 and 120 days, respectively. The male:female ratio maintained in the nomadic flock is 1:20–25, with which the nomadic farmers achieve up to 90% fertility and good hatchability. There are some flocks which do not have drakes. The average egg production is about 180 eggs with more than 60 g egg weight (Men 1997; Meyer et al. 2018). As the paddy harvesting season is different in different provinces of Mekong delta, the duck herds are moved between paddies using boats (66%), trucks (12%), or by foot (22%) throughout the year (Vergne et al. 2018). Some herds move to nearby Cambodia from Long An, Chau Doc, An Giang, and Kien Giang provinces. From the year 2000, the duck farmers were paying 4.3–20 USD per hectare of harvested paddy field for foraging their ducks; the present rate is 20–70 USD per hectare (Meyer et al. 2017). A flock of 1000 ducks needs 60–70 hectares of land to forage for 3 months period. They consume around 250 kg/ha in a day of foraging. During flood season (July to August), the ducks are kept in an elevated area near river or pond and given with commercial feed along with rice and or mollusks collected from rice fields and water sources. Farmers build simple night shelter with polythene sheets and wooden poles. The mortality percentage during laying period is 0–5%.

5.3.13.5 Breeder Flock Management

Selection of homebred birds for laying is based on traits like body weight, conformation, and appearance. The native breeds like Tau or Co and Sieu trung (Super egg), the exotic breeds like Khaki Campbell, Cherry Valley 2000 and their crosses at different levels like Vit Co, AnhDao crossbred (Cherry Valley crossbreeds), Agricultural crossbreds, and BachTuyet duck breed are reared for egg purpose. The local duck breeds like Ta or Bau, Super meat, local White Pekin ducks and their crosses with Cherry Valley White Pekin ducks are reared for meat purpose. Muscovy and Cherry Valley meat type Pekin ducks are not herded by the nomads as such. The state owned farms maintain grandparent and parent stocks for commercial duckling production.

The Department of Agriculture and Rural Development had evolved a new breed of saline water duck for the nomadic duck farmers of coastal provinces of Mekong delta (Cá Mau, Trá Vinh, and Tiên Giang) which can survive by drinking water with the salt content up to 15%, although the normal duck cannot survive even with 4% salt. These ducks reach 2.5–3 kg at 3 months of age and also lays about 240–250 eggs per year (ViêtNam-News 2016).

5.3.13.6 Flock Movement

About 55–67% of the duck flocks are herded within the province, about 32–45% are herded between provinces and 1% of the nomadic population move to bordering area of Cambodia via Vinh Hung district of Long An province, Hong Ngu district of Dong Thap province, Chau Phu district of An Giang province, and Ha Tien city of Kien Giang province. More than 50% of the farmers move to one location only and some follows two and three location movements. They return to the native village 3, 2, and one time per year, respectively. The distance of each movement is 30 km (1–85 km) and yearly distance is 125 km (21–763 km). About 59% of nomadic duck herders move back to their native place once or more but remaining 41% are moving all around (Henning et al. 2013; Meyer et al. 2017; Vergne et al. 2018). The herd movement is depended on the ability to obtain access to paddy field, familiarity with the location and inhabitants, cost effectiveness, distance, quality of paddy field, and easy access (Meyer et al. 2017). There are two main flock movements, namely direct line movement and circular pattern movement. Some individual farmers make irregular pattern of movements also.

Direct Line Movement

There is only a direct line movement recorded by Vergne et al. (2018) which starts from Ben Cau district of Tay Ninh province and enter into the Long An province where the flocks move through the eastern side of Duc Hue district and western side of Ben Luc district and reaches Tan Tru district. While returning back, duck flocks move through the western side of Thu Thua district and center of Duc Hue district of Long Province to the Ben Cau district of Tay Ninh province. Some flocks move from Tan Tru district to southern end of Chau Thanh district and move back before starting the return journey (Fig. 5.13). In this route, the ducks are reared along the Vam Co Dong river belt and its plain of reeds.

Circular Pattern Movement

Circular movements are recorded by two researchers.

(i) The first circular pattern recorded by Vergne et al. (2018) contains 5 direct line movements which altogether causes a circular pattern and a direct line movement (Fig. 5.13). This pattern covers both Vam Co Tay and Tien river belts. Here, duck herders take any point of circle as starting and ending point according to their convenience.
   - The 1st line starts from Tan An, the capital of Long An province and moves through north of Chau Thanh district and south of Than Phnoc district to Cai Lay district of Tien Giang province.
   - The 2nd line starts from Cai Lay district of Tien Giang province and moves through the western side of Tan Thanh district and center of Moc Hoa district to Vinh Hung district of Long An province.
   - The 3rd line starts from Vinh Hung district and moves through southwest of Tan Hung district of Long An province, Thap Muoi district of Dong Thap Province, Cai Be district of Tien Giang province and Vinh Long city to Long Ho district of Vinh Long province.
   - The 4th line starts from Long Ho district and goes through Mang Thit district of Vinh Long province, Cho Lach and Chau Thanh districts of Ben Tre province and Cho Gao district to northwest of Go Cong Tay district of Tien Giang province.
   - The 5th line starts from northwest of Go Cong Tay district of Tien Giang province and goes through the center of Chau Thanh district and ends in the starting point of 1st line, that is, Tan An the capital of Long An province.
   - Other than the circular movement, the 6th straight line starts from Go Cong Tay district of Tien Gian province and moves through the Chau Thanh, Tan

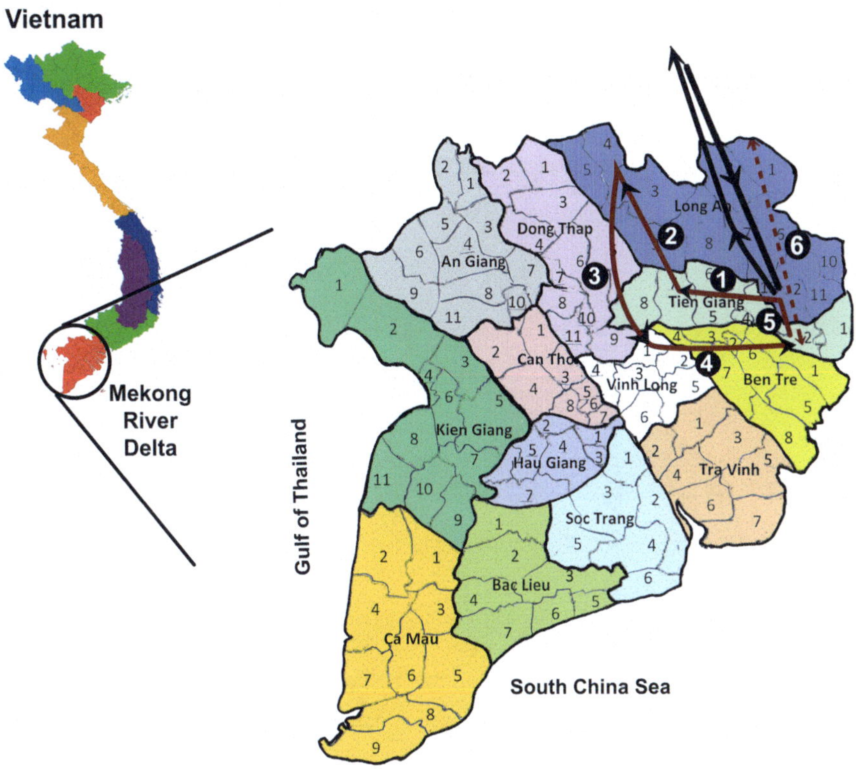

Fig. 5.13 Reference map of circular and direct line movement in Mekong delta region of Vietnam (Sourced from Vergne et al. 2018; Image courtesy: R. Richard Churchil)

Tru, and center of Ben Luc districts to northern end of Duc Hoa district of Long An province.

List of districts of different provinces of Mekong delta region mentioned in the Figs. 5.13 and 5.14.

A. **Long An province**: 1. Duc Hoa; 2. Duc Hue; 3. Moc Hoa; 4. Vinh Hung; 5. Tan Hung; 6. Ben Luc; 7. Thu Thua; 8. Thanh Hoa; 9. Tan Thanh; 10. Can Giuoc; 11. Can Duoc; 12. Tan Tru; 13. Tan An city (capital); 14. Chau Than.

B. **Dong Thap province**: 1. Tan Hong; 2. Hong Ngu; 3. Tam Nong; 4. Thanh Binh; 5. Thap Muoi; 6. Cao Lanh; 7. Cao Lanh city (capital); 8. Lap Vo; 9. Chau Thanh; 10. Sa Dec town; 11. Lai Vung.

(continued)

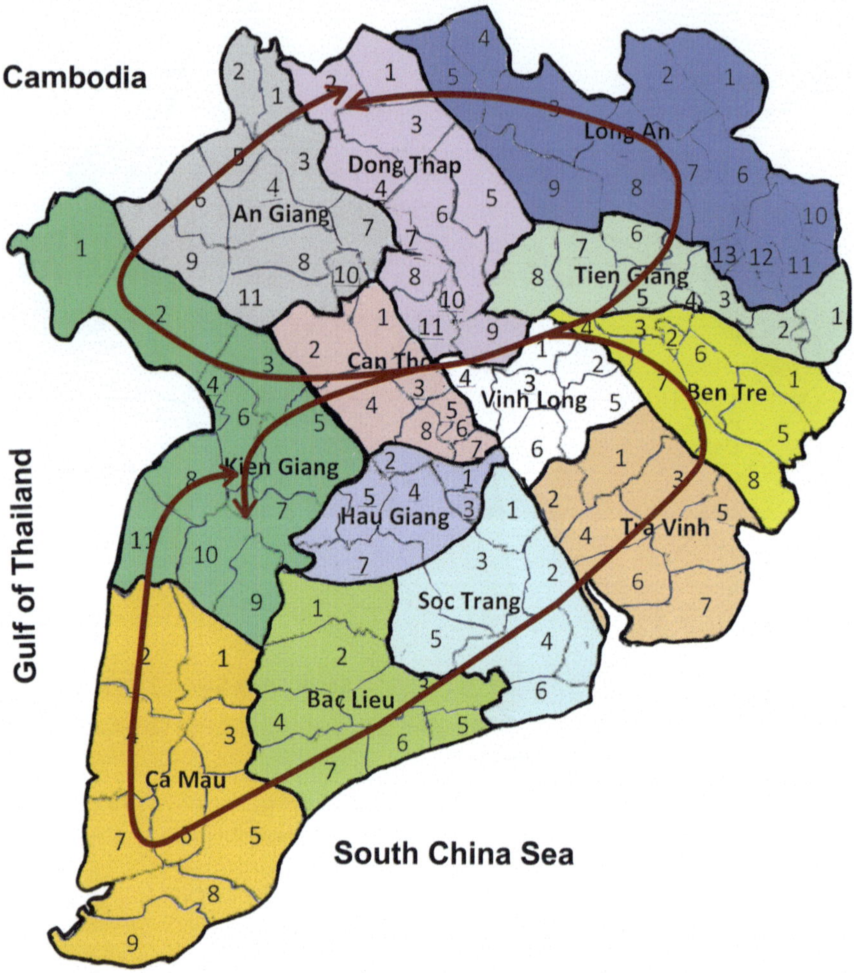

Fig. 5.14 Reference map of two circular movements in Mekong delta region of Vietnam. (Sourced from Men 2010; Image courtesy: R. Richard Churchil)

C. An Giang province: 1. Tan Chau (district level town); 2. An Phu; 3. Phu Tan; 4. Chau Phu; 5. Chau Doc (provincial city); 6. Tinh Bien; 7. Cho Moi; 8. Chau Thanh; 9. Tri Ton; 10. Long Xuyen (capital); 11. Thoai Son.

D. Kien Giang province: 1. Giang Thanh and Ha Tien city; 2. Hon Dat; 3. Tan Hiep; 4. Rach Gia (Capital); 5. Giong Rieng; 6. Chau Thanh; 7. Go Quao; 8. An Bien; 9. Vinh Thuan; 10. U Minh Thuong; 11. An Minh.

(continued)

E. Tien Giang province: 1. Go Cong Dong; 2. Go Cong Tay; 3. Cho Gao; 4. My Tho city; 5. Chau Thanh; 6. Tan Phuoc; 7. Cai Lay; 8. Cai Be.
F. Ben Tre province: 1. Binh Dai; 2. Ben Tre capital; 3. Chau Thanh; 4. Cho Lach; 5. Ba Tri; 6. Giong Trom; 7. Mo Cay; 8. Thanh Phu.
G. Vinh Long province: 1. Long Ho; 2. Mang Thit; 3. Tam Binh; 4. Binh Tan; 5. Vung Liem; 6. Tra on.
H. Can Tho province: 1. Thot Not; 2. Vinh Thanh; 3. O Mon; 4. Co Do; 5. Binh Thuy; 6. Ninh Kieu; 7. Cai Rang; 8. Phong Dien.
I: Tra Vinh province: 1. Cang Long; 2. Cau Ke; 3. Chau Thanh; 4. Tieu Can; 5. Cau Ngang; 6. Tra Cu; 7. Duyen Hai.
J. Hau Giang: 1. Chau Thanh; 2. Chau Thanh A; 3. Nga Bay Town; 4. Phung Hiep; 5. Vi Thuy; 6. Vi Thanh city; 7. Long My.
K. Soc Trang province: 1. Ke Sach; 2. Long Phu; 3. My Tu; 4. My Xuyen; 5. Thach Tri; 6. Vinh Chau.
L. Bac Lieu province: 1. Hong Dan; 2. Phuoc Long; 3. Vinh Loi; 4. Gia Rai; 5. Bac Lieu 6. Hoa Binh; 7. Dong Hai.
M. Ca Mau province: 1. Thoi Binh; 2. U Minh; 3. Ca Mau; 4. Tran Van Thoi; 5. Dam Doi; 6. Cai Nuoc; 7. Phu Tan; 8. Nam Can; 9. Ngoc Hien.

(ii) The second circular pattern reported by Men (2010) contains two opposite circles, both start from the same line (Fig. 5.14). Here, duck herders take any point of circle as starting and ending point according to their convenience. Both the ends of the base line make two semicircular movements; one upward and another downward in opposite directions and ends in same point to form two circles. Out of the two circles top one covers the northern provinces of Mekong delta and the bottom one covers the southern provinces of Mekong delta. The eastern side of the baseline starts from Long Ho district of Vinh Long province and passes through the southern part of Chau Thanh district of Dong Thap province, northern end of Binh Tan district of Vinh Long province, and ends at O Mon district of Can Tho province.

(a) **Top circle**

- From the western end of the baseline that is O Mon district of Can Tho province an upward semicircular route moves through the northern end of Co Do and southwest end of Vinh Thanh districts of Can Tho province, Tan Hiep and Hon Dat districts of Kien Giang province, TriTon, Tinh Bien, Chau Doc and northern border of Phu Tan districts of An Giang province and finally reaches the Hong Ngu district of Dong Thap province.
- From the eastern end of the baseline that is Long Ho district of Vinh Long province, the flocks pass through the northwest of Cho Lach district (Tien river belt) of Ben Tre province, southeast of Cai Lay, Chau Thanh and east of Tan Phuoc districts of Tien Giang province, southeast of Thanh Hoa, northwest of Thu Thua, north of Thanh Hoa,

center of Moc Hoa, Tan Hung districts of Long An province and finally move through south of Tan Hong district of Dong Thap province to south of Hong Ngu district and forms a circle with the western semicircle.

(b) **Bottom circle**

- From the western end of the baseline that is O Mon district of Can Tho province duck flocks pass through Co Do district of Can Tho province and north Giong Rieng district to south of Chau Thanh district of Kien Giang province.
- From the eastern end of the baseline that is Long Ho district of Vinh Long province, the flocks pass along the Chien river belt of Cho Lach district, east of Mo Cay district and go along the Ham Luong river belt, northwest border of Thanh Phu district of Ben Tre province and cross the Co Chien river and go through Chau Thanh district, southeast of Tieu Can, northwest of Tra Cu district of Tra Vinh province, then move to Long Phu district, pass along the Rach Nhu Gia river of My Xuyen district of Soc Trang province, pass through the Vinh Loi and Dong Hai districts of Bac Lieu province, then move through Dam Doi, South of Cai Nuoc, northeast of Phu Tan, Tran Van Thoi and U Minh districts of Ca Mau province, and finally move through An Minh, An Bien districts and reach the Chau Thanh district of Kien Giang province to form a circle.

Irregular Pattern Movement

Another pattern of flock movement is irregular movement pattern reported by Meyer et al. (2017). This pattern contains 219 unique sites; out of this 146 sites are located in An Giang province that means most of the farmers follow within province nomadicity. All the farmers herd back their flocks to their home land during the post-harvest season two to three times a year to flock the ducks in their own paddy field or hired paddy field.

5.3.13.7 Healthcare Management

The most common diseases are duck plague, duck cholera, and avian influenza. In Vietnam, all the ducks must be vaccinated against avian influenza. For small and medium scale farmers (<500 ducks), government offers free vaccination service and large scale nomadic farmers are charged for vaccination against avian influenza (Buranakanonda 2011).

5.3.13.8 Marketing

The price of 10 duck eggs is 1.3 USD in Vietnam. The spent ducks are sold in local markets and also to China and Cambodia illegally at the rate of 2.23–2.88 USD per duck. The marketing channel is different for drakes and spent hens, table eggs, balut eggs, and day-old ducklings (Meyer et al. 2018).

Marketing Channels for Drakes and Spent Ducks

- Farmer => Collector => Inter-province wholesaler/local wholesaler => Duck yard owner => Retailer cum slaughter point => Consumer
- Farmer => Collector => Retailer cum slaughter points => Consumer
- Farmer => Inter-province wholesaler/local wholesaler => Retailer cum slaughter point => Consumer
- Farmer => Retailer cum slaughter point => Consumer
- Farmer => Consumer
- Farmer => Inter-province wholesaler/local wholesaler => Duck yard owner => Inter-province wholesaler => Border wholesaler => China and Cambodia
- Farmer => Duck yard owner => Inter-province wholesaler => Border wholesaler => China and Cambodia

The retailers handle about 1000–1500 ducks per month and most of the retailers are women who travel 9 and 43 km in Red river and Mekong delta, respectively, for purchasing the ducks and also they travel 2 and 70 km, respectively, to sell the ducks. The local wholesalers handle 5000–20,000 ducks per month. The inter-provincial wholesalers handle 75,000–150,000 ducks per month. The duck yard owners who collect ducks alone from 100 s of kilometers away paddock them inside the fenced area with or without roof and feed the ducks for an average of 2.7 days to make the ducks to gain minimum marketing weight. The capacity of such yard is about 14,000 ducks with the handling capacity of 70,000 ducks per month.

Marketing Channel for Table Eggs

- Farmer => Hatchery => Inter-province wholesaler/local wholesaler => Retailer => Consumer
- Farmer => Hatchery => Retailer => Consumer
- Farmer => Collector => Hatchery => Inter-province wholesaler/local wholesaler => Retailer => Consumer
- Farmer => Collector => Hatchery => Retailer => Consumer
- Farmer => Collector => Inter-province wholesaler/local wholesaler => Retailer => Consumer
- Farmer => Collector => Retailer => Consumer
- Farmer => Inter-province wholesaler/local wholesaler => Retailer => Consumer
- Farmer => Retailer => Consumer

Most of the table eggs are handled by the hatcheries, followed by wholesalers and the contribution by the collectors is less than 10%. The farmer to consumer transfer of egg is almost non-existent. About 2% of the duck eggs are processed.

Marketing Channel for Balut Eggs

- Farmer => Hatchery => Wholesaler => Retailer => Consumer
- Farmer => Hatchery => Retailer => Consumer

Marketing Channel for Hatching Egg and Day-Old Ducklings

- Farmer => Hatchery => Farmer
- Farmer => Hatchery => itinerant sellers => Farmers

Only 16% of the farmers purchase the day-old duckling directly from hatcheries, remaining 84% of the farmers are supplied with day-old duckling at their premises by the itinerant sellers who not only supply day-old duckling but also supply feed and medicines to nomadic farmers.

5.3.14 Cambodia

The Cambodian duck farmers are following the backyard and free-range system of rearing only. There is no transhumance among the duck farmers, but there are some nomadic duck farmers of Mekong delta region of Vietnam who cross into the Cambodia illegally and rear their flocks.

5.3.15 Indonesia

Indonesia ranked fourth in world duck population till 2017. In the year 2018, the duck population increased to 60.01 million from the previous year's 49.71 million and now Indonesia ranks third (Hirschmann 2020) in the world. The duck population is mainly concentrated in West and Central Java, South Sulawesi Aceh, and South Kalimantan provinces. In Indonesia, ducks are reared for meat, egg, and fancy purpose. In West Sumatra, ducks are used for racing (Sudardjat 2003).

The extensive duck herders are classified into four groups (Petheram and Thahar 1983), namely

- ***Fully mobile***: The ducks move from one place to another according to harvesting time of rice field. The herders do not have any fixed place of living, their night shelters would be near the temporary duck confinements. They move to faraway places using transportation vehicles. They may use swamps during lean season and molting period.
- ***Semi-mobile***: It is similar to that of fully mobile system; the only difference is that the farmers have a permanent place for living. They use the home harvesting season as part of their nomadicity and also whenever the old flocks are sold or the ducks become non-layers due to molting or during lean season, the farmers move back to their home land or swamps and stay there till the new flock is purchased or egg laying starts.
- ***Home based***: The ducks are moved in, and around the village and not to the faraway places. Ducks are allowed to be herded around dikes, canals, ponds, paddy field, etc. Additional feed is given in the lean season with cassava or sago, corn, broken rice, or rice bran. As herding depends only on own harvest, egg production is low between harvests.

- ***Opportunist***: The farmers buy the ducks before the paddy harvest and sell them out when the paddy field is emptied.

In 1970s, the flock size of nomadic duck herd was in the range of 80–130 numbers. Petheram and Thahar (1983) reported that about 50% of the nomadic flocks were owned by herders themselves and others were herded for contract by receiving wage, cloths, and food allowances or percentage of egg production per day (10–50%). In Indonesia, only 10% of the farmers are following nomadic system and majority of the ducks are reared by free-range system and backyard system (86%) and very few (4%) go for intensive and semi-intensive system. Most of the nomadic herders are educated at elementary level (about 60%), followed by no formal education, junior high school, senior high school, and college education (about 1–2%). About 70% of the duck farmers have more than 5 years of experience and 85% are rearing ducks as main business. All nomadic duck farmers are males, while few of free-range farmers (about 4%) and half of the intensive and semi-intensive farmers are females. Most of the farmers (about 90%) are belonging to the productive age group between 20 and 60 years (CIVAS 2006; Yuyun et al. 2020). The districts with more paddy fields like Tegal, Pemalang, Brebes, Boyolali (Central Java), Subang, Cirebon (West Java), and Tangerang (Banten) are famous for nomadic duck rearing. In the past, coastal area duck herding was famous in Cirebon district of West Java but now it becomes scarce (CIVAS 2006).

There are three types of duck farmers, namely hatchery cum brooder farmers or brooder farmers, grower farmers, and layer farmers. Most of the duck farmers rear same age group of birds (about 60%) and remaining farmers rear 2–3 age groups such as ducklings, layer/breeder ducks, and ducks towards culling age.

5.3.15.1 Duckling Management

The nomadic duck farmers purchase the day-old ducklings and grow them in the cage made of nets which are located nearer to their home or made under the dome of the house. The ducklings are hand-fed up to 20–30 days by using commercial feeders and waterers (Hastang et al. 2020).

5.3.15.2 Grower Management

After 1 month, the ducklings are herded on the harvested paddy field from 6.00 am to 5.00 pm and rested at night in the Umbaran (a non-permanent cage that can be moved at any time) which has several basins for drinking water or open space confinement with bamboo or nylon net. At the time of entry into the Umbaran, ducks are counted and the shortfalls are assumed to be taken by predators or mingled with other flocks (Hastang et al. 2020).

Most of the farmers herd their flock within the district and others move out of the district but within the province and very few (about 13%) move out of the province. The herds are moved from one place to another by vehicle (52%) or by walk (11%) or by both (37%). Before long distance movement, one or two farmers make reconnaissance trip to assess the area of herding and if it is found suitable, a group of 4–10 farmers together rent a truck to move their flocks. The average foraging period in a location is about 2 months which varies from 26 to 169 days (Henning et al. 2016) and the average distance between two locations is 23 km. Most of the herders share their foraging area with other farmers and also with other species like chicken, Muscovy, and wild birds. Some farmers engage workers at the cost of 34.4 USD per month. Very few farmers hand-feed the growers after foraging depending on the quantity and type of feed available in the foraging area, herding time, herding area, and weather conditions.

5.3.15.3 Layer Management

Most of the farmers rear 100–500 ducks followed by herd sizes of 500–1000, less than 100 and only few farmers rear more than 1000 ducks. The flock size is decided by age of the herder, experience, land holding, education, and income level (Lestari and Siregar 2015). Only few males are herded along with females (sex ratio of 1:20 or less) to give secure feeling to the females and also to lead the flock. The duck herders feel that presence of male in the herd improves the egg production of females.

The places of herding of nomadic ducks are harvested paddy fields, fallowed paddy fields due to flood, water sources like ponds, canals, rivers, etc., dry paddy fields, and fully yarded places at the time of lean season. Whenever the feed is not adequate in the paddy field, supplementary feed is given. Some farmers partially hand-feed the ducks at the time of start of laying (21–30 weeks of age) and some others feed rice bran during peak yield.

As the duck herds contribute to the paddy farmers by eradication of golden snail from the paddy fields, most of the farmers (73%) do not pay for herding their ducks, instead they give 2–15 eggs to the land owners daily or at 2, 3, or 7 days interval according to the productivity of the ducks. Some farmers (4.9%) lease the land and others do both. The egg production is fluctuating in nomadic duck rearing due to movement stress, feed availability in the paddy field and its composition, efficiency of the farmer, and experience of the farmer (Setioko et al. 2014). The eggs are collected using bucket, cleaned and stored in egg crates inside the Dangau where the herders live (Hastang et al. 2020).

The nomadic farmers handover manure to the paddy field owner at free of cost for grazing their birds or throw away the manure near the night confinement or use or sell their manure as fertilizer or leave the manure as such inside the confinement.

5.3.15.4 Breeder Management

The egg type Indonesian ducks are Alabio from Kalimantan, Tegal from Central and West Java, Mojosari from east Java and Bali from Bali Island and other eastern islands. Among them, Tegal is used extensively for nomadic system. The duck breeds used for meat are Pekin, Muscovy, and Tiktok (Hinny) which are reared in intensive or semi-intensive systems. Other native breeds of duck in Indonesia are Cirebon and Tasikmalaya from West Java, Tangerang from Banten province, Magelang from Middle Java, Lombok from Lombok, Medan from Sumatra, Mengwi from Bali island, Pegagan from Sumsel, Maros from Sulsel, Tondano from Sulut, Kisaran (Hinny) from Sumatera, and also Bayang, Kamang, Rambon, Cihateup,

Pitalah, Kerinci, Metaram, Shadow and Damiaking (Tanabe 1992; Sudardjat 2003; Maharani et al. 2017, 2019; Suhaemi et al. 2019).

5.3.15.5 Healthcare Management

In Indonesian nomadic duck rearing system, loss due to insecticide poisoning is around 10% which goes up to 50% in some instances. Most of the nomadic herders (86.7%) are not vaccinating their birds against diseases like duck plague, duck cholera, and avian influenza and none is doing anthelmintic treatment to the birds although most of them are infested with parasites. Most of the nomadic farmers are giving antibiotics to their birds whenever they see paralysis (about 60%), diarrhea, cough, sneezing, off fed, weakness, twisted neck, and white droppings with the consultation of experienced farmers and no veterinarians are contacted by the nomadic farmers. Some farmers sell the sick birds for meat purpose at cheaper rate. Most of the farmers give vitamin supplements to their flocks. Some farmers give traditional medicines like papaya leaf, pace leaf, lamtoro leaf, kiomang leaf, kiareng leaf, peciplukan leaf, tamarine-brown sugar, salt, milk-soda, sugar-water coconut, ginger, kencur (*Kaempferia galanga*), and temulawak (*Curcuma xanthorrhiza*) to treat the sick birds and also to develop immunity of the birds and others sell their sick ducks or keep them untreated or consumes the sick ducks (CIVAS 2006). The dead birds are disposed by throwing into the water bodies (53.57%) or burial (32.14%) or burning (Yuyun et al. 2020).

5.3.15.6 Marketing

Following marketing channels are available in Indonesia for selling ducks and duck eggs.

- Farmer => Collector/middleman => Customer
- Farmer => Farmer group head => Customer
- Farmer => Farmer group head => Collector/middleman => Customer
- Farmer => Customer
- Farmer => Hatchery

Although there are several marketing channels, the nomadic duck farmers sell their eggs and ducks through the collector/middleman who supplies all needs of the farmers and also lend money to the farmers. So the price of the produce of nomadic farmers is decided by the collector. The collector/middleman sell them to the local market within sub-district or district. Most of the collector/middleman collect the eggs daily and some collect 2 or 3 or 4 or 5 days once and very few collect once in a week or once in 2 weeks.

5.4 Ethno-veterinary Practices in Duck Production

The farmers of Kerala state of India use indigenous medication for their duckling and adult ducks. There is a practice of providing medicated water for the ducklings during the first week. The formula for medicated water used by most of the duck farmers in Alappuzha region is presented in Table 5.5.

The ingredients are pulverized or smashed and added in required quantity of water (25 liters) and boiled for 15 min. The medicated water is believed to afford the ducklings more stamina and resistance to various ailments like cough, sneezing, fever, and paralysis (Ravindran 1983). A traditional medication practice adopted against fever and respiratory infection is given in Table 5.6.

The first three ingredients are pulverized or smashed, added with neem and coconut oils, and mixed well. Farmers administer 25 g of this mixture to sick birds orally for fast result (Annual Report 2001). Indigenous medicines are being practiced by duck farmers since long back. On screening the literature, it could be seen that the ingredients used in the medication had so many medicinal properties as listed under Table 5.7.

Gajendran and Karthickeyan (2011) reported certain ethno-veterinary practices followed in duck production in India. They reported common ailments like Coryza and respiratory distress due to exposure to chill weather during heavy mist, duck plague, swelling of joints, gasping for breath, etc. during summer in ducks. To treat these diseases, farmers use a decoction made of *Poduthalai leaves* (*Lippa nodiflora*), roots of paragrass, *Omum* (*Trachyspermum ammi*), and *Vasambu* (Sweet flag or *Acorus calamus L.*). These materials are ground well, mixed in water, and boiled. *Vasambu* is the underground stem of the aromatic marsh herb and has beneficial effect on the body as a stimulant tonic and antispasmodic (Kirtikar and Babu 1935).

Table 5.5 Composition of indigenous medication for ducklings followed in Kerala state of India (Joseph et al. 2004)

Sl. no.	Common name	Botanical name	Quantity
1.	Sweet flag	*Acorus calamus L.*	100 g
2.	Turmeric	*Curcuma longa*	20 g
3.	Pepper	*Piper nigrum*	20 g
4.	Jaggery	*Borassus flabellifer*	300 g
5.	Water	–	25 L

Table 5.6 Composition of indigenous medication for growers and layers followed in Kerala state of India

Sl. no.	Common name	Scientific name	Quantity (kg)
1.	Pepper	*Piper nigrum*	0.25
2.	Turmeric	*Curcuma longa*	2.50
3.	Lime	*Calcium hydroxide*	0.75
4.	Neem oil	*Azadirachta indica*	1.50
5.	Coconut oil	*Cocos nucifera*	1.00

Table 5.7 Medicinal properties of different ingredients used in medication for ducks

Sl. no.	Scientific name	Medicinal properties
1.	*Acorus calamus L.*	Analgesic, carminative, antipyretic, anthelmintic (David and Vasant 1994)
2.	*Curcuma longa*	Hepatoprotective, anti-inflammatory, antibacterial (Ammon and Walh 1991)
3.	*Piper nigrum*	Antipyretic, analgesic, anti-inflammatory, etc. (Lee et al. 1984)
4.	*Borassus flabellifer*	Hepatoprotective, remedy for cough and pulmonary complaints (Morton 1988)
5.	*Azadirachta indica*	Anti-inflammatory, antifungal, antibacterial, antiviral, immunostimulant, and hepatoprotective (Khosla et al. 2000; Singh and Sastry 2005)

5.5 Social and Environmental Impacts of Nomadic Duck Farming

The livelihood of almost all the nomadic duck farmers is dependent on the duck egg and spent duck. The duck vendors exploit the poor duck herders by advancing loan, which anyhow later recovered through collection of egg and spent ducks. There has been no distinct change in the socio-economic status of the duck farmers since many generations. The duck herders being less educated are not aware of the entire process of duck marketing other than rearing the ducks. They are not aware of the prices of the ducks supplied by the vendors, eggs collected by them, and other transportation charges. It is the high time for documenting the contribution of these duck farmers to the poultry economy and take appropriate measures to improve their livelihood.

Ducks help in biological control of pests. When ducks are used between the paddy plants of the rice field, they feed upon the larvae and insects of all the pests of rice like brown hopper, case worm, etc. This kind of control maintains the ecosystem intact without pollution by chemical pesticides. Sholev (1995) also opined that the duck could be used as scavenging birds utilizing large amount of insects, thus having a two-fold benefit of improving feed utilization efficiency and deducing insects' problem in the field. In China, ducks are specially trained to ingest grasshoppers which otherwise would destroy agricultural lands.

5.6 Constraints in Nomadic System of Duck Rearing

1. Low productivity of indigenous ducks and less availability of high yielding exotic ducks (Rahman et al. 2017).
2. Regional variation on acceptability of duck egg and meat. For example, consumption of duck egg and meat is very less in all Indian states except Kerala, West Bengal, and Assam, only where duck eggs fetch higher price than chicken eggs.

3. Getting bank loan is difficult as the nomadic farmers do not possess immovable property to pledge as bank security. High interest rates of microcredit providers (Dadon) (Rahman et al. 2017).
4. Lack of technical knowledge among the nomadic farmers (Rahman et al. 2017).
5. Inadequate veterinary service and vaccine supply and also non-availability of duck feed (Rahman et al. 2017).
6. Loss of feed resources and also loss of life due to disease spread by migratory birds.
7. Reduction in foraging area due to change in cultivation from paddy to cash crops like sugarcane, coconut, oil palm, and plantain. Urbanization of rural areas and encroachment of lakes and ponds.
8. High yielding short period paddy varieties reduce the duration of foraging in the harvested paddy field and makes the nomadic system less feasible (Men 1997).
9. Conversion of paddy field to aquaculture as happened in Krishna delta of Andhra Pradesh state and Pokkali paddy fields of Kerala state of India.
10. More incidence of avian influenza outbreak in nomadic duck population and associated restrictions on flock movement imposed in some countries.
11. High mortality level in nomadic ducks due to diseases, nutritional, and transit stress.
12. Wide spread use of pesticides in the paddy fields.
13. Inability to get technical guidance from the poultry experts as they are nomadic in nature.
14. Youngsters of new generations are not willing to take over the transhumance from the elderly herders. A case report of a nomadic duck farmer T.A. Varghese alias Manali from Ernakulam district of Kerala state of India is noteworthy. He is continuing the transhumant duck farming even at his 68 years of age, as none from the next generation of his family is willing to inherit nomadic duck rearing.
15. As nomadic duck rearing has received little or no attention from planners, bankers, and researchers, duck eggs contribution is reduced to 1% of the total egg production of India in this decade as compared to about 4% in the previous decades (BAHS 2019).
16. There is no facility to collect and market the duck eggs, drakes, and spent hens from the place of duck herding, so the nomadic farmers are fully dependent on intermediaries or egg traders cum financiers. The nomads have no bargaining power.
17. Lack of marketing guidance and facility. Non-existence of government sponsored marketing and extension services for value addition and transfer of technology.
18. Involvement of number of middlemen like hatchery owner, local wholesaler, inter-province wholesaler, retailer, itinerant sellers, collector, and duck yard owners reduces the profit of the traditional nomadic farmers.

5.7 Conclusion

The nomadic system of duck rearing is popular in rice growing countries like India, Bangladesh, Vietnam, and Indonesia. Although China is number one country in world duck population and also has more paddy fields, mostly free-range system of duck rearing is followed in China. The ducks are herded in the paddy fields at day-time and returned back to the fixed shelter every night. Most of the nomadic farmers of above mentioned countries are elementary educated or illiterates and almost 100% herders and middlemen are males except Bangladesh where some herders and traders are females.

The rich to moderate duck farmers hold acres of land of their own, move within the district or to nearby districts only, and do not move to faraway places. At the same time, poor and landless farmers with large flock size move to faraway places of nearby states or provinces and very rarely to nearby countries. The number of ducks and the wealth of the farmer decide the distance to be traveled, the mode and route of flock movement. Nomads who move for long distance hire vehicle together to reduce transport cost. Before moving to new location, one to two farmers make reconnaissance trip to the particular place for its suitability and availability of food materials. During lean seasons like summer and rainy season, the ducks are herded on the water bed and also handfed with energy source of feed ingredients. Most of the nomads move back to their native place at the time of paddy harvest and also between batch intervals.

Although nomadic system of duck rearing is most profitable business due to no cost or less cost on feed, it is decreasing day-by-day due to non-availability of manpower from the next generation to replace the elderly herders, reduction in scavenging areas, natural feed resources, drying up of natural water bodies, excessive use of pesticides in the crop fields, cultivation of improved and short duration varieties, shift from paddy cultivation to cash crop cultivation, industrialization, urbanization, etc.

Lender–debtor relationship and exploitation by the lender and collector/middleman is more common in nomadic duck rearing system of all the above mentioned countries. In all countries, there is financial bonding between hatchery, middleman/collector/itinerant sellers, and the nomadic farmers. Finally, the actual herders are getting small amount as profit and also few duck eggs and culled ducks in their dining table. The scheduled banks are not interested to give loan to these landless nomadic farmers because of the less traceability and non-availability of immovable property as security. So, the nomadic farmers are forced to get money from the microcredit institutions which levy heavy interest per year (50–100% of the loan amount).

In all countries, the flock size is in 1000's except Bangladesh where most of the flocks are having less than 500 ducks (some herds are having even less than 200 ducks). Most of the times, the landless poor farmers are forced to sell their flock for meat at the age of 2½–3½ months for getting quick return and come out of heavy debt.

The egg production of the nomadic ducks ranges from 100 to 200 per production year except in Bangladesh, where the same is 70 or even less due to the poor producing non-descript duck population, nutrient deficiency, and poor management.

The nomadic farmers of all countries are facing the problem of outbreak of diseases like duck plague, duck cholera, and avian influenza which cause heavy mortality and economic loss. Although there is vaccine for all the above mentioned diseases, the veterinary officials are finding difficulty in tracing out these nomadic herds. Other than India, all other countries are advocating vaccination against avian influenza. As the nomadic flocks are herded in new pasture daily, the occurrence of parasitic infestation is less in nomadic flocks compared to the birds which are reared under semi-intensive and intensive system of rearing. In some areas, nomadic flocks are abundantly fed with snails, slugs and earthworms in the paddy fields, ponds and river shore. This leads to heavy endoparasitic infestation but they are not treated for the same by nomads.

References

Ammon HPT, Wahl MA (1991) Pharmacology of Curcuma longa. Planta Med 57:1–7

Anitha P, Jalaludeen A, Peethambaran PA, Leo J (2009) Effect of crude fibre and enzyme supplementation on production performance of layer ducks. Indian J Poult Sci 44(2):265–267

Anitha P, Jalaludeen A, Peethambaran PA, Usha PTA, Leo J (2011) Study on organochlorine residue in foraging ducks. Int J Curr Res 3:305–308

Annual Report (2001) Annual Report of national agricultural technology project on 'productivity enhancement of ducks' for the year 2000–2001. Centre for Advanced Studies in Poultry Science (Lead Centre), Kerala Agricultural University, Mannuthy, Thrissur, Kerala

Annual Report (2002) Annual Report of national agricultural technology project on 'productivity enhancement of ducks' for the year 2001–2002. Centre for Advanced Studies in Poultry Science (Lead Centre), Kerala Agricultural University, Mannuthy Thrissur, Kerala

Asian Agribiz (2020) Duck, white broiler prices in Vietnam rebound. Asian Agribusiness Media Pvt Ltd, 25, International Business Park, Singapore. https://www.asian-agribiz.com/?s=Duck%2C+white+broiler+prices+in+Vietnam+rebound. Accessed 05 June 2020

BAHS (2019) Basic Animal Husbandry Statistics-2019. Department of Animal Husbandry & Dairying, Ministry of Fisheries Animal Husbandry & Dairying, Government of India, New Delhi, p 132. http://dahd.nic.in/circulars/basic-animal-husbandry-statistics-2019. Accessed 08 May 2020

BER (2019) Bangladesh economic review – 2019. Agriculture Finance Division, Ministry of Finance, Government of the People's Republic of Bangladesh, Dhaka, pp 93–106

Bhat PN, Bhat PP, Khan BU, Goswami OB Singh B (1980) Animal genetic resources in India. In: Proceedings SABRAO workshop on animal genetic resources in Asia and Oceania, pp 119–187

Buranakanonda A (2011) Small scale breeder farms still thriving in Vietnam. Poult World, 3rd March 2011. Available https://www.poultryworld.net/Other-Poultry-Species/Ducks/2011/3/Small-scale-breeder-farms-still-thriving-in-Vietnam-WP008575W/. Accessed 18 June 2020

Carlén E, Lansfors B (2002) Duck production in the Mekong Delta of Vietnam: a survey with emphasis on the development and the sustainability. Swedish University of Agricultural Sciences International Office

CIVAS (2006) Centre for Indonesian Veterinary Analytical Studies. A review of free range duck farming systems in Indonesia and assessment of their implication in the spreading of the highly pathogenic (H5N1) strain of avian influenza (HPAI). A report from the Centre for Indonesian

Veterinary Analytical Studies for the Food and Agriculture Organization of the United Nations, 94 p
David F, Vasant L (1994) In: David F, Vasant L (eds) The yoga of herbs, New Delhi, p 106
Desvaux S, Ton VD, Thang PD, Hoa PTT (2008) A general review and description of the poultry production in Vietnam. Published by PCR PRISE, National Institute of Veterinary Research (NIVR), No 86, Truong Chinh Street, Dong Da district, Hanoi, 34 p
Doan EZ (2020) Duck production in Vietnam (2005–2018). Statista. https://www.statista.com/statistics/660761/vietnam-duck-production/. Accessed 28 May 2020
Edan M (2006) Review of free-range duck farming systems in Northern Vietnam and assessment of their implication in the spreading of the highly pathogenic (H5N1) strain of Avian Influenza (HPAI) – final report, 31st March 2006. A report from Agronomes et Vétérinaires sans Frontières for the Food and Agriculture Organization of the United Nations, 101p
Edwards P (1986) Duck/fish integrated farming system. In: Farrell DJ, Stapleton P (eds) Duck production science and world practice. University of New England, pp 267–291
FAOSTAT (2018) Food and Agriculture Organization Corporate Statistical Database, Food and Agriculture Organization of the United Nations, Rome, Italy. http://www.fao.org/faostat/en/#data. Accessed 09 May 2021
Furuno T (2012) Traditional Asian duck paddy field grazing. In: The one duck revolution: why Asia does not do European farming, 1st edn. Self-published on Lulu.com, pp 60. ISBN: 9781329202078
Gajendran K, Karthickyan SMK (2009) Status of indigenous ducks in Tamil Nadu, India: a traditional perspective. Livestock Research and Rural Development, vol 21, Article #175. http://www.lrrd.org/lrrd21/10/gaje2110cit.htm
Gajendran K, Karthickyan SMK (2011) Indigenous technical knowledge in duck production in Tamil Nadu. Indian J Tradit Knowl 10:307–310
Gopinathan A, Murugan M, Sivakumar T (2015) A farming system perspective of rural duck farming of Tamilnadu, Shanlax. Int J Vet Sci 2(3):10–14
Halder G, Ghoshal TK, Samanta G (2007) Socio-economic background of duck owners and status of duck rearing in West Bengal. Indian Res J Ext Educ 7(2&3):55–59
Hastang, Sirajuddin SN, Asnawil A, Kadir S, Mahyuddin (2020) Characteristics of duck farmers moving from Pinrang regency to Sidrap regency Indonesia. IOP Conf Ser Earth Environ Sci 492(1):012139
Henning J, Henning KA, Long NT, Ha NT, Vu LT, Meers J (2013) Characteristics of two duck farming systems in the Mekong Delta of Viet Nam; stationary flocks and moving flocks and their potential relevance to the spread of highly pathogenic avian influenza. Trop Anim Health Prod 45:837–848
Henning J, Pfeiffer DU, Stevenson M, Yulianto D, Priyono W, Meers J (2016) Who is spreading avian influenza in the moving duck flock farming network of Indonesia? PLOS One 11(3): e0152123. https://journals.plos.org/plosone/article?id=10.1371/journal.pone.0152123. Accessed 06 July 2020
Hirschmann R (2020) Duck production in Indonesia 2005–2018. Statista, Agriculture Farming. https://www.statista.com/statistics/659226/indonesia-duck-production/. Accessed 05 July 2020
Hossain ME, Hoque MA, Fournie G, Das GB, Henning J (2019) Management and health of stationary and nomadic ducks in the coastal and Haor areas of Bangladesh. In: Proceedings of the 30th Australian poultry science symposium, 17–20 February 2019, Sydney, New South Wales, Australia, pp 143–146
Huang HH (1973) Management and feeding. In: The duck industry of Taiwan Chinese-American Joint Commission on Rural Reconstruction, Animal industry series, No 8, Taipei, Taiwan, pp 22–29
Huang JF, Hu YH, Hsu JC (2008) Waterfowl production in hot climates. In: Daghir NJ (ed) Poultry production in hot climates, 2nd edn. CAB International, Wallingford, pp 330–375
Islam R, Mahanta JD, Barua N, Zaman G (2002) Duck farming in North-Eastern India (Assam). Worlds Poult Sci J 58:567–572

Islam MM, Bhuiyan MNK, Harun MY (2012) Development of value chain: an effective way of profitable duck farming in Hoar areas of Bangladesh INFPD good practices of family. Poultry production note No 04, 15 p
Jalaludeen A (2002) Characterization and evaluation of native ducks of Kerala, India. Paper submitted for the 2nd world waterfowl conference, 7–9 October 2003 at Alexandria, Egypt
Jalaludeen A, Peethambaran PA, Joseph L, Lal JJ (2003) Socio-economic status and indigenous knowledge of duck farmers of Kerala, India. Paper presented at 2nd world waterfowl conference, 7–9 October 2003, Alexandria, Egypt
Jalaludeen A, Peethambaran PA, Joseph L, Manomohan CB (2004) Duck production in Kerala. ICAR, New Delhi, p 44
Joseph L, Jalaludeen A, Peethambaran PA (2004) Indigenous technical knowledge in duck farming. NATP on Productivity Enhancement of Ducks, Centre for Advanced Studies in Poultry Science, Kerala Agricultural University, Mannuthy, Thrissur, Kerala, India, 20p
Khalequzzaman M, Mahdi SHA, Rahman MM (2006) Effect of the traditional parched rice husk incubation techniques on the hatchability of duck eggs. Univ J Zool: Rajshahi Univ 25:37–40
Khan MAS, Raju KS, Eshwaraiah Reddy VR (1994) Economics of duck farming in Krishna Districts of Andhra Pradesh. Indian J Poult Sci 29(1):101
Khosla P, Bhanwara S, Sing J, Sethi S, Srivastava RK (2000) A study of hypoglycaemic effects of Azadirachta indica (neem) in normal and alloxan diabetic rabbits. Indian J Physiol Pharmacol 44:69–74
Kirtikar KR, Babu BD (1935) Indian medicinal plants, vol II. Lalit Mohan Publication, Allahabad
KSBB (2019) Poultry Breeds of Kerala, Pamphlet published by Kerala State Biodiversity Board, Thiruvananthapuram, Kerala, India. http://www.keralabiodiversity.org/images/2019/brochures/poultry.pdf. Accessed 05 June 2020
Lee EB, Shin KH, Woo WS (1984) Pharmacological study on piperine. Arch Pharm Res 7:127–132
Lestari VS, Siregar AR (2015) Some factors affecting to farm size of duck farming, In: Proceedings of 38th the IIER international conference, 26 September 2015, Zurich, Switzerland, pp 60–63
Livestock Census (2012) 19th Livestock Census – 2012. All India report Department of Animal Husbandry, Dairying and Fisheries, Ministry of Agriculture, Krishi Bhavan, New Delhi, 120p
Mahanta JD, Sapcota D, Mili DC, Chakraborty A (2000) A survey on the status of duck farming in Lakhimpur and Dhemaji district of Assam. Indian Vet J 78:531–532
Maharani D, Hariyono DNH, Putra DDI, Sidadolog JHP, Lee J (2017) Phenotypic characterization of Indonesian local ducks based on body measurements. In: The proceedings of the 7th international seminar on tropical animal production, 12–14 September 2017, Yogyakarta, Indonesia, pp 775–777
Maharani D, Hariyono DNH, Putra DDI, Lee J (2019) Phenotypic characterization of local female duck populations in Indonesia. J Asia-Pacific Biodivers 12:508–514
Men BX (1997) The role of scavenging duck duckweed and fish in integrated farming systems in Vietnam. In: 2nd FAO electronic conference on tropical feeds livestock feed resources within integrated farming systems, pp 407–416
Men BX (2010) Duck farming systems and avian influenza in the Mekong Delta of Vietnam, Smallholder poultry production. Food and Agriculture Organization, Rome, Paper No. 1, pp 1–14
Meyer A, Dinh TX, Nhu TV, Pham LT, Newman S, Nguyen TTT, Pfeiffer DU, Vergne T (2017) Movement and contact patterns of long-distance free-grazing ducks and avian influenza persistence in Vietnam. PLoS One 12(6):e0178241
Meyer A, Dinh TX, Han TA, Do DV, Nhu TV, Pham LT, Nguyen TTT, Newman S, Häsler B, Pfeiffer DU, Vergne T (2018) Trade patterns facilitating highly pathogenic avian influenza virus dissemination in the free-grazing layer duck system in Vietnam. Transbound Emerg Dis 65: 408–419
Morton JF (1988) Notes on distribution propagation and products of Borassus palms (arecaceae). Econ Bot 42(3):420–441

Nambi VA (2001) Modern technology and new forms of nomadism: duck herders in Southern India, Nomadic Peoples. New Ser 5(1):155–167
Narahari D (2009) Housing and management of ducks. In: Proceedings of 4th world waterfowl conference, 11–13 November 2009, Thrissur, India, pp 48–52
NATP 2004 National Agricultural Technology Project on 'Productivity enhancement of ducks' – final report. Kerala Agricultural University and Indian Council of Agricultural Research, India, 74 p
Nind L, Tu TD (1998) Traditional system of duck farming and duck egg incubation in South Vietnam. Worlds Poult Sci J 54(12):375–384
Palanisami K, Balasubramanian R (1998) Common property and private prosperity: tanks vs private wells in Tamil Nadu. Indian J of Agric Econ 53(4):600–613
Pervin W, Chowdhury SD, Hasnath MR, Khan MJ, Ali MA, Raha SK (2013) Duck production strategy and profile of duck farmers in the coastal areas of Bangladesh. Livest Res Rural Dev 25 (7). http://www.lrrd.org/lrrd25/7/perv25129.htm. Accessed 09 May 2021
Petheram RJ, Thahar A (1983) Duck egg production system in West Java. Agric Syst 10:75–86
Pingel H (2009) Waterfowl production for food security. In: Jalaludeen A (ed) Proceedings of 4th world waterfowl conference, 11–13 November 2009, Thrissur, India, pp 3–13
Pradeep B (2017) Pastoral nomads take to duck rearing like a duck to water. The Hindu, Nalgonda, 10 November 2017. https://www.thehindu.com/news/national/telangana/pastoral-nomads-take-to-duck-rearing-like-a-duck-to-water/article20111783.ece. Accessed 19 Apr 2020
Rahman SM, Fouzder KS, Sarkar NR (2017) Nomadic herded duck production as a livelihood tool in some selected areas of Bangladesh: a case study. IOSR J Agric Vet Sci 10:16–20
Ramachandran CK, Ramakrishnan A (1982) A survey on the status of duck farming in suburbs of Trissur town. Poult Guide 19(9):65–68
Ramakrishnan A (1982) Brief resume of research work on ducks carried out at Kerala Agricultural University. Poult Guide 19(10):27–29
Ramakrishnan A (1996) Duck production scenario in India – scope for improvement. In: Proceedings of the 20th world's poultry congress, 2–5 September 1996, New Delhi, India, vol 3, pp 589–595
Rao A (1981) Steps need for growth of duck farming industry in Andhra Pradesh. Poult Guide 18(9):49–52
Rao PC, Rao RV (1981) Studies on the pathology and prevalence of helminthic parasites of ducks. Indian Vet J 58(12):934–937
Rao STV, Preetam VC, Hafeez M, Swarna CL, Raju CS (2009) A study on migratory duck farming in delta areas of coastal Andhra Pradesh – India. In: Proceedings of 4th world waterfowl conference, 11–13 November 2009, Thrissur, India, pp 362–366
Ravindran TK (1983) A survey on the status of duck farming in Kerala state. MVSc thesis submitted to Kerala Agricultural University, Mannuthy, Thrissur, Kerala, India
Ravindran TK, Venugopalan CK, Ramakrishnan A (1984) A survey on the status of duck farming in Kerala state. Indian J Poult Sci 19(2):77–80
Rithamber V (1984) A study on duck farming in rural areas of Andhra Pradesh. MVSc thesis, Andhra Pradesh Agricultural University, Rajendra Nagar, Hyderabad, 120p. https://krishikoshegranthacin/displaybitstream?handle=1/5810027683. Accessed 17 May 2020
Rithamber V, Reddy R, Rao PV (1986) A survey study of duck farming and hatcheries in Andhra Pradesh. Indian J Poult Sci 21(3):180–185
Sankaralingam S (2017) Prospects of duck farming in India. In: Training manual on 'Prospects of diversified poultry and its impact in enhancing the rural economy'. Centre for Advanced Faculty Training, Veterinary College and Research Institute, Namakkal, India, 08–28 November 2017, pp 123–141
Sarkar S, Khan SU, Mikolon A, Rahman MDZ, Abedin J, Zeidner N, Sturn-Ramirez K, Luby SP (2017) An epidemiological study of avian influenza A (H5) virus in nomadic ducks and their raising practices in Northeastern Bangladesh 2011-2012. Influenza Other Respir Viruses 11(3):275–282

Setioko AR, Iskandar S, Raharjo YC, Soedjana TD, Murtisari T, Purba M, Estuningsih SE, Sunandar N, Promono D (2014) Business models of duck farming in agricultural systems with the rice planting index three times per year (Rice IP 300): 1 the effect of reciprocity between farmers and farmers. Indonesian J Vet Anim Sci 19(3):276–279

Sholev BA (1995) Comparative growth and efficiency of various avian species. In: Hunton P (ed) Poultry production, world animal science no 9. Elsevier, Amsterdam

Singh PK, Sastry VRB (2005) Contribution of neem in livestock production. Vet 29(6):19–21

Sivakumar T, Murugan M, Gopinathan A (2009) Indigenous feeding technologies prevalent in rural duck farming of Tami Nadu. In: Proceedings of the IV world waterfowl conference, 11–-13 November 2009, Thrissur, Kerala, India, pp 212–216

Srinivas P (2018) Andhra family makes duck-rearing a money-spinner. Telangana Today, Nalgonda, 10 May 2018. https://telanganatodaycom/andhra-family-makes-duck-rearing-a-money-spinner. Accessed 19 Apr 2020

Sudardjat DS (2003) National report on animal genetic resources in Indonesia: a strategic policy document directorate general of livestock services ministry of agriculture, Jakarta, Indonesia, 59 p

Suhaemi Z, Fati N, Manullang JR (2019) Production potential of local duck in West Sumatera for human nutrition and biodiversity conservation. J Sci Eng Res 6(12):196–200

Tamizhkumaran J, Rao SVN, Natchimuthu K (2013) Nomadic duck rearing in and around Puducherry region – an explorative study. Indian J Poult Sci 48:377–382

Tanabe Y (1992) Production evolution and reproductive endocrinology of ducks. Asian-Australas J Anim Sci 5(1):173–181

Veeramani P, Prabakaran R, Sivaselvam S, Sivakumar T, Karthickeyan SMK (2016) Social status of duck farmers in Northern Districts of Tamil Nadu, India. Indian Vet J 93(07):24–27

Vergne T, Meyer A, Willgert K, Tung DX, Thu NV, Long PT, Newman S, Thuy NTT, Pfeiffer DU (2018) Nomadic movements and infectious disease transmission: tackling highly pathogenic avian influenza viruses along the free-grazing duck movement network in Vietnam. In: Proceedings of Society for Veterinary Epidemiology and Preventive Medicine – 'analysis of joint effects between biosecurity production vaccine and antimicrobial use', 21–23 March 2018, Tallinn, Estonia, pp 116–127

ViêtNam-News (2016) Mekong delta to create animal husbandry zones. Viêt Nam News – The National English Language Daily, 04 June 2016. http://vietnamnews.vn/agriculture/297748/mekong-delta-to-create-animal-husbandry-zones.html. Accessed 05 June 2020

Yousuf MA, Das PM, Anisuzzaman Banowary B (2009) Gastro-intestinal helminths of ducks: some epidemiologic and pathologic aspects. J Bangladesh Agric Univ 7(1):91–97

Yuyun I, Wibawa H, Setiaji G, Kusumastuti TA, Nugroho WS (2020) Determining highly pathogenic H5 avian influenza clade 2331c seroprevalence in ducks, Purbalingga, Central Java, Indonesia. Vet World 13:1138–1144

Dr. S. Sankaralingam, MVSc (Poultry Science), is Assistant Professor in Department of Poultry Science, College of Veterinary science, Mannuthy, Thrissur, Kerala, India. He studied his BVSc from Madras Veterinary College, Vepery, Chennai, MVSc from College of Veterinary and Animal Sciences, Kerala Agricultural University, Mannuthy. He has 3 years of experience as Scientist in Department of Science and Technology Core Support Scheme, Government of India implemented through Peermade Development Society, Idukki district, Kerala and 11 years of experience as Veterinary Surgeon in the Department of Animal Husbandry, Government of Kerala and working as Assistant Professor for the past 11 years. He has published 25 research articles in the national and international journals and conference proceedings besides several abstracts in conference proceedings. He has co-authored two books on poultry production in Malayalam language. He has guided 4 MVSc students as major advisor and 8 MVSc students as minor advisor. He is the Course Director of Diploma in Poultry Production program of Kerala Veterinary and Animal Sciences University for the past several years and also Associate NCC officer of 1(K) R&V Sqn

NCC, Mannuthy since 2011. He has organized Kerala Veterinary Science Congress 2016 as Organizing Secretary and also was the Editor of Journal of Indian Veterinary Association, Kerala from 2016 to 2018. Presently, he is doing his PhD in Poultry Science (duck genetics).

Dr. J. D. Mahanta, PhD (Poultry Science), a Professor in Poultry Science is an inhabitant of Guwahati, Assam (India). He earned his Master's degree in Poultry Science from C.S. Azad University of Agriculture and Technology, Kanpur and PhD from Kerala Agricultural University, Mannuthy. He has 32 years of experience in teaching, research, and extension activities in the field of Poultry Science. He has published more than 80 research papers (including 39 papers on ducks) in national and international journals besides co-author of one textbook on poultry production. He has guided more than 35 students for MVSc and PhD degree program as Major Advisor/Member of Advisory committee.

Integrated Duck Farming

6

D. Sapcota and Kashmiri Begum

Contents

Abstract

Integrated duck farming systems are practised in many parts of the world because of its benefits and improvement of the economic status of the farmers. Four types of integration are practised, namely duck–fish farming, duck–rice farming, duck–

D. Sapcota (✉) · K. Begum
Department of Poultry Science, College of Veterinary Science, Assam Agricultural University, Guwahati, India
e-mail: debensapcota@aau.ac.in; kashmiri.begum.vdk19@aau.ac.in

A. Jalaludeen et al. (eds.), *Duck Production and Management Strategies*,
https://doi.org/10.1007/978-981-16-6100-6_6

rice–fish farming and duck–rice–fish–azolla farming. In duck-cum-fish farming, droppings of ducks act as feed to fishes, and conversely, some fishes become food for ducks, whereas in duck-cum-rice farming, ducks are raised in rice field and duck feeds on the insects, fishes, etc. and also cause continuous stirring of the water in the field which results in weed control and better growth of the rice and thereby yield. In duck–rice–fish farming, which is primarily practised in China, the fishes are bred in terrace fields and the glutinous rice seedlings used are of traditional varieties. This type of farming system has significance in fixing the problems of the world's agricultural ecological environmental degradation in addition to farmland and water pollutions. Duck–rice–fish–azolla farming is another system of farming, where floor of the duck house is constructed above the pond allowing the manure and spilled feed to fall directly into the pond. The duck manure serves as organic fertilizer for plankton production and the spilled feed particles as food for the fishes. Simultaneously, nutrients from the ponds are dispersed to the rice field by irrigating water.

Keywords

Integrated · Duck · Rice · Fish · Farming · Azolla · Polyculture

Abbreviations

cm	Centimetre
CML	Centre for Microfinance & Livelihood
DM	Dry matter
FAO	Food and Agriculture Organization
g	Gram
ha	Hectare
INR	Indian rupee
kg	Kilogram
m	mMetere
Wt	Weight
Yr	Year

6.1 Introduction

Integrated farming system can be defined as 'new knowledge based intensive production system for household food, nutritional and livelihood security of farm families through efficient management of human, crop and animal resources with regular income generation activities without degradation and endangering the natural resource base, life, soil, water and environment' (Sarkar et al. 2011). It is a multidisciplinary whole farm approach and very effective in solving the problems of small and marginal farmers. The approach aims at increasing income and employment from small holding by integrating various farm enterprises and recycling crop residues and by-products within the farm itself (Soni et al. 2014). The integrating farming system encompasses a general set of agronomic activities that invariably include livestock. Integrated livestock farming system mostly represents an appropriate combination of livestock farming with fishery, poultry/duck and crop production systems. In the integrated duck farming system, crop or fish are incorporated to exploit the maximum potentiality of the various sub-systems in terms of profitability and sustainability. It is an excellent resource management strategy, in which nothing is considered as waste; rather, all the materials are wealth and energy. In the integrated farming system, several enterprises are interlinked to boost up the productivity of quality products for maximizing the net income of the farmer. It is a holistic approach of agro-ecosystem considering the entire farm as a single unit that concerns with balanced nutrient cycle and welfare of all species.

In integrated duck farming system, different forms of production complement each other and the farmer will have better production and more profit. It has been proved that combining various integrated agricultural production methods helps in enhancing food production, increasing incomes and conserving biodiversity.

Mainly four major integrated duck farming systems are practised:

A. Duck–fish farming system.
B. Duck–rice farming system.
C. Duck–rice–fish farming system.
D. Duck–rice–fish–azolla farming system.

6.2 Duck–Fish Integrated Farming System

The integration of fish farming with agriculture and animal husbandry is considered as sustainable farming system. Since the ducks are highly adaptable with cultivated fishes so raising ducks over fishponds fits well with the fish polyculture system. Duck–fish integration is the common integrated farming system practised in China, Hungary, Germany, Poland, Russia and India. A fishpond provides an excellent disease free environment for ducks as it has a semi-closed biological system with various aquatic animals and plants (Anonymous 2020). Further, ducks consume juvenile frogs, tadpoles and dragonfly, hence creating a secure environment for fish. The embankments of the ponds are partly fenced with net to form a wet run. The integration of fish with duck becomes very productive to meet demand and supply and bring a significant productivity from a unit area, mostly for small holding farmers.

6.2.1 Advantages of Duck-Cum-Fish Farming (CML 2019)

1. Duck droppings fall directly into water containing necessary nutrients to surge the biomass of natural food organisms and to act as feed for cultured fish in the pond.
2. Ducks collect some portion of their feed naturally from the pond, hence requirement of additional feed will be less.
3. Ducks keep the waterbed clean and increase dissolved oxygen by dabbling action.
4. Duck house can be constructed at the embankment or over the water surface. Hence no additional land is required for keeping ducks.
5. Leftover feed for ducks may be used as supplementary feed for fish.
6. Production of duck eggs, meat and fish becomes possible from the same unit area. It ensures more income.
7. Ducks feed on predators and help the fingerlings to grow.
8. Survival of ducks increases due to the clean environment of fishponds (Majhi 2018).
9. Ducks keep aquatic plants in check.
10. The nutritional elements of soil get spread in water encouraging plankton production, by virtue of the digging action of ducks in search of organisms for benthic zone.
11. Reduces greenhouse gas emission.

6.2.2 Selection of Fish Species

Polyculture is the major fish farming system predominantly found in China. The duck-cum-fish farmers use some 20 fishes in their production system—Silver carp, Bighead carp, Grass carp, Crucian carp, Chinese bream, Common carp, freshwater crayfish, Snakehead and a few unidentified species (Weimin 2010). Selection of fish species is one of the main criteria where plankton feeders have to be 60% and omnivorous 40%. The optimum composition of fish species is catla (20%), silver carp (20%), rohu (20%), grass carp (10%), common carp (15%) and mrigal (15%). It should be stocked at a density of 8000–8500 fingerlings/ha for the targeted production level of above 3500 kg/ha. This type of integration is appropriate only for rearing and stocking ponds where fishes are exceeding 12 g (Biswas 2015). The ponds should be stocked in June–September months and harvested after rearing the fish for a year, in the south, coastal and northeastern states of India, as in this region the winter season is mild. Whereas in the northern and northwestern states of India, the best periods for stocking are in the month of March and harvestings in the months of October–November, due to severe winter, which affects the development of fishes. The fish for stocking the pool must be at least 10 cm in length; otherwise, ducks will eat them.

6.2.3 Ideal Housing

In this farming system, the duck house may be constructed with the help of locally available resources such as bamboo, cane, thatches, etc. above the pond surface such that the excreta and leftover excess feed go directly to the pond and serve as feed for the fishes. In general, duck only needs shelter for resting. Two small bamboo bridges may be constructed to the duck house one for feeding the ducks and collecting eggs and another bridge from the duck house to the pond surface for assisting the ducks to ascend or descend to pond water. The ducks are kept in duck house providing a floor area of about 0.3–0.5 m^2/ bird. For breeding purpose, one male duck should be kept for every 6–8 female ducks. The periphery of the pond should be fenced for protection of ducks and for ensuring that all manure is deposited in the water. Ducks can be housed in a variety of ways. A pen can be built which floats on the water or resting on stilts above the water or on the bank of the pond (van der Meulen and den Dikken 2004).

6.2.4 Management of Pond

The pond should not be in a flood prone area and the water level should be 1.5–3.0 m. Seasonal ponds, which can retain 8–9 months water, can be considered for this type of farming (Chakrabarti et al. 2014).

The soil pH should be within the range of 6.5–7.5 and this can be adjusted with the application of lime. Lime helps in maintaining pH, kills and decomposes parasites. The basal dose of lime and cow dung application per hectare of water bodies is 1200 kg and 5000 kg, respectively. The pond should be cleaned regularly from aquatic plants which prevent sunlight penetration and oxygen circulation in water. To kill predator fishes mahua oil cake (*Bassia latifolia*) may be applied at the rate of 2500 kg/ha of water bodies. Ammonia, tea seed cake and bleaching powder also can be applied to remove predator fishes.

Managing the water quality is most important aspect in duck–fish farming system. Oxygen plays an important role in determining the quality of water. Fish needs oxygen. Water plants especially algae produce oxygen with the help of sunlight. A simple test to check the water quality is to put your arm into the water until the elbow (van der Meulen and den Dikken 2004). If you can still see your hand, there are not enough algae and the pond needs more fertilizer. If you can see about half of your arm then there are enough algae in the water and the water quality is good. If you can hardly see your arm at all, then there are too many algae in the water, you should stop adding manure to the water, possibly add freshwater or aerate the water through stirring.

6.2.5 Plankton Diversity

The plankton increases after the application of duck droppings. Among phytoplankton, the most common genera found are *Chlorella, Spirogyra, Ulothrix, Volvox, Anabaena, Microcystis, Nostoc, Oscillatoria, Navicula, Pinnularia, Euglena* and *Trachelomonas*. Whereas *Daphnia, Moina, Cyclops, Diaptomus, Asplanchna, Brachionus* and *Keratella* were among the most abundant zooplankton genera (Kalita et al. 2015).

6.2.6 Selection of Duck and Their Maintenance

The kind of duck to be reared needs to be chosen carefully since not all the domesticated races are productive. Pekin breeds are used in Central Europe, China, the Philippines, Africa and Latin America. In Thailand Khaki Campbell strain is raised, mule duck in Taiwan and Muscovy ducks in Africa. Nageswari, Sylhet mete and Kuttanad ducks are used in India. Indian Runner ducks are also used in some countries. Out of these varieties, Khaki Campbell is widely used because of its better performance (Das et al. 2003). Ducks are stocked at 200–300 ducklings/ha of fishpond. The climatic conditions influence the stocking density and ratio of ducks as well the density of fish species polycultured in the pond. The stocking size should be lower if meat duck is raised in view of the huge amount of excreta. Annually duck manure production is 45–55 kg/duck/yr., which fertilizes the fishponds and is utilized as fish feed. The manurial value of duck excreta is given in Table 6.1. Apart from this, about 10–20% feed is wasted from farming which is utilized in ponds (Bhagaban 2006). Since the ducks are less efficient in the utilization of dietary protein, the unutilized nutrients in droppings can be utilized by the fishes. According to Sasmal et al. (2010) the mean dry matter loading rate of duck excreta was 6.2 kg/ha/day at a stocking density of 300 ducks/ha and they reported improved dissolved oxygen state, pH, alkalinity of water and plankton volume significantly.

Table 6.1 Nitrogen–phosphorus–ash value of duck manure (Sasmal et al. 2010)

Sl. No.	Proximate composition of manure	Fresh basis (%)
1	Moisture (DM basis)	76.6
2	Nitrogen	4.5
3	Phosphorus	1.8
4	Ash	12.1
	Mean loading rate	**Kg dry wt/ha/day**
1	Total input	6.20
2	Total nitrogen	0.29
3	Total phosphorus	0.10
4	Total ash	1.12

6.2.7 Feed for Ducks in Integrated Fish Production System

Mostly fine rice bran and poultry feed (layer mash) are used as duck feed at the rate of 100–120 g/duck/day. The feed should be stored at cool and dry environment to avoid the growth of mould producing aflatoxin. Alternatively, duck feed can be prepared by mixing broken rice, rice bran, finger millets, sesame cake, minerals and vitamins depending on the availability, cost and quality of feed materials. Apart from that, Duckweeds (Lemna, Wolfia, Azolla, etc.) can also be fed to the ducks. They also consume tadpoles, juvenile frogs, dragonfly larvae and various other organic materials. The leftover feed from ducks and duck droppings fulfils 60–65% of feed requirements of fishes. Sometimes algal bloom may increase, in that case plastic sheets are used to stop dropping mixed into the water. In order to ensure that the manure supply remains constant it is better to keep different age groups of ducks at the same time. A removable fence may separate different age groups.

6.2.8 Harvesting

Harvesting can be done partially with the demand for table size fish in the local market followed by restocking with the same species and same number of fingerlings. Final harvesting is done after 12 months of rearing. The duck eggs are collected every morning and after 1.5–2 years of age ducks can be sold out for table purpose in the market.

Pandey et al. (2016) conducted an experimentation on improving the productivity in fish-cum-duck farming unit as well as studying the economic viability. The duck variety taken in this study was Khaki Campbell and fishes were freshwater carps. The economics of fish-cum-duck farming revealed the total variable cost of INR 269,000 per hectare in 1 year, a total income of INR 658,000 per hectare in 1 year and a net calculated profit of INR 389,000 per hectare in 1 year. The benefit–cost ratio was INR 2.44 (Table 6.2).

Table 6.2 Economic analysis of integrated fish-cum-duck farming (Pandey et al. 2016)

Sl. No.	Particulars	Amount in INR (per hectare for 1 year)
A.	Variable cost	269,000.00
	Fish (fish seed/fingerlings, fish feed, lime, miscellaneous)	
	Duck (ducklings, duck feed, medicine, drinker/feeder, electricity, one labour, miscellaneous)	
B.	Returns	
	Sale of fish (INR 85 per kg)	414,000.00
	Sale of eggs (320 females; 48,000 eggs; INR 3 per egg)	144,000.00
C.	Gross return	558,000.00
D.	Net return (C-A)	289,000.00
	Benefit–cost ratio	2.07

6.2.9 Socio-Economic Status

This kind of farming practices in numerous forms, principally within the East and Southeast Asian countries is one in all the vital ecological balanced property technologies (Majumdar 2018). On farm waste use, integrated fish farming is very beneficial to the farmers since it improves the economy of production status and decreases the adverse environmental effect on agriculture field. In Nigeria, fish production is an important turnaround for empowerment of women and youth through integrated fish farming system. This farming system is a strong tool for job creation and poverty alleviation/hunger eradication (Zira et al. 2015). According to Zaman et al. (2005), duck farming was a primary source of income for 26.8% of the population of Northeast India, whereas it was a secondary source of income along with the agriculture for the rest of the farmers.

The fish and duck play a significant role in controlling weeds, cultivation, fertilization, oxygenation and controlling pests, etc. Ducks feed on the insects and continuous stirring of water check the weed growth in the field, thus the pond ecosystem is maintained by the duck which results in growth of the fish in pond and crop in field (Fig. 6.1).

6.3 Duck–Rice Integrated Farming System

It is a common practice in Asian countries with ducks being raised in the waterways around the rice fields and feeding in the rice fields after harvest. In Japan, this method is popularly known as 'Aigamo-rice cultivation'. In this type of farming technology, Aigamo ducklings of 7–14 days old are introduced into paddy field within two weeks of rice transplantation till the time of flowering. (Aigamo ducks are cross-breed of a wild male with domesticated female ducks). In any case, only sturdy seedlings are transplanted and small ducklings are introduced so that the seedlings are not damaged. The integration of ducks in rice field creates symbiotic relationship between rice and ducks. Rice productivity is increased due to the beneficial

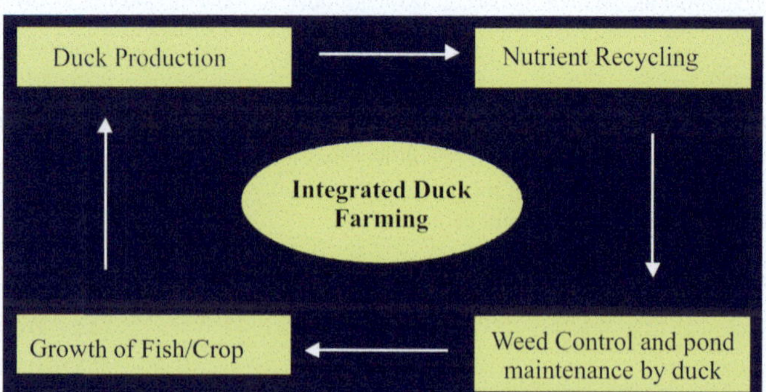

Fig. 6.1 Integrated Duck farming System

relationship between ducks and rice crop. Integrated rice–duck farming system has a mutually beneficial relation between duck and rice to increase productivity. It is suitably beneficial for poor farmers to produce organic rice at a very low cost. Various countries including Japan, China, Korea, Bangladesh, Indonesia, Philippines, India and Vietnam have showed the integration of ducks with rice field as an effective and productive farming technology. These countries had adopted the rice–duck farming system as one of the tools of organic farming where weeds and insects could be controlled by the ducks. In the Kerala State of India, duck farming is concentrated in low-lying areas and they feed on snails, small fishes, crabs and oyster shells and thereby help to maintain a balanced environment. By feeding snails and mosquitoes, which act as intermediary hosts of certain parasites, ducks take up the role of natural biological agent in the control of pests (Jalaludeen et al. 2004).

Muscovy ducks are weak to water. Hence they are not suitable. Similarly, geese are also not suitable as they eat the leaves of rice plant. In the integrated rice and duck farming, the ducks have several combined effects on rice, namely weeding effect, pest control effect, manuring effect, full time ploughing and muddying effect (Furuno 2001). Conversely, the paddy field has benefits for the ducks, viz. utilization of unused resources such as weeds, insects and water plants as food, utilization of unused space in the paddy field as duck habitat, abundant water and places for the ducks to hide under rice leaves. The approximate number of ducks to be released in the paddy field is 15–30 per 10 a in Japan and 45–60 ducks per 10 a in Vietnam (a = one or = 1/100 hectre). Usually the number of ducks are decided based on the amount of weeds, insects and other natural duck feed available in the field.

The paddy field is enclosed using a bamboo fence, a net, an electric fence or other available materials. This enclosure is used to protect the ducks from predators and to prevent the ducks from escaping. The main predators are dogs, foxes, cats, weasels, etc. Crows can also attack ducklings. While using electric fences, permission from the regulatory authority may be sought. Install the electric fence above the water surface inside the paddy field. In Japan, the ducks are raised in the paddy fields both day and night. However, in some other Asian countries they are released into the paddy fields in the daytime and taken out at night since they are liable to be stolen.

It is a low cost, organic farming method (Satnet Asia 2015). This form of mixed farming facilitates the poor farmers to acquire not only rice, as a main crop, but also other subsidiary products such as duck meat and eggs, from the same land simultaneously with equal efforts. The continuous movement of the ducks in the rice field enhances the aeration of the soil and prevents accumulation of harmful gases in the rhizosphere. Their activities also enhance the rice roots, stalk and leaf development, thereby accelerating rice growth.

6.3.1 Advantages of Duck–Rice Farming

1. Ducks eat harmful insects and weeds averting the use of chemical pesticides and manual weeding in the rice field.

2. Reduce cost of inputs and cost of labour.
3. Ducks get nutritious diet from eating insects and weeds in rice field.
4. The droppings of ducks act as a natural fertilizer to the rice crop.
5. The continuous movement of ducks, in the rice field, provides natural stimulation and aeration, which increases the availability of nutrients like nitrogen, phosphorus and potash to the rice crop.
6. Generate more income.
7. Rice–duck farming causes the reduction of emission of methane gas from rice field and contributing less to global warming.
8. Duck rearing serves as bio-shield against malaria and dengue diseases (Rizvi 2018).

6.3.2 Beneficial Effects of Rice–Duck Farming

1. **Effect on yield of rice**

The number of tillers per hill, number of grains per panicle and average grain weight are increased due to rice–duck farming system (Hossain et al. 2005). The yields of the rice–duck plots were 20% more as compared to the other normally grown rice plots. In all seasons and locations, the superiority of the rice–duck system was consistent. Similar findings were found by Choi et al. (1996) and Ahmed et al. (2004). Three per cent higher grain yield in plots with ducks raised with rice cultivation system was reported by Soon et al. (1995).

2. **Effect on insect population**

Introducing ducklings into the paddy field after the seedlings has been planted resulted in drastic reduction of insect population. The insect populations of green leafhopper, zigzag leafhopper, brown plant hopper, rice bug, short-horned grasshopper and long-horned grasshopper were significantly lower in rice–duck plots compared to without ducks (Hossain et al. 2005). Insects catching were observed efficiently in the rice–duck plots by the ducklings. Furuno (2001) and Hossain et al. (2002) made similar observations.

3. **Effect on weed population**

Weeds affect the yield of rice and showed that the weed plants per square metre of land were significantly lower in the rice–duck plots than to the sole rice plots (Hossain et al. 2005). Kang et al. (1995) reported lower weed growth ranging from 92 to 96% when ducks were reared along with rice. Ducks were reported to feed on young weed plants and weed seeds (Hossain et al. 2005). Additionally, their trampling motion could control the weed 90%, thus oxygenating the water and boosting the roots of the rice plants to grow vigorously. Furuno (1996) also observed that weed biomass was controlled more effectively in rice–duck plots compared to agrochemicals applied plot in Japan. Based on on-station study in the Philippines, Cagauan (2000) reported that in rice field weed biomass was lowered by ducks at rates from 52 to 58%.

4. **Effect on soil properties**

The movement of ducks in the rice field boosts the aeration of the soil and avoids building up of harmful gases in the rhizosphere leading to stimulation of growth of rice plants. Nitrogen, phosphorus, potash, calcium and sulphur levels were found to be high in cultivated land (Hossain et al. 2005). Feeding activity of the duck and its movement in the rice–duck plots disturbed the soil causing improvement in physical property of soils (Furuno 1996).

5. **Economic gains**

Hossain et al. (2005) obtained 50–60% higher net returns per hectare in the integrated rice–duck system compared to sole rice farming. Rice–duck system would be able to generate higher income in two ways; that is, by higher rice yield with low production cost and further income from the ducks. Additionally, these could lessen the insecticide and chemical fertilizer requirements, thus ensuring a safe environment and organic products.

6. **Rice provisioning ability**

The additional grain yield per unit area was obtained for higher grain yield per unit area due to rice–duck farming system.

Fig. 6.2 describes cycling and efficient utilization of nutrients under oxidation reaction by increasing dissolved oxygen, strengthening oxidation–reduction potential and improving electrical conductance. These effects are resulted from ducks' activities such as touching, pecking or shaking rice plants, eating weeds, pests, apple snails and other consumable organisms, depositing faecal matter and agitating and stirring paddy water. In this rice–duck co-culture system, less water drainage and no addition of chemical pesticide and herbicide help to reduce nutrients loss and hence may alleviate agricultural nonpoint source pollution.

6.4 Duck–Rice–Fish Integrated Farming System

For thousands of years Dong people, an ethnic minority of China practised this rice-fish-duck symbiotic system in China (Dai and Xue 2015). In this system of farming, traditional varieties of glutinous rice seedlings are used for cultivation in terrace fields along with that fishes are also bred. As the fish grow up to 10 cm long, the ducklings are released in the terrace field to be raised. The glutinous rice gives shade and organic food for the fish, ducks and other aquatic animals. 'Rice–fish–duck symbiotic system' of Dong people has turn out to be the protection pilot of FAO

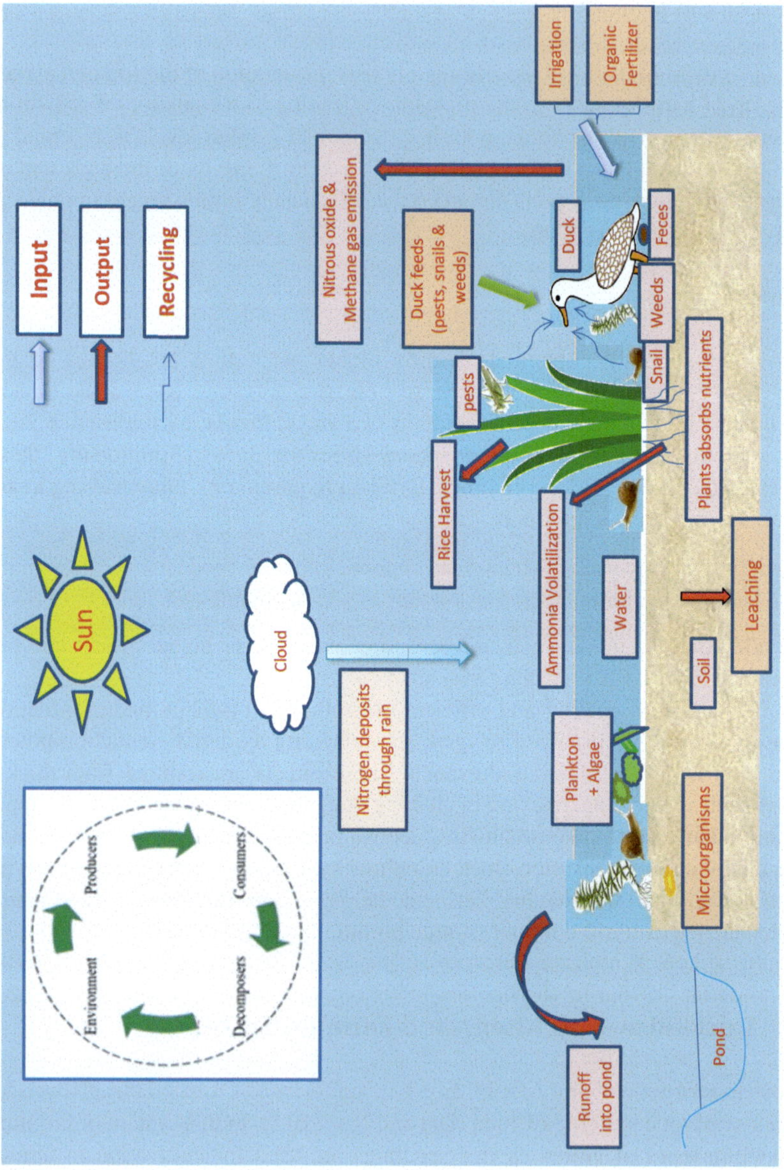

Fig. 6.2 Outline of nutrients flows in rice–duck farming co-culture paddy fields [NH_3, ammonia volatilization; N_2O, nitrous oxide; CH_4, methane]

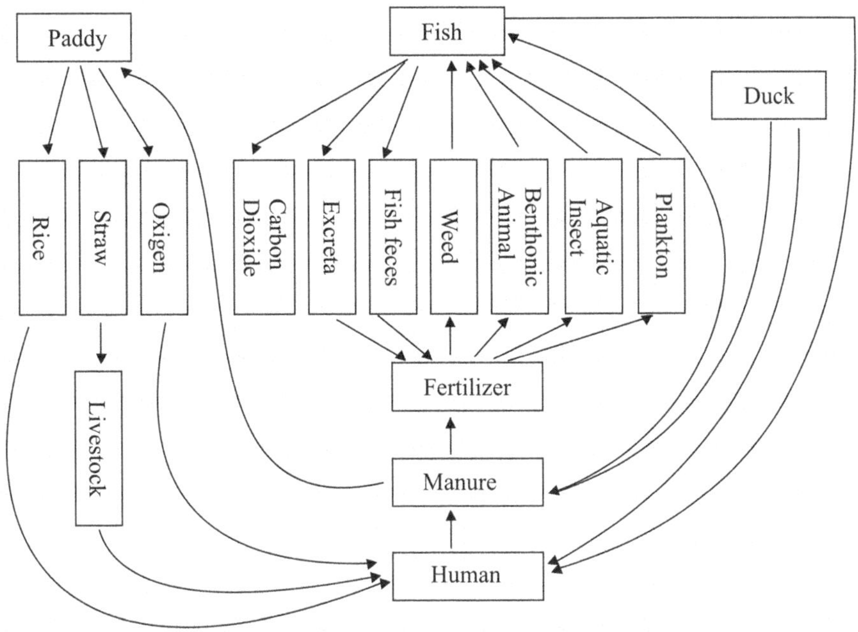

Fig. 6.3 Substance recycling and energy flowing in rice–fish–duck symbiotic system (Dai and Xue 2015)

Globally Important Agricultural Heritage and as one part of China's traditional knowledge and agriculture. This rice–fish–duck symbiotic system has practical importance and positive value for solving the problems of the world's agricultural ecological environmental deterioration, farmland and water pollution and agricultural products safety. Some of the benefits of this system described by them are given below.

1. **Economic benefits**

The glutinous rice yield is slightly lower than the hybrid rice and price of rice produced in this system is 3–4 times more than hybrid rice. The fish and ducks are typical 'green food', in this system as the prices of the glutinous rice are two times the ordinary varieties. The manure of fish and ducks is good organic fertilizer for glutinous rice, which can significantly decrease the input prices of pesticide, fertilizer, feed and labour.

2. **Ecological benefits**

The ecological cycle significantly improved farmland ecosystem stability, increased biodiversity (Fig. 6.3), to attain sustainable development and utilization of land.

3. **Social benefits**

This system is a natural three-dimensional way of agricultural production, which efficiently save the land and ease the contradiction between land and people as the Dong people lack arable land area and dominated by terrace fields.

4. **Cultural benefits**

The system is one of the surviving production modes for the Dong people and has a history of ages. It not only supports the traditional varieties of planting and breeding in the region, but it also promotes the persistence of Dong people's farming culture.

6.5 Duck–Rice–Fish–Azolla Integrated Farming System

Azolla, a weed floats on the water and absorbs essential nutrients from the water and it fixes nitrogen from the atmosphere. With the introduction of azolla, integrated rice and duck farming became a sustainable recycling system. Traditionally, in some Asian countries, azolla is used in rice fields as a green manure for the rice or as a weed supplement. Azolla was used in southern China and northern Vietnam from the time immemorial. In Vietnam, the azolla was ploughed under before seedling transplantation. However, in China, ploughing to add azolla in the soil before seedling transplantation as well as cultivation of azolla between rice tussocks after transplantation and ploughed under in later stage are being practiced.

The Japanese farmer and entrepreneur Dr. Takao Furuno has developed rice–duck–azolla–loach cultivation as an integrated bio-system, by combining duck raising into organic rice cultivation which excludes the requirement for fertilizers, herbicides and pesticides. This system is now being replicated through significant success entirely over Southeast Asia as an actual approach to improve farmers' incomes, lessen the impact on environmental influence and increase food security. His integrated bio-system has spread to more than 75,000 farmers in Japan, Korea, China, Vietnam, Philippines, Laos, Cambodia, Malaysia, Bangladesh, Iran and Cuba. Azolla–rice–duck–fish integration system is the most comprehensive combinatory use of azolla.

Azolla fixes nitrogen from the atmosphere and the ducks eat it. The duck excreta act as feed for microorganisms, water fleas and plankton, which flourish and eaten by the fish. The rice plants grow by making use of the nutrients provided by the decomposition of the excreta. The atmospheric nitrogen fixed by the azolla in the paddy field is used as a nutrient source for the natural growth of rice plants, ducks and fish. In duck–rice–fish–azolla integration, duck shed is constructed over the fishpond refuge that is contiguous to the rice field. The floor of the duck house has some spaces to allow the manure and spilled feed to fall directly into the fishpond. The duck manure serves as organic fertilizer for plankton production and the fish can directly consume the spilled feed. Nutrient from the fishpond refuge is dispersed to the rice field by irrigation water (Fig. 6.4).

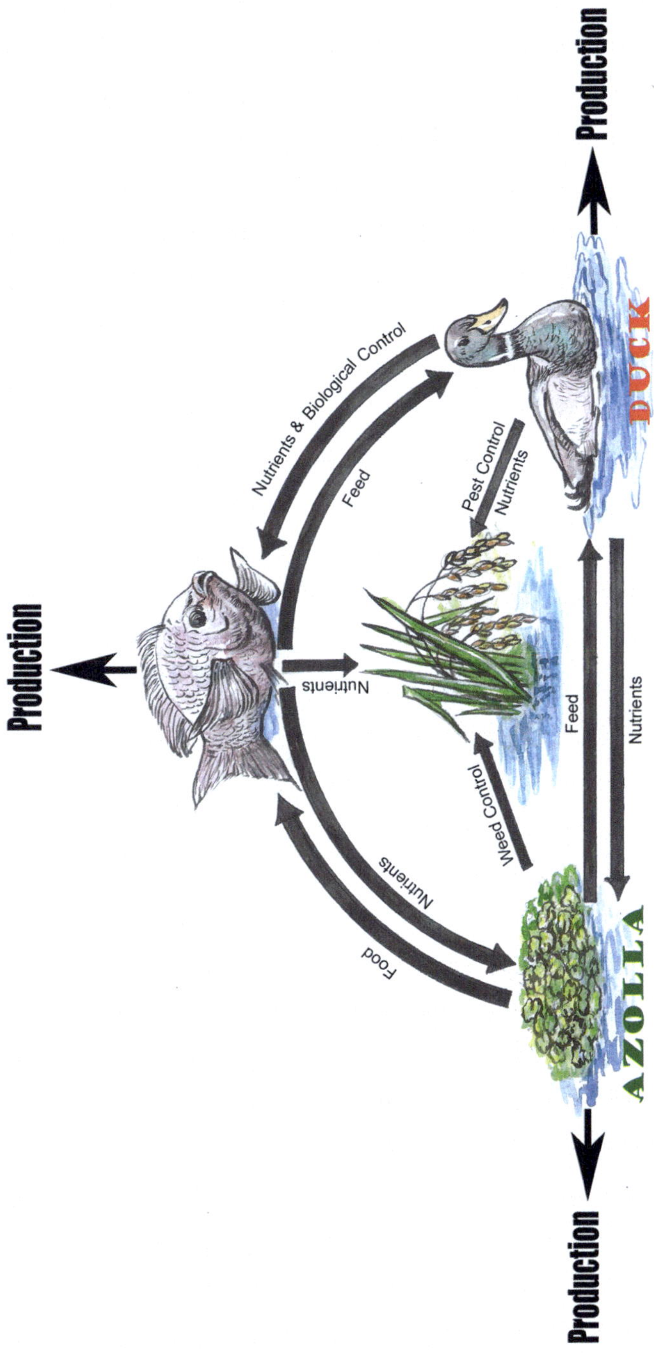

Fig. 6.4 Schematic presentation of the interrelationships among rice, fish, azolla and duck in an integrated farming system (Cagauan et al. 2000)

6.6 Scope of Integrated Duck Farming System

1. Due to small size of landholding of the farmers, integration will be the only way to enhance production of diversified products.
2. The modern intensive farming system has put threat to the human health and wealth due to use of chemical fertilizers, pesticides, insecticides and drugs in monoculture practices and it cannot sustain for a longer period.
3. Integrated duck farming system in contrast to monoculture may be an excellent tool to curb down the apprehension of environmental degradation, greenhouse gas emission and global warming.

6.7 Conclusion

Integrated duck farming system is the most suitable, eco-friendly and best system for controlling economic loss adopting crop diversification and optimum use of both land and water and managing waste resources at sustainable level. The profit with the integration is due to recycling of organic wastes which reduces feed costs to the fishes as well as extra revenue from the sale of birds. Duck feeds on the small vegetation present on the pond causing reduction in feed cost. The use of duck manure to increase fish production is viable and cost-effective than fish farming alone. It is an organic form of production system which is compatible with the environment and increases the income of the farmers with no extra efforts.

References

Ahmed GJU, Khan MA, Biswas JP, Islam MR, Islam MA (2004) Evaluation report integration of rice cum duck farming for resource poor farm households. A research project funded by Department for International Development (DFID) through PETRRA Agronomy Division Bangladesh Rice Research Institute

Anonymous. (2020) Kerala Agricultural University Agro-Infotech Portal: http://www.celkau.in/. Accessed 1 Jan 2020

Bhagaban K (2006) A training manual on polyculture and integrated fish farming. Assam Agricultural Competitiveness Project (AACP), Govt of Assam. College of Fisheries, Assam Agricultural University, Nagaon, Assam, India. Portal: https://docplayer.net/50822918-Polyculture-and-integrated-fish-farming.html. Accessed 2 July 2019

Biswas A (2015) Fish duck integrated farming. Portal: https://www.researchgate.net/publication/287595864_Fish_Duck_Integrated_Farming. Accessed 2 Apr 2019

Cagauan AG, Branckaert RD, Van Hove C (2000) Integrating fish and azolla into rice-duck farming in Asia. Naga, The International Center for Living Aquatic Resources Management, 23, 4–10. Portal: http://www.worldfishcenter.org/Naga/na_2359.pdf. Accessed 27 Sept 2019

Chakrabarti A,Dey A, Kumar D (2014) Livestock cum fishery integrated farming system. Kriwhisewa, E-Journal www.krishisewa.com. Portal: https://www.krishisewa.com/articles/livestock/402-livestock-fishery-integrated-farming.html. Accessed 27 Sept 2019

Choi CS, Shin BW, Kim DH, Yoo SJ, Soo JD, Rhee GS (1996) Rice growth and improvement of soil properties following rice-duck farming in a paddy field. RDA J Agric Sci Soil Fertilizer 38: 382–388

CML (2019) Centre for Microfinance & Livelihood, Guwahati, Integrated fish cum duck farming. Portal: http://cmlnortheast.com/wp-content/uploads/2020/05/CML_AR_2018.pdf. Accessed 25 Mar 2019

Dai, Xue D (2015) Dong people's rice-fish-duck symbiotic system in China. Case Study submitted to College of Life and Environmental Sciences, Minzu University of China; Nanjing Institute of Environmental Science, Ministry of Environmental Protection of China. Portal: https://satoyama-initiative.org/old/dong-peoples-rice-fish-duck-symbiotic-system-in-china/. Accessed 27 Sept 2019

Das GB, Haque ME, Ali ML, Chanda GC, Das C (2003) Performance of Indian runner Zending and Khaki Campbell in integrated duck-cum-fish farming system. Pak J Biol Sci 6:198–201

Furuno T (1996) Significance and Practice of Integrated Rice Cultivation and Duck Farming-Sustainable Agriculture, Kyushu International Center, Japan International Cooperation Agency and Kitakyushu Forum on Asian Women, pp 12

Furuno T (2001) The power of duck: integrated Rice and duck farming. Tagari Publications, Tasmania, pp 8–28

Hossain ST, Ahamed GJU, Islam MR, Mahabub AA (2002) Role of ducks in controlling weeds and insects in integrated rice-duck farming. Bangladesh J Environ Sci 6:424–427

Hossain ST, Sugimoto H, Ahmed GJU, Islam MR (2005) Effect of integrated rice duck farming on rice yield farm productivity and rice provisioning ability of farmers. Asian J Agric Dev 2:79–86

Jalaludeen A, Peethambaran P, Joseph L, Manomohan CB (2004) Duck production in Kerala. NATP on Productivity Enhancement of Ducks, Centre for Advanced Studies in Poultry Science, Kerala Agricultural University, Mannuthy, Thrissur, Kerala, India. pp 1–2

Kalita HK, Talukdar B, Baishya RA, Basumatary S, Das J, Dutta A, Barman RC, Sarma D (2015) Role of duck droppings on pond productivity through fish-integrated farming system in agro climatic condition, Assam. Nat Environ Pollution Technol 14:541–543

Kang YS, Kim JG, Park KH, Kang YS, Kim JI, Park JH (1995) Influence of Rice-duck farming system on yield and quality of Rice. Korean J Crop Sci 40:437–443

Li MJ, Li RH, Zhang JE, Guo J, Zhang CX, Liu SW, Hei ZW, Qiu SQ (2019) Integration of mixed-cropping and rice-duck co-culture has advantage on alleviating the nonpoint source pollution from rice (Oryza sativa L.) production. Appl Ecol Environ Sci 18:1281–1300

Majhi A (2018) Impact of integrated duck cum fish farming in improving the socio economic condition of rural fisherman at some selected region of Purulia district. Int J Adv Eng Res Dev 5: 1018–1027

Majumdar P (2018) Future aspects of integrated fish farming. Acta Sci Agric 212:45–47

Pandey A, Dhawan A, Srivastava A, Singh A (2016) Integrated fish-cum-duck farming: productivity enhancement through technological interventions. Indian J Ecol 43:894–896

Rizvi F (2018) Duck rearing serves as bioshield against malaria, dengue. Urdu Point. 11th December 2018. Portal: https://www.urdupoint.com/en/pakistan/duck-rearing-serves-as-bioshield-against-mala-503034.html. Accessed 23 Mar 2019

Sarkar AK, Singh RS, Singh MS, Yadav MS, Singh CS (2011) Integrated farming Systems for Sustainable Production. Agrotech Publishing Academy, Udaipur, pp 45–58

Sasmal S, Chari MS, Vardia HK (2010) Role of duck dropping on pond productivity through fish-duck integrated farming system. Livestock Research for Rural Development,22,172. Portal: http://www.lrrd.org/lrrd22/9/sasm22172.htm. Accessed 10 May 2021

Satnet Asia (2015) Integrated rice-duck farming. Satnet Asia, Technology fact sheet, PP 1–4 Portal: https://www.scribd.com/document/355634418/SATNET-FS14-Rice-duck-Farming-pdf. Accessed 23 Mar 2019

Soni RP, Katoch M, Ladohi R (2014) Integrated farming system- a review. IOSR J Agric Veterinary Sci 7:36–42

Soon KY, Kim JG, Park KH, Kang YS, Kim JI, Park JH (1995) Influence of rice-duck farming system on yield and quality of rice. Korean J Crop Sci 40:437–443

Van der Meulen SJ, Den Dikken G (2004) Duck keeping in the tropic. Agribook-series No 33, 2nd edn. Agromisa Foundation, Digigrafi, Wageningen, The Netherlands
Weimin M (2010) Distribution and characteristics of duck-fish farming systems in Eastern China. Smallholder poultry production. Food and Agriculture of the United Nations (FAO), Rome, p 28
Zaman G, Goswami RN, Aziz A, Nahardeka N, Roy TC, Mahanta JD (2005) Farming system of Nageswari ducks in North-Eastern India (Assam). Worlds Poult Sci J 61:687–693
Zira JD, Jaafaru A, Badejo BI, Ghumdia AA, Ali ME (2015) Integrated fish farming and poverty alleviation/ hunger eradication in Nigeria. J Agric Veterinary Sci 8:15–20

Dr. Deben Sapcota, a Doctor of Veterinary Medicine hails from Guwahati, Assam (India). He earned his Master's degree (Poultry Science) from APAU, Hyderabad and Ph.D. (Poultry Science) from IVRI, Bareilly. He joined AAU in 1987 as a Farm Manager (Assistant Professor) and is currently as Professor. He got post-doctoral research experience at North Carolina State University, USA. His main areas of interest include feed enzymes, probiotic, prebiotic, aflatoxin studies, gut microbiota and unconventional feed resources in poultry. He has supervised more than 23 postgraduate students and published over 200 papers and a textbook on poultry science.

Dr. Kashmiri Begum is native to Mangaldai, Assam (India). She earned her Master's degree (Poultry Science) in 2018 from College of Veterinary Science, Assam Agricultural University, Guwahati, Khanapara. Currently, she is pursuing PhD from the same university. She is also a Junior Research Fellow under National Agricultural Higher Education Project (Poultry Components).

Intensive Duck Rearing

7

K. Gajendran and P. Veeramani

Contents

K. Gajendran
Centre for Animal Production Studies, Tamil Nadu Veterinary and Animal Sciences University, Chennai, Tamil Nadu, India
e-mail: gajendrank@sugunainstitute.com

P. Veeramani (✉)
Department of Poultry Science, Madras Veterinary College, Tamil Nadu Veterinary and Animal Sciences University, Chennai, Tamil Nadu, India
e-mail: Veeramani.p@tanuvas.ac.in

A. Jalaludeen et al. (eds.), *Duck Production and Management Strategies*,
https://doi.org/10.1007/978-981-16-6100-6_7

Abstract

The traditional duck farming has been transmogrified into intensive duck rearing especially in Asian countries in the recent past, due to its numerous advantages over chicken farming. Better productivity in ducks is achieved even with less expensive, simple housing facilities resulted in less investment for commercial duck farming. Apart from the indigenous duck varieties, breeds like White Pekin and Aylesbury, improved varieties like Vigova are best suited for intensive duck rearing and make this as a lucrative venture. Increasing demand for duck eggs, meat and other duck products paved the way for intensive duck rearing as a profitable business. The scientific advancements in intensive management of chicken are well and easily adapted in duck farming also. With this background, different systems of duck rearing with special reference to intensive duck rearing, different management protocols, possible automations in housing and management are discussed exhaustively in this chapter. Apart from this, the topics like behaviours of ducks and specific managements for commercial and breeder ducks are also deliberated. The new scientific advancements in automation of intensive duck rearing such as rearing in environmentally controlled sheds, automation in feeding, watering, disease monitoring with internet/web based technology, etc. will improve labour efficiency and product quality. The intensive duck rearing with recent advanced technologies will satisfy the future demand for animal protein by improving productivity of ducks with lowered production cost.

Keywords

Intensive duck rearing · Systems of rearing · Management · Housing · Behaviour of ducks

List of Symbols/Abbreviations

COA	Council of Agriculture
RFID	Radio-frequency identification
PVC	Polyvinyl chloride
HEPA filter	High-efficiency particulate air filter
CSIRO	Commonwealth Scientific and Industrial Research Organisation

7.1 Introduction

Asia is considered as homeland for ducks, which comprises nearly 90% of world's duck population. Among the poultry species, ducks are hardy and prolific in nature, can easily and rapidly adapt to adverse environments and are resistant to many common poultry diseases. Ducks convert low quality feed efficiently into quality human food. In many parts of Asian countries, indigenous ducks are reared traditionally by poor farmers to earn their livelihood. Small-scale duck production makes a significant contribution to household economics and food security (Veeramani et al. 2016).

Next to chicken production, duck farming assumes importance among farming community in Asian countries as a lucrative business. There are numerous meat and egg productive duck breeds available throughout the world. The major advantages of commercial ducks compared to chicken farming are less investment in housing requirements. Ducks possess innate resistance to most of the common avian diseases causing heavy infection in other poultry. Intensive rearing of ducks can also be integrated with paddy, fish and other vegetable farming to improve the productivity in both ducks and agriculture. Moreover, the demand for duck products like meat and egg registered a positive trend in the local and international market. There are a lot of intensive duck farming enterprises in some of the Asian countries as that of chicken industry and still it has to be improved further to cater to the needs of the public. Indoor reared Peking ducks attain more bodyweight than free-range ducks (Knust et al. 1995). Further, intensive system of duck rearing attached with outside activity is the best system in terms of duck welfare and growth (Erisir et al. 2009).

In this chapter, an overview of duck farming under different systems of management with a special focus on intensive rearing of ducks, management and production is discussed.

7.2 Intensive Duck Farming

Under intensive duck farming, an all-in-all-out or batch system of rearing may be followed. In all-in-all-out only one age group of ducks will be reared at a particular period, whereas, in batch system, different age groups of ducks are reared at any point in time in the farm. The different age groups may be maintained separately in other sheds or in the same shed with partitions. Intensive system includes deep litter, pen and run and cage systems. Under these systems, ducks are reared in closed doors by providing required environment and ventilation as that of chicken (Rodenburg et al. 2005). Under intensive system, duck house with yard and swimming pool improves the productivity, carcass traits, meat quality, blood lipid profile and immunity (Ghanima et al. 2020).

Intensive duck farming is being practised in variants like pen and run, deep litter, cage, slat floor and wire floor systems and some in environmentally controlled houses. Importance of each system, management protocol followed along with its advantages and disadvantages are discussed in detail in this chapter.

7.3 Layout of Duck Farm

The farm layout needs to be systematically planned in such a way that different age groups of ducks are housed separately with proper biosecurity measures and to make sure that there is no drainage from the adult housing area to growers. The duck sheds for different age groups are placed in the farm layout by observing the wind direction. Deep litter system is ideal for commercial ducks, whereas litter cum slatted or wire floor is suitable for breeder ducks for improved productivity. A combination of litter and slats prevents leg damage in heavy breeding ducks compared to complete wire or slat floor.

7.3.1 Selection of Farm Site

The selected farm site should have easy operational convenience to carry out various activities of the farm. The farm site should be located away from the city but not too far away that it becomes strenuous to bring in and take out materials. A breeder duck farm should be located in an isolated place, away from commercial duck farm operation to prevent the spread of disease. A commercial farm should be located nearer to the marketing centres.

7.3.2 Location

A commercial duck farm should not be located in a congested area or in close proximity to existing farms. There should be a minimum distance of 1 km from existing farms. It should be away from a highway to avoid noise and dust pollution but it should have access to connecting roads. It may be nearer to the place where sufficient area is available for foraging of ducks, if necessary. The available area may be of fertile land to cultivate agricultural crops, so that the ducks may be allowed in the fertile land for foraging at the same time the fertility of the soil will also be increased.

7.3.3 Topography

The farm should be located in an elevated area, preferably on even land with proper facilities for drainage. The area should be well protected against strong winds as it could easily damage the duck house; therefore, windbreaks in the form of trees can be maintained at the farm boundaries. This would also provide shade during hot seasons.

7.3.4 Water

Water is the important necessary nutrient for efficient physiology of the bird. Abundant water is required for various activities. The water should be potable and free from any contamination. Before starting the farm, water available in that locality must be subjected to various tests for impurities like total solids, microbial contaminations and high percentage of salts of sodium, potassium and chlorides. Water should be tested for nitrites and nitrates whose presence will indicate organic pollution.

7.4 Distribution of Various Buildings in a Farm

The open ventilated houses, normally seen in tropics, where the airflow and light solely depends on the natural air currents and the rotation of earth in different seasons. The detailed plan of a model layout of various buildings of intensive duck farming unit is depicted in Fig. 7.1.

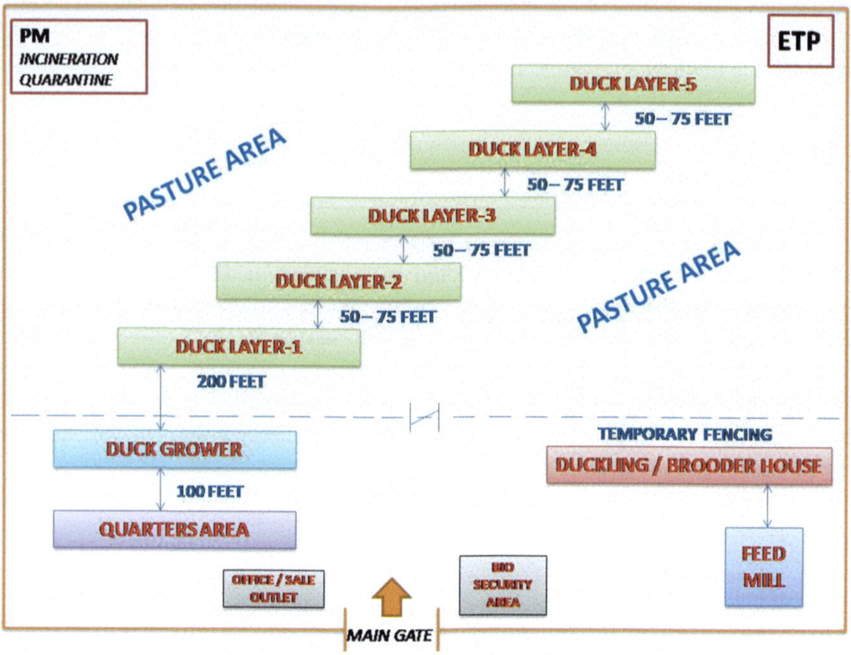

Fig. 7.1 Model layout of intensive duck layer farm

7.4.1 Duckling or Brooder Houses

This could be a cage house or a deep litter house. Normally chicks are reared in litter floor or brooder cages. The brooder house should be leeward in direction, i.e. away from the direction of wind flowing from the layer and grower houses.

7.4.2 Grower and Layer/Breeder Houses

Grower and layer houses should be located away from the duckling house and it could be of deep litter or cage house. The minimum distance between two houses should be 30–50 feet. The longitudinal axis of the shed should be east-west oriented and distance between houses can be utilized for pasture development or any cultivation. This will facilitate for foraging of ducks and also provide green environment inside the farm location.

7.4.3 Hatchery Unit

Hatchery building should be located away from the duck sheds. The hatching eggs collected from the breeder houses should be properly cleaned and stored in the egg storeroom and then to the hatchery. Accordingly, the egg storeroom and hatchery buildings are planned. Hatchery building should be covered with proper fencing and strict biosecurity should be adopted to prevent spread of diseases. The hatchery building should be located in leeward direction, i.e. away from the direction of wind flow from the duck sheds.

7.4.4 Feed Mill

Feed mill building should be located near the main entrance of the farm, which facilitates easy unloading of the inputs without entering into the farm premises. If possible, two entries may be planned for the feed mill, one for unloading the inputs from the outside vehicle and another one for loading the finished feed to transport to the sheds.

7.4.5 Other Buildings

Other buildings such as storeroom for egg and feed, office room, quarters for the workers should be located away from the duck sheds to prevent disease transmission. The egg storeroom and office may be located near the entrance of the farm premises to facilitate easy disposal of eggs and to minimize the people movement around the sheds. The another extreme end of the farm site should be earmarked for the construction of other structures such as post-mortem room, carcass disposal pit, room for sick birds and effluent treatment plant to have proper bio-security.

7.5 Housing for Ducks

7.5.1 Duck Shed Design and Construction

A duck house should provide maximum comfort to the bird. It should have a healthy atmosphere with good ventilation. It should be cooler during summer and comfortably warm during winter. It should have enough light and a comfortable microclimate. While planning for constructing a poultry house, provisions should be made for future expansion.

7.5.2 Orientation

The orientation of all duck houses in a tropical climate must be East-West. i.e. its long axis running East-West and the width extending North-South. This could be ascertained by placing a straight long pole perpendicular to a level ground at 12 noon and trace the falling shadow of this pole. The open-sided East-West oriented houses will prevent direct sunlight falling into the poultry house and also to minimize the splashing of rainwater into the house during monsoon.

7.5.3 Width

For a tropical open type house, the width preferred is 24–26 feet (7–8 metres) extending up to a maximum of 30 feet (9.5–10 metres). If the width goes beyond 10 metres, there will not be adequate cross ventilation, particularly during hot weather. Dampness inside the duck house is common, if cross ventilation is not adequate, dampness inside the shed will be a predisposing factor for most of the bacterial diseases.

7.5.4 Length

Depending on the topography and soil condition, the length can be extended to 150 feet (50 metres) preferably 100–150 feet (30–50 metres) and this can be partitioned suitably with a provision of separate doors for each compartment.

7.5.5 Height

From the floor level, the height of the house at the centre must be 10–12 feet. At the eaves level, it should be 7–10 feet (2–3 metres) and at the ridge level 10–13 feet (3.5–4.5 metres) depending on the type of roof material used, width and type of poultry house.

7.5.6 Foundation

The foundation should be strong enough to support the wall and other structures. The type and depth of the foundation will depend on the nature of soil, height and width of the building and the type of roofing material used. The height of the floor from the outer ground level depends on the water table of the soil and the extent of water seepage inside the shed during rainy season. Accordingly, the floor height ranges from 0.5 to 1 foot (0.15–0.31 m) from the ground level.

7.5.7 Floor

A cement concrete floor that allows easy cleaning and disinfection is advantageous in curtailing disease problems. This will also reduce problems due to insects, rodents, worms and water seepage. The floor should be constructed with slight slope for easy cleaning and free flow of water. Cemented floor is impermeable to parasites and easy to clean. The cemented floor should not be too rough or abrasive so as to prevent injury to the ducks. The constant contact of rough flooring with the duck skin leads to irritation, injury and abrasions resulting in damaged skin. Hence, smooth flooring may be advisable under deep litter system of management. If ducks are kept confined in rough flooring for longer duration, the detrimental effect of flooring also increases. This also depends on the age and size of the ducks. Similarly, floors with slats, wire or cages may also cause feet and leg injuries if they are not smooth, non-abrasive and free from sharp edges. In case of wire floor, if it occupies lesser floor space, the likelihood of injury is greatly reduced. Compared to slats and cages, properly constructed wire floors are the better choice and cause lesser injuries to the ducks. For proper wire flooring, 12-gauge welded wire with a mesh size of 1.9 cm for ducks below 3 weeks age and 2.5 cm for ducks of above 3 weeks of age should be used. Vinyl coated or smooth galvanized wires are preferable for making floors for ducks,

7.5.8 Advancements in Housing System

In recent decades, the indoor/confinement raising models are more popular over traditional duck in China, Taiwan, Vietnam, Cambodia and other Asian countries especially for the production of meat type ducks. Environmentally controlled housing as that of chicken industry, slatted floor and cage rearing are the recent advancements in duck farming. These developments are rapid due to industrialization, increased awareness on bird welfare and environment protection. The duck welfare depends on the microenvironment like temperature, humidity, litter moisture and ammonia inside the house (Jones and Dawkins 2010). Good ventilation with better litter quality is essential to maintain the health of body and plumage of ducks and to reduce mortality.

Indoor mesh bed and thick cushion grass models received better appreciation among the large scale duck farmers due to ease in management and reduced cost of production. These advancements are well received by the farmers as they enable proper waste disposal. Duck mesh bed with fermentation bed and duck cage-raising models are undergoing rapid development, the purpose of which is to transform wastes into organic fertilizers under aerobic fermentation conditions and therefore to reduce environmental pollutions (Hou 2016). In recent years, the practices for duck management incite more public concern on food safety and security and animal welfare. Under deep litter system, wood shavings are widely used; conversely, a raised plastic floor is a common flooring system used in duck production. Each flooring type requires different management practice that influences physical characteristics of growing ducks (Karcher et al. 2013).

At the same time, access to open water for swimming or in any form will be beneficial in improving eye, feather cleanliness and food pad condition. Muscovy ducks in hot climatic conditions recorded improved body weight, better immunity, decreased mortality rate and low body temperature when reared with access to a swimming pond (Farghly et al. 2017).

In intensive rearing of ducks, it is advisable to design the floor of the houses with a combination of slatted (30%) and litter (70%) from 14 days to slaughter for better performance. Slatted flooring, mesh-wire or synthetic floors and grass bed and plastic bed models are the recent advancements in flooring of intensive duck houses. These flooring have to be coupled with proper draining of wastewater and recycling of water which ensures environment friendly duck farming.

Although the comfort zone for mature ducks is about 8–26 °C, in commercial practice, 21 °C is considered optimum after 21 days of age. In enclosed housing system, good ventilation is more important to keep ammonia concentration below 15 ppm. Further reduction in inside temperature and ammonia level during hot climates can be achieved by installing cooling pads or fan ventilation system. Wood shavings and raised plastic flooring did not reveal any advantages of one system over the other as reported by Fraley et al. (2013). However, controlling the ducks' microenvironment with effective ventilation system is crucial to duck welfare (Jones and Dawkins 2010).

Plastic slats are developed specifically for the rearing and growing the ducks as per the standard requirements by different companies. Slats are designed specifically to suit the design of duck houses for complete or partial laying as in the case of slat house or slat cum litter floor house, respectively. The major advantages of slat flooring are it improves hygiene, reduces the cost on litter material, easy drainage of splashed water from nipple drinkers and improves the microclimate inside the house.

7.5.9 Automation in Operations

Automation of management system, recent advancement in production practices are highly essential for increasing the production efficiency. The experience from intensive chicken farming practices provides a good model for the automation in intensive duck production. Automation can increase overall productivity of ducks and reduce labour as well. An efficient economical automated animal house is provided to increase the health and longevity of ducks. The technologies involved in automated commercial poultry house such as automated feeding, watering, egg collection, bird weighing and automatic removal of contaminated bedding and replacement with fresh or recycled bedding are applicable to intensive duck farming. This automation in management practices greatly reduces workforce, ensures flock uniformity, reduces production cost and provides cleaner and healthier environment for the birds and workers. The increasing awareness of the health conscious consumers will also be properly addressed by the automation in intensive duck farming practices for the supply of hygienic duck products.

The level of automation with wireless multimedia sensor network technologies by combining different farm equipment into an internet enabled integrated and automated system is the recent advancements in intensive poultry and duck farming. This technology was demonstrated by several authors in poultry and integrated farming system; the same can also be applicable to intensive duck farming. Through the wireless sensor devices, feeding, watering and egg collection may be automated in duck farms and thereby reduce the workforce and production cost.

The most of the present feeding devices used in intensive duck farming are simple and less automated. An intelligent feeding system integrated with appropriate technologies like sensor detection, data processing and remote monitoring to accurately control the feeding quantity and monitor the feed consumption, now widely used in chicken production systems, can be applied in duck farming also to improve the production efficiency. In order to arrive at the requirement of feed in an intelligent feeding system, the parameters like age, type and stage of birds and the epidemic situation are analysed through the computer system. This system offsets the shortages of the traditional machine such as poor automation, low efficiency and inaccurate feeding. Besides, it is featured with lower cost, better feeding accuracy and working efficiency compared to manual feeding (Lijia et al. 2013).

The automatic feeding system for poultry can very well be applicable to intensive duck rearing. This automation in feeding should meet out the standards depending on feed distribution, age and type of birds. Easy access to the feed and avoidance of feed wastage are of great importance. Hence, this automation in feeding should be coupled with better conveying system.

Under the supervision of the Council of Agriculture (COA), Taiwanese researchers have developed the world's first-ever automated collection system of duck eggs with combined cloud computing with radio-frequency identification (RFID) technology to create a mechanism for harvesting and safekeeping fresh duck eggs. This is the smart, high-tech yet practical method of automation in duck egg collection in comparison to the old 'hands-on' method, which required farm helpers to spend 6 h per day in sorting alone, the automated system also produces cleaner eggs with unstained shells; a trait highly conducive to the food processing or the hatching process that follows.

The automated collection system of duck eggs is a pioneering achievement for the waterfowl industry. The Taiwan-developed technology is not only resource-friendly in terms of time and workforce, it is also well suited for further digital and wireless integration. Whether it is remotely accessing the system's software through the digital cloud or receiving real-time data through a mobile app in one's hands, this engineering feat is a solid case of tapping into smart technology to reinvigorate traditional farming industries.

7.5.10 Automation in Drinker System

The biological requirements of water to ducks can be satisfied by providing cup drinkers. Ducks are able to drink water from the cup and wash their bill and immerse their entire head to feel comfortable. These are important prerequisites to keep the ducks healthy and ideal for intensive broiler ducks. Furthermore, the ducks are able to carry out their natural behaviour such as dabbling, straining and preening. Combination of nipple drinkers may provide better results. In slat cum litter floor, drinker line should be installed above the plastic slats. Nipple drinkers with and without drip spark cup, round drinkers for open water supply and water connection with medicator are the various other automatic drinking systems, which can be very well used for intensive duck production.

7.6 Systems of Rearing

7.6.1 Pen and Run System

Deep litter system of rearing has a roofed house and an open run. Under this system, floor space allocation of 2 sq. ft. per duck in the pen area and 10 sq. ft. per duck in the run area should be provided till they attain 16 weeks of age. Local feeds may be prepared and provided to the birds to curtail the production cost. Cheaper vegetable feeds, household wastes and fodders as available under local conditions may replace one-third of the ration. Locally available litter materials can also be used but the use of litter material as that of chicken farming is limited, because of the wet litter problem. Concrete flooring with proper slopping to ensure draining of wastewater is advisable for intensive duck farming, since waterfowl drinks and excretes more water than chicken.

7.6.2 Deep Litter System

The ducks can also be maintained successfully in total intensive system on deep litter without provision for swimming (Fig. 7.2). In deep litter system of management watering of birds and litter management need special care. Since, ducks drink more water than other species of poultry; their dropping contains more than 90% moisture leads to wet litter problem. Extra care should be given to maintain litter floor in dry condition. Regular addition of fresh litter material, removal of old or wet litter and use of super phosphate may be followed for proper management under this system. In deep litter system, provision for proper ventilation inside the house should be planned to maintain better microenvironment inside the houses.

7.6.3 Mesh Bed and Cushion Grass Floor Models

Mesh bed or cushion made up of grass, rubber or PVC material may be used effectively on the concrete floor. This favours easy cleaning of floors, draining of wastewater and prevents moist microenvironment inside the duck house. Different sizes of mesh bed or cushion floors are available and can be used according to the size of the duck sheds.

Fig. 7.2 Deep litter system of management. (Photo courtesy: Big Dutchman)

7.6.4 Slat Floor System

Slats made up of PVC material plastic are being used for floor covering in duck houses (Fig. 7.3). Slats may be fixed 2–3 feet above the ground level for easy cleaning of the shed. The size of holes in the slats should be 0.8–1.2 cm for ducklings and it may increase in size as the age advances. The PVC slat materials should be tough, strong, resistant to acid, alkali and also resistant to corrosion. Different sizes and colours of slats are available in the market, it can also be custom prepared according to the need. Few breeder duck farms are having slat flooring for easy management and cleaning of the sheds. Slat flooring also favours feather hygiene and cleaner feathers than litter floor as described by Karcher et al. (2013). They reported physical condition of ducks remained good regardless of flooring system.

Advantages of Slats

- Easy to install, uninstall, clean and disinfect the floor
- Improves health and productivity of the bird
- Keeps the duck shed cleaner and drier
- Reduces cleaning time and labour cost
- Non-corrosive and resistant to acid and alkali
- Custom made available, hence, suitable for any dimensions of duck house and the panels can be easily modified and assembled

Fig. 7.3 Slatted floor for ducks. (Photo courtesy: Big Dutchman)

7.6.5 Wire Floor System

The wire floor designing may be of either all-wire mesh or a combination of litter and wire mesh. In the full wire mesh floors, mesh size of less than 1.5–2.0 cm is preferred for ducks. This wire mesh material may be of metal, plastic or plastic coated metal and should tough and strong enough to bear the weight of birds and equipment inside the sheds.

In combination of litter and wire mesh flooring, normally 2/3 of the floor area in lengthwise is covered with wire mesh and 1/3 with litter flooring. Waterers inside the shed are fixed on the wire mesh flooring and the feeders are placed on the litter flooring. This prevents wet litter problem and feed wastage. The combination of litter and wire mesh flooring is mainly practised for intensive rearing of breeder ducks. Whereas all-wire floors are not suitable for growers and laying breeder ducks (Scott and Dean 1991). Pure breeds with breeder duck farms and breeds of high productivity are being reared under intensive system of rearing, wherein automation in feeding and watering is practised as that of chicken.

7.6.6 Cage System

Ducklings and adult ducks may be reared under cage system as that of chicken. Colony cages are mostly used for brooding and rearing ducklings. In wire colony cages, the floor space of 1250 cm^2/bird should be provided. The standard pen size of colony cage for 10 ducklings is 1.25 m × 1.0 m. Generally ducklings raised in wire cages attain market weight very early than raised on ground due to confinement in a particular area. Grower and adult ducks can be reared in individual cages or in multiple bird cages under different tier system as practised in chickens. The advantage of cages is that less floor space is occupied. Cage system of rearing will also ensure better feeding efficiency, reduced wastage of feed, easy culling and medication. A study with Kuttanad ducks of Kerala reared in individual cages (size of 15″ front X 18″ depth X 21″ front height and 18″ rear height) resulted in increased productivity of ducks (Sankaralingam et al. 2017). They recorded duck day egg production of 224.80 and duck housed egg production of 218.72 up to 72 weeks period with better feed efficiency.

7.6.7 Environmentally Controlled Houses

This necessitates a completely enclosed house with no window. Common in most of the temperate countries, it is fully automated to have control over the microenvironment in the house by mechanical means/devices. The air speed, temperature and humidity are controlled mechanically. Automated ventilation system delivers fresh air and removes excess heat, moisture and undesirable gases from the duck house. A typical ventilation system consists of fans, air inlets, evaporative cooling system and controller/thermostats. Houses are designed to deal with both cold and hot weather extremes. Is it possible to indicate recommendations for air speed, temperature and humidity?

Advantages of Environmentally Controlled Houses

- More number of birds can be reared in a lesser space (floor space requirement is reduced compared to open-sided houses)
- Productivity of ducks is increased due to better management system
- Less chance of disease due to better microenvironment and better biosecurity
- Easy diagnosis and medication of ducks ensures better health status of birds
- Less labour required when compared to open shed
- Increased productivity of ducks results in increased profit

7.6.8 Space Requirements for Different Age Groups

7.6.8.1 Floor Space

Providing required and adequate floor space for ducks based on their stage of development is the basis for scientific management and successful duck farming. The floor space requirement at different age is given in Table 7.1. Overcrowding of ducks may lead to reduced growth, egg production and detrimental to their health, hence it should be avoided. However, overcrowding, is not usually a problem, but during cold weather condition, recommended stocking density will help in warmth through their metabolic heat when the ducks are confined in the shed.

If ducks are being reared under semi-intensive system, allow outdoors of at least twice as much space of indoors. Sufficient floor space is extremely important to

Table 7.1 Floor space requirement for ducks

Age (weeks)	Space (sq. ft.)/duck
2	0.60
4	1.50
6	2.25
8	2.50
Grower breeders	2.75
Laying breeders	3.00

obtain optimum growth with better economic performance, good welfare, better health and hygiene, particularly for the ducks above 17 days of age (Linden 2015). Even though, ducks are being reared under intensive system of rearing, welfare of ducks need not be compromised. Accordingly the stocking densities have to be planned. The maximum recommended stocking densities for ducks in full confinement houses are described by CSIRO (2002) as model code of practice for welfare of ducks is as follows:

- Duckling to 10 days: 50 birds/m^2
- Duckling to 8 weeks: 8 birds/m^2
- Breeders/adults: 5 birds/m^2

7.6.8.2 Type of Feeders and Space for Feeding

Different type of feeders used for other species of poultry, especially chicken, under intensive system, may be utilized for ducks satisfactorily. Special care is given to design the feeders for ducks to provide sufficient space for feeding by allowing the larger bill inside the feeders. Round type hanging feeders, trough feeders and feed hoppers are different types of feeders effectively used for ducks under intensive system. Since wet mash feeding is practised for ducks, automatic feeders or feeder hoppers are designed in such way that feed will move freely into the bottom of the feeder when it is consumed. Ducklings consume feed very frequently, like chicken, and it reduces as age advances. Ducks have efficient feed storage capacity as they grow older. Feeder space of 1 inch (2.5 cm) per duck up to 3 weeks of age has to be provided and gradually it may be reduced to half so long as no crowd at feed hoppers. For breeders in case of restricted feeding programme, sufficient linear feeding space of 4 inches (10 cm) has to be provided for uniformity and better production.

7.6.8.3 Waterers and Watering Space

Different types of waterers like trough, can or jar-type used for chicken and turkey are effectively utilized for watering the ducks, provided the watering space is sufficient enough to sub merge its bill. Minimum of 4 cm of drinking area is required for each duck. This requirement is also same for automatic water trough, cup or Plasson waterers. Nipple drinkers with cup are satisfactorily used for ducks by properly adjusting duck's height.

Minimum linear water space of 1 inch (2.5 cm) for starting and growing ducks and 2 inches (5.0 cm) for adult ducks is to be provided. In case of nipple drinkers, provide minimum of 15 nipples per 100 ducklings or growers and it should be increased to 20 nipples for 100 adult ducks. If nipple drinkers are used for ducklings, they should have access to water troughs until they learn to drink through nipple drinkers.

7.7 Duck House Equipment

7.7.1 Slat and Cages

Plastic slats and metal cages are being used for duck housing. Different sizes of slats and customized designs of slats are available in the market. According to the size of slats and dimension of cages, the housing specifications have to be planned. Slats and cages are fixed 3–4 feet above the ground level for easy cleaning and washing the floors. If it is multiple row cages, the working area between the rows of cages should be of concrete structure. For ducks plastic slat floor with big hole: 2.2 × 2 cm (length: 0.5 m * width: 0.5 m * thickness: 0.04 m) are available in the market and can be used effectively for duck house construction.

7.7.2 Indoor Mesh Bed and Thick Cushion Grass Floor

Duck house flooring may also be constructed with mesh bed or thick cushion grass models. These are made up of thick plastic, rubber or grass material with non-corrosive and resistant to acid and alkali. Custom designs of these models are available; accordingly, the duck house flooring could be planned. These models of flooring will favour easy cleaning and provide sufficient comfort to the birds.

7.7.3 Feeders and Waterers

Most of the feeders and waterers used for chicken are satisfactorily utilized for ducks also. In intensive rearing of ducks automatic feeding and watering equipment like circular feeder, augur feeder, watering channel with nipple and deep-drawn cup drinking systems are effectively utilized, since these automatic system facilitates improved productivity (Fig. 7.4). Wateres are placed on wire mesh floor instead litter area inside the duck house to reduce wet litter problem.

7.7.3.1 Deep-Drawn Cup Drinkers

Recently innovative cup drinkers for ducks are introduced. These drinkers not only provide the ducks the drinking water, but also satisfy characteristic behavioural needs (Fig. 7.5). The cup's special design allows the ducks to wash their bill and immerse their entire head and preen extensively in these drinkers. This activity keeps

Fig. 7.4 Automatic feeding and watering systems. (Photo courtesy: Big Dutchman)

Fig. 7.5 Deep-drawn cup drinkers. (Photo courtesy: Big Dutchman)

the nostrils of the ducks clean and eyes clear. The cup drinkers facilitates soaking of head and bill into the water, then duck are able to shake/spray water over the body without difficulty. This results in healthy and behaviourally satisfied birds leading to good growing performance. These drinkers have a special brim to return water with an overflow protection so as to avoid spilling of water and deterioration of litter quality. A mobile pendulum attached in the drinker ensures that the cup is constantly replenished with fresh water.

7.7.4 Handling and Transport Equipment

The equipment such as cages and crates made up of plastic used for handling and transportation chicken can satisfactorily be used for ducks also, but the size of these equipment should be large enough to hold the ducks. One crate with a size of 1.3 × 0.7 × 0.25 m can hold ten birds (Mousa-Balabel and Mohamed 2011). For transporting ducks to a longer distance, specialized vehicles with tier system is being utilized by the duck farmers (Veeramani 2013). Partitions are made in the truck by using wooden reapers or wire/slats with a height of 2–2.5 feet (0.61–0.76 m) to hold the ducks (Fig. 7.6). If provisions are made by using any ladder like materials for the ducks, they can climb into the vehicle without any assistance. Transport of ducks by unethical methods such as tying the legs and without proper transport equipment should be avoided.

7.7.5 Walls/Air Filters for Positive Pressure Houses

Under intensive system of rearing of ducks, houses have to be constructed with proper ventilation system. Normally, the automatic ventilation system is based on the principle of creating negative pressure inside the house and allows fresh air to enter from outside to reduce in-house temperature. To achieve this air inlets and exhaust fans are fitted along the side wall of the house. Exhaust fans are used to expel air from the shed and it creates negative pressure inside the house. This will also allow unfiltered air inside the duck houses through the cracks in the wall and gaps in the doors. Conversely, if the inlet air is filtered with HEPA filters, pushed and distributed through duct system with small exhaust fans for outlet, it will create positive pressure ventilation inside the house. The positive pressure ventilation house is effective in preventing unfiltered air from entering the house.

7.7.6 Tunnel Ventilation

In tunnel ventilation system, exhaust fans are installed at one end to expel the internal air which comes through the large openings located on the opposite end of the house. The outside air is drawn through the large openings, cooled at its entry by use of evaporative cooling pads or foggers. This cooled air reduces the temperature inside the house and makes the microenvironment conducive for improved performance of the ducks. The exhaust fans act as pumps to move the air through outlets. Pumping the air out of house leads to lowered air pressure inside the house, which in turn attracts the outside air through air inlets due to variation in the air pressure.

The major disadvantage of tunnel ventilation is extra investment and operation cost on exhaust fans compared to typical curtain-sided houses. This disadvantage may be offset by getting increased weight gain with reduction in mortality, especially during extremely hot weather. Cooling produced through air movement is

a

b

Fig. 7.6 Transport of ducks (**a**) Weld mesh cages for adult bit birds. (**b**) Trays for ducklings

commonly referred to as the 'wind chill effect'. Tunnel ventilation results in rapid air movement along the length of the house (Bucklin et al. 2018).

7.8 Management

7.8.1 Brooding Management of Ducklings

7.8.1.1 Extensive System

Under extensive system of management, where there is no provision for providing artificial heat, cold brooding of ducklings is widely practised. The tent like structures is made with the use of plastic sheets (Fig. 7.7) and the ducklings are allowed inside. The own metabolic heat helps in brooding of ducklings. Nomadic duck farmers are not practising artificial brooding by using electricity or coal, as practised in chicken (Veeramani 2013).

7.8.1.2 Intensive System

The scientific practices followed on brooding chicks can be applied to ducklings also. Brooding of ducklings may be carried out under deep litter or on wire floor or in battery cages. The brooding period of 3–4 weeks and 2–3 weeks is sufficient for layer and meat type ducklings, respectively. The brooding period may be extended 1–2 weeks during winter season than regular period. The required brooder temperature has to be maintained within the brooder. Under intensive system, bulbs, gas brooder or electrical brooder may be utilized as source of heat to maintain proper

Fig. 7.7 Tents for rearing ducklings (Veeramani 2013)

temperature inside the brooder. Hover type brooders are also used to maintain heat by providing 90–100 sq.cm per duckling as hover space.

Brooding temperature of 32 °C needs to be maintained inside the brooder during first week and it has to be reduced slowly by 3 °C per week until it reaches 24 °C at the time of fourth week. A 250-watt bulb is sufficient to brood 30–40 ducklings, under intensive system, other type brooders are also used. During brooding, the floor space of 1.0 sq. ft. is recommended for deep litter system and for wire floor, it is 0.5 sq. ft. per bird up to 3 weeks of age. Water level in the drinkers should be 5–7.5 cm deep; just sufficient to drink and not to dip themselves. The thickness of the litter recommended is 3 cm or more in order to absorb excess moisture in droppings of the ducklings. While releasing ducklings inside the brooder dipping their bills in the drinking water may sometimes be necessary to make them start drinking.

In case of floor brooding, clean dry litter materials like wood shavings, chopped straw and paddy husk may be used as bedding material. Brooder guards of sufficient size are used to confine the ducklings near heat, feed and water source, at the same time, it should be sufficient enough for the ducklings to move away from the heat source, if it is too warm inside the brooder. After 14 days of age, when outside room temperature is more than 70 °F (21.1 °C), ducklings may be allowed outdoors during part of the day.

7.8.1.3 Special Care for Ducklings in Intensive System

Under intensive system of management, ducklings are hatched artificially in incubators, as that of chicken. Newly hatched ducklings should be removed from the hatcher, when 95% of ducklings are dry to prevent dehydration. Day old ducklings are very sensitive and hence, they have to be handled carefully. For effective transport of ducklings from hatchery to the farm, fresh disposable boxes or properly disinfected, clean plastic boxes can be used. These boxes should give sufficient ventilation to ducklings to prevent huddling and heat stress. Before loading, the ducklings are kept in a holding room with the temperature of around 75 °F (23.9 °C) and relative humidity of 75% to prevent dehydration. In the same way, the transport vehicles should also provide adequate ventilation, temperature and humidity levels until the ducklings reach the farm.

Compared to newly hatched chicks, ducklings are at higher risk of dehydration and starvation during transport from hatchery to farm (Siregar and Farrell 1980). Therefore, the ducklings need to be monitored carefully for the signs of dehydration, heat stress and huddling while transport. The vehicles used for transport and equipment used for loading and unloading of ducklings should be properly disinfected before put into use.

7.8.2 Growing Management

Ducks with age group of 9–20 weeks are referred as growers, at this age they have to be maintained properly, since the reproductive organs undergo rapid growth and attain maturity during this grower phase. The grower ducks maintained with appropriate care will be metamorphosed into a good producer of eggs. Growers may be reared in separate grower houses or continued to be reared in brooder-cum-grower house. During grower stage, the floor space of 3 and 2.5 sq. ft. per bird is adequate under deep litter and slatted or wire flooring, respectively. Depth of water inside the drinkers should be 10–12 cm deep; this allows the ducks to immerse their head in the water. Feed them with grower mash up to 18–19 weeks of age and at during this stage, the daily feed consumption will be approximately 80–100 g. It is advisable to provide restricted feeding at this age to get better egg production during laying stage. Wet mash feeding is always preferable to ducks than dry mash feeding for better feed intake. Fresh, cool, potable water may be always made available to the birds. Follow proper litter management procedures. No artificial light is given during growing period. Vaccination against duck plague (duck viral enteritis) and duck viral hepatitis can be given during 8–9 and 16 weeks of age, respectively. Policy culling of poorly grown, injured and lame bird should be carried out regularly and to keep mortality level below 2–3% during growing phase for better productivity. Proper records on number of ducklings received, feed intake, body weight, mortality, culling, medicines and vaccines given, etc. must be maintained for analysing production efficiency and failures.

7.8.3 Layer Management

Ducks at the end of grower phase, when they are normally 18–19 weeks of age are shifted to layer houses. In deep litter system of management, floor space of 4 sq. ft. per bird is essential during laying. Feeding space of 10 cm is to be provided for wet and dry mash feeding and 7.5 cm for pellet feeding. Ad libitum feeding with good quality of duck layer mash has to be practised for better egg production. Normally, one nest box (30 × 30 × 45 cm dimension) is sufficient for three ducks during laying stage and it favours the production of clean and hatching eggs. Nest box with dimension of 40 × 40 × 45 cm is recommended for 3–5 meat type ducks. Nest boxes may be arranged at one side of the pen in a separate area so that females can be kept out of the nest area once all eggs have been laid. In order to maintain cleanliness in nest boxes fresh litter materials like straw or wood shavings may be regularly added. Lighting duration of 14–16 h is necessary for optimum egg production. Based on the availability of natural light source, the artificial lighting schedule may be planned accordingly. For every 200 sq. ft. area providing one 60-watt bulb gives adequate artificial light for laying ducks and the day length should never be decreased during laying period. Normally ducks are prolific layers and in most of the egg-laying breeds, the age at first egg and 50% egg production will be around 120 and 140 days, respectively.

It is important to look for the signs of sexual maturity in females, called nodding, which is a distinct forward and downward movement of the neck to the left and right side when the bird is walking through the house. In Pekin breed, egg production begins at 23 weeks of age. At about 25–26 weeks of age, about 10% egg production will be reached and peak egg production of 90% will be attained around 32 weeks of age (Klein-Hessling 2007).

Good quality clean, potable water is essential for better performance of ducks and it should be provided ad libitum to the ducks throughout the laying period. If it is slat cum litter floor, waterer or the water lines should be placed on the slat area to avoid water spillage on the litter and wet litter problem. The feeders and waterers must be arranged alternatively at equal distance. Their heights must be adjusted to avoid feed wastage and filled only to 2/3rd of their capacity at any time.

Collection of eggs: The ducks lay eggs during early morning unlike hens; which lay their eggs during early part of the day. Eggs have to be collected twice, in the morning itself; this facilitates lesser problems of dirty and cracked eggs. Plastic/cardboard egg trays especially designed to hold duck eggs may be used to collect eggs. Usually 30 egg capacity trays/filler flats, as that of collecting chicken eggs are used. In large intensive duck farms, cold storage is preferable to avoid quality deterioration of eggs.

7.8.3.1 Cage Rearing of Layers

Under intensive rearing, the management is easier for the ducks that are reared in cages. It requires less floor space. Egg collection is easier with clean eggs and better feed efficiency. In Europe, the use of individual cages for layers has been banned.

Cages of various sizes are available and it is ideal to house 3–5 ducks in a single cage. Conventional or reverse cages are used for housing ducks. According to the width of the duck layer shed, the individual cage size is determined. Fixing the longer side of the cage in front is preferable for easy feeding of ducks. Based on the height of the shed, the cages are organized into two or three rows one over the other in slanting arrangement on either side. This sort of arrangement of cages is referred as Californian cages. In cages, the floor space of 1 sq. ft. per duck is provided. The feeder and water channels are fixed on the front side of the cages. The preferable height of duck layer cage is 55 cm in front and 47.5 cm at the back with a slope of 1/16 towards front side to facilitate easy rolling of eggs. Sufficient gap should be provided under feed tray for the egg to roll out and reach front side of the cage. The weld mesh size of 2.5 × 5.0 cm size with 14-gauge thickness for floor and 7.5 × 7.5 cm size mesh of 16-gauge thickness for sides is sufficient. Water channel is fitted above the feeder; nipples drinkers, if used, are installed inside the cages. Cage rearing under controlled environment with unrestricted feeding favours increased egg production with better feed conversion efficiency in laying ducks (Avens et al. 1980).

Raised platform or high-rise or elevated cage house as that of chicken layer house can also be constructed for ducks. The cages are fitted at 5–7 feet above the floor level with walking platforms constructed between cage rows and sides. This arrangement widens the distance between birds and their droppings, which facilitate swift drying. The other advantages include less fly nuisance and ammonia problem, better ventilation, clean egg production and easy removal of droppings.

7.8.4 Breeder Management

Most of the breeder ducks are being maintained under slat cum litter floor or wire floor system with automatic feeding and watering facilities as that of chicken. Good quality breeder mash fortified with essential nutrients should be provided to the breeder ducks for the production of good quality hatching eggs. Feeding twice daily with restricted feeding facilitates better production and maintains flock uniformity. Other managements like lighting and watering management are practised as that of layer ducks. To stimulate adequate sexual response, light intensity of 10 lux (1-foot candle) at the duck's eye level is sufficient for both drakes and ducks. But, in practice, light intensity of 20–30 lux to be provided at duck level for breeder ducks.

7.8.4.1 Sexual Maturity

Even though the egg production starts by 120–140 days in egg-laying breeds of ducks, it is not desirable to bring ducks into egg production before 7 months of age because of problems with small egg size and low hatchability. Sexual maturity in ducks depends on breed and season. Egg type breeds attain sexual maturity earlier than meat type breeds. Hatching schedule is planned according to the season; so that ducks will sexually mature at proper age. This will facilitate better productivity of breeder ducks in terms of fertility and hatchability. Hatching season coincides with the April through July period which is appropriate for better productivity. Nest boxes can also be provided for breeder ducks if they are being maintained under litter floor. Individual nest box, unlike colony nest box in case of layer ducks, is provided for breeder hygiene. Normally nest box with dimension of 12 × 18 × 12 inches is sufficient for single breeder duck and in breeding farms trap nest can be used. Free access to feed and water in adequate points has to be provided to meet the daily nutrient requirements.

7.8.4.2 Fertility and Hatchability

Maintaining the proper sex ratio in the breeding flock is important to achieving high levels of fertility and hatchability. Sex ratio of 1:8–10 is optimum for light breeds and one male for every five females for heavy breeds for better fertility. Some extra males may be kept as replacement stock to compensate mortality. Normally fertility will increase rapidly during the first few settings of eggs, but will reduce as the age advances or at the end of the egg production cycle. Handling of the hatching eggs indicates the hatchability percentage, so the hatching eggs should be collected frequently, cleaning and storing of hatching eggs should be carried out properly to get maximum hatchability. Artificial insemination is not common in ducks except in case of mule duck production.

Reproductive performance of Orvia CKM Pekin duck breeder females through 42 weeks of lay (Klein-Hessling 2007).

Parameter	Performance result
Age at first egg (weeks)	: 23
Age at peak of lay (weeks)	: 32
Average duration of lay (weeks)	: 42
Total no. of eggs per female	: 230
Total no. of eggs per female (hen-housed)	: 220
Total no. of settable eggs per female	: 210
Average egg weight (g)	: 86
Average fertility (%)	: 94–96
Hatchability of fertile eggs (%)	: 85–87
No. of ducklings per female housed	: 165–175

7.8.5 Litter Materials and Litter Management

The litter materials such as paddy husk or straw, wood shavings, coconut coir and saw dust, used in chicken farming, can well be utilized effectively for rearing ducks under intensive system of management. Duck droppings contain more moisture than other species of poultry, it is necessary to provide extra care to keep the litter floor in dry condition. Regular addition of fresh litter material, replacing of wet or soiled litter with fresh new litter material and addition of disinfectant like super phosphate are the key points in better management of litter floor. If it is a soil floor, upper few inches of soil are removed and replaced with clean soil to keep duck house floor in a clean condition. Improved weight gain and better feed conversion ratio were observed in deep litter floor system than in cage system as observed by Sari et al. (2013). This result could be explained by the fact that feeding, lying, walking and object pecking activities were better in the deep litter floor system than in the cage system (Fouad et al. 2008). Stress is much higher in the cage system (Duncan et al. 1992).

7.8.6 Lighting Management

As that of chicken, the persistency of laying can be increased by providing proper lighting schedule. To achieve better productivity, combination of natural and artificial lighting regimen (14–17 h total light period per day) has to be provided for laying ducks for better continuous laying from 7 to 12 months. In a confined environment, artificial lighting is planned before sunrise or after sunset to maintain a constant light period of 16 h (and a dark period of 8 h) each day. Automation to switch on or off the lights through photo-sensors depending on the natural light or at a particular time using timers may also be planned. Fitting forty watt light bulbs at 8–10 feet (2.4–3 m) height, spaced 14 feet (4.3 m) apart will provide sufficient light (one foot candle light intensity at duck eye level) to stimulate egg production. For growing ducks artificial light is not required. However, totally confined growing ducks, to prevent stampeding, when the flock is disturbed, providing dim light with low wattage bulbs at night is beneficial. For ducklings, after brooding period, some supplemental light is recommended for at least the first 3 weeks.

7.8.7 Feeding and Watering Management

For ducklings during brooding period, feed can be provided on egg case flats or other rough paper for the first few days. After 2–3 weeks, ducklings can be fed with a balanced chick mash with more protein to improve the growth rate. Ad libitum feeding of ducklings should be followed for better growth performance. Wet mash feeding is being practised to avoid feed wastage as well as to overcome respiratory problems due to dry mash feeding. Less feed wastage and better feed efficiency are resulted by using crumbled or pelleted feeds. Since, mash feed forms sticky cake on the bill, ducks prefer pelleted or crumbled feed and a firm pellet size is important (Hall 2008).

7.8.7.1 Green Feed

For ducklings, providing greens along with feed fortified with essential vitamins and minerals increases bulkiness and reduces feed consumption which favours less possibility of cannibalism. During brooding stage, chopped greens are placed in the water trough, instead of placing on the floor, which will facilitate better feeding and avoid soiling of the greens. Moreover, the greens remain succulent and clean and the ducklings spend time in dabbling for the bits of greenery.

7.8.7.2 Creep Feeding

Creep feed is an exclusive feed for the young ones and creep feeder is designed specially to feed them. Sometimes ducklings are being maintained along with its parents, during that period it is preferable to feed the ducklings separately through creep feeder, where ducklings can freely eat without competing with adult birds. The creep feeder is intended to provide the duckling better access to the feed.

7.8.7.3 Feeding Laying Ducks

Layer rations should contain a higher level of calcium than other duck rations since additional calcium is required for eggshell formation. A level of 3% of the diet is adequate for most breeds of ducks including high egg producing breeds. When enough calcium is included in the ration, it is not necessary to feed oyster shells in addition. Some commercial duck producers have found it beneficial to restrict feed during growing stage in breeder ducks. The main advantages of food restriction in breeder ducks are lower lifetime feed intake, lower carcass fat content, delay in sexual maturity, higher egg production, lower mortality, higher fertility and hatchability (Olver 1995). Many breeder companies are adopting feed restriction during the rearing period; it may start between 4 and 6 weeks of age. Juvenile feed restriction has shown to increase egg production substantially and improved fertility (Klein-Hessling 2007). Breeders follow a pre-determined growth profile using a feeding regime. Emphasis is given on maintaining a high degree of flock uniformity. This requires that all ducks have the opportunity to eat simultaneously, hence sufficient feed and water space should be provided. Recommended feeder space allocations during rearing are 5 cm and 10 cm per duck, respectively, on circular type pans and on linear troughs, while one bell type drinker is sufficient for 100–200 ducks, depending on the stage and intensity of feed restriction. It is important to ensure a gradual release from feed restriction.

7.8.7.4 Watering of Ducks

Ducklings should have access to ad libitum drinking water, when they are started under the brooder. Optimum size of waterers should be used in such a way that the ducklings cannot get into the waterer, since ducklings are easily chilled when they become wet, which leads to mortality due to drowning. Pans or troughs with wire guards are satisfactory. Location of placing waterers is also important. Waterers are placed over low, wire-covered frames, the spilled water can drain under floor and also it helps reduce wet litter problems. Change waterers or adjust size as birds grow. The waterer should be wide and deep enough for a bird to dip its bill and head.

In ducks, the water intake is more in summer than in winter. Average water to feed ratio is 3.3:1 and 2.7:1 in summer and winter, respectively (Downing and Taylor 2010) with small differences between breeds and the sexes. They also stated that ducks prefer bell drinkers over nipple drinkers because it facilitates their behavioural activities like dabbling and head dipping. Siregar and Farrell (1980) reported a water to feed ratio of 4.1:1 and 2.3:1 in ducklings aged 5–22 and 11–23 days, respectively.

7.8.8 Bill and Claw Trimming

Beak trimming is a usual technique to prevent feather pecking and cannibalism in poultry. Among various breeds of ducks, feather pecking is more common in Muscovy ducks. Pecking generally begins as early as 13 days of age when the adult plumage begins to appear (Knierim et al. 2005). Bill trimming causes stress and pain to the birds. During trimming the upper bill is slightly shortened over lower and it is best practice to control pecking, especially in Muscovy ducks.

Bill trimming should be performed by the trained personnel and care should be taken to ensure least stress or discomfort to ducks. This may be carried out before 3 weeks of age and it can also be performed at hatchery by searing the nail of the upper bill with an electric beak trimmer. But bill trimming at early age is not ideal or advisable, because the regrowth of bill needs repetition and accuracy is also more difficult in duckling. During bill trimming, upper bill is cut at the mid-point of the nail (about ¼ inch is removed). Ducks are bill trimmed by cold cutting with scissors (Gustafson et al. 2007), hot-blade cutting with cautery, or tip searing which involves holding the end of bill against a cautery, blade for 2–3 s.

Prior to trimming, anti-stress medication along with vitamin K may be supplemented through drinking water to overcome the stress and to favour quick blood clotting. Proper feeding, watering and good ventilation are the main perquisites to prevent the stress during bill trimming. Due to stress, feed intake may be reduced, but will reach to normal within 24 h. Bill trimming by scissor and hot blade efficiently reduced aggressive pecking behaviour without affecting welfare, whereas trimming by hot blade significantly reduced body weight than trimming by scissor (Elshafaei et al. 2017).

In Muscovy ducks, the claws are very strong and sharp; they cause injury to other ducks. The claws can be trimmed to reduce scratching of pen mates or risk of injury to workers. This can be performed as early as 10 days of age or at the same time they are bill trimmed (Clauer and Skinner 2007). However, both claw and bill trimming when performed together cause less stress and more economical when compared to doing the operations separately in two different times. The claw should be trimmed close to the base but should not be trimmed too closely.

Clauer and Skinner (2007) stated that claw-induced injuries could be reduced by lowering stocking densities, providing environmental enrichment, improving the relationship between farm staff and the birds and more careful transportation in Muscovy ducks.

7.9 Behavioural Characters

Most ducks confine their displays to the water (or land) surface, since their heavy weight relative to their wing area ('high wing loading') dictates continuous flapping and makes complex manoeuvres, such as hovering and soaring, difficult or impossible. Aerial communication is thus largely restricted to short, ritualized flights (ordinarily close to the water surface) and vocalizations, including contact calls that help maintain flock coherence in these rapid fliers that often go long distances between landings (Ehrlich et al. 1988). The following are some of the displays or behavioural characters of ducks.

7.9.1 Special Characters of Ducks

Following are some of the special characters of ducks distinct from other poultry species:

Oily feathers: Prevents entering of water inside the feathers to reach the body and remains dry

Webbed foot: The front portion of three toes of legs interconnected with thin flexible skin referred as web. It helps the ducks for easy swimming by acting as a paddle of boat.

Bill: Strong and specially shaped to eat moss, insects, fish egg, hard snail, etc.

Time of lay: Ducks start egg laying at night or early in the morning unlike chicken. Therefore, the ducks are not allowed to go out from their house before 9 AM.

7.9.2 Broodiness

Broodiness is a condition, in which females stop laying and show tendency for nesting. It is considered as an ideal character for the propagation of species in wild conditions, but highly undesirable for domesticated species of ducks because of economic reasons. Most of the domestic breeds have lost their brooding instinct, almost entirely during the long process of domestication and selective breeding that took place for the development of utility duck breeds. However, some of the ducks particularly belong to heavy type of breeds, may sometime develop the urge to sit on eggs, especially Muscovy duck (Ye et al. 2019), but they are always found to be unreliable and abandon the eggs after a few days. Because of the unreliability of most breeds of domestic duck at sitting on their eggs and raising their young, and it has been the custom on farms for centuries to put duck eggs under a broody chicken for hatching.

7.9.3 Preening

Preening is a special behavioural display exerted by ducks and they preen especially after bathing. After swimming, ducks tend to perform of shaking movements to remove the water on their bodies. To maintain proper feathering conditions and the insulating property of its feathers, ducks do preening very carefully. The uropygial gland located at the base of the tail secretes oil and this oil is distributed evenly on the feathers by preening behaviour. This also favours interlocking of the feathers and helps in preventing the entry of water in the skin. Preening is often followed by sleeping for a short period.

Due to intensive rearing of ducks, due to availability of less or no water to swim, duck gets no opportunity to preen effectively. Even though the character of preening is bred-out due to intensive rearing, it will be quickly revert to this natural behaviour, when given chance to swim. Preening is essential for keeping plumage in good condition and waterproof. This is carried out by rubbing the bills with fine layer oil secreted by the uropygial gland at the base of the tail. Ducklings get this oil from their mother until they have developed their own glands. Providing open access to water increases the behavioural opportunities mainly preening in ducks as suggested by Rodenburg et al. (2005) at the same time it should not lead to hygiene or health problems. Whereas Rice et al. (2014) observed equal numbers of ducks preening regardless of association with water, and the flooring type had no effect on the amount of preening. Water lines do not allow for effective preening in Pekin ducks as suggested by Waitt et al. (2009).

7.9.4 Temperament

Ducks are not pugnacious in nature and rarely fight among themselves. They are quite calm in temperament except when incubating eggs or raising a flock. However, fighting may evoke when a new duck is introduced into an established flock and such fighting does not last for a long time and hardly results serious injury. High proportion of drakes in excess than optimum sex ratio leads to conflict or fight among themselves but is not of very serious problem. Drakes seldom fight among themselves when there is no female around for mating. Cannibalism and antagonistic behaviour is not usually encountered in a duck flock. Ducks are not aggressive towards humans also. It is most unlikely for a duck to inflict injury to even to children. However, one should be careful while approaching a broody incubating duck. The hostile posture of ducks, characterized by a pronounced uplifting of the bill, is seen when two males are chasing, fighting or threatening or when two males are competing for an attempt to mate a female. During the winter months, many species of wild ducks live together in the same land or water without being aggressive to each other. It could be due to low hormonal activity during that period. Similarly, the aggressiveness of domestic ducks is always low during the winter period. Thus, it is advantageous to introduce new ducks to an established flock during these months.

7.9.5 Social Behaviour

Ducks are social type of bird and gregarious in nature. Gregarious behaviour refers to the flocking or herding instinct. This behaviour arises out of social attachment and makes them stay together. In wild, they live in large flocks and forage together. Ducks gather in flocks mainly because it offers mutual protection from predators and easy exploitation of feed. Feeding together in flocks provides some safety from predators, particularly when searching for invertebrates below the surface of water, or when 'up ending' and searching for weed and crustaceans in the silt along the margins of lakes and pools. They communicate with each other by using voice and various types of actions. Nearly the same type of social behaviour is observed in the domesticated flocks. This social quality makes them suitable for keeping in a flock under captivity. As ducks live in flocks and coexist peacefully it is thus possible to keep ducks of different breeds together. It is because of their gregarious behaviour that makes ducks going out for scavenging to walk in single file, one behind the other.

7.9.6 Cannibalism

Ducks can develop the vice of cannibalism at any age but ducklings over 4 weeks old are more prone to develop cannibalism. Even though, the reason for cannibalism in ducks is clearly not known, but it may be associated with boredom and is aggravated by overcrowding, lack of ventilation and faulty nutrition. As that of debeaking in chicken, removal of the rim at the front of bird's upper bill is the only way to control cannibalism. It can be performed by using commercially available bill-trimming machines as that of automatic debeaker. Bill trimming should be performed only by a competent operator and only when it is essential to reduce damage and suffering in the flock. The reduction of light intensity inside the shed and use of coloured light source (red or blue bulbs) was found beneficial in controlling cannibalism (Rodenburg et al. 2005).

7.9.7 Duck Courtship Behaviour

The males will attract the females with their colourful plumage or feathers. Both sexes act out elaborate courtship behaviour. Domestic drakes are polygamous and not selective in their mating behaviour. They mate indiscriminately with the females in the flock. Hence, different varieties should not be mixed together particularly during the breeding season to maintain purity in each variety. In modern farming systems, the sex ratio of 1:8 to 1:10 for light breeds and 1:5 for heavy breeds are being maintained for better fertility and mating occurs throughout the year with the use of artificial lighting for 16–17 h per day. Ducks prefer to mate in water while swimming though they can also mate successfully on land. Mating on land is not much of a problem for light breeds kept for egg production. It is difficult for heavy meat type breeds to mate successfully on land because of their huge size. Following are some of courtship behaviours of ducks described by Barry (2015).

7.9.7.1 Head Pumping

This is a type of courtship behaviour in which males and females shake their heads in rhythm. It is quite often repeated to attract each other for mating and it also followed after mating (Fig. 7.8).

7.9.7.2 Head and Tail Up

During this behaviour, drake produces a loud whistle sound and pulls its wings and tail up exhibiting the purple-blue secondary feathers. This often happens as a quick gesture by males to attract the female in the group (Fig. 7.9).

Fig. 7.8 Head Pumping Behaviour

Fig. 7.9 Head and Tail Up Behaviour

Fig. 7.10 Grunt-Whistle Behaviour

Fig. 7.11 Nod-swimming Behaviour

7.9.7.3 Grunt-Whistle

It is a one-second behaviour exhibition by the group of drakes to draw attention of females. During this behaviour, male raise out of the water, elevates its head up and produces a sensational whistle followed by murmuring while coming back into a normal posture (Fig. 7.10).

7.9.7.4 Nod-swimming

Nod-swimming is a frequent ostentatious courtship behaviour exhibited by the ducks. During this display, both male and female swim swiftly for a short distance by keeping its neck low, just on the surface of the water (Fig. 7.11). Usually females express this behaviour as if it is interested in courtship and encouraging the nearby males to display this behaviour. Males exhibit this display happily after mating (Barry 2015).

7.9.8 Peck Order

The social type birds such as ducks tend to develop a social order. The order of rank and dominance in ducks is commonly referred as 'peck order'. Peck orders are useful in reducing conflicts in a group. Every duck house has a pecking order, which is established early on and then again, when a new duck is introduced to the flock. This ritual can be quite vicious but they will very quickly settle down. One should not interfere unless the new ducks are being denied food or water. If this does happen, it will be essential to put out an extra food and water supply away from the primary source. For this reason, it is essential to introduce new ducks slowly in the yard or range where there is ample space.

7.9.9 Herding or Flocking Behaviour of Ducks

Ducks tend to display flocking behaviour when they are in group, and foraging, similar to the herding behaviour in sheep. This social behaviour helps the birds to forage in a group and one bird will lead the others. This will help the duck handler to have close contact with the birds and control the duck flock by handling the duck, which is leading the whole flock.

7.10 Carrying and Handling of Ducks

Ducks can be carried safely to a lesser distance by holding both the wings or holding them under arm or holding top of their neck or by holding one wing and leg from the same side.

Ducks are mainly handled to check weight gain, medication, vaccination and sometimes to examine for diseases. They should be caught and handled by causing minimum stress to the ducks. The specially designed catching hook (Fig. 7.12) may be used for catching the ducks with minimal stress to the ducks. Ducks are caught around their neck by applying constant gentle pressure on the hook and should not caught by their legs, since it may lame the birds. Normally 1.5 m length of hook is convenient for all catchers and nets (like fishing net) attached to long handle works out satisfactorily for catching the ducks.

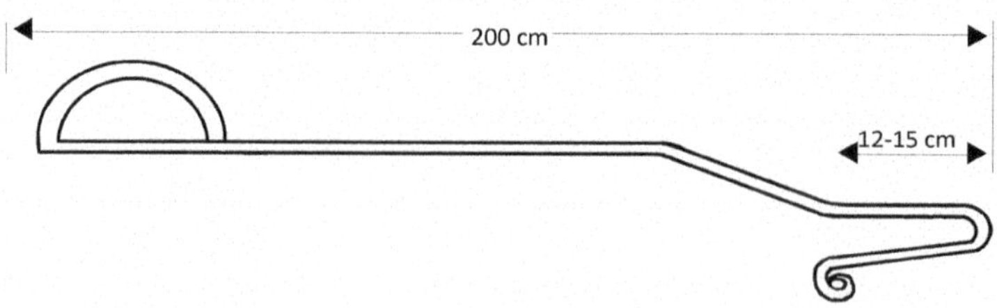

Fig. 7.12 Duck catching hook

7.11 Conclusion

Among different poultry, waterfowl especially duck rearing gained momentum during recent decades. Duck production is mainly concentrated in Asian countries and China is the major player in duck rearing. Even though, ducks are being reared under nomadic, semi-intensive system of management in several countries including China, intensive system of duck rearing attracts many entrepreneurs and local farmers to meet the growing demand and hygiene in production. Hence, the nomadic system of rearing ducks has been metamorphosed into intensive duck farming around the world, mainly in Asian countries.

Nomadic system of rearing is focused on indigenous duck germplasm available in local area, whereas the duck breeds like Pekin with improved productivity are mainly used for intensive rearing. The scientific management practices coupled with systematic genetic selection, proper feeding and advancement in refinement of environmental conditions paved way for the improved productivity.

Intensive rearing of ducks mainly concentrated on the increased productivity without compromising the hygiene in production. The intensive system of rearing underwent tremendous improvements in housing designs such as slat flooring, litter cum slat flooring, indoor mesh bed and cushion grass models in the recent past are best suited to meet the requirements of improved duck germplasm.

The new directions in automation of intensive poultry such as rearing under environmentally controlled sheds, automation in feeding, watering, disease monitoring with internet/web based technology, etc. largely cater the need for labour efficiency and product quality. These advancements in technology result in increased production per unit area with lowered production cost. The intensive rearing of ducks with the adoption of recent technologies and automation will complement the conventional production systems to cater the growing need for animal protein in the future.

References

Avens JS, Athearn RM, McNeal JH (1980) Egg production and efficiency of food conversion of Khaki Campbell ducks under different management systems. Br Poult Sci 21(5):333–337

Barry J (2015) How to recognize duck courtship displays. All about birds – Cornell University. https://www.allaboutbirds.org/news/what-to-watch-for-duck-courtship-video/. Accessed 09 June 2020

Bucklin RA, Jacob JP, Mather FB, Leary JD, Naas IA (2018) Tunnel ventilation of broiler houses IFAS Extension. University of Florida, Publication PS# 46. https://edisifasufledu/ps041. Accessed 20 Apr 2020

Clauer PJ, Skinner JL (2007) Raising waterfowl (Publication A3311). University of Wisconsin-Extension. https://www.canr.msu.edu/uploads/234/43487/livestock/raising_waterfowl.pdf. Accessed 20 Apr 2020

CSIRO (2002) SCARM report-83. Primary Industries Standing Committee – Model Code of Practice for the Welfare of the Animals – Domestic Poultry, 4th edn. CSIRO Publishing Australia

Downing JA, Taylor W (2010) The effect of strain and season on the performance of commercial ducks under Australian conditions. In: Proceedings of the Australian poultry science symposium, vol 10, pp 182–185

Duncan ET, Appleby MC, Hughes BO (1992) Effect of perches in laying cages on welfare and production of hens. Br Poult Sci 33:25–35

Ehrlich P, Dobkin DS, Wheye D (1988) The Birder's handbook. https://web.stanford.edu/group/stanfordbirds/text/essays/Visual_Displays.html. Accessed 20 Apr 2020

Elshafaei HE, Sharaf MM, Rashed RR, El-Kazaz SE (2017) Consequences of bill trimming on behavior welfare and performance of Muscovy ducks Alexandria. J Vet Sci 55(2):124–128

Erisir Z, Poyraz O, Onbasilar EE, Erdem E, Kandemir O (2009) Effect of different housing systems on growth and welfare of Pekin ducks. J Anim Vet Adv 8(2):235–239

Farghly MFA, Mahrose KM, Ullah Z, Rehman ZU, Ding C (2017) Influence of swimming time in alleviating the deleterious effects of hot summer on growing Muscovy duck performance. Poult Sci 96:3912–3919

Fouad MA, Razek AHA, Badawy EM (2008) Broilers welfare and economics under two management alternatives on commercial scale. Int J Poult Sci 7:1167–1173

Fraley SM, Fraley GS, Karcher DM, Makagon MM, Lilburn MS (2013) Influence of plastic slatted floors compared with pine shaving litter on Pekin duck condition during the summer months. Poult Sci 92:1706–1711

Ghanima MMA, El-Edel MA, Ashour EA, Abd El-Hack ME, Othman SI, Alwaili MA, Allam AA, Khafaga AF, Abd El-Aziz AH (2020) The influences of various housing systems on growth carcass traits meat quality immunity and oxidative stress of meat-type ducks. Animals (Basel) 10(3):410, p 11

Gustafson L, Cheng HW, Garner J, Pajor E, Mench J (2007) The effects of different bill-trimming methods on the well-being of Pekin ducks. Poult Sci 86:1831–1839

Hall AD (2008) Improvement in duck performance. Int Hatchery Pract 22(6):7–9

Hou S (2016) Duck production in China. In: Proceedings of XXV world's poultry congress- invited papers, 5–9 September 2016, Beijing, China, pp 209–212

Jones TA, Dawkins MS (2010) Environment and management factors affecting Pekin duck production and welfare on commercial farms in the UK. Br Poult Sci 51:12–21. https://doi.org/10.1080/00071660903421159

Karcher DM, Makagon MM, Fraley GS, Fraley SM, Lilburn MS (2013) Influence of raised plastic floors compared with pine shaving litter on environment and Pekin duck condition. Poult Sci 92: 583–590

Klein-Hessling H (2007) Pekin duck breeders require special management. Worlds Poult Sci J 23: 14–18

Knierim U, Bulheller M, Kuhnt K, Hartung J (2005) Minimum requirements for keeping Muscovy ducks (*Cairina moschata dom*). Final report of the research project #01HS039. Federal Office for Agriculture and Food, Bonn, Germany

Knust U, Wicke M, Pingel H, Lengerken G, Salomon V (1995) Muscle structure and meat quality of ducks reared on pasture and under intensive conditions respectively. In: Proceedings of XII European symposium on the quality of poultry meat, Zaragoza, Spain, pp 189–193

Lijia XU, grad Chaolin LU, grad Xianping ZU, grad Nan J, Xinbao W (2013) Research and development of intelligent feeding system. INMATEC Agric Eng 40(2):53–62

Linden J (2015) Recent advances in the production, management and nutrition of intensively-farmed domestic ducks. The Poultrysite. https://www.thepoultrysite.com/articles/recent-advances-in-the-production-management-and-nutrition-of-intensivelyfarmed-domestic-ducks. Accessed 20 Apr 2020

Mousa-Balabel TM, Mohamed RA (2011) Behavioral and physiological changes of Pekin ducks after transportation. Kafrelsheikh Vet Med J 9(2):19–44

Olver MD (1995) Effect of restricted feeding during the rearing period and a 'forced moult' at 40 weeks of production on the productivity of Pekin breeder ducks. Br Poult Sci 36(5):737–746

Rice M, Meelker A, Fraley SM, Fraley GS (2014) Characterization of Pekin duck drinking and preening behaviors and comparison when housed on raised plastic versus pine litter flooring. J Appl Poult Res 23:735–741

Rodenburg TB, Bracke MBM, Berk J, Cooper J, Faure JM, Guemene D, Guy G, Harlander A, Jones T, Knierim U, Kuhnt K, Pingel H, Reiter K, Serviere J, Ruis MAW (2005) Welfare of ducks in European duck husbandry systems. Worlds Poult Sci J 61(4):633–646

Sankaralingam S, Ashitha K, Neethu KP, Azhaguraja M, Srinivasan C (2017) Production performance of Kuttanad duck (*Anas platyrhynchos*) under cage system of rearing. Kerala Veterinary Science Congress, pp 210–213. ISSN 2394-255X

Sari M, Önk K, Isik S, Tilki M, Tufan T (2013) Effects of housing system, slaughter age, and sex on slaughter and carcass traits of Native Turkish Ducks. Turk J Vet Anim Sci 37:694–700

Scott ML, Dean WF (1991) Nutrition and management of ducks. ML Scott of Ithaca Publisher, New York

Siregar AP, Farrell DJ (1980) A comparison of the energy and nitrogen metabolism of fed ducklings and chickens. Br Poult Sci 61:213–227

Veeramani P (2013) Genetic characterization and phylogenetic analysis of indigenous ducks of Tamil Nadu. PhD thesis submitted to Tamil Nadu Veterinary and Animal Sciences University Chennai Tamil Nadu India, 177 p

Veeramani P, Prabakaran R, Sivaselvam SN, Sivakumar T, Karthickeyan SMK (2016) Social status of duck farmers in northern districts of Tamil Nadu, India. Indian Vet J 93(07):24–27

Waitt C, Jones T, Dawkins MS (2009) Behaviour synchrony and welfare of Pekin ducks in relation to water use. Appl Anim Behav Sci 121:184–189

Ye P, Li M, Liao W, Ge K, Jin S, Zhang C, Chen X, Geng Z (2019) Hypothalamic transcriptome analysis reveals the neuroendocrine mechanisms in controlling broodiness of Muscovy duck (*Cairina moschata*). PLoS One. https://doi.org/10.1371/journalpone.0207050. PMID: 31071089; PMCID: PMC6508920

Dr. K. Gajendran, Ph.D. (Poultry Science)

Dr. K. Gajendran, retired Director, Centre for Animal Production Studies, TANUVAS and presently heading the Suguna Institute of Poultry Management, Udumalpet, Tiruppur District, Tamil Nadu as the Principal of the Institute. He is native of Villupuram District of Tamil Nadu (India). He earned his Master's degree and Ph.D. in Poultry Science from Tamil Nadu Agricultural University. His research work on 'Characterization duck farming practices in Tamil Nadu' was the pioneer work in ducks. He served as Professor and Head of Livestock Research Station, Kattupakkam, the prestigious research institute of TANUVAS, for more than a decade. He has handled several external funded projects as Principal Investigator sponsored by national and international agencies like Institute Village Linkage Programme and National Agricultural Technology Project for 'Productivity Enhancement in Ducks' sponsored by ICAR, New Delhi, India. He has published more than 60 research publications in peer-reviewed reputed national and international journals. He received many national awards and international recognitions.

Dr. P. Veeramani, Ph.D. (Poultry Science)

Dr. P. Veeramani is native of Thirukoilur, Villupuram District of Tamil Nadu (India). He earned his Master's degree and Ph.D. in Poultry Science from TANUVAS, Chennai. He started his academic career from Kerala Veterinary and Animal Sciences University and presently working as Associate Professor at Madras Veterinary College, Chennai. He has completed the genetic characterization and phylogenetic analysis of native ducks of Tamil Nadu as his Ph.D. research work. Based on this work proposal was sent for breed registration of 'Arni Ducks of Tamil Nadu' at National Bureau of Animal Genetic Resources (NBAGR), Karnal, India. He has wide experience in duck farming. He has participated in many national conferences and received medals for his presentation. He has published more than 30 peer-reviewed research publications in national and international journals and also editor for some of the reputed journals.

Feeding and Nutrient Requirements of Ducks

8

A. B. Mandal

Contents

A. B. Mandal (✉)
Division of Avian Nutrition & Feed Technology, ICAR-Central Avian Research Institute, Izatnagar, Uttar Pradesh, India

A. Jalaludeen et al. (eds.), *Duck Production and Management Strategies*,
https://doi.org/10.1007/978-981-16-6100-6_8

Abstract

Ducks, when raised under grazing system, search their feed by scavenging. A combination of feed sources of floral and faunal origin along with fallen grains provides a near balanced feed for the ducks. However, when reared under semi-intensive and intensive farming systems, scientific feeding and nutrition are very important. Ducks have very fast growth rate during early juvenile life compared to chicken. Feed or energy restriction is followed during growing period to produce lean carcass with more breast and leg muscles, as they have tendency to deposit body fat. The factors like growth rate, composition of body, and production status determine nutrients requirements. There is a scope to reduce dietary protein through supplementation of critical amino acids lysine and methionine. Ducks are more susceptible poultry species to aflatoxins. The poor quality and quantity of protein and energy deficiency exacerbates the deleterious effect of aflatoxins. Although ducks can be fed dry or wet mash or pellets, they prefer wet mash due to difficulties in swallowing. The ducks are having habit of eating and drinking alternatively. The droppings of ducks are watery; therefore provision of water and feed must be in tandem to manage wet litter conditions. Ducks are sensitive to imbalances of nicotinic acid and available phosphorous in the feed. They are also sensitive to sudden changes in their diets and the quality of drinking water.

Keywords

Duck nutrition · Feed additives · Feeding of ducks · Nutrient requirements

Abbreviations

ADP	Adenosine diphosphate
AMP	Adenosine monophosphate
ATP	Adenosine triphosphate
CFU	Colony forming unit
CP	Crude protein
DNA	Deoxyribonucleic acid
FAD	Flavin adenine dinucleotide
FCR	Feed conversion ratio
FMN	Flavin mononucleotide
IU	International unit
ME	Metabolizable energy

MJ	Mega joule
NAD	Nicotinamide adenine dinucleotide
NADP	Nicotinamide adenine dinucleotide phosphate
RNA	Ribonucleic acid

8.1 Introduction

Wild ducks have long been hunted for food and sport in many countries. Thereafter, several species of ducks have been domesticated for eggs, meat, and feathers. Ducks have been the second major alternate birds after chicken for meat and egg production and have also become popular in certain parts of Asia. The availability of huge amount of snails, oysters, small fishes, algae, fungi, waterweeds, earth warms, etc. favoured domestication of the ducks for the first time in the world. There are certain advantages of farming ducks over chickens. Duck farming is easier since ducks need little attention and less rearing space; feeding procedures are also simple. Duck farming is suitable in places where there are plenty of ponds, lakes, backwaters, coastal areas, and paddy fields. In developing countries like India, ducks are mainly kept by the people belonging to lower strata of the society as a source of income but presently the scenario has changed. Even large farmers are adopting the duck farming as an alternative readymade source of animal protein. Social inhibition also exists in rearing chickens, but no such inhibition exists in duck farming. The ducks have natural tendency for foraging hence they are widely reared in rice growing area. They lay 40–50 eggs more than native chicken when reared under backyard. The duck egg is also heavier by 10–20 g and has fishy smell that has led to fetch more prices in some places, while in some parts that limits its consumption. Ducks generally lay during second year (even during third year) without considerable loss of production, whereas in case of chicken the peak egg production period is only 1 year. They do not require any elaborate housing like chicken and management practices are easy. Ducks lay 95–98% of their eggs before 9 AM thus egg collection is easier. Ducks are more hardy, easily brooded, and resistant to common avian diseases. Ducks move far for grazing/scavenging from the house of owner but they come back to their nest at evening without any mistake. Cannibalism and agnostic behaviour, very common in chicken, are not normally encountered in ducks.

Although little care is given on nutrition for ducks being reared in backyard production system, but nutrition plays an important role in semi-intensive and intensive duck farming systems. Therefore, to understand and apply principles of nutrition in practical feed formulation and feeding of ducks, it is necessary to have knowledge on comparative digestive system development of ducks compared to other poultry species, feeds and its constituents, nutrient requirements, and special aspects of duck feeding including feeding habits and precautions to be taken.

8.2 Comparative Digestive System Development

The digestive system in poultry consists of the alimentary tract starting from mouth to cloaca through which the ingesta passes and undigested feeds are eliminated, salivary glands, liver, and pancreas. The nutrients consumed are digested into their smaller units suitable for absorption; absorbed into the blood stream in the digestive tract. Salivary glands, liver, and pancreas help the process of digestion. Moreover, liver plays a significant role in nutrient metabolism, nutrient storage, and many other functions. Similarly, the function of the pancreas is the production of digestive enzymes as well as hormones. The digestive tract is more or less similar in different poultry species. However, ducks and geese have glands in the crop mucous membranes. In ducks, the small intestine had lower weight per unit metabolic body weight than broiler chickens (Jamroz et al. 2002a, b) but higher than turkey (Applegate et al. 2005). The digestive system in different poultry species develops at different rates. The rapid digestive tract development is associated with higher growth rate. The intestinal length or intestinal surface area to body weight ratio is relatively more in the first week of life, decreasing with age in chickens (Soriano et al. 1993) and in Pekin ducks (King et al. 2000). The small intestine is 3.7 times heavier and 1.6 times longer in 7-day-old ducklings than turkey poults of the same age (Konarkowski 2006). The morphological and functional growth of the digestive tract in Pekin ducks ceases after 7 weeks of age (Watkins et al. 2004). The factors affecting growth of digestive tract are body size, species, breed, age, sex, and physiological status of the birds, quantity and quality of ingested food (Gille et al. 1999). Genetic background of the ducks did not alter significantly the length of intestine and its segments, the diameter of different intestinal segments, and the weight and proportion of the gizzard, liver, heart, and spleen (Wasilewski et al. 2015). The avian caeca was reviewed by Clench and Mathias (1995). The caeca of ducks are generally moderate to long (4–38 cm). The caecal length depends upon the fibre content in feed. Anti-peristaltic motility of the colon moves fluid from the cloaca through the colon and into the caeca but not to the small intestine. This colonic antiperistalsis exists in ducks and geese. Anaerobic bacteria present in caeca (less than 10^{10} per gram weight mass) are responsible for caecal fermentation, yielding short chain fatty acids and certain vitamins. The short chain fatty acids provide energy to the birds. Moreover, propionic and butyric acids also stimulate growth and healing of intestinal epithelium. The ducks produce wetter droppings than the chicken. Again with the increase in environmental temperature, the water content in droppings is also increased. Caeca also helps in maintaining water balance in the body.

8.3 Principles of Nutrition

Nutrients are the chemical substances that are required by the living organisms to support and maintain life. These are water, carbohydrates, proteins, fats, minerals, and vitamins. The inorganic nutrients are water and minerals, while the organic nutrients are carbohydrates, proteins, lipids or fats, and vitamins. These nutrients are present in different feedstuffs in variable quantity. Therefore, nutrient requirements are better met through a mixture of feedstuffs and certain supplements, i.e. feed. All the nutrients are essential for growth, egg production, and good health of the ducks. Some nutrients especially water-soluble vitamins are synthesized in minor quantity in the lower part of digestive tract by the microorganisms (mainly bacteria) normally present there.

8.3.1 Energy

Energy is not nutrient, but it is the effect of carbohydrates, fat, and protein present in feed. In feed, energy is present in the form of chemical energy. When feed is consumed, these nutrients (carbohydrates, fat and protein) are digested to smaller units, absorbed and transported to different organs. The units of these nutrients (amino acids from proteins, glucose, and other single unit sugars from carbohydrates, and glycerol, fatty acids, and monoglycerides/diglycerides from fat) undergo various chemical reactions in the body and are either utilized for the synthesis of proteins, glycogen, and fat and/or broken down further to carbon dioxide, water, and ammonia to liberate energy. The released energy is required to maintain all biological activity (movement, walking, heartbeat, respiration, panting, etc.), vital processes (consumption, digestion, absorption, transportation, etc.), and chemical reactions occurring in the body for synthesis of proteins, fats, and glycogen, formation of egg, etc. The energy is also deposited in the body in the form of protein as structural component, and fat and glycogen as readymade available source of energy whenever required for vital activities and processes and chemical reactions. Thus, the gain in body weight is the gain of energy and egg production is conversion of dietary energy to egg energy. The dietary energy concentration influences feed intake and performance of ducks. Feed intake is increased with the decrease in dietary energy concentration up to a certain extent and vice versa. Thus dietary energy forms base for setting requirements of various nutrients for practical feed formulation. Increased dietary energy also improves the feed conversion efficiency. A positive correlation was also observed between dietary energy concentrations (MJ/kg diet, x) and FCR (Cherry and Morris 2008) in ducklings as follows:

$$\text{FCR} = 6.203 - 0.549x + 0.018x^2$$

Ducks have tendency to deposit body fat. Feed or energy intake is restricted during growing period to have lean carcass with more breast and leg muscles. However, feed restriction reduces growth.

8.3.2 Water

Water is an essential constituent of animal body. The water content is about 70% at hatch, 60–68% at 6 weeks of age, and as low as 52–57% when adult. The water content of body decreases with age and fat deposition in the body. Water plays a significant role in the body. It is an essential constituent of all tissues and cells; gives the rigidity and shape to the body. It helps to maintain body temperature by absorption, distribution, and elimination of heat, in digestion, absorption, transportation of nutrients and chemical reactions because of its solvent and ionizing properties. It is required for excretion of waste products and helps in respiration. It is an essential constituent of egg, and one egg of 65 g contains about 44 g of water. One gram of meat protein can hold 3 g of water and provides gain of about 4 g in live weight. Whatever the source of drinking water, it should be clean and free from salts, hardness, clay, and germs. Temperature of water is also an important factor affecting water intake. Drinking of hot water (water temperature above 30 °C) reduces egg weight, shell weight and shell thickness, and egg production. Warmer water is also an ideal environment for bacteria to multiply. Otherwise, the ducks require more water than chickens each day.

8.3.3 Carbohydrates

Carbohydrates constitute a major portion (70–75%) of dry matter of duck feed and serve as a major source of energy. They also provide bulk to the diet required for filling of the gut (hunger satisfaction) and normal movement of gastro-intestinal tract. Structural carbohydrates are hygroscopic and, therefore, facilitate easy passage of faeces (laxative effect). Glucose and glycogen supply energy required for synthesis of tissues. Some monosaccharides are involved in energy transformations and tissue synthesis. Ribose is an essential component of adnenosine triphosphate (ATP), adnenosine diphosphate (ADP), and adnenosine monophosphate (AMP) which is required for energy transformation and being a component of RNA it helps in protein or tissue synthesis. Proteins and fat metabolism require carbohydrates.

8.3.4 Protein and Amino Acids

The body contains 16–19% protein. However, when expressed on fat-free basis, the body contains more or less constant protein (21.6%). Proteins and amino acids are essential structural constituent of living organisms. Proteins play important role in body structural functions, e.g. collagen and elastin are the components of connective tissues; and glycoproteins and lipoproteins are the components of cell membrane. Protein is also the constituent of feathers.

Proteins also play important role in muscle contraction and transportation of nutrients and oxygen, e.g. haemoglobin and myoglobin transport oxygen in blood and muscle respectively; serum albumin transports fatty acids and transferrin transports iron. Proteins have buffering capacity and help to regulate acid-base balance in body. All enzymes are protein in nature and these enzymes play key role as catalyst in chemical reactions. Enzymes are involved in metabolism (breakdown and synthesis) of nutrients. Some hormones are proteins and act as chemical regulator of body activities. Some important protein hormones are insulin (regulates blood-glucose level), growth hormone (regulates body growth), parathyroid hormone and calcitonin (regulates blood-calcium level), and gonadotrophic hormone (stimulates primary sex organs). Antibodies are proteins, which perform vital role in protecting animals from specific diseases. Proteins are involved in production. The growth (production of meat) is a function of protein deposition. The egg contains many proteins (albumen, globulin, ovomucin, avidin, etc.). One duck egg of 70 g contains 8.4 g of protein. Prothrombin is a protein required for clotting of blood and opsin is a protein involved in dim-light vision.

The ducks require all the 20 amino acids for protein synthesis and other biological functions. However, some of the amino acids are synthesized in the body, but there are few amino acids, which are either not synthesized or synthesized at a rate in the body not sufficient to meet the requirement of the birds for normal growth and production. Essential amino acids are those that are not synthesized in the animal body at a rate required for normal growth and other production functions, hence must be supplied through diet. These are histidine, isoleucine, leucine, lysine, methionine, arginine, phenylalanine, threonine, tyrosine, and valine. In addition glycine and proline are also essential for rapidly growing birds. The limiting amino acids are those dietary essential amino acids, which are usually deficient in diet/feed. For example, in maize-soybean meal diets, the first, second, and third limiting amino acids are methionine, lysine, and threonine, respectively. The most limiting amino acid in maize-fish meal based diet is tryptophan. Both lysine and methionine are most limiting in maize-ground nut cake based diet. Lysine is limiting in rapeseed meal, sesame meal, and maize gluten meal.

The quality of protein depends upon the composition/assortment of essential amino acids. A protein can be said of ideal quality if it supplies all the essential amino acids in appropriate quantity and proportion conforming nearer to the amino acid composition of body plus excess nitrogen in the form of non-essential amino acids for the synthesis of other non-essential amino acids required for the synthesis of body protein. The protein quality can be improved by using more than one protein source and supplementation of synthetic amino acids which is (are) deficient or limiting. Proteins are digested to amino acids that are absorbed and utilized for synthesis of body protein or egg protein. Poultry species are unable to synthesize some of amino acids. Therefore, the birds should be supplied with all the essential amino acids in appropriate quantity and non-essential amino acids to meet their requirements.

In growing ducks partial deficiency of protein or one or more amino acid(s) may result in poor growth, which is directly proportional to the degree of deficiency. Protein deficiency may also be caused due to excess of energy through reduced intake of protein. A protein deficiency causes an increased deposition of fat in body tissue. Severe deficiency of protein or even of a single amino acid results in immediate cessation of growth and striking weight loss. Deficiency of some of the amino acids may lead to symptoms like feather pecking, cannibalism, ruffled feathers (due to arginine deficiency), and perosis or slipping of gastrocnemius tendon (due to deficiency of sulphur containing amino acids like methionine). The protein sources are classified into animal and vegetable protein supplements depending upon origin. The animal protein sources include fishmeal, meat meal, meat-cum-bone meal, blood meal, etc. The oil cakes and meals such as soybean meal, groundnut cake, rapeseed meal, sunflower seed meal, maize gluten meal, sesame meal, etc. are some of the common vegetable sources of protein.

8.3.5 Lipids

Lipids are the group of substances, which are insoluble in water but soluble in organic solvents such as ether, chloroform, or benzene. Fat content in body depends upon the age and nutritional status of the bird. Essential fatty acids are not synthesized in the body and thus are dietary essential, i.e., must be supplied through diet. The essential fatty acids are linoleic and arachidonic acids in poultry. However, if sufficient linoleic acid is available, the later acid is synthesized in the body. Symptoms of essential fatty acid deficiency include retarded growth, loss of feather, scaly skin, diarrhoea in young chicken, reduced egg size, reduced egg production, and marked decrease in hatchability in layers/breeders. Fats are rich sources of energy. Fats provide 2.25 times more available energy than proteins or carbohydrates, thus fats/oils are used in diet to increase energy level of feed. Oils are the richer sources of essential fatty acids. Fats increase the palatability and reduce the dustiness of feeds, and reduce wear and tear of feed-processing machineries. Fats are the carrier of fat-soluble vitamins and their precursors and are essential for the absorption of fat-soluble vitamins. The adipose (fatty) tissues of animals are the storage houses of energy and help in storing the fat-soluble vitamins. Fats decrease the heat increment (wasteful heat production) in the body. Phospholipids are the constituent of lipoprotein complexes of cell membrane. Glycolipids (cerebrocides) are constituent of brain and nerve fibres. Cholesterol is an essential constituent of cells, and 7- dehydrocholesterol, derived from cholesterol, is the precursor of vitamin D_3. Bile acids are also synthesized in liver from cholesterol. The steroid hormones (glucocorticoids, oestrogen, etc.) are synthesized from cholesterol. The essential fatty acid, arachidonic acid, is the precursor of prostaglandins.

8.3.6 Vitamins

Vitamins are organic compounds present in food in minute quantities, necessary for the birds in minute quantities for growth and maintenance of life through diet and must be supplied through the diet since the body is unable to synthesize them. The vitamins are regulatory substances, perform specific function and on deficiency produce certain disorder. The diseases or syndromes are specific for each vitamin. Vitamins are classified depending upon their solubility in different solvents, viz., fat or fat solvents and water. The fat-soluble vitamins include vitamins A, D, E, and K, while the water-soluble vitamins are all the members of vitamin B complex (thiamin, riboflavin, pantothenic acid, pyridoxine, folacine, niacin/ nicotinamide, biotin, cyanocobalamin, choline, etc.) and vitamin C. However, Vitamin C is synthesized in the body of poultry birds. The water-soluble vitamins are distributed in every living tissues and are not stored in body, except vitamin B_{12}, thus require daily supply through diet. Fat-soluble vitamins are stored if they are adequately supplied through diets. The fat-soluble vitamins are involved in metabolism of body structural components like vitamin A in epithelium and eye, vitamin D in bones, vitamin E in connective tissues, and vitamin K in blood. Water-soluble vitamins are components of coenzymes involved in nutrient metabolism.

8.3.6.1 Vitamin A

Vitamin A exists as such in animals and animal products, while in plant kingdom it is present as carotene, which is converted to vitamin A in gut epithelium. There are three forms of vitamin A, viz., retinol, retinal, and retinoic acid. First two are active and functional. Retinoic acid helps in maintenance of the body and promotes growth in chicken but ineffective in preventing night blindness or supporting normal reproduction. Carotenoids, the precursors of vitamin A, are present abundantly in leaves in association with chlorophyll and lower concentration in roots, seeds, fruits, and flowers. The different carotenoid compounds include alpha, beta, and gamma-carotenes and xanthophylls or oxygenated carotenoids. Carotene with at least one beta-ionone ring is converted to retinal. Only one molecule of vitamin A is produced from each of one mole of alpha and gamma-carotene, while two from beta –carotene. Other pigments, zeaxanthin and xanthophylls, have no vitamin A activity and the vitamin A activity is expressed in international units (IU). One IU is equivalent to 0.3 μg crystalline all-trans retinal or 0.344 μg all-trans retinal acetate or 0.55 μg of retinyl palmitate or 1.0 μg of beta-carotene in poultry. Vitamin A is essential for the formation of rhodopsin (visual purple) needed for vision in dark. Retinoic acid has no role in vision. Deficiency causes structural damage to the eyes. Retinoic acid cannot support reproduction. Its deficiency in males leads to failure of spermatogenesis. The damage can, however, be reversed in both the cases by retinal. Drying of germinal epithelium in laying birds and testicular degeneration in male receiving vitamin A deficient diet takes place. Eggs contain 200–1500 mg retinyl esters per kg mass (shell-free). Breeding hens must get sufficient quantity of vitamin A to get its adequate concentration in eggs for hatching. Vitamin A helps in synthesis of mucopolysaccharides. In its absence, keratinization of epithelial cells takes place that affects respiratory tract, alimentary tract, urinary tract, reproductive tract, and corneal epithelium. Vitamin A controls the activity of osteoclasts and osteoblasts of the epithelial cartilage and thus plays role in the development of bones. Vitamin A is involved in growth and differentiation Carotenoids and xanthophylls impart yellow colour of egg yolk. Foraging birds like geese and ducks are less prone to vitamin A deficiency as they take good amount of green. The deficiency signs are delayed growth, weak and emaciated with ruffled plumage, staggering gait, ataxia, droopiness, nervousness, water discharge from eyes and nose (also in chicks hatched from vitamin A deficient eggs), appearance of cheesy material under the eyelids, xerophthalmia or dry eye (all young one may not show), corneal hyperkeratosis, deposits under vent, nodular lesion (hyperkeratosis) of the exits of mucous glands of oesophagus and mouth. Ducklings exhibit a nasal discharge and may be paralysed later. Laying birds show symptoms of vitamin A deficiency slowly, but eye disorder is more acute along with decreased egg production, and increased incidence of blood spot in eggs. Hatchability is decreased due to abnormal development of circulatory system and skeletal system.

8.3.6.2 Vitamin D

Vitamin D is the collective name for a family of compounds, which have antirachitic activity. There are two forms of vitamin D, ergocalciferol (D_2) and cholecalciferol (D_3). Ergocalciferol is produced from ergosterol, which is present in plants, and cholecalciferol is derived from 7-dehydrocholestrol found in animals and birds. Green and fresh plant has no vitamin activity. One IU of vitamin D activity is equivalent to 0.025 mg of crystalline vitamin D_3 in poultry (or 1 mg = 40 IU). Vitamin D is involved in metabolism and utilization of calcium and phosphorus required for formation of bones, beaks, claws, and eggshell. It reduces oxidation of citric acid and stimulates the activity of enzymes phytase which hydrolyze phytic acid, yielding inorganic phosphate. Vitamin D stimulates incorporation of phosphorus into phospholipids and promotes absorption of Zn, Mg, Fe, and other elements. The birds exposed to sunlight generally do not exhibit signs of vitamin D deficiency. However, under intensive production system deficiency is more common.

8.3.6.3 Vitamin E

Compounds with vitamin E activity are called tocopherols (Tokos - child birth and Phero-to bear) and toco-trienols. Alpha–tocopherol is the most active form. One IU of vitamin E activity is equivalent to 1 mg of synthetic dl-tocopherol. Vitamin E is a potent anti-oxidant and functions at least in part on protecting cell membrane as well as vitamin A and polyunsaturated fatty acids (cell membrane components) from oxidation. It prevents free radical induced oxidation damage by trapping reactive radicals. It protects lipid layer of mitochondria from oxidation and aids in phosphorylation of creatine phosphate and ATP. Vitamin E has great role in reproduction. It protects the embryo against free radical generating stress conditions. It is essential for normal hatchability under stress condition. An increased level of dietary vitamin E is associated with enhanced level in egg. There is no adverse effect on hatchability in laying ducks due to deficiency. Green grass, lucerne meal, vegetable oils, wheat germ, synthetic vitamin E, rice bran oil, sunflower oil, and safflower oil are excellent sources of vitamin E (> 300 mg per kg), whereas alfalfa, orchard grass, distillers' dried grains, rice, soybean, wheat germ, etc. are fairly good sources (30–300 mg per kg) of vitamin E.

8.3.6.4 Vitamin K

Vitamin K is essential for the synthesis of prothrombin (factor II), a precursor of thrombin, needed for normal blood coagulation. Vitamin K is also involved in synthesis of factor VII (tissue thromboplastin component), factor IX (plasma thromboplastin component), and factor X (Stuart factor). Normally, vitamin K is not deficient in duck mash. Green forages or leafy materials (spinach, cauliflower, cabbage), soybeans, wheat bran, liver meal, fishmeal, and egg yolk are the good sources of vitamin K. The commercial source is Vitamin K_3 IP or menadione.

8.3.6.5 Thiamine

Thiamine (Aneurin, vitamin B_1) is required for energy utilization and prevention of stargazing or polyneuritis condition. Thiamine functions in all cells as the coenzyme cocarboxylase or thiamine pyrophosphate (TPP) and lipothiamide pyrophosphate (LTPP). TPP is involved in removal of carbon from keto acids. When thiamine is deficient, pyruvic and lactic acids accumulate in blood and tissues that give rise to deficiency symptoms. Thiamine is involved in the synthesis of acetylcholine that transmits nerve impulse. Thiamine deficiency is associated with accumulation of pyruvate and lactate in blood, reduced synthesis of fatty acid and energy metabolism. Yeast, wheat germ, rice polish, whole grain and their by-products, lucerne meal and liver meal are the natural sources of thiamine. Synthetic vitamin B_1 is aneurin hydrochloride or thiamine hydrochloride.

8.3.6.6 Riboflavin

Riboflavin (vitamin B_2) is required to promote growth and increase hatchability. Riboflavin, in the form of flavin mononucleotide (FMN) and flavin adenine dinucleotide (FAD), acts as coenzymes of several enzymes involved in biological oxidation–reduction reactions. They help in regulation of cellular metabolism and primarily involved in carbohydrate metabolism. However, they also help in protein and fat metabolism. Liver meal, yeast, rice bran, molasses, fishmeal, lucerne, and milk products are natural sources of this vitamin. Yeast is the richest natural source (125 μg/ g), whereas leaves and leafy forages are also good sources of riboflavin.

8.3.6.7 Niacin

Niacin functions as a constituent of two important coenzymes (NAD and NADP) responsible for transfer of hydrogen from substrates to molecular oxygen, resulting in water formation. These coenzymes are involved in carbohydrate (glucolysin, oxidative pathway, and TCA cycle), protein (degradation and synthesis of amino acids), and fat (fatty acid oxidation and synthesis, glycerol synthesis, and steroid synthesis) metabolism. It is also required for rhodopsin synthesis. Rice polish, molasses, wheat bran, groundnut cake, and lucerne meal are natural sources of niacin.

8.3.6.8 Pyridoxine

Pyridoxine (Vitamin B_6) is found largely in products of vegetable origin. Pyridoxal phosphate (Co-decarboxylase) is the active form of this vitamin. Vitamin B_6

functions as part of coenzyme systems in the body that aid in the metabolism of amino acids, fatty acids, and in the release of energy. It is responsible for conversion of linoleic acid to arachidonic acid. It helps in synthesis of porphyrin, epinephrine, and nor-epinephrine. It helps in incorporation of iron into haemoglobin and transportation of amino acids. Vitamin B_6 is distributed among the plants and animal foods of the diet. Yeast, cereal grains and their by-products, liver meal, and synthetic vitamin (vitamin B_6) are the best sources of this vitamin. Generally the food supply is sufficient to meet the physiological requirement of pyridoxine.

8.3.6.9 Pantothenic Acid

Pantothenic acid, being a part of coenzyme A (CoA), has importance in cellular metabolism of carbohydrates, amino acids, and fat. It mediates acetylation reactions (the acceptance or transfer of acetyl group) and helps in synthesis of steroid and porphyrin part of the haemoglobin. Yeast, liver meal, groundnut meal, green gram, lucerne meal, milk products, fermentation residues, cereal seeds and by-products, cane molasses, and synthetic vitamin (calcium pantothenate) are the good sources of this vitamin.

8.3.6.10 Folacin

Folacin (folic acid) functions as co-enzyme in the transfer of one carbon unit, which is important in the metabolism of many body compounds. It is required for the biosynthesis of purines, pyrimidines, glycine, serine, and creatine. Folic acid plays a dominant role in the haematopoiesis along with vitamin B_{12} and ascorbic acid. Anaemia, leukopenia, and thrombocytopenia along with reduced growth rate are the signs of deficiency but it is unlikely to be deficient in common feeds of birds. However, antagonists create deficiency, if present in the diet. It is widely distributed in nature especially in green foliage of plants.

8.3.6.11 Biotin

Biotin is an essential part of certain coenzymes. Biotin is involved in carbon dioxide fixation. Carboxylation reactions are involved in the synthesis of fatty acids. Yeast, groundnut cake, green leaves, molasses, sunflower meal, and liver meal are good sources of biotin. Wheat and corn products are poor sources. It is available commercially as vitamin H or biotin (2%).

8.3.6.12 Choline

Choline occurs in the body in very large amounts. Choline may be synthesized in the body when sufficient amounts of methionine are available. Choline is needed for protein and fat metabolism. It is a source of labile methyl groups, which are used in the synthesis of vital substances, e.g. epinephrine, phospholipids like lecithin and sphingomyelin. It is involved in the transmission of nerve impulses as a component of acetylcholine. Choline prevents accumulation of fat. Fish meal, liver meal, soybean meal, groundnut meal, wheat by-products (wheat germ), legumes, and synthetic choline chloride are the good sources of choline.

8.3.6.13 Cyanocobalamin

Cyanocobalamin (vitamin B_{12}), also called animal protein factor, is a cobalt containing vitamin. It is required for growth (thus called chick growth factor) and to prevent anaemia (antipernicious anaemia factor). It functions as a coenzyme in the metabolism of body cells specially the cells of bone marrow, the nervous system, and the gastro-intestinal tract. The vitamin participates in nucleic acid synthesis. Factors responsible for pernicious anaemia may be related to the function of this vitamin. Animal protein sources like fishmeal, meat meal, liver meal, and cow dung are very good sources of vitamin B_{12}. The origin of the vitamin in nature is probably the result of microbial synthesis. Feedstuffs of plant origin are devoid of this vitamin.

8.3.6.14 Vitamin C

Vitamin C (L-ascorbic acid) is highly soluble in water. This vitamin acts as antioxidant. It is synthesized in the body to meet the requirement but its rate of synthesis is reduced under stress, therefore, supplementation of this vitamin during stress is beneficial. L-ascorbic acid has an important metabolic role because of its reducing properties and function as an electron carrier. This vitamin is not stored in body or transmitted to eggs. Vitamin C is essential for the collagen formation, the main supportive protein of all connective tissues. It is involved in the metabolism of lipids and synthesis of steroid hormones. It helps in conversion of folic acid to tetra hydro folic acid (active form) and tryptophan to serotonin, and hydroxylation of proline, lysine, and aniline. It keeps the iron in reduced state. Vitamin C is universally distributed. Fresh fruits specially citrus and leafy vegetables are good sources of this vitamin.

8.3.7 Minerals

Animal body contains 3–5% minerals on empty body weight basis. The mineral content increases with age. About 30–40 minerals are present or found in animal body in varying amounts. The contents also vary in different organs depending upon the functions they perform. Among the mineral matter of the body, calcium ranks first and phosphorus second. All the minerals present in the body of birds may not be essential. Minerals give rigidity and strength to skeleton (Ca, P, and Mg). Different organic molecules in the body contain several minerals as essential component like proteins (sulphur and phosphorus), phospholipids (P), mucopolysaccharides (S), enzymes (Zn, Mn, Se, Cu, Fe), hormones (I, S), and vitamins (S, Co and P). Minerals act as activator of many enzymes responsible for digestion and metabolism of organic nutrients. As soluble salt, minerals help in maintaining acid-base balance (Na, K, Ca, P, and Cl), fluid balance (Na, K, and Cl), osmotic tension (Na, K, and Cl), muscular contraction and relaxation (Ca, Na, and K), and transmission of nerve impulse (Na and K). Minerals also help in synthesis of haemoglobin and thus oxygen transport (Fe and Cu). Thus minerals are required for growth, production, and reproduction performance of the birds.

Mineral deficiencies under field conditions are not uncommon. The minerals which are apt to be deficient in diet of ducks include calcium, phosphorus, manganese, zinc, copper, iodine, and sodium.

8.3.7.1 Calcium

Calcium (Ca) is the major element (1.3–1.5% of body) present in the body. About 99% of body calcium is present in bones as structural component of skeleton, while the remaining 1% is present in soft tissues and body fluids. Ca serves many physiological functions like coagulation of blood and liberation of insulin, regulation of acid-base balance, excitability of nerves, permeability of cell membrane, heart beat and muscle tone, and activates digestive enzymes. It is an essential component of eggshell. Green forages, snails, oyster, small fishes, fish meal, meat-cum bone meal, sesame or gingili cake, finger millet, rapeseed meal are very good sources of calcium. The different supplemental sources are limestone (34%Ca), oyster shell (34%Ca), marble grits or powder (34%Ca), bone meal-steamed or sterilized (24% Ca), bone ash (36%Ca), di-calcium phosphate (22–28%Ca), and rock phosphate (33%Ca).

8.3.7.2 Phosphorus

Phosphorus (P) is an essential mineral required for normal growth, development, egg production, and health. It is an essential component of bone. Being component of many organic molecules (creatine phosphate, ATP, ADP, AMP, NAD, NADP, FAD, FMN, nucleic acids, phospholipids, phosphoproteins, etc.), it is involved in metabolism of carbohydrates, proteins, and lipids, storage and transfer of energy in body, gene expression, cell membrane structure, etc. The deficiency of P causes health problems and decreased egg production. All cereals and their by-products are good sources of P but availability of P is less from those ingredients (average 30% is available). Bone meal (steamed 12% P), bone ash (17% P), di-calcium phosphate (19% P), rock phosphate (18% P), sodium acid phosphate (22% P), etc. are the supplementary sources. The requirement of P is expressed as total and available phosphorus. The concentrate feeds of plant origin are quite good sources of P, but about 50% or more of phosphorus present in feeds of plant origin is in the form of phytic acid (ester of inositol and six phosphates) or phytate (salts of Ca and Mg, phytin). This bound form of P is not available to the poultry birds unless bound P is released. The phytin P is about 80% of total P in bran, 40–65% in oilseed meals, 50–70% in grains, and 30–40% in beans. The available phosphorus is calculated by thumb rule method as plant P $\times$ 0.30 + supplemental P + P from animal feed origin.

8.3.7.3 Magnesium

Magnesium (Mg) regulates neuromuscular irritability, activates enzymes transferring phosphate from ATP involved in carbohydrates and lipid metabolism, acts as co-factor for decarboxylation of some peptidases and alkaline and acid phosphatases. Generally all the feed ingredients contain sufficient amount of magnesium to meet its requirement. Oilcakes and leguminous fodders are rich sources of

Mg. The supplements are magnesium oxide and magnesium sulphate. Magnesium is not usually deficient in diet.

8.3.7.4 Sodium

Sodium (Na) and chlorine (Cl) help in acid-base balance along with bicarbonate ions, electrolyte balance, fluid balance, and regulation of osmotic tension. Sodium is involved in transmission of nerve impulse, action potential in skeletal muscle, and active process of absorption. Sodium is a limiting mineral in feedstuffs of plant origin. Use of fishmeal containing high salt or saline water may increase the incidence of toxicity. Common salt is the cheap source of sodium and chloride.

8.3.7.5 Sulphur

Sulphur(S) is an essential component of certain amino acids (methionine, cystine and cystein), glutathione, enzymes, hormones, lipoic acid, bile acids, certain vitamins, and many other organic compounds. It is also required for the growth of feather.

8.3.7.6 Manganese

Manganese (Mn) is an important mineral for reproduction as hatchability is improved with dietary supplementation of manganese. Therefore, it needs special care in formulation of diets for breeding ducks. Moreover, it is required for normal growth, egg shell formation, and egg production. It is involved in several enzyme systems required for metabolism of amino acids and nucleic acids, oxidative phosphorylation and synthesis of fatty acids, cholesterol, and mucopolysaccharides. Grains are usually deficient in manganese. Borderline deficiency in feeds exists in certain parts of tropical countries. Lucerne meal and grain by-products are good sources of manganese. The supplemental sources are manganese salts (sulphate, 25% Mn, and oxide, 60% Mn).

8.3.7.7 Zinc

Zinc (Zn) is required for growth, feathering, egg production, and hatchability in birds. It is associated with many enzymes as an essential component or activator, responsible for metabolism of carbohydrates, lipids, proteins, and nucleic acids. It is also involved in absorption, transportation, and metabolism of vitamin A. Being component of carbonic anhydrase; it is involved in bicarbonate metabolism required for different functions including removal of CO_2, reabsorption of sodium, regulation of buffering system, and egg shell formation, etc. It is a component of superoxide dismutase along with copper. Zn is not stored sufficiently in the body for longer period. Other dietary minerals such as Ca, Cu, and Fe decreases the availability of Zn. Feedstuffs are mostly deficient in zinc, thus its supplementation is required. Supplemental sources are zinc oxide (73% Zn), zinc carbonate, or zinc sulphate (36% Zn).

8.3.7.8 Copper

Copper (Cu) helps in utilization of iron, haemoglobin synthesis, maturation of red blood cells, pigmentation of feather, formation of connective tissues, myelination of spinal cord, myoglobin synthesis, bone formation, and reproduction. It is also a component of super oxide dismutase involved in anti-oxidative process, thus protecting cells from oxidative, damage. Copper is normally not deficient in feed. But deficiency of copper has been reported in chicks. Too much molybdenum or iron may also decrease its availability. Dietary supplementation of Cu is thus required. Distillery by-products, fishmeal, liver meal, oil cakes and dried whey are the natural feed sources of copper. The supplementary sources of copper are copper sulphate (25% Cu), cupric carbonate (53% Cu) or cupric oxide (75% Cu).

8.3.7.9 Iron

Iron (Fe) is an essential component of haemoglobin and myoglobin. It also plays key role in enzyme systems involved in oxygen transport and oxidative process. The deficiency symptoms are anaemia, depressed body weight gain, and loss of appetite. The feed ingredients of tropical countries contain sufficient iron and thus its supplementation is seldom required except in secondary anaemia caused by parasitic infection or fluid/blood loss. Supplementary sources are ferrous sulphate and ferric sulphate.

8.3.7.10 Iodine

Iodine (I) is a component of thyroid hormones, responsible for regulation of basal metabolism. It is supplied in the trace mineral mixture. Fishmeal and aquatic weeds are good sources of iodine. Feeds of plant origin are deficient in iodine. Certain goitrogens present in certain feedstuffs affects its utilization. The different supplements of iodine are potassium iodide (59% I), potassium iodate (76% I), and calcium iodate (65% I).

8.3.7.11 Selenium

Selenium (Se) as component of enzyme glutathione peroxidase, it plays an important role as antioxidant. It acts in cytoplasm and helps to remove free radicals. Usually Se deficiency does not occur in chickens. It is closely associated with vitamin E and one can spare the other. Deficiency causes exudative diathesis and white muscle disease or muscular dystrophy. Its excess intake causes toxicity.

8.4 Nutrient Requirements of Ducks

All the nutrients required for chicken or other poultry birds are also required for ducks but the dietary concentration and ratio of different nutrients may vary. Whatever the system of rearing (extensive, semi-intensive, and intensive), the ducks should receive all the nutrients through their feeds as per their requirement for optimum growth and reproductive performance. The nutrient requirements of ducks are given in Table 8.1.

Table 8.1 Nutrient requirements of ducks

		White Pekin ducks[a]			Suggested for Indian ducks[b]			
Nutrients	Units	0–2wk	2–7wk	Breeding	0–2wk	2–9wk	9–20 wk	Breeding/ layer
ME	Kcal	2900	3000	2900	2750	2750	2800	2650
CP	%	22.0	16.0	15.0	20	16–18	15–16	18–19
Arginine	%	1.10	1.00	–	1.05	0.90	0.70	0.70
Lysine	%	0.90	0.65	0.60	0.85	0.60	0.56	0.56
Methionine	%	0.40	0.30	0.27	0.38	0.27	0.25	0.25
Met+Cys	%	0.70	0.55	0.50	0.65	0.52	–	–
Tryptophan	%	0.23	0.17	0.14	0.23	0.23	–	0.21
Calcium	%	0.65	0.60	2.75	0.65	0.60	0.60	2.75
Non phytate P	%	0.40	0.30	–	0.40	0.35	0.30	0.30
Sodium	%	0.15	0.15	0.15	–	–	–	–
Chloride	%	0.12	0.12	0.12	–	–	–	–
Vitamin A	IU	2500	2500	4000	4000	3000	3000	4000
D_3	ICU	400	400	900	1500	–	–	–
E	IU	10	10	10	20.0	0.20	–	0.38
K	Mg	0.50	0.50	0.5	2.0	0.20	–	1.0
Riboflavin	Mg	4.0	4.0	4.0	10.0	8.0	–	9.6
Pyridoxine	Mg	2.5	2.5	3.0	4.0	4.0	–	6.0
Niacin	Mg	55	55	55	50.0	50.0	–	25.0
Pantothenic acid	Mg	11.0	11.0	11.0	20.0	10.0	–	28.5

[a]NRC (1994) as percent or unit per kg of feed (90% DM), [b]Khaki Campbell and other breeds (Mandal et al. 2004; Fouad et al. 2018)

8.4.1 Protein and Amino Acids

Similar to chickens, the ducks require quality protein in diet with assortment of essential amino acids matching to the requirements of birds. The limiting amino acids in diets vary depending upon feed resources used in mixed feed. However, in maize or wheat and soybean meal based diets methionine seems to be the first limiting, lysine is second limiting, and threonine is the third limiting amino acids. As different type of ducks have different growth rate, development of breast muscle, and production status, the protein and amino acid requirements also vary. Again, protein constitutes about 40% of feed cost, and proteinic ingredients are costlier in terms of price, besides excess proteins are metabolically costlier when converted to energy. Thus calorie to protein ratio in the diet must be considered for efficient use of both protein and energy. Diets low in protein delays growth rate. However, with increasing protein content in diet from 14% to 24% improved body weight gain linearly (Cherry and Morris 2008). FCR was also improved by 1.3% for each 1%

increase in protein (Cherry and Morris 2008). However, there is scope for reduction of dietary protein through supplementation of limiting amino acids.

Pekin ducks require 22 and 16% protein during 0–14 days and 14–49 days of age, respectively, while breeders require 15% protein (NRC 1994). A diet with 16.5% protein was suggested for Khaki Campbell layers (Thongwittaya and Tasaki 1992). Otherwise, the protein levels of 20–22% for starter, 17–19% for grower, and 16% for breeders are recommended. The information on the amino acid requirements of ducks is scanty. A level 0.90% lysine, 0.40% methionine, 0.70% methionine+ cysteine, and 1.10% arginine during 0–14 days of age and 0.65% lysine, 0.30% methionine, 0.55% methionine plus cysteine, and 1.0% arginine during 14–49 days of age was recommended for Pekin ducks (NRC 1994). A three phase feeding (0–14, 14–35, and 35–49 days of age) was recommended by Hou (2007). Slightly higher values for lysine (1.10%) and methionine (0.50%) and methionine +cysteine (0.82%) was specified for Pekin ducklings during 0–14 days of age by Hou (2007). The values for 15–35 days of age were 0.85, 0.40, 0.70, and 0.60%, and for 36–49 days of age, they were 0.65, 0.30, 0.60, and 0.45% for lysine, methionine, methionine plus cysteine, and arginine, respectively.

On the other hand, levels of 1.16% lysine, 0.42% methionine, 0.76% methionine pluscysteine, 0.84%threonine, and 0.94% arginine have been suggested during 0–21 days of age of Pekin ducks (Amino News 2010). The corresponding requirements during 21–49 days of were 0.90%, 0.42%, 0.77%, 0.66%, and 0.76%, respectively. The requirements of tryptophan were 0.21% to 0.23% during 0–14 and 0.16–0.20% during 15–49 days of age. The research information on nutrition of native duck is practically lacking but scanty reports are available on protein requirement of Khaki Campbell ducks. Growing Pekin ducks required 16% protein during 9–16 weeks of age (Mohanty et al. 2016). Indian Council of Agricultural Research has suggested requirements of 20.5, 16.5, 15.0, and 16.5% of protein, 1.0, 0.75, 0.60, and 0.75% of lysine, 0.45, 0.35, 0.30, and 0.30% of methionine and 0.85, 0.65, 0.60, and 0.75% of methionine plus cysteine during 0–8, 8–16, 16–20, and above 20 weeks of age, respectively (ICAR 2013), for Indian native ducks. Zhang et al. (2016) suggested dietary threonine levels ranging from 0.86% to 0.95% based on growth, FCR, and breast yield in White Pekin ducklings (1–14 days of age). Like broiler chickens, higher level of threonine was required for feed conversion ratio. However, Xie et al. (2014) recommended lower level (0.67%) of threonine in diets of White Pekin ducklings (1–21 days of age) based on body weight. White Pekin ducks required diets with 0.73–0.75% threonine during 14–35 days of age for maximum body weight gain, feed conversion, and breast meat yield (Zhang et al. 2014). Laying ducks require 0.57% threonine in the diet from 17 to 45 weeks of age. Wang et al. (2013) suggested dietary levels of 0.95% arginine for body weight gain, 1.16% for FCR, and 0.99% for breast yield in male White Pekin ducks from 0 to 21 days of age. Between 17 and 31 weeks of age, laying ducks require diets containing 1.13% arginine for the development of reproductive system, and 1.46% arginine for maximum egg production and eggshell quality (Xia et al. 2017). The protein supplements that can be used in duck rations include soybean meal, mustard cake, corn gluten meal, rice gluten meal, rice dried distillers grains with solubles, fish meal, meat-cum-bone meal, etc. Groundnut cake and de-corticated cottonseed meal (solvent extracted, free from free-gossypol) can also be used. However, it is necessary to ensure that these ingredients are free from aflatoxins. The fish meal gives very good result but the level may be limited to 5% to avoid fishy flavour in meat and eggs.

8.4.2 Energy

The dietary energy plays an important role in duck nutrition like other poultry birds. The maintenance requirement of ducks was calculated through regression as 139.3 kcal per $kg^{0.75}$at 10 °C and 125 kcal per $kg^{0.75}$ at 26 °C (Cherry and Morris 2008). The requirements of other nutrients also depend on the concentration of dietary energy. NRC (1994) suggested 2900 kcal ME per kg for White Pekin ducks from 0 to 2 weeks of age, 3000 kcal MEper kg from 2 to 7 weeks of age, and 2900 kcal ME per kg for breeders. Reports on energy requirement of Khaki Campbell x Thai Native ducks indicated its requirement as 2700 kcal ME per kg diet with 16.5% CP from 18 to 37 weeks of age (Thongwittaya and Tasaki 1992). The native ducks are fed starter ration with 2750 kcal (0–2 weeks of age), grower ration (3–8 weeks of age) with 2750 kcal ME per kg, and 9–20 weeks with 2700 kcal ME per kg of diet. ICAR (2013) has suggested energy requirements of Indian native ducks as 2800, 2650, 2500, and 2650 kcal ME per kg diet during 0–8, 8–16, 16–20, and above 20 weeks of age, respectively. The energy supplements are maize, wheat, pearl millet, sorghum, rice bran, rice bran (solvent extracted), small millets, etc. Wheat is preferred over maize as earlier one is less susceptible to *Aspergillus flavus* growth. Increasing dietary energy level from 2600 to 3100 kcal ME per kg diet during 2–6 weeks of age improved production performance; in addition, increased fat deposition also in the body of Pekin ducks (Fan et al. 2008). Accordingly, 3000 kcal ME per kg diet at 18% protein levels from 2 to 6 weeks of age was suggested. Xie et al. (2010)recommended 2900 kcal of ME per kg at 20.5% protein diet for male White Pekin ducks during 0–3 weeks of age based on the growth performance and carcass quality. Thongwittaya et al. (1992) suggested 2700 kcal of ME per kg with 16.5% CP during 18–37 weeks of age for Khaki Campbell and native crossbreds. Generally, diet with 2500 kcal of ME per kg and 17% CP is used for egg-laying ducks under commercial practice (Fouad et al. 2018).

8.4.3 Minerals

Minerals play important roles in formation of bone, enzymes (as component or catalyst), vitamins (as component), blood cells, egg shell formation, and many other organic molecules, blood clotting, nutrient metabolism, proper muscle, and nerve function and immunity. The starting and growing ducklings require 1.0% calcium, while breeders require 3% calcium. The requirements for total phosphorus range from 0.6% to 0.7%. The requirements of available phosphorus are 0.40%, 0.35%, and 0.40% for starting, growing and laying or breeding ducks. NRC (1994) has suggested 0.65% calcium during 0–2 weeks of age, 0.60% during 2–7 weeks of age, and 2.75% in the breeder ration with 0.40% and 0.30% available P during 0–2 and 2–7 weeks of age, respectively, for Pekin ducks. Slightly higher values of calcium 2.90% and available phosphorus 0.47% have also been suggested. Dietary calcium and available phosphorus requirements of indigenous layer ducks reared in cages were 4.0% and 0.6%, respectively (Ravi et al. 2005). The requirements of zinc and manganese are 60 and 50 mg per kg, respectively, for all phases of growth and production. Scanty information is available on requirements of trace minerals in ducks. Male Pekin ducklings (0–8 weeks of age) require 30 mg zinc per kg of diet for body weight gain, while 120 mg Zn per kg diet was beneficial to reduce body fat deposition and improve meat quality (Attia et al. 2013). For laying ducks, 30 mg Zn per kg in the diet was adequate during peak laying period (Chen et al. 2017). Attia et al. (2012) recommended 7 mg copper per kg diet, for male White Pekin ducklings (0–8 weeks of age). A higher level of manganese (90 mg per kg diet) was specified for laying ducks from 17 to 36 weeks of age in order to optimize their antioxidant defence system(Fouad et al. 2016). White Pekin ducklings (0–21 days of age) need 0.236 mg Se per kg of diet to protect tissues against oxidative damage (Wang 2006). Cherry Valley hybrid ducklings (0–7 weeks of age) require 0.4 mg Se per kg diet for improved growth rate (Baltić et al. 2017). The levels of 0.18 and 0.24 mg Se per kg of diet was recommended for egg production during early and peak laying phases, respectively, while 0.38 mg Se per kg of diet was recommended to enhance the activity of the glutathione redox system (Chen et al. 2015). A dietary level of 0.3% common salt is sufficient for all categories of ducks. The requirements prescribed for chicken may be used in practical feed formulation. The feedstuffs of vegetable origin are generally low in minerals. Therefore, mineral supplements are added to commercial duck feed. Limestone, marble, and oyster shells are common sources of calcium, whereas di-calcium phosphate and mono-calcium phosphate are common sources of phosphorus and calcium. The micro-minerals are usually supplemented through trace mineral premix. Salt is good and cheap source of sodium and chlorine.

8.4.4 Vitamins and Linoleic Acid

Vitamins are required in small quantities and associated with tissue (fat-soluble vitamins) as well as nutrient metabolism (water-soluble vitamins). Vitamins are essential for normal body growth and functions and reproduction. A deficiency of one or more vitamins may cause deficiency syndromes and diseases. The most critical vitamins are vitamin A, riboflavin, vitamin D_3, vitamin K, and niacin. The ducks are comparatively sensitive to deficiency of vitamin E and selenium. NRC (1994) has suggested requirements of 2500 IU of vitamin A from 0 to 7 weeks of age and 4000 IU for breeders. Slightly higher values (4000 IU of vitamin A during 0–2 weeks, 3000 IU during 2–20 weeks, and 4000 IU during laying phase) have also been used in diet (Mandal et al. 2004; Fouad et al. 2018). ICAR (2013) recommends vitamin A requirements as 3200, 2250, and 4000 IU during 0–8, 8–20, and above

20 weeks of age, respectively. However, Wei et al. (2009) reported that a level of 2500 IU vitamin A per kg diet improved the productive performance in White Pekin ducklings during 0–21 days of age. Wang et al. (2016) observed that maize-soybean meal diet could satisfy the requirement of vitamin A for performance and egg quality in laying ducks (20–36 week of age). However, addition of 400 IU vitamin A per kg diet improved the antioxidant defence system. The requirements of vitamin D_3 suggested for Pekin ducks are 400 IU from 0 to 7 weeks of age and 900 IU for breeders (NRC 1994). However, Wang et al. (2009) reported that diets containing 1000 IU vitamin D_3 per kg were sufficient to meet the needs of White Pekin ducks during the first 2 weeks of age. A level of 550 IU vitamin D per kg was sufficient for the ducklings of egg type ducks during 0–4 weeks of age (Xie and Wang 2012). For optimum egg production and better egg shell quality dietary requirement of vitamin D_3 was about 1000 IU per kg (Chen 2000). Chen and Hsu (2004) suggested dietary supplementation of 400 mg α-tocopherol per kg to enhance egg production, egg weight, and feed conversion ratio in laying ducks during 30–36 week of age. ICAR (2013) recommended 400, 350, and 650 IU of vitamin D_3 for 0–8, 8–20, and above 20 weeks of age, respectively. The requirement of thiamine was suggested as 1.55 mg per kg diet for laying ducks from 22 to 42 weeks of age (Fouad et al. 2018). The vitamin niacin requires (45 mg per kg diet minimum) special care to prevent leg weakness in ducks. Use of Brewer's yeast in duck ration (5–7%) prevents weakness and promotes growth since yeast is a good source of niacin. The requirements of riboflavin and pantothenic acids are 3–5 mg and 11 mg per kg of feed, respectively. Choline supplementation improves growth of ducklings. Wen et al. (2014a, b) reported that diets containing 810 and 1182 mg choline per kg were optimum for improved growth rate and prevention of perosis, respectively, in Pekin ducklings from 0 to 3 weeks of age, while 980 mg choline per kg was required to enhance average body weight gain during 3–6 weeks of age. The requirement of biotin is 0.10 mg per kg body weight for all the phases. Folic acid requirement is 0.60 mg for starting (0–8 weeks) and laying period (>20 weeks of age), while it is 0.40 mg per kg during growing (9-20 weeks of age) period. Linoleic acid is an essential fatty acid and acts as component of organic molecules. It also influences egg size. The requirement of linoleic acid is 0.8–1% in diet for all the purposes. The sources of vitamin A under extensive rearing system are fishes, snails, green grasses, distiller's grain, and leaves, etc. Certain water-soluble vitamins are synthesized by the microorganisms present in the digestive tract. Vitamin D can be produced, if the birds' skin is exposed to sunlight. However, vitamins must be supplied to the birds reared under intensive production system.

8.5 Economizing Dietary Protein Supply

Extensive studies have been conducted on minimizing dietary protein levels through dietary supplementation of limiting amino acids in commercial broilers, layers, and turkeys. Similar scope exists in ducks also (Baéza and Leclercq 1998). Feeding low protein diet (lowered by 12.4% of dietary protein), balancing dietary concentration of methionine, lysine, threonine, and tryptophan through supplementation, had equal performance in finishing Pekin ducks (Baéza and Leclercq 1998). However, there is a need for extensive studies for determination of levels of limiting amino acids in diets of different strains of ducks besides minimum protein levels required by ducks of different strains for optimum performance. The limiting amino acids like methionine, lysine, threonine, tryptophan, etc. are available commercially at reasonable rates.

8.6 Amino Acid Digestibility Values in Feed Formulation for Ducks: Limitation

Generally, the practical feeds of chicken (commercial broilers and layers) are formulated based on digestible amino acids. Lot of works have been conducted in India and elsewhere on requirement of digestible amino acids for different categories of commercial chickens. Similarly, amino acid digestibility values of different feedstuffs have been determined in chicken. The values of requirements and amino acid digestibility cannot be used in ducks as lot of differences exist in amino acid digestibility between chicken and ducks. The average values of amino acid digestibility (apparent ileal) were determined as 76, 69, and 59% in chickens, duck and geese, respectively (Jamroz et al. 2001). The comparative digestibility values of methionine and lysine were determined as 70 and 72% for chicken, 44 and 57% for ducks, and 52 and 41% for geese, respectively (Jamroz et al. 2001; Helmbrecht 2012). The experiment was further repeated by the same authors (Jamroz et al. 2002b) and results were more or less similar. The average digestibility values of amino acids were 72% in chicken, 55% in ducks, and 63% in geese. The comparative digestibility values of methionine and lysine were determined as 74 and 62% for chicken, 55 and 41% for ducks, and 60 and 41% for geese, respectively. Both the experiments indicated higher digestibility values for chicken than ducks. In another experiment (Rodehutscord et al. 2004) with soybean meal and rapeseed meal as test ingredients in corn starch and wheat gluten based basal diets in broiler chicks, turkey poults, and Pekin ducklings of 3 weeks of age also indicated similar trend. The average digestibilities of amino acids were lower for ducklings than chicks or poults. The methionine and lysine digestibility values were also significantly lower in duckling than in the other two species.

8.7 Feeding and Watering

Feeding practices of ducks depend on the number of ducks raised and rearing system. There are three systems of rearing of ducks, i.e. backyard, semi-intensive, and intensive. Under backyard production system only a few ducks are raised by the households. The ducks are reared mostly in backyard or free range system in Asia and other developing countries. Little attention is given to feed requirement of such ducks. Under backyard or free range system, duck feeds comprise of fallen paddy grains, small fishes, frogs, snails, insects, earthworms, green tender leaves/grasses, aquatic weeds, etc. Anitha et al. (2012) assessed the nutritional and physiological status of foraging ducks in harvested paddy fields in three locations in Kerala, India. Crop of ducks contained paddy grains, crabs, snails, worms, and weeds. Chemical analysis of crop contents were carried out and it contained 41.0–53.0% moisture, 10.2–17.5% CP, 11.3–19.8% CF, 1.8–3.7% EE, and 9.5–17.0% total ash. The mean fibre fractions, viz., Neutral Detergent Fibre (NDF) and Acid Detergent Fibre (ADF) ranged between 28.1 to 39.5 and 12.7 to 23.4%, respectively. After returning to home at evening, those are given with some amount of home grown concentrates such as paddy grains, broken rice, rice polish, rice bran, wheat bran, thrashed fishes/snails, kitchen waste/garbage, table scraps, etc. depending upon the availability. During early age, the ducklings are fed boiled rice, ground rice, earthworms, etc. After 1–2 weeks of age those are also allowed to scavenge with adult ducks. Ducks are voracious eaters and good foragers. The growth and egg production under this system depends upon their access to areas for foraging and feed availability. Although extra care is not required in feeding and nutrition when the birds are being reared under conventional backyard rearing system with very few birds, but for intensified farming even under scavenging/backyard rearing system with sizable number of birds definitely require supplementary feeding to augment their production level.

Under semi-intensive system, birds are allowed for foraging as well as supplementary feed. The amount of supplementary balanced feed depends upon the foraging area, availability, and nutritional quality of natural feedstuffs as in backyard system of rearing. Generally, during rainy seasons, ducks get more protein but less energy. During winter, ducks get more energy and less protein. On the other hand, during lean period both energy and protein are deficient. Under intensive production system any number of birds can be raised on balanced feed. Adult ducks should not have free access to feed as they become fatty. Ducks may be fed dry mash, wet mash or pellets. They can successfully be reared on dry mash if practiced from day-old. Precautions are needed to reduce the wastage of feed. Due to difficulties in swallowing of dry mashes, ducks prefer wet mash. Feeding of wet mash improves digestion of feed as intestinal juice penetrates easily into the wet feed particles and thus feed intake is increased. Fresh wet mash need to be prepared each time. However feeding wet mash needs precaution so that fungi should not grow in feed and feeders.

Ducks under commercial operation, in other countries, are fed pelleted mash. Pellet feeds have certain advantages such as improvement in palatability, feed conversion efficiency, uniform mixing and no segregation of feed ingredients, uniform body weight, higher density, destruction of certain anti-nutritional factors and pathogen, if any. Under commercial intensive production system any number of birds can be raised. The birds are fed balanced feed. The carcasses of ducks have more fat than broiler chickens. The ducks have tendency to consume more energy and deposit more fat from hatching to 7 weeks of age. Qualitative or quantitative feed restrictions may reduce the fat deposition. As the degree of feed restriction is increased, the sexual maturity is also delayed. Feed restriction also increases the fertility and hatchability of eggs, decreases mortality, and increases egg size. The ducklings generally consume up to 10 kg feed per bird up to 8 weeks of age, and thereafter, 160–230 g feed per day. Khaki Campbell ducks eat about 12.5 kg of feed up to 20 weeks of age. Thereafter, the feed intake is 120–160 g per bird per day depending upon the rate of production and availability of greens. The diameters of pellets are 3 mm for starter (0–2 weeks of age) and 4.5 mm after 14 days of age. Ducks are generally fed three to four rations. However, having more feeding phases is better for economic production as with age and increased body weight, the requirements are changed. Ducks are exposed to feed as soon as kept in brooder but always better to allow for sufficient drinking water at least 1–2 hours before offering feed. Properly designed feeders and drinkers should be used to minimize feed wastage. Drinkers should be placed near feeders.

The ducks reared under backyard system in ponds or water bodies are always exposed to drinking water. However, high producing ducks, reared under intensive production system, require consistent supply of clean drinking water. Such ducks should not be allowed to go thirsty frequently to obtain optimum growth and egg production. Ducks do not require elaborate drinker but drinker should be four to six inches depth to allow the ducks to clean their eyes and bills. Performance of ducks is improved if they have enough water to dip their heads and necks. If the ducks are not allowed to do so, their eyes seem to get scaly and crusty leading to blindness. They also like to clean their bills. Although ducks are fond of water, but young ducks should not be allowed to be too wet. Ducklings need a consistent supply of clean water. Chick fountains are recommended for watering ducklings. Ducklings play in water, making a mess and splashing out their drinking water. Thus there is a need to clean and refresh drinkers often. Water temperature should not be too low or too high.

8.8 Precautions in Duck Feeding

Ducks are more susceptible to aflatoxins than other poultry birds due to difference in rate of bio-transforming aflatoxin to less toxic epoxide in the liver due to varying liver enzymes with other poultry species. Moreover, due to increased sensitivity to aflatoxins, ducks are more likely to show liver lesions and hepatic tumours even on exposure to lower level of aflatoxins for a long period. The safe level of aflatoxin is below 30 ppb in duck mashes. The toxicity signs are reduced feed intake and growth rates, poor feed efficiency, increased susceptibility to infectious diseases, and enlarged liver and kidney. The low quality protein or less protein and or dietary energy in the diets exacerbates the deleterious effect of aflatoxins. Thus, fresh and good quality ingredients, free from aflatoxins, are used in duck ration. Being maize and groundnut cake, more susceptible to *Aspergillus* growth, these ingredients are replaced totally or partially by coarsely ground wheat and toasted soybean meal in the diet of ducks considering tolerance level of aflatoxins. Therefore, methionine and cystine are the major limiting amino acids in mashes of ducks. The formulations, however, are not static and change depending upon the availability and cost of feed ingredients. The ingredients used for compounding duck feed should not be stored for longer period to prevent growth of Aspergillus. Again, moisture content of stored ingredients should not contain more than 12%. Adverse effects on the gastro-intestinal tract are probably the major cause of economic losses resulting from trichothecene. T-2 toxin can cause caustic injury to the mucosa, destroying cells on the tips of villi, and affect rapidly dividing crypt epithelium. Histopathology of GI tract lesions due to acute intoxication by purified T-2 toxin is characterized by haemorrhage, necrosis, and inflammation of the intestinal epithelium, which occur before transient shortening of villi and reduction in the mitotic activity in crypt epithelium. Necrosis also occurs in the mucosa of the proventriculus and gizzard. Use of suitable toxin binders reduces the toxicity of mycotoxins that affect gut health also.

Nicotinic acid concentration in diets of meat type ducks needs to be considered. Niacin is an essential vitamin and it is required for synthesis of organic molecules (NAD and NADP) involved in energy transfer, duck health, development, and bone growth. Ducks are unable to convert tryptophan into niacin. Moreover, meat type ducks grow at faster rate thus the dietary requirement of niacin is also increased. Deficiency signs of niacin deficiency in ducklings vary depending on the dietary supply. Most common signs are bowed legs, enlarged hock joints, crippled, weak, and unable to walk. Moreover, ducks are sensitive to available phosphorous deficiency in the diet. They are also very sensitive to sudden changes in their diets. The ducks are having habit of eating and drinking alternatively.

8.9 Use of Feed Additives

Poultry strains, employed in commercial production system, are more prone to various categories of stressors including physical (debeaking, vaccination, handling, transportation, etc.), nutritional (unbalanced diet, poor feeding management, feed quality, intrinsic and extrinsic toxins, starvation, etc.), physiological (high growth, egg production, and accumulation of waste products and metabolites), social stress (overcrowding, multi-age, and multi-species rearing system), pathological (diseases, infection, and toxins), and environmental (extreme ambient temperature, wet litter, ammonia, and bad aroma and poor ventilation) stressors. Therefore, they require specialized management practices including dietary care. Improvement and maintenance of physical appearance, consistency, nutritive quality, shelf life, and texture of diet are also deserves utmost attention. Pre-harvest losses arising from mortality and morbidity are significant. Maintaining gut health is also an issue to improve nutrient utilization, and minimizing nutrient wastage (Mandal et al. 2006), and thus to reduce the cost of production and environmental pollution. Keeping this in view, the use of appropriate feed additive(s) in poultry feeding has become obligatory as feed additives have made it possible to provide more eggs and meat and better feed utilization efficiency.

8.9.1 Antibiotics

Antibiotic compounds have been employed in low concentrations as feed additive throughout the world for more than 60 years to promote growth, egg production, and feed conversion efficiency. Antibiotics act through one or more of the following mechanisms: keeping the animals free from sub-clinical infection; enhancing intestinal absorption capacity through preventing thickening of intestinal wall, altering metabolic activity, sparing micro-nutrients by preventing destruction, and suppressing competition for critical nutrients between host and pathogenic organisms reducing harmful bacterial metabolite production and thus reducing stress. There are also evidences that some antibiotics when used as feed additive help the birds to overcome from low level of infection normally present continuously even at some of the well-organized poultry units. The effect is more pronounced when the animals are kept under unhygienic condition and are fed relatively vitamins and/or amino acid imbalanced/deficient diet. However, antibiotics are not the substitute for hygienic condition or balanced diets. However, antibiotics are losing public image. In recent years, the public has become increasingly alarmed by the emergence and spread of antibiotic-resistant bacteria, which have serious consequences for disease control in humans and animals. Therefore, efforts are going on to replace feed antibiotics through the use of direct fed microbials (probiotics), prebiotics, synbiotics, feed enzymes, essential oils, acidifiers, etc.

8.9.2 Direct Fed Microbials (DFM)

Direct fed microbials (DFM, Probiotics) refer to 'a source of live (viable), naturally occurring microorganisms' which includes bacteria, fungi, and yeasts. FDA has accepted 42 organisms as generally accepted as safe (GRAS) and as being appropriate for use in animal feeds. Each of these organisms is naturally occurring in the gut of normal, healthy animals, and is not genetically engineered. A well-balanced gut microflora is necessary to establish normal intestinal flora and to protect the gut from invasion of enteric pathogens (*Escherichia coli, Enterobacteria, Streptococci, Staphylococci, Pseudomonas, Clostridia,* etc.) and secondarily to serve the function of antibiotic feed additives. Probiotics especially the lactobacilli and *Bacillus cereus* are important in the development of immuno-competence against enteric infections. The crop and intestinal microbial counts (total plate count, coliforms, *Staphylococcus aureus,* and yeast and mould) were significantly lowered on probiotic supplementation (Elangovan et al. 2011). Supplementing probiotics (*Gallipro*) to broiler chicken improved growth performance, immunity, and gut microbial population. The addition of probiotics to the diet has been found to improve growth performance and feed conversion in meat type birds, and egg mass and egg weight in laying birds. Certain probiotics reduce egg yolk cholesterol, serum cholesterol, and total lipids. A commercial probiotic preparation containing lactic acid bacteria (120,000 million CFU per kg), live yeast cells(5000 billion CFU per kg), and traces of multienzymes at 0.05% in White Pekin ducks from day-old to eight weeks of age resulted in higher body weight and weight gain and superior feed conversion efficiency(Thomas et al. 2005).

8.9.3 Prebiotics

Prebiotics are certain non-digestive feed components that benefit the host by selectively accelerating growth rate and/or proliferation of one or more of a limited number of bacteria in the colon such that the health of the host can be improved, have been referred to as the prebiotics. The prebiotics (inulin, galacto-oligo-saccharides, fructo-oligo-saccharides, mannan-oligo-saccharides, and lactose derivatives) act by lowering gut pH, facilitated by lactic acid production, preventing colonization of pathogens (exclusion), modifying metabolic activity of normal intestinal flora, and stimulating immune system. Bifidobacteria and Lactobacillus population increase due to use of prebiotics (as substrate for those organisms) in feed, and they dominant in the colon. The prebiotics have been tried in chickens, rabbits, and other monogastric animals to combat colonization of pathogens (*Salmonella spp.*, *Escherichia coli*, *Vibrio cholera*, *Salmonella. typhimurium*, *Salmonella. enteridis*, etc.). The effective prebiotic has been mannan-oligosaccharides.

8.9.4 Synbiotics

Synbiotics are the combination of both probiotics and prebiotics. The efficacy of probiotic preparations may be potentiated by simultaneous administration of prebiotics. Synbiotics are more effective than when used probiotics and prebiotics alone.

8.9.5 Enzymes

Enzymes have become the most vital nutritional discovery since vitamins and minerals. All of the processes of life depend on enzymes. They are involved in the function of every system in our bodies. Extra enzymes can add to the animals' own digestive capacity. Feeding high viscosity cereal grains to broilers result in larger microbial populations in the ileum. Viscosity reduces the passage rate of the feed leading to overall reductions in consumption and decreased performance, sticky droppings, and dirty eggs. A viscous environment slows down digestion processes, and encapsulates nutrients, making them inaccessible to digestive enzymes. Viscous gels are formed in the digesta by the soluble non-starch polysaccharides (NSP), which are not digested by the animals own enzymes. Thickening of unstirred water layer due to gel formation also inhibit absorption. Thus those non-starch polysaccharides form a major target substrate for feed enzymes. The addition of enzymes to address NSP viscosity can improve gut health, feed efficiency, improve manure quality, and facilitate the use of lower cost feed ingredients. However, it is necessary to make the enzyme to work to obtain maximal benefits through effective formulation.

In many studies, bacterial count from the small intestine of broilers fed wheat-based diets became lower in xylanase-supplemented birds than the not supplemented ones, because dietary enzyme supplementation reduces the microbial population in the small intestine and hence the entire gut ecosystem changes. Dietary enzyme supplementation changes the animal's gut microflora and decreases some of the adverse effects of microbial fermentation. Dietary cereals producing high intestinal viscosities lead to increased overall bacterial activity in the small intestine. The supplementation of xylanase to cereal based diets producing high intestinal viscosity, changes the composition and metabolic potential of bacterial populations and may specifically influence fat absorption. There are several factors that influencing degree of improvement obtained by adding enzymes to the diet, including the type and amount of cereal in the diet, the level and type of anti-nutritive factor present in cereal, the composition and quantity of enzymes used, and the age of the birds, type of gut microflora present and the physiology of the bird. Incorporation of a multi-enzyme preparation (containing endoxylanase, beta glucanase, cellulose, alpha amylase, pectinase, protease, phytase, alpha galactosidase, and beta galactosidase) in 12% fibre diet improved the production performance of indigenous layer ducks (Anitha et al. 2009).

8.9.6 Organic Acids

Organic acids are used in food sanitation programme, acting as a food additive and preservative. These are used to reduce the risk of survival of salmonellae in feed. Van Immerseel et al. (2005) reviewed the physiological action of organic acids and the reduction of Salmonella in supplemented feeds. Acids (formic, acetic, propionic, butyric, fumaric, and lactic acids) or their salts (Ca, Mg, Na, and K) are used most widely as feed additives and feed preservatives. Europe continues to be the largest regional market with high demand of acidifiers for use in pig and poultry, supported largely by legislation that bans use of antibiotics in feed (Menconi et al. 2014) as organic acids have gained popularity due to their high nutritional value and anti-microbial properties. By preventing the growth of pathogenic bacteria it prevents the feed deterioration and extent the shelf life of perishable food ingredient. Organic acid commonly used to reduce the pathogenic microbial load (like Salmonella and *Escherichia coli*) includes short chain fatty acids, volatile fatty acids, and weak carboxylic acids that are only partly disassociated. Organic acids also reduce the colonization of pathogens on intestinal wall, preventing damage to epithelial cells. Dietary acidification lowers down the intestinal pH below 6.0 preventing the growth of unfavourable organisms and promotes the normal microflora growth. Inhibition of undesirable microbes not only prevents the accumulation of toxic metabolites, but also spares more nutrients available for the host, ensuring higher feed utilization efficiency. Stabilization of intestinal pH also increases the efficiency of all digestive enzymes. Daily application of short chain fatty acids increases epithelial cell proliferation; quick repairing of the intestine, increased villous height and increased absorptive capacity. Propionic acid is an effective mould inhibitor and can completely inhibit feed mycotoxin. Thus addition of vinegar to poultry feeds or the drinking water may reduce the bacterial load to which the birds are exposed or reduce the spread of bacteria within a flock. Addition of vinegar (%?) in water is a common practice. Butyrate plays several functions in the lower gut.

8.9.7 Phytogenics

Herbs, essential oils or phytogenic compounds, parts of many plants/ weeds such as bark, fruit, seed, leaf, flower, roots and their extracts are utilized to cure different diseases/ ailments in men and animals in traditional systems of medicine. Phytogenic compounds have anti-microbial, anti-oxidative, and sedative properties attributed to the presence of polyphenols, especially flavonoids. Inclusion of phytogenic compounds in feed results in a destruction of pathogenic bacteria without affecting Lactobacilli or Bifidobacteria. Active principles of different spices, e.g. curcumin (turmeric), capsaicin (red chillies), eugenol (cloves), linalool (coriander), piperine (black pepper), zingerone (zinger), cuminaldehyde (cumin), and cinnamon oil (cinnamon) inhibit lipid peroxidation. The essential oils consist of a mixture of hydrocarbons (terpenes and sesquiterpenes), oxygenated compounds (alcohol, esters, aldehydes, and ketone), and small amount of non-volatile residues. They are known to promote feed intake (because of appetizing smell), digestive enzyme production, regulate gut microflora, enhance the immune system, act as antioxidant, and destroy pathogens. Herbal liver tonics are beneficial for stimulating liver activity and safeguarding liver from toxins and fibrosis. Some of the extracts from spices or plants/herbs exhibit antioxidants activity when added to diet. The saponins help to enhance the nutrients absorption by improving the permeability of intestinal wall. Glycocomponents bind ammonia and other toxic metabolites in the digestive tract thus neutralize their harmful effects. Essential oils could be used to control *Clostridium perfringens* and coccidiosis.

8.9.8 Metabolic Modifiers

Metabolic modifiers are a group of compounds that modify animal metabolism in specific and directed ways. They have the overall effect of improving productive efficiency (weight gain or egg production) and improving carcass composition (lean: fat ratio) in growing birds. Two classes of compounds have received major focus, one the somatotropins (STS) and the other adrenergic agonists. Both the metabolic modifiers are already in use in the USA and are also approved by several countries. Commercially these compounds are produced by the use of recombinant DNA technology to selectively produce specific components for a species.

8.9.9 Bacteriophages

Bacteriophages are much more specific than antibiotics and thus can be used to treat pathogenic infection. Phages are currently being used therapeutically to treat bacterial infections that do not respond to conventional antibiotics, particularly in Russia and Georgia. They tend to be more successful than antibiotics where there is a bio-film covered by a polysaccharide layer, which antibiotics typically cannot penetrate. It is likely that interest in the use of phage in the poultry industry will increase with increased consumer demand to reduce or eliminate use of antibiotics and chemical treatments.

8.9.10 Other Additives

Natural betaine increases lean meat yield and reduces the need for choline chloride and methionine in feed formulations. It also alleviates the negative effects of osmotic stress and helps to maintain gut integrity, thus improves birds' performance. It is also claimed that broilers exposed to coccidial challenge, natural betaine improved nutrient digestibility compared to synthetic anhydrous betaine and betaine-HCl products. Moreover, a significant improvement in feed conversion was observed in heat stressed broilers fed with natural betaine supplementation, while synthetic anhydrous betaine and betaine-HCl products only showed numerical improvement compared with the control. Beta-glucans, known as biological response modifiers, have ability to activate the immune system. It has been established that a receptor on the surface of innate immune cells (Complement Receptor 3, CR3 or CD11b/CD18) is responsible for binding to beta-glucans, allowing the immune cells to recognize them as 'non-self'. Beta-glucans increase the quantity of phagocytic cells and their activity, improving the immunity, preventing infection, without causing inflammation. Maintaining sound gut health is an effective tool for improved production, feed conversion and health, provides scope for minimizing use of antibiotic growth promoters, and facilitates safe food production, eco-friendly environment, better profitability, birds' welfare, and organic poultry production. A sound gut health also improves immunity through ensuring absorption of the raw materials required structure and function of immune system.

8.10 Crude Fibre Digestibility in Ducks

The ducks under scavenging production system consume significant amount of fibre. Disruption of cell wall during the process of digestion plays important role for the availability of the highly digestible cell contents. Moreover, fibre, if digested provides significant quantity of volatile fatty acids to meet some amount of energy requirements. The fibre digestibility was estimated as 21.1% and 22.1% in Pekin and Mallard ducks, respectively, fed rye grass (Jeroch et al. 1997). Similarly, feeding of diet with barley and oat fibre with 20.5 and 28.2% fibre contents, respectively, to young geese resulted digestibility coefficients of 23% and 6%, respectively (Vetesi et al. 2000). Other reports (Nehring et al. 1961; Schubert et al. 1982) indicated crude fibre digestibility up to 36% in ducks. In an another study (El Beeli et al. 2002) fibre digestibilities in scavenger ducks were 40.8, 48.1, and 52.8% in ducks of 4–5, 15–18, and 36–40 months of age, respectively. Indigenous layer ducks could utilize a crude fibre level of 12% without affecting their performance (Anitha et al. 2009). The factors affecting digestibility of crude fibre are quality of fibre, lignification of fibre, content of dietary fibre, feed particle size, rate of passage, microbial profile, dietary energy–protein ratio, and age of the birds.

8.11 Conclusion

The ducks reared in backyard production system require little attention on feeding and nutrition. However, scientific feeding and nutrition plays important role for ducks reared under semi-intensive and intensive farming systems. Therefore, to understand and apply principles of nutrition in practical feed formulation and feeding of ducks it is necessary to have knowledge on comparative digestive system development of ducks compared to other poultry species, feed and its constituents, nutrient requirements, special aspects of duck feeding and watering. The feeding principles of ducks are similar to those of other poultry birds but the requirements of different nutrients differ due to variation in growth rate, composition of body and production status. There is scope for improved utilization of nutrients, performance, and gut health through addition of various feed additives. Ducks have tendency to deposit body fat. Feed or energy intake is restricted during growing period to have lean carcass with more breast and leg muscles. However, feed restriction reduces growth. Protein is the most costly nutrient. There is scope to have reduction of dietary protein through supplementation of critical amino acids lysine and methionine. Ducks are more susceptible to aflatoxins than other poultry birds. The low quality protein or less protein and or dietary energy in the diets exacerbates the deleterious effect of aflatoxins. Thus, fresh and good quality ingredients, free from aflatoxins, are used in duck ration. Maize and groundnut cake being more susceptible to *Aspergillus* growth, these ingredients are replaced totally or partially by coarsely ground wheat and toasted soybean meal in the diet of ducks considering tolerance level of aflatoxins. The ingredients used for compounding of duck feed should not be stored for longer period to prevent growth of Aspergillus. Moisture content of stored ingredients should not contain more than 12% moisture. Ducks may be fed dry mash, wet mash or pellets. Due to difficulties in swallowing of dry mashes, ducks prefer wet mash. Feeding of wet feed improves digestion of feed as intestinal juice penetrates easily into the wet feed particles and thus feed intake is improved. However, feeding wet feeds needs precaution so that fungi should not grow in feed and feeders. The deep litter gets moist because of more moisture content in droppings, hence appropriate supply of water and feed must be taken care of. Nicotinic acid concentration in diets of meat type ducks needs to be considered. Moreover, ducks are sensitive to an available phosphorous deficiency in the diet. Ducks are also very sensitive to sudden changes in their diets and the quality of drinking water. The ducks are having habit of eating and drinking alternatively. Extensive works have been conducted on feeding and nutrition of chickens but less attention has been given on duck feeding and nutrition.

References

Amino News (2010) Revised amino acid recommendations. Evonik Degussa GmbH, Germany 14(01)

Anitha P, Jalaludeen A, Peethambaran PA, Leo J (2009) Effect of crude fibre andenzyme supplementation on production performance of layer ducks. Indian J Poult Sci 44(2):265–267

Anitha P, Jalaludeen A, Peethambaran PA, Leo J (2012) Nutritional and physiological status of foraging ducks. Indian J Poult Sci 47(1):64–66

Applegate TJ, Karcher DM, Lilburn MS (2005) Comparative development of the small intestine in the Turkey Poult and Pekin duckling. Poult Sci 84:426–431

Attia YA, Qota EM, Zeweil HS, Bovera F, Abd al-Hamid AE, Sahledom MD (2012) Effect of different dietary concentrations of inorganic and organic copper on growth performance and lipid metabolism of white Pekin male ducks. Br Poult Sci 53:77–88

Attia YA, Abd Al-Hamid AE, Zeweil HS, Qota EM, Bovera F, Monastra G, Sahledom MD (2013) Effect of dietary amounts of inorganic and organic zinc on productive and physiological traits of White Pekin ducks. Animals 7(6):895–900. https://doi.org/10.1017/S1751731113000050

Baéza E, Leclercq B (1998) Use of industrial amino acids to allow low protein concentrations in finishing diets for rowing Muscovy ducks. Br Poult Sci 39:90–96

Baltić MŽ, Starčević MD, Bašić M, Zenunović A, Ivanović J, Marković R et al (2017) Effects of dietary selenium-yeast concentrations on growth performance and carcass composition of ducks. Anim Prod Sci 57(8):1731–1737

Chen TF (2000) Effects of vitamin D_3 supplementation on laying performance and eggshell quality of laying Tsaiya ducks. Asian-Australas J Anim Sci 13:131–133

Chen TF, Hsu JC (2004) Effects of n-3 polyunsaturated fatty acids-enriched diet supplemented with different levels of alpha-tocopherol on lipid metabolism in laying Tsaiya ducks. Asian-Australas J Anim Sci 17:1562–1569

Chen W, Zhang HX, Wang S, Ruan D, Xie XZ, Yu DQ, Lin YC (2015) Estimation of dietary selenium requirement for Chinese egg-laying ducks. Anim Prod Sci 55:1056–1063

Chen W, Wang S, Zhang HX, Ruan D, Xia WG, Cui YY, Zheng CT, Lin YC (2017) Optimization of dietary zinc for egg production and antioxidant capacity in Chinese egg-laying ducks fed a diet based on corn-wheat bran and soybean meal. Poult Sci 96:2336–2343

Cherry P, Morris TR (2008) Nutrition and body composition. In: Cherry P, Morris TR (eds) Domestic duck production: science and practice. CABI, Oxfordshire

Clench MH, Mathias JR (1995) The avian cecum: a review. Wilson Bull 107:93–121

El Beeli MY, Musharaf NA, Abdolla HO, Bessei W (2002) Crude fibre digestibility in scavenger ducks. Arch Geflügelk 66(4):169–172

Elangovan AV, Mandal AB, Shrivastav AK, Yadav AS (2011) Supplementing probiotics (Gallipro) to broiler chicken on growth performance immunity and gut microbial population. Anim Nutr Feed Technol 11:169–176

Fan HP, Xie M, Wang WW, Hou SS, Huang W (2008) Effects of dietary energy on growth performance and carcass quality of white growing Pekin ducks from two to six weeks of age. Poult Sci 87:1162–1164

Fouad AM, Li Y, Chen W, Ruan D, Wang S, Xia W, Zheng CT (2016) Effects of dietary manganese supplementation on laying performance egg quality and antioxidant status in laying ducks. Asian J Anim Vet Adv 11:570–575

Fouad AM, Ruan D, Wang S, Chen W, Xia W, Zheng C (2018) Nutritional requirements of meat-type and egg-type ducks: what do we know? J Anim Sci Biotechnol 9:1
Gille U, Salomon FV, Rönnert J (1999) Growth of digestive organs in ducks with considerations on their growth in birds in general. Br Poult Sci 40:194–202
Helmbrecht A (2012) Amino acid nutrition in ducks. Evonik Nutrition & Care GmbH Germany, XXIV World's Poultry Congress, 5–9 Aug, 2012, pp 1–7
Hou SS (2007) Evaluation of amino acid requirement for ducks. Evonik Degussa GmbH, AminoNews™ 9(3)
ICAR (2013) Nutrient requirements of poultry. Indian Council of Agricultural Research, New Delhi
Jamroz D, Jakobsen K, Orda J, Skorupinska J, Wiliczkiewicz A (2001) Development of the gastrointestinal tract and digestibility of dietary fibre and amino acids in young chickens ducks and geese fed diets with high amounts of barley. Comp Biochem Physiol Part A 130: 643–652
Jamroz D, Jakobsen K, Knudsen KEB, Wiliczkiewicz A, Orda J (2002a) Digestibility and energy value of non-starch polysaccharides in young chickens, ducks and geese fed diets containing high amounts of barley. Comp Biochem Physiol Part A 131:657–668
Jamroz D, Jakobsen K, Orda J, Wiliczkiewicz A, Skorupinska J (2002b) Aspects of development of digestive activity of intestine in young chickens ducks and geese. J Anim Physiol Anim Nutr 86: 353–366
Jeroch HR, Timmler R, Guy G (1997) Comparison between geese Pekin and mule ducks in order to their ability to digest common rye-grass, 11th European Symposium on Water-fowl, France
King D, Asem E, Adeola O (2000) Ontogenetic development of intestinal digestive functions in white Pekin ducks. J Nutr 130:57–62
Konarkowski A (2006) Development of small intestine in chicks. Polish Poult 7:21–23
Mandal AB, Yadav AS, Johri TS, Pathak NN (2004) Nutrition and disease management of poultry. International Book Distributing Agency, Charbagh
Mandal AB, Chatterjee PN, Elangovan AV (2006) Augmenting gut health for improved nutrient utilization and welfare of poultry. Proceedings of XII Animal Nutrition Conference (7–9 Jan 2006) held at Anand, Gujarat, India, pp 131–135
Menconi A, Kuttappan VA, Hernandez-Velasco X, Urbano T, Matte F, Layton S, Kallapura G, Latorre J, Morales BE, Prado O, Vicente JL, Barton J, AndreattiFihlo RL, Lovato M, Hargis BM, Tellez G (2014) Effect of a commercially available organic acid product on body weight loss carcass yield and meat quality during preslaughter feed withdrawal in broiler chickens: a poultry welfare and economic perspective. Poult Sci 93:448–455
Mohanty SR, Babu LK, Sahoo SK, Pradhan CR, Panigrahi B, Joshi SK (2016) Effect of feeding different levels of proteins on growth feed consumption and mortality in growing khaki Campbell duck. Scholars J Agric Vet Sci 3(1):58–61
Nehring R, Richter G, Knabe B (1961) Die Verdaulich-keiverschiendenerFutterstoffebeiEnten. Arch Geflügelk 25:263–289
NRC (1994) Nutrient requirements of poultry, 9thedn. National Research Council, National Academy Press, Washington, DC
Ravi S, Pethambaran PA, Jalaludeen A, Leo J (2005) Influence of dietary calcium available phosphorus on production performance of indigenous layer ducks in cages. Egyptian Poult Sci J 25(4):1031–1036
Rodehutscord M, Kapocius M, Timmler R, Dieckmann A (2004) Linear regression approach to study amino acid digestibility in broiler chickens. Br Poult Sci 45:85–92
Schubert R, Richter G, Gruhn K (1982) Comparative investigation of the digestive performance of *Cairina,* Pekin ducks and laying hens. Arch Tierenahr 32:306–320
Soriano ME, Rovira N, Pedros N, Planas JM (1993) Morphometric changes in chicken small intestine during development. Z Gastroenterol 31:578
Thomas KS, Jalaludeen A, Peethambaran PA, Amritha V, Leo J (2005) Influence of probiotic supplementation on the performance of white Pekin ducks. Egyptian Poult Sci J 25(4):1037–1043

Thongwittaya N, Tasaki I (1992) Energy and protein requirements of Khaki Campbell x Thai Native laying ducks. Asian-Australas J Anim Sci 5:365–368

Thongwittaya N, Pleusamran P, Choktaworn N, Tasaki I (1992) Energy and protein requirements of khaki Campbell x Thai native growing ducks. Asian J Anim Vet Adv 5:357–363

Van Immerseel F, Boyen F, Gantois I, Timbermont L, Bohez L, Pasmans F, Haesebrouck F, Ducatelle R (2005) Supplementation of coated butyric acid in the feed reduces colonization and shedding of salmonella in poultry. Poult Sci 84:1851–1856

Vetesi MS, Orosz S, Mezes M (2000) Influence of feeding diets with barley or out to growing geese on performance of digestibility of nutrients and concentration of VLDL triglyceride and cholesterol in blood plasma. Arch Geflügelk 64(4):1–5

Wang Z (2006) Studies on the requirement of selenium and its mechanism of action for Peking ducks between 0 w and 3 w of age. MS Thesis, Guangdong Ocean University, Guangdong, China

Wang S, Hou S, Xie M, Huang W, Yu J (2009) Effects of different concentrations of vitamin D on growth performance blood biochemistry and tibia parameters for White Pekin duclings. Chin J Anim Vet Sci 41:1414–1420

Wang C, Xie M, Huang W, Xie JJ, Tang J, Hou SS (2013) Arginine requirements of white Pekin ducks from 1 to 21 days of age. Poult Sci 92:1007–1010

Wang S, Zhang X, Chen W, Ruan D, Xia W, Li Y, Wang SL, Lin YC, Zheng CT (2016) Effect of different concentrations of vitamin a on production performance egg quality and antioxidant indicators in ducks. Chin Poult 38:1004

Wasilewski R, Kokoszyński D, Mieczkowska A, Bernacki Z, Górska A (2015) Structure of the digestive system of ducks depending on sex and genetic background. Acta Vet Brno 84:153–158

Watkins E, Butler P, Kenyon B (2004) Post-hatch growth of the digestive system in wild and domesticated ducks. Br Poult Sci 45:331–341

Wei L, Hou S, Huang W, Xie M, Zhang T, Liu F (2009) Effects of vitamin A on performance and intestinal development of Beijing ducks from 0–3 weeks of age. Chin J Anim Nutr 21:462–467

Wen ZG, Hou SS, Tang J, Feng YL, Huang W, Guo YM, Xie M (2014a) Choline requirements of male white Pekin ducks from 21 to 42 days of age. Br Poult Sci 55:548–552

Wen ZG, Tang J, Hou SS, Guo YM, Huang W, Xie M (2014b) Choline requirements of white Pekin ducks from hatch to 21 days of age. Poult Sci 93:3091–3096

Xia W, Fouad AM, Chen W, Ruan D, Wang S, Fan Q, Wang Y, Cui Y, Zheng C (2017) Estimation of dietary arginine requirements for Longyan laying ducks. Poult Sci 96:144–150

Xie J, Wang A (2012) Effects of dietary vitamin D level on immune and antioxidant functions of egg-type ducks. Chin J Anim Nutr 24:819

Xie M, Zhao JN, Hou SS, Huang W (2010) The apparent metabolizable energy requirement of white Pekin ducklings from hatch to 3 weeks of age. Anim Feed Sci Technol 157:95–98

Xie M, Zhang L, Wen ZG, Tang J, Huang W, Hou SS (2014) Threonine requirement of white Pekin ducks from hatch to 21 days of age. Br Poult Sci 55:553–557

Zhang Q, Xu L, Doster A, Murdoch R, Cotter P, Gardner A, Applegate TJ (2014) Dietary threonine requirement of Pekin ducks from 15 to 35 days of age based on performance yield serum natural antibodies and intestinal mucin secretion. Poult Sci 93:1972–1980

Zhang Q, Zeng QF, Cotter P, Applegate TJ (2016) Dietary threonine response of Pekin ducks from hatch to 14 days of age based on performance serology and intestinal mucin secretion. Poult Sci 95:1348–1355

Dr. A. B. Mandal, Ph.D. (Animal Nutrition) Born on 06.10.1957 in West Bengal, India, did B. V.Sc. & A.H from Bidhan Chandra Krishi Vishwavidyalaya (BCKV), Mohanpur, M.V.Sc. (Animal Nutrition) from Haryana Agricultural University, Hisar and Ph.D. (Animal Nutrition) from Indian Veterinary Research Institute, Izatnagar. Served as Veterinary Surgeon, Assistant Professor, Associate Professor, Principal Scientist, and Director of Central Avian Research Institute. Contributed significantly towards development of feeding standards of livestock and poultry, area-specific mineral mixture, utilization of alternate feed resources, least-cost feed formulation, clinical nutrition, designer poultry products, in-ovo feeding and nutritional biotechnology, welfare, and clean and green poultry production. Have seven books, 383 research papers in reputed international (92) and national (291) journals, 93 book chapters and reviews, 27 manuals and technical bulletins, 35 lead papers, and 78 extension/popular articles to his credit. Received several prestigious awards and recognition. He along with his colleagues has developed two feed formulation software (MakeFeed Poultry and MakeFeedDairy), being widely used by farmers and industry. He is a Fellow of National Academy of Veterinary Sciences (NAVS) & National Academy of Agricultural Sciences (NAAS).

Incubation and Hatching of Duck Eggs

9

Stella Cyriac and Leo Joseph

Contents

S. Cyriac (✉)
Department of Poultry Breeding and Genetics, College of Avian Sciences and Management, Kerala Veterinary and Animal Sciences University, Palakkad, Kerala, India
e-mail: stella@kvasu.ac.in

L. Joseph
Centre for Advanced Studies in Poultry Science, Kerala Veterinary and Animal Sciences University, Wayanad, Kerala, India

A. Jalaludeen et al. (eds.), *Duck Production and Management Strategies*,
https://doi.org/10.1007/978-981-16-6100-6_9

Abstract

Hatching egg production as well as its incubation are specialized enterprises, which need utmost care and attention. At the time a fertile egg is laid, there is already a small embryo floating on the yolk. The vitality of this embryo must be preserved during storage until the incubation process starts. To achieve this, the eggs have to be handled carefully. Specific management procedures also help to minimize the loss of hatchability during extended egg storage. Fertility and hatchability are affected by factors such as breeder management, egg quality, egg storage and incubation conditions. Production of fertile eggs, proper handling and accurate incubation process are the key factors in the production of good quality ducklings. In this chapter, the production of hatching eggs, their selection and storage conditions are discussed. The physical requirements for incubation and the various hatchery operations including candling, fumigation, hatch day activities, disinfection and waste management are dealt in this chapter. Embryonic mortality, the common symptoms shown during hatching, its causes and corrective measures, and the various factors affecting fertility and hatchability are also explained in this chapter.

Keywords

Hatching eggs · Artificial insemination · Incubation · Fertility · Hatchability · Hatchery hygiene

List of Symbols/Abbreviations

%	Per cent
^{0}	degrees
°C	Degree Celsius
°F	Degree Fahrenheit
AI	Artificial Insemination
AM	Ante Meridiem
BPSE	Beltsville Poultry Semen Extender
cm	centimetre
CO_2	Carbon dioxide
DIC	Differential Interference Contrast
EGR	Ejaculatory Groove Region
ft^3	Cubic feet
g	gram
H_2O_2	Hydrogen Peroxide
LPSE	Lake Poultry Semen Extender
m.Osm./kg H_2O	Milliosmol per kilogram water
m^3	Cubic metre
ml	millilitre
mm	millimetre
mm^3	cubic millimetre
NATP	National Agricultural Technology Project
NS	Normal Saline
PBS	Phosphate Buffer Saline
ppm	Parts per million
QAC	Quaternary Ammonium Compounds
RH	Relative Humidity
TES	N-tris [hydroxymethyl] methyl-2-aminoethane sulfonic acid
µl	microlitre
µm	micrometre

9.1 Introduction

Ducks are waterfowl that belong to the *Anatidae* family. They are distributed all over the world, except in Antarctica (Idahor et al. 2015), because of their ability to adapt to different environment. The domestication of Mallard ducks happened in Southeast Asia at least 4000 years ago, during the Neolithic age. This species of poultry was also farmed by the Romans in Europe and the Malays in Asia during Neolithic era. All domestic duck breeds except Muscovy ducks originated from Mallard ducks (*Anas platyrhynchos*). Domestic ducks have an incubation period of 28 days and are devoid of broodiness. Unlike the domestic duck breeds, Muscovy ducks have a long incubation period and strong broodiness. It is not known at which point of time the domestic ducks lost their broodiness. Hatching of duck eggs have been practiced since human domesticated ducks. The process of hatching changes a microscopic germ into a downy duckling capable of walking, eating and expressing its needs by its voice and actions over a period of 28 days (Banerjee 1991). Chickens were domesticated before ducks, so natural incubation using broody hens was quite familiar and it was practiced during the earlier period. The natural method of incubation of duck eggs using the broody hens appears to be the most popular method. In Kerala, broody hens have been used for production of large number of Kuttanad ducklings in Alappuzha region. They were called '*chatti*' ducklings since earthen pots (called '*chatti*' in local language) were used to place the hens for incubation. Similar practice has been adopted in Tamil Nadu, India (NATP 2004). However, this traditional method is having many limitations and novice farmers are mainly practicing artificial incubation of eggs nowadays. Success of duck farming depends mainly on the regular availability/supply of day old ducklings. The reproductive performance in ducks is measured by two major economic factors; namely, fertility and hatchability. These factors are mainly influenced by environment and genetics. Healthy and vigorous ducklings could be produced only through proper care of hatching eggs before and during the incubation. This chapter elaborates the details on the production and care of hatching eggs, incubation of eggs, fertility and hatchability.

9.2 Production of Hatching Eggs

The ducks are oviparous animals that lay eggs with little or no embryonic development, within the body of the mother. In the process of egg formation, the ovary releases ova, otherwise called 'yolk' that is fertilized in the female reproductive tract by the spermatozoa, thereafter the layers of albumen, shell membrane and shell are secreted over the yolk and the resultant fertile egg is laid. An infertile egg results if the yolk is not fertilized before the formation of peripheral layers. Therefore, production of fertile egg, which can be used for the production of ducklings, is the major activity in a breeder farm. A good breeder flock of ducks has to be maintained for production of eggs suitable for hatching purpose. The hatching eggs can be produced either through natural mating or through artificial mating. Natural mating is the most popular method adopted by farmers.

9.2.1 Natural Mating

In natural mating, the females should be at least 24 weeks of age for collection of hatching eggs. The males should be introduced in a flock of females at least 2 weeks before the intended collection of hatching eggs. Males of the same hatch/batch is preferable than older males. The male: female ratio (also called as sex ratio) should preferably be 1:10 to 15 for lighter breeds like Kuttanad and Khaki Campbell and 1: 8 to 10 for heavier breeds like White Pekin and Aylesburry. Giri et al. (2014) reported that in Khaki Campbell ducks the ideal sex ratio for production of fertile eggs under intensive rearing is 1:5. Sex ratio of one drake to five ducks was found optimum in Muscovy ducks according to Idahor et al. (2015), whereas one male to four females as per Nickolova (2005) and Alonso-Alvarez (2006). However, Banerjee (2013) reported that even a wider sex ratio of one drake to six ducks can also be used with satisfactory results if the number of males is limited. Less number of drakes in a breeding flock will result in reduced fertility. On the other hand, larger number of drakes, in the excess of sex ratio, makes the whole flock restless and that will increase the feed cost also. The hatching egg collection may be started 1–2 weeks after introduction of males. In Muscovy ducks, the drakes have to be introduced 1 month before the intended collection of hatching eggs since their frequency of mating is lesser compared to other waterfowls.

9.2.2 Artificial Mating

Artificial mating is a technique that involves collection of semen from the male, its evaluation, processing and finally deposition within the female's reproductive tract. This helps in utilizing the superior males effectively and can be used as a tool to reduce inbreeding. It also helps to promote outbreeding by transporting semen from a distant population. Also, the problems associated with the transport of birds, viz. the risk of introducing diseases can be kept minimum if only semen is transported from a distant population. The number of drakes required for breeding can also be reduced by employing artificial insemination (AI). Besides these, artificial insemination technology helps to save the cost of breeding pen, trap nests and labour. AI is the only method of mating in species hybridization, especially for production of mule ducks generated by crossing Muscovy drakes with females of common ducks. Artificial Insemination (AI) technique will facilitate rearing ducks in cages for breeding purpose and selection programs. This system has the advantage of generating accurate data and clean hatching eggs.

9.2.2.1 Semen Collection

Even though, there are methods like electro ejaculation and artificial vagina method for collection of semen from drakes, the ideal, easy and cost effective method of semen collection is the massage method. The procedure is the same as that outlined by Lake and Stewart (1978a) in chicken. The following modifications are recommended while attempting semen collection. The drake has to be kept on a table and should be held in position by one assistant. Hold the glass funnel used for collecting semen, in between the fingers of the technician's hand used for massaging the abdomen. With the other hand, gently stroke the back toward the tail ending with the thumb and forefinger on each side of the cloaca. The cloacal region has to be gently pressed to extrude the phallus. If complete extrusion of phallus is allowed, the semen might be lost as it flows down the spiral shaped phallus. Hence, it is preferable to collect the semen directly to the funnel from the engorged base of the phallus before complete extrusion (Fig. 9.1).

Semen collections should be done after clipping the feathers around the vent region in order to avoid contamination. The birds should be trained at weekly

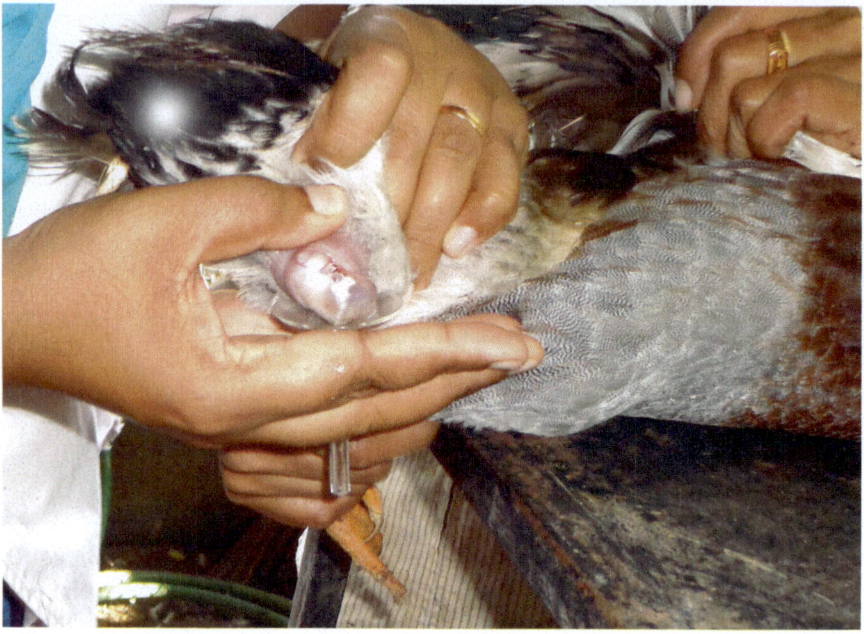

Fig. 9.1 Semen collection from a drake

intervals for a month prior to collection, which includes handling and attempting to ejaculate the birds. The trained males readily ejaculate good quality semen. It is better to collect semen in early morning hours before feeding, as this will avoid faecal contamination of semen. Artificial insemination should also be carried out in the early morning hours after egg laying.

A volume of 0.16 ml and 0.18 ml of fresh semen with a concentration of 3.03 and 3.22 billion spermatozoa/ml was reported in White Pekin and Kuttanad drakes, respectively, through the manual massage method by Cyriac et al. (2013). The copulatory organs of drakes are much larger than those of gallinaceous birds, and the extended phallus – like folds could protrude several centimeters from cloaca in these species. Sturkie (1986) reported that the phallus of drake is a vascularized and spiral shaped sac with an ejaculatory groove that can be protruded by muscle and lymph infiltration, and retracted by a cartilage like ligament. In spite of its anatomical similarity to a mammalian penis, however, the copulatory organ of waterfowl is functionally similar to that of gallinaceous birds because semen is transported in an open groove that spirals down the outside surface of the protruding organ (Etches 1996). This was clearly observed in Muscovy drakes, whereas in White Pekin and Kuttanad breeds, the phallus was not always extruded out during collection as reported by Humphreys (1972) who collected semen from the base of engorged phallus. According to the findings of Fujihara and Nishiyama (1976), this ejaculatory groove region (EGR) of the drake can also be considered as an androgen dependent accessory sex organ in addition to the other accessory sex organs, viz. paracloacal vascular bodies and dorsal proctodeal gland.

A transparent fluid is also secreted during semen collection in drakes, which might be the lymphatic exudates from the phallic folds. Fujihara et al. (1976) concluded that the ejecting mechanism of the transparent fluid from the EGR in drake may be different from that in the cock and the lymph produced by the vascular bodies may be causing the swelling of EGR and erection of phallus. In chicken, the seminal fluid is derived from the testes and the excurrent ducts; the lymphatic exudates from the phallic folds often contribute to the ejaculate when semen is collected by abdominal massage (Etches 1996).

During molting period, the testes appeared to be completely regressed and so the functions of testes, viz. the production of spermatozoa and secretion of testosterone ceased during this period.

9.2.2.2 Semen Evaluation

Semen evaluation includes macroscopic as well as microscopic evaluation. Semen can be macroscopically examined for volume, colour, consistency, appearance, presence of contaminants and pH and microscopically for motility, concentration, percentage of live and dead spermatozoa as well as abnormal spermatozoa.

The volume of semen can be measured by drawing semen from the glass funnel into a tuberculin syringe. The colour and consistency are usually judged by visual examination. The samples can be classified as yellowish white, white or chalky white based on the colour and based on the consistency, the samples can be watery (lowest score), medium thick, thick or very viscous (highest score). The pH of fresh semen can be estimated by using pH paper or pH metre. The semen sample has to be examined for the presence of possible contaminants such as blood, faecal matter or uric acid crystals.

In order to examine the sperm motility, place a drop of undiluted semen on a dry, clean glass slide immediately after collection of semen. Take samples with the help of a micropipette. Spread it uniformly. Place a cover slip over it and observe quickly under low power objective of the microscope. Based on the activity of swirls, semen can be graded and thus motility of spermatozoa can be scored. The scoring system suggested by Wheeler and Andrews (1943) is usually followed where '0' (zero) score indicate no activity and '5' indicate extremely rapid eddies and movement.

The sperm concentration can be estimated as per the procedure suggested by Taneja and Gowe (1961). Add 10 μL of undiluted semen to 1990 μL of formal saline (saline with few drops of formalin) and agitate it sufficiently to mix the contents. Place a thick cover glass over the well-cleaned counting chamber of 'Neubaur' haemocytometer and add a drop of diluted semen to the counting chamber. Allow the sperm cells to settle, focus the chamber under high power objective and focus five large squares each consisting of 16 small squares in order to count the number of sperms. In each square, count all the sperms in the centre and those touching the upper and right hand borders and ignore those straddling the other two edges. Divide the number of sperm cells counted in 80 small squares by 100 to give the concentration of spermatozoa in million/mm^3 or $\times 10^9$/ml of semen.

Table 9.1 Semen quality characteristics of White Pekin, Kuttanad and Muscovy ducks (Cyriac et al. 2013)

Characteristics	White Pekin	Kuttanad	Muscovy
Semen volume (ml)	0.16	0.18	0.15
Semen appearance score	3.29	3.29	2.46
Semen pH	7.48	7.33	7.27
Initial motility of spermatozoa (%)	57.50	58.33	60.83
Concentration (billions/ml)	3.03	3.22	1.94
Live spermatozoa (%)	96.45	94.27	94.74
Dead spermatozoa (%)	3.55	5.73	5.26
Normal spermatozoa (%)	88.78	87.96	89.54
Abnormal spermatozoa (%)	11.22	12.04	10.46

The percentage of live and dead spermatozoa of duck semen can be estimated using staining method described by Lake and Stewart (1978b) with some modifications. Mix one drop of 2% Eosin and four drops of 10% Nigrosin on a clean glass slide. Add one drop of semen sample to it and mix it gently and uniformly. Allow to stand for 2–3 min. From this mixture, make a moderately thick smear on a clean glass slide. Air-dry and examine under oil immersion objective of the microscope. Count about 300 spermatozoa (50 spermatozoa in each of 6 microscopic fields). The unstained spermatozoa can be categorized as live and stained or partially stained spermatozoa can be counted as dead ones and calculate the percentage of live and dead spermatozoa.

The smears prepared for live sperm estimation can be used to study the percentage of abnormal spermatozoa (Lake and Stewart 1978b) also. Count 200 spermatozoa under the oil immersion objective of the microscope and classify as either normal or abnormal and calculate the total percentage of abnormal spermatozoa. The semen quality characteristics of Kuttanad, White Pekin and Muscovy ducks are given in Table 9.1.

9.2.2.3 Duck Spermatozoa

The spermatozoon of the domestic duck resembles that of other non-passerine birds, corresponding to a basic type of spermatozoon similar to that of reptiles, commonly referred as sauropsid type (Simoes et al. 2012). The duck spermatozoa are characterized by the presence of an elongated head, short mid-piece and a long principal piece (Majhi et al. 2016). They observed a total length as around 67.04 μm. The sperm head as well as the principal piece is elongated in nature with a size of approximately 13.56 μm and 50.10 μm, respectively. Mid-piece is coiled in shape and is about 3.35 μm in length. The duck sperm head is elongated and cylindrical in shape with a tapering tip and the head tip contains mitochondria with high oxidative potential.

The head of the spermatozoon consists of acrosome; a reduced and moderately electron-dense homogenous material. The mid-piece contains electron-dense material associated with the proximal centriole and nuclear membrane, and a long distal

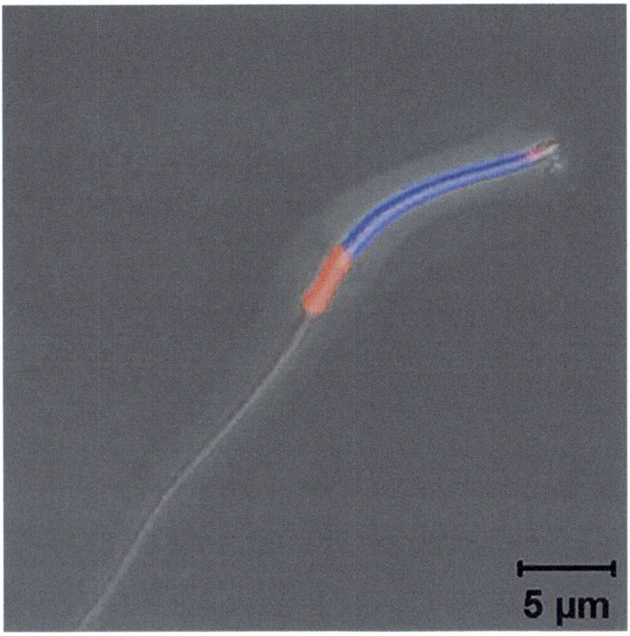

Fig. 9.2 Sperm cell of White Pekin duck (Majhi 2020; personal communication)

centriole surrounded throughout its length by 11–12 elliptical mitochondria. The mid-piece is separated from the principal piece by dense annulus. Posterior to the annulus, the axoneme is surrounded by a dense fibrous sheath, representing the principal piece or flagellum, which is a long segment with a smooth surface but smaller in diameter than the mid-piece. The differential interference contrast (DIC) image of a single sperm cell of White Pekin duck is given shown in Fig. 9.2.

9.2.2.4 Semen Diluents

The semen of ducks is highly concentrated and the volume is too low to handle conveniently. Many poultry semen extenders have been developed and their efficiency has been evaluated in chicken and other avian species. However, research work on storage and preservation of drake semen is limited and the viability of drake sperms after dilution in different extenders is least studied. Cyriac et al. (2014) reported that common poultry semen diluents namely, Normal Saline (NS), Phosphate Buffer Saline (PBS), Lake Poultry Semen Extender (LPSE) and Beltsville Poultry Semen Extender (BPSE) could be used to maintain the motility of drake spermatozoa also. Dilution of semen with LPSE at 1: 1 ratio is better in maintaining the motility. Dilution with PBS at 1:1 ratio is recommended for dilution of drake semen that is to be used for insemination immediately after collection of semen. Readymade PBS powder is available in markets and this make the preparation of PBS easier when compared with LPSE. But, if the semen is to be used later, dilution with LPSE in 1: 1 ratio under refrigeration conditions is recommended. Normal

Table 9.2 Composition (g/100 ml double distilled water) of Lake Poultry Semen Extender and Beltsville Poultry Semen Extender

Sl. no.	Component	LPSE	BPSE
1.	Sodium glutamate	1.350	0.867
2.	Tripotassium citrate	0.128	0.064
3.	Sodium acetate	0.510	0.430
4.	Magnesium acetate tetra hydrate	0.080	–
5.	Glucose	0.800	–
6.	Potassium diphosphate	–	1.270
7.	Fructose	–	0.500
8.	TES[a]	–	0.195
9.	Potassium monophosphate	–	0.065
10.	Magnesium chloride	–	0.034
11.	pH	7.2	7.55
12	Osmotic pressure (m.Osm./kg H_2O)	310	333

[a]N-tris [hydroxymethyl] methyl-2-aminoethane sulfonic acid

saline can be prepared by dissolving 0.9 g sodium chloride in 100 ml distilled water and PBS by dissolving 1.0078 g PBS powder in 100 ml double distilled water. The composition (g/100 ml double distilled water) of Lake Poultry Semen Extender as described by Lake (1960) and Beltsville Poultry Semen Extender as described by Sexton (1977) are given in Table 9.2.

9.2.2.5 Insemination of Females

The oviduct of the duck may be exposed from the cloaca before insemination. The birds in the cages should be caught by holding on their legs with head down and the bird's abdominal region against the body of the holder. The oviduct is everted by pressing the tail towards the back and by gently pressing the abdomen with the other hand. Since, only the left oviduct is functional in ducks, as in case of other poultry, the left hand side of the abdomen has to be pressed.

The semen in thin plastic straws (cannulas) connected to a syringe through an adapter tube (Fig. 9.3) should contain the correct volume of semen to give minimum 100 million spermatozoa. This cannula is inserted deep into the everted oviduct; preferably to a depth of about 60 mm in ducks (20 mm in smaller domestic fowls) so that the semen is deposited near the sperm storage glands in the females. At the time of injection of semen, the pressure around the cloacal region was gradually released until it returned to its normal position (Fig. 9.4). The eversion of oviduct in ducks is a little complicated, and the extent of eversion is not as much as that occurs in chicken. Semen loss and backflow of semen out of the oviduct has to be avoided by blowing the semen in the cannula before the oviduct reverts (Rose 1997).

Either fresh semen or diluted semen can be used for insemination. Normal saline, phosphate buffered saline, etc. can be used as diluents. Since ducks complete egg laying in the early morning hours, the preferred time for insemination is morning hours, as the reproductive tract is free from any obstructions during this time. Two to

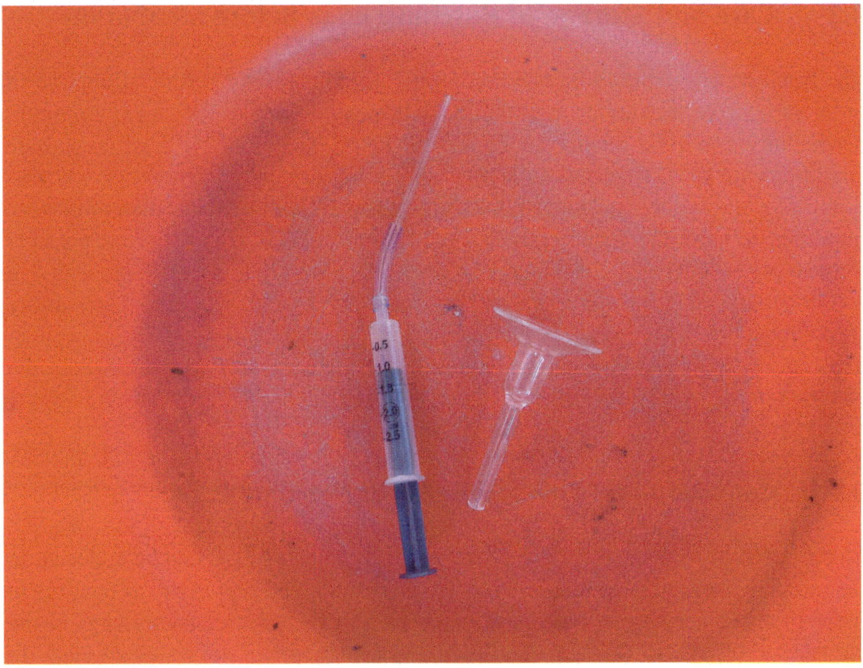

Fig. 9.3 Semen Collection funnel and AI syringe

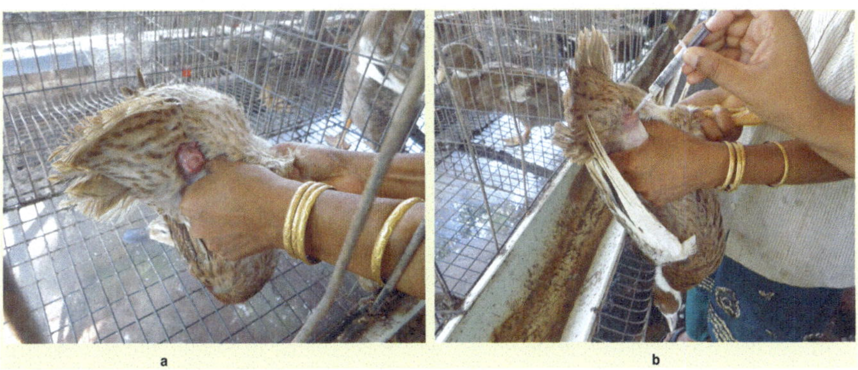

Fig. 9.4 Artificial insemination in ducks. (**a**) Eversion of Vagina (**b**) Insemination

three days after the first AI, hatching egg collection can be started. Once collection of hatching eggs is started, AI has to be repeated once in every 5 days for getting optimum fertility (Cyriac et al. 2018). According to Etches (1996), 9.2 inseminations per ejaculate (assuming 100×10^6 cells per insemination) were possible with 0.23 ml Pekin semen having a concentration of 4.0×10^9 cells/ml. The number of inseminations can be increased further if using diluted semen.

9.3 Care and Storage of Hatching Eggs Before Incubation

The hatching eggs should be handled properly and carefully before incubation in order to achieve maximum hatchability with healthy viable ducklings. Optimum hatchability and good quality ducklings can be produced only if most favourable conditions are provided between laying and setting of eggs. A fertile egg contains 1-day old embryo of around 20,000 cells when it is laid. It is, therefore, very important to store the fertile eggs in a cold room (12–18 °C) as soon as possible after laying in order to maintain the viability of the embryo. It is well documented in earlier studies that storing eggs reduces viability of embryos and hatchability (Proudfoot 1969; Meijerhof 1992; Fasenko et al. 2002). Mishandling as well as exposure to high and fluctuating environmental temperature will deteriorate the hatching potential. Hence, extreme care is required while handling hatching eggs.

9.3.1 Collection and Transportation of Hatching Eggs

Since, ducks lay eggs in the early morning hours, hatching eggs can be collected in the morning itself. In deep litter system, the litter should be kept as dry as possible in order to avoid dirty eggs. Since, ducks are having the habit of laying eggs in the corners, additional fresh litter may be added in the corners to get clean eggs. In addition, nest boxes can be provided. Nest box of size 30 × 30 × 45 cm is recommended for every 3–5 egg type laying ducks and that with dimensions of 40 × 40 × 45 cm is recommended for every 3–5 broiler ducks for getting clean eggs. The nest boxes can be protected with portable guards for preventing the birds from soiling the nests during the day. Fresh litter can be added in the nests before removing the guards in the late afternoon and this will reduce the incidence of soiled eggs. Nest boxes should be kept clean in order to ensure that the eggs are free from *Salmonella* contamination.

The eggs should be collected on clean, sanitized plastic filler flats or in clean paper filler flats. Measures should be taken to keep the eggs at a temperature below the physiological zero (17 °C) as quickly as possible. The hatching eggs should not be shaken or jarred and if any such cases occur while transit, such eggs should be allowed to settle for 24 h before setting in incubator.

In extensive system of rearing ducks, which is popular in Kerala, the farmers are housing ducks in double net enclosures in such a way that one encircles the other during the night hours. In the early morning hours, farmers remove the inner circle so that ducks move adjacent to the outer circle and lay eggs there and this facilitates to get comparatively cleaner eggs, less soiled with mud and droppings. Eggs are collected at 5 AM and a second collection is made at 6 AM. The farmers will move the hatching eggs to a store house immediately without exposing them to sunlight. This is because when eggs are exposed to sunlight, the cuticle may get dried up and this will open the pores and it favours the entry of microorganisms. The albumen and yolk quality will be deteriorated which will affect the viability of embryos (Jalaludeen et al. 2004).

9.3.2 Selection of Hatching Eggs

The eggs collected from the farm should be regularly screened and the following points have to be considered while selecting eggs for hatching purpose.

(a) *Size of eggs:* Even though the size of eggs varies with breed and strain of ducks, medium sized eggs are preferred over too small or too large eggs as they create hindrance in setting in incubation trays and also do not hatch well. The optimum size of hatching eggs is 65–70 g. El-hack et al. (2019a) opined that heavier eggs hatch better than those with low weight.
(b) *Shape of eggs:* Oval shaped eggs are preferred for hatching purpose. Elongated and spherical eggs do not hatch well.
(c) *Shell quality:* The eggs should have clean, sound and thick shell. Any cracks in the shell can be detected at the time of candling and should be discarded.
(d) *Interior quality:* Hatching eggs should have good albumen and yolk quality and should be free from blood spots, meat spots or other defects.

Egg samples may be tested for the above parameters before starting hatching egg collection from a new flock.

9.3.3 Cleaning and Sanitizing of Hatching Eggs

The dirty eggs can be cleaned with damp cloth dipped in warm water (preferably at 40 °C) or with a dry cloth or with sand paper depending on the nature of the dirt and severity of soiling. Certain chemicals can also be used for cleaning dirty eggs and the recommended ones are 0.3% glutaraldehyde solution at 40 °C and 2500 ppm sodium hypochlorite solution at 40 °C. Be sure that the eggs are dry before storing. Only clean eggs should be kept in the incubator, as there is chance for the soiled eggs to nurture bacterial growth and produces gas inside and this will finally explode inside the incubator contaminating other eggs. Harikrishnan et al. (2013) collected hatching eggs from ducks reared in semi-intensive system with deep litter in housing. They reported that eggs subjected to dry cleaning with muslin cloth gave better hatchability and good quality day old ducklings than eggs washed with water or with glutaraldehyde solution or sodium hypochlorite solution at 40 °C.

9.3.4 Candling of Hatching Eggs

The cleaned eggs have to be candled individually before storage. The shell quality and interior quality can be easily assessed by this method. A box type table top egg candler (Fig. 9.5) is ideal for this purpose. The following details can be verified.

1. Position and size of air cell – Air cell should be at the broad end with 1–3 mm size (fresh eggs may not have air cell).

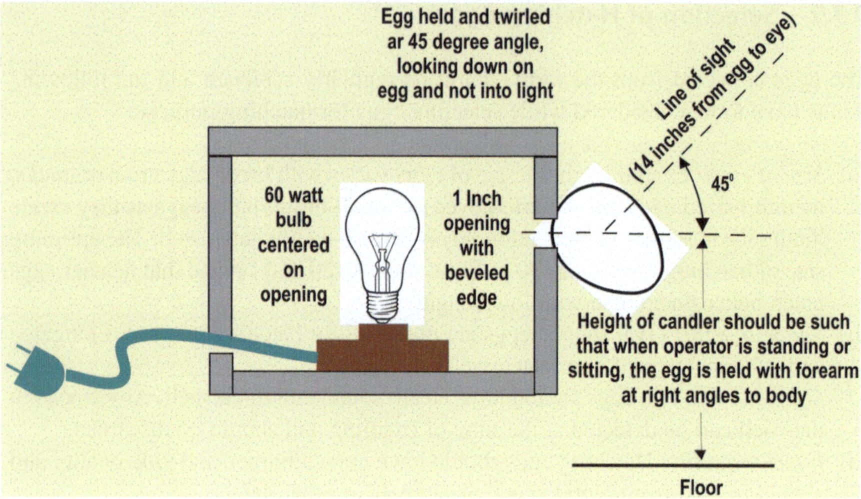

Fig. 9.5 Table top box type egg candler

2. Presence of cracks in the shell – Cracks appear as a light corrugated line and such eggs should not be used for hatching purpose.
3. Uniformity of shell intensity and thickness – The thin shells and patches appear brighter (Fig. 9.6).
4. Presence of inclusion bodies – Eggs having blood spot, meat spot should be discarded.

9.3.5 Fumigation

Disinfection of hatching eggs prior to incubation helps to limit the bacterial load and ensures better hatchability. Excessive bacterial contamination can also lead to poor chick quality, poor growth and performance (Scott and Swetnam 1993) and increased mortality (Reid et al. 1961). Pre-incubation fumigation of hatching eggs should be done before storage. If the harmful bacteria penetrate the shell, they reach the shell membrane and get protected from the fumigant. Thus, fumigate the eggs as soon as possible, preferably on arrival at the hatchery. It is better not to fumigate after egg handling, right before setting. This is because the bacteria might have already entered the egg by this time. Another reason is that any residue of the formalin on the eggs at the time of setting can cause the embryo to die, as the young embryonic germ is very susceptible to certain disinfectants such as formalin. Potassium permanganate and formalin fumigation are recommended for sanitizing hatching eggs at 1:2 ratio. This is done in a closed chamber with proper air circulation and ventilation facilities. The potassium permanganate should be taken in a clay or mud pot into which formalin is added. The formaldehyde fumes generated has to be circulated inside the chamber with the help of fans and it has to be eliminated outside through the

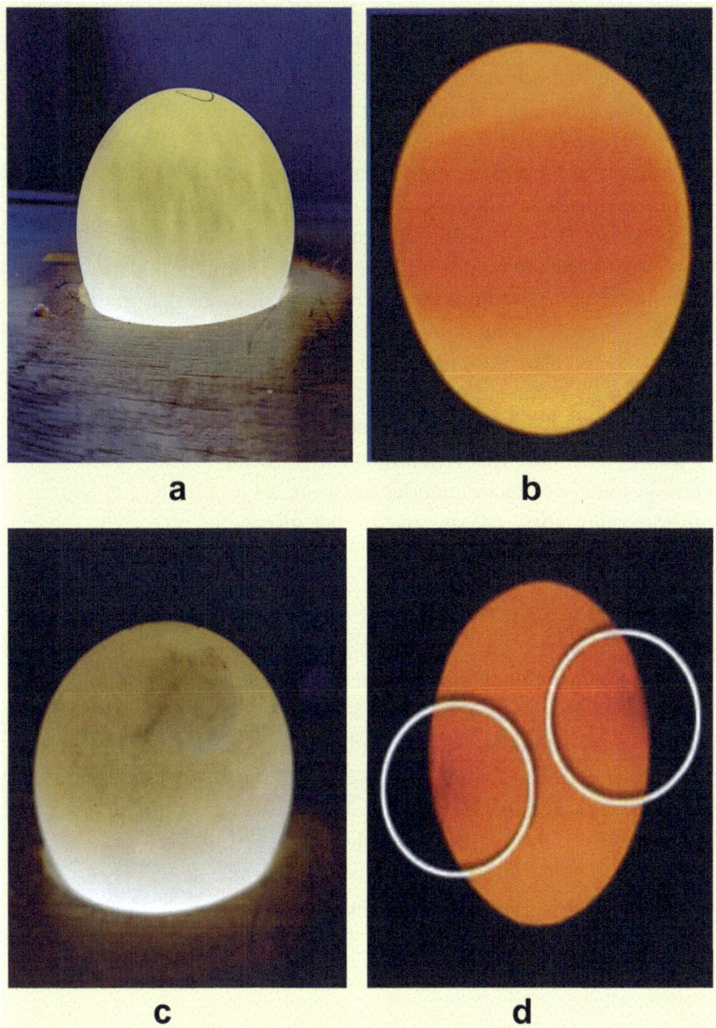

Fig. 9.6 Appearance of various conditions during candling prior to incubation. (**a**) Good quality hatching egg (**b**) Poor quality, large air cell (**c**) Cracked egg (**d**) Double yolk

ventilation after a prescribed contact time. Airflow is one of the key factors deciding the efficiency of fumigation. The entire surface of the hatching egg should be exposed to the fumigant for getting good results.

A preparation with 20 g potassium permanganate and 40 ml formalin (1X concentration) will generate formaldehyde gas sufficient to disinfect a chamber with volume 2.8 m^3 or 100 ft^3. Similar amount of gas can also be produced by 10 g paraformaldehyde. The hatching eggs after cleaning have to be fumigated with 3X concentration for 30 min before storing or setting. Formaldehyde is a surface disinfectant and it will not sanitize underneath the dirt. Therefore, it is important to destroy microorganisms on the surface of the eggshell and ensure that the shell is not covered in dirt. The setter/hatcher/rooms should also be fumigated with 3X concentration for 30 min before incubation. If needed, fumigation can be carried out in hatcher after transfer of eggs with 1Xconcentration for 5 min. In such cases, the formaldehyde gas has to be neutralized with ammonium hydroxide, and by this, the formaldehyde gas will be converted into harmless hexamine. For this, a vessel containing ammonia (26–29%) or cloths saturated with ammonium hydroxide may be introduced into the incubator at the end of fumigation period. Ammonium hydroxide solution can be applied at the rate of not more than half of the volume of formalinused. Spread on the floor of the incubator and close the doors. Formaldehyde fumigation can also be carried out by cheesecloth method wherein, clean cheese cloth dipped in formalin (15 ml for 2.8 m^3) has to be hanged inside the incubator and fan may be turned on. The soaked cloth may be allowed to stay there for 3 h. A higher dose than that recommended for the purpose will lead to negative results. Fumigation is not recommended between 24 and 96 h after setting as it can cause embryonic mortality. Always use rubber gloves for protection while fumigation and the doors should not be opened before the prescribed time.

9.3.5.1 Fumigation with H_2O_2

Formaldehyde is an irritant gas, which causes irritation of eyes and nose. Therefore, alternative disinfectant like Hydrogen Peroxide can be used for fumigation of hatching eggs without affecting the hatching potential. H_2O_2 has no odour and poses minimal safety problems to the handlers. It produces nascent oxygen, which is a very potent disinfectant. It can be used directly or preparations with nascent oxygen may be used. Usually, H_2O_2 at 5% concentration is suggested as comparable to formaldehyde for fumigation of hatching eggs (Samberg and Meroz 1995). Commercial preparations are available under the trade name ‘Huwasan’ in India (Joseph 2011, personal communication) and a mist machine/fogger (Fig. 9.7) can be used to spray mist like droplets into the space for disinfection (incubator, storage room, etc.).

9.3.6 Storage of Hatching Eggs

The cleaned eggs should be cooled to a temperature below the ‘physiological zero’ as quickly as possible. *Physiological zero* refers to the temperature below which the embryonic growth will be suspended and this temperature range is from 12 to 18 °C. Hence, the temperature of the cold room for storing eggs should be in the range of 12–18 °C (55–65 °F). The embryo remains in a state of embryonic diapause when the eggs were held at temperatures below their physiological zero. Embryo development still occurs at a minimum rate during storage (Bagliacca et al. 2005). The embryonic growth will be re-initiated in a temperature above the physiological zero. Fluctuation in temperature is not recommended in storage room. The temperature gradient inside the storage room results in variation in time at which the eggs reach

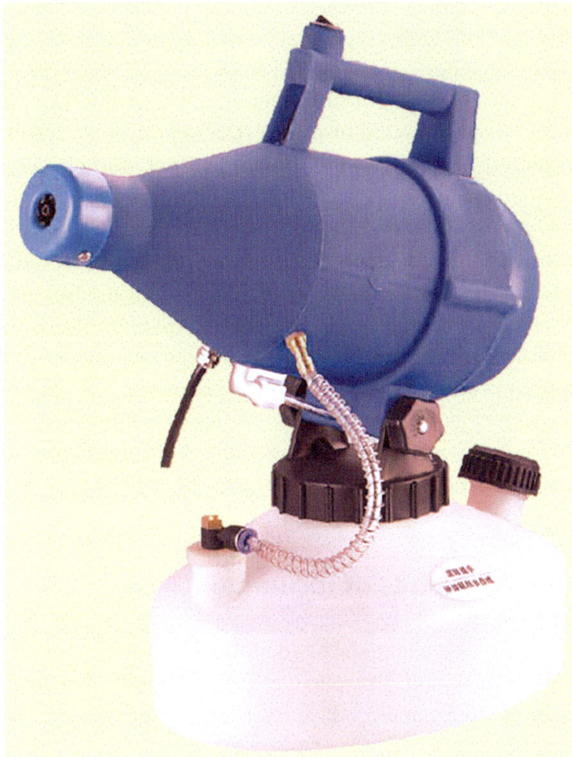

Fig. 9.7 Mist machine/fogger

incubation temperature when placed in the incubator. This will later lead to hatching of eggs in a prolonged period of window. The best way to ensure uniformity of temperature is to install many thermometers at different points inside the egg holding room. Cormic (2013) recommended the use of minimum-maximum thermometer instead of standard one as it can record the extreme temperatures during previous 24 h. Also, it is better to avoid placing the thermometers against walls as the reading can be influenced by the temperature of the wall when compared to free hanging ones. The moisture loss from the egg occurs through the eggshell to the atmosphere of the egg holding room during the storage of eggs. One way of preventing the moisture loss from the eggs is by maintaining higher humidity in the egg storage room; so that, the air cannot attract more moisture from the eggs. The humidity of the egg storage room should be maintained at 60–70% in order to prevent moisture loss from the eggs before they are incubated.

The hygiene of the room should be well maintained by keeping it clean and tidy. The eggs should be kept in clean and fumigated filler flats in the storage room and the room should be free from vermin. In order to avoid loss due to reduced hatchability, maximum storage period should be limited to 7 days. Prolonged storage can reduce the hatchability and lengthen the incubation time. Sometimes, it may lead to morphological abnormalities of the embryo (Arora and Kosin 1966) and poor quality ducklings. Prolonged storage period decreases hatchability with increased early and late embryonic deaths; mainly due to moisture loss and albumin degradation in eggs during storage (El-hack et al. 2019a). The longer period of storage also increases the spread of time over which hatching take place, reduces total hatchability and overall quality of chicks (Decuypere et al. 2001). The hatching eggs have to be placed with small end down position during storage. This position helps to maintain the air space into original position and provide highest rate of survival of embryos. For better results, the egg position has to be altered periodically by turning to a new position once daily during storage, if stored for more than 7 days. It is possible to store the eggs upside down with the pointed end up unlike setter trays. Mayes and Takeballi (1984) found that in eggs stored with narrow end up position, the yolk and the embryo are placed in the central position of the egg and thereby the embryo is protected during storage. However, this position is not recommended for hatching eggs during transport, as this may cause loose air cells (Schulte-Drüggelte 2011). Lima et al. (2012) also pointed out that the adverse effects of long storage period of 14 days may be lessened if the eggs are stored with the small end up.

9.4 Methods of Incubation

There are two methods of incubation: natural method and artificial method. The incubation period, irrespective of the method of incubation is 28 days for eggs of common ducks and 35 days for Muscovy duck eggs (Table 9.3).

9.4.1 Natural Method of Incubation

This is the traditional system of incubation in which the eggs are incubated with the help of broody hens. Depending on the size of broody hens, 12–15 eggs can be set under each hen. The broody hens should be healthy, quiet, good sitters and have good body size. A saucer shaped nest made of bamboo or wood filled with bedding materials should be provided at a safe and comfortable place. The hens should be provided with feed and water at least twice a day, by taking them out of the nest. In order to maintain the humidity, lukewarm water can be sprinkled on the eggs. Candling can also be practiced after 7 days of incubation to remove infertile eggs. Among the exotic duck breeds, Muscovy ducks are having maternal instinct and they are good sitters, they can be used to incubate eggs from common ducks also. If ducks

Table 9.3 Incubation period of eggs of different waterfowl species

Sl. no.	Species	Incubation period	No. of days in setter	No. of days in hatcher
1	Common duck	28	25	3
2	Muscovy duck	35	32	3
3	Mule duck	31	28	3
4	Geese	42	39	3

are provided with water ponds, they may go to the ponds and get wet, which will meet the increased humidity requirement for duck eggs.

The custom hatching of duck eggs using broody hens was practiced in the Kuttanad area (Alappuzha district) of Kerala, India before 1990s. In this system of hatching, eggs were mainly incubated in three definite seasons, namely January–February, August–September and November–December. The local housewives engage in this activity as a cottage industry to custom hatch the duck eggs as a replacement hatch for the farmers rearing large flocks under extensive system of management. The housewives had been paid a fixed amount as remuneration (hatching cost) by the duck farmers. A hatchability of about 80–85% was obtained by this method. The main advantage was the low cost of incubation as the only expenditure involved was the feeding cost of broody hens. The broody hens were fed mainly with cooked rice soaked in water overnight; as the farmers believed that the hens remain broody for a long period with this feeding. The same hen were used continuously for hatching a second set of eggs also after the first hatch; while the ducklings hatched were collected by the farmers after paying hatching cost to the housewives (Jalaludeen et al. 2004). Due to the various limitations for this natural method of incubation, farmers resort to the artificial method of incubation after 1990s.

9.4.2 Artificial Method of Incubation

The artificial method of incubation of eggs started more than 2000 years ago in China and Egypt. In ancient times, the egg chambers were simply heating compartments. In modern hatcheries, the machines are equipped to maintain a perfect control of temperature, humidity, ventilation and turning of eggs under extremely hygienic condition to produce large number of chicks/ducklings. Charles A. Cyphers designed the first large sized incubators for duck eggs in 1895 (Brown 1979).

9.4.2.1 Parched Rice Husk Hatching

This is one of the traditional artificial incubation techniques developed in China for hatching duck eggs. The temperature during incubation is maintained by the heat supplied by the heated paddy rice and the heat generated by growing embryo. This method is still used in China, Bali and Indonesia with hatchability up to 80% (Smith 1990). Usage of parched rice or rice husks can give a hatchability of 65–75%. As duck egg shells are less brittle when compared to chicken eggs and hence this system is more suited for incubation of ducks eggs.

The procedure of parched rice or rice husk incubation was described by Sonaiya and Swan (2004). The hatching eggs are first heated by placing under the sun or in a heated room. Simultaneously, the unhusked rice (paddy) is also heated with continuous stirring until it reaches to a temperature of 60 °C. The heated paddy is used to provide warmth for the eggs during first 2 weeks of incubation. The paddy pillows are prepared by packing about 2.5–3.0 kg of the heated rice. The pillows are of the same diameter (about 40 cm) as the cylindrical incubation basket and about 8 cm thick. It is better to make pillows with black-coloured material, as they can easily absorb the sun's heat while reheating. The temperature of the pillows is allowed to drop to 40 °C. A warm paddy pillow is placed into the egg incubation basket first and a loose pack of 40 pre-warmed duck eggs is placed on top of the warm pillow. The egg packs can be made using a square piece of cloth or soft duster cloth with pinholes of about 45 cm on each side. The warm pillows and egg packs are placed alternatively up to the brim of the basket finishing with a paddy pillow at the top. The basket is then covered with a blanket to conserve the heat. This procedure is repeated until all the eggs are placed in baskets.

The filled in egg baskets are then placed in large bamboo frame setting boxes. The walls of the setting boxes are lined with rice husk bags to conserve heat and rice husk material is also placed between the cylinders for better insulation.

Reheated paddy rice (or rice husk) is added thrice a day at regular intervals for first 3 days and twice a day during the next 4–6 days. This will ensure optimum temperature to the eggs, which is most suitable for embryo development. The egg packs are restacked in separate empty baskets but with reheated paddy or rice husk. Thus the top layer of eggs goes to the bottom and the bottom layer comes to the top in the new cylinder. The emptied basket is used to place eggs from the next basket and the cycle is thus continued.

Heat is to be provided for 13–14 days during summer and for 18–19 days in the cold season. Afterwards, the heat emitted by the hatching eggs is sufficient to maintain the temperature inside the incubation basket. Incubating eggs during their first 6 days are called 'new eggs', those from 7 to 13 days are called 'in-between eggs' and after day 13, they are called 'old eggs'. If a basket contains old and in-between eggs, the embryo generated heat is sufficient to incubate the new eggs. The new eggs are usually introduced at intervals and placed between the layers of old eggs. The eggs in the baskets should be regularly turned and they should be exposed to air three to five times daily during the process of restacking.

9.4.2.2 Hatching with Incubators

Incubators can be classified into still air incubators and cabinet or forced draft type depending on the provision for controlling temperature, humidity and ventilation.

(i) ***Still air incubators***

In this, a temperature gradient occurs across the machine. The heat supplied at the top of the machine passes down slowly to the bottom through the eggs as the air cools. There can be as much as 18 °F (7.7 °C) difference between the temperature at the top and bottom and 5–6 °F (2–3 °C) across the eggs. The thermometer should be correctly positioned above the eggs in order to get accurate reading, and proper airflow has to be ensured to maintain optimum temperature inside the machine (Fig. 9.8).

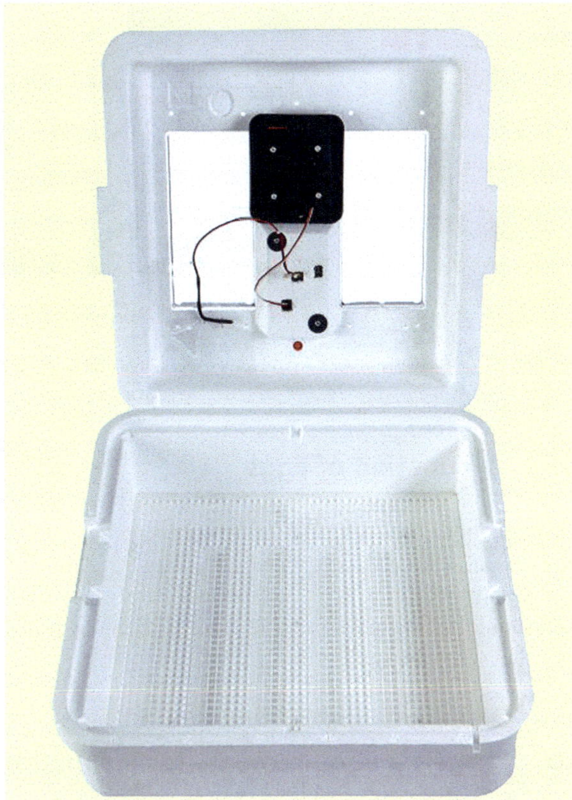

Fig. 9.8 Still air incubator

(ii) ***Cabinet incubators/Forced draft incubators***

Smith patented forced draft type of incubator in 1918; and in 1923, Petersime Incubator Co. introduced all electric incubator. In cabinet/forced draft type of incubators, the air is moved round by fan or paddles and hence, the same temperature is maintained in every part of the machine. The cabinet incubators (Fig. 9.9) can be of two types:

(a) Combined setter and hatcher: In this machine, the hatching trays are placed in a part of the machine where the temperature is 1 or 2 °C lower than that of the setter trays, in order to compensate for the excess heat generated during hatching. Along with this, some alterations are needed in the humidity, temperature or both.
(b) Separate hatchers: In this, the high humidity requirement as well as the low temperature requirement can be adjusted in the hatcher compartment.

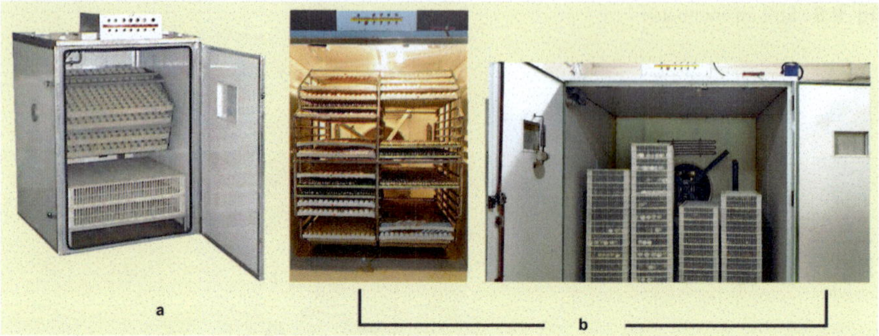

Fig. 9.9 Forced draft incubator (**a**) Combined setter and hatcher (**b**) Separate setter and hatcher

9.5 Incubation of Eggs

The eggs should be moved from the cooler room to a well-ventilated room about 6 h prior to the setting. If eggs and equipment are not fumigated before storage, the fumigation procedures have to be completed before setting. The eggs will sweat and shell should be dry before setting. The eggs can be transferred in to setter trays from filler flats after the shell is dry.

9.6 Physical Requirements for Incubation

Physical conditions such as temperature, humidity, ventilation and turning of eggs are essential for proper incubation of eggs. While the broody ducks/hens naturally create these conditions, in natural method of incubation, these conditions should be created manually or by automatic devices in artificial method of incubation. Table 9.4 shows the optimum conditions required for incubation of domestic duck eggs and Muscovy duck eggs.

The optimum conditions for hatching duck eggs in incubators with separate setter and hatcher compartments are detailed in the following. The eggs of common ducks are kept in the setter compartment for the first 25 days and that of Muscovy ducks are kept in setter compartment for 32 days. The eggs with live embryos are transferred to the hatcher compartment on 25th and 32nd days, respectively, for common and Muscovy ducks and are kept there for the last 3 days.

9.6.1 Temperature

The temperature is considered one of the most critical factors in the process of incubation of eggs. Temperature is having an effect on duration of hatch, size of embryos, embryonic mortality and viability of ducklings after hatching (EL-hack

Table 9.4 Optimum conditions required for incubation of duck eggs

Parameters		Optimum conditions
Incubation period (days)	Common ducks	28
	Muscovy ducks	35
Temperature (°F)	In Setter	99.5
	In Hatcher	98.5
Relative humidity (%)	In Setter	75
	In Hatcher	85
Gaseous concentration (%)	Oxygen	21
	Carbon Dioxide	0.4
Position	In Setter	Broad end up
	In Hatcher	Horizontal
Turning	In Setter	Hourly at 45 degrees
	In Hatcher	No turning
Transfer to Hatcher	Common ducks	On 25th day
	Muscovy ducks	On 32nd day

et al. 2019b). Even though slight variation from the optimum may affect the above parameters, the intensity of adverse effects depends on the extend and duration of temperature deviation from the optimum and the stage of incubation at that time. Hence, the thermostat controlling the temperature as well as the wet and dry bulb thermometers installed in the machine should be checked regularly with thermometers of known accuracy.

In forced draft incubators with different setter and hatcher compartments, the optimum temperature requirement for ducks eggs in the setter is 38 °C (99.5 °F) and that in the hatcher is 37.5 °C (98.5 °F).

9.6.2 Humidity

Relative Humidity (RH) is referred as the percentage of saturation of the atmospheric air with moisture. RH is a ratio (%) of the amount of moisture in the atmosphere to its fully saturated level. Since the saturated level is dependent on temperature, RH is a function of both moisture content and temperature. RH is measured by comparing dry and wet bulb thermometer readings based on Table 9.5. The relative humidity in the setter should be 75% and that in the hatcher should be 85% for better hatching results of duck eggs. However, the relative humidity requirement for chicken eggs during incubation is only 60% and 70% in setter and hatcher compartments, respectively, which is much lesser than that of duck eggs. During incubation, the hatching eggs must loose certain proportion of weight through moisture loss in order to produce strong and viable ducklings. The weight loss is about 14–17% for Kuttanad duck eggs at 24th day of incubation according to Harikrishnan et al. (2013). The shells of duck eggs are highly porous and therefore the rate of water loss is higher with duck eggs due to evaporation. Hence, high humidity is necessary inside the

Table 9.5 Relative Humidity chart using Wet and Dry Bulb Thermometer readings in Degree Fahrenheit

		DRY BULB TEMPERATURE									
		85	**90**	**95**	**96**	**97**	**98**	**99**	**100**	**101**	**102**
	68	41	31	23	22	21	19	18	17	16	15
	69	44	34	25	24	23	21	20	19	18	17
	70	47	36	28	26	25	23	22	21	20	18
W	**71**	50	39	30	28	27	25	24	23	21	20
E	**72**	53	41	32	30	29	27	26	25	23	22
T	**73**	56	44	34	33	31	30	28	27	25	24
	74	60	47	37	35	33	32	30	29	27	26
B	**75**	63	50	39	37	36	34	32	31	29	28
U	**76**	66	53	42	40	38	36	34	33	31	30
L	**77**	70	55	44	42	40	38	37	35	33	32
B	**78**	73	58	47	45	43	41	39	37	36	34
	79	77	62	49	47	45	43	41	39	38	36
T	**80**	80	65	52	50	48	46	44	42	40	38
E	**81**	84	68	55	52	50	48	46	44	42	40
M	**82**	88	71	57	55	53	51	48	46	43	43
P	**83**	92	74	60	58	55	53	51	48	46	45
E	**84**	96	78	63	61	58	56	54	51	49	47
R	**85**	100	81	66	64	61	59	56	54	52	50
A	**86**		85	69	67	64	61	59	57	54	52
T	**87**		89	72	70	67	64	62	59	57	55
U	**88**		92	76	73	70	67	65	62	60	57
R	**89**		96	79	76	73	70	67	65	62	60
E	**90**		100	82	79	76	73	70	68	63	63
	91			86	82	79	76	73	71	68	65
	92			89	86	83	79	76	74	71	68
	93			93	89	86	83	80	77	74	71
	94			96	93	89	86	83	80	77	74
	95			100	96	93	89	86	83	80	77

incubator compartments to reduce the evaporative moisture loss from the eggs. The high humidity requirement for duck eggs can be met by sprinkling lukewarm water from 2nd day of incubation until the eggs are transferred to hatcher.

Extreme variation in humidity during incubation should be avoided. Low RH in setter can cause a greater loss of moisture and this will result in smaller sized ducklings. On the other hand, high RH can lower the weight loss and this will result in large sized ducklings. Too low humidity in the hatcher will lead to high rate of pipped eggs, with dried embryos or dead embryos in shell. Too high humidity in the hatcher will delay the hatch and the navels will not be healed properly. The embryos will not be dried properly, shell will stick to the ducklings and the hatchability will be reduced (Vedder et al. 2017).

9.6.3 Ventilation

Like all living creatures, hatching eggs also require oxygen. Ideal ventilation is essential to ensure the entry of oxygen and the removal released carbon dioxide through the pores in the egg shell. There are three to six thousand pores are present in the eggshell, according to Gildersleeve and Boeschen (1983). Optimum oxygen content in the incubator should be as that in the atmosphere, that is, 21%. The hatchability is reduced when the oxygen content is reduced or increased. It is generally believed that for every 1% reduction in the oxygen content, hatchability will reduce by 5%. However, the increase in oxygen content is not as detrimental as that of decreased oxygen content. The developing embryos give off the carbon dioxide which concentration should not exceed 0.5% in the setter and 0.75% in the hatcher for maximum hatchability. The increased carbon dioxide concentration will reduce the hatchability. The fans in the incubator will help to dissipate the heat generated by the developing embryos and to keep the CO_2 levels within the limit. Hence, power failure for a long time should be viewed seriously, as this can lead to the death of embryos.

9.6.4 Position

In the setter trays, eggs are set in the small end down position. The air cell being on the broad end, this position will permit the embryos to develop with head towards the large end of the egg near the air cell. This position will also help the embryos to break the air cell and initiate the pulmonary respiration at the time of hatch. In the hatcher, the eggs are placed horizontally. This will allow free movement of the ducklings at the time of hatching. This will also assist in maintaining hygiene by limiting the contamination of hatchery through the spread of fluff generated during hatch.

9.6.5 Turning

Proper rotation and turning of the hatching eggs is essential to ensure the accurate distribution of temperature, relative humidity and ventilation inside the incubator (El-hack et al. 2019b). Eggs are turned gently during incubation at regular intervals in order to avoid adhesion of the embryos to the shell/shell membranes (Elibol and Brake 2008). The egg trays are turned at 45 degrees in opposite directions, which will give a total turning of 90 degrees to the eggs. In forced draft incubators, provision is given for turning at hourly intervals in the setter. No turning is required for the last 3 days when the eggs are in the hatcher.

Incubators both setters and hatchers should be run for a day or two before use in order to ensure that all the above conditions are optimum inside the machine, especially temperature. The thermostats, thermometer and pilot lights should be checked carefully. There should be provisions for continuous power and water supply. Adequate spare parts should be kept in hand for repair/replacement in case of emergency. Manufacturer's directions should be followed closely while operating the machines. The cleanliness should never be compromised. A detailed analysis of the common symptoms of various hatching problems, probable causes and corrective measures is given in Table 9.6.

9.7 Assessing Embryonic Growth

Candling is the common practice adopted in hatcheries to assess the embryonic growth in the incubated eggs. Light is passed through eggs keeping them in dark room in order to assess the status of embryo. Eggs do not need to be removed from the setter trays for candling. Candling on 5th to 7th day helps to remove infertile eggs wherein the infertile eggs appear clear as a fresh egg and the developing ones show spider like radiating lines of blood vessels and signs of growth. Eggs with small spot fixed in a position, and those with a red line running right across them (referred as blood ring) are signs of ceased development (early dead germs) and should be discarded. Eggs are again candled on the day of transfer (25th and 32nd day for common and Muscovy duck eggs, respectively) and those with development appears dark and show a pulsating movement at the edge of the air cell and others will be dead (mid-dead germs). Candling is not practiced in some hatcheries to save time and labour and in some other hatcheries, candling is practiced only on fifth or seventh day. Candling also helps to remove contaminated/infected eggs, which may explode inside the machine and contaminate other eggs. Candling process has to be completed as quickly as possible and cooling of eggs should not occur by placing the eggs outside for a long time, to avoid reduction in hatchability. The appearance of various conditions during fifth day candling is presented in Fig. 9.10.

9.8 Transfer of Eggs to Hatcher

The hatcher room as well as hatcher trays should be cleaned, disinfected/fumigated and dried before re-traying of eggs from the setter trays. The temperature and other requirements inside the hatcher room should be optimum prior to transfer. Eggs containing only live embryos are transferred to hatcher. Eggs containing live embryos can be differentiated easily with the help of an egg candler. Infertile eggs will appear transparent while embryonated eggs will be opaque during candling. Embryo movement/heart beat can be observed to differentiate eggs with live embryos and dead embryos. The appearance of egg during 25th day candling is presented in Fig. 9.11. The transfer operation should be done gently and as quick as possible to avoid cooling of eggs as this may delay hatching. The eggs with live embryos are transferred to the hatcher compartment on 25th and 32nd day, respectively, for common and Muscovy ducks and are kept there for the last 3 days. The transfer of eggs too early or too late will result in lower hatchability as the developing embryos are subjected to sub-optimal conditions. The eggs are placed in

Table 9.6 Common symptoms, probable causes and corrective measures during hatching

Sl. no.	Conditions	Probable reason(s)	Remedial measure(s)
1.	Eggs appear clear on candling without any visible embryonic development	Infertile eggs Inadequate male nutrition	Provide a balanced breeder ration Replace underweight males
		Less number of males. Improper ratio of males to females	Increase the number of males Follow correct male to female ratio
		Competition among males for mating	Reduce the number of males Rear all males together Divide large pens
		Sterility of males	Identify the sterile males by semen evaluation and exclude them from breeding
		Selective mating in pens	Replace all males in the pen.
		Old age of males	Spiking of the flock (introduce younger males replacing old ones)
		Wrong method of artificial insemination and over-dilution of semen	Use proper techniques. Evaluate semen quality characteristics
		Diseased flock	Screen the flock for diseases using standard procedures
		Damaged eggs	Collect eggs early in the morning and thereafter at frequent interval to reduce the contact time with ducks
		Eggs stored for more than 7 days or incorrect storage conditions	Store eggs at 15–18 ° C and 60% RH for maximum 7 days
2.	Blood rings – Indication of early embryonic death	Improper storage conditions	Practice correct duration and conditions for egg storage
		Incorrect incubation temperature	Ensure the accuracy of thermometer and incubator temperature Check temperature settings
		Unbalanced nutrients in breeder duck feed	Use breeder ration with balanced nutrients
		Improper fumigation	Practice correct fumigation techniques Never fumigate between 24 and 96 h after setting
3.	Large number of early dead embryos	Incorrect incubator temperature (usually high)	Practice recommended incubation temperature
		Lack of turning of eggs or turning angle	Turn at least 3 times daily, 45° to one side (total 90°)

(continued)

Table 9.6 (continued)

Sl. no.	Conditions	Probable reason(s)	Remedial measure(s)
		Faulty ventilation	Increase ventilator opening in the incubator and/or room
		Bacillary white diarrhoea or other infectious diseases	Use eggs from disease-free flock.
		Faulty nutrition of breeders	Provide a balanced breeder ration to breeders
4	Pipped eggs fail to hatch	Low humidity	Increase the humidity during end stage of incubation
		Faulty ventilation	Increase ventilation rate in machine by opening air outlet
		Malposition of embryos due to improper setting of eggs	Set eggs with board end up position Turn eggs at regular interval in setter Turning should not be done in the hatcher
		Faulty nutrition of breeders	Provide well-balanced breeder ration to breeders
5.	Early hatching (chicks with unhealed navels)	Marginally high incubation temperatures	Adhere to optimum temperatures in incubators Check the machine frequently to ensure the maintenance of temperature Also check the temperature of incubator room
		Improper storage of eggs	Ensure temperature of 15–18 °C and RH of 60–70% in egg storage room
6.	Late hatch or not hatching uniformly	Marginally low incubator temperature	Double check incubator temperature
		Drafty incubator (cool spots in the machine due to faulty design)	Contact manufacturer of the machine to rectify the fault
		Old eggs or improper storage	Collect the eggs frequently, transfer them to hatchery immediately and store properly Storage period should not be more than 7 days
7.	Sticky embryos (embryos may be smeared with egg contents)	High humidity inside the incubator	Double check incubator humidity Check the size of air cell and egg weight loss as indicator of humidity inside the incubator
		Low temperature inside the incubators	Ensure the maintenance of optimum temperature all the time
		Inadequate ventilation	Increase air vent opening

(continued)

Table 9.6 (continued)

Sl. no.	Conditions	Probable reason(s)	Remedial measure(s)
		Improper fumigation of eggs	Fumigate using correct quantity of chemicals
8.	Embryos adhering to shell Embryos dried too much.	Low humidity in hatcher (especially at the time of hatching)	Increase humidity inside the incubator
		Excess ventilation	Reduce the opening of air vents without affecting ventilation
9.	Crippled and malformed ducklings	Incorrect incubation temperatures (mostly on higher side)	Use recommended incubation temperature setting Reduce the temperature by a fraction if condition persists
		Lower incubator humidity	Ensure optimum relative humidity all the time
		Wrong position of egg setter trays or inadequate turning	Set the eggs with broad end up position and turn eggs at least once in 8 h
		Lethal genes in flock	Follow proper culling practices breeding flock Cull all ducklings/ducks with defects Do not use them as breeders.
		Improper breeder nutrition	Provide a well-balanced breeder ration to breeders
10.	Abnormal, weak or small ducklings	Higher setter or hatcher temperature	Practice recommended incubation temperature Reduce temperature by a fraction if condition persists
		Using small eggs	Set only standard or large sized eggs Sort the hatching eggs before storage Discard small eggs
		Lower incubation humidity	Check humidity periodically Weigh a sample of eggs regularly for weight loss
		Improper ventilation in hatcher	Increase ventilator opening little more
		Diseased or unhealthy flock	Follow regular screening and disease control programmes
		Improper nutrition of breeders	Ensure balanced nutrients in the breeder ration (especially vitamins and trace minerals)
		Excessive fumigation in hatcher	Use measured quantity of fumigants based on the volume of the incubator
11.	Ducklings with laboured breathing	Use of excessive fumigant	

(continued)

Table 9.6 (continued)

Sl. no.	Conditions	Probable reason(s)	Remedial measure(s)
			Use measured quantity of fumigants based on the volume of the incubator
		Respiratory diseases	Check health condition of breeder flock Carry out cleanup and disinfection of incubator and hatchery
12.	Large, soft-bodied mushy ducklings	Low incubation temperature.	Use recommended incubation temperature Reduce the temperature setting by a fraction if the condition persists
		Poor ventilation	Increase ventilator opening of incubator
		Navel infection (Omphalitis)	Clean and disinfect setter compartment frequently Hatcher to be cleaned thoroughly and fumigated between hatches. Store the eggs properly
13.	Unhealed navels	Improper incubator temperature	Follow optimum incubation temperature
		Higher hatching humidity	Open up air inlet and outlet. Maintain adequate humidity
		Navel infection (Omphalitis)	Clean and disinfect setter compartment between settings of eggs. Use dry sanitized hatching trays Store and fumigate the eggs properly
14.	Short down	High incubation temperatures	Follow optimum incubation temperatures
		Low incubation humidity	Increase water flow Place a disinfected damp cloth in incubator
		Excessive ventilation	Reduce air vent openings to restrict air outflow
		Delay in taking duckling out of hatcher	Remove ducklings as soon as they are fluffy but not later than 24 h after hatching

horizontal position in the hatcher trays. Care must be taken to avoid breakage, rupture and haemorrhages since the shells are more brittle at this stage due to the absorption of calcium from the shell by the embryos for skeletal development. It is better to spread corrugated newspapers, shredded paper pieces or jute gunny bags in the trays for providing grasp to the ducklings and this will also prevent leg problems/

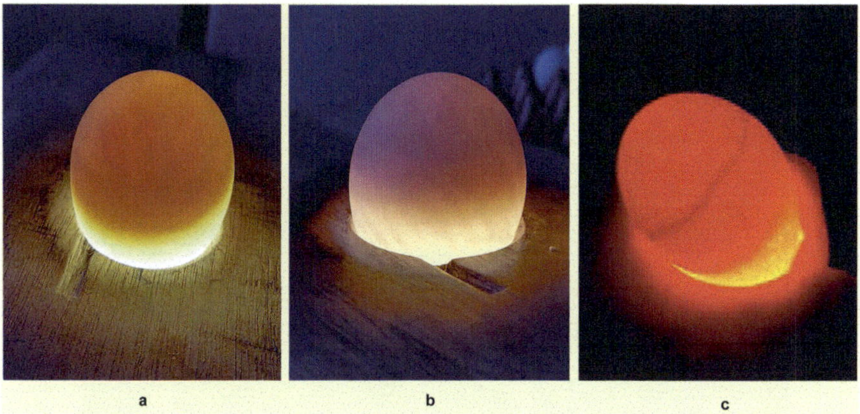

Fig. 9.10 Appearance of egg during 5th day candling (**a**) Developing embryo (**b**) Infertile egg (**c**) Presence of blood ring

Fig. 9.11 Appearance of egg during 25th day candling (**a**) embryonated egg (**b**) Infertile egg

splayed legs. Generally, the hatcher compartment requires slightly lower temperature and slightly higher humidity than setter for facilitating the piping, hatching process and drying of down feathers of the newly hatched ducklings. Automated transfer equipment is available and this will carry out transfer process more gently and quickly than manual transfer (Vimal and Babu 2013). For getting information on pedigree, the eggs are transferred to pedigree hatching boxes placed in the hatcher trays (Fig. 9.12). The various stages of embryo development are depicted in Fig. 9.13.

Fig. 9.12 Hatched ducklings in pedigree hatching boxes

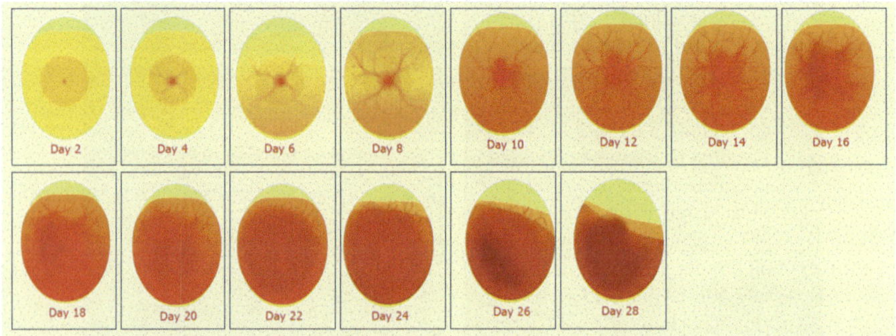

Fig. 9.13 Stages of embryo development

9.9 Taking the Hatch

The ducklings are ready to be taken out when they are dry and fluffy. A few number (about 5%) may have some moist feathers on the back. Excessive drying of ducklings in the hatchery should be avoided as it may cause dehydration. Usually there exists a gap of 36 h between the first and last hatch and this gap is referred as 'hatch window'. Immediately after taking the ducklings from the hatcher, they should be kept in a well-ventilated holding room with 24 °C temperature and 75% humidity to reduce chilling and dehydration.

9.10 Hatch Day Operations

The following operations may be carried out on the day of hatch:

9.10.1 Sexing

Vent sexing also called as venting is a method popularized in 1930s by a Japanese professor, Kiyoshi Masui. Although, vent sexing is one of the most accurate methods of separating sexes, it is labourious and time consuming. The ducklings' intestinal contents are evacuated first by applying slight pressure on both sides of the abdomen in a downward motion (Purchase 1978). Then, hold the duckling on its back in the right hand and its legs are free up in the air pointing away and head towards the wrist of the examiner. Hold the vent with the tail up and press in and up between the duck's legs with the thumb and forefinger of right hand with very slow, gentle pressure to open the vent. Sex of ducklings is then determined by examining of the interior of cloaca. The genitalia in female ducklings is a cone-shaped organ, while males have a smaller, longer and pointed organ inside the cloaca. However, the male's organ may not even appear for several seconds after the vent is opened (Nagmote et al. 2009).

9.10.2 Grading

Ducklings can be graded as good, average and poor by looking into certain characteristics. For grading ducklings, Onbaşılar et al. (2011) gave scores to various parameters according to their importance within a total scale of 100. The parameters used for grading ducklings were activity, down and appearance, eyes, legs, navel, remaining membrane and retracted yolk. The assessment for these parameters was done as per Tona et al. (2003) in broiler chicks.

9.10.3 Wing Banding

This refers to the application of metal bands on the wing web region. Specific numbers encrypted on the wing bands help in the identification of birds (Fig. 9.14).

9.10.4 Packing

The weak ducklings should be discarded and the healthy ones can be placed comfortably in boxes made of corrugated cardboard or plastic (partitioned into compartments) for transportation. These boxes should have holes for air exchange. Plastic boxes can be washed and fumigated for reuse. The number of ducklings that can be placed in each box depend upon season, distance to be travelled and mode of

Fig. 9.14 Wing banding

transportation. Ducklings should be delivered to farm about 12 h after the entire group is removed from the hatcher. The yolk reserve in the newly hatched ducklings enables them to survive without feed and water for 72 h. However, it is better to put them on feed and water as early as possible.

9.11 Embryonic Mortality

The infertile eggs as well as embryonic mortality at various stages are discarded and the weaklings are counted in order to analyse the fertility and hatchability. Embryonic mortality can be classified in to three depending on the stage at which the embryo has died during the incubation. This information is necessary to determine the cause of death and to correct the same. Break out study may help to find out the type of embryonic mortality.

Early embryonic mortality: This occurs in the first week of incubation and is caused by faulty fumigation and poor pre-incubation storage conditions, improper incubation conditions, genetic factors, etc. This should not be confused with infertile eggs.

Mid embryonic mortality: The embryonic mortality, which occurs in the 2nd and 3rd week of incubation comes under this category. This occurs due to poor genetic, nutritional and management problems in the parent stock.

Late embryonic mortality: The embryonic mortality, which occurs in the last 3 days of incubation, comes in this category and is also called as dead in shell. The main reason is poor incubator conditions (Singh 1990).

9.12 Analysing Fertility and Hatchability

The following parameters are estimated to assess the efficiency of the breeding farm as well as the hatchery:

$$\text{Fertility percentage} = \frac{\text{Number of fertile eggs}}{\text{Total number of eggs set}} \text{ x } 100$$

$$\text{Hatchability percentage (on total eggs set)} = \frac{\text{Number of ducklings hatched}}{\text{Total number of eggs set}} \text{ x } 100$$

$$\text{Hatchability percentage (on fertile eggs set)} = \frac{\text{Number of ducklings hatched}}{\text{Number of fertile eggs set}} \text{ x } 100$$

Even though fertility and hatchability are two independent factors, a decrease in one or both will reduce the economic returns. For good returns, the fertility, hatchability on total eggs set and hatchability on fertile eggs set should be 90%, 75% and 80%, respectively. Both genetic and non-genetic factors affect fertility and hatchability. The non-genetic factors include management of breeder (age, season, feed, etc.), egg quality, incubation conditions, etc. (El-hack et al. 2019a)

9.13 Factors Affecting Fertility

The term fertility refers to the percentage of eggs, which are fertile and have life in them. Fertility can be assessed by break open analysis to check the changes in blastoderm after 2 days of incubation. The infertile eggs and early dead embryos appear clear during candling. Therefore, break out examination is necessary to distinguish. In break open analysis, the infertile eggs should be distinguished from those having fertile blastodisc, but died before oviposition, or blastoderm died after oviposition but before setting, or those died during very early stage of development. Abd El-Hack et al. (2019) reviewed the factors influencing the fertility and hatchability of duck eggs. The important factors, which affect the fertility, are given in the following.

9.13.1 Male: Female Ratio

Male: Female ratio should be preferably 1:8 for heavier breeds and 1:10 for lighter breeds, in natural mating. The quality of the breeder stock as well as its management also affects fertility. For males, the quality of semen has to be evaluated. Breed, age, nutrition, climatic condition, method of mating, etc. can affect semen quality. Brillard (2003) and Mohan et al. (2018) opined that the quality and quantity of semen deposited by the male influence the fertility. Brillard (2003) described that fertility depends on the ability of females to ovulate, store sperms and provide an appropriate environment for the formation and development of the egg. Age, method of mating, rate of lay, nutrition and climate may affect the reproductive health of females, which in turn can lead to infertility.

9.13.2 Management

Housing system influences the fertility of ducks. The ducks by nature and prefer to mate in water. Therefore, the access to swimming significantly increases the fertility rate (Ojewola 2006).

9.13.3 Age and Rate of Lay

Females should be at least 24 weeks of age for collection of hatching eggs. Males should be introduced at least 2 weeks before the collection of hatching eggs. Breeding males should be 6–7 months old for good results. Insko et al. (1947) pointed out that fertility declines with age. As the age of the female increases, the functioning of the sperm storage tubules decreases, which influences egg quality. A low rate of lay can also lead to low fertility. Hence, fertility will be higher in the first year of laying than in the subsequent years.

9.13.4 Heredity

Fertility is one of the poorly heritable traits with great individual variations. The presence of lethal genes is also considered as one of the factors contributing to poor fertility (King'ori 2011).

9.13.5 Environment

Temperature and photoperiod are two important factors that affect fertility. Fertility declines during severe hot and cold conditions. Kamar (1961), Chowdhury et al. (2004) and Widiyaningrum et al. (2016) reported high values of fertility during the cooler seasons. Similarly, Lora (1959) reported that fertility is adversely affected by high environmental temperature. On the other hand, fertility can be improved with increasing day length when the environmental temperature is relatively low. The low fertility in Khaki Campbell ducks can be improved by supplemental artificial light so as to provide 14 h photoperiod when the environment temperature was relatively low (Lora and Carranza 1959). The high environmental temperature affects reproductive efficiency of both males and females by hampering the production of germ cells, release of the egg, fertilization and embryo survivability (King'ori 2011).

9.13.6 Nutrition

The diet of breeders should be of good quality and quantity. Feed should be regulated to prevent excessive weight gain, which will ensure the quality of semen and number and eggs (King'ori 2011). Nutritional deficiencies especially that of vitamin E, selenium can lead to poor fertility.

9.14 Factors Affecting Hatchability

Hatchability is the percentage of fertile eggs that hatch into chicks (King'ori 2011; Taplah et al. 2018). It can also be expressed as the percentage of eggs hatched out of the total eggs set for incubation. The important factors, which affect the hatchability, are given in the following.

9.14.1 Breeding Flock

Breeds with good genetic potential at their optimum ages can ensure good hatchability. The age of the breeding flock also affects the hatchability. As the age advances, the external as well as internal quality of eggs are reduced (El-hack et al. 2019a). The ratio between yolk and albumin decreased with increasing age of the birds (Peebles et al. 2001). Age affects the formation of egg shell by influencing the deposition of calcium and minerals in the uterus (Onbaşılar et al. 2014). The shell permits controlled gaseous exchange and loss of moisture from the egg. The poor shell quality allows excessive loss of water from the eggs during incubation (Peebles et al. 2001). Inbreeding tends to reduce hatchability. The hereditary defects such as chromosomal aberrations and lethal genes transmitted by sire and dam can also cause high early embryonic mortality (Liptoi and Hidas 2006).

9.14.2 Season

El-hack et al. (2019a) reported decreased reproductive efficiency in both male and female ducks in the summer months due to high environmental temperature compared to other seasons and this leads to lower hatchability.

9.14.3 Nutrition

Nutrients stored in albumen and yolk are utilized by the embryo for tissue formation, heat generation and muscle activity (Uni et al. 2012; Onbaşılar et al. 2014). Vitamin deficiency especially that of B complex vitamins, mineral deficiency especially that of calcium, phosphorus and manganese can reduce hatchability considerably and will occur more frequently than deficiency of major nutrients like protein, fat, carbohydrates, etc. Nutrient deficiencies hamper proper development of embryos, cause disorders of the musculoskeletal, immune and cardiovascular systems, reduce hatchability and increase embryonic deaths (Uni et al. 2012).

9.14.4 Egg Quality

Egg quality has a significant effect on hatchability, since the microenvironment conditions during storage early stage of incubation alter both the external and internal qualities of eggs (Narushin and Romanov 2002). Heavier eggs are more likely to hatch than smaller ones. The shell thickness and internal contents are more in heavier eggs, providing more reserves of nutrients and energy to the growing embryo (Toro et al. 2015). Demirel and Kirikci (2009) found a greater increase in yolk compared to albumen as the egg size increases and hence this could be considered as a major factor influencing hatchability. Hatchability will be reduced in contaminated/infected eggs. Infection by *Salmonella*, *Pseudomonas*, etc. will considerably reduce the hatchability.

9.14.5 Pre-incubation Storage Conditions

Mishandling of hatching eggs, poor pre-incubation storage conditions, poor cleanliness and hygiene in hatchery can reduce the hatchability. Onbaşılar et al. (2007) reported that higher early and late embryonic deaths due to water loss and degradation of albumin during storage. The hatchability of eggs declines when incubated in hatcheries located at an altitude beyond 750 m from the mean sea level. Low oxygen pressure and lower level of haemoglobin in growing embryos may be the reasons for the death of embryos.

9.14.6 Incubation Conditions

Marked reduction in hatchability will occur when there is deviations in the physical conditions inside the incubator such as temperature, humidity, ventilation, position and turning from the optimum conditions (Archer et al. 2017) as well as when the hygiene is not proper.

9.15 Hatchery Hygiene and Sanitation

Hatchery can be regarded as the greatest potential source from where disease can spread. Diseases that originate in hatchery can be either egg transmittable or that are transmitted from sources other than eggs (e.g. Coliform infection). Infection can occur through eggs, egg boxes, vermins, flies, clothing and hand of hatchery staff, dead or ailing birds. In order to avoid the spread of diseases, use clean nest boxes, clean egg collection equipment and incubate only clean eggs. The hatchery should be at a safe distance from poultry farms and other hatcheries. The visitors should not be allowed. The building must be designed for ease of sanitation. The movement of hatching eggs/ducklings in a hatchery must be unidirectional starting from the receiving area to storage room, to setter room, to hatcher room, to the duckling packing room and finally to dispatch area. Secondary rooms like area for fumigation, washing area for hatcher trays and movable equipment and storage room for chick boxes and other equipment should be located subsidiary to the main flow of hatching eggs/ducklings.

The poor hygienic condition of the hatching eggs arriving at the hatchery is a major source of contamination in the hatchery. All eggs entering the incubator should be fumigated. The equipment, setter and hatcher rooms should be cleaned and fumigated between the hatches. Before the start of hatching season, the entire hatchery building with all equipment should be thoroughly cleaned and disinfected. The flow of eggs from the receiving area to the chick despatch area should be unidirectional. The ventilation inside the hatchery should be proper to prevent recirculation of air. The walls, ceiling and floor should be constructed with materials that can be washed easily. The downs as well as dust should be removed from fans, air conditioner, ventilator, etc. The staff should be trained to make sure that cleaning and disinfection are not compromised.

Effective cleaning and disinfection programs are important in controlling microorganisms such as *Salmonella*, *Pseudomonas*, *Proteus*, *E. coli*, *Staphylococcus*, *Streptococci*, *Aspergillus*, etc. Soiled eggs, unclean hatchery surfaces, air-borne contaminants and movable equipment and personnel are considered as the four key factors for hygiene. The presence of organic matter such as soil, dust, feathers and litter material protects harmful organisms from the action of chemical disinfectants. Therefore cleaning and thorough washing are necessary prior to disinfection. In some instances, this organic matter can even inactivate certain groups of disinfectants. Water under high pressure can be used for removing dirt. Soaking the whole area with soap for about 15 min and washing with water will remove the dirt as 'suspended dirt'. Apply the disinfectant after drying as presence of water will dilute the disinfectant. The machines should be cleaned and disinfected thoroughly after removing eggs in case of setters and ducklings in case of hatcher. This can be accomplished by scraping, vacuuming and mopping the floors and wiping the wall areas and fan blades. The exterior surfaces should be cleaned by damp mopping at least once a week. The top surfaces of incubators should never be used for storing equipment or other items. Each machine should be emptied and thoroughly cleaned at least once in a year. Eggs should be transferred to hatcher before egg pipping starts in order to avoid contamination of setter compartment.

Formaldehyde fumigation is a very powerful method of destroying microorganisms from eggs, egg cases, setters, hatching machines and fibre chick boxes. As the use of this product can cause hazards for human health and safety, the other possible alternatives are:

Chlorine Dioxide: It does not have harmful effects on the eggshell cuticle, the natural barrier that prevents the entry of microbes and other contaminants (Patterson et al. 1990). As the concentration of chlorine is low, this product is not unpleasant to handle.

Phenolic Compounds: These are effective against bacteria and fungi, but their efficacy against spores and viruses is highly dependent on the concentration of the products used. Even though these chemicals are relatively inexpensive, they are toxic to humans. Phenolic compounds are best used in footbaths/foot dips in the buildings and as floor disinfectants.

Quaternary Ammonium Compounds (QAC): Products based on these compounds are effective only against bacteria; but the action against fungi and viruses is mainly dependent on the concentration. These compounds have little or no effect on spores. Products based on these compounds are good detergents and are non-toxic to man but they are relatively expensive. The quaternary ammonium compounds are best used in disinfection of hatchery floors, walls and incubator trays and in fogging. The application of a 3.0% concentration reduces aerobic bacteria counts on the egg surface (Brake and Sheldon 1990).

Iodophors, glutaraldehyde and peracetic acid: These compounds are all highly effective against bacteria, fungi, viruses and microbial spores. These products are relatively non-toxic, but the cost limits their use in large-scale operations.

Ozone: Ozone is an effective hatchery disinfectant (Sheldon and Brake 1991). Both gaseous and aqueous ozone are capable of inactivating many poultry pathogens from the surfaces of eggshells and incubators (Whistler and Sheldon 1989a, b, c). However, over-exposure of hatching eggs to ozone will lead to high embryo mortality (Whistler and Sheldon 1989b).

Hydrogen peroxide: This can be used as a surface decontaminant; but unlike formaldehyde, the hydrogen peroxide readily decomposes into water and oxygen and quickly destroyed. This chemical has no unpleasant lingering odour and poses minimal safety problems for workers if handled properly. But it should be kept in mind that this chemical is a strong oxidizing agent that can irritate the skin, eyes and mucous membranes; therefore, should be handled with caution. It can also discolour clothes and hair. Hydrogen peroxide is less expensive than ozone, as it does not require on-site generation. Sheldon and Brake (1991) reported that hydrogen peroxide at 5% concentration is comparable to formaldehyde as a disinfectant for hatching eggs, without any adverse effect on the hatchability.

9.16 Hatchery Waste Management

Hatchery wastes are inevitable by-products of the industry and it should be managed in sustainable, hygienic and cost effective ways. The rising costs for waste disposal, environmental regulations and awareness have created a need for hatcheries to find suitable sustainable alternative waste management strategies. Hatchery waste includes solid wastes like eggshell debris and fluff, infertile eggs, dead embryos, culled chicks, egg fluids and wastewater. A first way towards more effective hatchery waste management is reducing the amount of waste produced. A second option is turning hatchery waste into a more valuable by-product. The solid wastes can be converted into value added products such as compost, fertilizer, biogas and animal feed. Tacon (1982) detailed the nutritive value of dead day old chicks and eggshells. Chick shell meal was described as a mineral-based product equivalent to a trace element enriched limestone or a phosphorus deficient steamed bone meal. Abiola et al. (2012) showed that 10% of fishmeal could be replaced with whole hatchery waste meal in broiler diets without adverse effects on growth and carcass traits. Recycling is another option wherein the wastewater produced may be recycled and used for irrigation or other purposes. Glatz et al. (2011) detailed in a review, the different methods being used to handle and treat hatchery waste. A good start is to increase the efficiency in the hatchery by optimizing hatchability (improved procedures, better incubation programs, etc.) and to reduce the use of resources (mainly energy and water) and thereby reducing the level of waste. However, the scale of operations at many hatcheries is too small and development of treatment systems may not be viable.

References

Abd El-Hack ME, Hurtado CB, Toro DM, Alagawany M, Abdelfattah EM, Elnesr SS (2019) Fertility and hatchability in duck eggs. Worlds Poult Sci J 75(4):599–608. https://doi.org/10.1017/S0043933919000060

Abiola SS, Radebe NE, Westhuizen CVD, Umesiobi DO (2012) Whole hatchery waste meal as alternative protein and calcium sources in broiler diets. Arch Zootec 61(234):229–234

Alonso-Alvarez C (2006) Manipulation of primary sex-ratio: an updated review. Avian Poult Biol Rev 17:1–20

Archer GS, Jeffrey D, Tucker Z (2017) Effect of the combination of white and red LED lighting during incubation on layer broiler and Pekin duck hatchability. Poult Sci 96:2670–2675

Arora K, Kosin I (1966) Changes in the gross morphological appearance of chicken and Turkey blastoderms during pre-incubation storage. Poult Sci 45:819–825

Bagliacca M, Paci G, Marzoni M (2005) Effect of egg weight categories storage time and storage temperature on incubation length in duck eggs (*Cairina moschata L* and *Anas platyrhynchos domesticus L*). J Poult Sci 42:205–214

Banerjee GC (1991) A text book of animal husbandry, 7th edn. Oxford/IBH Co Publishing Ltd, New Delhi, 747p

Banerjee S (2013) Morphological traits of duck and geese breeds of West Bengal India. Anim Genet Resour 52:1–16

Brake J, Sheldon BW (1990) Effect of a quaternary ammonium sanitizer for hatching eggs on their contamination permeability, water loss and hatchability. Poult Sci 69:517–525

Brillard JP (2003) Practical aspects of fertility in poultry. Worlds Poult Sci J 59:441–446

Brown AFA (1979) In: Batty J, Batty M (eds) The incubation book. Spur Publications Saiga Publishing Co Ltd, Royal Prade, 246p

Chowdhury MMI, Ashraf A, Mondal SP, Mondol NMAAM, Hasan MM (2004) Effect of season on the hatchability of duck eggs. Int J Poult Sci 3:419–421

Cormic J (2013) Egg storage: good practices. https://wwwpetersimecom/hatchery-development-department/article-1/. Accessed 21 Aug 2020

Cyriac S, Joseph L, Peethambaran PA, Narayanankutty K, Karthiayini K (2013) Semen quality characteristics of White Pekin Kuttanad (*Anas platyrhynchos domesticus*) and Muscovy (*Cairina moschata momelanotus*) drakes. Indian J Anim Sci 83:29–33

Cyriac S, Joseph L, George GT (2014) Comparative efficiency of four poultry semen extenders for dilution and storage of drake semen. J Vet Anim Sci 45:32–37

Cyriac S, Joseph L, Anitha P (2018) Effect of natural mating and frequency of artificial insemination on fertility and hatchability parameters of Kuttanad (*Anas platyrhynchos domesticus*) duck eggs. J Vet Anim Sci 16:42–47

Decuypere E, Tona K, Bruggeman V, Bamelis F (2001) The day old chick: a crucial hinge between breeders and broilers. Worlds Poult Sci J 57:127–138

Demirel S, Kirikci K (2009) Effect of different egg storage times on some egg quality characteristics and hatchability of pheasants (*Phasianus colchicus*). Poult Sci 88:440–444

El-hack MEA, Hurtado CB, Toro DM, Alagawany M, Abdelfattah EM, Elenesr SS (2019a) Fertility and hatchability in ducks: a comprehensive review. Worlds Poult Sci J 75:4–9

El-hack MEA, Hurtado CB, Toro DM, Alagawany M, Abdelfattah EM, Elenesr SS (2019b) Impact of environmental and incubation factors on hatchability of duck eggs. Biol Rhythm Res. https://doi.org/10.1080/09291016.2019.1628394

Elibol O, Brake J (2008) Effect of egg position during three and fourteen days of storage and turning frequency during subsequent incubation on hatchability of broiler hatching eggs. Poult Sci 87: 1237–1241

Etches RJ (1996) Reproduction in poultry, 1st edn. CAB International, Wallingford, 318p

Fasenko GM, Robinson FE, Christensen VL (2002) Effects of long-term storage on the egg embryo and chick. In: Deeming DC (ed) Practical aspects of commercial incubation in poultry. Lincoln University Press, pp 33–39

Fujihara N, Nishiyama H (1976) Studies on accessory reproductive organs in the drake 4 effects of androgen on the ejaculatory groove region of drake. Poult Sci 55:1324–1330

Fujihara N, Nishiyama H, Nikashima N (1976) Studies on accessory reproductive organs in the drake 2 macroscopic and microscopic observations on the cloaca of the drake with special reference to the ejaculatory groove region. Poult Sci 59:927–935

Gildersleeve RP, Boeschen DP (1983) The effects of incubator carbon dioxide level on Turkey hatchability. Poult Sci 62:779–784

Giri SC, Kumar P, Jaiswara R, Bais RKS, Sahoo SK, Sastry KVH, Saran S (2014) Ideal mating sex ratio in duck houses for optimum production of fertile eggs. Indian J Poult Sci 49:106–107

Glatz P, Miao Z, Rodda B (2011) Handling and treatment of poultry hatchery waste: a review. Sustainability 3:216–237

Harikrishnan S, Narayanankutty K, Chacko B, Anitha P, Jalaudeen A (2013) Comparative assessment of egg sanitizing agents on the hatchability of Kuttanad duck eggs. Int J Curr Res 5:3987–3988
Humphreys PN (1972) Brief observations on the semen and spermatozoa of certain passerine and non- passerine birds. J Reprod Infertil 29:327–336
Idahor KO, Akinola LAF, Chia SS (2015) Egg colour weight and shape: possible indices in the predetermination of duckling sex. J Rec Adv Agric 3:37–344
Insko WM Jr, Steele DG, Whuteman ET (1947) Reproductive phenomena in ageing hens. Kentucky Agric Exp Stat Bull 498:1–23
Jalaludeen A, Peethambaran PA, Joseph L, Manomohan CB (2004) Duck production in Kerala, first edition, NATP on ducks. Centre for Advanced Studies in Poultry Science, Kerala Agricultural University, Mannuthy, Thrissur, Kerala, p 44
Kamar GAR (1961) Seasonal variation in fertility and hatchability of duck eggs. In: Proceedings of 56th annual convention of national poultry butter and egg association, September 13–15, 1961, Miami, Beach, Florida
King'ori AM (2011) Review of the factors that influence egg fertility and hatchability in poultry. Int J Poult Sci 10:483–492
Lake PE (1960) Studies on the dilution and storage of fowl semen. J Reprod Fertil 1:30–35
Lake PE, Stewart JM (1978a) Artificial insemination in poultry. No. 213 Bulletin of Ministry of Agriculture Fisheries and Food, Her Majesty's Stationary Office, London, p 34
Lake PE, Stewart JM (1978b) The staining of spermatozoa for differential count of live and dead spermatozoa. Br Poult Sci J 19:187–194
Lima JCS, de Silva PL, Coelho LR, Borges MS, Freitas AG, de Fonseca BB (2012) Effects of inverting the position of layers eggs during storage on hatchery performance parameters. Braz J Poult Sci 14:245–248
Liptoi K, Hidas A (2006) Investigation of possible genetic background of early embryonic mortality in poultry. Worlds Poult Sci J 62:326–337
Lora RP (1959) Study of the seasonal changes of the fertility of Khaki Campbell ducks. Arch Zoot 8:225–241
Lora RP, Carranza J (1959) Influence of supplementary artificial light and gonadotropic hormone on the fertility of Khaki Campbell ducks. Arch Zoot 8:277–284
Majhi RK, Kumar ARK, Kumar A, Yadav M, Kumar P, Maity A, Giri SC, Goswami C (2016) Light and electron microscopic study of mature spermatozoa from White Pekin duck (*Anas platyrhynchos*): an ultrastructural and molecular analysis. Andrology 4:232–244
Mayes FJ, Takeballi MA (1984) Storage of eggs of the fowl (*Gallus domesticus*) before incubation. Worlds Poult Sci J 40:131–140
Meijerhof R (1992) Pre-incubation holding of hatching eggs. Worlds Poult Sci J 48:57–68
Mohan J, Sharma S, Kolluri G, Dhama K (2018) History of artificial insemination in poultry its components and significance. Worlds Poult Sci J 74:1–14
Nagmote SG, Murthy HNN, Isloor S, Rao NR (2009) Approaches in sexing of ducks. In: Proceedings of the fourth World Waterfowl Conference, Centre for Advanced Studies in Poultry Science College of Veterinary and Animal Sciences Kerala Agricultural University, and World's Poultry Science Association (India branch), 11–13 November, 2009, Mannuthy, Thrissur, pp 407–410
Narushin VA, Romanov MN (2002) Egg physical characteristics and hatchability. Worlds Poult Sci J 58:297–303
NATP (2004) Final report. National Agricultural Technology Project (NATP) on productivity enhancement of ducks. Kerala Agricultural University, Thrissur, p 74
Nickolova M (2005) Effect of the sex ratio on the egg fertility of Muscovy duck (*Cairina moschata*). J Cent Eur Agric 5:367–372
Ojewola FEG (2006) Effect of management systems on semen quality of Muscovy drakes. Int J Poult Sci 5:482–484

Onbaşılar E, Poyraz Ö, Erdem E (2007) Effects of egg storage period on hatching egg quality and hatchability, chick quality and relative growth in Pekin ducks. Arch Geflügelk 71:187–191
Onbaşılar E, Erdem E, Poyraz O, Yalem S (2011) Effects of hen production cycle and egg weight on egg quality and composition hatchability duckling quality and first week body weight in Pekin ducks. Poult Sci 90:2642–2647
Onbaşılar E, Erdem E, Hacan Ö, Yalçin S (2014) Effects of breeder age on mineral contents and weight of yolk sac embryo development and hatchability in Pekin ducks. Poult Sci 93:473–478
Patterson PH, Ricke SC, Sunde ML, Schaefer DM (1990) Hatching eggs sanitized with chlorine dioxide foam: egg hatchability and bactericidal properties. Avian Dis 34:1–6
Peebles ED, Doyle SM, Zumwalt CD, Gerard PD, Latour MA, Boyle CR, Smith TW (2001) Breeder age influences embryogenesis in broiler hatching eggs. Poult Sci 80:272–277
Proudfoot FG (1969) In: Carter TC, Freeman BM (eds) The handling and storage of hatching eggs: the fertility and hatchability of the hens' egg. Oliver and Boyd, Edinburgh, pp 127–141
Purchase HG (1978) Raising geese. USDA Farmers' Bull 2251:1–14
Reid WM, Maag TA, Boyd FM, Kleckner AL, Schmittle SC (1961) Embryo and baby chicks mortality and morbidity induced by a strain of Escherichia coli. Poult Sci 40:1497–1502
Rose SP (1997) Principles of poultry science. CABI Publishing, CAB International, Wallingford, p 134
Samberg Y, Meroz M (1995) Application of disinfectants in poultry hatcheries. Sci Tech Rev Office Int Epiz 14:365–380
Schulte-Drüggelte R (2011) Recommendations for hatching egg handling and storage. Lohmann Inf 46:55–58
Scott TA, Swetnam C (1993) Screening sanitizing agents and methods of application for hatching eggs: 1. Environmental and user friendliness. J Appl Poult Res 2:1–6
Sexton TJ (1977) A new poultry semen extender: 1. Effect of extension on the fertility of chicken semen. Poult Sci 56:1443–1446
Sheldon BW, Brake I (1991) Hydrogen peroxide as a hatching egg disinfectant. Poult Sci 70:1092–1098
Simoes K, Orsi AM, Artoni MB (2012) Ultrastructure of the spermatozoa of the domestic duck (*Anasplatyrhynchos* sp.). Anat Histol Embryol 41:202–208
Singh RA (1990) Poultry production, 3rd edn. Kalyani Publishers, New Delhi, p 324
Smith AJ (1990) Poultry-tropical agriculturist series. CTA/Macmillan, London, pp 179–184
Sonaiya EB, Swan SEJ (2004) Small scale poultry production-technical guide. Food and Agriculture Organisation of the United Nations, Rome
Sturkie PD (1986) Avian physiology, 4th edn. Springer, New York, p 516
Tacon AGJ (1982) Utilisation of chick hatchery waste: the nutritional characteristics of day old chicks and eggshells. Agric Wastes 4:335–343
Taneja GS, Gowe RS (1961) Spermatozoa concentration in the semen of two breeds of fowl estimated by three different methods. Poult Sci 40:608–615
Taplah AJ Jr, Suministrado DC, Amongo RMC, Paras FO Jr, Elauria JC, Torkpah DP (2018) Economic analysis of duck eggs incubation using hot spring as heat source. J Dev Agric Econ 10(2):38–44
Tona K, Bamelis F, De Ketelaere B, Bruggeman V, Moraes VMB, Buyse J, Onagbesan O, Decuypere E (2003) Effects of egg storage time on spread of hatch chick quality and chick juvenile growth. Poult Sci 82:736–741
Toro DM, Aguilar YM, Bertot RR, Hurtado CB, Nava OR (2015) Effect of dietary supplementation with *Morinda citrifolia* on productivity and egg quality of laying hens. Rev Cienc Agric 12:7–12
Uni Z, Yadgary L, Yair R (2012) Nutritional limitations during poultry embryonic development. J Appl Poult Res 21:75–184
Vedder O, Kürten N, Bouwhuis S (2017) Intraspecific variation in and environment-dependent resource allocation to embryonic development time in common terns. Physiol Biochem Zool 90: 453–460

Vimal AM, Babu BA (2013) Managing hatcheries, 1st edn. Centre for Advanced Studies in Poultry Science, Kerala Agricultural University, Mannuthy, Thrissur, India and Department of Animal Husbandry, Kerala, India, p 24

Wheeler NC, Andrews FN (1943) The influence of season on semen production in the domestic fowl. Poult Sci 22:361–367

Whistler PE, Sheldon BW (1989a) Biocidal activity of ozone versus formaldehyde against poultry pathogens inoculated in a prototype setter. Poult Sci 68:1068–1073

Whistler PE, Sheldon BW (1989b) Bactericidal activity eggshell conductance and hatchability effects of ozone versus formaldehyde disinfection. Poult Sci 68:1074–1077

Whistler PE, Sheldon BW (1989c) Comparison of ozone and formaldehyde as poultry hatchery disinfectants. Poult Sci 68:1345–1350

Widiyaningrum P, Lisdiana L, Utami N (2016) Egg production and hatchability of local ducks under semi intensive vs extensive managements. J Indones Trop Anim Agric 41:77–82

Dr. Stella Cyriac, an Assistant Professor at Kerala Veterinary and Animal Sciences University hails from Kottayam, Kerala (India). She earned his Master's degree and Ph.D. in Poultry Science from Kerala Veterinary and Animal Sciences University. She joined as an Assistant Professor, Department of Poultry Breeding and Genetics, College of Avian Sciences and Management, Thiruvazhamkunnu, Palakkad, in 2015. She did her Master's and Doctoral research projects in ducks. She is currently leading a DBT funded research project on 'Genetic up-breeding of duck production to strengthen livelihood security in NER of India by converging conventional and molecular techniques' at College of Avian Sciences and Management, Thiruvazhamkunnu, Palakkad, Kerala.

Dr. Leo Joseph, a Doctor of Veterinary Medicine hails from Kerala, India. He joined as a faculty member in the College of Veterinary and Animal Sciences, Kerala in 1979 in the Poultry Science Department. He worked in the area of Poultry Breeding and was instrumental in the development of Gramasree, a synthetic breed for backyard rearing and the genetic improvement of Kuttanadducks of Kerala. He was associated with the Poultry Breeding Project as a Geneticist for the development of a hybrid layer, ILM 90. He retired from Kerala Veterinary and Animal Sciences University as Director of Poultry Science in 2016.

Nutritive Value of Duck Meat and Eggs 10

Elisabeth Baéza and Jeng-Fang Huang

Contents

Abstract

Duck egg and meat production and consumption are mainly concentrated in Asia. Different species or breeds and different production systems are used all over the world. This will affect the growth rate, body weight and slaughter age of ducks and thus, the muscle development and further meat quality. As with other poultry species, ducks can be sold as whole carcasses, cut and processed products. Owing to genetic selection and to improvements in bird management, particularly in nutrition, the meat yield of ducks has increased and carcass fattiness has decreased. Duck meat is red, and duck muscles have a higher oxidative energy

E. Baéza (✉)
Animal Physiology and Rearing Systems, The French National Research Institute for Agriculture, Food and the Environment, Nouzilly, France
e-mail: elisabeth.baeza-campone@inrae.fr

J.-F. Huang
Ilan Branch, Livestock Research Institute, Council of Agriculture, Executive Yuan, Ilan, Taiwan
e-mail: huangajf@tlri.gov.tw

A. Jalaludeen et al. (eds.), *Duck Production and Management Strategies*,
https://doi.org/10.1007/978-981-16-6100-6_10

metabolism than those of chickens or turkeys. Lipids are, therefore, an important component of duck meat quality. Finally, duck meat displays the characteristics of poultry meat (high level in unsaturated fatty acids representing around 60% of total fatty acids) and red meat (high level of haeminic pigment and higher level of phospholipid and total lipid contents than chicken and turkey breast meat). Owing to the varieties in laying duck species used in the world, the weight range of duck egg is 60–90 g. The egg composition is mainly influenced by the age. It is quite difficult to modify the chemical composition of eggs except the fatty acid composition and micronutrient content that can be manipulated by the way of feeding laying ducks.

Keywords

Proteins · Lipids · Fatty acids · Minerals

Abbreviations

ANSES	French National Agency for food, environment and work health safety
CLA	Conjugated Linoleic Acids
CP	Crude Proteins
DHA	Docosahexaenoic Acid
EPA	Eicosapentaenoic Acid
FA	Fatty Acids
ME	Metabolizable Energy
MUFA	Mono Unsaturated Fatty Acids
PUFA	Poly Unsaturated Fatty Acids
SFA	Saturated Fatty Acids

10.1 Introduction

Although chicken eggs are typical in the Western diet with a worldwide production number of 1.577 trillion hen shell eggs in 2019, worldwide production of shell eggs from other poultry species saw a 18% increase from 2009 exceeding 95 billion in 2019 with 95% of production concentrated in Asia (FAO 2021). China is the largest production country of duck eggs (71 billion in 2019). The duck eggs have been consumed in many Asian countries for a very long time. In America and Europe, duck eggs are usually not consumed due to the potential *Salmonella* risk.

From 2009 to 2019, the world duck production strongly increased, and meat production rose from 3.939 to 4.858 million tonnes (FAO 2021). Asia is the leading production region and in this area, the duck production is expanding most rapidly. Asia accounted for 82.1 and 84.8% of world's duck meat output in 2009 and 2019, respectively. In Europe, France is the leading producer with 221,400 tonnes in 2019. In Asia, China's contribution (3.389 million of tonnes in 2019) represents 82% of the total region. The trade in fresh and frozen duck meat largely increased between 2000 and 2018 from 170,000 to 306,000 tonnes. Two-thirds of duck meat exports came from Europe.

The main species used to produce duck meat are Pekin, Muscovy (mostly in France) and mule ducks (mostly in France for the production of fatty liver and Taiwan). Other duck species such as Jinding and Shao ducks in China, Tsaiya ducks in Taiwan, Khaki Campbell, Indian runner and Desi ducks in India, Vietnam, Cambodia and Indonesia are used mainly for egg production and meat is a by-product (Baéza 2006).

Different production systems are used in the world. In developed countries, ducks are mainly reared intensively in total-confinement housing or with access to pasture (free range, organic or 'label rouge' systems) and sometimes to water for swimming. In Asia, extensive systems such as fish-duck farming are largely used, ducks being housed above fishponds. To decrease feed costs, ducks can also be reared in rice fields. Indeed, the growth rate, body weight and slaughter age of different species and genotypes used can vary strongly, affecting the muscle development and further meat quality.

Duck eggs can be used fresh but also as salted eggs and preserved eggs (also known as 'century eggs', 'thousand years of eggs' or 'pidan eggs'). These processed eggs were originally developed in China, centuries ago to extend the shelf life of duck eggs in the absence of refrigeration (Hou 1981).

As with other poultry species, ducks can be sold as whole carcasses, cut and processed products. Owing to genetic selection and to improvements in bird management, particularly in nutrition, the meat yield of ducks has increased and carcass fattiness has decreased. At the same time, interest in the technological quality of duck meat has increased. Duck meat is red, and duck muscles have a higher oxidative energy metabolism than those of chickens or turkeys. Lipids are therefore an important component of duck meat quality. Finally, duck meat displays the characteristics of poultry meat (high level in unsaturated fatty acids representing around 60% of total fatty acids) and red meat (high level of haeminic pigment and higher level of phospholipid and total lipid contents than chicken and turkey breast meat).

10.2 Duck Meat

10.2.1 Chemical Composition

10.2.1.1 Proteins

Duck meat displays a high protein content of around 19.5 to 22.3% in the breast and 19.3 to 21.7% in the thigh (Table 10.1).

The main amino acids are glutamic acid, aspartic acid, lysine, leucine, arginine and alanine with values comprised between 2.4 and 18.8 g/ 100 g dry matter in the breast and between 2.3 and 14.2 g/ 100 g in the thigh (Kokoszynski et al. 2017; Table 10.2). The amino acid composition of duck meat is relatively stable. However,

Table 10.1 Proximate composition of breast and thigh muscles of ducks (% fresh meat)

Piece	Duck type	Dry matter	Proteins	Lipids	Minerals	References
Breasts	Male Pekin 7 weeks	24.7	22.3	1.5	0.9	Koci et al. (1982)
	Female Pekin 7 weeks	24.3	21.0	2.1	1.1	Paquin (1988)
	Muscovy	26.0	21.0	2.0	1.4	Paci et al. (1993)
	Male Muscovy 11 weeks	23.5	20.2	2.5		Paci et al. (1993)
	Female Muscovy 9 weeks	23.7	19.6	1.6		Paci et al. (1993)
	Male Pekin 8 weeks	23.4	20.8	2.1		Paci et al. (1993)
	Mule 10 weeks	25.0	21.1	2.2		Paci et al. (1993)
	Male Muscovy 12 weeks	24.0	21.0	1.7	1.0	Salichon et al. (1993)
	Pekin 7 weeks	22.3	19.5	2.3	1.1	Smith et al. (1993)
	Male mule 10 weeks	24.4	21.8	2.1	1.1	Baéza et al. (2000)
	Female mule 10 weeks	23.7	21.7	1.7	1.1	Baéza et al. (2000)
	Pekin 7 weeks	24.1	20.9	1.7	1.4	Mazanowski et al. (2003)
	Pekin 9 weeks	24.5	20.1	1.8	0.9	Ali et al. (2007)
	Pekin 6 weeks	23.1	20.2	1.8	1.1	Muhlisin et al. (2013)
	Pekin 7 weeks	24.1	21.7	1.3	1.1	Qiao et al. (2017)
	Pekin 7 weeks	24.7	22.1	2.0		Kokoszynski et al. (2017)
Thighs	Male Pekin 7 weeks	25.2	21.7	2.7	0.8	Koci et al. (1982)
	Female Pekin 7 weeks	24.9	20.5	3.4	1.0	Paquin (1988)
	Muscovy	26.0	20.5	5.0	1.4	Paci et al. (1993)
	Pekin 6 weeks	22.9	19.3	2.7	0.9	Muhlisin et al. (2013)
	Pekin 7 weeks	22.7	20.0	3.0	1.1	Qiao et al. (2017)
	Pekin 7 weeks	26.4	20.4	5.1		Kokoszynski et al. (2017)

it is possible to make it vary marginally. Kokoszynski et al. (2017) obtained light modifications of amino acid profile in breast and thigh muscle by providing to Pekin ducks a commercial diet (85%) plus whole wheat grains (15%) during the last 13 days of rearing (Table 10.2).

Table 10.2 Amino acid profile (g/100 g dry matter) of breast and thigh muscles of Pekin ducks fed with a commercial complete diet (C) or commercial complete diet (85%) plus whole wheat grain (15%, WW) for the final 13 days and slaughtered at 49 days of age (n = 10; derived from Kokoszynski et al. 2017)

	Breast			Thigh		
Amino acids	C	WW	P value	C	WW	P value
Aspartic acid	11.1	12.8	0.179	10.6	9.6	0.099
Threonine	3.9	5.1	0.027	3.7	3.6	0.637
Serine	2.4	3.1	0.110	2.4	2.3	0.478
Glutamic acid	15.5	18.8	0.117	14.2	12.5	0.019
Proline	4.5	3.8	0.078	3.5	3.2	0.040
Glycine	3.7	5.1	0.079	3.8	3.2	0.257
Alanine	2.9	3.3	0.519	3.0	2.5	0.013
Valine	4.0	5.0	0.042	3.4	3.4	0.878
Isoleucine	3.8	5.0	0.184	3.8	3.5	0.234
Leucine	5.7	6.3	0.545	6.0	5.4	0.053
Tyrosine	2.6	2.9	0.604	2.6	2.5	0.220
Phenylalanine	3.4	3.9	0.240	3.0	2.8	0.093
Histidine	3.0	3.2	0.265	2.5	2.5	0.467
Lysine	9.8	11.2	0.512	7.3	8.6	0.350
Arginine	6.5	7.1	0.721	6.5	5.6	0.040

The protein content of duck meat is mainly influenced by slaughter age. The protein content of breast of mule duck increased from 20.6 to 22.4% when the animals were slaughtered at 8 or 13 weeks of age (Baéza et al. 2000). The protein content of breast and thigh muscles was higher for spent layers slaughtered at 500 days of age than for Cherry Valley ducks slaughtered at 38 days of age: 24.34% and 22.14% vs. 21.73% and 19.99% (Qiao et al. 2017).

The age also has an effect on the collagen content of meat. The collagen content was two-fold lower in breast muscle of Muscovy duck at 12 weeks of age compared to ducks at 8 weeks of age (Baéza et al. 2002; Fig. 10.1).

Finally, the haeminic pigment content, particularly myoglobin, also increases with age. For example, iron content was two-fold higher in the breast muscle of Muscovy duck at 12 weeks of age when compared with Muscovy duck at 8 weeks of age (Baéza et al. 2002; Fig. 10.1).

10.2.1.2 Lipids

The most variable fraction concerns lipids. The lipid content of duck meat is around 1.3 to 2.5% in the breast and 2.7 to 5.1% in the thigh (Table 10.1). Among the lipids, the most variable fraction is that of triglycerides, content of which is positively correlated with the lipid content; duck breast contains 0.5–0.8% triglycerides (Baéza 2000). In breast meat, the phospholipid content is around 1.1% and the cholesterol content is comprised between 0.07 and 0.12% (Baéza 2000).

The feed composition can have an effect on the intramuscular content of lipids. This will depend on the energy content and the ratio of energy to proteins or the

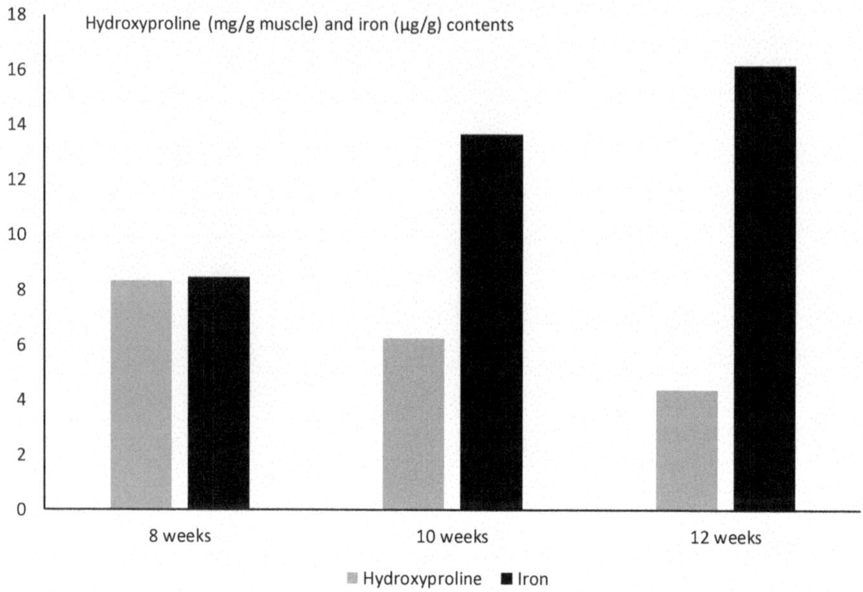

Fig. 10.1 Hydroxyproline and iron contents of breast muscle from male Muscovy ducks slaughtered at 8, 10 and 12 weeks of age (n = 10, derived from Baéza et al. 2002)

essential amino acid (lysine, methionine) content. Unfortunately, the effects of these factors, in relationship with lipid content of duck meat, were not studied. Only, the meat yield and the carcass fatness were examined. For example, increasing crude protein content of feed from 15 to 19% decreased the subcutaneous fat relative to eviscerated carcass weight from 6.78 to 6.17% and increased the breast meat yield relative to eviscerated carcass weight from 17.93 to 20.10% in Pekin ducks slaughtered at 35 days of age (Zeng et al. 2015a). Increasing dietary methionine concentration from 0.30 to 0.68% between 15 and 35 days of age decreased the subcutaneous fat relative to eviscerated carcass weight from 6.95 to 6.51% and increased the breast meat yield relative to eviscerated carcass weight from 17.9 to 19.6% (Zeng et al. 2015b). The effect of the dietary energy source (lipids vs. starch) on the carcass composition and lipid deposition in muscles was not investigated in ducks. The effect of feed restriction during rearing on carcass composition was investigated. Leclercq and De Carville (1978) tested slight (5%) and severe (20%) feed restriction between 8 and 12 weeks of age in male Muscovy ducks. The slight feed restriction had no effect on slaughter body weight or carcass composition, whereas the severe feed restriction had a negative impact on slaughter body weight (3453 vs. 3629 g) and breast yield (15.8 vs. 16.2%). It also decreased carcass fatness. Another way to restrict feed intake is to dilute diets with feedstuffs containing high fibre content. Wu et al. (2012) tested 40 and 60% inclusion of rice hulls in the basal diet of Pekin ducks between 8 and 14 days of age. Ducks were fed the same starter diet (12.13 MJ ME/ kg and 210 g CP/ kg) from hatching to 14 days of age and the

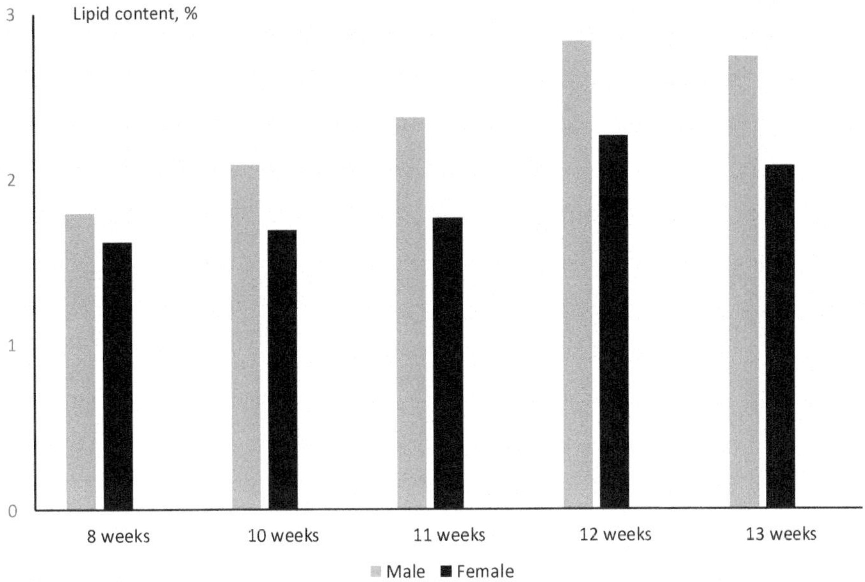

Fig. 10.2 Effect of age and sex on lipid content of breast meat of mule duck (n = 10, derived from Baéza et al. 2000)

same grower diet (12.55 MJ ME/kg and 180 g CP/kg) between 14 and 42 days of age. The use of 40% rice hulls had no effect on body weight, feed conversion ratio or meat yield at 42 days of age. Carcass fatness and lipid content of breast muscle were decreased, whereas the protein content of carcass and breast muscle was increased. Han et al. (2016) also found an increase in breast meat yield and a decrease in abdominal fat percentage of Pekin ducks by testing different levels of dietary cellulose supplementation from 0 to 1500 mg/kg during the whole rearing period. The practice of overfeeding ducks has the reverse effect, with the intramuscular lipid content being two-fold higher than that of ad libitum fed ducks (Baéza et al. 2013).

The lipid content of duck meat is also influenced by age, sex, genotype and production system. In mule duck (males and females), the lipid content in breast meat increased from 1.79 to 2.74% when ducks were slaughtered at 8 or 13 weeks (Fig. 10.2; Baéza et al. 2000). The lipid content of breast and thigh muscles was higher for spent layers slaughtered at 500 days of age than for Cherry Valley ducks slaughtered at 38 days of age: 2.42% and 5.92% vs. 1.29% and 2.95% (Qiao et al. 2017).

For a given age, the females usually display higher carcass fatness and intramuscular lipid content than males. However, the sex effect on muscle lipid content is more or less pronounced. In mule ducks slaughtered at different ages between 8 and 13 weeks, the intramuscular lipid content of breast meat was higher for males compared to females (Fig. 10.2; Baéza et al. 2000). In Chaohu ducks reared indoor on litter (rice hulls) or slats and with or without access to outdoor, the females

Table 10.3 Fatty acid composition (% total FA) of dietary oils and breast meat from female Muscovy ducks selected or not for body weight and slaughtered at 63 days of age (n = 3, derived from Schiavone et al. 2004)

	Fatty acid composition of dietary oils		Fatty acid composition of duck breast meat					
			SO groups		FO groups		P values	
Fatty acids	SO	FO	Selected	Rural	Selected	Rural	Diet	Strain
C14:0	0.1	ND	0.36	0.27	0.42	0.44	*	Ns
C16:0	10.5	ND	23.23	21.29	23.09	22.73	Ns	Ns
C16:1	0.1	ND	1.02	0.68	0.97	0.96	*	Ns
C18:0	4.4	ND	14.76	16.23	15.21	15.94	Ns	Ns
C18:1	24.2	10.9	21.36	17.50	18.62	16.95	Ns	Ns
C18:2 n-6	52.4	2.2	17.96	18.29	15.85	15.03	Ns	Ns
C18:3 n-3	6.9	1.4	0.37	0.35	0.27	0.24	*	Ns
C20:0	0.4	0.2	0.12	0.15	0.14	0.14	Ns	Ns
C20:4 n-6	ND	0.4	9.97	12.88	7.97	9.57	*	*
C20:5 n-3	ND	6.6	ND	ND	1.76	1.67	*	Ns
C22:5 n-3	ND	ND	0.57	0.62	1.42	1.46	**	Ns
C22:6 n-3	ND	8.4	0.70	1.21	6.56	7.36	**	Ns
SFA	16.1	22.2	38.76	38.93	39.23	39.72	Ns	Ns
MUFA	24.6	28.1	25.70	21.23	23.27	21.68	Ns	Ns
PUFA	56.3	19.3	35.61	38.98	36.44	37.84	Ns	Ns
n-6 FA	52.4	2.9	30.36	33.76	25.28	25.78	*	Ns
n-3 FA	6.9	16.4	1.64	2.18	10.01	10.73	*	Ns
n-3 FA/n-6 FA	7.6	0.2	18.51	15.49	2.53	2.40	**	Ns

SO = Soybean oil; FO = Fish oil
*$P < 0.05$; **$P < 0.01$; ns = not significant; ND = not detected
SFA, MUFA, PUFA = saturated, monounsaturated and polyunsaturated fatty acids

displayed a higher abdominal fat percentage at 63 days of age but the intramuscular lipid content of breast muscles was similar for both sexes (Zhang et al. 2018). Among different species used for duck meat production, the Muscovy duck displayed lower intramuscular lipid content than Pekin duck and the mule duck has intermediate values to its parental species (Chartrin et al. 2006). Matitaputty et al. (2015) compared the thigh meat quality of Alabio and Cihateup strains and their reciprocal crossbreds. The Cihateup ducks displayed higher lipid content in thigh meat than the Alabio ducks and the hybrids. There are few studies investigating the effect of production system on the duck meat quality. Zhang et al. (2018) compared two rearing systems previously mentioned and found a higher abdominal fat percentage and a higher intramuscular fat in breast of ducks reared indoor on litter with access to outdoor than that of ducks always reared indoor on slats.

The fatty acids (FA) from total lipids of duck meat consist approximately one-third part of saturated fatty acids (SFA), one-third part of monounsaturated fatty acids (MUFA) and one-third part of polyunsaturated fatty acids (PUFA, Baéza 2000). The main SFA are palmitic (2/3) and stearic (1/3) acids (Table 10.3).

The oleic acid is the main MUFA (5/6) followed by palmitoleic acid (1/6). The main PUFA are linoleic and arachidonic acids. The total lipids of duck meat also contain linolenic acid and long chain PUFA from n-6 and n-3 series. As for all poultry species, the main variation factor of FA composition in duck meat is the FA composition of feed. Palm and coconut oils increase the short chain SFA percentages and the animal fat enriches the duck meat lipids with palmitic and stearic acids. On the reverse, the vegetable oils increase the linoleic and linolenic acid percentages. The marine oils significantly increase the percentages of long chain n-3 PUFA (Table 10.3, Schiavone et al. 2004).

In Europe, replacing animal fat (tallow and lard) with vegetable oils (rapeseed, soya, linseed) in poultry feed, since 2001, induced an increase in PUFA percentage in poultry meat. In western countries, the daily intake of FA is not satisfactory because the n-6 FA/ n-3 FA ratio is around 15 whereas ANSES (French National Agency for food, environment and work health safety) recommends a value of 5. In order to increase the daily intake of n-3 FA, different strategies have been undertaken. All had the aim to increase n-3 FA in the different food daily consumed (bread, dairy products, fresh and processed meats, eggs) by avoiding large and definitive modifications of eating habits, such as a large increase in fat fish consumption to the detriment of meat and dairy products. The enrichment of poultry meat with n-3 FA has been, therefore, largely studied. The use of fish oils rich in long chain PUFA is the more effective way (Schiavone et al. 2004; Huang et al. 2006). Microalgae or algae can replace the fish oils (Baéza et al. 2017; Islam et al. 2014). It is also possible to use linseed or rapeseed oils which are rich in linolenic acid even if, in this case, the percentage of long chain n-3 PUFA deposited in muscles remains low (Li et al. 2017). Numerous studies have also investigated the poultry meat enrichment with CLA (conjugated linoleic acids). It is a group of 16 isomers of the linoleic acid, which have beneficial health effects in humans in particular during cancers, atherosclerosis, obesity and diabetes. It would also stimulate the immune functions (Brown and Mac Intosh 2003). The CLA are natural compounds found in the lipid part of cattle meat and milk. As for chickens and hens, it is possible to enrich duck meat with CLA. Halle et al. (2012) tested 0, 1 and 2 g CLA/kg feed in Pekin ducks reared until 49 days of age. The CLA content in breast meat increased from 0 to 0.41 and 0.68, respectively. The brake in the CLA use is their cost and sometimes their negative effect on the poultry growth performances.

The dietary energy source can also influence the FA composition of duck meat. A diet rich in carbohydrates stimulates the hepatic lipogenesis and by way of consequence the synthesis of SFA and MUFA, whereas a diet rich in lipids stimulates mainly the direct deposition of dietary FA in the peripheral tissues. This is particularly the case of overfeeding. The daily intake of carbohydrates (corn starch) is quite high. In the 'magret' (breast muscle of overfed duck), the percentages of MUFA and PUFA represent 50 and 16% of total FA, respectively, vs. 36 and 31% in the fillet from ad libitum fed duck (Girard et al. 1993).

Other factors that can influence the FA composition of duck meat are age, sex, species and genotype having a direct effect on the lipid content of meat and by way of consequence on FA composition. Actually, in ducks, an increase of meat lipid content is correlated with an increase of the percentage of neo-synthesized MUFA to the detriment of the PUFA percentage and conversely (Baéza 2000). For example, in the breast meat of mule duck slaughtered at 8 or 13 weeks of age, the percentage of MUFA increased from 29.6 to 33.0%, whereas that of PUFA decreased from 35.5 to 32.1% and that of SFA remained stable (35.2 and 34.3%, Baéza et al. 2000).

10.2.1.3 Minerals

The mineral content of duck meat (calcium, phosphorus and potassium essentially) is around 1.1% (Table 10.1). The micronutrient contents of raw duck meat are given in Table 10.4 (Favier et al. 1995). It is relatively stable if the amount of minerals provided by feed covers the duck requirements.

10.2.1.4 Micronutrients

The duck meat contains also different micronutrients (vitamins, carotenoid pigments, trace elements, Tables 10.4 and 10.5) which content mainly depends on the feed intake.

Many studies have been undertaken to enrich poultry meat with vitamins E, A and C. In ducks, only the enrichment of meat with vitamin E was investigated. Russell et al. (2004) distributed to Pekin ducks, from 0 to 42 days of age, diets supplemented with 20, 400 or 1000 mg vitamin E/ kg. At slaughter age, the α-tocopherol content was 0.4, 4.2 and 11.2 mg/ kg, respectively, in breast meat, and 0.5, 3.8 and 6.5 mg/kg in thigh meat.

The enrichment of poultry meat with different trace elements such as selenium, magnesium, zinc and copper was also tested. Baltic et al. (2015) distributed to Pekin ducks, diets supplemented with four increasing selenium levels (0, 0.2, 0.4 and 0.6 mg/kg) from 0 to 49 days of age. At slaughter age, the selenium content increased from 0.05 to 0.87 mg/kg in breast meat and from 0.04 to 0.64 mg/kg in thigh meat. The selenium can be used under various forms among which the organic

Table 10.4 Micronutrient contents of raw duck meat (derived from Favier et al. 1995)

Micronutrients (mg/ 100 g)	Contents
Calcium	11
Phosphorus	202
Potassium	262
Sodium	90
Magnesium	19
Iron	2.1
Retinol (μg)	24
Thiamine	0.33
Riboflavin	0.43
Niacin	5.4
Pantothenic acid	1.6
Vitamin B6	0.34
Vitamin B12 (μg)	1.3
Folate (μg)	30

Table 10.5 Mineral content (mg/100 g dry matter) of breast and thigh muscles of Pekin ducks fed with a commercial complete diet (C) or commercial complete diet (85%) plus whole wheat grain (15%, WW) for the final 13 days and slaughtered at 49 days of age (n = 10; derived from Kokoszynski et al. 2017)

	Breast		Thigh			
Minerals	C	WW	P value	C	WW	P value
Sodium	86.0	90.5	0.238	107.8	94.4	0.063
Potassium	375.4	371.8	0.763	323.1	322.6	0.963
Phosphorus	46.7	47.0	0.859	40.3	40.3	0.991
Magnesium	22.1	20.8	0.182	18.8	18.3	0.660
Zinc	1.2	1.8	0.256	2.6	2.8	0.314
Iron	2.9	3.1	0.259	2.7	2.7	0.648
Copper	0.3	0.4	0.742	0.2	0.1	0.321

forms (selenized yeasts, seleno-methionine) are more efficient to deposit selenium in muscles than the mineral form (sodium selenite). Selenium and vitamin E are mainly used for their antioxidant activity.

10.2.2 Processed Duck Meat Products

In their review, Huda et al. (2011) detailed the various processed duck products, which can be found in different countries such as boiled salted duck, charcoal grilled duck, roasted duck, smoked duck sausages, patties, nuggets and meatballs. In France, many products have been developed from overfed ducks: thighs and gizzards 'confits' (cooked with fat), fillets salted and dried or smoked, 'rillettes', stuffed necks. There is a large variety of recipes. Very often, these recipes are confidential for the processing industry. It is, therefore, very difficult to have access to all details of processing procedures such as cooking time and temperature, use of spices or other ingredients (flour and eggs). Some studies detailed the chemical composition of processed products such as Lorenzo et al. (2011) which described that of dry-cured duck breast.

10.3 Duck Eggs

10.3.1 Physical Composition

Owing to the varieties in laying duck species used in the world, the weight range of duck egg is 60–90 g (Huang and Lin 2011). The percentages of eggshell, egg white and egg yolk relative to egg weight account for 11–13%, 45–58% and 28–35%, respectively (Huang and Lin 2011). The egg composition is mainly influenced by the age. Increasing the age of laying ducks from 26 to 42 weeks increased the weight of egg from 77.5 to 85.3 g and the yolk percentage from 28.4 to 33.3%, whereas the albumen percentage decreased from 58.8 to 54.9% (Applegate et al. 1998). The genotype, nutrition, laying age, raising environment, ambient temperature and relative humidity, and egg storage conditions will mainly affect the laying performance, relative percentages of eggshell, egg white and egg yolk, eggshell quality and technological and sanitary quality of eggs (Huang and Lin 2011).

10.3.2 Chemical Composition

10.3.2.1 Proteins

The amount of proteins in duck egg white is around 8.8 to 10.5% (Table 10.6). The major proteins in duck egg white are ovalbumin (40%), ovomucoid (10%), ovomucin (3%), conalbumin (2%) and lysozyme (1.2%) (Lin 2000). The amount of proteins in duck egg yolk is around 15.8 to 16.7% (Table 10.6). These proteins are livetins, phosvitin, vitellin and vitellenin mainly associated with lipids to form lipoproteins. The major amino acids in the egg white and egg yolk of ducks are glutamate, serine, aspartate, leucine, alanine, valine and lysine (Ganesan et al. 2014;

Table 10.6 Proximate composition of white and yolk of ducks (%)

Parts	Dry matter	Proteins	Lipids	Minerals	References
White	11.7	8.8	0.13	0.53	Chang (1992)
	12.3	10.5	0.03	0.74	Kaewmanee et al. (2009)
Yolk	56.5	16.0	37.3	1.6	Kaewmanee et al. (2009)
	55.3	15.8	36.8	1.7	Huang and Lin (2011)
	55.7	16.7	36.7	2.3	Sinanoglou et al. (2011)

Table 10.7 Amino acid contents of white and yolk of duck fresh eggs (expressed as residues/1000 residues; derived from Ganesan et al. 2014)

Amino acids	White	Yolk
Aspartic acid/asparagine	91	89
Threonine	67	60
Serine	97	99
Glutamic acid/glutamine	123	110
Glycine	61	54
Alanine	68	77
Cysteine	1	4
Valine	70	64
Methionine	54	26
Isoleucine	41	50
Leucine	81	86
Tyrosine	33	34
Phenylalanine	58	36
Lysine	61	70
Histidine	17	37
Arginine	6	5
Tryptophan	33	52
Proline	37	45

Table 10.7). Egg proteins are noteworthy because of their essential amino acid content, which is perfectly suited to human requirements, particularly due to the presence of a large quantity of lysine and sulphur amino acids (Seuss-Baum et al. 2011). The digestibility of raw egg differs substantially from that of cooked egg. This is usually attributed to the action of protease inhibitors contained in egg white (Seuss-Baum et al. 2011). Structural changes due to cooking would also facilitate the action of digestive enzymes. Moreover, raw egg white proteins weakly stimulate secretion of pancreatic enzymes and bile salts, thus reduce their digestibility. In contrast, raw yolk proteins are highly digestible, while overcooking reduces the digestibility of lipoproteins (Seuss-Baum et al. 2011). Many egg proteins are implicated in the development of human allergy. The major egg allergens contained in the egg white are lysozyme, ovomucin, ovalbumin and ovomucoid (Mine and Yang 2011).

10.3.2.2 Lipids

The lipids in duck eggs are mainly stored in the yolk, which content is about 37% (Table 10.6). The main lipids are triglycerides (591 mg/g lipids), diglycerides (9.5 mg/g lipids) and phospholipids (399 mg/g lipids, Kaewmanee et al. 2009). The cholesterol content of duck egg is about 46 mg/g lipids (Huang and Lin 2011). Among the phospholipids, the phosphatidylcholine is the main component (75.6%) followed by the phosphatidylethanolamine (16.0%), the lysophosphatidylcholine (2.7%) and the sphingomyelin (2.4%) (Sauveur 1988). The digestibility of egg triglycerides and phospholipids is excellent (Seuss-Baum et al. 2011). The main fatty acids are oleic (C18:1, 47.5%), palmitic (C16:0, 27.2%), linoleic (C18:2, 8.08%), stearic, (C18:0, 6.19%), arachidonic (C20:4, 2.62%) and palmitoleic (C16:1, 2.25%) (Kaewmanee et al. 2009). As that of hen eggs, it is easy to modify the fatty acid composition of duck egg yolk by using different dietary lipid sources. Huang and Lin (2011) reported various studies concerning the enrichment of egg yolk with long chain n-3 polyunsaturated fatty acids (PUFA) such as eicosapentaenoic acid (EPA) and docosahexaenoic acid (DHA) by including fish oil or fish meal in diets. The increase in EPA and DHA of egg yolk was obtained after 1 week of feed distribution when diets contained 5 and 6% refined cod liver oil and 2 weeks when diets contained 2, 3 or 4% refined cod liver oil (Chen and Hsu 2003). However, with the inclusion of 5 and 6% refined cod liver oil in diet, the score for fishy flavour of hard-boiled eggs increased significantly. By increasing dietary level of rice bran from 0 to 30%, Ruan et al. (2015) observed decreased contents of saturated and monounsaturated fatty acids (SFA and MUFA) and increased content of PUFA in egg yolk. By increasing the dietary level of corn-dried distillers' grains with solubles, from 0 to 30%, Ruan et al. (2018) observed decreased contents in MUFA and increased content in PUFA in egg yolk.

10.3.2.3 Minerals

In egg, the minerals are mainly distributed in the yolk (Table 10.6) which is rich in phosphorus mainly associated with proteins and lipids, calcium and potassium (Table 10.8). In comparison with hen eggs, there are few studies on the enrichment

Table 10.8 Mineral content of white and yolk of duck fresh eggs (mg/100 g; derived from Ganesan et al. 2014)

	White	Yolk
Calcium	23.12	158.22
Magnesium	79.89	18.23
Sodium	754.00	72.3
Potassium	604.40	131.42
Zinc		3.95
Copper	0.79	0.14
Iron	1.11	8.51
Sulphur	1.70	0.28

of duck eggs with micronutrients. By increasing dietary level of selenium from 0.04 to 0.40 mg/kg feed, Chen et al. (2015) observed a linear increase in selenium content of white and yolk of the egg.

10.3.2.4 Micronutrients

The vitamins are mainly distributed in the yolk, particularly those that are lipid-soluble (vitamins A, D, E, K). The yolk is also rich in carotenoid pigments, that give yellow-orange colour; they are, zeaxanthin (35.18 μg/g), lutein (98.79 μg/g), canthaxanthin (47.65 μg/g) and neolutein (14.96 μg/g) (Sinanoglou et al. 2011). In practice, xanthophylls, shrimp meal or some food-graded pigment are added in the duck diet for deeper yolk colour (Huang and Lin 2011).

It is quite difficult to modify the chemical composition of eggs except the fatty acid composition and micronutrient content that can be manipulated by way of feeding laying ducks.

10.3.3 Processed Duck Egg Products

Duck eggs are consumed as table eggs in many countries. There are many varieties of recipes, for example, boiled eggs, stuffed eggs, or deviled eggs, coddle eggs, roasted eggs, scrambled eggs, poached eggs and so on (Jalaludeen et al. 2009). Duck eggs in some countries are processed into salted eggs and pidan or alkalized eggs. Balut (embryonic eggs) is a nutritious delicacy in some South East Asian countries, such as the Philippines, Vietnam and Laos (Jalaludeen et al. 2009). Ganesan et al. (2014) compared the nutritional value of fresh, Pidan and salted eggs. Proteins, lipids and mineral contents are enhanced during pickling and salting process. Kaewmanee et al. (2009) also reported the changes in chemical composition of duck eggs during salting for up to 14 days. Zhao et al. (2014) detailed the nutritional characteristics of pidan duck eggs and showed that alkali treatment destroyed some amino acids and vitamins in egg white.

10.4 Conclusion

Feed is the main variation factor of duck meat and egg nutritional quality as it can directly modify the chemical composition and particularly the micronutrient content and fatty acid composition. Indirectly, feed will also have an effect on muscle growth and egg weight and by way of consequence on the physico-chemical characteristics of muscles such as water, protein, haeminic pigments and glycogen contents with consequences on colour and pH of meat, and on the weights of white and yolk of eggs. However, other factors can have a leading role on certain characteristics of duck meat and egg such as the genotype or the rearing system. Duck meat and egg production is well adapted to consumer demand thanks to an important segmentation of products induced by the diversity of genotypes, production systems (free range or standard) and the modalities of product presentation (whole carcasses, cut or processed meat, table or processed eggs).

References

Ali MS, Kang GH, Yang HS, Jeong JY, Hwang YH, Park GB, Joo ST (2007) A comparison of meat between duck and chicken breast. Asian Australas J of Anim Sci 20(6):1002–1006

Applegate TJ, Harper D, Lilburn MS (1998) Effect of hen production age on egg composition and embryo development in commercial Pekin ducks. Poult Sci 77:1608–1612

Baéza E (2000) Carcasses et viandes de canard: des facteurs de variation de mieux en mieux connus. Viande et Produits Carnés 21:159–167

Baéza E (2006) Major trends in research into domestic ducks and recent results concerning meat quality. 12th European Poultry Conference, Verona (Italy), 10–14/09/06, pp 8

Baéza E, Salichon MR, Marché G, Wacrenier N, Dominguez B, Culioli J (2000) Effects of age and sex on the structural chemical and technological characteristics of mule duck meat. British Poult Sci 41:300–307

Baéza E, Dessay C, Wacrenier N, Marché G, Listrat A (2002) Effect of selection for improved body weight and composition on muscle and meat characteristics in Muscovy duck. Br Poult Sci 43: 560–568

Baéza E, Fernandez X, Marie-Etancelin C (2013) Qualité des carcasses et de la viande des palmipèdes gavés INRA. Prod Anim 26:425–434

Baéza E, Chartrin P, Bordeau T, Lessire M, Thoby JM, Gigaud V, Blanchet M, Alinier A, Leterrier C (2017) Omega-3 polyunsaturated fatty acids n-3 provided during embryonic development improve the growth performance and welfare of Muscovy ducks (Cairina moschata). Poult Sci 96(9):3176–3187

Baltic MZ, Starcvic MD, Basic M, Zenunovic A, Ivanovic J, Markovic R, Janjic J, Mahmutovic H, Glamoclijaa N (2015) Effects of dietary selenium-yeast concentrations on growth performance and carcass composition of ducks. Anim Product Sci 57(8):1731–1737

Brown J, Mac Intosh M (2003) Conjugated linoleic acid in humans: regulation of adiposity and insulin sensitivity. J Nutr 133:3041–3046

Chang HS (1992) Introduction. In: Egg processing, 2nd edn. Hua Shiang Yuan Publishing Company, Taipei, pp 1–20

Chartrin P, Bernadet MD, Guy G, Mourot J, Duclos MJ, Baéza E (2006) Effects of genotype and overfeeding on fat level and composition of adipose and muscle tissue in ducks. Anim Res 55: 231–244

Chen TF, Hsu JC (2003) Incorporation of n-3 long-chain polyunsaturated fatty acids into duck egg yolks. Asian Australas J Anim Sci 16(4):565–569

Chen W, Zhang HX, Wang S, Ruan D, Xie XZ, Yu DQ, Lin YC (2015) Estimation of dietary selenium requirement for Chinese egg-laying ducks. Anim Production Sci 55(8):1056–1063
FAO (2021) FAO Statistics Division Primary livestock production data Total world production (1000 no): eggs (hen) in shell and other bird eggs in shell 2006 to 2016 by geographic region. FAOSTAT http://www.fao.org/faostat. Accessed February 2021
Favier JC, Ripert JI, Toque C, Feinberg M (1995) Répertoire général des aliments, Editeurs INRA et Tec Doc, Paris, 2ème edition, p 891
Ganesan P, Kaewmanee T, Benjakul S, Baharin BS (2014) Comparative study on the nutritional value of Pidan and salted duck egg. Korean J Food Sci Anim Resour 34(1):1–6
Girard JP, Culioli J, Denoyer C, Berdagué JL, Touraille C (1993) Discrimination de deux populations chez deux espèces de volailles sur la base de leur composition en lipides. Eur Poult Sci 57:9–15
Halle I, Jahreis G, Henning M, Köhler P, Dänicke S (2012) Effects of dietary conjugated linoleic acid on the growth of chickens and ducks for fattening and fatty acid composition of breast meat. J Consum Prot Food Saf 7:3–9
Han H, Zhang K, Ding X, Bai S, Luo Y, Wang J, Peng H, Zeng Q (2016) Effects of dietary nanocrystalline cellulose supplementation on growth performance carcass traits intestinal development and lipid metabolism of meat ducks. Anim Nutrition 2:192–197
Hou HC (1981) Egg preservation in China Food Nutr Bull 3 (2) http://archive.unu.edu/unupress/food/8F032e/8F032E03.html. Accessed June 2018
Huang JF, Lin CC (2011) Production composition and quality of duck eggs. In: Nys Y, Bain M, Van Immerseel F (eds) Improving the safety and quality of eggs and egg products, vol 1. Woodhead Publishing Series in Food Science Technology and Nutrition, Cambridge, UK, pp 487–508
Huang JF, Huang CC, Lai MK, Lin JH, Lee CH, Wang TY (2006) Effects of dietary fish oil on the contents of eicosapentaenoic acid and docosahexanenoic acid and sensory evaluation of the breast meat in mule ducks. Asian Australas J Anim Sci 19:231–235
Huda N, Putra AA, Ahmad R (2011) Potential application of duck meat for development of processed meat products. Curr Res Poult Sci 1(1):1–11
Islam MM, Ahmed ST, Kim YJ, Mun HS, Kim YJ, Yang CJ (2014) Effect of sea tangle (Laminaria japonica) and charcoal supplementation as alternatives to antibiotics on growth performance and meat quality of ducks. Asian Australas J Anim Sci 27(2):217–224
Jalaludeen A, Churchil RR, Joseph L, Anitha P (2009) Duck meat egg and their products, 4th World Waterfowl Conference, Thrissur, Kerala (India) 11–13 November 2009. pp 57–64
Kaewmanee T, Benjakul S, Visessanguan W (2009) Changes in chemical composition physical properties and microstructure of duck egg as influenced by salting. Food Chem 112:560–569
Koci E, Baumgartner J, Illes V, Palanska O (1982) Carcass and nutritive value of the small white broiler duck and its crosses. Eur Poult Sci 46:157–161
Kokoszynski D, Kotowicz M, Brudnicki A, Bernacki Z, Wasiewski PD, Wasilewski R (2017) Carcass composition and quality of meat from Pekin ducks finished on diets with varying levels of whole-wheat grain. Anim Prod Sci 57:2117–2124
Leclercq B, De Carville H (1978) Intérêt du rationnement du caneton mâle de Barbarie entre les âges de 8 et 12 semaines. Ann Zootech 27(1):1–7
Li M, Zhai S, Xie Q, Tian L, Li X, Zhang J, Ye H, Zhu Y, Yang L, Wang W (2017) Effects of dietary n-6:n-3 PUFA ratios on lipid levels and fatty acid profile of Cherry Valley ducks at 15-42 days of age. J Agric Food Chem 65:9995–10002
Lin CW (2000) Egg science and utilization in other avian species. In: Chemistry and utilization of eggs. Hua Shiang Yuan Publishing Company, Taipei, pp 555–583
Lorenzo JM, Purrinos L, Temperan S, Bermudez R, Tallon S, Franco D (2011) Physicochemical and nutritional composition of dry-cured duck breast. Poult Sci 90:931–940
Matitaputty PR, Wijaya CH, Bansi H, Laudadio V, Tufarelli V (2015) Influence of duck species and cross-breeding on sensory and quality characteristics of Alabio and Cihatup duck meat. J Food 13(4):522–526

Mazanowski A, Kisiel T, Gornowicz E (2003) Carcass quality meat traits and chemical composition of meat in ducks of paternal strains A44 and A55. Anim Sci Pap Rep Inst Genet Anim Breed Jastrzebiec Poland 21(4):251–263

Mine Y, Yang M (2011) Egg allergy. In: Nys Y, Bain M, Van Immerseel F (eds) Improving the safety and quality of eggs and egg products, vol 2. Woodhead Publishing Series in Food Science Technology and Nutrition, Cambridge, UK, pp 254–271

Muhlisin DSK, Yeong RS, Hong RK, Hyung JK, Byoung KA, Chang WK, Hak KK, Sung KL (2013) Comparison of meat characteristics between Korean native duck and imported commercial duck raised under identical rearing and feeding condition. Korean J Food Sci Anim Resour 33(1):89–95

Paci G, Baguacca M, Marzoni M, Avanzi CF (1993) Quality of Italian strains of Muscovy common and Muscovy x common ducks bred under two different technologies. In: Proceedings 11th WPSA European Symposium on the Quality of Poultry Meat, Tours (France) 4–8/10/93, p 66–73

Paquin J (1988) Valeur nutritionnelle des viandes de volailles, In: L'aviculture française informations techniques des services vétérinaires, Ed Rosset, Paris, pp 743–748

Qiao Y, Huang J, Chen Y, Chen H, Zhao L, Huang M, Zhou G (2017) Meat quality fatty acid composition and sensory evaluation of Cherry Valley spent layer and crossbred ducks. Anim Sci J 88:156–165

Ruan D, Lin YC, Chen W, Wang S, Xia WG, Fouad AM, Zheng CT (2015) Effects of rice bran on performance egg quality oxidative status yolk fatty acid composition and fatty acid metabolism-related gene expression in laying ducks. Poult Sci 94:2944–2951

Ruan D, Fouad AM, Fan QL, Chen W, Xia WG, Wang S, Cui YY, Wang Y, Yang L, Zheng CT (2018) Effects of corn dried distillers' grains with solubles on performance egg quality yolk fatty acid composition and oxidative status in laying ducks. Poult Sci 97:568–577

Russell EA, Lynch PB, O'Sullivan K, Kerry JP (2004) Dietary supplementation of α-tocopheryl acetate on α-tocopherol levels in duck tissues and its influence on meat storage ability. Int J Food Sci and Technol 39:331–340

Salichon MR, Leclercq B, Rémignon H, Marché G, Blum JC (1993) Composition biochimique des filets de canard de Barbarie, In: Proceedings of 11th WPSA European Symposium on the Quality of Poultry Meat, Tours (France) 4–8 October 1993, pp. 66–73

Sauveur B (1988) Structure et composition de l'œuf. In: Sauveur B (ed) Reproduction des volailles et production d'œufs. INRA, Paris, pp 347–375

Schiavone A, Romboli I, Chiarini R, Marzoni M (2004) Influence of dietary lipid source and strain on fatty acid composition of Muscovy duck meat. J Anim Physiol Anim Nutr 88:88–93

Seuss-Baum I, Nau F, Guérin-Dubiard C (2011) The nutritional quality of eggs. In: Nys Y, Bain M, Van Immerseel F (eds) Improving the safety and quality of eggs and egg products, vol 2. Woodhead Publishing Series in Food Science Technology and Nutrition, Cambridge, UK, pp 202–236

Sinanoglou VJ, Strati IF, Miniadis-Meimaroglou S (2011) Lipid fatty acid and carotenoid content of edible egg yolks from avian species: a comparative study. Food Chem 124:971–977

Smith DP, Fletcher DL, Buhr RJ, Beyer RS (1993) Pekin duckling and broiler chicken pectoralis muscle structure and composition. Poult Sci 72:202–208

Wu LY, Guo X, Fang Y (2012) Effect of diet dilution ratio at early age on growth performance, carcass characteristics and hepatic lipogenesis of Pekin ducks. Braz J Poult Sci 14(1):43–49

Zeng QF, Cherry P, Doster A, Murdoch R, Adeola O, Applegate TJ (2015a) Effect of dietary energy and protein content on growth and carcass traits of Pekin ducks. Poult Sci 94:384–394
Zeng QF, Zhang Q, Chen X, Doster A, Murdoch R, Makagon M, Gardner A, Applegate TJ (2015b) Effect of dietary methionine content on growth and carcass traits and feather growth of Pekin duck from 15 to 35 days of age. Poult Sci 94:1592–1599
Zhang C, Ah Kan Razafindrabe RH, Chen K, Zhao X, Yang L, Wang L, Chen X, Jin S, Geng Z (2018) Effects of different rearing systems on growth performance carcass traits meat quality and serum biochemical parameters of Chaohu ducks. Anim Sci J 89:672–678
Zhao Y, Tu YG, Xu MS, Li JK, Du HY (2014) Physicochemical and nutritional characteristics of preserved duck egg white. Poult Sci 93(12):3130–3137

Dr. Elisabeth Baéza obtained first a diploma of Engineer in Agronomy (ENSFA, Rennes, France) specialized on Animal Productions, then a Ph.D. in Animal Sciences from the University of Montpellier (France). She was recruited at INRAE Nouzilly Centre (France) in 1993 to work on waterfowl production and poultry meat quality. She has more than 121 peer-reviewed research publications in national and international journals and written three book chapters. She is an Associated Editor for *INRAE Productions Animales* and *European Poultry Science* journals. She has an expertise activity for the French National Agency on Food and Environment Safety (ANSES).

Dr. Jeng-Fang Huang obtained his bachelor degree on Animal Husbandry from the National Taiwan University (Taiwan). He was granted Master's and Ph.D. degrees from the Department of Animal Science, Michigan State University (USA), and the Department of Animal Production, National Taiwan University (Taiwan), respectively. He had worked in the research of duck production and management for more than 23 years at the Ilan Branch Institute (also known as Duck Research Center, DRC), Livestock Research Institute, Taiwan. Presently, he is the Director General of the Livestock Research Institute in Taiwan and is serving as the President of the Chinese Society of Animal Science.

Duck Slaughter Processing and Meat Quality Measurements

11

Wen-Shyan Chen

Contents

W.-S. Chen (✉)
Animal Products Processing Division, Livestock Research Institute, Council of Agriculture, Tainan, Taiwan
e-mail: wschen@mail.tlri.gov.tw

A. Jalaludeen et al. (eds.), *Duck Production and Management Strategies*,
https://doi.org/10.1007/978-981-16-6100-6_11

Abstract

The duck egg and meat are highly nutritious just like chicken and full of flavor. Duck meat is very lean because most of the fat associated with it is found under the skin. From farm to fork, totally mechanized systems are needed in production, catching, transporting from duck house to vehicle through conveyors, assembling in coops and trucks, transporting to the processing plants and processing into products and by-products. The slaughtering procedure for ducks is slightly different from that of chicken. The different steps in duck processing such as assembling, pre-slaughter handling, slaughtering, bleeding, scalding, feather release, inspection and evisceration, chilling and pinfeather removal, cutting and product packaging, and storage management are described in this chapter. The carcass characteristics of duck meat, physicochemical properties such as shear value, color, pH value, water-holding capacity, cooking loss and drip loss, oxidation susceptibility, microbiological detection, and sensory evaluation are discussed in this chapter.

Keywords

Duck · Slaughter processing · Meat quality

Abbreviations

MFI	Myofibrillar fragmentation index
MUFA	Monounsaturated fatty acid
PUFA	Polyunsaturated fatty acid
RDF	Rendered duck fat
SFA	Saturated fatty acid
TBA	Thiobarbituric acid
TBARS	TBA-reactive substances
WHC	Water-holding capacity

11.1 Introduction

Ducks had long association with man in Asia, especially in far-east Asian countries such as China, Korea, Taiwan, Indonesia, Thailand, Vietnam, and Malaysia. The waterfowl, especially domestic ducks are the source of food and livelihood for people in many parts of the world, particularly Asian countries, since ancient times. Apart from meat and eggs the waterfowl also provide down-feathers for making bedding and warm jackets. Compared to chicken eggs, duck eggs are more stable during storage at room temperature (Jalaludeen et al. 2009). Their albumen and yolk indices and Haugh unit score are not severely affected with storage. Compared to chicken eggs, duck eggs have many positive attributes; the important ones are greater stability against deterioration, less pronounced chalazae, less darkening around the yolks on boiling, low sulfur odor, and albumin devoid of yellow coloration. Duck meat is an excellent source of high quality protein containing a well-balanced array of amino acids. Duck meat also contains generous amounts of minerals and vitamins. The fat contained in duck meat is similar to olive oil that can lower cholesterol. The bezoar acid, linoleics, and a-linoleic acid contained in duck meat are able to lower blood diseases and to increase immune system performance, prevent diabetes and skin diseases (Jalaludeen et al. 2009).

Many special products such as pressed salt duck, roasting Pekin duck, duck roll, smoked duck steak, tea-smoked duck, crispy skin duck, herb duck, etc. are prepared from duck meat. Thousand-year duck eggs and salted duck eggs are the popular processed duck egg products. The professional bakers prefer duck eggs because of their rich yolks and higher protein content to get better texture and shape for the baked goods (Jalaludeen and Churchil 2006). Besides meat and eggs, visceral parts of ducks are also processed into salted duck gizzards and marinated duck tongues, which are famous appetizers. The duck slaughter processing and duck meat quality measurements are elaborately discussed in this chapter.

11.2 Duck Slaughter Processing

11.2.1 Introduction

According to the survey of livestock and poultry in the second quarter of the 2018 of Council of Agriculture, Taiwan (NAIF 2018), the number of meat duck farms in the Taiwanese region was 2181 and the total number of raising ducks was almost near six million. The species of ducks reared was mainly mule ducks, with a proportion of 75.10%, followed by Pekin duck (19.27%) and Muscovy duck (5.62%). The raising area of duck is mainly in the central and southern parts of Taiwan, and most of the slaughterhouses are concentrated in this area. The geo-relationship between duck raising location and duck slaughterhouse contributes to the transportation and slaughtering of ducks.

It is estimated that there are about 5–10% damage among the waterfowl containing ducks and geese when transported from the farm to the slaughterhouse (Chen et al. 2012). Duck collection and transportation from farm to slaughterhouse is an extremely stressed process, the loss of ducks includes economic losses caused by death and poor appearance of carcasses. Actually, there are not many published literatures related to duck slaughter process. Chen et al. (2012) investigated the stresses before slaughter on changes of the physicochemical characteristics of duck muscle and found many stressors such as temperature, noise, and water shortage having damage effect on carcass quality, and lower the prices of ducks. Studies have pointed out that the new type of chicken catching system seems to be changed to the collection system for ducks. For example, Hale et al. (1973) pointed out that the birds (poultry) catching methods in various stages include the following: transporting birds from the farm house to vehicle by special conveyor, enlarging and redesigning coops and trucks, and designing completely total mechanized systems for producing, harvesting, and transporting birds.

The duck carcass should be slaughtered according to the specifications of the duck slaughtering operation in Taiwan. It must be inspected by veterinarian and subjected to rapid cooling treatment. The operation procedure of the duck slaughtering operation specifications are as follows;

- the carcass rapidly cools down to a temperature of 4 °C or less in the meat center,
- the internal organs, especially the lungs need to be removed completely,
- carcass appears clean without residual feathers,
- no fractures and blood stasis,
- no antibiotics, antibacterial agents, and pesticide residues,
- the total bacterial count should meet the food hygiene standards (frozen duck meat below 3×10^6 CFU/g, refrigerated duck meat below 3×10^7 CFU/g) (Huang 1998).

The slaughtering procedures for ducks in different countries may be slightly different, but Chen et al. (2012) have shown the flow chart for reference (Fig. 11.1).

11.2.2 Duck Assembling

The ducks are not raised in large size in Taiwan. Whenever possible, ducks for slaughter are caught at night for loading, the time when the weather is cooler, the ducks are easier to catch, struggle less, and settle down in the coops faster. Usually, ducks are assembled by hand and carried to the transport vehicle by holding one or two legs. At present, there is usually a metal cage in the duck farm or the holding pen of slaughterhouse (Fig. 11.2), which is conducive to the concentration and shipment of the ducks. It is particularly effective in reducing the labor force and increasing efficiency of collecting the ducks.

11.2.3 Pre-Slaughter Handling

The conditions of poultry transport from the farm to the slaughterhouse and pre-slaughter processing will result in a 3–4% weight loss (Mountney and Parkhurst 1995). Another study pointed out that improper handling of waterfowl before they are slaughtered can result in 10% of carcasses being listed damage or lost (Chen et al. 2012). For example, continuous noise, vibration and movement, rise in temperature, confinement in a narrow environment, and continued exposure to upset stressors tends to have a compounding effect, which with time can lead to undesired levels of stress. It is obvious that the stress plays a vital effect on the meat quality of the ducks.

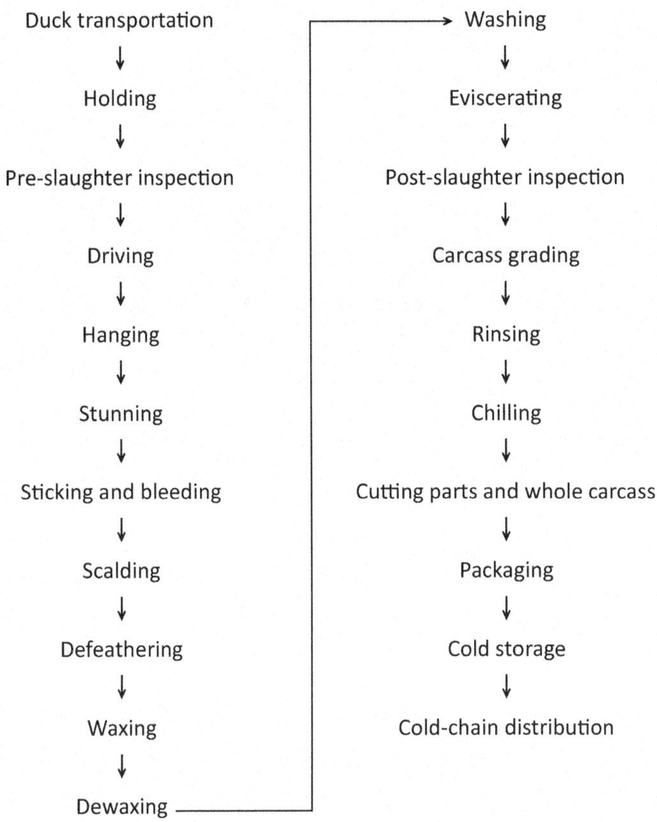

Fig. 11.1 Flow chart of duck slaughter

Chen et al. (1991) studied the effects of fasting and enforced exercise on the meat quality of duck muscle. The ultimate pH was increased and the lactate content of breast and thigh muscles was lower in stressed birds compared to control ones. The activity of lactic dehydrogenase was significantly increased by the stress, and the activities of creatine phosphokinase and alkaline phosphatase were also slightly increased. After transport, ducks are placed in a holding pen to rest and to reduce the effect of transport stress (Fig. 11.3).This contributes to preserve the meat quality of ducks after slaughter.

11.2.4 Slaughtering

The slaughtering procedure includes electric stunning and bleeding. Generally, stunning with an electric shocker is frequently used in slaughtering ducks to prevent

Fig. 11.2 Metal cage can be used to collect duck easier

Fig. 11.3 Ducks stay in holding pen before slaughter to reduce stresses after transportation

Fig. 11.4 The ducks are stunned by slipping through the electrified water

struggling and beating. Mountney and Parkhurst (1995) demonstrated that the shockers can be a knife which has an electric current in the blade or an electric plate which touch the heads of the birds as they move along the conveyor. The local slaughter plants in Taiwan use the electric water shock method to carry out the stunning of the ducks (Fig. 11.4). The voltage used was about 90 volts. This kind of voltage can make the ducks coma in a state of no struggle and maintain the best quality of meat. Stunning procedure also relaxes the duck muscles that will lead to loosened feather follicles enabling easy plucking of feathers. The feathers from the duck wings are to be sold for making badminton in Taiwan (Fig. 11.5).

11.2.5 Bleeding

There are several ways of cutting ducks so that they can be bled, such as "modified kosher method" of slaughter is the most widely used in processing operation because it is easier to gain better bleeding and leaves the head and neck intact for use in suspending the duck carcass for further eviscerating operations (Mountney and Parkhurst 1995). In local slaughterhouses, then the duck is stunned by electric water shock, hooks are used to lift the duck feet and manual bloodletting of the neck is performed by knife (Fig. 11.6). The stainless steel metal grooves are used under the slaughter line to collect the duck blood, because duck blood is a quite important food material in Taiwan's cuisine culture.

Fig. 11.5 Pulling out the major duck feathers for making the badminton

Fig. 11.6 Ducks are bled by knife

11.2.6 Scalding

Scalding is the process of immersing poultry carcasses in hot water to loosen the feathers for easier plucking (Barbut 2002). The combination of hot water temperature and immersion time produce different effects on the processing and the processed carcasses (Sams and McKee 2010). The scalding process significantly determines the appearance, color and visual quality of the duck carcass. It can also influence yield and shelf life of the duck product. The feathers can be easily removed by the roller machine if the hot water reach feather follicles and the feather follicles get soaked (Fig. 11.7). The temperature and time should be properly operating according to the species and age of the duck. Industrial production uses mechanical scalding tanks that generally use steam to heat the water and keep the water temperature within the specified range continuously. Generally, the proper scalding temperature for duck is about 61–62 °C for 60 s. This temperature and time can effectively remove the feathers and preserve the quality of the duck carcass. Mechanical defeathering equipment is generally equipped with a number of rubber squeegees on the opposite drum or a number of opposing shafts mounted on a flat-type hair-dripping machine. In the photograph (Fig. 11.8), a model of defeathering equipment with a number of squeegees on the shaft, and a metal disc is shown. When driven by the motor, the rubber spurs on the two sides of the defeathering machine are rapidly rotated. When the scalded duck passes through the middle gap, it is in close contact with the feathers of the duck body and rubs against each other. This frictional force exceeds that of the duck hair follicles on the feathers. The grip force allows the feathers to be smoothly removed within few seconds without mechanical damage to carcass.

Fig. 11.7 The ducks are immersed into hot water to facilitate easy mechanical plucking

Fig. 11.8 Mechanical defeathering equipment

11.2.7 Feather Release

Waterfowl animal carcass such as duck and goose contains many pinfeathers and are not easily removed during slaughter process. It must take a lot of time to remove the pinfeathers in the processing plants. Rosin, a depilatory aid for livestock and poultry industry, is allowed to use in duck and goose slaughter industry in Taiwan's slaughter plants. Rosin (Fig. 11.9), also called colophony or Greek pitch (Latin: *pix græca*), is a solid form of resin obtained from pines and some other plants (Wikipedia 2018a). The rosin block is dissolved in hot water; the slaughtered duck is completely immersed in the rosin solution and then immersed in cold water (Fig. 11.10). This facilitates easily removal of duck pinfeathers. The duck carcass in figure (Fig. 11.11) shows the clean appearance of the duck skin treated with rosin, which is smooth with the feathers removed completely. Feather-containing rosin colloid can be recycled after heating and melting, and thus the cost of the rosin block can be reduced. The rosin colloid must be heated to kill microorganisms before reuse, because it may serve as a serious contamination. In Western countries, slaughterhouses use mainly paraffin (synthetic wax).

11.2.8 Inspection and Evisceration

Most duck processing plants in Taiwan now operate under the government inspection. Each slaughterhouse should be equipped with a public veterinarian to check whether the viscera of the duck have any lesions and whether the appearance of the carcass is normal (Fig. 11.12). If the duck viscera are found to have lesions, the carcass and internal organs will be indwelled for further examination of the disease

Fig. 11.9 Solid rosin used in scalding

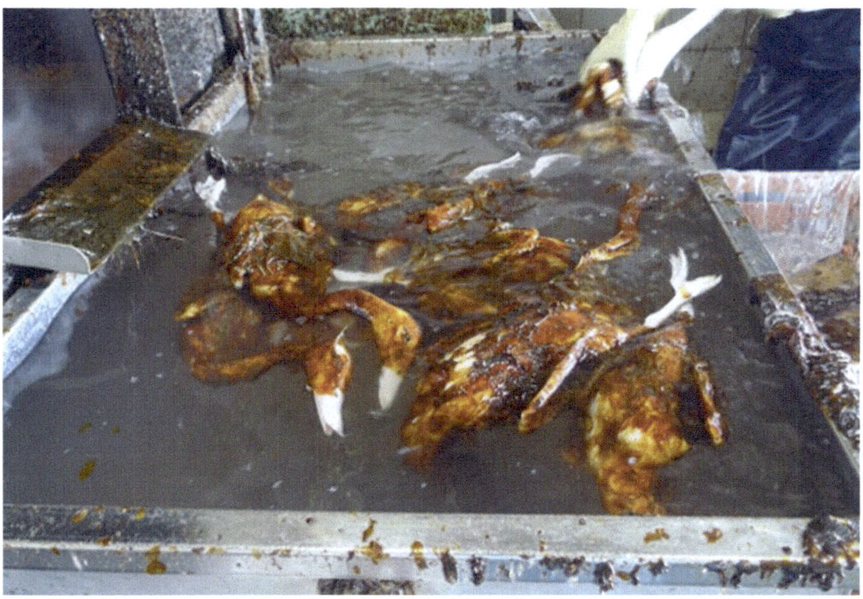

Fig. 11.10 The duck carcasses are first immersed in the rosin solution, and then in cold water to harden the rosin

or lesions. It is important that the viscera is taken out carefully and maintained intact to prevent contaminating the carcass. In processing plants, the head of the duck are hung on the conveyor line, the hook is made of metal, which is relatively strong and

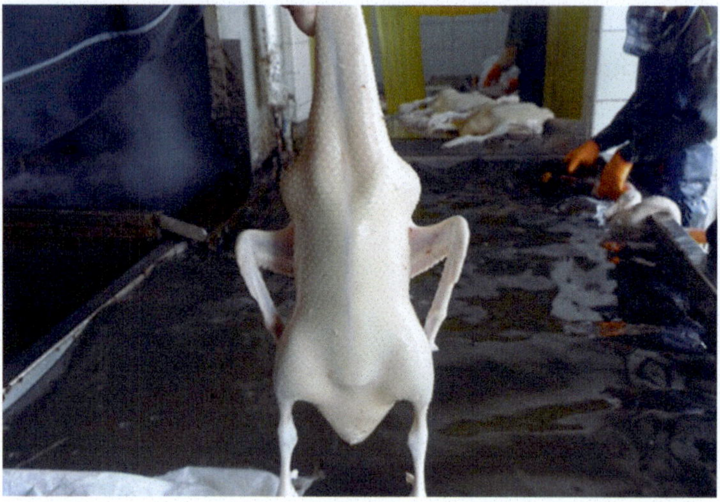

Fig. 11.11 The duck skin treated with rosin solution

Fig. 11.12 The veterinarian is inspecting the duck carcass

washable (Fig. 11.13). The size of the hook varies depending on the type and weight of slaughtered duck. For example, the Muscovy duck has a larger body size and therefore requires a larger gap. However, the mule duck is smaller and requires a smaller gap to facilitate the tight hanging of the duck head. It is necessary to use a

Fig. 11.13 Duck carcass hanging on conveyor

knife to cut from the tail of the duck, to pull out the internal organs of the duck and to hang it outside, and to maintain the integrity of the internal organs to avoid damage. The visceral damage will cause microbes to contaminate not only the duck carcass but also the working places. Usually, a vacuum pipette is used to absorb the internal organs, which can quickly suck out the internal organs and keep the internal organs intact and hygienic (Fig. 11.14).

11.2.9 Chilling and Pinfeather Removal

Carcass chilling is an essential step in today's commercial poultry processing. The core body temperature of ducks after slaughter is often higher than 40 °C. Keeping duck carcass with this high body temperature will easily damage its quality. Especially, the meat protein quality is sensitive to high temperature; the higher the carcass temperature, lower the protein quality and functionality of muscle. Therefore, during the slaughter process, soaking the carcasses in cool water at about 7 °C can quickly decrease the body temperature and keep the good quality of duck meat (Fig. 11.15). The temperature of this cooling water is close to that of Ali et al. (2007a) who also suggested that it is possible to chill duck carcasses in processing plants at 10 °C without any problem of toughening the meat. The carcass-cooling tank has the function of rotating and generating bubbles, which can effectively reduce the temperature and clean the duck carcass. The pinfeather of the duck skin has a

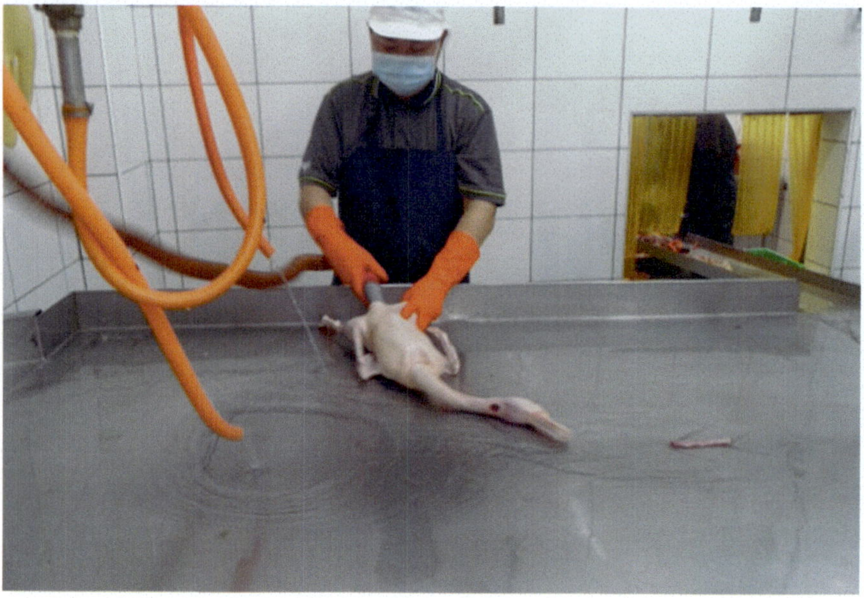

Fig. 11.14 Use a vacuum pipe to suck the internal organs of the duck

Fig. 11.15 Rotary sink for cooling and moving ducks

great influence on the quality of the carcass. The consumer prefers the duck meat without residue of black pinfeather. In slaughter plant, when the duck is fully chilled, the pinfeather is carefully removed manually (Fig. 11.16), to ensure the quality of the duck.

Fig. 11.16 Manual pulling of feathers

11.2.10 Cutting and Product Packaging

The room temperature of the cutting chamber should be below 15 °C according to the hygiene regulatory standards of Bureau of Animal and Plant Health Inspection and Quarentine (BAPHIQ 2018) of Taiwan. The metal table top has holes allow water droplets to drip easily and the carcass surface dry quickly. The whole duck with bones, the duck without head or the cut meat should be drained before packaging (Fig. 11.17). This is necessary to reduce the formation of ice crystal during the frozen storage period. The safety of packaging materials should be in accordance with hygiene regulations, and different products should be packaged with different packaging materials (Fig. 11.18). The workers should wear sanitary gloves and strictly abide by various sanitary regulations to ensure the quality of duck meat.

11.2.11 Storage Management

After the duck product is packaged, it should not be placed indoors for a long time. It should be moved to the refrigerated (4 °C) or freezer storage (−18 °C) room as soon as possible (Fig. 11.19). The shelf life of different duck products should be tested to

Fig. 11.17 Duck cutting room

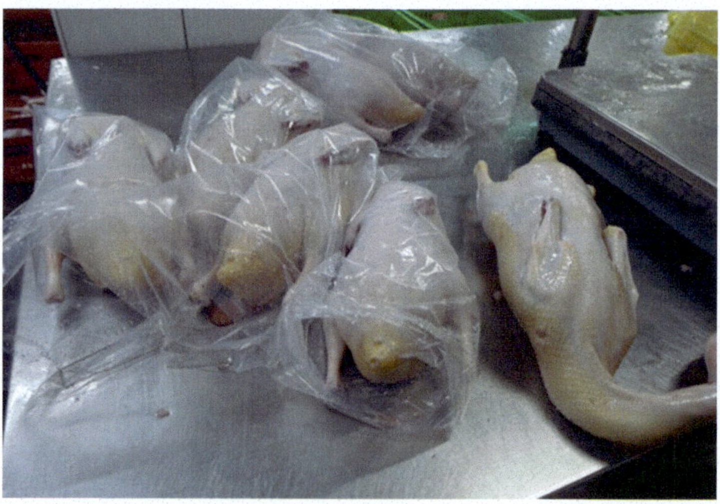

Fig. 11.18 Packaged duck

avoid the quality variation and bacteria multiplying during freezing or refrigeration. The product should be transported to a sales place using a refrigerated truck, or transported after being covered with crushed ice, but it is necessary to prevent the ice water from penetrating into the meat product.

Fig. 11.19 Packed products placed on the shelves in cold storage room

11.2.12 Supplementary

The slaughtering equipment and sanitation inspection regulations for ducks are basically the same as the operation methods of poultry slaughterhouses. For the basic operations of duck slaughtering is given in the instructions of the BAPHIQ (2000). The authorities are responsible for the slaughter inspection of poultry in Taiwan. Poultry (chicken, duck, and goose) slaughtering operations should meet the summarized requirements as follows:

1. Containers for the transport of live poultry should be kept clean and should be cleaned and disinfected after use;
2. Each carcass should be able to submerge in water and continuous flow of water will keep the water clear;
3. Water spray cleaning system in the feather removal machine should be able to clean the sludge and feathers attached to the surface of the carcass;
4. The carcass dropped in the feather removal area should be properly cleaned and disinfected before being hung back in the conveyer;
5. Feathers container should be placed in the contaminated operation area and moved out at the appropriate time;
6. The operation of taking the internal organs should not be carried out when the feather removal is not complete;
7. The carcass must be cleaned before taking out the internal organs;
8. Remove the internal organs gently and smashing of the internal organs should be avoided. The tool that has come in contact with the smashed organs should immediately disinfect in hot water above 83 °C to avoid cross-contamination;
9. The carcass should be cleaned before entering the cooling tank.

11.3 Duck Meat Quality Measurements

11.3.1 Introduction

There are certain standards for the determination of meat quality in meat industries very successfully, such as the method of measuring pork carcass and meat quality for international pork trade or meat quality determination as prescribed by National Pork Producers Council (NPPC 1991). However, there are only few units or institutes that conduct research and establish standards for measuring the quality of duck meat. Duck meat quality characteristics such as carcass appearance, chemical composition, physical determination, microbiological examination, and sensory evaluation are described as follows.

11.3.2 Carcass Characteristics

Duck carcass with full fattening skin, attractive color, and full meat percentage. No residue of pinfeathers in duck skin indicates that the ducks are slaughtered at correct age and attained good maturity, in other way, the meat has better flavor due to the prolonged feeding period. Muscovy duck has some unique features include its bare face and the red caruncles positioned around the eyes and beak. The duck farmers and consumers are like Muscovy faces with abundant red caruncles, because more and abundant the red caruncles indicate adequate fattening of carcass and good maturity of the duck (Fig. 11.20). The duck rearing system is gradually changing to a slat floor housing system in Taiwan and some other countries to avoid the infection

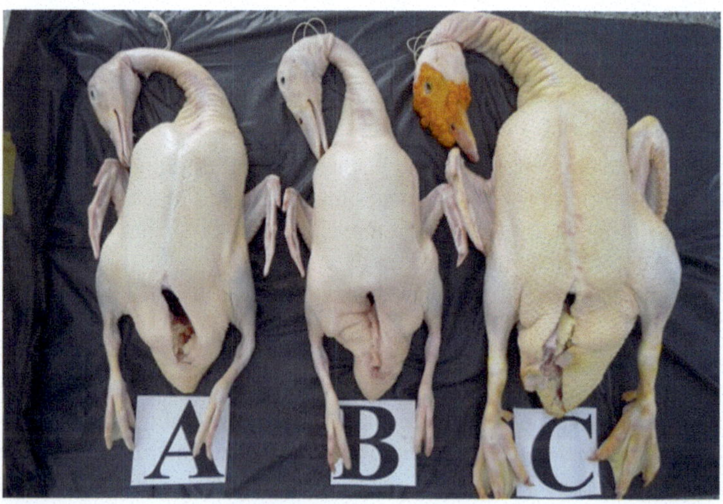

Fig. 11.20 The visual appearance of duck skin and color (A: mule duck; B: Pekin duck; C: Muscovy duck)

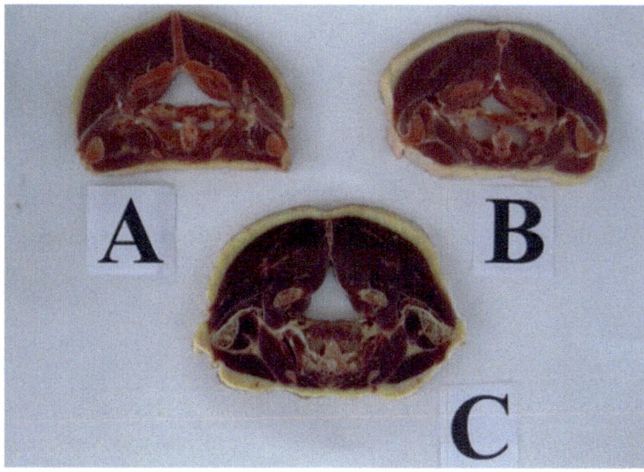

Fig. 11.21 The cross section of duck breast meat (A: mule duck; B: Pekin duck; C: Muscovy duck)

of avian influenza, and the feed formula is mainly based on corn and soybeans. Fernandez et al. (2003) reported that paler breast muscle of overfed mule duck and geese contained higher lipid levels.

The cross section of breast from duck carcass shows that different duck species have different muscle color and skin fat color (Fig. 11.21). Among them, the muscle color of Muscovy duck is dark brown, while that of mule duck and Pekin duck is dark red. Muscovy duck breast is fuller and thicker, and the thickness of breast from Pekin duck and mule duck is thinner. In the aspect of skin thickness, that of Muscovy duck and Pekin duck is the thickest, while that of mule duck is thinner. Because of variations in skin types among breeds, the breed preferred for different duck processed items are also different. Generally, the color of duck breast is darker than other poultry meat, this is because muscles doing work need more oxygen, and the oxygen is delivered to those muscles by the red cells in the blood. One of the proteins in meat, myoglobin, holds the oxygen in the muscle, and gives the meat a darker color. The studies in Taiwan Livestock Research Institute (TLRI 2009) revealed that duck breast and leg meat has higher myoglobin content (3.7 mg/g and 2.3 mg/g, respectively) than those of geese breast meat (1.5 mg/g) and chicken leg meat (0.6 mg/g). The low level of myoglobin in chicken meat and high level of that in duck meat can explain the dark coloration in duck meat compared to chicken meat. In terms of carcass characteristics, it can be found even after fattening, the fat accumulation of poultry is mostly concentrated in the skin and the abdomen, but intramuscular fat is rare. According to the Council Science and Technology of Japan, the melting point of duck fat is only 14 °C, much lower than the temperature of the chicken fat (37 °C), and the low melting point of fat makes duck meat still very delicious even when served cold (CSTJ 2018).

TLRI (1992) published a manual, which describes the duck carcass portioning and grading specifications. The procedure of cutting and cut-up parts of the duck carcass are shown in Fig. 11.22. Ducks are small animals, and do not need to be cut too finely. Irrespective of species, the procedure of making cut up parts described in this manual is suitable for Muscovy or mule or Pekin ducks. Making cut up parts is

Whole duck

Whole duck without wings and feet

Whole duck without neck, wings and feet

Breast halves without drummette

Whole thigh

Part thighs with back portion

Half-drummette, mid-joint and tip

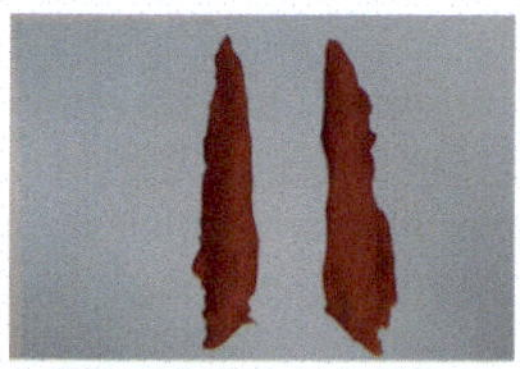

Tenderloins

Fig. 11.22 Sliced meat from mule duck

necessary for the sale at various markets and for the manufacture of processed duck products.

The percentage of each part of the duck carcass is a very important economic character that aids in breeding improvement and the price of marketed duck meat. Generally, the consumer likes duck meat with thick breasts and a high percentage of leg meat. In particular, the processing industry pays attention to the percentage of lean meat and the percentage and quality of duck skin, because the taste of many processed duck products is closely related to the crispness of duck skin. Chen et al. (1984) analyzed the percentages of different portions of mule duck carcass (Table 11.1). The breast meat, thigh meat, and wings having higher proportions are also the most costlier and popular parts of the duck carcass. The head and neck, wings, and feet are also very popular edible by-product in Taiwan. Duck by-products, such as tongue, head, heart, and gizzard, are processed to produce consumer's favorite duck by-product preparations.

Table 11.2 showed the percentage of carcass cuts from mule and Cherry Valley (Pekin) ducks. Regardless of male or female, the weight of Pekin duck is heavier

Table 11.1 Percentage yields of carcass and cut-up parts of mule duck (Chen et al. 1984)

Portion of carcass	Weight (g)	Percentages relative to carcass weight (%)
Live weight	2574	–
Carcass weight	2004	77.9 (% relative to live weight)
Head and neck	308	15.4
Breast	418	20.9
Thigh	390	19.5
Wings	277	13.8
Back part	468	23.4
Feet	66	3.31

Table 11.2 Yield and cut-up parts of Mule and Cherry Valley (Pekin) ducks (Chen et al. 1984)

	Mule duck (8 weeks old)	Cherry Valley (8 weeks old)	
Portion of carcass	(Male and female)	Male	Female
Live weight, g	2491	3541	2891
Dressing percentage, %	63.3	61.7	66.0
Breast, %	22.6	23.8	23.3
Back, %	17.9	22.0	21.2
Thigh, %	15.0	14.1	14.0
Wing, %	12.4	9.4	9.9
Head and neck, %	14.1	14.1	13.1
Feet, %	2.9	2.7	2.5
Abdominal fat, %	0.51	2.7	2.2

than that of mule duck and Pekin duck's breast and back muscle percentage are higher than that of mule duck. Pekin duck had higher percentage of abdominal fat than mule duck (Chen et al. 1984). Pan et al. (1985) also found that Pekin duck had heavier abdominal fat, which was as 1.9 to 3.6 times than that of mule duck. Even now, the high abdominal fat percentage of Muscovy duck and Pekin duck has a negative economic impact on the duck industry because removal the abdominal fat in duck processing factory is time-consuming and the recovered fat is a useless waste. However, development of value-added utilization of duck fat is an important issue in the future in the duck processing industry.

11.3.3 Physicochemical Properties

11.3.3.1 Proximate Composition

The nutritional properties of duck meat are detailed in the previous chapter. To analyze the proximate composition of poultry meat, the World Poultry Science Association published a guide (Petracci and Bąéza 2011).

11.3.3.2 Fat Oxidation

Meat oxidation susceptibility depends mainly on fat and protein oxidation. Compared to chickens, ducks have high fat depot in the abdominal cavity and subcutaneous tissue. Rendered duck fat (RDF) is prepared from abdominal fat. RDF can be used as a substitute for other cooking fats and oils. The unique flavor attributes of RDF create a potential for increased use in various food applications. The off-flavor, off-odor, and warmed-over flavor appears to be related to lipid oxidation in meat, which is undesirable. Lee et al. (2012) investigated the fat from different animals, and stated that the duck fat had the lowest peroxide value (1.2 meq/ kg) and the highest unsaturated fatty acid content (74%) (Table 11.3). The major components of the unsaturated fatty acid in duck fat are oleic acid, linoleic acid, and linolenic acid. Lee et al. (2015) examined the effect of rendering and purification temperature on fatty acid composition, and found that the carbon chain lengths of duck fat are almost between 12 and 20. The rendering time of animal fat will obviously affect the quality of the oil. Generally, long-term rendering time can produce more percentage of the oil. However, the quality of the oil will decrease with increased rendering temperature and time, resulting in cracking or auto-oxidation of the oil. Lee et al. (2012) showed that the rendering rate of lard (back and abdominal lard) increased with increasing rendering temperature, but that of duck and geese fat decreased with increasing rendering temperature (Table 11.4). Gong et al. (2007) compared rendered duck fat (RDF) to commercial sources of soybean oil, lard, tallow, butter, and

Table 11.3 Fatty acid composition of different animal fats (Lee et al. 2012)

	Animal fat sources				
Type of fatty acid	Back lard	Abdominal lard	Chicken fat	Duck fat	Goose fat
Monounsaturated fatty acids (MUFA), %	46.3^{c}	38.1^{d}	46.9^{c}	52.2^{a}	50.7^{b}
Polyunsaturated fatty acids (PUFA), %	16.6^{d}	18.4^{c}	22.8^{a}	21.5^{b}	18.2^{c}
Saturated fatty acids (SFA), %	37.1^{b}	43.5^{a}	30.2^{c}	26.3^{d}	31.0^{c}
n-6 fatty acids, %	1.15^{c}	1.25^{b}	1.43^{a}	1.02^{d}	0.76^{c}
n-3 fatty acids, %	0.28^{c}	0.41^{b}	0.51^{a}	0.30^{c}	0.10^{d}
Free fatty acids, %	0.35^{d}	0.32^{d}	0.64^{b}	0.52^{c}	0.74^{a}

Table 11.4 The rendering rate (%) of various animal fats at different temperature

Temp. °C	Back lard	Abdominal lard	Duck fat	Goose fat
80	14.6^{d}	54.7^{b}	82.0^{a}	85.1^{abc}
100	51.1^{c}	74.6^{a}	80.0^{a}	86.3^{ab}
120	66.9^{ab}	74.9^{a}	81.3^{ab}	86.7^{a}
140	63.2^{b}	76.9^{a}	81.0^{ab}	83.2^{c}
160	69.0^{a}	73.7^{a}	79.5^{bc}	84.8^{abc}
180	68.9^{a}	76.3^{a}	79.1^{c}	83.5^{bc}

Table 11.5 Changes on TBA values (mg/ kg) of breast meat form different duck species during storage at 4 °C (Lee et al. 2018)

Species	Storage period		
	Day 1	Day 4	Day 7
Mule duck	0.62 ± 0.08[a]	0.93 ± 0.06[a]	1.32 ± 0.17[a]
Pekin duck	0.62 ± 0.09[a]	0.72 ± 0.11[b]	1.09 ± 0.23
Muscovy duck	0.36 ± 0.09[b]	0.43 ± 0.15[c]	0.52 ± 0.19

olive oil and found that the RDF of duck fat was highly susceptible to lipid oxidation during storage compared to the other fats and oils.

Thiobarbituric acid (TBA) is the most widely used test for evaluating the extent of lipid peroxidation in foods or meats (Witte et al. 1970) due to its simplicity and its results highly correlate with sensory evaluation scores. The basic principle of the method is the reaction of one molecule of malonaldehyde and two molecules of TBA to form a red malonaldehyde-TBA complex, which can be quantitated spectrophotometrically (532 nm). This method has been criticized as being nonspecific and insensitive for the detection of low levels of malonaldehyde. Other TBA-reactive substances (TBARS) including sugars and other aldehydes could interfere with the malonaldehyde-TBA reaction. Therefore, the TBA test is applicable for comparing samples of a single material at different states of oxidation. Despite high contents in lipids, unsaturated fatty acids and iron, duck meat is in the intermediate position between chicken and turkey meat for the susceptibility to oxidation during storage, the chicken displaying the lowest and turkey the highest (Gong et al. 2010). By combining the duck genotype and the nutritional level, Baéza et al. (2018) obtained duck fillets exhibiting a large variation in lipid content from 2.0 to 7.9%. Storage at +4 °C or − 20 °C of raw meat or at +4 °C of dry-cured fillets did not affect its susceptibility to lipid and protein oxidation. Lee et al. (2018) investigated the change of TBA value from three types of duck meat during storage at 4 °C for 7 days. They pointed out that the TBA value of all duck meat increased with the extended storage time (Table 11.5). On the fourth and seventh day, the Muscovy duck had the lowest TBA value, followed by Pekin and mule ducks. Therefore, it may be necessary to pay more attention to the preservation technology of mule duck meat during the process of refrigeration, freezing, and storage. By taking advantage of good chilling technology during slaughter processing and good packaging materials, the speed of fat oxidation may be delayed to preserve the quality of duck meat during storage period. Suryanti et al. (2014) marinated duck meat with ginger extract at 0% (control), 5% and 10% (w/v) and then aged for 72 hours at 10 °C. The results showed that the TBA value in duck meat marinated with ginger extract at the end of aging was lower than that of the control. Marinating with ginger extract can also arrest the increase of TBA value in duck meat, which in turn has implications on the inhibition of the oxidation process.

To evaluate the protein oxidation, it is usual to measure the meat content in carbonyl and thiol groups. He et al. (2019) exposed Pekin ducks to a chronic heat stress (32 °C from 10.00 am to 6.00 pm and 25 °C for the rest of the day vs. 25 °C for control ducks). Ducks were slaughtered at 9 weeks of age. The breast meat of stressed ducks had a higher carbonyl content and a higher TBARS value than that of control ducks. During dry-cured processing of duck meat, Du et al. (2018) showed that the carbonyl content increased, whereas the content in SH groups decreased to the expense of S-S groups suggesting oxidation of the myofibrillar proteins and resulting in aggregation and decreased digestibility of proteins.

11.3.3.3 Shear Value

The texture is the major crucial factor for the consumer's acceptability and perception of quality of meat and meat products. Whole muscle and processed meats are often investigated to assess tenderness and juiciness. Typical methods for determining physical properties of the meat **include** firmness, toughness, and tensile strength.

The most common biting or shearing type system is the Warner-Bratzler shear during the past few decades. For the evaluation of tenderness in meat and meat products, some institutes recommends use of the Warner-Bratzler shear force machine (AMSA 1978). The Warner-Bratzler shear device is relatively cheap and provided that the methodology is standardized, results should be reasonably comparable between laboratories. There are some devices such as texture detector (Fig. 11.23) could be more convenient, less time consuming and accurate equipment for testing the softness or juiciness of meat and meat products. Lee et al. (2018) detected the firmness and toughness of breast meat of mule, Pekin and Muscovy ducks by texture analyzer, and the results showed that the breast meat of Muscovy duck had the highest firmness and toughness followed by mule and Pekin ducks (Table 11.6). The same researchers also showed that breast meat of Pekin duck was the softest quality, while that of Muscovy duck had the highest hardness. From the muscle shear value point of view, it should be noted that the softening or cooking

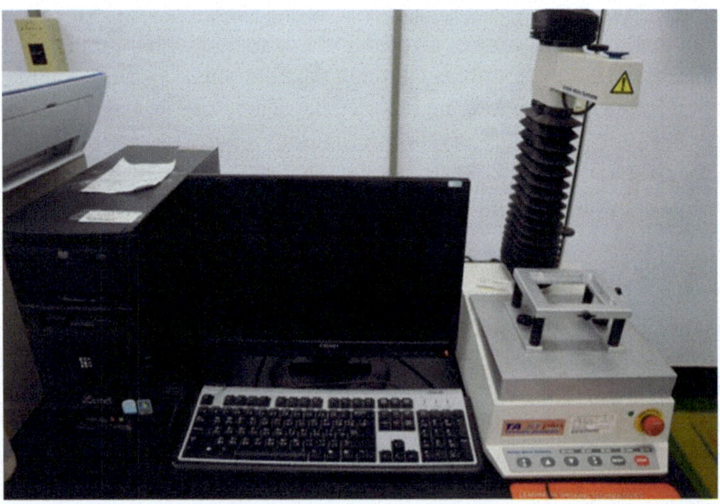

Fig. 11.23 Texture Analyser (TA.XT Plus)

Table 11.6 The shear value of breast meat from different duck species (Lee et al. 2018)

	Mule duck	Pekin duck	Muscovy duck
Shear value, kg	3.91 ± 0.44^{b}	3.00 ± 0.35^{c}	6.64 ± 0.57^{a}
Toughness, kg × sec	7.58 ± 1.18^{b}	5.55 ± 0.82^{c}	11.64 ± 1.37^{a}

Table 11.7 Chilling temperature on Myofibrillar Fragmentation Index (MFI) and shear value of mule duck breast (Yang 2018)

	Mule duck breast chilling temperature	
Physicochemical property parameters	5 °C	15 °C
MFI (%)	88.28 ± 3.8^{a}	108.37 ± 11.5^{b}
Shear value (N)	103.31 ± 10.8^{a}	63.46 ± 7.3^{b}

method of Muscovy meat is different from other species duck meat. Treating Muscovy duck meat with high temperature and long-term heating can effectively soften the muscle texture. Yang (2018) determined the shear value of breast meat of Mule duck, which is dipped into cold water at 5 °C and 15 °C for 60 min, the results showed that the meat immersed at 5 °C had higher shear value than the samples immersed at 15 °C for 60 min (Table 11.7). The shear value is highly correlated with the tenderness of the muscle (Veiseth et al. 2004); according to the outcome of Yang (2018), it is recommended that the slaughtered duck breast meat should be immediately immersed in 15 °C-chilled water for 60 min., after slaughter, and then moved to a 5 °C refrigerator for 24 hr. This can effectively improve the shear value or tenderness of the muscle.

11.3.3.4 Instrumental Evaluation of Color

The color of meat is an important factor affecting consumers' preference for choosing raw meat or meat products (Jeremiah et al. 1972). Lynch et al. (1986) reported that 74% of consumers use the color of meat as an indicator for purchasing the meat, and they usually use redness as an indicator for measuring the freshness of the meat. The Hunter colorimeter (Hunter Association Laboratory, Inc., Reston, VA) is a commonly used objective determination of muscle color (Fig. 11.24). The Hunter colorimeter determines the meat color and expressed by L, a, and b or L*, a*, and b* values. L-value, which range in value from 0 to 100, indicates lightness with 100 equal to white and 0 equal to black. Positive a-values quantitate red color and negative a-values are an indication of green. Positive b-values are a determination of yellow and negative b-values indicate blue color (Miller 1994). The Hunter colorimeter is easy to standardize and to obtain measurements. Lee et al. (2018) measured the duck breast meat color by Color Difference Meter and showed that breast meat of Pekin and mule ducks had higher "L" value than that of Muscovy duck. Muscovy and mule ducks had higher "a" value than Pekin duck. However, breast meat of Muscovy duck had the highest "b" value among all breeds (Table 11.8). Haraf et al. (2009) determined color parameters include L-, a-, and b-values of 6 conservative duck flocks in Poland, and found that "L" value ranged from 42.56 to 45.27, "a" value ranged from 16.35 to 17.65, and "b" value ranged

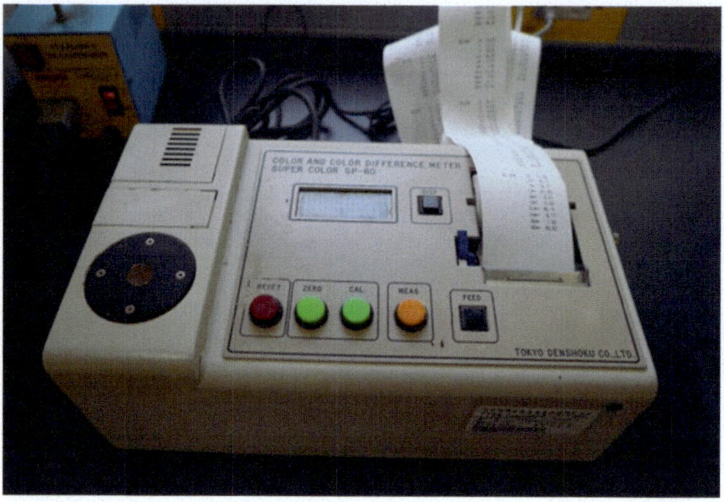

Fig. 11.24 Color and Color Difference Meter

Table 11.8 Lightness (L-value), redness (a-value), and yellowness (b-value) values of breast meat from different duck breeds (Lee et al. 2018)

Color parameters	Mule duck	Pekin duck	Muscovy duck
L-value	32.01 ± 0.82^{b}	32.79 ± 1.00^{a}	28.48 ± 1.07^{c}
a-value	15.68 ± 0.49^{a}	14.67 ± 0.93^{b}	15.76 ± 0.65^{a}
b-value	2.62 ± 0.35^{b}	2.63 ± 0.33^{b}	3.35 ± 0.62^{a}

from 2.67 and 4.56. From this study, the main factor affecting the color of duck meat was genotype, but other elements such as age, feed, feeding period, and feeding methods can also affect the color of duck meat.

11.3.3.5 pH Value

The pH of muscle/meat is an indicator for measuring the acidity of meat (Klont 2005). In a normal living animal, the pH is approximately 7.2. Animal muscles contained some glycogen and it is broken down to lactic acid when muscle turns into meat after slaughter. The ultimate pH is determined by the extent of the pH decline at 24 hours after slaughter. The variation in ultimate pH influences factors such as color and the ability of the meat to retain water. A low ultimate pH results in meat proteins having decreased water-holding capacity and a lighter color. For slaughterhouses and further processors, a lower ultimate pH will lead to less saleable product, due to increased drip losses during the production processes of fresh meat and/or cooked products. Meat with a higher ultimate pH, which retains more water during storage, will also keep more juice after preparation of the meat. Ali et al. (2007b) found that the pH of chicken meat was higher than that of duck meat at 30 min post-mortem. They also showed that the rate or pattern of pH drop immediately after death is different between the meats of two poultry species, but the final pH values after

Table 11.9 pH, Water-Holding Capacity (WHC), and cooking loss of breast and leg meats from local native ducks and imported ducks in Korea (Kim et al. 2013)

	Local native ducks		Imported ducks	
Quality parameters	Breast	Leg	Breast	Leg
pH value	5.67 ± 0.02^{c}	6.75 ± 0.09^{a}	5.73 ± 0.06^{c}	6.63 ± 0.12^{b}
Water-holding capacity (%)	48.17 ± 4.42^{b}	59.58 ± 10.55^{a}	36.54 ± 2.99^{c}	65.90 ± 9.85^{a}
Cooking loss (%)	31.52 ± 1.35^{a}	32.21 ± 4.57^{a}	31.80 ± 2.12^{a}	28.73 ± 4.33^{a}

24 hours were similar. Kim et al. (2013) analyzed the duck breasts and legs from Korean native duck and imported duck, and found that the legs had higher pH value (6.63–6.75) than the breast meat (5.67–5.73), but no significant difference in pH was observed between the breasts of local ducks and imported (Table 11.9). The pH values of breast meat of Korean native duck and imported duck were similar to that of Beijing duck (Kim et al. 2012). Mazanowski et al. (2003) found that the pH of A44 and A55 duck strains were 6.0 and 6.4, respectively. The variability in pH values may be attributed to the factors such as species, feeds, and slaughter conditions.

11.3.3.6 Water-Holding Capacity

Meat quality is a complex concept because it is influenced by a multiple factor of physicochemical parameters. Among the related physical characteristics, water-holding capacity (WHC) is one of the most important factor affecting economic traits. WHC is generally defined as the capability of meat to maintain its own/or additional water under an external force such as heating, pressing, and centrifuging (Sakata et al. 1993). For consumers, it affects the appearance before cooking and the mouthful during consumption and it is also important for the meat industry because it affects the processing yields and the final product attributes (consistency, color, and saltiness) (Roseiro et al. 1994).

Muscle/ meat contains 70–75% water, which is bound primarily to the muscle proteins within the muscle cell. The high pH of about 7.0 in the muscle cell and its physiological salt concentration allows the muscle proteins to bind about 90% of the intracellular water. For assessment of WHC under field and laboratory conditions, there are several methods available such as filter paper press (Ockerman 1972), drip loss (Honikel 1987), and centrifugation methods (Bouton et al. 1971), but standardization of methods is essential if investigations carried out by different groups or companies are to be directly comparable. Lee et al. (2018) adopted the drip loss method to measure WHC in mule, Pekin and Muscovy ducks, the results showed that Pekin duck had the highest drip loss (%) during the 48 hours storage period, and the Muscovy duck had the lowest drip loss during the same period (Table 11.10).

Table 11.10 The drip loss and cooking loss of breast meat from different duck species (Lee et al. 2018)

	Mule duck	Pekin duck	Muscovy duck
Drip loss, %	5.07 ± 0.91^a	5.31 ± 2.91^a	4.73 ± 0.52^a
Cooking loss, %	31.29 ± 0.88^a	28.20 ± 1.56^b	29.20 ± 1.15^b

11.3.3.7 Cooking Loss

The procedure for measuring cooking loss of duck meat is described in the following. Duck breast meat sample (about 200 g wrapped in a plastic bag under vacuum) is immersed in water bath temperature of 80 °C for 30 min, and then surface dried meat is weighed. The final internal temperature of breast meat was about 70 °C under water heating. Cooking loss of meat is a very important meat quality and economic traits. In particular, the slaughter industry and meat processing industry are striving to maintain or reduce cooking loss. Joseph et al. (1992) reported that duck meat have relatively lower WHC than chicken meat, resulting in less emulsion stability and greater cooking loss in duck meat. Biswas et al. (2006) also found lower emulsion activity and cooking yield in patties from duck compared to broiler meat. Ali et al. (2007b) observed that breast meat from Cherry Valley ducks had 34.5% of cooking loss; while, White Pekin ducks deboned meat had 31.26% of cooking loss. Lee et al. (2018) reported that breast muscle of mule duck had higher cooking loss (31.29%) than those of Muscovy (29.20%) and Pekin (28.20%) ducks. Choi et al. (2016) showed that when storage periods after slaughter increase from 15 min to 24 hr., the cooking loss in meat stored at 0, 15, and 30 °C also increased significantly up to 2 h but it was not significantly different thereafter. They also found that the storage temperature did not affect cooking loss in duck breast muscle.

11.3.3.8 Drip Loss

More recently, cut-up parts such as breast and legs are becoming more available and popular in Taiwan; the cut-up part duck offers more options for meat industries and consumers, especially duck steak is very popular in consumption markets. When the duck carcass is cut into pieces, if the storage temperature or the transportation conditions are not good, the drip loss of the duck meat will increase due to the increase in the meat surface area. Generally, duck muscle has more red muscle fibers compared to chicken muscle (Smith et al. 1993). The experiment results from Lesiak et al. (1996) showed that turkey thigh muscle stored at 0 °C resulted in greater drip loss than stored at 12 °C, but the greatest drip losses were observed at 30 °C conditions.

11.3.4 Microbiological Detection

The microbiological safety and quality of poultry meat are equally important to producers, retailers, and consumers, and both contribute to microbial contaminants on the processed product. Contamination of poultry meat with foodborne pathogens remains an important public health issue, because it can lead to illness if there are malpractices in handling, cooking or post-cooking storage of the product.

The poultry meat such as chicken, duck, and goose that we eat every day are polluted by many physical, chemical, or biological contaminants during slaughtering process. Therefore, many advanced countries have established relevant regulations for slaughter hygiene inspections to make sure of the sanitation of the meats. The microorganisms we detect contain the so-called indicator microorganisms and specific pathogens. The indicator microorganisms include three items: total plate count, *Escherichia coli* and coliforms. The detection rate of these microorganisms and the number of bacteria per square centimeter carcass or per ml of immersion water can reflect the cleanliness of the entire slaughter process. When these values increase significantly, it is necessary to review carefully which part of the entire slaughter process is problematic. In terms of specific pathogenic microorganisms, there are several microorganisms associated with human food poisoning such as *Staphylococcus aureus, Salmonella spp., Clostridium perfringens, Escherichia coli* O157: H7, and *Campylobacter spp.* In a pathogen detection project in Taiwan, the above-mentioned six specific pathogenic microorganisms were detected (Yeh 2012). For example, *Salmonella* is a pathogenic microorganism common to humans and animals. In addition to causing infections in many animals, it can also cause food poisoning when a poultry meat or its products, contaminated with the bacteria, is ingested. Detection of *Salmonella* contamination on carcass surfaces is an important indicator of the effectiveness of slaughtering. In Taiwan (Fig. 11.25), *Salmonellas* survive well in the environment, but *Campylobacters* appear less well adapted for survival outside the alimentary tract of warm-blooded animals. In addition, growth of *Salmonellas* only occurs under conditions of high moisture, reduced oxygen and an environmental temperature above 30 ° C. The organisms are particularly sensitive to drying and the effects of freezing and thawing, which can cause a 1 to 2 log reduction in the level of contamination on poultry meat. Taiwan's Food and Drug Administration (TFDA 2019) declared that frozen foods must not be spoiled, badly discolored, smelly, contaminated or contain foreign matter or parasites. In addition, the number of microorganisms needs to comply with government health regulations must meet the following regulations (Table 11.11).

Immersion in chilling water is the commonly used operation for duck and geese carcasses chilling in Taiwan. In addition to lowering carcass temperature, immersion of carcass in chilling water also reduces a part of microorganisms inside and outside the carcass. In the USA, where water-immersion chilling is the norm and super-chlorination of process water is permitted, there is also an option to use a chemical decontamination treatment for carcasses, which may involve substances such as trisodium phosphate, acidified sodium chlorite or peroxyacetic acid (Russell 2003). However, there is currently a very different situation in the EU, because super-chlorination is not allowed, immersion chilling has been largely replaced by air chilling or evaporative cooling, and any form of chemical decontamination is unacceptable. Therefore, in case of fresh carcasses that are air chilled, there is no marked reduction in carcass contamination (Allen et al. 2000; Fluckey et al. 2003).

Raw and undercooked poultry meat has high risk of causing food illness. This is mainly due to two types of bacteria, *Campylobacter* and *Salmonella*, which are

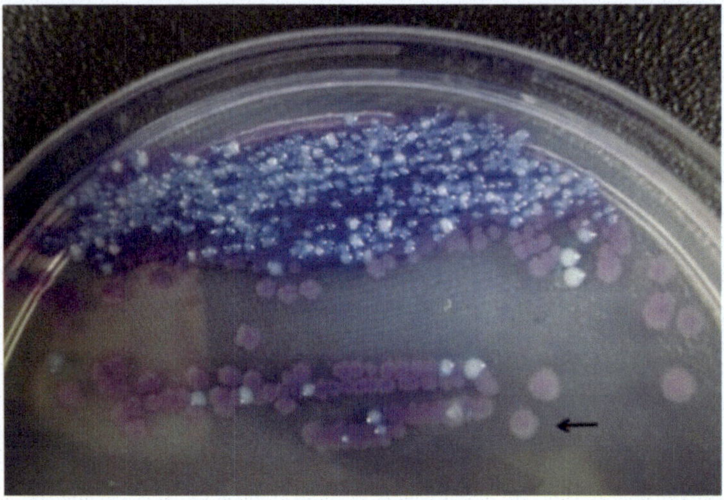

Fig. 11.25 *Salmonella* is cultured on a Chromogenic Salmonella agar. Purple colonies appear (arrows) (Yeh 2012)

Table 11.11 Frozen food hygiene standards in Taiwan (TFDA 2019)

	Total plate count (cfu/g)	Coliform (MPN/g)	*Escherichia coli* (MPN/g)	Volatile base nitrogen (mg/100 g)
Frozen foods that can be eaten without heat conditioning	Less than 100,000	Less than 10	Negative	–
Frozen foods that need to be heated and prepared for consumption	Heated before freezing: Less than 100,000	Less than 10	Negative	Less than 15
	Unheated before freezing: Less than three million	–	Less than 50	Less than 15

commonly found in the guts and feather of these birds. Berndtson et al. (1992) isolated Campylobacters from chickens and found contaminations in 89% of the neck skins and 93% of the peritoneal cavity swab samples and 75% of subcutaneous samples; but muscle samples were only very sparsely contaminated. Rahimi et al. (2011) pointed out that the most prevalence *Campylobacter* species isolated from duck meat was *Campylobacter jejuni* (88.5%); the remaining isolates were *Campylobacter coli* (11.5%). The consumption of contaminated duck meat or meat products, like other types of meat, poses the risk of foodborne diseases. Incidences of *Campylobacter* spp., *Escherichia coli*, *Listeria Monocytogenes*, and *Enterococcus* contamination in duck meat have been reported (Adzitey et al. 2013; Hu et al. 2011; Jamali et al. 2015). Kim et al. (2016) detected the levels of total aerobic bacteria were in the range of 3.53–6.19 and 3.62–6.85 log CFU/g in raw and smoked duck products, respectively. The most common species of microorganisms which

Table 11.12 Storage periods for duck and goose meat (USDA 2018)

	Method of storage	
Type of meat	Refrigerator (4 °C)	Freezer (−18 °C)
Fresh duck and goose	1–2 days	6 months
Cooked duck and goose	3–4 days	2 to 3 months
Smoked duck breast: Vacuum sealed	2 weeks	1 to 2 months
Smoked duck breast: After opening	7 days	1 to 2 months

Table 11.13 Changes in total plate counts (log CFU/g) of breast meat of different duck species during storage period at 4 °C (Chen et al. unpublished data)

	Duration of storage		
Species	Day 1	Day 4	Day 7
Mule duck	$3.78 \pm 0.57^{a*}$	3.69 ± 0.80^{a}	4.11 ± 0.82^{a}
Pekin duck	3.91 ± 0.88^{a}	3.58 ± 0.47^{a}	4.59 ± 0.60^{a}
Muscovy duck	3.79 ± 0.66^{a}	3.56 ± 0.58^{a}	4.87 ± 1.06^{a}

are identified as responsible for spoilage in raw and smoked duck products are *Aeromonas* spp. or *Pseudomonas* spp. and *Leuconostoc mesenteroides*, respectively.

Salmonella may be found in the intestinal tracts of chicken and duck. The raw poultry or duck must be handled carefully to prevent cross contamination. This can occur if raw duck, goose or their juices contact cooked food or foods that will be eaten raw such as salad. Salmonellosis is a foodborne illness characterized by stomach pain, diarrhea, and nausea, therefore cooking whole duck or goose meat to a safe minimum internal temperature of 74 °C as measured using a thermometer is recommended (USDA 2018). Check the internal temperature in the innermost part of the thigh and wing and the thickest part of the poultry breast. USDA (2018) suggested proper storage period for processed duck products at different freezing or refrigerating temperatures (Table 11.12). Chen et al. (unpublished results) found no difference in the total plate counts in refrigerated ducks of three different genotypes until fourth day with the values ranging from 3.56 to 3.91 log CFU/g. After extending refrigeration to seventh day, no difference was observed in total plate counts within different duck species, but total plate count in Muscovy duck was slightly higher than those of mule and Pekin ducks (Table 11.13).

11.3.5 Sensory Evaluation

Sensory analysis is a technical discipline which applies the principles of experimental design and statistical analysis, using human senses (sight, smell, taste, and hearing) to analyze consumer products or food (Wikipedia 2018b). By applying statistical techniques to the results, it is possible to make inferences and insights about the products under test. Perry (2009) mentioned that the methods for testing sensory evaluation can be divided into discriminatory, descriptive, and affective tests. The definition for each as follows;

1. Discriminatory tests: indicate whether a difference between samples that humans can detect.
2. Descriptive analysis tests: these involve the detection and description of both the qualitative and quantitative aspects of a product. Panelists need to be trained to detect and describe the perceived sensory attributes of a sample. The qualitative aspects of a product combine to define the product, and include all of the appearance, aroma, flavor, and texture properties of the product, which differentiate it from others. Descriptive tests are used to obtain detailed description of the aroma, flavor, and oral texture of food and beverages.
3. Affective tests: these assess the personal response to a product by consumers. Do consumers like or dislike a product? It is based solely on preferences, and does not measure attributes of the product per se. Companies when developing new products, maintaining market share of current products and assessment of market potential, use them.

Cooking meat or processed meat products usually require checking of color, aroma, and taste, especially whether sweet, salty, sour or spicy, and this sensory attribute is quite different among different consumers. For example, Geldenhuys et al. (2014) conducted that descriptive analysis in order to establish the sensory attributes of different poultry breast includes Egyptian goose, guinea fowl, Pekin duck, ostrich, and broiler chicken. Results showed that Egyptian goose meat had a very intense game aroma, game flavor, and metallic aftertaste, mainly attributable to the muscle's high percentage of polyunsaturated fatty acids and iron. The advantage of sensory evaluation is that you get a direct quantification of a human's perception of meat attributes. The disadvantages are that it is time-consuming, expensive, and there is a huge variation. There are three major attributes of meat, which contributes to the sensory perception of eating quality: tenderness (or texture), juiciness, and flavor. Tenderness is primarily determined by the myofibrillar and connective tissue components of muscle. A simplistic definition of tenderness is the psychological response to physicochemical stimuli caused by chewing (Bailey 1964); juiciness is depends on the amount of water retained in the cooked meat, the amount of melted lipids in the sample, and the degree of salivation occurring during chewing. Judge et al. (1989) described that the principal sources of detected juiciness are the intramuscular lipids and water. In combination with water, melted lipids constitute a broth that, when retained in meat, is released upon chewing and this broth may stimulate saliva flow and thus improve "apparent" juiciness. For the purpose of practical sensory analysis, flavor can be defined as the impressions perceived via the chemical senses from a product in the mouth (Caul 1957). This includes the aromatics, the tastes caused by soluble substances in the mouth (sweet, bitter, sour, and salty), and the chemical feeling factors which stimulate nerve ends. Miller (1994) described the rating for different descriptive attributes, namely, juiciness, tenderness, connective tissue, flavor intensity, off-flavor characteristics, and off-flavor intensity (Table 11.14). The rating scale of each item is divided into 8 points except off-flavor characteristics, for which alphabets are used.

Table 11.14 Meat descriptive attributes for whole muscle meat products (Miller 1994)

Rating	Juiciness	Tenderness	Connective tissue	Flavor intensity	Off-flavor intensity	Off-flavor characteristic	
						Rating	Description
8	Extremely juicy	Extremely tender	None	Extremely intense	Extremely intense	A	Acid
7	Very juicy	Very tender	Practically none	Very intense	Very intense	F	Fish-like
6	Moderately juicy	Moderately tender	Traces	Moderately intense	Moderately intense	L	Liver
5	Slightly juicy	Slightly tender	Slight	Slightly intense	Slightly intense	M	Metallic
4	Slightly dry	Slightly tough	Moderate	Slightly bland	Slightly bland	O	Old (freezer burned)
3	Moderately dry	Moderately tough	Slightly abundant	Moderately bland	Moderately bland	U	Rancid
2	Very dry	Very tough	Moderately abundant	Very bland	Very bland	B	Bitter
1	Extremely dry	Extremely tough	Abundant	Extremely bland	Extremely bland	SO	Sour
						X	Other

Table 11.15 Hedonic rating scale (Jones et al. 1955)

Scale value	Description
1	Dislike extremely
3	Dislike moderately
5	Neither like nor dislike
7	Like moderately
9	Like extremely

An historical landmark in sensory evaluation development was the hedonic scale developed at the Army Food and Container Institute in the late 1940s (Jones et al. 1955). This method provided a balance 9-point scale for liking with a centered neutral category and attempted to produce scale labels with adverbs that represented psychologically equal steps or changes in hedonic tone. An example of the 9-point scale is shown in Table 11.15. The test would involve several alternative versions of the product and can be conducted in some central location or sensory test facility. The reliability, validity, and discriminative ability of the scale was proven in food acceptance tests with soldiers in the field and in the laboratory, as well as in large-scale food preference surveys (Peryam and Pilgrim 1957).

Many studies have analyzed the effects of poultry strain on meat quality, mainly focusing on sensory characteristics and consumer acceptability of cooked meat and on lipid oxidation during meat storage. Because lipid levels in poultry meat are low, for example, chicken and turkey breast meat contained only 1–2% lipid. The relationship between lipid levels and sensory characteristics has received little attention. Conversely, the lipid level is higher in duck meat than in chicken and turkey (Baéza et al. 2002). Chartrin et al. (2006) also reported that the development of duck meat flavor is affected by the intramuscular fat level. Different genotypes of ducks are used to produce meat, including common ducks such as Pekin, Muscovy, and mule ducks. There have been few reports focus strain and crossbreeding effects on meat quality and sensory parameters in poultry, especially in ducks. Larzul et al. (2002) compared the breast meat from Pekin, mule and Muscovy ducks and found significant additive effects on breast meat quality, as well as significant heterosis effects on color parameters.

The most important facility for sensory evaluation is the panel booth. The panel booth area should contain individual compartments where panelists can assess samples without influence by other panel members (Figs. 11.26, 11.27, 11.28 and 11.29). This area may contain as few as 4 individual sections but 5–10 are most common. Each booth should be equipped with a counter, a stool or chair, a pass-through opening to the food preparation area and individual lighting and electrical outlets. The sensory testing facility consists of individual testing booths with controlled lighting. Room temperature water is also served. The detailed physical structure design of the evaluation room can refer to American Society for Testing and Materials (ASTM 1986) evaluation laboratories design guideline. When those sensory panelists evaluate meat, providing independently and individual booths are

Fig. 11.26 Panel booths with individual sections for each panelist

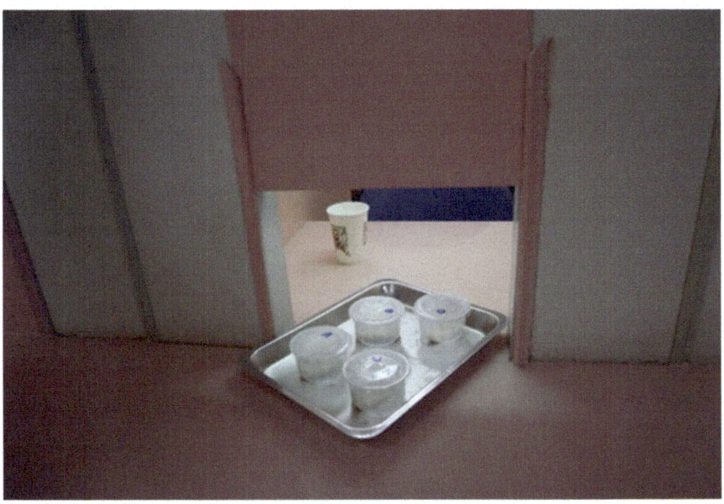

Fig. 11.27 Pass-through opening to introduce food item from preparation room to panel booth

necessary in order to control test conditions and provide privacy for panelists. In this way, panelists can concentrate on the meat sample with a minimum of distraction from the surrounding environment or other panelists.

Fig. 11.28 Panelist evaluating the meat at individual booth

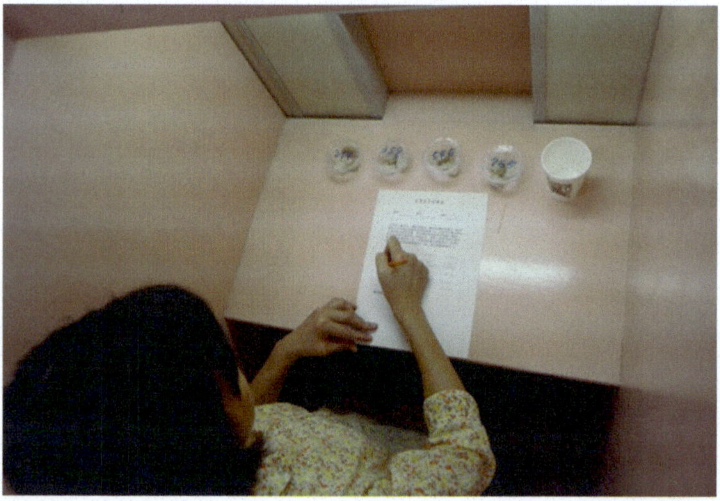

Fig. 11.29 An evaluation booth provided with water for gargling between analyses of different samples

References

Adzitey F, Ali GRR, Huda N, Cogen T, Corry J (2013) Prevalence antibiotic resistance and genetic diversity of *listeria monocytogenes* isolated from ducks their rearing and processing environments in Penang Malaysia. Food Control 32:607–614

Ali MS, Yang HS, Jeong JY, Moon SH, Hwang YH, Park GB, Joo ST (2007a) Quality of duck breast and leg meat after chilling carcasses in water at 0, 10 or 20°C. Asian-Australas J Anim Sci 20:1895–1900

Ali MS, Kang GH, Yang HS, Jeong JY, Hwang YH, Park GB, Joo ST (2007b) A comparison of meat characteristics between duck and chicken breast. Asian-Australas J Anim Sci 20:1002–1006

Allen VM, Corry JEL, Burton CH, Whyte RT, Mead GC (2000) Hygiene aspects of modern poultry chilling. Int J Food Microbiol 58:39–48

AMSA (1978) Guidelines of cookery and sensory evaluation of meat. American Meat Science Association, National Livestock and Meat Board, Chicago

ASTM (1986) Physical requirement guidelines for sensory evaluation laboratories. In: Eggert J, Zook K (eds). ASTM Publication Code Number (PCN) 04-913000-36, Race Street, Philadelphia, pp. 6-20

Baéza E, Dessay C, Wacrenier N, Marché G, Listrat A (2002) Effect of selection for improved body weight and composition on muscle and meat characteristics in Muscovy duck. Br Poult Sci 43: 560–568

Baéza E, Chartrin P, Bernadet MD (2018) Effect of lipid contend on oxidation susceptibility of duck meat. Europ Poult Sci 82(241):10

Bailey ME (1964) Meat tenderness. Agric Sci Rev 2:29

BAPHIQ (2000) Slaughtering Practices, Bureau of Animal and Plant Health Inspection and Quarantine (BAPHIQ), Council of Agriculture, Taiwan

BAPHIQ (2018) Electric poultry slaughterhouse setting standards. Bureau of Animal and Plant Health Inspection and Quarantine (BAPHIQ). Council of Agriculture, Taiwan

Barbut S (2002) Poultry products processing-An industry guide. CRC Press, Boca Raton

Berndtson E, Tivemo M, Engvall A (1992) Distribution and numbers of campylobacter in newly slaughtered broiler chickens and hens. Int J Food Microbiol 15:45–50

Biswas S, Chakraborty A, Sarkar S (2006) Comparison among the qualities of patties prepared from chicken broiler spent hen and duck meats. J Poult Sci 43:180–186

Bouton PE, Harris PV, Shortthose WR (1971) Effect of ultimate pH on the water holding capacity and tenderness of mutton. J Food Sci 36:435–439

Caul JF (1957) The profile method of flavour analysis. Adv Food Res 7:1–40

Chartrin P, Meteau K, Juin H, Bernadet MD, Guy G, Larzul C, Baéza E (2006) Effects of intramuscular fat levels on sensory characteristics of duck breast meat. Poult Sci 85:914–922

Chen MT, Lee YP, Huang MC, Liu DC, Huang HH (1984) Studies on characteristics of the duck carcass and meat I. Carcass quality of the duck. J Chin Soc Anim Sci 13:109–116

Chen MT, Lin SS, Lin LC (1991) Effect of stresses before slaughter on changes to the physiological biochemical and physical characteristics of duck muscle. Br Poult Sci 32:997–1004

Chen MC, Jiang A, Ockerman HW (2012) Waterfowl slaughter process. In: Chen MC, Jiang A, Ockerman HW (eds) Duck and goose meat product processing technology. Fu-Lin Publisher Inc, Taichung City, pp 7–13

Choi JH, Choi YS, Kim HW, Song DH, Kim CJ (2016) Effects of postmortem temperature on the physicochemical characteristics of prerigor Pekin duck breast muscles. Poult Sci 95:645–650

CSTJ (2018) Council of Science Technology of Japan, A comparison of the unsaturated fatty acids in duck meat and other meat Available at: http://www.gerchean.com.tw/ English/Comparison.htm. Accessed 4 Oct 2018

Du X, Sun Y, Pan D, Wang Y, Ou C, Cao J (2018) Change of the structure and the digestibility of myofibrillar proteins in Nanjing dry-cured duck during processing. J Sci Food Agric 98:3140–3147

Fernandez X, Auvergne A, Renerre M, Gatellier P, Manse H, Babile R (2003) Preliminary observations on the colour variability of breast meat ('magrets') in force-fed ducks. Anim Res 52:567–574

Fluckey WM, Sanchez MX, McKee SR, Smith D, Pendleton E, Brashears MM (2003) Establishment of a microbiological profile for an air-chilling poultry operation in the United States. J Food Prot 66:27–29

Geldenhuys G, Hoffman LC, Muller M (2014) Sensory profiling of Egyptian goose (*Alopochenaegyptiacus*) meat. Food Res Int 64:25–33

Gong Y, Weber PF, Richards MP (2007) Characterizing quality of rendered duck fat compared to other fats and oils. J Food Qual 30:169–186

Gong Y, Parker RS, Richards MP (2010) Factors affecting lipid oxidation in breast and thigh muscle from chicken, Turkey and duck. J Food Biochem 34:869–885

Hale KK, Thompson JC, Toledo RT, White HD (1973) An evaluation of poultry processing – a special report. University of Georgia, Athens

Haraf G, Ksiazkiewicz J, Woloszyn J, Okruszek A (2009) Characteristics of meat colour of different duck populations. Archive Tierzucht 52:527–537

He J, Xia C, He Y, Pan D, Cao J, Sun Y, Zeng X (2019) Proteomic responses to oxidative damage in meat from ducks exposed to heat stress. Food Chem 295:129–137

Honikel KO (1987) How to measure the water-holding capacity of meat? Recommendation of standardized methods. In: Tarrant PV, Eikelenboom C (eds) Evaluation and control of meat quality in pig. MartinusNijhof, Dordrecht, pp 129–142

Hu Q, Tu J, Han X, Zhu Y, Ding C, Yu S (2011) Development of multiplex PCR assay for rapid detection of *Riemerellaanatipestifer*, *Escherichia coli* and *Salmonella enterica* simultaneously from ducks. J Microbiol Methods 87:64–69

Huang JC (1998) Process and quality control of processed duck products. In: Duck meat processing and utilization seminar, I-Lan Branch, Taiwan Livestock Research Institute, I-Lan County, Taiwan

Jalaludeen A, Churchil RR (2006) Duck eggs and their nutritive value. Poult Line 6(10):35–39

Jalaludeen A, Churchil RR, Joseph L, Anitha P (2009) Duck meat, egg and their products, In: Souvenir of IV World Waterfowl Conference, Thrissur, Kerala, pp 57–64

Jamali H, Ghaderpour A, Radmehr B, Wei KSC, Chai LC, Ismail S (2015) Prevalence and antimicrobial resistance of campylobacter species isolates in ducks and geese. Food Control 50:328–330

Jeremiah LE, Carpenter AL, Smith GC (1972) Beef color related to consumer acceptance and palatability. J Food Sci 37:476–479

Jones LV, Peryam DR, Thurstone LL (1955) Development of a scale for measuring soldier's food preferences. Food Res 20:512–520

Joseph JK, Balogun OO, Famuyiwa MA (1992) Carcass evaluation and organoleptic assessment of some selected Nigerian birds. Bull Anim Health Prod Africa 40:97–102

Judge MD, Aberle ED, Forrest JC, Hedrick HB, Merkel RA (1989) Principles of meat science, 2nd edn. Kendall/Hunt Publishing Company, Iowa

Kim HW, Lee SH, Choi JH, Choi YS, Kim HY, Hwang KE, Park JH, Song DH, Kim CJ (2012) Effect of rigor state thawing temperature and processing on the physicochemical properties of frozen duck breast muscle. Poult Sci 91:2662–2667

Kim DS, Song YR, Kim HR, Kwon HJ, An BK, Kang CW, Kim HK, Lee SK (2013) Comparison of meat characteristics between Korean native duck and imported commercial duck raised under identical rearing and feeding condition. Korean J Food Sci Anim Resour 33:89–95

Kim HJ, Yong HI, Lee HJ, Jung S, Kwon JH, Heo KN, Jo C (2016) Identification of microorganisms in duck meat products available in Korea and the effect of high hydrostatic pressure. Korean J Food Sci Anim Resour 36:283–288

Klont R (2005) Influence of ultimate pH on meat quality and consumer purchasing decisions. www.thepigsite.com/articles/1506/influence-of-ultimate-ph-on-meat-quality-and-consumer-purchasing-decisions. Accessed 18 Oct 2018

Larzul C, Imbert B, Bernadet MD, Guy G, Rémignon H (2002) Qualité du magret dans un croisement factoriel Barbarie X INRA 44. In: Proc 5èmes Journées de la Recherche sur les Palmipèdes à Foie Gras, Pau, ITAVI, France, Paris, pp 29–32

Lee HJ, Lee MR, Wu HY (2012) Application of poultry fat in cosmetics researches. J Taiwan Livest Res 45:227–238

Lee HJ, Lee MR, Chen WS (2015) The comparison of fatty acid composition in duck fat with temperature partition. J Chin Soc Anim Sci 44(suppl):321

Lee MR, Tu RJ, Chen WS (2018) Analyzing physical traits of mule, Pekin and Muscovy duck breast meat in Taiwan. J Chin Soc Anim Sci 47(suppl):347

Lesiak MT, Olson DG, Lesiak CA, Ahn DU (1996) Effect of post-mortem muscle temperature and storage time on the water-holding capacity of Turkey breast and thigh muscles. Meat Sci 43:51–60

Lynch NM, Kastner CL, Kropf DH (1986) Consumer acceptance of vacuum packaged ground beef as influenced by product color and educational materials. J Food Sci 51:253–272

Mazanowski A, Kisiel T, Gornowicz E (2003) Carcass quality meat traits and chemical composition of meat in ducks of paternal strains A44 and A55. Anim Sci Pap Rep Inst Genet Anim Breed Poland 21:251–263

Miller RK (1994) Quality characteristics. In: Kinsman DO, Kotula AW, Breidenstein BC (eds) Muscle foods, meat, poultry and seafood technology. Chapman and Hall, New York, pp 325–327

Mountney GJ, Parkhurst CR (1995) Processing fresh poultry. In: Mountry GJ, Parkhurst CR (eds) Poultry products technology, 3rd edn. The Haworth Press Inc, New York

NAIF (2018) National animal industry foundation waterfowl industry status. https://www.naif.org.tw/industrialContentaspx?param=frontMenuID=13%EF%BC%86sDate=%EF%BC%86eDate=%EF%BC%86key1=%EF%BC%86frontTitleMenuID=12%EF%BC%86forewordTypeID=0%EF%BC%86pn=1&frontTitleMenuID=12&frontMenuID=13&forewordID=2437. Accessed 27 Oct 2018

NPPC (1991) Procedures to evaluate market hogs, 3rd edn. National Pork Producers Council, Des Moines

Ockerman HW (1972) Quality control of post-mortem muscle tissue. Animal Science Department, The Ohio State University, Columbus

Pan CM, Lee SR, Lin CY, Kan CL, Chen BJ (1985) Measurement on growth and carcass traits of meat ducklings. Taiwan Livest Res 18:167–174

Perry D (2009) Sensory measurement of meat quality MEAT418/518 Meat Technology, 14-1-8. The Australian Wool Education Trust, Licensee for education activities, University of New England

Peryam DR, Pilgrim FJ (1957) Hedonic scale method of measuring food preferences. Food Technol September 1987:9–14

Petracci M, Baéza E (2011) Harmonization of methodologies for the assessment of poultry meat quality features. World's PoultSci J 67:137–151

Rahimi E, Alian F, Alian F (2011) Prevalence and characteristic of campylobacter species isolated from raw duck and goose meat in Iran, vol 9. International Conference on Food Engineering and Biotechnology, International Proceedings of Chemical Biological and Environmental Engineering, IACSIT Press, Singapoore, pp 171–175

Roseiro LC, Santos C, Melo RS (1994) Muscle pH60, colour (L a b) and water-holding capacity and the influence of post-mortem meat temperature. Meat Sci 38:353–359

Russell SM (2003) Disinfection of poultry carcasses during scalding and immersion chilling. J Turkey 51:5–8

Sakata R, Deguchi T, Nagata Y (1993) Effectiveness of the filter paper press method for determining the water holding capacity of meat. Fleischwirtsch 73:1399–1400

Sams AR, McKee SR (2010) First processing-slaughter through chilling. In: Owens CW, Alvarado CZ, Sams AR (eds) Poultry meat processing, 2nd edn. CRC Press, Boca Raton, FL, pp 25–50

Smith DP, Fletcher DL, Buhr RJ, Beyer RS (1993) Pekin ducklings and broiler chicken pectoralis muscle structure and composition. PoultSci 72:202–208

Suryanti U, Bintoro VP, Atmomarsono U, Legowo AM (2014) Antioxidant activity of Indonesian endogenous duck meat marinated in ginger (*Zingiberofficinale Roscoe*) extract. Int J PoultSci 13:102–107

TFDA (2019) Taiwan's food and drug administration sanitation standard for the microorganism in frozen foods, amended date 2019-08-15. Ministry of Health and Welfare, Taiwan

TLRI (1992) Taiwan Livestock Research Institute, manual for classification and cutting specifications of mule duck carcasses, Yilan branch, Taiwan Livestock Research Institute, Yilan County, Taiwan

TLRI (2009) Taiwan livestock research institute, poultry meat and egg processing. Taiwan Livestock Research Institute, Tainan city, Taiwan

USDA (2018) Duck and goose from farm to table. https://www.fsis.usda.gov/wps/portal/fsis/topics/food-safety-education/get-answers/food-safety-fact-sheets/poultry-preparation/duck-and-goosefrom-farm-to-table/ct_index. Accessed 17 Oct 2018

Veiseth E, Shackelford SD, Wheeler TL, Koohmaraie M (2004) Indicators of tenderization are detectable by 12 h postmortem in ovine logissimus. J AnimSci 82:1428–1436

Wikipedia (2018a) Rosin. https://en.wikipedia.org/wiki/Rosin. Accessed 25 Oct 2018

Wikipedia (2018b) Sensory Evaluation. https://en.wikipedia.org/wiki/Sensory_analysis. Accessed 24 Oct 2018

Witte VC, Krause GF, Bailey ME (1970) A new extraction method for determining 2-thiobarturic acid values of pork and beef during storage. J Food Sci 35:582–585

Yang SU (2018) Effects of chilling treatment on postmortem proteolysis and tenderization in mule duck breast muscle, Master Thesis, National Chiayi University, Chiayi County, Taiwan

Yeh GS (2012) Importance of microbial detection on carcass surface. Agric Ext Newslett Bimonthly 93:1–3. (College Bio-ResoAgri Nat Taiwan Univ)

Dr. Wen-Shyan Chen holds Master's (Animal Nutrition) and Ph.D (Meat Science) degrees of National Chung Hsing University, Taichung, Taiwan. He began his career in Animal Products Processing Division, Livestock Research Institute in 1993 and involved in pig and poultry (chicken, geese and duck) slaughter and meat processing research for almost three decades. He has 150 scientific publications in domestic and foreign journals and academic conferences. With sound theoretical knowledge, he involved in dissemination of research findings to the stakeholders in meat processing industry. He is the recipient of Agricultural Contribution Award of Taiwan Agricultural Society in 2020.

Further Processing of Duck Meat and Egg 12

George T. Oommen, T. Sathu, and Wen-Shyan Chen

Contents

G. T. Oommen (✉) · T. Sathu
Meat Technology Unit & Department of Livestock Products Technology, College of Veterinary and Animal Sciences, Kerala Veterinary and Animal Sciences University, Thrissur, Kerala, India
e-mail: sathu@kvasu.ac.in

W.-S. Chen
Animal Products Processing Division, Livestock Research Institute, Council of Agriculture, Tainan, Taiwan
e-mail: wschen@mail.tlri.gov.tw

A. Jalaludeen et al. (eds.), *Duck Production and Management Strategies*,
https://doi.org/10.1007/978-981-16-6100-6_12

Abstract

The production of duck meat has risen steadily in Asian and European countries and has become the third most widely produced poultry meat in the world next to chicken and turkey. It is not considered either a seasonal dish or with any cultural and religious restrictions on its consumption. The duck meat and egg products and their style of cooking vary from country to country and even among regions or provinces within a country. Further processing of duck meat and egg into value-added and convenient ready-to-eat (RTE) and ready-to-cook (RTC) products and their standardization of processing are inevitable for commercialization, uniformity of products, and marketing. With a view to introducing the vast culinary and industrial scope of duck dishes, a few popular RTE duck meat and egg products, their methods of preparation and ingredients with respect to each style of cooking in the various States of India, Bangladesh, Sri Lanka, Thailand, Malaysia, Singapore, Indonesia, Vietnam, China, Taiwan, France, and England are discussed in this chapter. The scope of further processing of duck meat and egg are also reviewed in order to impart a comprehensive idea to students, researchers, product developers, meat processors, and entrepreneurs. Further processing of duck meat and egg needs innovation to increase the market demand of the products.

Keywords

Duck products · Kerala duck · RTE duck · Cooking duck · Duck cuisine · Duck egg

Abbreviations

cm	centimetre
g	gram
h	hour
kg	kilogram
l	litre
lb.	pound
mg	milligram
min	minute
ml	millilitre
tbsp.	tablespoon
tsp.	teaspoon

12.1 Introduction

Since ancient times duck meat and eggs have been eaten as food around the world, especially in Asian countries. In meat technology terms, duck refers to the meat of domesticated fresh water mallard ducks. Duck meat generally refers to the breast and legs of the bird. Duck and their products are getting popular and in strong demand all over the world, especially in China, India, other Southeast Asian countries, Australia, France, Germany, other European countries, South Africa, UK, and American continent. Among the various meat type duck breeds, Pekin, Rouen, and Aylesbury are popular in most of the countries. Commercially many crosses have been made because of the demand for fast growing, table ducks for meat. Mule (mulard/ moulard) ducks are produced by crossing Muscovy ducks with common ducks. These sterile hybrids are bred predominantly for meat and fat liver production. Health conscious consumers prefer the lean meat of mule ducks. Muscovy ducks have suitable size for cutting and further processing, especially for cured duck.

In many Asian countries, indigenous duck breeds are reared for both egg and meat production. Moreover, spent ducks are also utilized for preparing various products specific to those countries and regions. Duck meat is used in the preparation of many ready-to-eat (RTE) and ready-to-cook (RTC) curries with different spices and styles in various parts of the world. The method of cooking varies from country to country and even among regions/provinces within a country. However, duck meat and egg products in different countries are highly diverse and worth investigating scientifically, especially, considering duck more than as a delicacy like turkey and goose during New Year, Easter, and Christmas. Therefore, in this chapter, an attempt has been made to document a few popular RTE duck meat and egg products and their methods of preparation in the States of India, Sri Lanka, Bangladesh, Thailand, Malaysia, Singapore, Indonesia, Vietnam, China, Taiwan, France, and England. A brief overview of further processing of duck meat and egg are also given with a view to imparting a comprehensive idea to students, researchers, product developers, food processors, and entrepreneurs.

12.2 Further Processing of Duck Meat

In earlier days duck was sold as whole live bird only. But later, after processing on-line in large numbers in processing plants, they are being sold as whole fresh/ chilled/frozen duck. Further processing of poultry is the transformation of raw carcass into value-added, convenient products by the addition of different spices, additives, curing agents and subjecting it to mechanical deboning, comminution, grinding, chopping, mixing, emulsification (frankfurter, bologna, meat balls), restructuring (burger, patties, ham), marinating, curing, tumbling, massaging, coating (nuggets), cooking, portioning, and slicing (Keeton 2001).

12.2.1 Scope of Developing Duck Meat Products

The top five duck meat producing countries in the world are China, France, Myanmar, Vietnam, and Malaysia and more than 50% of the world duck meat is produced by China (FAOSTAT 2017). Ramadhan et al. (2010) reported that the production of duck meat has risen steadily and has become the third most widely produced poultry meat in the world next to chicken and turkey. Moreover, duck meat is not generally considered a seasonal dish as in the case of turkey and goose and there are no cultural and religious restrictions on its consumption.

Now, duck carcasses are further processed into cut-up parts, portions, mechanically deboned duck meat, and several other further processed/value-added RTE/RTC products. Although the functional and sensory characteristics of spent duck meat are inferior, it can be improved by processing techniques into emulsion-type products. This can maximize its utilization and add value to the products. Therefore, Ramadhan et al. (2010) opined that the increased availability of spent egg laying ducks is an opportunity to expand production of further processed meat products. An advantage of duck meat is its unique combination of red meat and white poultry meat characteristics with comparatively higher fat content, especially with skin, good meat flavour, and rich unsaturated fatty acid content (Chartrin et al. 2006).

In a comparative study between duck and chicken sausages (Bhattacharyya et al. 2007) and duck and chicken meat patties (Biswas et al. 2006), lower emulsion stability, cooking yield, and water retention capacity were observed for both duck sausages and patties. Comminuted emulsion products permit inclusion of edible by-products such as skin, gizzard, and heart from the spent birds, which would reduce the cost of production and improve the yield and quality of finished product. Yang et al. (2009) prepared low fat duck sausage by replacing 10% of meat with hydrated cereal flours, viz., rice, wheat, corn, millet, and barley and reduced the off-flavours associated with pure duck meat desirably. Kim et al. (2020a) optimized the quality characteristics of semi-dried restructured jerky by the addition of 1% duck skin gelatin and 0.3% carrageenan. Kim et al. (2020b) studied the physico-chemical properties and sensory characteristics of duck jerky formulated with different combinations of collagen and konjac. Their study revealed that a combination of collagen and konjac at a ratio of 60:40 significantly improved the quality of duck jerky products. Shim et al. (2018) observed that a 1:1 ratio of duck skin and pre-emulsified skin could improve the quality characteristics of restructured duck ham.

Ramadhan et al. (2010) in his review article on duck meat utilization and the application of surimi-like material in further processed meat products suggested that further research might improve the quality of further processed products made with duck meat. He reviewed many studies in which the functional properties of surimi-like materials made from low value meat from beef, pork, mutton, mechanically deboned chicken, broiler, and spent hen have been investigated. Jin et al. (2007) reported that spent hen surimi can replace around 40–60% of the meat in sausage formulations. Therefore, the processing of duck meat into a surimi-like material and its incorporation into various further processed meat products enhances its scope for formulating new products.

Chen and Lin (1997) reported an increase in duck meat consumption in Taiwan and boiled salted duck, charcoal grilled duck, roasted duck, and smoked duck are the famous processed duck meat products there. Tai and Tai (2001) also listed a few most popular duck meat products in Asia, viz., roasted Peking duck, crispy duck skin, ginger root duck, herb duck, pressed salt duck, smoked duck steak, duck roll, tea smoked duck, salted duck gizzards, and marinated duck tongues. In his review on the potential use of duck meat in the development of processed meat products, Huda et al. (2011) reported a few traditional duck meat products popular in China, France, Korea, and Indonesia. They are Peking roasted duck, Nanjing cooked duck, Zhangcha duck in China, Canard à l'Orange in France, Oritangin Korea, and Gulai itiak lado mudo in Indonesia. Boiled salted duck, charcoal grilled duck, roasted duck, and smoked duck are other famous processed duck meat products in Taiwan. Biswas et al. (2019) also enlisted several popular products manufactured by the top five market players in duck meat industry in China, New Zealand, Australia, Thailand, USA, Europe, Africa, and Asia with a forecast of their enhanced production in the future. Also they suggested diversification of duck meat products in many forms apart from the traditional products, to attract more consumers and to increase production.

Different types of cured duck products are available in Southeast Asian countries. Nanjing cooked duck and Nanjing dry-cured duck are two famous Chinese duck products that commonly use Gaoyou ducks in their formulation as reported by Wang et al. (2009). The delicious final taste Nanjing cooked ducks is enhanced due to the brining process used before boiling as observed by Liu et al.(2007). In Nanjing dry-cured duck processing, there are different steps, such as dry curing, marinating, piling, and drying processes. Chinese and Southeast Asian consumers accept this product due to its good flavour and texture (Xu et al. 2008). Another popular traditional Chinese duck meat product, Nanjing water-boiled salted duck, is prepared by cooking meat at low temperature to produce a savoury taste and tender texture (Liu et al. 2006).

12.2.2 Methods of Cooking Duck

Before cooking, frozen duck has to be thawed overnight in refrigerator and drain well. The natural gamy and muddy odour of fresh or frozen duck is due to its special feeding habits in swamp, rice fields and other muddy waters. This odour can be got rid of by massaging duck with a mixture of vinegar/lime juice and turmeric or crushed ginger and white wine or vinegar or salt and vinegar. Keep aside the duck after massaging for 15 min and wash under running cold water before chopping into pieces of required size. Rice/coconut/cider vinegar is preferable to synthetic vinegar for washing, marinade, and cooking. Duck covered with butter milk in a glass bowl and chilled overnight would not only remove the odour but tenderize meat, as well.

Duck is cooked into various dishes or products around the world. The selection of the breed of duck and the method of cooking vary widely all over the world. For each method of cookery, the type, age, and size of the duck must be considered. In USA, UK, France, other European countries, China, Southeast Asian countries, and Australia, duck is generally roasted, braised, barbecued or grilled as a popular product. Indian products include a wide variety of regional and traditional duck preparations native to states of Kerala, Tamil Nadu, Andhra Pradesh, West Bengal, and Assam. As a result of the wide diversity in the food habits of the people from state to state in the Indian sub-continent, the cuisine or style of cooking of duck also varies. In Indian food culture, duck is a side dish rather than a main dish. Accordingly, in Indian style of cooking of meat, especially for whole or cuts of duck are moist cooking. It includes stewing, boiling, and braising. In almost all products in Indian or neighbouring Southeast Asian style of cooking, meat after seasoning with spices is cooked in water with sautéed vegetables and forms a product with thick or semi-thick or dry gravy. Meat will be well done, as well. On the contrary, in East Asian, European, and American style of cooking, meat is mildly seasoned before roasting or grilling with stuffing to rare or medium doneness. The sauce is prepared separately with roast juice or otherwise.

Stewing is generally recommended for tougher meat cuts of uniform size by fully submerging in liquid for even cooking slowly in low heat. The liquid with spices form the gravy. Stewing includes cooking for a longer period on simmering. In a simmer, small bubbles of steam will occasionally rise from the bottom of the pan, but the liquid as a whole is below the boiling point. Boiling usually means a rolling boil where all the liquid is at the boiling point and is applied to older poor quality birds. Braising is for whole duck or cuts in the least amount of liquid and to cook meat in a very low heat. For roasting, duck may be placed on a rack, in a roasting pan in an oven at 160 °C to 200 °C to ensure even application of heat and may be rotated on a rotisserie. Roasting is the preferred method of cooking whole duck. Roasting and grilling of poultry are generally resorted to in English and French cuisines. Most of the duck preparations involve roasting for at least part of the cooking process to aid in crisping the skin.

In various cuisines, whole spices are dry roasted in a frying pan or wok, without oil or water as a carrier. This changes the chemistry of proteins in the food and enhances the flavour and taste of spices. On the other hand, in tempering they are roasted shortly in oil or ghee (butter oil). The essential oils are liberated from the cells in spices and enhance their flavour when poured with oil into a dish. The latter is better as the oil will adsorb the flavour and during sautéing it will be evenly transferred to meat and other ingredients.

In comparison to chicken, duck is darker with a distinct strong flavour and legs are fattier than that of the breast. Being a waterfowl and due to the higher subcutaneous fat, meat on cooking is juicer, tender, and flavourful. Duck is mostly cooked with skin-on and takes longer time for cooking unless pressure cooked. The skin of duck on moist cooking gelatinizes and improves tenderness, texture and imparts flavour for most duck products. On roasting, skin becomes crispier as a result of rendering of the subcutaneous fat. The rendered duck fat enables in reducing cooking oil during sautéing of the ingredients besides imparting the unique flavour.

In cookery, accurate measurements and use of uniform good quality spices and condiments are essential to get good results each time you make a product. All measurements of ingredients in the products are metric or its equivalent in cup or spoon level measures due to pragmatic convenience in measuring. Glass or graduated plastic measuring jugs/cups and weighing scales enable in accurate measurements. A set of metric measuring spoons would be ideal for measuring spices and condiments, rather than scales.1 metric cup = 250 ml, 1 table spoon (tbsp) = 15 ml; 1 teaspoon (tsp) = 5 ml.

12.2.3 Duck Meat Products in Indian Style

A cuisine is a style of cooking with ingredients and techniques associated with a specific culture or geographical region. Regional traditions, customs, procedure of preparation, and ingredients create dishes unique and particular to a region.

In India, ducks are mostly concentrated in the Southern, Eastern, and North-eastern States, mainly coastal region rice fields, with non-descriptive indigenous desi stocks, which are light weight type and poor layers. These Indian breeds of ducks are sold for slaughter at the end of their egg laying period and as a result eating qualities become less desirable and meat yield will also be lesser. Nowadays, surplus drakes of 4–5 months of age of indigenous varieties are slaughtered for preparing many products. In many Indian states, people have not tried eating duck products from outside or never used it in domestic cooking, probably due to unawareness on its cookery and nutritive value. Duck is perceived to be fit for a special dish during Easter, Christmas, New Year, wedding feasts, in restaurants and catering establishments. The duck meat products in various cuisines in India vary from state to state due to variations in the food habits and availability of ingredients. The most popular RTE duck meat products in the Indian cuisines, viz., Kerala, Tamil Nadu (Chettinadu), Andhra (Telugu), Bengali, and Assamese are discussed in the following.

12.2.3.1 Kerala

Kerala, with high number of non-vegetarian population, is highly diverse and distinct from other parts of the country due to the culinary influence of Portuguese, Arab, Dutch, and English traders who visited the port towns of Kerala during pre-independence period. The Syrian Christian and Malabar Muslim non-vegetarian preparations are famous. Although there are diversities in the preparations of non-vegetarian dishes between the pre-independent Central Travancore and Malabar in the north Kerala, both use similar spices and condiments. Almost all of the products, irrespective of different communities, have coconut either in the form of shredded coconut, coconut oil or coconut milk imparting the unique flavour and taste.

The popular traditional duck products in Kerala style are Duck Curry, Duck Stew, Duck Korma, Duck Pappas/Mappas, Duck Roast, and Duck Fry. Their formulations, ingredient composition, and processing steps are detailed in the following (Cicy 2020, personal communication).

Duck Curry of Kerala

Duck curry (duck in ground roasted coconut with spices in thick or semi-thick seasoned gravy) is the signature duck preparation in Kerala, especially one of the most loved duck dishes of the Christians in the Kuttanad region (Central Kerala) and elsewhere during Easter, Christmas, New year, and wedding feasts as a side dish with rice. It is served for breakfast with vellayappam (flat fermented rice flour pan cake with ground coconut), milk hopper (palappam - bowl-shaped thin fermented rice flour pan cake with coconut milk baked in a special wok by swirling dough), string hopper (idiyappam – rice noodles curled into flat spirals), chappathi/ rotti, wheat bread, nan, kulcha, paratha/ porotta, boiled tapioca, etc. Kuttanad region in the Central Travancore is referred to as the rice bowl of Kerala with large expanse of rice fields, backwaters and Kuttanad ducks. So Kuttanad/ Kumarakom Duck Curry is synonymous with this preparation is an indication of its geographical popularity and significance.

Cooking time: 50–60 min.
Ingredient composition.

Sl. No.	Ingredients	Quantity
1	Duck with skin-on (cut into medium size pieces)	1 kg
2	Thinly sliced coconut	50 g (1/4 cup)
3	Coconut oil	As required
4	Turmeric powder	1 g (a pinch)
5	Mustard	2.5 g (1/2 tsp)
6	Large, skinned, and sliced onion	4 no
7	Ginger paste	15 g
8	Garlic paste	15 g
9	Green chilli (split lengthwise)	3–4 no
10	Curry leaves	4–5 sprigs
11	Vinegar (cedar/rice/coconut)	5 ml (1 tsp)
12	Water	As required
13	Salt	To taste
14	Freshly ground black pepper	5 g (1 tsp)
15	Fresh spice blend (garam masala)*	2.5 g (1/2 tsp)
For coconut spice/masala paste		
1	Grated coconut	200 g (1 cup)
2	Fennel (anise)	5 g (1 tsp)
3	Cardamom	2 pods
4	Cinnamon	2.5 cm piece
5	Cloves	3 buds

(continued)

Sl. No.	Ingredients	Quantity
6	Kashmiri chilli powder	25 g (1 ½ tbsp)
7	Coriander powder	15 g (1 tbsp)
8	Turmeric powder	4 g (3/4tsp)

Processing steps.

- Heat coconut oil in a wok or pan; add coconut slices, a pinch of turmeric powder and salt. Fry in medium flame, stir until it turns golden brown. Drain and keep aside.
- To prepare coconut spice/masala paste, dry roast grated coconut, fennel, cardamom pods, cinnamon, and cloves in a wok until evenly browned. Stir continuously over medium low flame. Reduce to low flame and add Kashmiri chilli powder, coriander powder, and turmeric powder. Mix well and cook for a minute, stir continuously.
- Transfer the coconut mixture to a plate and allow it to cool without over roasting in woke. Grind this mixture to a fine paste and keep aside.
- Add mustard to heated oil in the wok and let it splutter. Add sliced onion and sauté until it turns golden brown.
- Add ginger and garlic paste and sauté for a few minutes or until the raw odour of ginger and garlic subside or mellow.
- Then add green chilli, curry leaves and duck pieces, cook for about 5 min, stir well in between. Add the prepared coconut spice paste and mix well.
- Add water 250 ml (1 cup or as required), vinegar, and salt to taste. Cover the wok and stew until duck is well done in medium flame. If the duck is not meat type or tough, it has to be pressure cooked at 15 lb. pressure for about 20 min. Accordingly, sautéing and further cooking can be done in a pressure cooker.
- Add fried coconut slices, black pepper powder, and spice blend (garam masala) to taste. Mix well and switch off the flame. Serve hot as side dish with any main dish.

Notes

- Spice blend (garam masala)* freshly ground: cinnamon – 2.5 cm stick, cloves – 6 buds, cardamom – 4 pods, star anise – 1, black pepper – 2.5 g, fennel – 5 g, mace – 2 g. It is preferable to use freshly ground spice blend to retain the aroma.
- If hotter variety chilli is used instead of Kashmiri chilli, reduce the quantity accordingly. Kashmiri chilli imparts an appealing red colour to the product, but not hot.

Duck Curry of Kerala: Alternate Method

The above duck curry with slight variations in the preparation would make another dish of choice. The curry can be prepared without roasted coconut spice paste and ingredients one to five for the paste. But increase coriander powder to 30 g (2 tbsp) while retaining the same quantity of Kashmiri chilli powder and turmeric. Increase the quantity of spice blend to 10 g (2tsp). Use 1 g (1/4 tsp) asafoetida also for a distinct flavour. Replace vinegar with 250 g (three) sliced ripe tomato or equivalent tomato puree.

Skip the steps two and three in the instructions for duck curry. Follow the steps four and five. In step six, after sautéing green chilli and curry leaves in woke, add Kashmiri chilli powder, coriander powder, turmeric powder, and spice blend made into a paste with a little water. Stir continuously over low flame and add asafoeitida, tomato and continue sautéing for 2 min. Then add duck pieces and cook for 5 min. Thereafter, follow step seven and eight without adding vinegar. Pepper powder can also be added as paste with chilli, coriander, and turmeric in step six.

Duck Stew of Kerala

Duck stew (mildly spiced duck stewed with potato and coconut milk in seasoned loose gravy) is yet another widely accepted side dish served with vellayappam, palappam, string hopper, bread, etc. as in the case of duck curry. Use of coconut milk gives a desirable colour and flavour besides high calorie.

Cooking time: 50–60 min.
Ingredient composition.

Sl. No.	Ingredients	Quantity
For marinade		
1	Duck with skin-on (cut into medium size pieces)	1 kg
2	Freshly ground black pepper	10 g (2tsp)
3	Vinegar (cedar/ rice/ coconut)	15 ml (1tbsp)
4	Turmeric powder	3 g (1/2 tsp)
5	Salt	10 g (2 tsp)
For gravy		
1	Coconut oil	120 ml (1/2 cup)
2	Chopped onion	200 g
3	Chopped ginger	15 g
4	Chopped garlic	10 g
5	Curry leaves	4–5 sprigs
6	Green chilli (split lengthwise)	5
7	Crushed black pepper	5 g (1 tsp)
8	Cinnamon	2.5 cm stick
9	Cardamom	2 pods
10	Cloves	4
11	Turmeric powder	2.5 g (1/2 tsp)
12	Coriander powder	15 g (1 tbsp)
13	Kashmiri chilli powder	5 g (1 tsp)
14	Spice blend (garam masala)*	10 g (2 tsp)
15	Fresh semi-thick second extract coconut milk	500 ml (2 cups)
16	Fresh thick first extract coconut milk	250 ml (1 cup)

(continued)

Sl. No.	Ingredients	Quantity
17	Salt	To taste
18	Cubed potato	250 g
19	Hot water	As required
For seasoning		
1	Mustard	2.5 g (1/2 tsp)
2	Chopped shallot (small red onion)	15 g (1 tbsp)
3	Broken red chilli pieces	2–3
4	Curry leaves	2 sprigs

Processing steps.

- Marinate the duck pieces with the marinade in a stainless steel or glass bowl by rubbing and keep aside for about 2 h at room temperature.
- Heat coconut oil in a wok; add chopped onion and sauté in medium flame until it turns golden brown by caramelisation. Add chopped ginger and garlic and sauté until the raw odour of ginger and garlic subside. Add green chilli, curry leaves, black pepper, cinnamon, cardamom, and cloves and sauté further for 2 min with stirring.
- Then, add turmeric, coriander, Kashmiri chilli powder, and spice blend made into a paste with water and sauté in reduced heat with constant stirring. Add marinated duck pieces and continue sautéing for 2–3 min.
- Add the coconut milk, semi-thick second extract, and salt to taste to the wok and mix well with required quantity of hot water, if necessary. Cover the wok and stew till half cooked and add diced potato. Bring to boil till potato and duck are well done. If the duck is not meat type or tough, it maybe pressure cooked at 15 lb. pressure for about 20 min. Accordingly, sautéing and further cooking can be done in a pressure cooker.
- Then, add coconut milk, thick first extract, simmer for 2 min in low heat without boiling. Check for salt and spice level. The gravy should cover the duck pieces fully. The thickness of gravy may be adjusted with hot water. Remove the wok from the flame and keep aside for seasoning.
- For seasoning, add mustard to heated coconut oil in another frying pan. As mustard splutter, add chopped shallot, broken pieces of red chilli, and curry leaves till it becomes brown. Pour it into the prepared stew in the wok and serve hot as a side dish.

Notes

- Spice blend (garam masala)* - see the notes in Duck curry.
- The consistency of gravy can be increased by adding ground cashew nut or poppy seeds (khuskhus) or corn flour.
- Depending on the choice of consumers, in addition to potato, carrot, beans, green peas, and tomato can also be added.

Duck Korma of Kerala

Korma is prepared by marinating and then braising a meat in curd/ yoghurt or cream base or meat stock. The flavour of a korma is based on a mixture of spices, especially ground coriander and cumin, combined with yoghurt kept below curdling temperature and incorporated slowly and carefully with the meat juices. A korma can be mildly spiced or hot. When the Mughals came to India from Central Asia, they introduced their own style of cooking, including the slow braised, yoghurt-based or cream-based dish, korma. Like most Mughlai dishes, this one uses nuts, dried fruits and spices in addition to cream. Fried onions give the korma a slightly sweet flavour, yoghurt adds acidity, spices give a warmth and nut paste add richness or creamy texture. Most South Indian kormas use either ground coconut or coconut milk, which imparts a unique flavour.

Duck Korma (mildly spiced duck braised with curd/yoghurt, ground coriander leaves and grated coconut, nuts in seasoned gravy) is usually served as a side dish with chappati/ rotti, nan, kulcha, paratha, pathiri (non-fermented flat thin rice flour pancake), palappam, vellayappam, pulaav (pilau), etc. as in the case of duck curry and stew.

Cooking time: 50–60 min.
Ingredient composition.

Sl. No.	Ingredients	Quantity
For marinade		
1	Duck with skin-on (cut into medium size pieces)	1 kg
2	Freshly ground pepper	10 g (2 tsp)
3	Turmeric powder	2.5 g (1/2 tsp)
4	Ground ginger-garlic paste in equal quantities	30 g (2 tbsp)
5	Curd/yoghurt plain	45 g (3 tbsp)
6	Salt	10 g (2 tsp)
For onion paste		
1	Coconut oil	30 ml (2 tbsp)
2	Sliced onion	250 g (1 cup)
3	Curd/yoghurt plain	60 g (1/4 cup)
For nut paste		
1	Cashew nut	10
2	Almond (optional)	8
For korma gravy		
1	Coconut oil	30 ml (2 tbsp)
2	Sliced onion	250 g (1 cup)
3	Ground ginger-garlic paste in equal quantities	30 g (2 tbsp)
4	Green chilli (sliced lengthwise)	4–5
5	Coriander powder	45 g (3 tbsp)
6	Kashmiri chilli powder	5 g (1 tsp)
7	Turmeric powder	5 g (1 tsp)
8	Spice blend (garam masala)*	10 g (2 tsp)

(continued)

Sl. No.	Ingredients	Quantity
9	Fenugreek powder	2.5 g (1/2 tsp)
10	Cumin powder	5 g (1 tsp)
11	Salt	To taste
12	Hot water	As required
13	Finely ground coconut to smooth paste	125 g (1/2 cup)
14	Ground coriander leaves	30 g (2 tbsp)

Processing steps.

- Marinate the duck pieces with the marinade containing curd and other ingredients in a glass or stainless steel bowl by rubbing. Keep aside the marinated duck pieces covered for about 2 h at room temperature.
- Heat coconut oil in a frying pan over medium high heat, add sliced onion, and fry evenly until golden brown. If the onion slices sizzle, then the oil is ready for frying. Excess oil on fried onion, if any, is to be moped with paper towel. Grind the fried onion with curd in a blender or food processor in to a paste and keep aside.
- Grind cashew nut and almond in a blender and make a fine paste with a little water and set aside.
- Heat coconut oil in a wok; add sliced onion and sauté in medium flame until it turns golden brown. Add ginger-garlic paste and sauté until the raw odour of ginger and garlic subside. Add sliced green chilli and sauté further for 2 min with stirring. Then, add coriander, Kashmiri chilli, turmeric powder, and spice blend made in to a paste with a little water. Sauté for a minute with constant stirring under a low flame. Finally, while stirring, add fenugreek and cumin powder.
- After sautéing, when the oil starts separating, add the marinated duck pieces followed by onion-curd paste, water, and salt to taste. Mix well, cover the wok, and braise over medium high flame till the duck is well done. If the duck is not meat type or tough, it maybe pressure cooked at 15 lb. pressure for about 20 min. Accordingly, sautéing and further cooking can be done in a pressure cooker.
- Remove the lid and add cashew nut–almond paste, coconut paste, and ground coriander leaves. Stir well. Check salt, spice level, and consistency. If it is too thick, pour the required quantity of hot water. Simmer for another few minutes without boiling. Remove from heat. Garnish with coriander leaves and serve hot.

Notes

- Spice blend (garam masala)* - see the notes in Duck curry.
- Nuts shall be avoided for those having allergy. Instead add more curd/plain yoghurt.

- Coconut, finely ground to smooth paste can be replaced with thick first extract coconut milk.
- In Korma, do not use any kind of starch to thicken gravy.

Duck Pappas of Kerala

Pappas/mappas is a typical Kerala Syrian Christian delicacy, with thick and creamy coconut milk, potato, and tomatoes. Pappas means potato in Portuguese and it simply refers to a curry with potatoes.

Duck pappas is stewed duck in thick coconut milk with distinct flavour of coriander, black pepper, tomato, and potato but without red chilli. As a side dish, it goes well with all rice flour based fermented and non-fermented pan cakes for breakfast mentioned earlier in this chapter.

Cooking time: 40–50 min.
Ingredient composition.

Sl. No.	Ingredients	Quantity
For marinade		
1	Duck with skin-on (cut into medium size pieces)	1 kg
2	Turmeric powder	2.5 g (1/2 tsp)
3	Freshly ground pepper	10 g (2 tsp)
4	Spice blend (garam masala)	2.5 g (1/2 tsp)
5	Vinegar (cedar/rice/coconut)	15 ml (1 tbsp)
6	Salt	10 g (2 tsp)
For sautéing and gravy		
1	Coconut oil	45 ml (3 tbsp)
2	Black peppercorns	3–4
3	Cardamom	3–4 pods
4	Cloves	3–4 buds
5	Cinnamon	2.5 cm stick
6	Sliced onion	100 g
7	Julienned ginger	10 g (2 tsp)
8	Crushed garlic	5–6 cloves
9	Green chilli (split lengthwise)	5
10	Curry leaves	2–3 sprigs
11	Coriander powder	30 g (2 tbsp)
12	Turmeric powder	5 g (1 tsp)
13	Spice blend (garam masala)*	5 g (1 tsp)
14	Chopped tomato	100 g
15	Cubed potato	200 g
16	Salt	To taste
17	Fresh semi-thick second extract coconut milk	250 ml (1 cups)
18	Fresh thick first extract coconut milk	120 ml
For seasoning		

(continued)

Sl. No.	Ingredients	Quantity
1	Coconut oil	10 ml (2 tsp)
2	Mustard	2.5 g (1/2 tsp)
3	Finely chopped shallot	10 g (2 tsp)
4	Curry leaves	2 sprigs

Processing steps.

- Marinate the duck pieces with the marinade in a glass or stainless steel bowl by rubbing. Keep aside the marinated duck pieces for about 2 h at room temperature.
- Heat coconut oil in a wok and temper the whole spices; black pepper, cardamom, cloves, and cinnamon, by roasting to release the essential oils from the cells to enhance their flavour.
- To the tempered spices in the wok, add sliced onion, ginger, garlic, green chilli, and curry leaves and sauté for some time on low flame. Sauté till the raw odour of ginger and garlic tone down and onion turns golden brown.
- Add turmeric, coriander powder, spice blend, and sauté further with stirring on low flame. Then, add chopped tomato and sauté for 2 min on medium flame.
- Put the marinated duck pieces and cubed potatoes to the sautéed mixture and stir followed by the addition of coconut milk, semi-thick second extract. Add salt to taste. If the duck is not meat type or tough, it may be pressure cooked at 15 lb. pressure for about 20 min. Accordingly, sautéing and further cooking can be done in a pressure cooker.
- Cover woke and cook on medium flame till the duck is well done and gravy thickens. Then, add coconut milk, thick first extract, mix and simmer for a minute with occasional stirring. Check salt and spice level. Remove from heat.
- For seasoning, add mustard to heated coconut oil in another frying pan. As mustard splutter, add chopped shallot and curry leaves till it becomes brown. Add it into the prepared duck pappas in the wok and serve hot as a side dish.

Notes

- Spice blend (garam masala)* - see the notes in Duck curry.
- Onion can be replaced with shallot 250 g which would be more flavourful.
- Vinegar in marinade can be replaced with lime juice, or curd/yoghurt.

Duck Roast of Kerala

Duck Roast is a marinated low fat fried, braised and seasoned duck in thick gravy in Kerala. In this, duck is not roasted in an oven or tandoori but cooked for a longer time so that the liquid grey thickens and gets coated over the duck pieces. In particular, for thickening coconut milk is added.

Cooking time: 50–60 min.
Ingredient composition.

Sl. No.	Ingredients	Quantity
For marinade		
1	Duck, cut-up parts with skin-on, (meat type duck is preferred)	1 kg
2	Garlic paste	15 g (1 tbsp)
3	Ginger paste	15 g (1 tbsp)
4	Vinegar (cedar/rice/coconut)	15 ml (1 tbsp)
5	Salt	10 g (2 tsp)
Whole spices for tempering		
1	Cardamom	5–6
2	Cloves	6–8
3	Cinnamon	2.5 cm stick
4	Star Anise	1
5	Fennel (Anise)	8 g (1 ½ tsp)
6	Mace	3
7	Black pepper	15 g (1 tbsp)
For sautéing and gravy		
1	Coconut oil	60 ml (4 tbsp)
2	Kashmiri chilli powder	20 g (1 ½ tbsp)
3	Coriander powder	30 g (2 tbsp.)
4	Turmeric powder	4 g (3/4 tsp)
5	Mustard	5 g (1 tsp)
6	Onion	100 g
7	Sliced shallot	100 g
8	Crushed garlic	10–15 cloves
9	Crushed ginger	15 g
10	Curry leaves	3 sprigs
11	Tomato, chopped	100 g
12	Fresh thick first extract coconut milk	200–250 ml
13	Hot water	As required
14	Salt	To taste
15	Black pepper powder	10 g (2 tsp)

Processing steps.

- Heat a little coconut oil in a wok on low medium flame and temper the whole spices one to seven by roasting to release the essential oils from the cells to enhance their flavour. Remove from heat and after cooling grind to a fine powder. Mix with Kashmiri chilli powder, coriander powder, and turmeric powder and kept aside.
- In a large glass or stainless steel bowl the duck pieces are marinated with the marinade along with half of the above spice mixture. Score diagonally skin-on cut-up parts for quicker penetration of marinade and rendering of fat during

cooking. The marinated duck pieces are kept chilled for overnight or for at least 4 h prior to further cooking.

- The chilled marinated duck pieces are to be kept at room temperature for 30 min prior to further cooking.
- The marinated duck pieces are low fat fried until brown in hot coconut oil in a wok. The pieces are fried on medium heat in batches and kept aside.
- In the same oil, put mustard to splutter. Add sliced onion, shallot, and curry leaves and sauté until onion turns light brown.
- Add chopped ginger, garlic, and sauté until the raw odour subside followed by chopped tomato till mashed well.
- Then add the remaining half of spice mixture saved after marination and sauté for 2–3 min. on medium heat.
- Add fried duck pieces and mix well. Then, add coconut milk and approximately 500 ml hot water, salt to taste and mix well.
- Cover the wok and braise on medium heat until well done. Stir occasionally.
- Once the duck is well done, open the wok and cook until the gravy is nicely thick.
- Add curry leaves, black pepper powder, and mix well. Check salt level. Remove from heat, cover and serve after 2–3 h so that the thick gravy will evenly coat the pieces. Serve hot as a side dish.

Notes

- Vinegar in marinade can be replaced with lime juice or curd/yoghurt.
- Coconut milk can be replaced with khuskhus (poppy seeds) and coriander leaf paste in curd.

Duck Fry of Kerala

Duck fry in Kerala is duck marinated in thick spice mix and vinegar/curd/lime juice, sautéed and low fat fried in the rendered fat from duck skin and meat juice during cooking.

Cooking time: 50–60 min.
Ingredient composition.

Sl. No.	Ingredients	Quantity
For marinade		
	Duck, cut-up parts with skin-on (meat type duck is preferable)	1 kg
1	Ginger paste	15 g (1tbsp)
2	Garlic paste	15 g (1 tbsp)
3	Turmeric powder	5 g (1 tsp)
4	Pepper powder	15 g (1tbsp)
5	Vinegar (cedar/ rice/ coconut)/ lime juice/ curd	15 ml (1tbsp)
6	Salt	10 g (2 tsp)

(continued)

Sl. No.	Ingredients	Quantity
For sautéing		
1	Coconut oil	75 ml (5 tbsp)
2	Chopped onion	150 g
3	Chopped tomato	150 g
4	Turmeric	2.5 g (1/2 tsp)
5	Kashmiri chilli powder	10 g (2tsp)
6	Coriander powder	15 g (1tbsp)
7	Black pepper power	5 g (1 tsp)
8	Cumin powder	5 g (1 tsp)
9	Spice blend (garam masala)*	10 g (2 tsp)
10	Curry leaves	4 sprigs
11	Salt	To taste
12	Hot water	As required

Processing steps.

- Marinate the duck pieces with the marinade in a glass or stainless steel bowl by rubbing. Score diagonally skin-on cut-up parts for quicker penetration of marinade and rendering of fat during cooking. Keep aside the marinated duck pieces for about 2 h at room temperature.
- Heat coconut oil in a wok; add sliced onion and sauté in medium flame until it turns golden brown.
- Then, ingredients four to nine for sautéing are made into a paste with a little water and mixed with the onion in the wok. Sauté for a minute and add tomato and curry leaves with constant stirring under a low flame.
- Add the marinated duck pieces, salt to taste, and mix well. Close the wok and simmer on medium low flame with intermittent stirring till the meat is well done. As the fat renders, duck skin and meat get fried in fat and meat juice/stock.
- If the meat stock is insufficient to cook, add the required quantity of hot water. Check salt level.
- Complete the cooking till the duck pieces are evenly coated with the dry gravy. Remove from the flame and serve hot after 30 min as a side dish.

12.2.3.2 Assamese

Assamese style of cooking is characterized by very little use of spices, cooking over fire and strong flavour due to the use of various fruits and vegetables, fresh, dried or fermented. Assamese meat preparations are similar to those of Southeast Asian and East Asian countries. They are characterized by low amount of spices and oil, higher quantity of ginger, garlic, curry leaves, green chilli, fermented bamboo shoots, and lemon juice. The preferred oil for cooking is the pungent smelling mustard oil. Although, there are many breeds of duck in Assam and North East India, Pati Hanh breed (a type of Mallard duck) is mostly preferred for cooking. Duck dishes combined with vegetables, especially ash gourd, pumpkin, and green papaya are very popular in Upper Assam. Generally a cast iron woke is used for frying, sautéing, and stewing. In Assam, the best season to eat duck is considered to be from late October to April. It tastes best during December–January, the rice harvest season, during when ducks are full of fat. Slight ingredient composition variations exist in the products for the same dish among various communities in Assam. The important Assamese duck products are Hanh Aloo Aru Kumura (Assamese duck curry) and Hanh Guta Bhaji (Assamese duck dry fry).

Hanh Aloo Aru Kumura of Assam

Assamese duck curry with potato and ash gourd is an ingenious very popular delicacy served as a side dish with rice on special occasions like Bihu or any celebrations or festivals. There are many variations of this dish, though basically the same.

Cooking time: 60 min.
Ingredient composition.

Sl. No.	Ingredients	Quantity
For marinade		
1	Duck with skin-on, cut into small pieces	1 kg
2	Turmeric powder	10 g (2 tsp)
3	Grated raw papaya	250 g (1 cup)
4	Mustard oil	10 ml (2 tsp)
5	Salt	5 g (1 tsp)
Whole spices for tempering		
1	Coriander	10 g (2 tsp)
2	Cumin	5 g (1 tsp)
3	Black pepper	10 g (2 tsp)
4	Cinnamon	5 cm piece
5	Cardamom	7–8 pods
6	Cloves	15–20 buds
7	Bay leaf	2
8	Red chilli	4–5
For sautéing		
1	Mustard oil	100 ml
2	Sliced onion	300 g
3	Ground ginger-garlic paste in equal quantities	30 g (2 tbsp)
4	Red chilli powder	10 g (2 tsp)
5	Cubed potato	300 g
6	Cubed ash gourd	300 g

Processing steps.

- Marinate the duck pieces with the marinade in a glass or stainless steel bowl by rubbing. Keep aside the marinated duck pieces for about 1 h at room temperature.

- Dry roast or temper the whole spices in a pan, grind into powder and keep aside.
- Cube the potatoes fresh without turning brown.
- Heat mustard oil in a wok for sautéing other ingredients; add sliced onion, a little salt and sauté on low flame for a few minutes.
- Then, add ginger-garlic paste and red chilli powder to onion and continue sautéing.
- Once the onion is fried, add marinated duck pieces and potatoes. Cover the wok and cook on medium to low flame for 20–25 min, stir intermittently without sticking to the bottom of the wok.
- When the oil start separating, add a little more salt to taste. Then, add ash gourd cubes and cook for 10–15 min more.
- Add the already prepared roasted and ground spice powder to the curry and mix. Cook for another 10 min and check whether potato and ash gourd are done. Check salt level.
- Remove the wok from the flame, keep the curry for some time, and serve hot with Joha rice. Garnish with fresh chopped coriander leaves.
- Note: For marinade, lemon juice can also be used. Ash gourd can be replaced with bottle gourd or pumpkin. The gravy should neither be too watery nor thick.

Hanh Guta Bhaji of Assam

Hanh guta bhaji (Assamese duck dry fry) is popular Christmas special dish in upper Assam. Duck is cooked slowly in a pan and not pressure cooked.

Cooking time: 60 min.
Ingredient composition.

Sl. No.	Ingredients	Quantity
For marinade		
1	Turmeric powder	5 g (1 tsp)
2	Salt	5 g (1 tsp)
For sautéing		
1	Duck with skin-on (cut into small pieces)	500 g
2	Mustard oil	60 ml
3	Chopped onion	100 g
4	Ground ginger-garlic paste in equal quantities	30 g (2tbsp)
5	Green chilli (split lengthwise)	4–5
6	Turmeric powder	5 g (1 tsp)
7	Coriander powder	10 g (2 tsp)
8	Cumin powder	5 g (1 tsp)
9	Salt	To taste
10	Cubed potato	100 g
11	Chopped coriander leaves	10 g

Processing steps.

- Marinate the duck pieces with the turmeric powder, salt, and mustard oil in a glass or stainless steel bowl by rubbing. Keep aside the marinated duck pieces for about 1 h at room temperature.
- Heat mustard oil in a wok and sauté sliced onion till it is fried and golden brown. Stir and add green chilli, ginger-garlic paste, turmeric powder, coriander powder, cumin powder, and salt to taste on medium flame. Stir to avoid sticking to the bottom of wok.
- Add marinated duck pieces with constant stirring followed by the addition of cubed potato. Cover and cook slowly on medium flame for 25–30 min until duck is well done. Stir intermittently. The fat renders and meat gets fried in that.
- Remove the wok with dry fried duck from the flame. Garnish with chopped coriander leaves before serving with rice.

12.2.3.3 Bengali

Bengali food preparations are a blend of sweet and spicy flavours. The staple food of people in Bengal is rice and fish. A Bengali lunch or dinner generally comprise of rice, lentil, vegetables, and fish curry. Bengali meal follows a multi-course tradition in a specific sequence as in French or English cuisine. Fish, mutton, chevon, chicken, duck and their eggs are usually consumed across Bengal. The non-vegetarian preparations are of four different types based on spiciness/ hotness, thickness of gravy and flavour such as a light stew like preparation (jhol), a more spicy and hot one (jhal), a moderately spicy rich gravy preparation (kalia), and a more sweetish one (korma).

A variety of spices and their mixes are used in preparing Bengali food items, viz., turmeric, cumin, coriander, black onion seeds (black cumin), red chilli, mustard, bay leaves, fenugreek, fennel, wild celery seeds, onion, ginger, coconut and a combination of five spices called panchphoron (five spices) which comprises of black cumin, cumin, fenugreek, fennel, and thymol seed (ajwain/ ajowan/ caraway/ carom). Phoron in Bengali means, frying the spices singly or in combination either at the beginning of preparing a product or adding later to it is a unique style in Bengali preparations. Mustard is also used either as an ingredient of panchphoron or as a paste in Bengali dishes is another specialty. The principal medium for cooking is mustard oil. Bengali spicy duck curry is one of the popular duck meat products.

Bengali Spicy Duck Curry

Cooking time: 60 min.

Ingredient composition.

Sl No	Ingredients	Quantity
1	Duck with skin-on (cut into small pieces)	1 kg
For tempering		
2	Mustard oil	60 ml (4 tbsp)
3	Cardamom	2 pods
4	Black pepper	4
5	Cinnamon	5 cm stick
6	Cloves	4 buds
7	Bay leaf	1
For sautéing		
1	Sliced onion	250 g
2	Ground ginger-garlic paste in equal quantities	25 g (1 1/2 tbsp)
3	Sliced medium sized tomato	1
4	Salt	To taste
5	Cumin powder	15 g (1 tbsp)
6	Turmeric powder	5 g (1 tsp)
7	Kashmiri chilli powder	2.5 g (1/2 tsp)
8	Red chilli powder	2.5 g (1/2 tsp)
9	Coriander powder	15 g (1 tbsp)
10	Hot water	500 ml (2 cups)
11	Spice blend (garam masala)*	10 g (2 tsp)

Processing steps.

- Heat mustard oil in a wok and temper the spices Sl. No. three to seven with stirring.
- Add chopped onion and sauté till it turns golden brown.
- Then add ginger-garlic paste and stir well followed by the addition of sliced tomato and a pinch of salt. Stir and sauté tomato till it is soft.
- Add ingredients for sautéing duck pieces Sl. No. five to nine and stir well. Then, add 250 ml hot water and continue sautéing.
- Transfer the duck pieces into the sautéed mix in the wok, stir and mix well followed by the addition of salt to taste. Cook for 20 min and add water to cover the duck pieces.
- Cover the wok and continue boiling for 30 min with intermittent stirring.
- After 30 min of cooking, add spice blend (garam masala). Check for salt and spice level.
- Cover and cook till the duck well done and the gravy gets slightly thicker.
- After 2 min, remove wok from heat and serve hot with rice or pulaav.

12.2.3.4 Andhra

Andhra/ Telugu style of cooking is seen among the people of Andhra Pradesh and Telangana states of India. All the three regions, viz., Costal Andhra, Rayalaseema, and Telangana have distinctive preparations. Generally it is tangy, very hot, and spicy as in Andhra spicy duck curry. The staple food is rice, millet, and ragi (finger millet). Most of the populations cook non-vegetarian dishes but popular dishes are served in Andhra style. The gravy base is usually onion, tomato, coriander, tamarind, and coconut. Pepper is usually used for fried preparations.

Andhra Spicy Duck Curry

Cooking time: 50–60 min.
Ingredient composition.

Sl No	Ingredients	Quantity
1	Duck with skin-on (cut into small pieces)	500 g
For marinade		
1	Ground ginger-garlic paste in equal quantities	10 g (2 tsp)
2	Red chilli powder	10 g (2 tsp)
3	Salt	2.5 g (1/2 tsp)
4	Refined oil	60 ml (4 tbsp)
5	Turmeric powder	1 g (¼ tsp)
6	Chopped coriander leaf	
For sautéing and gravy		
1	Cumin seed	10 g (2 tsp)
2	Chopped onion	150 g
3	Chopped green chilli	2
4	Hot water	250 ml (1cup)
5	Coriander powder	5 g (1 tsp)
6	Spice blend (garam masala)*	5 g (1 tsp)
7	Curry leaves	4 sprigs

Processing steps.

- In a glass or stainless steel bowl the duck pieces are marinated with ginger-garlic paste, red chilli powder, salt, turmeric powder, and chopped coriander leaves. Mix well with 30 ml oil and keep aside at room temperature for 30 min.
- To heated oil in a wok, add cumin seed and stir. Add chopped onion, green chilli, and mix thoroughly. Cover the wok and sauté.
- Add remaining ginger-garlic paste, mix, cover, and sauté followed by the addition of marinated duck pieces. Stir well and mix.
- Add 250 ml hot water and stir. On boiling, add coriander powder, cover, and continue cooking till the duck is well done and gravy thickens.
- Add spice blend, curry leaves, and mix well. Check salt level. Cover and cook for a minute.

- Remove from flame and transfer to serving dish. Serve hot with rice or other main dish.

12.2.3.5 Tamil Nadu

In Tamil Nadu, there are different styles of cooking in the various geographical regions in the State and eating habits of the community, as in Kerala. Among those, the most popular one is of the Nattukotai Chettiars of Chettinadu, the enterprising trading community, which is known as Chettinadu style. They use a variety of fresh ground spices. Most of the products are eaten with rice and rice based Dosa, Idli, Appam, Idiyappam, etc. Major spices used in this cuisine is non-vegetarian, which include, star anise, kalpasi (black stone flower which is a lichen), tamarind, chilli, fennel, cinnamon, cloves, bay leaf, pepper, cumin, and fenugreek. More coriander than chilli is used and so dishes are not that spicy. The major duck products of this region are Chettinadu duck curry and Duck Chettinad are discussed in the following.

Chettinadu Duck Curry of Tamil Nadu

Duck without skin cooked in sautéed spices, shallot, garlic, tomato puree, and coconut milk with a thick mild spiced gravy.

Cooking time: 40–50 min.
Ingredient composition.

Sl No	Ingredients	Quantity
1	Eight pieces duck without skin (cut-up parts)	1 kg
For dry roasting of spices		
1	Coriander	15 g (1 tbsp)
2	Cumin	15 g (1 tbsp)
3	Fennel	15 g (1 tbsp)
4	Poppy seeds (khuskhus)	15 g (1 tbsp)
5	Kalpasi (black stone flower) (optional)	5 g (1 tsp)
For sautéing and gravy/sauce		
1	Kashmiri chilli	15 g (1 tbsp)
2	Turmeric powder	2.5 g (1/2 tsp)
3	Coconut oil	30 ml (2 tbsp)
4	Cinnamon stick	2.5 cm
5	Cloves	4 buds
6	Star anise	½ flower
7	Sliced shallot	3
8	Chopped garlic	6 cloves
9	Curry leaves	3 sprigs
10	Chopped and pureed tomato	4 nos
11	Thick first extract coconut milk	150 ml
12	Salt	To taste

Processing steps.

- Dry roast the spices coriander, cumin, fennel, kalpasi and poppy seeds in a frying pan on medium heat for 2–3 min.
- Grind the roasted spices finely and add Kashmiri chilli powder and turmeric powder. Mix well and set aside.
- Heat coconut oil in a wok over medium heat. Add cinnamon stick, cloves, and star anise and sauté for 30 sec.
- Add shallot, garlic, and curry leaves and cook, stir occasionally until lightly browned.
- Add spice powder kept aside and salt and mix well. Put it in high flame and add duck pieces and sauté for 2–3 min till all the duck pieces get seared with formation of brown crust on all sides.
- Add tomato puree, stir well and when the curry starts simmering, lower the heat. Cover and continue cooking till the duck is well done.
- Add the thick coconut milk, mix well, and simmer for a minute. Check the salt and spice level.
- Transfer to a serving dish and garnish with chopped coriander leaves before serving with rice.
- Note: As coconut milk is used in the preparation and to retain its flavour, it is preferable to use coconut oil itself for cooking, though it can be replaced with sunflower oil or any other refined oil.

Duck Chettinadu of Tamil Nadu

Duck breast with skin-on is pan-fried to render fat, marinated with ground roasted fresh spices paste, oven roasted, and served with rich sauce made of coconut milk, tomato, onion, and spices sautéed in coconut oil.

Cooking time: 50–60 min.
Ingredient composition.

Sl No	Ingredients	Quantity
1	Duck breast with skin-on (meat type)	500–600 g
For dry roasting of spices		
1	Coriander	8 g(1 ½ tsp)
2	Black pepper	5 g (1 tsp)
3	Cinnamon	5 cm stick
4	Cloves	5–6 buds
5	Star anise	3
6	Fennel (anise)	15 g (1 tbsp)
7	Cardamom	8–10 pods
8	Red chilli	1
9	Poppy seeds (khuskhus)	30 g (2 tbs)
10	Kalpasi (black stone flower)	7–8
11	Marathi moggu (kapok buds)	3 pieces
12	Roasted chana dal	15 g (1 tbsp)

(continued)

Sl No	Ingredients	Quantity
13	Turmeric powder	8 g (1 ½ tsp)
For marinade		
1	Ginger-garlic paste	8 g (1 ½ tsp)
2	Lemon juice	10 ml (2 tsp)
3	Red chilli powder	5 g (1 tsp)
4	Spice powder (prepared and set aside)	10 g (2 tsp)
5	Coconut oil	15 ml (1 tbsp)
6	Salt	To taste
For sautéing and gravy/sauce		
1	Coconut oil	30 ml (2 tbsp)
2	Star anise	½
3	Cardamom	2 pods
4	Black peppercorn	4
5	Red chilli	1
6	Kalpasi (black stone flower)	2
7	Marathi moggu (kapok buds)	1 bud
8	Fennel	2.5 g (½ tsp)
9	Chopped onion	100 g
10	Curry leaves	3 sprigs
11	Salt	To taste
12	Turmeric powder	10 g (2 tsp)
13	Chopped tomato	100 g
14	Water, hot	250 ml (1 cup)
15	Thick first extract coconut milk	125 ml

Processing steps.

- To make spice powder, dry roast all ingredients Sl. No. 1–12 in a frying pan on medium heat for 2–3 min with stirring.
- Grind the roasted spices finely and add turmeric powder. Mix well and set aside.
- In order to prepare the marinade paste, mix all the ingredients, one to six in a glass or stainless steel bowl and keep aside.
- Prior to marinating, duck breasts are pan-fried for 10–15 min. For pan-frying, score the skin of the breast diagonally and season with salt. Care is taken not to cut through meat.
- Place the breast with skin side down in a cold frying pan over a medium high heat.
- On pan-frying, skin gets crisp and the subcutaneous fat rendered. The rendered fat is decanted out of the pan. Though rendered fat is delicious, it will make the breast soggy.
- Meanwhile, sauce/gravy for the dish is prepared. Heat 30 ml coconut oil in a wok and temper the spices sl. no. two to eight with constant stirring.
- When it starts to sizzle, add chopped onion and salt. Stir well and add curry leaves. Continue sautéing till onion becomes golden brown.

- Add spice powder followed by turmeric powder. Stir well, add chopped tomato, and sauté till tomato gets cooked.
- Add water and salt to taste. Add coconut milk, with stirring and simmer in low heat for 5 min and then to medium heat. When the sauce thickens remove from flame.
- Meanwhile, place the pan-fried duck breasts with skin side down in an oven tray and marinate the flesh side with the marinade paste.
- Finish the cooking of the marinated duck in an oven preheated to 200 ^{0}C. After a few minutes of roasting on skin side, flip it over and finish on skin side.
- In 15–20 min duck will get cooked to medium and in 20–25 min, well done.
- Take the pan out of oven; rest for a few minutes for setting before cutting the duck breast.
- Serve in serving dish or plate with the prepared sauce and rice.

Note: Before adding coconut milk, 15 g desiccated coconut can be added to the sautéing mixture to improve thickness and flavour of sauce. This is optional

12.2.4 Bangladeshi Products

Bangladeshi style of cooking has been largely influenced by Mughalai cuisine, the legacy left behind by Persian rulers. Later on, Turkish and Arabic cuisines also influenced and the dishes are becoming popular. However, the cuisine is closely related to that of neighbouring Bengali and North-East Indian traditional cuisines with rice, fish curry, and lentil. Generally, Bangladeshi dishes are spicy. Bangladesh also developed a multi-course tradition which is known as Bengali Keta, as in Bengali cuisine. Two popular duck dishes Bangladeshi duck curry and Bangladeshi hasher kalia (spicy duck curry) are given in the following.

12.2.4.1 Bangladeshi Duck Curry

Cooking time: 45–50 min.
Ingredient composition.

Sl No	Ingredients	Quantity
1	Duck with skin-on (cut into small pieces)	1 kg
For spice blend (garam masala)		
1	Bay leaf	1
2	Black peppercorn	10
3	Cinnamon	5 cm piece
4	Cardamom	6 pods
5	Grated nutmeg	1.25 g (1/4 tsp)
6	Cloves	4 buds
7	Garlic	2 cloves
For sautéing		
1	Mustard oil	15 ml (1 tbsp)

(continued)

Sl No	Ingredients	Quantity
2	Chopped onion	250 g (3 nos)
3	Ground ginger-garlic paste in equal quantities	15 g (1 tbsp)
4	Red chilli powder	5 g (1 tsp)
5	Coriander powder	5 g (1 tsp)
6	Cumin powder	5 g (1 tsp)
7	Salt	To taste
9	Hot water	750 ml (3 cups)
10	Quartered tomato	8–10
11	Chopped coriander leaves	30 g (2 tbsp)

Processing steps.

- Dry roast the ingredients for spice blend in a frying pan, grind and make into a paste with 50 ml water. Keep aside.
- In a hot non-stick pan, fry duck pieces with skin side down. The skin gets fried in the fat rendered from skin. When the skin is brown and crisp while meat is still pink, remove from flame and keep aside. Meat side can also be seared, if required. Save the rendered fat for sautéing other ingredients which would impart the unique flavour.
- Sauté chopped onion in the rendered fat saved, till onion is lightly brown. Add ginger-garlic paste and continue sautéing for 2–3 min till the raw flavour subsides.
- Then add spice blend and salt to taste with stirring. After 34 min of sautéing, add chilli powder, coriander powder, and cumin powder. Stir and mix well.
- Pour hot water and boil on low heat till the gravy gets thickened. Meanwhile, stir and add tomatoes and chopped coriander, cook for 2–3 min.
- Remove from flame and serve hot.

12.2.4.2 Bangladeshi Hasher Kalia

Cooking time: 50–60 min.
Ingredient composition.

Sl No	Ingredients	Quantity
1	Duck with skin-on (cut-up parts)	1.5 kg
For spice blend		
1	Cinnamon	2.5 cm x 2 sticks
2	Cardamom	3 pods
3	Cloves	2.5 g (½ tsp)
4	Black pepper	2.5 g (1/2 tsp)
5	Fenugreek	2.5 g (½ tsp)
6	Coriander	2.5 g (½ tsp)
7	Cumin	5 g (1 tsp)

(continued)

Sl No	Ingredients	Quantity
For marinade		
1	Plain curd/yoghurt	125 ml (1/2 cup)
2	Ginger paste	10 g (2 tsp)
3	Garlic paste	10 g (2 tsp)
4	Cumin paste	5 g (1 tsp)
5	Coriander powder	5 g (1 tsp)
6	Turmeric powder	5 g (1 tsp)
7	Red chilli powder	5 g (1 tsp)
8	Onion paste	125 g (1/2 cup)
For sautéing and cooking		
9	Mustard oil	62 g (1/4 cup)
10	Whole garam masala—Cinnamon, cardamom, bay leaf	As required
11	Salt	10 g or to taste
12	Water	250 ml (1 cup)
13	Fried onion	125 g (1/2 cup)
14	Fried garlic	30 g (2 tbsp)

Processing steps.

- Temper all the ingredients for spice blend, one to seven, in a frying pan, grind into fine powder and keep aside.
- Mix well all the ingredients one to eight for marinade. Marinate duck pieces by massaging and keep aside for 15 min at room temperature in a glass or stainless steel bowl.
- Heat mustard oil in a pressure cooker and add whole garam masala and fry a little.
- Add marinated duck pieces and sauté for 5 min, until oil is separated from spices. In Bengali this process is koshano.
- Add 250 ml water, salt to taste, cover and pressure cook for 20 min.
- Transfer meat from pressure cooker to a wok and add fried onion, fried garlic, and spice blend. Mix and cook for 5 min more.
- Turn off the flame and transfer to a serving dish and garnish with fried onion.

12.2.5 Sri Lankan Products

Rice is the staple food of the Sri Lankans with meat, fish, and vegetables prepared as curries. Sliced onions, green chilli, black pepper, cinnamon, cardamom, cloves, nutmeg, and saffron are used to add flavours. Although the ingredients used are similar to those in Indian dishes, pepper and chilli are added a little excess to give an extra hotness. Sri Lankan black curry spice blend is generally used in curries with coconut milk and cashew either during the preparation or as a garnish prior to serving. This whole spice mix is a mixture of several spices like cumin, coriander, mustard, coconut, mace, black pepper, clove, cardamom, curry leaves, cinnamon, and rice. Similarly, there is white curry spice blend as it does not contain red chilli or turmeric. It is usually paired with coconut giving a white colour to sauce. Lemongrass and pandanus leaves (pandan) give sweet flavour to Sri Lankan and Southeast Asian dishes. Pandanus leaves are used in briyani in some parts of Kerala, as well. The same dish is made differently in north and south Sri Lanka. Dishes from northern Sri Lanka have great influence of South India.

12.2.5.1 Sri Lankan Black Curry Duck

Cooking time: 1 h 15 min.
Ingredient composition.

Sl No	Ingredients	Quantity
1	Duck with skin-on, medium sized pieces (meat type)	1 kg
2	Coconut milk	400 ml
3	Soaked Goraka (*Garcinia gummi-gutta*/Malabar tamarind/ kudampuli)	3 slices
4	Chopped garlic	3 cloves
5	Chopped ginger	15 g (1 tbsp)
6	Chopped green chilli	2
7	Sugar	30 g
8	Lemon grass	10 cm piece
9	Black pepper ground	15 g
10	Green cardamom	4 pods
11	Salt	To taste
12	Ground Sri Lankan black curry blend	60 g
13	Scotch whisky	150 ml
14	Coconut oil	60 ml
15	Curry leaves	2 sprigs
16	Sliced onion	250 g
17	Water	As required

Processing steps.

- In a wok, place the duck pieces, coconut milk, soaked goraka, chopped garlic and ginger, green chilli, sugar, lemon grass, black pepper, black curry spice blend, salt, and enough water to cover all components.
- Bring to boil and let simmer for 1 h and add Scotch whisky. Stir occasionally.
- Meanwhile, heat coconut oil in a pan, add curry leaves and as it sizzles, add onion. Sauté on low heat till onion turns golden brown due to caramelisation. Set aside.
- Once the duck is well done, add onion and simmer for 10 min.
- Transfer to serving dish and serve with rice.

12.2.5.2 Sri Lankan Duck Curry

Cooking time: 30 min.
Ingredient composition.

Sl No	Ingredients	Quantity
1	Chunks of roasted duck with skin	500–600 g
2	Ground Sri Lankan black curry blend	2 g (1/2 tsp)
3	Finely chopped birds eye chilli	2
4	Roughly chopped onion	50 g
5	Halved cherry tomatoes	4
6	Zucchini cut into small chunks	1
7	Tomato ketchup	15 ml (1 tbsp)
8	Raspberry vinegar	15 ml (1 tbsp)
9	Salt	To taste
10	Desiccated coconut	60 g (1/4 cup)
11	Lime juice	15 ml (1 tbsp)
12	Warm water	60 ml
13	Peanut oil (groundnut oil)/ rape seed oil/sunflower oil	15 ml (1 tbsp)

Processing steps.

- Heat oil in a wok on medium flame, add chopped onion, and stir well till it becomes golden brown.
- Add spice blend, zucchini, chopped chilli, and stir well until the zucchini gets softened.
- Then add duck chunks and mix together followed by the addition of tomatoes and salt. Cover and cook for a few minutes. With stirring add tomato ketchup and raspberry vinegar. Reduce the heat, stir and add desiccated coconut, lime juice, and warm water.
- Cover the wok and cook for few more minutes. Remove from flame.
- Transfer from wok to serving dish, garnish with chopped coriander leaves, and serve with rice.

12.2.6 Southeast Asian Duck Meat Products

The products of a few popular duck dishes in Malaysian, Singaporean, Thai, Indonesian, and Vietnamese products are discussed to show the popularity of duck delicacies in their diet. Malaysian cuisine is an amalgamation of the culinary traditions and practices of the three major ethnic groups, viz., Malay, Chinese, and Indian those make up its population. At the same time, it is strongly influenced by

Malayan Peranakan, Thai, Portuguese, Dutch, Arab, and British cuisines. Malaysia shares culinary practices in Minangkabau, in west Sumatra in Indonesia and Thailand also. Therefore, it is highly complex with spices from all these places. In all meat dishes, the natural fresh spices from the Malayan peninsula are generally used rather than dry whole or ground spices as in Indian way of cooking. This makes the chicken and duck dishes a specialty and popular compared to other regions. Generally, all meat, poultry, and seafood dishes are prepared with coconut milk, lemon grass, kaffir lime leaves, ginger and its flower, fresh turmeric and its leaves, candlenuts, pandan leaves, galangal, tamarind, palm sugar, shallots, onion, red, green, yellow and bird's eye chilli, belachan(sundried smooth paste of fermented and salted tiny shrimp), fish sauce, rempah (ground spice paste) like Indian wet masala or Thai curry paste or kerisik (grated coconut dry fried or toasted and ground to a paste). The staple food is rice, as in any other Asian countries, and the duck dishes are served along with that at any meal. Satay has been the most popular spiced roasted meat in Southeast Asia, especially in Indonesia and Malaysia. It is prepared from beef, chicken or goat meat, which can be substituted with duck meat. Duck meat Satay offers a variation in its taste because of the fat content in duck meat which gives a unique stronger aroma during roasting (Mathew 2020, personal communication).

Because of the proximity of Singapore to Malaysia and the mutually related cultural influences, both the cuisines have a lot of similarities but with certain regional differences. Singapore cuisine has also the same influence as that of Malayan. In the case of Singapore, being a cosmopolitan city, their method of cooking is influenced by Japan, Korea, and Thailand also.

In Thai cuisine, curry refers to both the dish as well as the curry paste for that dish. The colour of the chilli and other ingredients give the distinct red, green, and yellow colour to the paste and the term. Usually, the green is the mildest in hotness, red the hottest, and yellow falling in between. Thai green curry paste contains coriander (cilantro) leaves, kaffir (makrut) lime leaf and peel, basil, lemon grass, garlic, ginger, and shallots besides green chilli. With red curry paste, dry red chilli, tomato, and chilli powder are also added. Yellow curry is mild sweet with yellow chilli, turmeric, coriander seeds, cumin, galangal, ginger, garlic, lemon grass, and fish sauce. Both red and yellow curry use coconut milk also.

The products of a few typical traditional duck dishes like Malay duck curry, Rendang itiak, Singapore Nyonya duck curry, Thai green curry with roasted duck, Bebek goring Madura or Indonesian fried duck-Madura style and Vietnamese roast duck or Vit quay are given below with variations.

12.2.6.1 Malay Duck Curry

Although there are different ingredient composition for the dish Malaysian duck curry, the one with kerisik, rempah spice paste, and coconut milk is found more flavourful, delicious, and popular. Kerisik is grated coconut dry fried to golden brown in colour and ground to a paste. This ingredient is usually added in Malay, Singaporean, Peranakan, and Indonesian curries to give aroma and thickness to gravy. The Malay word rempah means a blanched spice paste made with a blend of herbs and spices and traditionally used in all chicken, duck, and seafood curries. It can be stored for a month in chiller or for 3 months in freezer. The formulary and processing steps described by Moonshadow (2019) with variations are as follows.

Cooking time: 1 ½ to 2 h.
Ingredient composition.

Sl No	Ingredients	Quantity
1	Duck with skin-on, cut-up parts or 6–8 duck legs, jointed	~ 1. 2 kg
For Kerisik		
1	Desiccated coconut (grated coconut – 250 g)	40 g
For Rempah (Malaysian spice/curry paste)		
1	Fresh, finely sliced red chilli	5
2	Dried red chilli soaked in hot water till softened	8–10
3	Roughly chopped candle nuts	10–12
4	Finely chopped, sliced shallots (equivalent quantity onion,)	250 g
5	Chopped garlic	3 cloves
6	Chopped ginger	15 g (1 tbsp)
7	Fresh chopped turmeric	10 g (2 tsp)
8	Finely chopped lower 1/3 portion lemon grass	3 stalks
9	Sliced galangal	15 g (1 tsp)
10	Belachan	10 g
11	Cumin powder	10 g (2 tsp)
12	Coriander powder	15 g (1 tbsp)
13	Palm sugar	a pinch
14	Coconut oil or any vegetable cooking oil	90 ml
For curry/sauce		
1	Curry leaves	6 sprigs
2	Thick coconut milk	500 ml
3	Salt	~ 5 g (1 tsp)
4	Coriander leaves for garnishing	As required

Processing steps.

- In order to prepare kerisik, dry fry desiccated or grated coconut in a frying pan with occasional stirring, till it is golden brown. Grind the roasted coconut into a smooth paste and keep aside.
- Grind coarsely all the ingredients for rempah Sl No 1 to 9 in a food processor with three tablespoon full of coconut oil. To the coarsely ground paste, add the rest of the ingredients Sl No 10 to 13 in to a fine paste and kept aside. The soaked red chilli can be halved and deseeded before making the paste, if only less spice and hotness is required.
- Sauté the prepared rempah in hot coconut oil in an open wok with stirring. Sauté over low heat until the oil starts to separate from rempah, about 20 min. Stir continuously to ensure that the paste does not burn. If necessary, add a little coconut milk to prevent burning or sticking to the bottom of the pan.

- Add curry leaves and stir for half-a-minute followed by the addition of duck pieces. Stir well so that all the pieces get evenly coated with the sautéed paste.
- Add coconut milk with stirring and bring to simmer. Cover the wok and cook the duck until it is well done, about an hour.
- Add kerisik and salt to taste, with stirring. Simmer in high flame till the desired consistency for the curry is achieved. Remove from heat, garnish with fresh coriander leaves, and serve hot with steamed plain white rice.

12.2.6.2 Rendang Itiak of Malaysia

Rendang Itiak (Rendang bebek) is also known as Duck Rendang. Rendang is a spicy meat dish from beef, buffalo meat, mutton, lamb, chevon, chicken, duck, or vegetables and originally from the Minangkabau highland region in west Sumatra in Indonesia. It spreads throughout Indonesia, Malaysia, and Singapore and is a highly popular dish as it is served during ceremonial occasions and festivals to honour guests. This is a meat cooked slowly for longer time and braised in coconut milk and spice mixture containing lemongrass, galangal, turmeric, fish paste, and palm jaggery. Therefore, the shelf life of rendang at room temperature is 3 to 4 weeks and in freezer for 6 months. The colour and liquid content indicate the type of rendang as gulai (wet and light yellow due to lesser red chilli and more turmeric), kalio (brown and medium wet), and rendang (very dark brown, dry, and too spicy). Rendang duck is more like kalio and is cooked for shorter periods unlike that of beef rendang. The formulary and processing steps of Rendang Itiak of Malaysia (Bok 2019) with variations are discussed.

Cooking time: 1 h 15 min.
Ingredient composition.

Sl No	Ingredients	Quantity
1	Duck with skin-on, cut-up parts (half duck) or 7 duck legs	1.5 kg
For marinade		
1	Salt	2.5 g (1/2 tsp)
2	Turmeric powder	15 g (1 tbsp)
For rendang paste		
1	Dry red chilli	10
2	Garlic	2 cloves
3	Shallots	50 g
4	Chopped galangal	10 g (5 cm)
5	Fresh chopped turmeric in small pieces	10 g (thumb size)
6	Lower 1/3 portion finely chopped lemon grass	4 stalks
7	Chopped ginger	10 g
8	Cumin powder	5 g
9	Coriander powder	5 g
10	Water	200 ml

(continued)

Sl No	Ingredients	Quantity
Spices		
1	Rendered duck fat/coconut oil or vegetable cooking oil	60 ml (4 tbsp)
2	Cardamom	4 pods
3	Star anise	2
4	Cloves	5–6 buds
5	Cinnamon (broken in to small pieces)	5 cm
6	Kaffir lime leaves (tear in to halves)	6
7	Finely sliced turmeric leaf	1
8	Kerisik	50 g
For seasoning		
1	Coconut water	400 ml
2	Hot water	300 ml
3	Salt	3 g (1/2 tsp)
4	Palm sugar	30 g
5	Chicken powder/ fish sauce	15 g
6	Tamarind paste mixed with water	15 ml (1 tbsp)
7	Thick coconut milk	30 ml

Processing steps.

- Score the skin diagonally on the duck cut-up parts or legs. Marinate the duck pieces with the marinade containing salt and turmeric in a stainless steel or glass bowl by rubbing and set aside.
- For rendang paste, soak dry red chilli in boiling water for 5 min to soften. Halve, deseed, and chop chilli before adding with other ingredients in a food processor. Blend well to a smooth paste. Add water, mix again, and keep aside.
- Pan fry the marinated duck with skin side down in a frying pan over a medium high heat until the subcutaneous fat has almost rendered. Keep the fried duck on a plate and keep aside. The rendered duck fat is saved for frying the spices to impart more flavour.
- The rendered fat saved in the frying pan is decanted into a wok with sufficient quantity of coconut oil to fry cardamom, cinnamon, star anise, and cloves for a few seconds till the aroma evolves. Then add the rendang paste and sauté until the oil separates.
- Transfer the fried duck from the frying pan into a wok and add kaffir lime leaves, turmeric leaves, kerisik and mix so that duck gets evenly coated with the paste.
- Pour the coconut water and bring the mixture to a boil. Cover the wok, reduce the heat, and simmer for 15 min. After 15 min, add 400 ml hot water and simmer for further 45 min. With occasional stirring or till the required doneness.

- Add salt, palm sugar, and tamarind juice to taste followed by coconut milk and stir. Bring to simmer and remove from heat. Serve with coconut duck and pineapple salad.

12.2.6.3 Singapore Nyonya Duck Curry

Nyonya or Peranakan cuisine is of Peranakans, the descendants of original Chinese migrants in Penang, Malacca, Singapore, and Indonesia. A female Peranakan whose father is Chinese is known as nyonya/nonya. The cuisine is influenced mainly by Chinese, Malay, and Javanese. Serve with rice, bread or paratha/ porotta. The formulary and processing steps of this Singaporean duck dish of Kho (2019) with variations are described.

Cooking time: 50–60 min.
Ingredient composition.

Sl No	Ingredient	Quantity
1	Duck with skin-on (cut-up parts)	2 kg
2	Meat curry powder (spice blend) #	45 g (3 tbsp)
3	Turmeric powder	15 g (1 tbsp)
4	Coconut oil	15 ml (1 tbsp)
5	Rempah Kuning paste ##	250 g
6	Cinnamon	5 cm stick
7	Coriander powder	15 g (1 tbsp)
8	Cumin powder	5 g (1 tsp)
9	Finely chopped lower 1/3 portion lemon grass	4 stalks
10	Halved kaffir lime leaves	2
11	Coconut milk	50 ml
12	Water	250 ml (1 cup)
13	Coconut milk	50 ml
14	Cubed potato	400 g
15	Salt	5 g (1 tsp)

Processing steps.

- Marinate the duck cut-up parts with spice blend, turmeric, and a pinch of salt and keep aside in a glass or stainless steel bowl.
- To hot coconut oil in a wok, add Rempah Kuning paste, cinnamon stick, coriander powder, and cumin powder. Mix well and sauté for 5 min until the aroma of spices come. Add chopped lemon grass and Kaffir lime leaves and fry for 1 min. Add marinated duck pieces with stirring and fry for 5 min.
- Add water and coconut milk and bring to boil. Simmer for 1 h or till the duck is just cooked. Add potato cubes and simmer for further 30 min. Stir well and check salt level.

- Remove lemon grass and Kaffir lime leaves and skim off rendered fat on the top of the curry. Garnish with fried Kaffir lime leaves before serving.

Note:

- # For ingredient of spice blend: see notes of Kerala Duck Curry.
- ## For Rempah paste: see Malay Duck Curry

12.2.6.4 Thai Green Curry with Roasted Duck

Thai green curry uses the same ingredients as red curry but red chilli is replaced with fresh green chilli. The green curry is generally used for white meat such as chicken and duck. Because of the slight sweetness of the dish this is pleasing and popular to the Western palate. The original recipe is from a 1926 Cook Book (Hanuman et al. 2016).

Cooking time: 1 h 45 min.
Ingredient composition.

Sl. No	Ingredients	Quantity
1	Roasted duck breast or duck breast with skin-on	500 g
2	Green curry paste	250 g
3	Coconut cream	750 ml
4	Coconut milk	750 ml
5	Green chilli	100 g
6	Roasted and ground coriander	5 g (1 tsp)
7	Roasted and ground cumin	5 g (1 tsp)
8	Palm sugar	15 g (1 tbsp)
9	Fish sauce	75 g (5 tbsp)
10	Kaffir lime leaves, julienned	30 g (2 tbsp)
11	Thai basil leaves	30 g (2 tbsp)
Green curry paste		
1	Fresh bird's eye chilli	60 g
2	Salt	15 g
3	Thinly sliced galangal	25 g
4	Finely chopped lower 1/3 portion lemon grass	15 g
5	Chopped coriander roots	30 g
6	Kaffir lime zest	10 g
7	Chopped chilli leaves	15 g
8	Finely chopped shallots	45 g
9	Finely chopped Thai garlic	45 g
10	Fermented shrimp paste	15 g
11	Roasted and ground coriander seed	5 g
12	Roasted and ground cumin	5 g
13	Thai cardamom	4 pods

Processing steps.

- Either a restaurant prepared Thai style roasted duck or fresh duck breast with skin-on can be used for the dish. Season the duck breast with 5 g salt, 5 g pepper and sweet soy sauce. Roast the duck breast with skin side down in the roasting pan in a preheated oven at 220 °C till it is golden brown, about 20 min. Remove from the oven and keep aside.
- All the ingredients for the green curry paste are ground into a smooth paste in a food processor and keep aside.
- Cook the coconut cream in a hot wok till it cracks and the oil separates. Add the green curry paste to the coconut cream, stir and sauté by gradually adding coconut cream.
- Add a handful of julienned kaffir lime leaves, coriander, and cumin powder with stirring followed by the addition of coconut milk, palm sugar, and fermented fish sauce. Stir well and add sliced roasted duck.
- Add chopped green chilli, julienned kaffir lime leaves, Thai basil leaves, and coconut cream. Remove from flame and serve hot with steamed white rice.

12.2.6.5 Bebek Goreng Madura

Bebek goreng madura or Indonesian Fried Ducks-Madura Style is one of the two famous fried duck delicacies in Indonesia because of its crispiness, tenderness, and juiciness. The other one is bebek goreng Bangkalan. Madura Island is located in the northeast of Java, Indonesia and their recipe for bebek goreng is popular. The duck is usually cut into pieces, boiled or steamed and then deep-fried until crispy. Before it is fried, the pieces are generously coated in spices such as garlic, ginger, turmeric, galangal, or coriander. Fried duck pairs well with sambal pencit (green mango sauce) and poyah (yellow shredded toasted coconut). The formulary and processing steps of Nasution (2019) with variations are followed here.

Cooking time: 1 h 20 min.
Ingredient composition.

Sl No	Ingredients	Quantity
1	Duck legs (Maryland duck)	2
2	Calamansi (Philippine lime) juice	15 ml (1 tbsp)
3	Coconut water (water can be used)	250 ml (1 cup)
4	Water	50 ml
5	Chopped lemongrass	1 stalk
6	Kaffir lime leaves, without midrib	4 leaves
Spices to be ground for spice blend		
1	Shallots	6
2	Garlic	3 cloves
3	Toasted coriander seeds	15 g (1 tbsp)
4	Toasted white pepper	3 g

(continued)

Sl No	Ingredients	Quantity
5	Ginger 2 cm	5 g
6	Galangal	3 cm
7	Turmeric powder	4 g
8	Kaffir lime leaves, without midrib	3 leaves
9	Chopped lower white portion lemon grass	3 cm
10	Candlenut	3
11	Salt	To taste

Processing steps.

- In a stainless steel or glass bowl place the two duck legs and marinate with calamansi juice and salt for 30 min. Cover and keep chilled. This is to eliminate the odour of duck.
- After chilling take the marinated duck legs out from fridge. Rinse and pat dry with paper towel.
- In a stainless steel pressure cooker marinate the duck legs with ground spice blend (bumbu, in Indonesian) Sl No 1 to 11 and keep aside for 30 min in a fridge.
- Take the cooker with marinated duck out from fridge and add water and coconut water.
- Stew the duck legs in the pressure cooker for 20 min under pressure. Once the duck is well done, release the pressure.
- Deep fry the cooked duck legs in coconut oil or in an air fryer.
- Serve with Sambal pencit, Poyah, and cooked rice. The left over yellow cooking fluid of duck is used as sauce.

12.2.6.6 Vietnamese Roast Duck-Vit Quay

This dish, Vit quay is a Vietnamese version of the Peking duck, which is succulent, flavourful, and tender. Normally, it takes 3 to 4 days for the preparation of Peking duck. Vit quay can be also prepared in 12 to 24 h. Duck is usually roasted in a clay oven or in an open charcoal oven. In Vietnam, roast ducks are prepared in street rotisseries on skewers over burning charcoal which imparts a desirable smoky flavour and colour. In a Vietnamese home, the duck is served with pickled vegetables or a salad, several dipping sauces and either fragrant steamed rice or stir-fried rice made using the duck offal. Vit quay is served in one course.

Duck meat is considered ‘cool’ and is served during the hot summer with ginger fish sauce which is ‘warm’. Conversely, chicken which is ‘warm’ and pork which is ‘hot’ are eaten in the winter. In Vietnamese dishes pepper is generally used for replacing red chilli and fish sauce is a commonly used condiment. Many Vietnamese dishes are a combination of five tastes. Vietnamese roasted duck is very popular world over, especially in Europe and America as it is less hot and spicy (Helen 2017; Pohl and Lister 2020).

Cooking time: 12–24 h.
Ingredient composition.

Sl No	Ingredients	Quantity
1	Whole duck	1.5 kg
Mix 1 rub/cleaning paste		
1	Chopped ginger	50 g
2	Cooking wine/ mai que lo wine*/rice wine/vodka	15 ml
3	Salt	5 g
Mix 2 marinade		
1	Vietnamese five spices powder (finely ground mixture of star anise, cloves, fennel, pepper, and cinnamon)	5 g
2	Salt	5 g
3	Sugar	10 g
4	Chicken stock powder/chicken bouillon/ mushroom seasoning	5 g
5	Pepper powder	2 g
6	Grated/ powder garlic	5 g
7	Cooking wine/cider vinegar/mai que lo wine	5 ml
8	Oyster sauce/hoisin sauce	30 ml
9	Annatto/paprika powder	2 g
10	Honey/maltose	15 ml
11	Scallion (spring onion/green onion) (cut in to 5 cm long and crushed)	4
12	Sliced ginger	3 slices
13	Sliced shallots	4
Mix 3 scalding/poaching liquid		
1	Water	2 L
2	Soy sauce	30 ml
3	Cider vinegar	30 ml
4	Honey/maltose	30 ml
5	Broken star anise	3 pods
6	Cinnamon	2 sticks
Mix 4 glaze		
1	Cedar vinegar	30 ml
2	Maltose/honey	30 ml
3	Annatto powder	3 g
Mix 5 dipping sauce		
1	Juice from duck drippings, without fat on the top of dripping	75 ml
2	Soy sauce	30 ml
3	Sugar	10 ml
4	Chilli garlic paste	10 g
5	Lime juice	10 ml
To serve		
1	Julienned cucumber and carrot	50 g
2	Chopped scallion (green onion/spring onion)	30 g
3	Rice paper	3

Note: *Mai Que Lo wine is a cooking wine prepared by roasting 2 star anise, 3 cm long cinnamon, 3 cloves, and ½ tsp. dried citrus peel in a pan for 3 minutes till fragrant. Insert in a cloth bag and keep submerged in 300 ml vodka or rice wine for 1 month. This infusion can be used instead of cooking wine

Processing steps.

- Remove the neck and giblet from the whole duck and trim off preen gland and excess fat from the cloacal region. Wash the duck under cold running water, drain and pat dry well with paper towel. Pass the fingers carefully, both from the vent side and neck, between the skin and meat to separate them but without removing and tearing skin. Keep the duck aside for cleaning with the cleaning paste to remove the gamy/muddy odour.
- To prepare the rub or cleansing paste, coarsely grind ginger with salt and cooking wine in a food processor.
- Rub well the paste all over the body and inside the body cavity of the duck. Keep in refrigerator for 1 h. Then wash inside and outside with running cold water. Drain well and pat dry with kitchen paper towel.
- Mix thoroughly all the ingredients for marinade Sl No 1 to 10 in a glass bowl till the honey/maltose is dissolved.
- Two third of the marinade is rubbed inside the body cavity and stuffed with sliced ginger, shallots, and scallion. Rub the rest of the marinade liberally all over the duck, inside and outside the skin.
- With a bamboo skewer the skins of both sides at the opening of duck cavity starting from the cloacal end to the sternum is sewed by passing the skewer through the skin. Care is taken not to leave any gap between the skins to prevent dripping of meat fluid from inside. Hang the duck horizontally by passing the arm of a suspension tool kept in a baking dish to collect the drip or marinade. This helps in the uniform marinating and drying of the skin. Even without the tool also the duck can be directly kept in the dish. Keep the uncovered duck with the dish in the refrigerator for at least 4 h or until the evening. This allows the skin to get dried faster so that it will be crispy on roasting. If any of the marinade drips into the tray, baste it back over the duck.
- In order to prepare poaching/scalding liquid, boil two litres of water in a wok or sauce pan. Add soy sauce, cider vinegar, honey, star anise and cinnamon. Simmer for two minutes till the flavour of spices evolve and water turns light brown.
- Take the marinated duck from the fridge. Keeping the duck on the suspension tool, ladle the hot poaching liquid all over the duck for 20 to 30 seconds for the skin to shrink. Or else, the duck can be bathed in the hot poaching liquid for about five minutes, as well.
- Once the duck is cool to touch, either the wings are cut or they are bend over and tucked under the body (trussing). This is to prevent obstruction to the rotation of the skewer in rotisserie. Pat dry the duck with paper towel and keep

in refrigerator for at least 5 h or overnight for drying the skin. This is essential to crisp the skin-on roasting.

- Prepare the glaze by mixing annatto powder, cedar vinegar, and honey in a glass bowl. Baste the glaze all over the duck with a brush. Remove the duck from the suspension tool and pass a metal skewer through the neck to the cloacal region.
- The duck on skewer can be roasted in a convection oven with horizontal rotisserie or on a rack or in a dish. Roast the glazed duck in the preheated oven at 180 °C for 25 min per 450 g, that is for about 1 h and 30 min for this duck. Baste every 20 min with the glaze and juices that drip into the pan. Wipe the excess melted fat from the skin surface with paper towel. Roast to an internal temperature of 75 °C and check with meat thermometer.
- If the glazed duck is kept on a rack or in a dish for roasting, keep the breast side down for first 45 min. Thereafter, turn the duck over with breast side up and continue cooking for further 45 min to an internal core temperature of 75 °C. Baste every 20 min with glaze.
- If more colour and crispness are required, increase the oven temperature to 200 °C during the last 5 min. The duck is done if the juices run clear when bird is pierced with a skewer or tooth pick into thigh. Remove from the oven and cover with aluminium foil during rest to cool for 20 min before slicing.
- Prepare the dipping sauce by mixing the juice from duck drippings in the oven tray/dish, soy sauce, chilli garlic paste, sugar, and lime juice.
- Spread a little dipping sauce on rice paper and keep a slice of roasted duck dipped in the sauce, julienned carrot, cucumber and chopped scallion, roll, and serve.

12.2.7 Taiwanese/ Chinese Products

The duck industry is an ancient livestock industry in China. China is the largest duck producer in the world. Due to the improvement in the processing technology, duck meat RTE and RTC are increasing. Many varieties of duck meat products in China are popular because they are a part of the Chinese people's daily diet. The preference for seasoning and cooking techniques of Chinese provinces depend on the differences in their historical background and ethnic groups. The dishes that originated from various parts of China, especially Fujian province have influenced heavily Taiwanese cuisine and other East Asian countries.

The Taiwanese people generally love duck meat due to its special flavour and taste. Due to the improvement of modern farming and processing technology, it provides good quality meat and a large variety of duck products such as roasted duck, duck steak, and a variety of novel and delicious duck processed products. The duck skin contains higher fat content than other poultry. Through high temperature baking or frying, the fat of the duck skin renders and then produce the flavour of the oil accompanied by the crispness of the duck skin, combined with the protein of duck meat. This special procedure of cooking, make the duck meat to generate a special taste. The processing technology of duck meat, not only increases the crispness of duck skin, but also improves the tenderness and flavour of duck meat.

The recipes and processing steps of Peking ducks, Cantonese roast duck, Roasted duck, Crispy fried duck, Eight treasure duck, Eight treasure duck: Alternate procedure, Tea smoked duck breast, Tea smoked duck, Smoked duck steak, Nanjing salted duck, Salted water duck, Pressed salted duck and ginger duck are discussed in the following.

12.2.7.1 Peking Duck

Peking duck is also known as Beijing roast duck and one of the oldest iconic dishes in China originally prepared for the imperial class. A dish consisting of thin pieces of tender roasted duck and crispy skin wrapped in small pancakes with julienned spring onion, cucumber, and Hoisin sauce or sweet bean sauce. This is one of the 15 most popular dishes in China. This dish is reserved for special occasions like Chinese New Year Day, Christmas, festivals, and other holidays.

Roast duck is a popular Chinese dish that comes in two forms: a northern Beijing style and a southern Cantonese style (Zhou et al. 2014). The roast duck is characterized by crispy skin, attractive taste and dark red colour, and tender meat shining with oil (Chen et al. 2009). Both are cooked in different types of ovens and vary in how they are processed, sliced, and consumed. Recently, hybrids of both have come with Peking duck more similar in style to Cantonese duck. Peking duck is traditionally roasted in a closed oven but in these days, both open and closed ovens are being used. High temperatures in the oven promotes the formation of desirable flavour compounds, generated through the lipid oxidation process, Maillard reaction and interactions between these reaction products which greatly contribute to its overall acceptance (Estévez et al. 2003).

A thin and crisp skin characterizes this product. Peking duck is served in three courses as the crispy skin, stir-fried meat, and a broth made from the carcass. Traditionally, the duck is sliced in front of the diners by the cook. The carving technique of the duck is unique with about 108 cuts to separate it for three different eating styles.

Free-range White Pekin ducks are overfed four times a day from 45th day for 15–20 days till they attain at least 3 kg, usually by 65 days. Such young fatty ducks are the best for the preparation of Peking duck and not skinny ducks. In the traditional method it takes 4 to 5 days for the processing of Peking duck and involves the following unique steps. They are the use of two tools, yadu and yacheng, filling duck cavity with hot water and inflating with air to separate skin and meat. Yadu is for plugging the vent to prevent loss of duck fluid from the cavity and the yacheng as a spacer at the chest opening. Similarly, duck is eviscerated through an incision on any one side of the chest under the wing. In the general dressing practice, it is through the incision on the ventral side near the vent. The other step to separate skin from the underlying meat of duck to keep the skin crisp on roasting is by pumping in air with the help of a hand pump or air compressor. Skin is not torn or removed. If with head-on, it is easy to hang in open oven and during inflation with air.

The following recipe for Peking duck is of Wong (2019) with alterations is a simplified method skipping the above steps but without compromise to other procedures, flavour, colour, and crispness of the duck.

Cooking time: ~2 days.
Ingredient composition.

Sl No	Ingredients	Quantity
1	Chilled or frozen 1 whole duck	2.6 kg
For marinade		
1	Hoisin sauce	30 ml
2	Soy sauce	45 ml
3	Salt	15 g
4	Sugar	15 g
5	Ground ginger	15 g
6	Grated garlic	30 g
7	White pepper powder	10 g
8	Chinese five spice powder	10 g
9	Bundled spring onion/green onion/scallion	2
Mix 2 for scalding liquid		
1	Water	2 L
2	Dried citrus/mandarin orange peel	2
3	Licorice	5 g
4	Broken star anise	5
5	Cinnamon	1 stick
6	Rice vinegar/dry sherry	30 ml
Mix 3 for glaze		
1	Sugar	50 g
2	Honey/maltose	50 g
3	Chinese red vinegar	180 ml
4	Shaoxing wine (Chinese rice wine)	30 ml
5	Water	350 ml
Mix 4 for dipping sauce		
1	Chopped garlic	5 g
2	Ground ginger	3 g
3	Sweet bean paste	80 g
4	Sugar	5 g
5	Water	45 ml
6	Hoisin sauce/plum jam	100 g
To garnish and serve		
1	Chinese pancakes/tortilla	6
2	Finely sliced spring onion/green onion/scallion	5 bulbs
3	Julienned cucumber	50 g

(continued)

Sl No	Ingredients	Quantity
4	Julienned carrot	50 g
5	Chopped parsley to garnish	15 g
6	Orange, sliced in rounds, to garnish	1

Processing steps.

- If the duck is frozen, thaw in refrigerator on the previous night. Remove excess fat from the vent region and any offal from the cavity. Wash the thawed/chilled/fresh duck under cold running water, drain and pat dry well with paper towel inside the cavity and outside. Pass the fingers carefully between the skin and meat to separate them, both from the vent side and neck, without removing or tearing skin. Massage skin gently. This simpler method is more ideal than inflating with air.
- Mix all the ingredients for the marinade in Hoisin sauce in a glass or stainless steel bowl, except spring onion.
- The marinade is rubbed inside the body cavity and stuff with bundled spring onion. Instead of bundling, spring onion can be chopped, as well. One or two ginger slices can also be placed in the cavity, if required. Rub the marinade liberally all over the duck, in the space between the skin and meat and outside.
- With a bamboo skewer the skins of both sides at the opening of duck cavity starting from the cloacal end to the sternum are sewed by passing the skewer through the skin. No care is taken to leave any gap between the skins to prevent dripping of meat fluid from inside.
- Refrigerate the marinated duck for 24 h with breast down and uncovered. It can be hung by two hooks passing under the wings also.
- To prepare scalding liquid, boil one and a half litre water in a wok or sauce pan. Add licorice, star anise, cinnamon, citrus peel, and rice vinegar. Simmer for five to ten min till flavour of spices evolve and water turns light brown.
- Meanwhile, prepare the glaze in a stainless steel bowl by mixing sugar, honey, vinegar, and wine. As maltose is very thick, first dissolve well in half litre boiling water in another vessel with constant stirring. Mix the maltose liquid with the previous solution.
- Take out the marinated duck from the refrigerator and scald by immersing in the boiling scalding liquid for 15 to 20 s and by ladling the liquid on the back and breast.
- Ladle the glaze all over the duck and when cold to touch, hang the duck in refrigerator for 12 to 24 h to dry the skin. Instead, hang the duck in a cool dry place for 5 h and use a fan to blow air on duck till the skin becomes dry and springy. Skin should be dry while touching.
- Preheat the oven to 240 °C. Place the duck on the rack with breast side down. Add a little water in the drip tray at the bottom to prevent the rendered fat from splattering. Then roast for 15 min. Reduce the heat to 180 °C and continue to roast for 1 h 15 min. Baste the duck with glaze every 20 min. After 45 min of roasting, turn the duck with breast side up. Roast to an internal temperature of 75 °C at the core of the thigh and check with a meat thermometer.

- Once it is well done and the skin is golden brown and crisp, switch off the oven and keep the duck to rest for 15 min before carving.
- Prepare the dipping sauce by mixing well all the ingredients in a serving bowl.
- Carve the duck hot just before serving only. With a cleaver make a cut along the sternum, then on the side of the duck and slice sideways diagonally. Place the thin slices on a warm plate and garnish with orange slices and parsley. Serve with pancake, julienned carrot, cucumber, and sliced spring onion.

12.2.7.2 Cantonese Roast Duck

Although Cantonese roast duck (CRD) and Peking duck (Beijing roast duck) look alike and equally delicious, two different styles of processing, cooking, slicing, and consumption exist. The former is a South Chinese and the latter a North Chinese style dish. This is common in Hong Kong also. Cantonese roast duck is highly popular for its taste as it is treated with aniseed, prickly ash peel, cassia, sweet grass, and licorice.

In CRD, skin is air dried before roasting as in Peking duck but not as much as the latter by hanging for 2 or 3 days. Similarly, duck is not inflated. A wet marinade is filled in the cavity before it is skewered in CRD which imparts juiciness and savoury to meat. The wet marinade is drained out before cutting. Only an expert cook can carve hot Peking duck in the desired style which is not required in the case of CRD. The comparatively easier process of CRD makes it more household friendly, popular, and cheaper than Peking duck without much compromise in crispness of skin, flavour, and juiciness. This roast duck is served with stir-fried vegetables and cooked rise. The recipe of Parkinson (2019) with alterations is furnished in the following.

Cooking time and to marinade: 1 day.
Ingredient composition.

Sl No	Ingredients	Quantity
1	One whole duck with head	2 kg
For preparation of duck with rub		
1	Salt	5 g
2	Roasted Sichuan peppercorn	1.25 g (1 pinch)
For marinade		
1	Sesame oil	5 ml
2	Finely chopped spring onion	15 g
3	Chopped or grated ginger	5 g
4	Minced/grated garlic	15 g
5	Sugar	15 g
6	Salt	2.5 g (1/2 tsp)
7	Shaoxing wine/dry sherry/sake	30 ml
8	Hoisin sauce	15 g
9	Dark soy sauce	10 g
10	Five spice powder	10 g
11	Sun dried tangerine/orange peel	1 piece
For glaze		

(continued)

Sl No	Ingredients	Quantity
1	Honey/maltose	60 ml
2	Rice vinegar	15 ml
3	Soy sauce	30 ml
4	Warm water	250 ml

Processing steps.

- Remove any excess fat from the cavity and tips of wings. Wash the duck thoroughly in cold water. Pat dry inside and outside of the duck with paper towel.
- For the preparation of the duck, rub salt and roasted Sichuan peppercorn on the duck and place in fridge for 2 h.
- To prepare the marinade, heat sesame oil in a hot frying pan and sauté ginger, spring onion, and garlic until the raw flavour subsides. Add sugar, salt, wine, sauces, five spice powder, and tangerine peel with constant stirring. Allow to boil and then reduce heat and simmer for 2–3 min. Stop heating and keep aside to cool.
- Remove duck from the fridge and tie its neck tightly. Hold the duck upside down by holding its legs and pour the cooled marinade into the cavity. Sew the skins of the cavity together tightly with a metal or bamboo skewer to prevent leak of marinade.
- After filling with the marinade and sewing, scald the duck in boiling water in a wok by holding upside down by its legs. Ladle boiling water all over the skin for 2 min until the skin is contracted. This is to make the skin crispier on roasting.
- Meanwhile, prepare the glaze in a glass bowl by mixing honey, vinegar, and sauce in warm water. If maltose is used, boil water in a saucepan and melt maltose with constant stirring followed by the addition of other ingredients for glaze and cool.
- Take the marinated duck from the fridge, pat dry the skin with paper towel, and brush the glaze on the skin. Repeat brushing with the glaze 3–4 times as it gets dried.
- In a preheated oven at 200 °C, hang the duck head down from the top rack. Place a tray with hot water at the bottom to collect the duck fluid.
- Roast the duck for 25 min. Baste the glaze all over the skin. Reduce the oven temperature to 180 °C and roast for further 30 min. Baste again with the glaze or the drip. Roasted till the internal temperature at the thigh reaches 80 °C measured with a meat thermometer.
- Remove from the oven and rest for 10 min. Open the cavity and drain the meat fluid into a tray. Keep duck on a cutting board and cut into the desirable serving size.
- Transfer the duck fluid in the tray to a saucepan. Skim the fat off and simmer for 2 min and add corn flour with stirring till it gets thickened. Season the sauce with salt and pepper, to taste.

12.2.7.3 Roasted Duck

Ingredient composition.

Sl No	Items	Quantity
Marinating ingredients		
1	One whole Pekin duck with head	~ 2 kg
2	Table salt	15 g
3	Brown sugar	30 g
4	Cinnamon powder	2 g
5	Citrus peel powder	2 g
6	Sand ginger/ galangal powder	1 g
7	Crushed garlic	6 g
8	Crushed red onion	5 g
9	Sand ginger	3 g
10	Star aniseed powder	2 g
11	Sliced ginger	6 g
12	Red wine	20 ml
Smearing ingredients (sharing)		
13	Maltose	100 g
14	Vinegar	110 ml
15	Water	100 ml

Processing steps.

- Pekin duck is used as the raw material for this preparation. The rich subcutaneous fat and the thick skin of Pekin duck impart crispness and flavour to Roasted duck.
- Carefully remove the organs of abdominal cavity, wings and feet (Fig. 12.1).
- Separate the duck skin and meat using a traditional blow pipe.
- The blow pipe is penetrated through the opening of the neck. The backflow of air is blocked by closing the entry point by hand and air is filled until the skin and meat of the duck are completely separated (Fig. 12.2).
- All the marinating ingredients (mixed and kept ready) are stuffed into the abdominal cavity and the cavity is sealed completely using a long metal tail needle (Fig. 12.3).
- The whole duck is immersed in hot water (80 °C) for 40 sec, and then air dried in windy area or under shady place for 3 h.
- After air-drying, the whole duck skin is smeared with the smearing ingredients completely three times.
- In an electric oven, the smeared whole duck is cooked at two-stage temperature: stage 1 at 130 °C for 60 min and stage 2 at 170 °C for 30 min (Fig. 12.4).

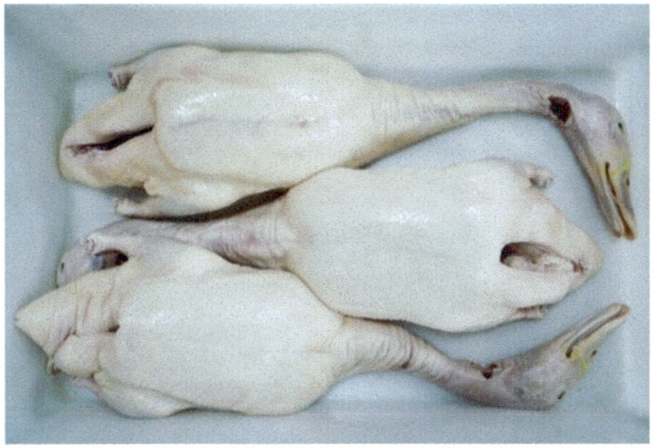

Fig. 12.1 Pekin ducks after removing wings and feet

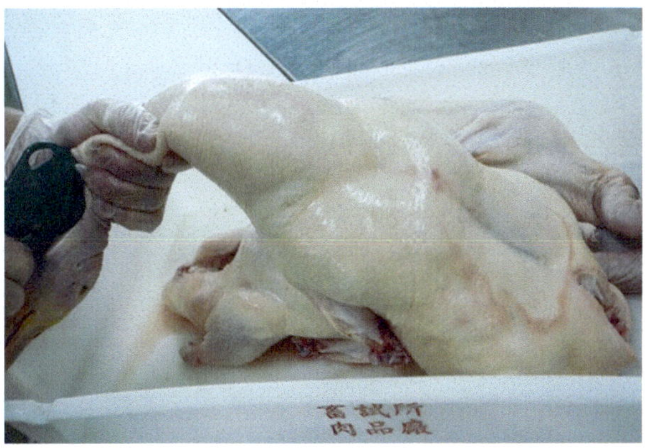

Fig. 12.2 Separation of skin from the meat using a blow pipe

Fig. 12.3 The ducks stuffed with marinating ingredients and sealed using a needle

Fig. 12.4 The final product of Roasted Duck

12.2.7.4 Crispy Fried Duck

Ingredient composition:

Sl No	Items	Quantity
1	One whole mule duck with head	2 to 2.5 kg
2	Table salt	20 g
3	Star anise powder	12 g
4	Ginger slices	30 g
5	Chinese red pepper powder	12 g
6	Green onion	25 g
7	Rice wine	50 ml

Processing steps.

- The wings and feet are removed from mule ducks and the surface of the carcass and the abdominal cavity are washed completely.
- Mix all the ingredients well (except rice wine), and spread it all over the duck skin and inside the abdominal cavity. Keep the duck in a metal utensil and keep it in the refrigerator for 8–12 h.
- Wash all the ingredients from the duck body and then smear inside and outside surface of the duck with rice wine and leave it for 30 min.
- Prepare the steamer with 1/3 full water and place the duck on the top and then steam for 90 min (Fig. 12.5).
- Heat the oil first to 140 °C and fry the whole duck in hot oil on both sides for 7–8 min at 160 °C (Fig. 12.6).

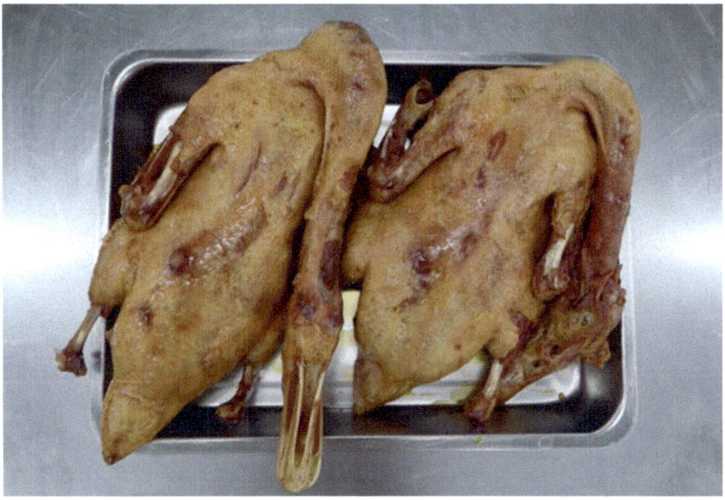

Fig. 12.5 Whole duck after steaming for 90 min

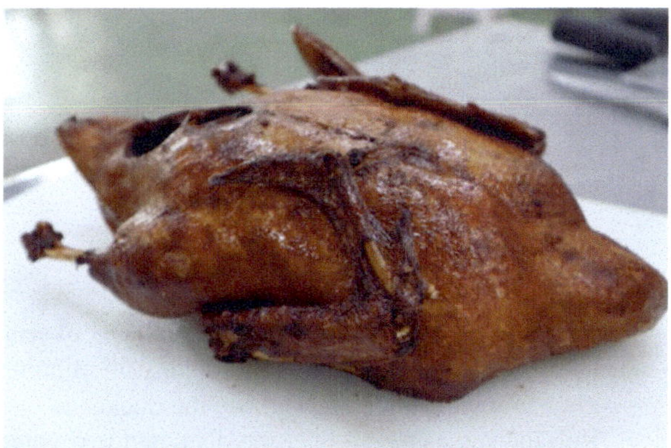

Fig. 12.6 The final product of Crispy Fried Duck

12.2.7.5 Eight Treasure Duck

Eight treasure duck is a Chinese traditional Shanghaies origin served on the eve of Chinese New Year. The duck is first fried to render out the fat, washed, pat dried, marinated, stuffed/filled with eight different ingredients (hence the name 'treasure'), steamed or braised till soft to serve. Each treasure ingredient signifies a particular attribute with medicinal property: 1) gingko nut – wealth, 2) lotus seed – fertility, 3) Chinese mushroom – longevity, 4) ginger – strength, 5) barley – abundance, 6) pork – blessings, 7) chestnut – unity, and 8) mandarin orange peel – wealth. This dish is gradually vanishing from the younger generation due to its complexity in the preparation. The recipe followed here is of Winston (2012) and Leng (2018) with variations.

Ingredient composition.

Sl No	Ingredients	Quantity
Preparation of duck		
1	Whole mule duck	2.5–3 kg
2	Salt	15 g
3	Soy sauce	30 ml
Preparation of 8 treasure for stuffing		
1	De-shelled gingko nut. Soak in hot water for easy peeling	30 g
2	Dried lotus seeds. Soaked 24 h in water to soften and then drained	30 g
3	Dried Chinese mushroom (soaked 24 h and drained)	5 pieces
4	Washed barley	30 g
5	Sliced ginger	4 slices
6	Lean and sliced pork	20 g
7	Salt	2.5 g
8	Chestnut (remove extra skin)	30 g
9	Broken mandarin orange peel	2 pieces
10	Dried and pitted red dates (optional)	5
11	Dried shrimp, rinsed, and drained	45 g
12	Oyster sauce	45 g
13	Shaoxing wine/any rice wine	5 ml
14	Hot water	60 ml
For sauce		
1	Corn starch	15 g
2	Water	45 ml
3	Salt and pepper	To taste

Processing steps.

- Rub salt all over the duck. Pluck remnants of feathers from follicles with the help of a tweezers. Wash thoroughly in running cold water. Pat dry with kitchen towel inside and outside the duck.
- Rub the soy sauce all over the duck and marinate in the fridge overnight.
- In order to render the fat from the duck, sear it in a dry hot wok flipping to all sides till the fat completely render out and duck turns lightly brown.
- Pour boiling water over the duck to remove excess rendered fat on the skin. Pat dry and keep aside for stuffing. Save the rendered duck fat for sautéing the 8 treasures.
- Pit the soaked lotus seeds and remove the green stem to avoid bitterness. Similarly, remove the bitter core from the gingko nut with the help of a toothpick.
- In a wok, heat 15 ml rendered duck fat and sauté sliced lean pork and ginger slices with stirring. The rest of the ingredients for the preparation of the 8 treasures for stuffing, except hot water, are also added with pork and ginger in the wok with stirring. Sautéed till the aroma comes.

- Stuff the cavity of the duck with the treasures and hot water. Close the cavity by sewing with a bamboo skewer or thread. Keep the stuffed duck in a ceramic dish and cover with aluminium foil.
- Keep the dish with duck on a metal platform over water in another larger and deeper wok. Cover the bigger wok and heat. Steam the duck under low heat for 6 h. Check water level in the wok for steam production.
- Once the duck is done, transfer any fluid in the duck cavity to the duck fluid in the ceramic dish. This is placed under low heat. Add corn starch to the fluid to prepare the sauce. When the fluid starts bubbling and thickens, season with salt and pepper to taste. Keep the duck covered with foil until the opening of the duck cavity at the time of serving. Drizzle the sauce over duck on opening, while serving with glutinous steamed rice. Garnish the duck with chopped spring onion or broccoli florets.

12.2.7.6 Eight Treasure Duck: Alternate Procedure

Ingredient composition:

Sl No	Ingredients	Quantity
1	One whole mule duck	2 to 2.5 kg
2	Rice	70 g
3	Ginkgo	4 g
4	Chestnut	4 g
5	Lotus seed	4 g
6	Pork ham	5 g
7	Ground garlic	3 g
8	Shrimp	5 g
9	Pea kernel	5 g
10	Soy sauce	30 g
11	Sesame oil	5 ml
12	Shaoxing wine	5 ml

Processing steps.

- Using the mule duck as the raw material, the surface of the duck is washed and then the two-sectioned wing and the feet are removed.
- Sharp knives are used to gradually remove the bones of the whole duck from the opening of the ducktail and the fat and lymph glands are also removed from the meat inside (Fig. 12.7).
- The inside and outside surface of the body are cleaned. Coarsely broken ginger (1%), spring onion (1.5%), and Japanese sake (1.2%) are mixed well, and then smeared on the inside surface of the abdominal cavity completely, while the outside of the duck body is smeared with salt (1.5%) and kept in the refrigerator for 1 h.

Fig. 12.7 Whole mule duck with removed bones

- Ducks are washed in tap water, then dipped into 80 °C hot water for about 30–40 s and then placed in ice water to cool down the temperature of the carcass for 10 min.
- Mix all ingredients well by hand (with the addition of soy sauce, sesame oil, and Shaoxing wine), and steamed for 1 h. After cooling, the ingredients mixed with soy sauce, sesame oil, and Shaoxing wine are filled in the abdominal cavity of the ducks and the cavity is closed with a long metal needle (Fig. 12.8).
- The ducks are first cooked in steam for about 1 h, then in oven at 200 °C for about 30 min (Fig. 12.9) and finally take out the product and cut it out.

12.2.7.7 Tea Smoked Duck Breast

Tea smoked duck in Chinese is known as Zhangcha duck. It is a classic Chinese dish of Szechuan cuisine prepared by hot smoking a marinated whole duck or breasts over tea leaves and other spices. The preparation is time consuming with three different steps as in Peking duck. Traditionally, whole duck is used for tea smoked duck. But for convenience in cooking at home and restaurants, breasts of fatty Pekin and mule ducks are used. The ducks can be smoked on a grill in a covered wok over gas stove rather than a smoke house or electric smoke generator. For smoke generation wood chips of pear, apple, maple, cherry, pecan, and hickory are preferable due to their sweet flavour of the smoke. Breasts with skin-on will keep breast moist and juicy during smoking. Ducks during smoking shall be cooked to an internal temperature of 73 °C to 77 °C, for food safety. However, general practice is to cook to medium rare doneness at 57 °C.

This is eaten more at banquets and festivals than in daily meal. Tea smoked duck is good in sandwiches, on salads, with stir-fried vegetables, steamed rice, and even as

Fig. 12.8 Duck after stuffing all ingredients inside the abdominal cavity

Fig. 12.9 Final product of Eight Treasure Duck

a snack. This goes well with sweet, sour, and spicy flavour of Southeast Asian foods. The recipe of Shaw (2017) with alterations is given in the following.

Cooking time: 16 h, including curing.
Ingredient composition.

Sl No	Ingredients	Quantity
1	Fresh/frozen mule duck breasts, skin-on, about 450 g each	4
For cure		
1	Salt	30 g
2	Sichuan peppercorns	15 g

(continued)

Sl No	Ingredients	Quantity
3	Black pepper	10 g
4	Sodium nitrite (optional), to give pink colour on cooking	1.25 g
5	Shaoxing wine or dry sherry	30 ml
Smoking ingredients		
1	Rice grain	125 g
2	Loose tea leaf	125 g
3	Brown sugar	125 g
4	Star anise	1 pod
5	Orange or tangerine peel, dried	5 g
6	Pear/cherry/maple wood chips (optional)	150 g
To finish and cook vegetables		
1	Sliced mushroom	240 g
2	Coarsely chopped Bok choy (Chinese cabbage)	240 g
3	Fresh minced ginger	15 g
4	Sugar	15 g
5	Soy sauce	15 g
6	Duck/chicken stock	80 ml
7	Fresh and hot thinly sliced red chilli	2–4

Processing steps.

- If the duck breasts are frozen, thaw in the refrigerator. Trim off the excess fat and skin. Score the skin diagonally with a sharp knife without cutting through meat to create a crisscross pattern in order to enable rendering of fat during cooking.
- Pat dry with paper towel. Coat the Shaoxing wine all over the duck breasts. Grind all the other ingredients for curing and coat the spice mixture on the meat side only.
- Put each breast in a Ziploc cover and keep chilled in refrigerator to cure for at least 4 h or preferably 12 h or overnight.
- After curing, rinse off cure and pat dry with paper towel. Keep the duck breasts with skin side up on a cooling rack to dry the skin for 2–3 h at room temperature.
- After bloating moisture from duck breasts, place them with skin side down in a cold unoiled cast iron skillet. Place the skillet over high flame and as the duck fat renders with sizzling sound, reduce flame to medium. Cook until skin is medium brown, usually by 8 min.
- Transfer the breasts to a plate and pour off the rendered fat from skillet completely. Save it for frying the skin side of breasts after smoking and stir frying vegetables.
- Keep the breasts again in the skillet with skin side up. Cover the skillet and cook for 3 to 5 min over medium heat. Flip them over with skin side down and cook for 1–2 min more over high flame. Total cooking time would come to approximately 15 min. With a meat thermometer check the internal temperature of 57 °C for medium rare doneness or 75 °C for well doneness. Leave the breasts on a cutting board to rest for 10 min before smoking.

- To smoke the duck breasts, put a double layer of aluminium foil in the bottom of an unoiled wok. About 5 cm aluminium foil shall extend beyond the round rim of the wok. This is to seal the wok after closure with lid.
- Mix all the ingredients for smoking and spread it over the foil in a thin layer in the wok. Place the duck with skin side down on a rack in the wok. Cover and seal the lid with the overhanging foil.
- Place the wok over high flame for 3–5 minutes till the tea mixture start smoking with snapping, crackling, and popping noise. The smoke will be initially white, then light yellow and then dark yellow. As soon as the smoke turns dark yellow, lower the heat to medium and start timing to smoke for 20 to 30 min depending on the size of the breast.
- Take the breast out of the wok and cool. It can be kept chilled for a week or can be frozen or finished by crisping the skin as follows.
- To finish the duck breasts by crisping the skin, heat the rendered duck fat already saved in the frying pan over medium high heat. Then place duck breasts with skin side down until skin is crisp. Do not cook meat side again as it is already cooked. Remove the breasts from the pan, slice and keep aside.
- The vegetables are cooked in the rendered duck fat already saved in the frying pan used to crisp the duck breast skin. Add mushrooms and toss to coat with the rendered duck fat and keep undisturbed for 2 to 4 min. Add the bok choy, ginger and stir-fry for a minute. Add sugar, soy sauce, and duck stock and toss to combine with vegetables. Boil for about 4 min till the liquid thickens. Remove the pan from the stove and add sliced duck breast and red chilli. Toss to combine and serve hot.

12.2.7.8 Tea Smoked Duck

Ingredient composition.

Sl No	Items	Quantity
1	One whole mule duck	1 kg
2	Table salt	30 g
3	Brown sugar	22 g
4	Sodium nitrite	150 mg
5	Cinnamon powder	2 g
6	Chinese red pepper	6 g
7	White pepper	2 g
8	Chicken soup powder	2 g
9	Cumin powder	1 g
10	Anise powder (0.2%).	2 g

Processing steps.

- Choose a well fleshed whole mule duck.
- Take the table salt and Chinese red pepper in a pot and fry them until the flavour of Chinese red pepper evolves.

Fig. 12.10 Ducks smeared with mixed ingredients

- After cooling the fried mixed ingredients, mix it with the other seasonings. Smear evenly over the outside and inside surface of the duck and store in refrigerator at 5–10 °C for 2 days (Fig. 12.10).
- Boiling water is poured into the abdominal cavity to wash away residual blood and making the skin bright.
- Ducks are hanged on the metal hook, and then take the black tea as smoking material, heat the black tea at 60 °C for 30 to 40 min (Fig. 12.11).
- Finally, ducks are steam heated for 60 min and fried at 180 °C for a few min until the skin becomes golden colour (Fig. 12.12).

12.2.7.9 Smoked Duck Steak

Ingredient composition:

Sl No	Items	Quantity
1	Mule duck breast with skin	1 kg
2	Table salt	14 g
3	Brown sugar	12 g
4	Polyphosphate (STPP)	3 g
5	Sodium nitrite	150 mg
6	Pepper powder	3 g
7	Isolated soy protein	15 g
8	Cinnamon powder	1 g
9	Ice water	150 ml

Processing steps.

- Frozen mule duck breast is used as raw material. The frozen duck breast is thawed at 4 °C overnight, and then excess duck skin and fat are trimmed off.

Fig. 12.11 Smoked ducks using black tea smoke

Fig. 12.12 Final finished product of Tea Smoked Duck

- The completely thawed duck breast is placed in a massage machine in the chilling room.
- All the ingredients and ice water are added into a massage machine. The massage machine should be run for 10 min on 10 min off-cycle. This mode of operation of 10 min massaging and 10 min rest cycle should be continued for 3 h and then the duck breast is marinated overnight (Fig. 12.13).
- Using stainless steel hooks, the breast meat is hung on the iron frame in the smoke house.
- The operation mode of the smoke house is, drying for 60 min at 50 °C followed by smoking for 40 min at 60 °C and finally cooking for 30 min at 80 °C or till the core temperature of 72 °C is reached (Fig. 12.14).

Fig. 12.13 The duck breast kept for curing overnight

Fig. 12.14 The original flavour duck breast steak (up); the black pepper flavour duck breast steak (down)

- The smoked duck steak is then cooled down to room temperature (Fig. 12.15), packaged, and stored at −18 °C.

12.2.7.10 Nanjing Salted Duck

Nanjing cuisine is a part of Huaiyang cuisine which is one of the four culinary heritages of China. The others are Cantonese, Shandong, and Sichuan cuisines. In Nanjing area, which is closer to Shanghai, people consume more duck than chicken. The salted duck is available in all restaurants in vacuum-packed pouches and gift boxed as souvenirs (Judy 2015).

Fig. 12.15 The final product of Smoked Duck Steak

Total processing time: 5 h.
Ingredient composition.

Sl No	Ingredients	Quantity
1	Whole duck	2 kg
2	Salt	60 g
3	Sichuan peppercorn	38 g (2½ tbsp)
4	Star anise	4 pods
5	Cumin seed	10 g
6	Bay leaves	3
7	Cinnamon	5 cm stick
8	Black pepper	10 g
9	Ginger	5 slices
10	Chopped scallions (spring onion)	3
11	Shaoxing wine	125 ml
12	Sesame oil	10 g

Processing steps.

- Wash the duck thoroughly in cold water. Set aside in a colander for draining water. Pat dry with paper towel.
- Pan fry salt and Sichuan peppercorn in medium heat with constant stirring for 5–8 min or till the salt turns slightly brown.
- Rub well the salt and Sichuan peppercorn all over the duck and inside. Use at least 1/3 of the marinade in the cavity. Marinate and chill the duck uncovered for 3 h.
- Meanwhile, prepare a spice pouch by tying star anise, cumin, bay leaf, cinnamon, and black pepper in a square cheese cloth and keep aside.
- Heat a wok big enough to contain the duck and fill it with enough water to submerge the duck. Add the spice pouch, ginger, and scallion and bring the water to boil. Add wine and simmer over low heat for some more time.
- Transfer the marinated duck from fridge into the simmering water in the wok. Bring water to boil. Lift the duck and empty the water from the cavity. Lower

the duck back into water and boil again. This is to ensure that no cold water is trapped inside the cavity of the duck.

- Cover the wok and simmer for 15 min without roll boiling. After 15 min turn off the stove and set aside the duck for 40–50 min. Transfer the duck to a cutting board.
- Lightly brush sesame oil over the duck and allow cooling before cutting and serving.

12.2.7.11 Salted Water Duck

Ingredient composition.

Sl No	Items	Quantity
1	Whole mule duck	2 kg
Marinating ingredients:		
2	Table salt	70 g
3	Brown sugar	36 g
4	Chinese red pepper	6 g
5	Garlic powder	2 g
6	White pepper powder	6 g
Cooking ingredients		
7	Water	3 L
8	Table salt	36 g
9	Cinnamon stick	6 g
10	Lemon leaf	2nos
11	Star anise	2 g
12	Shallot	44nos
13	Shaoxing wine	50 ml

Processing steps.

- Use one mule duck weighing about 2 kg as raw material. Remove the internal organs including the kidneys completely.
- The marinating ingredients listed in the table are mixed well and smeared over the outer surface of whole duck carcass and inside the abdominal cavity. The marinated duck is kept at 4 °C overnight (Fig. 12.16).
- Rinse the internal and external surface of the duck using 80 °C hot water to remove the marinating ingredients and blood.
- In a mud pot add all cooking ingredients and boil with the duck immersing in it. It is cooked in medium fire for about 60 min (Fig. 12.17).
- Take out the cooked duck, cool and cut into slices for consumption (Fig. 12.18).

12.2.7.12 Pressed Salted Duck

Ingredient composition.

Sl No	Items	Quantity

(continued)

Fig. 12.16 Mule duck and Shaoxing wine

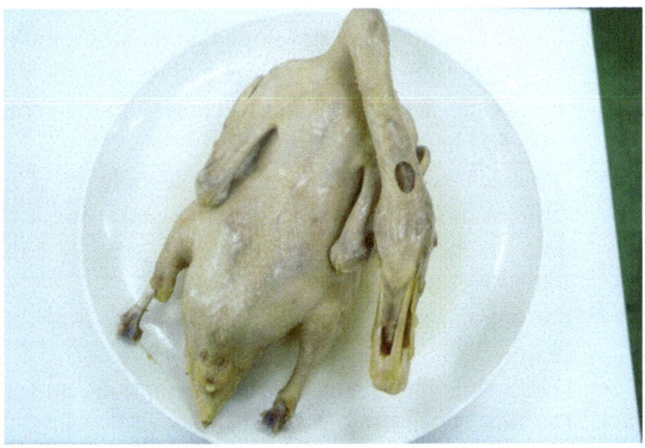

Fig. 12.17 Salted water duck

Sl No	Items	Quantity
1	Mule duck	1 kg
2	Table salt (3.0%)	30 g
3	Brown sugar	20 g
4	Sodium nitrite	150 mg
5	Chicken soup powder	4 g
6	Five spices powder	1 g
7	White pepper powder	2 g
8	Chinese red pepper	4 g
9	Potassium sorbate	2 g

Fig. 12.18 Final product of Chopped Salted Water Duck

Processing steps.

- Choose healthy, well developed, and mature mule duck as raw material.
- Cut the chest open with a knife and remove the kidneys. Wings connecting to chest bones, leg joints and ribs connect to the vertebrae bone are dislocated by knife and hand.
- The duck is then placed flat on a tabletop. The water inside and outside surfaces of the duck carcass is mopped using a cloth. Then spread the thick cloth on the inside of the abdominal cavity of the duck, and apply force to flatten the duck.
- The outer and inner surfaces of the duck carcass are completely smeared with mixture of ingredients given in the table.
- The duck is then hanged on the metal hook and stored at 4 °C for 4 days.
- The outside and the inside surfaces of the duck are rinsed and then dried at 45 to 48 °C for about 6 hr.
- Smoked at 50 °C for 2 h in a smoke house.
- The finished products are hanged on shady and well ventilated for about 1 week, and this procedure is necessary for promoting ageing process in the meat (Fig. 12.19 and 12.20).

12.2.7.13 Ginger Duck

Ginger duck is considered a medicated non-greasy warm food suitable for autumn and winter. It originated in Quanzhou in China and has become popular in Taiwan and other neighbouring countries. It is a traditional Han Chinese street food that contains one of the 50 fundamental Chinese herbal medicines, Astragalus. Majority of Taiwanese population are Han, which is an ethnic group. In the authentic recipe of

Fig. 12.19 Finished press salted duck

Fig. 12.20 Chopped press salted duck

ginger duck, instead of water, only Chinese cooking rice wine is used. Ginger duck is a special dish on Chinese New Year Day.

Ingredient composition.

Sl No	Items	Quantity
1	Cut Muscovy duck meat	1 kg
2	Rice wine	150 ml
3	Broken mature ginger	100 g
4	Black sesame oil	50 ml

(continued)

Sl No	Items	Quantity
5	Table salt	12 g
6	Angelica	12 g
7	Cinnamon stick	2
8	Wolfberry	1 g
9	Rehmannia	1.5 g
10	Chicken soup powder	3 g

Processing steps.

- Chopped Muscovy duck is used as a major ingredient and mature ginger and other ingredients are also used for producing the ginger duck.
- The mature ginger is washed completely and cut into short sections. Each section is half cut and slightly crushed (Fig. 12.21).
- Chopped duck meat is boiled for 3 min, washed in cold water and kept ready.
- In a low heated pan, the black sesame oil is poured. When the oil temperature is reached about 160 °C, the crushed ginger is added in the oil and fried moderately until the flavour of ginger emanates.
- When the ginger pieces turn dry yellow, add the chopped duck meat and stir-fry until it is half cooked (Fig. 12.22).
- After the duck meat is fried, pour rice wine and fire it for about 10 min to let the alcohol of the rice wine volatilize, leaving only the wine flavour (Fig. 12.23).
- Pour water into the pan to cover the meat completely and boil it. Add remaining ingredients, boiled for about 2 h (Fig. 12.24) and serve hot.

Fig. 12.21 Muscovy duck, mature ginger and rice wine

Fig. 12.22 Stir-fried chopped duck and mature ginger

Fig. 12.23 Firing to let the alcohol of the rice wine volatilize and leave only the wine flavour

Fig. 12.24 The final product of Ginger Duck

12.2.8 French Products

French cuisine embraces the cooking traditions and practices from France and those developed due to the influence of cultures of Spain, Italy, Switzerland, Germany, and Belgium. The most popular ducks for French cuisine are Muscovy and overfed mule ducks. *Foie gras* (Abitbol 2017), Duck à l'Orange, Roast duck, and duck confit are a few very popular French duck dishes discussed in the following.

12.2.8.1 Duck *Foie Gras*

Foie gras (Eng: pronounced as fwa:'gra: in French it means 'fat liver') is the most famous duck dish of French cuisine and heritage protected by law. This is a traditional specialty duck product made from the liver of duck or goose fattened by gavage or force feeding, twice daily for 12.5 days. Ducks are slaughtered at 100 day. Muscovy and mule ducks are generally used for this preparation. France is the largest producer and consumer of this dish followed by Hungary, Bulgaria, USA, Canada, China, and other countries. *Foie gras* has a rich, buttery, and delicate flavour unlike that from normally reared duck. There are two different methods of preparation, viz., traditional cold and current hot methods. Duck *foie gras* from southwestern France holds the red label (label rouge) a distinction of quality. It is either sold as whole or used in the preparation of mousse, parfait (a frozen dessert) or pate (paste).This is served as an appetizer in French cuisine on special occasions.

Cooking time: 1 h 20 min.
Ingredient composition.

Sl No	Ingredients	Quantity

(continued)

Sl No	Ingredients	Quantity
1	Duck liver lobe	450 g
2	Salt	5 g
3	Black pepper, ground	2.5 g (1/2 tsp)
4	Sweet white wine/brandy/cognac	50 ml

Processing steps.

- Remove the veins from the fresh liver lobes by pulling with fingers. If liver is frozen, thaw slowly by 24 h in refrigerator.
- Dissolve salt in wine and add pepper and mix. Pour it over foie gras in a ceramic baking dish and keep covered for 12 h in a refrigerator.
- Preheat oven to 120 °C. Lower the oven temperature to 100 ^{0}C and cook for 50 min to an internal temperature of 50 °C.
- Cool and remove excess fat from the top and keep in fridge with a weight on the top for 24 h.
- Foie gras may be served as slices immediately or within 5 days after reaching maturity. The slices can also be pan seared in butter and served.

12.2.8.2 French Duck à l'Orange

Duck à l'Orange is one of the most popular classic French dishes in which the seared duck breast is roasted and glazed with a sweet orange sauce. The plumpest duck breast is ideal due to thick subcutaneous fat. Orange pairs well with duck and is better than lemon. The sauce can be prepared ahead of time and sear the duck right before serving as hot.

Cooking time: 45 min.
Ingredient composition.

Sl No	Ingredients	Quantity
For orange sauce		
1	Granulated sugar	60 g
2	Water	30 ml
3	Sherry vinegar	30 ml
4	Orange juice	375 ml
5	Minced shallot	30 g
6	Chicken stock	375 ml
7	Oranges (sections cut)	4
8	Orange zest (for sauce and garnishing)	30 ml
For preparing and cooking duck breast		
1	Halved duck breast	500–600 g
2	Salt	To taste
3	Black pepper	To taste
4	Cold unsalted butter	60 g

Processing steps.

Preparation of orange sauce

- In a sauce pan, dissolve the sugar in cold water and boil for a few minutes until it is a dark brown caramel.
- Add vinegar, orange juice, shallots and chicken stock and simmer gently until the sauce is reduced to almost one third, a little less than a cup.
- Add butter to the sauce pan with orange zest. Stir and melt the butter in the pan over medium heat. Mix with the sauce well and add orange sections.
- Cool the sauce and set aside till the duck breast is prepared.

Preparing and cooking duck breasts

- Dry the two halves of breast with paper towels.
- Score the skin diagonally with a sharp knife to create a criss-cross pattern to enable rendering of fat during cooking. While scoring, take care not to cut through meat.
- Rub the breast of the duck with butter and sprinkle both the meat and skin sides with a little salt and pepper.
- Heat a frying pan over high heat and sear the sides of the breast quickly. Then cook the duck breast for about 9–11 min on each side.
- Remove duck from the pan, place on a warm plate, and leave them for 5 min rest covering with paper towels. This is to soften the duck after cooking.
- Place the duck on a hot plate, whole or sliced. Pour the warmed up sauce over the duck and spread.
- Garnish with the remaining orange zest and serve immediately.

Note: Zest of orange or any citrus fruit is a food ingredient that is prepared by scoring or cutting from the unwaxed rind or peel or skin. The exocarp (flavedo) of the rind is removed using a zester or peeler or scorer. Inner white pith, the endocarp (albedo) is bitter and not used

Variations in the Preparation of French Duck à l'Orange

Some variations in the preparation of French Duck à l'Orange are noticed, both in roasting duck and in sauce making (Edden 1980). Put duck breast rubbed with butter, salt, and pepper in a roasting tin with 150 ml white wine. Cook in the oven 30 min per 450 g at 190 °C, and baste occasionally. When cooked, remove the duck from pan to a serving dish, join it, and keep hot. In the preparation of orange sauce, replace shallot, orange zest, and chicken stock with finely shredded orange rind or peel, juice from one lime and 30 ml brandy, but retaining orange juice and sherry vinegar. After draining excess rendered duck fat from the roasting tin, the shredded orange rind and the sauce are added to the sediment and juices remaining in that.

Blend corn flour or arrowroot with a little water, stir it into the juices and return the tin to the heat. Boil for 2–3 min with stirring. Pour the sauce over the duck joints and garnish with orange slices.

12.2.8.3 Roast Duck with New Potatoes

This main French duck roast is served with classic peppercorn sauce. Comparatively this recipe takes a little more time as the sauce is to be prepared simultaneously. Pekin ducks are the ideal type of duck for this dish. Muscovy, mule or Long Island ducks are also used. They will have duck flavour without gaminess. The skin-on roasting gets crispy and in order to ensure that excess fat is to be trimmed off. Too much rendered fat will make the duck soggy. The tender new potatoes put along with the duck in the pan will get roasted in the fat drippings and adsorbing the flavour of the seasonings.

Cooking time: 1 h 35 min.
Ingredient composition.

Sl. No	Ingredients	Quantity
1	Ducklings - 2, fresh and oven ready, about 2 kg each	4 kg
2	Olive oil	As required
3	Salt	To taste
4	Black pepper	To taste
5	New potatoes	1 pack
6	Rosemary	3 twigs
7	Fresh cut parsley	Small handful
8	Largely sliced onion	100 g (1)
9	Carrot	100 g (2)
10	Port wine	250 ml (1 cup)
11	Raspberry vinegar	30 ml
12	Chicken stock	1 L (4 cups)
13	Salted butter	60 g (4 tbsp)
14	Plain flour	60 g (4 tbsp)
15	Green peppercorns in brine	300 g (1 can)

Processing steps.

- Preheat oven to 230 °C.
- Keep aside the neck, gizzard, heart, and wings of the duck for preparing the sauce. After trimming off excess fat, dry the duck thoroughly with a paper towel.
- Prick the skin all over the duck with a fork and rub lightly with olive oil followed by salt and pepper.
- Place washed new potatoes in the roasting pan. Put rosemary, parsley, and sprinkle a little salt on potatoes.
- Place the roasting rack above the potatoes and keep the duck on the rack.

- Place the roasting pan into the centre of the oven and roast at 230 °C for 20 min. Reduce the oven temperature to 180 °C and continue to cook for 45 min. Baste occasionally with the rendered fat.
- After roasting, remove the duck from the oven, transfer to a plate, and cover with aluminium foil. Allow the duck to rest for 30 min.
- Remove the roasted potatoes from the roasting pan and transfer to a dish, sprinkle salt, cover, and keep hot.

Preparation of peppercorn sauce

- While the duck is roasting, heat olive oil in a saucepan and sauté the chopped duck wings, neck, gizzard, onions, and carrot on medium high heat until browned. Stir frequently to prevent burning and sticking to the bottom of pan.
- When browned, remove the pan from the heat and add port wine and raspberry vinegar.
- Carefully return the pan to the heat for a minute without the port wine getting inflamed. Add chicken stock and simmer for 1 h.
- Melt butter in a wok over medium heat and add plane flour, whisk and make a roux. Strain the stock into the sautéing wok and whisk thoroughly to mix with the roux followed by the addition of canned green peppercorn. Continue whisking till the sauce thickens.
- Joint the roast duck to separate the legs and breast keeping the skin intact. Serve with the roasted new potatoes and the peppercorn sauce.

12.2.8.4 Duck Confit

In French this duck dish is known as Confit de Canard. This French preparation involves duck legs first curing in salt, partly or fully, then marinating and poaching the meat in its own fat and aromatic herbs and spices, especially garlic. The French word confit means preservation. Because it is so tender, duck confit can be shredded and served over salad, in stews, or on bread. It enhances the flavour with vegetables and potatoes. Before serving the Duck confit, warm it either with skin side down in a pan with duck fat or crisp it by broiling.

Cooking time: 3 h 20 min.
Ingredient composition.

Sl No	Ingredients	Quantity
1	Duck legs, skin-on bone-in, frenched*	4
2	Salt	45 g (3 tbsp)
3	Toasted and ground whole juniper berries	5 g (1 tsp)
4	Duck fat	6 cups
5	Thyme	4 sprigs
6	Sliced round lemon	2
7	Fresh Bay leaf	2
8	Split garlic	1 head

Processing steps.

- In a small bowl, mix salt and toasted, ground juniper berries. Season the duck by rubbing duck legs with salt and juniper mix. Keep on a butter paper lined tray, cover and chill overnight or for up to 2 days.
- Prior to cooking, the spices are rinsed off and pat dry. With a fork prick all over the duck skin. Melt duck fat in a wok over medium heat for deep frying at 200 °C. Place duck legs, thyme, lemon, bay leaves, and garlic in the wok. Cook and maintain an oven temperature of 190 to 210 °C, till meat is tender, approximately 2 to 2 ½ h.
- Remove the duck legs and set aside. Decant the fat into a large bowl leaving any cloudy liquid behind.
- Place the duck in a Pyrex dish with decanted duck fat until it covers the top of legs by 2.5 cm. Cover and chill. The duck legs can be used right away or stored for up to 1 month. The refrigerated duck fat can be used for sautéing vegetables, frying potatoes, and other culinary purposes.
- Before serving duck confit, remove the duck from the storing oil and place skin side up on a wire rack in an oven at 400 °C. Cook until the skin is golden brown and crisp, about 30–35 min. Serve with garlicky potatoes.
- Note: Frenching* duck legs is to cut around the skin-on the drumstick about 2.5 cm below the tip of the bone. Scrape down towards the meat to expose the bone. Discard any skin or bone.

12.2.9 English Products

English cuisine encompasses the cooking traditions and practices associated with England. The distinctive styles of English cooking shares many of those of British, Chinese, Indian, and American. This cuisine absorbed the cultural influence of the post-colonial territories, especially South Asia and the spices. Mashed potatoes with caramelized onions are a sweet and flavoured side dish, ideal for duck. Many duck dishes are there in English cuisine with fruit glazes such as cherry, cranberry and plum besides honey, rosemary, and garlic glazes. A glaze in culinary is a coating of glossy sweet or savoury substance applied by dipping, dripping or basting. A savoury glaze for meat can be prepared from reduced stock, as well. Two recipes are presented in the following (Illsleys 2004).

12.2.9.1 Duck Breast with Beetroot and Red Onion

Cooking time: 30–40 min.
Ingredient composition.

Sl No	Ingredients	Quantity

(continued)

Sl No	Ingredients	Quantity
1	Duck breasts, 170 g each	4
2	Oranges	3
3	Olive oil	40 ml (4 tbsp)
4	Thinly sliced red onion	450 g
5	Salt and pepper	To taste
6	Red wine	300 ml
7	Fresh and grated ginger	2.5 cm
8	Light Muscovado sugar (unrefined cane sugar with a little molasses)	10 g
9	Port wine	200 ml
10	Cooked and sliced beetroot	175 g

Processing steps.

- Grate the zest and squeeze the juice of two oranges and slice the other orange.
- Heat olive oil in heavy based frying pan and add onions. Sauté gently with frequent stirring till it is soft and caramelized.
- Meanwhile, score the duck skin-on breasts with a sharp knife and rub a little salt in to the cuts.
- In a roasting pan, put the breasts with skin side up and orange slices in between. Roast at 200 °C for about 15 min.
- Add the wine and orange juice to the onions in a wok. Boil and simmer for about 5 min until the liquid is totally reduced.
- Then add the grated ginger, orange zest, sugar, and port. Simmer for 5 min or until syrupy. Add the beet root and warm through. Season with salt and pepper to taste.
- Let the duck rest for 5 min to set, slice and serve with roasted orange slices and beetroot-red onion relish.

12.2.9.2 Roasted Duck with Cranberry Glaze

Cooking time 2 h 20 min.
Ingredient composition.

Sl No	Ingredients	Quantity
1	Duck, oven ready, 1.6 kg each	2
2	Water	175 ml
3	Salt to taste	To taste
4	Plain flour, for dredging*	15 g (1 tbsp)
5	Corn flour	30 g (2 tbsp)
6	Lemon juice	30 ml
7	Red wine	60 ml (4 tbsp)
8	Cranberry sauce, whole berry	400 g
9	Water cress, trimmed for garnishing	As required

(continued)

Sl No	Ingredients	Quantity

Processing steps.

Preparation of duck for roasting

- Prick duck skin with fork and rub well with salt. Instead the skin can be scored with a sharp knife without cutting meat.
- Place duck on a rack in a roasting tin and roast at 200 °C for 20 min per 450 g, basting occasionally. About 15 min before the end of cooking time, baste the duck, dredge with flour and finish cooking at 220 °C. After roasting, spoon the fat from the roasting juice and keep aside the juice for adding in the sauce. Keep the duck hot in the tin.

Preparation of stock and cranberry glaze

- Prepare the stock by simmering duck giblet in water in a covered sauce pan for 1 h. Strain the stock.
- Blend corn flour, lemon juice, and wine and mix with the strained giblet stock with stirring.
- Heat the cranberry sauce in a pan until softened followed by the addition of corn flour stock mixture in to it with stirring. Bring to boil and simmer for 3–4 min.
- Pour two thirds of the above sauce into a sauceboat and keep warm. Strain the remaining sauce in to a saucepan to remove the cranberries. Add the roasting juice without fat to the strained cranberry sauce. Boil rapidly until reduced to a rich cranberry glaze.
- Brush the complete glaze over the ducks in the tin. Halve the ducks and arrange them on a serving dish. Garnish with watercress and serve with cranberry sauce.

12.3 Duck Egg Products

Duck eggs, like chicken eggs, are a good culinary ingredient in all cuisines. In any form, eggs can be served in sweet and savoury courses and at any meal. Similar to chicken eggs, the duck eggs are also well suitable for all egg preparations (Jalaludeen and Churchil 2006). However, due to its larger size, it is unsuitable for pickling (Jalaludeen et al. 2009). Many people prefer duck eggs to chicken eggs due to their allergy to the latter. Select duck eggs with clean and unbroken shells only for cooking. Eggs should always be cooked at moderate to low temperatures. If the heat is too high, egg will become dry and rubbery. Even at low temperature they should not be cooked for long. Refrigerated eggs are to be thawed before cooking. It is ideal to eat hard or soft-boiled, fried or poached eggs as fresh as possible for safety.

Quan and Benjakul (2019) made a review on the composition and functional properties of duck albumen and how they are affected by processing conditions in comparison with hen albumen. Lower contents of ovalbumin, conalbumin, lysozyme, and ovoflavoprotein are found in duck albumen. However, duck albumen shows better gelling and foaming properties than hen albumen. During storage, duck albumen gel properties are enhanced, while foam volume and foam stability are decreased. Moreover, the changes in quality indices of duck egg including the thinning of the albumen, an increase in albumen pH, loss of water, and CO_2 occur as storage time is increased. Duck egg albumen contains protease inhibitors and can thus be used as an additive to prevent proteolysis in some foods, particularly surimi gel, preventing the degradation of muscle protein and improving the gel property. Nevertheless, the functional properties can be altered when duck eggs are stored or processed. Therefore, duck albumen should be considered as a potential protein additive to replace hen albumen in food products.

Duck eggs are richer with more albumen and fat than turkey, goose, and hens' eggs which make cakes and other pastries fluffier. The official substitution ratio is two duck eggs for every three chicken eggs. If duck egg is almost of the same size as hens' egg, substitute with one duck egg only. Duck eggs stay fresher longer, due to their thicker shell.

A few popular duck egg savoury dishes such as boiled egg, deviled egg, coddled egg, roasted egg, egg omelette, scrambled egg, poached egg, balut, salted duck egg, century eggs, spring eggs and a dessert dish, egg roll and egg roll cookies are discussed in the following, in brief. Other products include omega 3 fatty acid enriched egg, yolk pigment enhanced egg, duck egg pasta products (Alamprese 2017), and nutraceuticals (Tahergorabi and Jaczynski 2017).

12.3.1 Boiled Eggs

'Boiled' is a misnomer, as eggs should be simmered rather than boiled. Hard-boiled duck eggs are just like hard-boiled chicken eggs. Because of the extra albumen content, it is especially important not to overcook duck eggs. If allowed to overcook by keeping at higher temperature for longer time, egg will become tough and rubbery and a dark line may form between the yolk and the white all around. When boiled, the white of duck eggs turns bluish and yolk turns red-orange.

To hard-boil a duck egg, put it into boiling water, lower the heat and simmer for at least 10–12 min; for soft-boiled eggs allow about 7 min. As soon as cooking is complete, plunge egg into cold or chilled water for easy peeling of shell. Very fresh eggs are hard to peel cleanly. Hard-cooked eggs can be served whole or sliced evenly with an egg slicer. The white can be cut into wedges, as well. The slices and wedges can be used in salads, mousses or to make attractive garnishes.

12.3.2 Stuffed or Deviled Eggs

Hard-boiled duck eggs can also be cut in half, lengthwise so that the solid yolk can be removed from the two halves. The yolk is then blended with other ingredients such as mayonnaise, mustard, and seasonings and the mixture is stuffed back in to the yolk cavity of the egg halves and served.

12.3.3 Coddled Eggs

A coddled egg is cooked slowly than a boiled egg, but yields the same results, except that the egg is bit tender. Pierce the large end of the eggs with a pin. In a saucepan place the eggs in a single layer and add cold water and 7.5 g of salt and leave opened. Place the pan on medium heat and bring the water to a simmer, but not to a full boil. Remove the pan from the heat and cover it. The length of the time that the eggs remain in the covered pan determines the degree of firmness of the yolk: soft yolk - 4 to 6 min, medium yolk – 6 to 8 min, and hard yolk - 20 to 25 min. To stop the cooking process, run cold water over the eggs. It is best to use older eggs for coddling because they peel easier. Soft-coddled eggs are often served in an egg cup and eaten directly from the shell because they are difficult to peel.

12.3.4 Roasted Eggs

Roasted egg is prepared using two different cooking processes. It is first hard-boiled in simmering water and then roasted in an oven with shell. It is removed from the oven when the shell becomes brown. This method is used to prepare eggs traditionally served during the Jewish Passover. The eggs should not be overcooked in order to achieve good results, regardless of the cooking processes. Unless eggs are rapidly cooled, they continue cooking even after removal from the oven.

12.3.5 Egg Omelette

An omelette is usually made with two or three eggs and is cooked very quickly in a sauce pan. Salt is then rubbed on the pan and wiped clean after each use. The salt treatment helps to prevent eggs from sticking to the pan the next time it is used. Additional ingredients, such as curry leaves and other herbs, cheese or finely chopped meat can be added to the eggs before they are cooked without changing the way the basic omelette is cooked. The extra ingredients will probably require that a spatula be used to fold the omelette in the pan. Extra ingredients can also be placed on the top of the plain omelette when it is served.

12.3.6 Scrambled Eggs

In order to prepare scrambled eggs, break the eggs into a bowl, add a tablespoon of cold water, and whisk together so that the yolk and whites are blended. Salt and pepper can be added to the eggs or added when the eggs are served. Coat the pan lightly with oil or butter, heat it over a medium heat, and pour the egg mixture into the pan. As the mixture begin to set, use a spatula to scrap the eggs from the edge of the pan to the centre. It is best to remove the eggs from the heat source when they are still very moist, because the internal heat of the eggs will complete the cooking process. When two or three eggs are scrambled together, it usually takes a minute or less before the cooking process is complete. Eggs that become too dry indicate that they have been overcooked. For increasing the quality of duck egg scrambling, lemon extract may be helpful. The citric acid content of the lemon will adhere to the ovomucin factor, which will effectively reduce the duration of scrambling. The duck egg has a strong flavour than chicken's egg. Therefore, duck eggs either on scrambling or in omelette are well complemented by onions, pepper, mushroom, etc.

12.3.7 Poached Eggs

A poached egg cooked on the stovetop is one that is cooked in simmering water without the shell. Unlike a boiled or coddled egg that benefit from the use of an older egg, a poached egg is best when a fresh egg is used. This is because the fresh egg, when placed into the heated water, will not spread out like an older egg, yielding better results with the shape and texture of the egg. If an older egg is used, it can be simmered in the shell for a few seconds so that the white is just slightly congealed. When the egg is broken in to the simmering water, it will not spread out as much. One tablespoon of vinegar added to the water will also help coagulating the white to keep it from spreading too much. In another method, an oval, slotted metal container (poacher) with an attached handle is used. The metal poacher is convenient for use and creates a pleasing shape. Both fresh eggs and older eggs can be poached well using the metal poacher. If the egg is to be poached in a microwave, there are various poaching dishes that can be used, such as the cooking dish.

12.3.8 Balut

Balut (embryonic egg) is a fertilized developing egg embryo that is boiled and eaten from the shell. Consumption of fertilized duck eggs is familiar in the food customs of Chinese, Laotians, Cambodians, and Thais. It is commonly sold as street food in Philippines. The length of incubation before the egg is cooked is a matter of local preference, but generally ranges between 14 and 21 days. It is considered as a delicacy and is highly nutritious.

12.3.9 Salted Duck Eggs

The salted duck eggs are famous for many centuries as a traditional product popular in Asia, especially in China. Benjakul and Kaewmanee (2017) gave the details of sodium chloride preservation of duck eggs. There are two methods of production: 1) immersing in brine (brining method) and 2) packing each egg in damp salted charcoal or soil paste (coating method). In Philippines, the popular method of processing salted eggs is by the Pateros style. Fresh duck eggs are individually coated with a mixture of clay from ant hills or termite mounts, salt and water in the ratio 1:1:2. With some variations in the ingredients, eggs are dyed red to distinguish them from fresh duck eggs and they are known as red coloured eggs. The eggs are sold with the salt paste removed and vacuum packed.

Due to the salt curing process, salted eggs have a briny aroma, a gelatin-like egg white and a firm textured round yolk which is bright orange-red in colour. Egg albumen has a sharp salty taste. Yolk is fatty and less salty and so consumer demand is very high. The physico-chemical and rheological changes in albumen and yolk are due to dehydration. The yolk is used in Chinese moon cakes as a symbol of moon and in pastries. Salted duck eggs are steamed or boiled and peeled and eaten as a condiment to rice gruel for breakfast or cooked with other food as flavouring. As a result of brining, albumen gets liquefied and so generally discarded. The researchers suggested the following uses of the salted egg albumen, viz., 1) incorporation in frankfurters and tofu, 2) extraction of lysozyme, 3) desalination of the egg white to restore the functional properties of albumen such as foaming, emulsification, and gelation, 4) creating a protein hydrolysate that can be used to enhance the water holding capacity of meat and marine foods, and 5) using the protein hydrolysate as nutraceuticals.

12.3.10 Century Eggs

Century eggs are also known as hundred year eggs, thousand year eggs, preserved eggs, and millennium eggs. Traditionally, century eggs were made by preserving duck eggs in a mixture of salt, lime, and ash, then wrapping in rice husks for several weeks. During this preservation, the pH of the eggs rises and transform them. The eggs are cured for 4 to 5 weeks. It is a Chinese preserved food product and delicacy. Chicken eggs can also be preserved by this method.

12.3.11 Spring Eggs

Ingredient composition.

Sl No	Items	Quantity
1	Duck egg	1 kg
2	Soy sauce	120 g
3	Mirin	60 g
4	Brown sugar	30 g
5	Japanese sake	30 g

Fig. 12.25 The Spring Eggs

Processing steps.

- Raw shell duck eggs are used for the preparation of Spring eggs. The eggs are washed to clean the surface from dirt and faeces.
- If it is a refrigerated egg, then it needs to be thawed to room temperature.
- Put the eggs in the pot and add cold tap water to cover the eggs completely and heat the water until boiling, and then keep it in simmering for 5 min.
- Eggs are taken out, placed in ice water for about 20 min, and cracked to remove eggshells.
- Dip the eggs into the marinade for 24 h in refrigerator, and let egg white and yolk absorb the flavour of the marinade to add taste to the product.
- Cut the eggs before serving to show that the centre of the egg; which is gelatinous (Fig. 12.25).

12.3.12 Egg Rolls

Ingredient composition.

Sl No	Ingredients	Quantity
1	Duck egg or chicken egg	1 kg
2	Unsalted butter	600 g
3	Plain flour	550 g
4	Brown sugar	450 g
5	Custard powder	50 g
6	Baking soda	25 g

Processing steps:

- Sieve the plain flour, custard powder, baking soda, and then mix well (Fig. 12.26).
- Whisk the butter and sugar at low speed in a stainless steel bucket for 5 min until it turns light yellow in colour.
- Add three eggs at a time, whisk well before adding another three eggs. Make sure the ingredients are thoroughly mixed to make a dough.
- Carefully whisk at low speed while adding the flour mix (Fig. 12.27).
- The pan is coated with cooking oil and heat under low heat on both top and bottom sides.
- Place 2 tsp. dough on the bottom pan and gently press with the top pan for 5 sec. Cook for 5 sec.
- Lift the top pan up and the pancake should not stick to the pan. Roll the pressed and cooked dough onto the rolling pin (Fig. 12.28). Let it cool to become the final egg roll (Fig. 12.29).

12.3.13 Egg Roll Cookies

Egg roll cookies are wafer thin crispy cookies rolled on hollow SS tubes. They are sweet with a buttery and mild vanilla flavour. This is used as a dessert and is a very popular street food in Taiwan and China besides other Southeast Asian and East Asian countries. Both chicken and duck eggs are used. Sesame seeds and vanilla essence are optional ingredients. An electric hand mixer or whisk can be used for

Fig. 12.26 Eggs and other ingredients

Fig. 12.27 Egg dough

Fig. 12.28 Rolling the egg press-dough

making the egg roll batter. Similarly, a rectangle electric egg roll cookie machine for large-scale production or a gas non-stick egg roll maker pan with round press plates for domestic cooking can be used. Immediately after production the cookies may be stored in airtight containers without getting soggy and rancid. This formulary is of Kwa (2019).

Fig. 12.29 The final product of Egg roll

Ingredient composition.

Sl No	Ingredients	Quantity
1	Duck eggs	3
2	Unsalted butter	120 g
3	Castor sugar/icing sugar	85 g
4	Fresh thick first extract or canned coconut milk	60 ml
5	Plane/all-purpose flour	90 g
6	Corn flour	20 g
7	Salt	2 g (a pinch)
8	Vanilla essence	5 ml (1 tsp)
9	Sesame seed (optional)	~ 10 g

Processing steps.

Whisk butter in a glass bowl till it is light and fluffy. Add sugar and beat again.

- To the butter-sugar mixture, add egg one by one and whisk till fluffy, each time.
- Add thick coconut milk to the batter, whisk followed by the addition of plain flour, corn flour, and salt through a sieve. Whisk till light and fluffy.
- Add vanilla essence and sesame seed to the batter and mix well. Set aside for 15 min.
- Switch on the bottom and top plate heaters of the egg roll cookie machine and set time and temperature at low heat. To check the temperature the following thumb rule may be used: a drop of water placed on the cooking surface, should bubble and evaporate quickly, but not sizzle violently.

- Pour one tablespoon batter at the centre of the bottom plate of the machine and lightly press with the top plate in such a way that no batter squirts out of the plates. If at all a little batter comes out, it may be scraped off with a spatula.
- Toast the egg roll for 8–10 sec till it becomes golden brown. Roll on the SS hollow pipes and keep aside for cooling. Transfer the cold egg roll cookies carefully to airtight containers.

12.4 Conclusion

With a view to introducing the vast culinary and industrial scope of duck dishes, a few popular RTE duck meat and egg products, their methods of preparation and ingredients with respect to each style of cooking in the various States of India, Bangladesh, Sri Lanka, Thailand, Malaysia, Singapore, Indonesia, Vietnam, China, Taiwan, France, and England are discussed in this Chapter. The use of meat type duck would improve the eating quality of meat as well as its increased use in various dishes. But spent duck meat can also be used in comminuted meat products. Among Indian cuisines, highly diverse dishes are in vogue particularly, in Kerala in using the vast array of oriental spices and coconut in different forms. Assamese, Bangali, Telugu, and Chettinadu duck dishes are also popular. Bangladeshi dishes are more related to Bangali with the use of panchphoron spice blend as its specialty. Similarly, Sri Lankan dishes are related to that of Kerala, Singapore, Thailand, and Malaysia. The duck dishes in French and English cuisines with roasting, grilling, and separate preparation of the sauces and glazes make their dishes distinct from that of other dishes discussed. Strategy for the use of duck eggs in the production of nutraceuticals and other duck products with minimal cholesterol level, reduced saturated fat and rich omega 3 fatty acid content are to be formulated. Use of duck egg in pasta quality modulation is also to be investigated. Value-added duck products need innovation to increase their market demand. The development of new RTE and RTC duck meat products are required to introduce duck meat in modern food regime.

References

Abitbol V (2017) France *Foie gras*. https://www.196flavors.com/france-foie-gras/. Accessed 27 June 2020

Alamprese C (2017) The use of egg and egg products in pasta production. In: Hester PY (ed) Egg innovations and strategies for improvements. Academic Press, pp 251–259

Benjakul S, Kaewmanee T (2017) Sodium chloride preservation in duck eggs. In: Hester PY (ed) Egg innovations and strategies for improvements. Academic Press, pp 607–625

Bhattacharyya D, Sinhamahapatra M, Biswas S (2007) Preparation of sausage from spent duck – an acceptability study. Int J Food Sci Technol 42:24–29

Biswas S, Chakraborty A, Sarkar S (2006) Comparison among the qualities of patties prepared from chicken broiler, spent hen and duck meats. J Poult Sci 43:180–186

Biswas S, Banerjee R, Bhattacharya D, Patra G, Das AK, Das SK (2019) Technological investigation into duck meat and its products – a potential alternative to chicken. World Poult Sci J 75:1–12

Bok N (2019) Duck rendang. www.sharefood.sg/2019/11/14/duck-rendang-2/. Accessed 24 June 2020

Chartrin P, Méteau K, Juin H, Bernadet MD, Guy G, Larzul C, Rémignon H, Mourot J, Duclos MJ, Baéza E (2006) Effects of intramuscular fat levels on sensory characteristics of duck breast meat. Poult Sci 85:914–922

Chen BH, Lin YS (1997) Formation of polycyclic aromatic hydrocarbons during processing of duck meat. J Agric Food Chem 45:1394–1403

Chen G, Song H, Ma C (2009) Aroma-active compounds of Beijing roast duck. Flavour Fragrance J 24:186–191

Cicy MJ (2020) Personal communication. Mannuthy, Kerala State, India

Edden G (1980) The house keeping step-by-step cook book. Edbury Press, London, pp 273–276

Estévez M, Morcuende D, Ventanas S, Cava R (2003) Analysis of volatiles in meat from Iberian pigs and lean pigs after refrigeration and cooking by using SPME-GC-MS. J Agric Food Chem 51:3429–3435

FAOSTAT (2017) Food and agriculture organisation of the United Nations. FAOSTAT database on agriculture. Rome, Italy: 2017

Hanuman, Khun T, Lertviriyavik T (2016) Thai green curry with roasted duck and young chilies. https://thaifoodmaster.com/thai_food_recipes/thai_curry_recipes/6440. Accessed 24 June 2020

Helen (2017) Roast duck-vit quay, https://www.cooking.youtube.club/2017/11/the-ultimate-roast-duck-v-quay-html/. Accessed 26 June 2020

Huda N, Putra A, Ahmad R (2011) Potential application of duck meat for development of processed meat products. Current Res Poult Sci 1:1–11

Illsleys J (2004) Good house keeping cookery book. The cooks classic companion. Collins and Brown Ltd, London, p 156

Jalaludeen A, Churchil RR (2006) Duck eggs and their nutritive value. Poult Line 6(10):35–39

Jalaludeen A, Churchil RR, Joseph L, Anitha P (2009) Duck meat egg and their products. In: Souvenir of IV world waterfowl conference, Thrissur, Kerala, India, 11-13 Nov 2009, pp. 57–64

Jin SK, Kim IS, Jung HJ, Kim DH, Choi YJ, Hur SJ (2007) The development of sausage including meat from spent laying hen surimi. Poult Sci 86:2676–2684

Judy (2015) Nanjing salted duck. https://www.thewoksoflife.com/nanjing-salted-duck/#. Accessed 30 June 2020

Keeton JT (2001) Formed and emulsion products. In: Sams AR (ed) Poultry meat processing. CRC Press, Boca Raton, pp 195–226

Kho B (2019) Nonya duck curry. https://www.womensweekly.com.sg/recipe/food/asian-recipes/nonya-duck-curry/. Accessed 20 Jan 2020

Kim S, Kim T, Ku S, Kim MJ, Jung S, Yong HI, Choi Y (2020a) Quality characteristics of semi-dried restructured jerky: combined effects of duck skin gelatin and carrageenan. J Anim Sci Technol 62(4):553–564

Kim T, Kim H, Lee Y, Jang H, Kim Y, Choi Y (2020b) Quality characteristics of duck jerky: combined effects of collagen and konjac. Poult Sci 99(1):629–636

Kwa K (2019) Egg roll cookies. https://www.youtube.com/watch?v=xI_usckAI7c. Accessed 26 June 2020

Leng TB (2018) Eight treasure duck. https://www.youtube.com/watch?v=myAR7rrgvgs. Accessed 28 June 2020

Liu Y, Xu XL, Ouyang GF, Zhou GH (2006) Changes in volatile compounds of traditional Chinese Nanjing water-boiled salted duck during processing. J Food Sci 71:S371–S377

Liu Y, Xu XL, Xhou GH (2007) Changes in taste compounds of duck during processing. Food Chem 102:22–26

Mathew J (2020) Personal communication. Johor, Malaysia

Moonshadow (2019) Malay chicken curry with kersik. https://www.allspices.co.uk/recipe/33614 co.uk/recipe/33614/malay-chicken-curry-with-kersik.aspx. Accessed 24 June 2020
Nasution P (2019) Bebek Madura: Indonesian Fried Duck. https://www.indonesiaeats.com/bebek-madura-indonesian-fried-duck. Accessed 26 June 2020
Parkinson R (2019) Cantonese roast duck. https://www.thespruceeats.com/Cantonese-roast-duck-694866. Accessed 30 June 2020
Pohl A, Lister T (2020) Roast duck. https://www.cooked.com/uk/Andreas-Pohl/Hardie-Grant-Books/Real-Vietnamese-Cooking/Poultry/Roast-duck-recipe. Accessed 26 June 2020
Quan TH, Benjakul S (2019) Duck egg albumen: physicochemical and functional properties as affected by processing and storage. J Food Sci Technol 56(3):1104–1115
Ramadhan K, Huda N, Ahmad R (2010) Duck meat utilization and the application of surimi-like material in further processed meat products. J Biol Sci 10:405–410
Shaw H (2017) Tea smoked duck. https://www.honest-food.net/tea-smoked-duck-recipe. Accessed 28 June 2020
Shim J, Kim T, Kim Y, Jeon K, Ahn K, Paik H, Choi Y (2018) The ratios of pre-emulsified duck skin for optimized processing of restructured ham. Korean J Food Sci Anim Resour 38(1):162–171
Tahergorabi R, Jaczynski J (2017) Nutraceutical egg products. In: Hester PY (ed) Egg innovations and strategies for improvements. Academic Press, pp 271–280
Tai C, Tai JJL (2001) Future prospects of duck production in Asia. J Poult Sci 38:99–112
Wang D, Xu W, Xu X, Zhou G, Zhu Y, Li C, Yang M (2009) Determination of intra intramuscular phospholipid classes and molecular species in Gaoyou duck. Food Chem 112:150–155
Winston (2012) Treasure duck. https://www.thehungryescavator.com/2012/07/8-trasureduck-recipe.html. Accessed 28 June 2020
Wong A (2019) Pekin duck. https://www.greatbritishchefs.com/recipes/peking-duck-recipe. Accessed 28 June 2020
Xu W, Xu X, Zhou G, Wang D, Li C (2008) Changes of intramuscular phospholipids and free fatty acids during the processing of Nanjing dry-cured duck. Food Chem 110:279–284
Yang HS, Ali MS, Jeong JY, Moon SH, Hwang YH, Park GB, Joo ST (2009) Properties of duck meat sausages supplemented with cereal flours. Poult Sci 88:1452–1458
Zhou X, Xie F, Wang Y, Tang W, Zhou Y, Xiao Y (2014) Evaluation of antioxidant activities of extract from Beijing roast duck. Food Sci Technol Res 20:449–457

Dr. George T. Oommen, a native of Thrissur, Kerala, India, is Ph.D. in Meat Science and Technology from Madras Veterinary College, TANUVAS, Chennai and a postgraduate in Veterinary Public Health from Kerala Agricultural University. He was a Professor and the Head of Meat Technology Unit and Department of Livestock Products Technology, Kerala Veterinary & Animal Sciences University. During 34 years of service, he was involved in teaching, research, and entrepreneurship development activities in meat science and technology, public health, quality assurance, meat and poultry products development and abattoir management. He guided 50 Postgraduate and Doctoral research work, published 59 papers in journals, national and international conference proceedings, co-edited, and co-authored a book and compendia. He is a technical member/expert in BIS and FSMS audit team of several national organisations and institutions.

Dr. Sathu T hailing from Thrissur, Kerala did his Master's Degree (Livestock Products Technology) from IVRI, Izatnagar, India, and Ph.D. from Kerala Veterinary & Animal Sciences University (KVASU). He served the State Animal Husbandry Department of Kerala for six years from 2002 and now a Faculty in the Meat Technology Unit and Dept. of Livestock Products Technology, KVASU. He is involved in teaching and research in meat, poultry and egg, new products development, its commercialisation, supervision of meat and poultry processing and a national trainer for HRD. He supervised nine PG and one Doctoral research work and published more than 45 papers in national and international journals and conference proceedings and recipient of 10 best research paper awards. He is a Technical Expert to Quality Council of India.

Dr. Wen-Shyan Chen obtained his Master's Degree (Animal Nutrition) and Ph.D. (Meat Science) from National Chung Hsing University, Taichung, Taiwan. He joined Animal Products Processing Division, Livestock Research Institute, Taiwan in 1993 for research in meat processing. He was engaged in the slaughter and dressing of pig, chicken, geese and duck, meat processing and research for almost three decades. He published more than 150 papers in national and international journals and conferences. He assists domestic manufacturers in commercializing meat processing technology, solves their problems and establishes a liaison between academic research and Industrial applications. He received Agricultural Contribution Award from the Taiwan Agricultural Society in 2020.

Value Addition of Feather and Down 13

Yao K. Y. Chen

Contents

Y. K. Y. Chen (✉)
Technical Commission, International Down and Feather Bureau (IDFB), Höchst, Austria
e-mail: yao@hoplion.com.tw

A. Jalaludeen et al. (eds.), *Duck Production and Management Strategies*,
https://doi.org/10.1007/978-981-16-6100-6_13

Abstract

Geese and ducks are raised primarily for their meat; therefore, feather and down are the by-product of meat processing industry. Recycling of feather and down, otherwise go as waste, to turn into valuable products is an environmentally friendly business. Although the down of goose and duck is basically very similar in structure, it can be recognized under microscope. The goose down and feather tend to have a more valuable marketing image and is usually more expensive and more sought-after than duck down and feather. China is the leading country in down and feather production. The supply chain of this industry involves many entities such as farm, slaughterhouse, collector, processor, down suppliers, manufacturer, trader, and retailer. The raw and fine washing are the major protocols in the processing of down and feather. The indicators of down and feather quality are content analysis, fill power, specie, turbidity, fat and oil content, and oxygen number. The feather and down are used in bedding, furniture, and cushion, apparels such as jacket, vest, and fashion style, outdoor gears such as military, camping, and sports materials, and badminton and decoration items. International Down and Feather Bureau functioning from Austria is functioning with the objectives of protecting consumer safety, developing and promoting International Standards for feather and down, and re-assessing Testing Institutes annually and providing assurance. This organization contributes immensely to dissuade the apprehensions of consumers on animal welfare and zoonotic importance associated with this industry.

Keywords

Applications · Down · Feather · Processing · Quality control · Waterfowl

Abbreviation

IDFB International Down and Feather Bureau

13.1 Introduction

Down and feather are the biodegradable and environmentally friendly natural products for bedding, apparel, and outdoor gear. The use of down and feather assumes significance from the point of view of resource conservation of a by-product of the waterfowl meat processing industry from going to waste. More importantly, production of down and feather products require a lower carbon footprint compared to the manufacture of synthetic materials. Moreover, these materials are the best insulator of natural fibers. Comforters made of them provide the sleepers with the added warmth leading to less room heating. Down and feather are durable and keep the insulating ability long lasting for bedding, apparel, and sleeping bags.

The International Down and Feather Bureau (IDFB) is the global association of experts of collection and processing of down and feather material and manufacturing of its products from different countries in the world. The members are experts in the production of down and feather products from all over the world. Although this traditional business is more than 4000 year old, ignorance of actual process of feather harvesting and processing has raised many unnecessary controversies. This chapter dwelt with the history, basics, processing, products, marketing, and misconceptions on down and feather with a holistic idea of enlightening the readers on this subject.

13.2 What Is Feather and Down?

Feather and down only come from waterfowl, which are aquatic birds such as ducks and geese that are accustomed to spend a substantial amount of time in freshwater. Land fowl are birds that live only on land such as chickens and turkeys. Both waterfowl and land fowl have feathers, but only waterfowl have down.

Geese and ducks are raised for their meat. Therefore, feather and down are the by-product of the poultry meat processing business. When the birds are slaughtered, the feather and down are left as waste. The waste is recycled and turned into valuable products. Therefore, it is an environmentally friendly business.

The most valuable portion of the waste is "down," which is obtained from the breast area of the geese and ducks, that grow under the outer protective layer of feathers. Although the commercial products contain a mixture of both down and feathers, the value of the products containing higher proportion of the down are priced higher.

Feather is the principal covering of birds, has a flat, two-dimensional shape, and has a leaf structure with a hard central stiff quill. Feathers contain a quill shaft in the middle and series of softer fibers arising on either side (Fig. 13.1) (Anon 2010a). Feathers can protect the birds from cold and moisture and provide insulation.

Down is a light, fluffy coating cluster, which formed in three dimensions with down fibers (Fig. 13.2). It looks much like a pod of dandelion plant. The special structure entraps more air and therefore has more "loft" or insulation (Anon 2010a). It is the most efficient natural insulator in the world, and it can help birds to keep warm and protect them from other elements. Down is used in comforters, pillows, bedding articles, and garments.

Down and feather are structurally different. A down cluster has unique identity and it will never develop into a feather. It is not a young or small feather (Anon 2010a).

Fig. 13.1 Structure of a feather

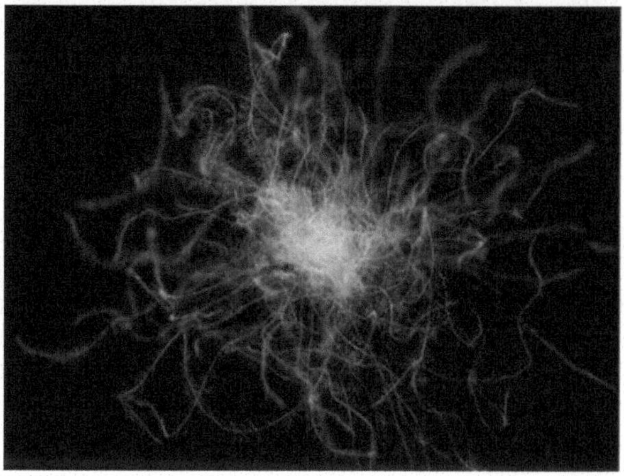

Fig. 13.2 Structure of down

13.3 Difference Between Feather and Down Materials

13.3.1 White and Gray

Down can be classified into white and gray. White down is generally preferred because it does not show through and darken light colored coverings. However, there is no connection between the color and the quality of down. The quality of down is determined mainly by the maturity and specie of the bird.

13.3.2 Goose and Duck

The two most common types of down and feather used in natural filling products are goose and duck, which are both waterfowl species. The down of goose and duck is basically similar in structure, and it can be recognized under microscope. However, goose down and feather tend to have a more valuable marketing image and is usually more expensive and more sought-after than duck down and feather. One of the major reasons for this is that goose down and feather have less odor problems. Because the diet of duck is usually more diverse compare with goose (goose is primarily herbivorous).

Furthermore, down clusters of geese are generally larger than duck and typically come from more mature birds. The down with larger clusters tends to have higher fill power performance, which is fluffier and more resilient, and can be used in high quality of down comforters with better insulation. Large and mature ducks have good quality down too.

The goose down is priced much higher compared to duck down with thc same down cluster content. However, in some situations, a high quality duck down can be better than a mediocre goose down. One notable exception is that the down from the Eider duck, known as Eiderdown, clusters of which are very dense with exceptional "cling" property. Eiderdown is the softest and best insulating of all downs, even the very best goose down, which means, that eiderdown makes the very best comforters and the softest pillows in the world. However, it is very costly due to its rarity (Anon 2019a).

13.3.3 Eiderdown

Eiders (/ˈaɪ.dər/) are the large sea ducks belonging to the genus *Somateria*. The scientific name of Eiders is derived from Ancient Greek words *somatos* meaning "body" and *erion* meaning "wool" (Jobling 2010). Eiderdown is the down of eider duck. It is considered the best quality and the most expensive of all downs.

The Eiders (*Somateria mollissima*) have unique features to adjust to the cold and the fluctuating climate in northern latitudes (Hanna et al. 2004). Eiders nest on coastal islands in colonies as colonial breeders with number of individuals ranging from less than 100 to even up to 15,000 in a single colony (Chapdelaine et al. 1986). The down is molted off from the breast portion of the female ducks, with which the duck makes its nest (Fig. 13.3). The molting of the down allows direct contact between skin in the breast region to eggs under incubation and to the young ducklings while brooding and the down around the nest insulates the eggs from the chilling.

The Eider's nests are found on the shores of arctic countries such as Alaska, Canada, Greenland, Norway, Finland, and Russia (Merkel and Gilchrist 2010). At present, eiderdown collected from Iceland forms the overwhelming majority of this product marketed in the world.

The Eider returns to the same nesting places year after year during hatching season (Peter et al. 2010). The eider farmers exercise great care every year to make the nesting area peaceful and attractive for the birds. They use all kinds of devices, including pennants and bells, to make the nesting areas very attractive. The farmers also guard the breeding colonies to protect from predatory animals and birds. The

Fig. 13.3 Eider duck. (Photo courtesy: Jonathan Cannon, Bangor, United Kingdom)

ducks reward the farmers with down for protecting their nesting areas. After hatching of the eggs, the duck abandons their nest with her nestlings, leaving the down behind for the farmer. Thus Eider and the farmers live together in perfect harmony and co-operation. This type of cohabitation of man and wild birds in the natural environment is unique (Anon 2019b). There are community-based management systems for commercial harvesting of eiderdown is also in place northern Canada or elsewhere since very long (McDonald and Fleming 1991).

The Young's moduli, a measure of stiffness of an elastic material for ultimate strength and strain at break, is found to have greater in eiderdown barbs compared to that of goose or duck. The individual eiderdown plumes are also more compression resistant than those from geese and ducks (Edward 2015). The Eiderdown's clusters cling together in a Velcro-like manner. Eiderdown forms a cohesive uniform mass, while other down falls apart easily into its single elements on handling. Eiderdown has unique thermal insulation qualities for regulating temperature and moisture far better than other materials.

13.4 Supply Chain of Feather and Down

The main place of origin of feather and down includes China, Southeast Asia, Europe, and North America.

China is one of the main places of feather and down origin because of the large area and population of the country combined with a long tradition of keeping waterfowl and that duck meat as one of the popular menu in Chinese cuisine.

In the world, more than 3.26 billion ducks were slaughtered during the year 2018 as per the Food and Agriculture Organization of the United Nations statistics with

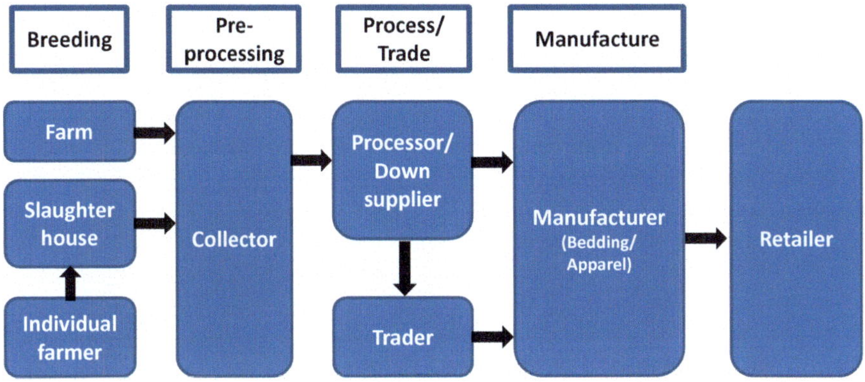

Fig. 13.4 Feather and down supply chain

the mainland China alone contributing 74.41 percent (2.42 billion ducks) of it. The next nine countries that follow China in descending order of share are Myanmar (4.44%), Viet Nam (2.32%), France (2.25%), Bangladesh (1.69%), Indonesia (1.55%), Taiwan (1.14%), Malaysia (1.11%), Hungary (1.04%), and Egypt (1.01%). These top ten countries contribute more than 90% of world's ducks slaughtered. In terms of quantity of duck meat produced, China alone produced 3.24 million tonnes, which is 67.5% of world's production of 4.80 million tonnes in the year 2018 (FAOSTAT 2021).

The supply chain of feather and down is depicted in Fig. 13.4.

13.4.1 Definitions in Supply Chain

Feather and down are the by-product of the meat industry. The slaughterhouses slaughter the birds for meat and collect the feather and down materials, and then sold to the collectors for pre-processing. There are certain entities/individuals involved in every step of this process.

Farm: Any location where birds are bred or raised for food production is referred a farm. Collector-based farms and industrial farms are two main types of farming system.

Slaughterhouse: Any location where birds are slaughtered for food production.

Collector: An entity, which collects down and feathers from small farms or other collections agents.

Processors/ down suppliers: Those who purchase the material from collectors. After proper washing and sorting process, the feather and down are sold to manufacturers to produce bedding/ apparel products. Then the products are sold to the retailers.

Manufacturer: Any location where down and feathers are filled into fabric shells for consumer bedding and apparel products.

Trader: Any entity, which purchases finished down, and feathers and/or finished consumer products for re-sale to other entities. Often traders organize production of finished products from one or more sewing factories and sell the products to brand names and retailers.

Retailer: Any entity, which offers finished consumer products to end consumers under their own brand name or other brand names.

13.5 Processing of Feather and Down

Raw washing is the process of washing and sorting of original feather (Fig. 13.5). After raw washing process, the feather and down are washed, sterilized, de-dusted, mixed, and packed by fine washing process (Anon 2010b) (Fig. 13.6).

Down and feather products must be washed and processed before making them into bedding, apparel, and outdoor gear. They are thoroughly washed, rinsed, and sanitized to remove the dirt, oil, and bacteria, dried at a high temperature and then separated based on quality before filling in the final products. When exporting down and feather products to other countries, the products must meet the quarantine requirements of the importing countries.

Besides washing process, down materials can also be separated into the different qualities. The sorting machines, usually a tall structure, are adjusted continually for best results. This sorting machine contains a long channel and the air current in the machine makes the best down drifting up and blown to the maximum length. There will be at least one, but up to six chambers in a sorting machine. The more chambers it has, the more space it needs. The air current can bring the lightest and best quality of down to the end, which means last chamber and are graded as the finest quality, while the down of lesser quality and feathers drop to the front chambers.

13.6 Applications of Feather and Down

The evidence of use of feather and down has been found from 4000 years ago (Nina and Gardener 1936). The images of geese/ ducks are present in ancient Egyptian murals and it is said that Egyptians have domesticated geese since 200 years before. By 1800s, the gypsies of Eastern Europe had started gathering feathers and selling them to small feather processors, founding the modern feather and down industry model.

During the eighteenth century Industrial Revolution, people began automating the production process, started mass production of feather and down. Different parts

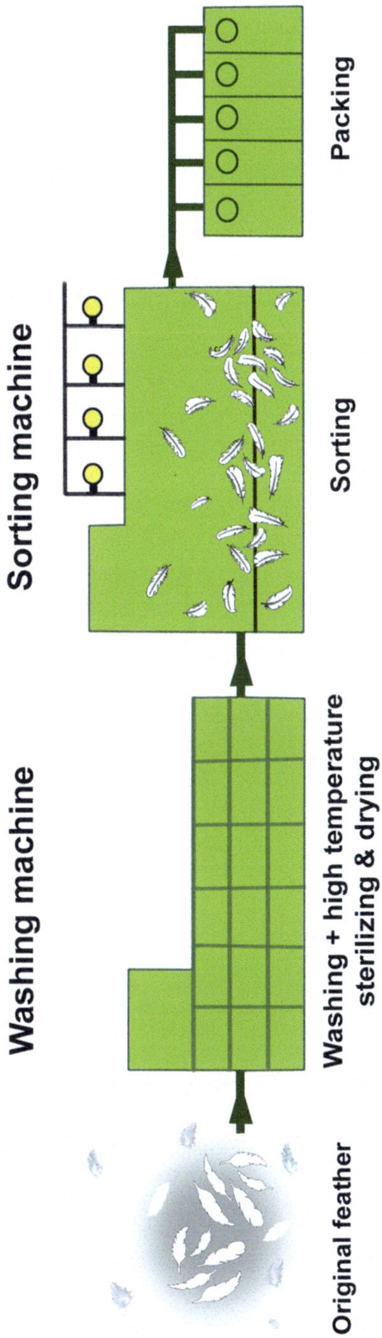

Fig. 13.5 Raw washing process of feather and down. (Recreated by: R. Richard Churchil)

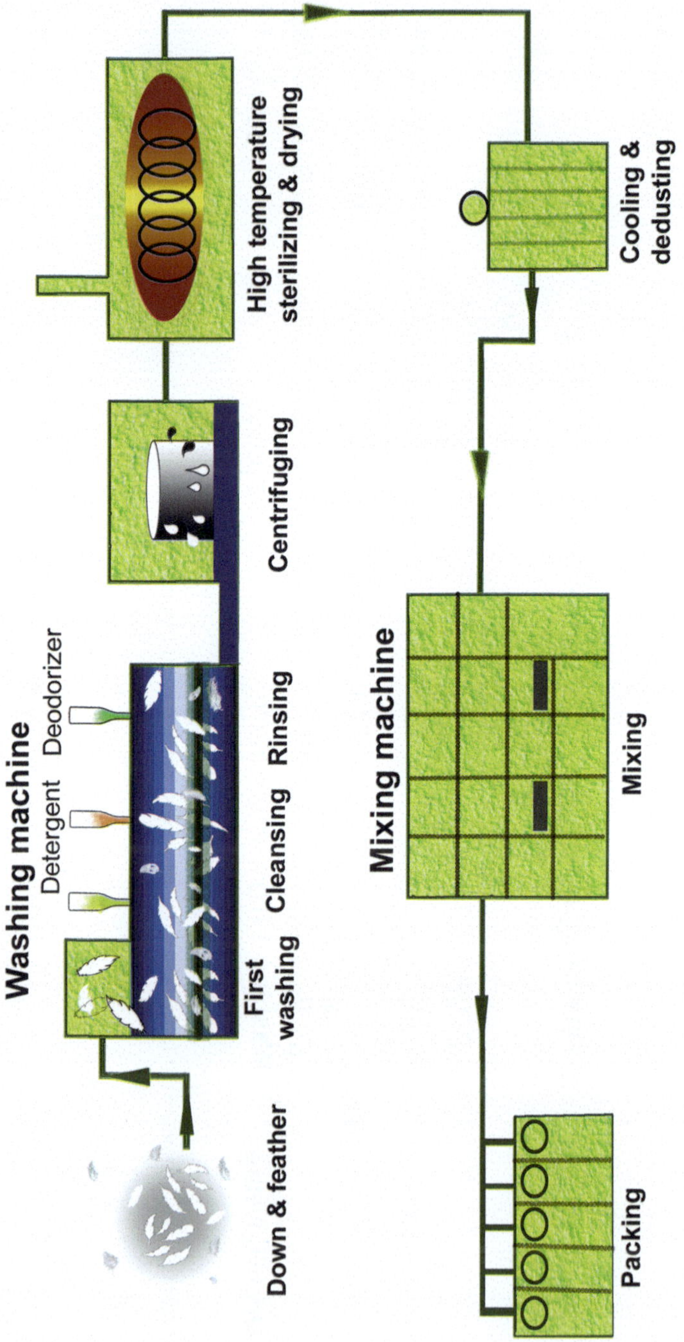

Fig. 13.6 Fine washing process of feather and down. (Recreated by: R. Richard Churchil)

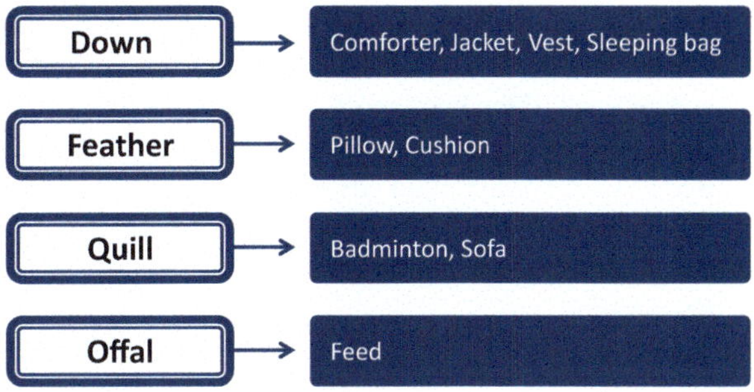

Fig. 13.7 Applications of feather and down

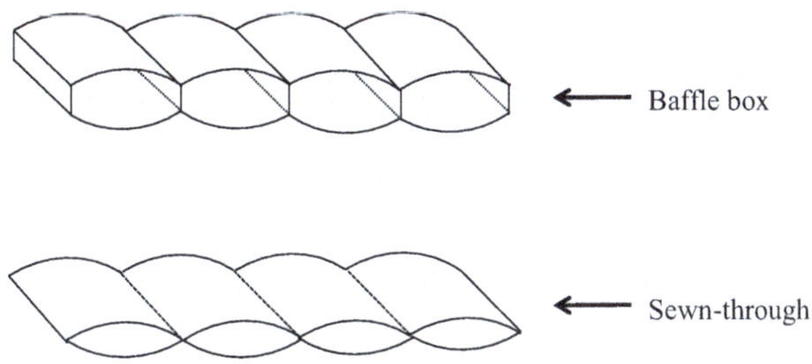

Fig. 13.8 Types of construction

of feather and down have different usage. It can be divided into a few categories as shown in Fig. 13.7.

- Home textile (Bedding, Furniture, Cushion).
- Apparel (Jacket, Vest, Fashion Style).
- Outdoor (Military, Camping, Sports).
- Other special quill products (Badminton, Decoration).

13.6.1 Home Textile (Bedding, Furniture, Cushion)

The bedding products made by down and feather, such as comforters, pillows, and featherbeds, are designed with the comfort of sleepers in mind. These products come in various types and styles for different purposes and usage.

13.6.1.1 Types of Construction (Anon 2019c)

1. ***Baffle Box Construction***: In this type of construction, the product has boxes sewn into it. The vertical strips of fabric are sewn at the top to bottom of the fabrics that create baffled walls between boxes as shown in Fig. 13.8. Although the baffles allow down to move within the boxes initially, the sidewise movement is prevented.
2. ***Sewn through Construction***: In this design, the top and bottom fabrics of the product are sewn together to help preventing the down from moving (Fig. 13.8). Products with sewn through construction may have boxes, channels, or other patterns. A multiple box sewn through construction generally prevents shifting of down more efficiently than an end-to-end channel construction.

13.6.1.2 Down and Feather Beddings (Anon 2019c)

1. ***Featherbeds***: A featherbed adds a layer of comfort, softness, and warmth to sleepers and their bed. Featherbeds are used on top of mattresses to make them softer, and to absorb some of the wear and tear that mattresses usually endure. Because one of the main advantages of down and feather is its natural breathability. Therefore, it is necessary to use only a cotton cover and not a waterproof cover. The cotton cover will protect the featherbed topper from dirtying and body perspiration.
2. ***Duvets and Comforters***: The duvets are large flat bags filled with down and feather used to cover the sleeper in bed. Duvets avoid the need for multiple sheets, blankets, and quilts. Duvets are similar to comforters, but when they are filled with down, they are commonly referred to as a duvet.
3. ***Pillows***: Pillows are more easily scrunched up and folded into the preferred sleeping position of the sleeper. The pillows made of only down are lighter in weight compared to those made of feathers. They may be (i) Firm density pillows (for side sleepers), (ii) Medium density pillows (back sleepers), and (iii) Soft density pillows (for stomach sleepers).

13.6.2 Apparel (Jacket, Vest, Fashion Style)

Down apparel can keep the user warm without the weight of heavy insulation. A properly maintained apparel is durable and can last many years. Apart from the filling weight and specie of down, the indicators of quality of apparel are fill power, feather ratio, shell material, and hood design.

13.6.3 Outdoor (Military, Camping, Sports)

Down is the best filling material for outdoor and military products. It provides warmth and lighter than other natural and synthetic materials, so it is easier to carry. Also, it can be compressed easily and make the total gear kit more compact.

13.6.4 Other Special Quill Product (Badminton, Decoration)

The quill feathers are usually separated from the other down and feathers at slaughterhouse. It will be removed by hands before the birds are put into plucking machine. It can avoid quill feathers being broken or damaged by machine. The quill collector could be different from down and feather collector.

13.7 Quality of Feather and Down

13.7.1 Main Factors in Determining the Quality of Down

13.7.1.1 Goose and Duck Growing Period

The longer the growing time, the more mature duck/goose down is, resulting in larger and stronger down cluster and higher fill power. Commonly, goose or duck are reared for meat purpose, and some are for producing *foie gras* or eggs. Depending on different eating habits in different areas, the growing period of goose and duck is also different. For example, the duck meat segment's growing period is around 30–40 days for roasted duck. Some Chinese like a soup called "Old Duck Soup." The duck used for this dish is the local Muscovy duck, and the growing periods can even reach 120 days. Duck for the purpose of *foie gras* have a relativity short growing period. On the reverse, the ducks dedicated to egg production have a long rearing time, sometimes more than 3 years.

13.7.1.2 Goose and Duck Growth Environment

The environment in which the goose and duck are raised is one of the important factor that affects the quality of down. Free-range geese and ducks near water have down that has good color and high cleanliness compared to indoor rearing. The geese and ducks living in cold weather areas develop large down clusters in order to keep them warm, which results in feathers and down with high fill power.

13.7.2 Origin and Quality of Down

It is often asked whether goose down from one geographical location is better than another, i.e. "Is Hungarian goose down better than Polish goose down?" or "Is Canadian goose down better than Hungarian goose down?," etc. The answer is an emphatic "No." Generally speaking, location has no association with the quality of down. In other words, down clusters of equal fill power will be the same size (fill power measures the size of the down cluster) regardless of origin. For example, an 800 cu-inch fill power goose down from Hungary will be the same quality as an 800 cu-inch fill power down from Poland or Canada. (Anon 2019d).

13.7.3 Quality Control and Testing

The down and feather products are to be tested frequently because of the fact that the natural raw material is highly non-homogeneous in nature. The products in one shipment may vary in terms of number of quality parameters from that of another consignment.

The quality testing of feather and down is time-consuming because the acceptable methods require tedious manual work. It is not possible to use machines for the task like separation based on specie and content analysis. Testing of each and every shipment of raw material is essential in order to ensure that the quality specifications are met, although all the shipments are coming from a single exporter/ farmers' cooperative. The raw material coming out of washing and processing is tested again to ensure the quality of finished products.

The feather and down in the manufactured finished products should also be verified whether correct fill material was used and also to ensure that the manufacturing process did not adversely affect the fill material. The government enforcement agencies perform the testing of finished products again at the point of retailing.

The major tests include content analysis, fill power, species, turbidity, fat and oil, and oxygen number.

13.7.3.1 Content Analysis

This is the weight percentage of down cluster which occupies down and feather filling material. Certain amount of the sample is separated by hand into the following components: down, feathers, down fiber, broken and damaged feathers, feather fiber, quill feather, residue, and land fowl feathers. Under the same volume, higher down cluster content usually means better insulation.

13.7.3.2 Fill Power

Fill power (also called "loft") is the most important index to determine down quality. It states which volume can be filled by certain amount of down under specific testing conditions of temperature and humidity. Most other properties such as its weight or insulation capacity depend on it. Longer growing period of birds means it will have more mature and larger down cluster, which will lead to the higher fill power result. Under the same down cluster content and the same filling weight, higher fill power means better insulation and more comfort.

13.7.3.3 Species

The species of down can be recognized by using microscope. Down filling material with stated specie, e.g. "goose down and feather" should receive species determination. Every country has its own standard of labeling. For example, in the USA, if the goose down and feather content is less than 90%, then it cannot be called as "goose down and feather."

13.7.3.4 Turbidity

The dirt and dust of both organic and non-organic origin in the down and feathers can be tested by turbidity test. The sample is soaked and agitated in pure water, and then the clarity of water is measured by eyes or machines.

13.7.3.5 Fat and Oil Content

Fat and oil are extracted from a down and feather samples. The amount of extractable fat and oil from the feather indicates its cleanliness and its susceptibility to odor problems. But natural down and feathers require some oil to function well.

13.7.3.6 Oxygen Number

The oxygen number indicates product cleanliness. It is usually done with turbidity testing because the organic materials have to be measured after soaking and agitating the down and feathers in a solution of pure water.

13.8 Controversial Issues Related to Feather and Down

13.8.1 Harvest Plucking

Some animal rights organizations such as FOUR PAWS and PETA have sought to apprehend the down and feather industry. They maligned feather and down industry, falsely implicating that most of feather and down comes from live plucking.

Live plucking originated from an old custom in Eastern European countries that helps the goose to molt during its molting season. For centuries, family farms have inherited this tradition, collecting down and little feathers from goose in its molting season. According to EU laws, the collecting of down and feathers from the specific European goose species during molting season permitted only during a specific time and only using a specific method. Unfortunately, some farmers do not abide by the EU animal protection law to practice live plucking of geese. Certainly, all animal protection organizations and down industrial organizations condemn such behavior which disregards European law and animal rights (CFDIA 2009).

The ducks are generally not live-plucked because they have short growth period (around 40 days) and high reproduction ability. Its hatching is not affected by seasonal interference. In addition, duck can be slaughtered when FCR (feed conversion ratio) is most reasonable. Therefore, there is no necessity for duck live plucking (CFDIA 2009).

The growing period for meat goose is short. Only a few specific breeding geese, which have long living time, may risk suffering live plucking in order to reduce production cost. However, the incidence of this practice is very limited (CFDIA 2009).

13.8.2 Bird Flu

There is some concern as to whether feather and down products contain bird flu virus. With proper washing and high-temperature sterilized process, the bird flu virus will be destroyed and not harmful to human beings at all. However, bird flu outbreaks are closely related to feather and down industry. It may cause the quantity of supply side to reduce and the price might rise rapidly.

13.9 International Down and Feather Bureau (IDFB)

13.9.1 Introduction

International Down and Feather Bureau (IDFB) was founded in 1953 in Paris, France. The current location is in Bregenz, Austria. IDFB holds meetings twice in a year. One internal meeting is being held in January every year in Frankfurt, and only presidential board and committee members are attending this meeting. The plenary meeting is being held in June in different countries. IDFB members from America, Europe, and Asia take turns as the organizer of the meeting. It is an international trade association of the feather and down industry. As a global association for feather and down industry, IDFB provides a platform for exchange of information on feather and down, testing and certification institutes and promoting the advancement of global feather and down industry. IDFB also provides consumers clear and accurate feather and down related expertise and develops universal feather and down testing standards.

13.9.2 Goals

- Provide safety to the consumers with clear and easily understandable definitions.
- Develop and promote International Standards for down/ feather fillings. (IDFB Standards contain definitions and testing regulations).
- Annual re-assessment of approved Testing Institutes to guarantee the maintenance of International Standards.
- Provide assurance: Annual check-up of the approved Testing Institutes that guarantees uniform testing methods worldwide.

Unit	Members
Executive committee (3 persons)	• president • vice presidents (two persons)
Presidential board (PB)	• four largest consumer countries and two largest producer countries have to be represented, that is, USA, China, Japan, Germany, France, and Hungary. • one member represents all of the “smaller” nations. • the chairman of technical commission and public relation committee.
Supervisory board	• the chairpersons of organizations on their representatives. • at least one person of each nation.
Technical commission (TC)	• chairman • members are the experts of feather and down testing and production.
Public relation committee (PRC)	• chairman • representatives taking care of public relationship and marketing in industry.

(continued)

Unit	Members
Members	• national associations of feather and down industry. • individual processors or traders of feather and down raw material. • individual producers or traders of finished products filled with feather and down materials. • independent testing institutes for feather and down as filling material.

Note: The author of this chapter is the present Chairman of Technical Commission of IDFB

13.9.3 Technical Commission (TC)

The IDFB TC develops and improves the testing instruments, updates the IDFB testing regulations, provides annual round robin assessment of independent testing institutes and research more accurate testing methods. The incumbent TC chairman, Mr. Yao Chen, had been elected in 2006 and was the first Asian-elected chairman. He successfully leads the Asian feather industry to step into the international scene.

This working group includes members with technical expertise who are tasked with the following:

- Further development and monitoring of quality testing method of feather and down materials.
- Organization of round robin tests, followed by certification by independent test institutes.
- Scientific research with the aim of adding new supporting evidence for the advantages of down and feathers.

References

Anon (2010a). Anonymous, Down & feather trivia, International Down and Feather Testing Laboratory, DFL Headquarters, 1455, South 1100 East, Salt Lake City, UT 84105, USA. http://www.idfl.com/media/pdfs/xpdfs/IDFL%20Info%20-%20Down%20and%20Feather%20Trivia.pdf. Accessed 30 May, 2019

Anon (2010b) Anonymous, Washing and sorting down and feathers (typical factory process), IDFL laboratory and institute, 1455 South 1100 East Salt Lake City, UT 84105, USA. http://www.idfl.com/media/pdfs/xpdfs/IDFL%20Info%20-%20Washing%20and%20Sorting%20Down%20and%20Feathers%20(Typical%20Process).pdf. Accessed 31 May 2019

Anon (2019a) Anonymous, About Eiderdown, Plumaria Bay, 37960, Theo Lane, Birds view, Washington, USA – 98237. https://www.plumeriabay.com/info/eider-down.aspx. Accessed 30 May 2019

Anon (2019b) Anonymous, High Quality Icelandic Eiderdown, King Eider Co. Ltd. Víkurgata 1, 340 Stykkishólmur, Iceland. http://kingeider.is/Eiderdown.html. Accessed 31 May 2019

Anon (2019c) Anonymous, For Your Home, International Down and Feather Bureau (IDFB), Josef-Huter-Strasse – 316900, Bregenz, Austria, http://www.idfb.net/down-feather-you/for-your-home/. Accessed 31 May 2019

Anon (2019d) Anonymous, Goose down - a warm & fluffy wonder, Plumaria Bay, 37960, Theo Lane, Birdsview, Washington, USA – 98237. https://www.plumeriabay.com/info/what-is-goose-down.aspx. Accessed 30 May 2019

CFDIA (2009). Finding the truth about 'live-plucking' & 'harvesting', China Feather and Down Industrial Association. https://www.idfl.com/documents/50/IDFL_Info_-_Live_Plucking.pdf . Accessed 30 May 2019

Chapdelaine G, Dupuis P, Reed A (1986) 'Distribution, abondance et fluctuation des populations d'eider à duvet dans l'estuaire et le golfe du Saint-Laurent' [Distribution, abundance and population fluctuations of the common eider in the estuary and gulf of St. Lawrence]. In: Reed A (ed) Eider ducks in Canada. Canadian Wildlife Service Report Series (in French), Ottawa, pp 6–11

Edward FM (2015) The structure and properties of down feathers and their use in the outdoor industry. PhD thesis, University of Leeds. http://etheses.whiterose.ac.uk/id/eprint/9268. Accessed 30 May 2019

FAOSTAT (2021) Food and Agriculture Organization Corporate Statistical Database, Food and Agriculture Organization of the United Nations, Rome, Italy.http://www.fao.org/faostat/en/#data. Accessed 14 July 2021

Hanna E, Jónsson T, Box JE (2004) An analysis of Icelandic climate since the nineteenth century. Int J Climatol 24:1193–1210

Jobling JA (2010) The helm dictionary of scientific bird names. Christopher Helm, London., ISBN 978-1-4081-2501-4, p 359

McDonald M, Fleming B (1991) The role of traditional knowledge in community-based management of an eiderdown industry developing in Northern Canada, In: Proceedings of the conference on 'Common Property Conference, the Second Biennial Conference of the International Association for the Study of Common Property, Winnipeg, Manitoba, Canada on 26–30 September 1991

Merkel F, Gilchrist G (2010) Seabirds – common eiders. Arct Biodivers Trends:38–40

Nina NM, Gardener AH (1936) Ancient Egyptian paintings. University of Chicago Press, Chicago, p 209

Peter LF, Fast H, Gilchrist G, Clark RG (2010) Nest-site materials affect nest-bowl use by common eiders (*Somateria mollissima*). Can J Zool 88(2):214–218. https://doi.org/10.1139/Z09-131. Accessed 30 May 2019

Yao K. Y. Chen was born in 1954. He has been in down and feather industry for 44 years since 1977. He is the fourth-generation President and incumbent Chairman of Hop Lion Feather Works Corporation, a family-run business founded in 1908. He became the first Asian-elected Chairman of the Technical Commission of International Down and Feather Bureau (IDFB) in 2006, and is also the Chairman of Taiwan Time-Honored Stores Club. Yao work on domestic to international operations, including sourcing, processing and marketing and established huge resources and connections to develop the Asian feather industry internationally.

Duck Diseases and Disease Management 14

Yen-Ping Chen, Chao-Fang Yu, and Yu-Hua Shih

Contents

Abstract

Commercial production of duck meat is a growing industry around the world. The change to more intensive production systems means biosecurity and disease control are becoming more important. Diseases described in this chapter are common infectious diseases of the ducks: duck virus hepatitis, duck virus enteritis, avian influenza, parvovirus infection, *Riemerella anatipestifer* infection, fowl

Y.-P. Chen (✉) · C.-F. Yu · Y.-H. Shih
Animal Health Research Institute, Council of Agriculture, New Taipei City, Taiwan
e-mail: ypchen@mail.nvri.gov.tw; yhshih@mail.nvri.gov.tw

A. Jalaludeen et al. (eds.), *Duck Production and Management Strategies*,
https://doi.org/10.1007/978-981-16-6100-6_14

cholera, reovirus infection, duck circovirus infection, and duck egg drop syndrome.

Keywords

Disease management · Infectious diseases · Prevention and control

Abbreviations

AI	Avian influenza
AIV	Avian influenza virus
CPE	Cytopathic effect
DAstV-1	Duck astrovirus type 1
DAstV-2	Duck astrovirus type 2
DEV	Duck enteritis virus
DHAV	Duck hepatitis A virus
DVE	Duck virus enteritis
DVH	Duck virus hepatitis
GPV	Goose parvovirus
HA	Hemagglutinin
HI	Hemagglutinin inhibition
HPAI	Highly pathogenic avian influenza
ICTV	International Committee on Taxonomy of Viruses
IgY	Immunoglobulin Y
IVPI	Intravenous pathogenicity index
LAMP	Loop-mediated isothermal amplification
LPAI	Low pathogenic avian influenza
MDPV	Muscovy duck parvovirus
NI	Neuraminidase inhibition
OIE	World Organization for Animal Health
OmpH	Outer membrane protein H
rDEV-AI	Recombinant duck enteritis virus
rFPV-AIV	Recombinant fowl pox virus
rHVT-AIV	Recombinant Herpes turkey virus
rNDV-AIV	Recombinant Newcastle disease virus

14.1 Introduction

In this chapter, the most common infectious diseases of ducks are listed and described. Similar to managing other avian diseases, management of duck diseases needs rapid identification and immediate and effective treatment, while removing any possible pathogens via solid biosecurity measures. Therefore, the first part of this chapter gives general recommendations on the control and prevention of duck diseases. Following the recommendations, comprehensive knowledge of each disease is shared, including the etiology, transmission, host range, clinical signs, pathology, and feasible prevention and control strategies specific to the disease.

14.2 Recommendations for the Control and Prevention of Diseases

The following recommendations to the control and prevention of diseases are provided.

- Always follow good biosecurity practices in the farm.
- Buy ducks from an established disease-free source and new ducks should be quarantined before being placed on the farm premises.
- Provide quality feed and water.
- Provide a stress-free environment for ducks; this includes providing proper housing, management, ventilation, and nutrition.
- If you have to introduce additional ducks from an outside source, it is better to establish a second flock.
- Never mix ducks of different ages because younger ducks may be more susceptible to diseases.
- Keep breeders away from growing ducks.
- Keep a close watch on your ducks and monitor them frequently.
- Immunize ducks against known infectious diseases at the proper time and give timely medications. Always use the correct vaccine or medication at the recommended dose.
- When inspecting the ducks, always go from the youngest to the oldest.
- Isolate sick ducks immediately. Removing sick ducks from the flock reduces the number of infectious organisms.
- Safely dispose dead ducks immediately by either incinerating or burying them. Obtain an early diagnostic report by sending samples/carcasses to a veterinary laboratory for diagnosis.
- Discourage visitors to the farms and do not allow visitors enter your farms without wearing proper personal protective clothing. Make sure that they have not contacted other poultry for at least 14 days prior to their visit.
- Do not allow a buyer to bring unclean crates and/or boxes on to the farm for transporting the ducks when selling ducks.
- Thoroughly clean and disinfect the buildings and equipment between flocks. This may not make the building sterile but it can reduce the number of infectious microorganisms.
- Feed and water the ducks indoors, where possible in order to reduce the risk of exposing your ducks to wild birds.
- Keep detailed records at all times.

14.3 Viral Diseases

14.3.1 Duck Virus Hepatitis

Duck virus hepatitis (DVH) is the highly contagious disease characterized by hepatitis and sudden death in ducklings. DVH is on the list of diseases notifiable to the World Organization for Animal Health (OIE). This disease is caused by duck hepatitis virus (DHV) types I, II, and III. DHV type I, also known as duck hepatitis A virus (DHAV), belongs to a genus *Avihepatovirus* in the family *Picornaviridae*. DHAV can be classified into three genotypes: DHAV-1, DHAV-2, and DHAV-3. DHAV-1 is the most virulent virus and widely distributed (Kim et al. 2006). DHAV-2 has been only isolated in Taiwan (Tseng and Tsai 2007), and DHAV-3 has been reported in South Korea, China, and Vietnam (Kim et al. 2007; Fu et al. 2008; Wen et al. 2014). DHAV-1 is a positive sense single-strand RNA virus and the genomic size is approximately 7700 nucleotides. DHAV is highly resistant to disinfectants and acidic environment (Kim et al. 2006). DHV type II is caused by duck astrovirus type 1 (DAstV-1) and DHV type III is caused by duck astrovirus type 2 (DAstV-2); both are members of the *Astroviridae* family (OIE 2017).

DHAV spreads rapidly to all susceptible ducklings by direct contact with infected ducklings or indirect contact with contaminated water, feed and equipment (Woolcock and Tsai 2013). DHAV can be transmitted through respiratory or oral route (Hanson and Tripathy 1976; Priz 1973). Infected adult ducks can become asymptomatic carriers and excrete the virus in their feces for up to 8 weeks after infection (Woolcock and Tsai 2013). Egg transmission of the virus has not been reported. Wild birds might act as mechanical carriers of this virus (Asplin 1961).

DVH mainly occurs in ducklings. DHAV-2 was isolated from goslings in Taiwan in 2004 with a mortality rate of over 70%. The infected goslings showed the similar hemorrhagic hepatitis lesions as DHAV infected ducklings (Tseng and Tsai 2007). In 2011, DHAV-3 was isolated from geese with hemorrhagic liver lesions. The morbidity of infected geese was 20–40% and the mortality was less than 5% (Liu et al. 2011).

The susceptibility of ducks for DHAV-1 infection is decreasing while age increases. The mortality may be up to 95% for ducklings under the age of 1 week, but mortality will be low when the age reaches over 4 weeks. Spread of DHAV-1 in a flock is very rapid, and the course of the disease usually occurring within 3–4 days. The clinical signs include reduced movement, opisthotonus, depression, lethargy, ataxia, and sudden death. DHAV usually occurred in ducklings, and breeders on infected premises do not become clinically ill and continue in full production (Woolcock and Tsai 2013). However, DHAV-1 strains isolated from egg-laying ducks in China in 2016 can infect adult ducks and result in decreased egg production. The infected ducks show decreased feed uptake, exterior feather detachment and egg drop syndrome (Zhang et al. 2018).

The main lesions of DVH are enlargement of the liver and diffuse hemorrhagic foci in liver. Sometimes, enlarged spleen and swollen kidneys with some congestion of renal blood vessels may be seen. The microscopic lesions in the liver include

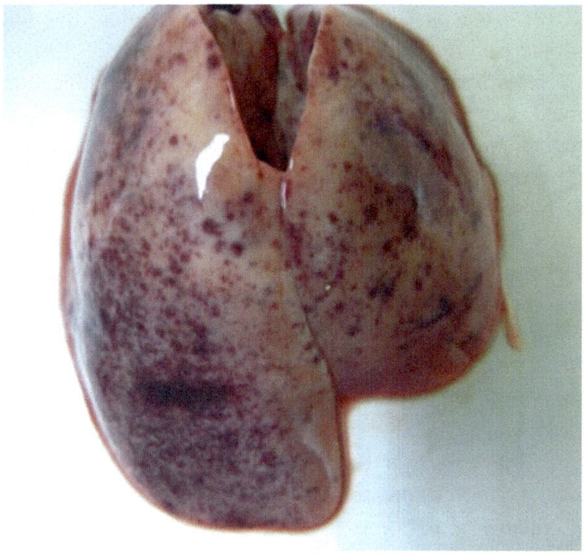

Fig. 14.1 Diffuse hemorrhagic foci in liver of DHAV-2 infected Tsaiya duck

hepatocyte necrosis and bile duct hyperplasia (Fig. 14.1). Infection of the new strain of DHAV-1 in adult ducks could induce ovarian hemorrhage, ovarian necrosis, and follicle atresia (Zhang et al. 2018).

Diagnosis of DVH is based on the clinical signs and gross pathological lesions in ducklings, but it is not possible to distinguish DHAVs' genotypes by clinical signs. The neutralization test is reliable for typing DHAV, but it seems time-consuming and needs higher technique (Sandhu et al. 1992). Molecular diagnostic methods such as RT-PCR and real-time RT-PCR can be used for DHAV typing (Chen et al. 2013; Kim et al. 2008). Direct fluorescent antibody test or immunohistochemistry of liver slides can also be used to rapid and accurate diagnosis of DVH (Zhang et al. 2012).

Strict isolation of duck flocks could prevent DVH, but it is very difficult in the epidemic area. DHAV usually causes lethal diseases of ducklings less than 4 weeks of age. There are some immune strategies to prevent DVH: injecting the immunoglobulin Y (IgY) in day-old ducklings for passive immunization; immunization of breeders 1 month before laying to ensure high level of maternal antibodies in hatched ducklings; and direct immunization of ducklings with a live attenuated vaccine. However, immune response in ducklings is not induced until 3–5 days after vaccination, meanwhile ducklings have a high potential risk of DHAV infection at 1–3 days old (Kim et al. 2009). IgY injection could provide good protection against DHAV. In DVH outbreak, IgY products are efficacious to prevent the disease from spread. Although immunization is useful, it does not contribute to eradication of the disease. The application of inactivated or attenuated vaccines to breeder ducks to produce maternal immunity against DHAV in their progeny is a standard method for prevention.

Live attenuated DHAV vaccine strains are produced by serially passaging the virus in chicken embryos or duck embryos (Kim et al. 2009; Woolcock and Tsai

2013; Wu et al. 2020a). In addition, a bivalent live attenuated vaccine has shown protective efficacy against DHAV-1 and DHAV-3 in ducklings (Kang et al. 2018).

Several recent studies demonstrate various molecular strategies for the development of candidate vaccines against DHAV. Zou et al. (2016) and Niu et al. (2020) demonstrated that attenuated duck enteritis virus strain was an efficient vector for developing bivalent and/or polyvalent live attenuated vaccine against DHAV-1 and DHAV-3. Wang et al. (2019a, b) indicated that the vaccination of the recombinant avian adeno-associated virus expressing VP1 and VP3 protein, respectively, of DHAV-1 induces protective immunity against DHAV-1 challenge. Moreover, a recombinant *Lactococcus lactis,* which can express VP1 protein of DHAV-3, was constructed and animals immunized with the recombinant showed effective protection efficacy against DHAV-3 infection (Song et al. 2019). In addition, an antigenic mimotope of DHAV is identified by phage display technology and might be a promising vaccine candidate for the prevention of both DHAV-1 and DHAV-3 infection (Zhang et al. 2020). Wang et al. (2018) indicate that the DHAV-1 virus-like particles expressed by baculovirus expression system in insect cells show a promising prospect as a candidate vaccine for DHAV-1. Furthermore, a suicidal DNA vaccine of the VP1 gene of DHAV-1 is also constructed and is able to elicit efficient immune response against DHAV-1 challenge (Fu et al. 2012).

14.3.2 Duck Virus Enteritis

Duck Virus Enteritis (DVE), also called as duck plague, is an acute, contagious, and fatal disease of waterfowl, resulting in significant economic losses in the waterfowl breeding industry. The disease has been reported in domestic and wild waterfowl in The Netherlands, USA, China, France, Belgium, India, Thailand, England, Canada, Hungary, Denmark, Austria, Vietnam, Germany, Bangladesh, Egypt, and Poland (Metwally 2013; Dhama et al. 2017).

The aetiological agent of DVE is Anatid alphaherpesvirus 1 or duck enteritis virus (DEV), a member of the genus *Mardivirus*, subfamily *Alphaherpesvirinae*, and family *Herpesviridae* (ICTV 2017). DEV has similar morphology to other herpesviruses and is composed of a DNA core, icosahedral capsid, tegument, and envelope (Yuan et al. 2005; Dhama et al. 2017). The virus particles are approximately 126–129 nm in diameter (Metwally 2013). The genome of DEV is a linear double-stranded DNA molecule with a length of approximately 158 kb and encodes 78 putative proteins (Li et al. 2009).

Field strains of DEV display differences in virulence, but all appear to be immunologically identical and antigenically related (Shawky and Sandhu 1997; Metwally 2013). The virus is sensitive to lipid solvents and heat (10 min at 56 °C). Significant reduction in titers is observed at pH 5, 6, and 10 and rapid inactivation occurs at pH 3 and 11 (Metwally 2013).

The virus can be transmitted by direct contact from infected to susceptible ducks or indirect contact with a contaminated environment. Since the waterfowl is dependent on an aquatic environment, water seems to be a natural route of viral transmission. Most outbreaks have occurred in duck flocks with access to open bodies of water cohabited with free-living waterfowl (Metwally 2013). The introduction of the virus in domestic ducks is usually via transmission from wild waterfowl (Gough 2002). Oral, intranasal, and parenteral administration of infected tissues can experimentally transmit DVE (Spieker et al. 1996). In addition, vertical transmission has been also reported in persistently infected waterfowl under experimental conditions (Burgess and Yuill 1981).

Migratory waterfowl has been reported as DEV carriers during many outbreaks. Recovered birds may become latent carriers and then may shed the virus periodically for a period. Latency and reactivation of the virus may be the cause of outbreaks in susceptible wild and domestic ducks. DEV may undergo latency as other herpesviruses, and trigeminal ganglion, lymphoid tissues, and peripheral blood lymphocytes seem to be the main latency sites for the virus (Shawky and Schat 2002; Metwally 2013; Dhama et al. 2017).

Members of the family *Anatidae* of the order *Anseriformes* such as ducks, geese, and swans are the natural hosts for the DEV. Some species such as domestic Muscovy ducks (*Cairina moschata*) exhibit higher susceptibility (Campagnolo et al. 2001). While others, such as mallards (*Anas platyrhynchos*), are more resistant to the lethal effects and are considered a possible natural reservoir of the infection (Metwally 2013; Dhama et al. 2017). The disease mostly affects older waterfowl but has been reported in birds ranging from seven days of age to mature breeders (Campagnolo et al. 2001; Metwally 2013). Field observations exhibit that recovered birds from natural infection are immune to re-infection with DEV (OIE 2018). The infection has not been reported in avian species other than anseriformes, mammals or humans and does not pose a zoonotic risk (Metwally 2013).

In domestic ducks, the incubation period for DEV is 3–7 days. After the onset of clinical signs, death usually occurs within 1–5 days (Metwally 2013). Mortality may range from 5 to 100%, depending on species, immune status, age, and sex of the affected birds and the virulence of the virus (Dhama et al. 2017; OIE 2018). Sudden, high, and persistent mortality is often the first sign of the disease in domestic ducks and mortality is often more severe in susceptible adult breeder ducks (Metwally 2013). In addition to sudden death, the clinical signs include weakness, depression, photophobia associated with half-closed pasted eye-lids, polydipsia, loss of appetite, thirst, ataxia, nasal discharge, watery diarrhea, ruffles feathers, and soiled vents (Metwally 2013; Dhama et al. 2017; OIE 2018). In 2–7 week-old ducklings, losses may be lower than in older ducks, and the infected ducklings frequently show dehydration, loss of weight, conjunctivitis, lacrimation, nasal exudate, blue coloration of the beaks and blood-stained vents (Metwally 2013; OIE 2018). In infected laying flocks, egg production may drop obviously (Metwally 2013).

The lesions of DVE are characterized by disseminated intravascular coagulopathy and necrotic degenerative changes in mucosa and submucosa of the gastrointestinal tract and lymphoid tissues. Vascular damage throughout the body results in hemorrhages in various tissues or the presence of free blood in the body cavities and intestinal lumen can be noticed. Petechiae and ecchymosis may be seen on the surfaces of heart ("paintbrush" appearance), liver, pancreas, intestine, lungs, and kidney. The "paintbrush" appearance of heart is often observed in adult ducks rather than in ducklings. In adult layers, hemorrhages may be observed in ovarian follicles, and sometimes a large amount of hemorrhages may exude to the abdominal cavity. Eruptions of specific digestive mucosal lesions, found in the oral cavity, esophagus, caeca, rectum, and cloaca, undergo progressive alterations during the course of the disease. Macular surface hemorrhages initially appear, then develop into elevated yellowish white hard plaques and organizes into green superficial scab, which may merge and form large, patchy, diphtheritic membranes. In the esophagus, macules arise parallel to longitudinal folds. Necrotic degenerative changes are obvious in the lymphoid and parenchymatous organs. Hemorrhagic annular bands can be seen from serous and mucous surfaces in the different portions of intestines. The spleen may be normal or smaller, dark, and mottled. The thymus may be atrophied and has multiple petechiae and necrotic foci. The bursa of Fabricius may be severely congested or hemorrhagic. In the ducklings, lymphoid lesions are more prominent than tissue hemorrhages, while in adult ducks, tissue hemorrhages and reproductive tract lesions are more prominent because of naturally regressed cloacal bursa and thymus. In geese, intestinal lymphoid disks are analogous to annular bands in ducks, and lesions resembled "button-like ulcers" can be seen. In swans, diphtheria esophagitis may be commonly seen. The liver exhibits enlarged, pale copper in color, and may have pinpoint hemorrhages and white necrotic foci. The pancreas may have petechiae and multifocal necrosis (Metwally 2013; Dhama et al. 2017; OIE 2018).

The microscopic lesions are characterized by vascular damage, resulting in the development of generalized hemorrhages and progressive degenerative changes of parenchymatous organs. Smaller blood vessels, venules, and capillaries are more significantly involved than larger blood vessels. Eosinophilic intranuclear inclusions and cytoplasmic inclusions may be seen in epithelial cells of the digestive, respiratory and reproductive tracts and in visceral organs (Metwally 2013; Dhama et al. 2017; OIE 2018).

The tentative diagnosis of DVE may be made based on disease history, clinical symptoms, and gross and histopathologic lesions. The confirmatory diagnosis of DVE should depend on isolation and identification of the virus. Suitable samples for isolation of DEV are tissues from liver, spleen, bursa and kidney, peripheral blood lymphocytes, and cloacal swabs. The virus isolation may be attempted by inoculation of sample homogenates into susceptible one-day-old ducklings, various cell cultures, and duck embryos. Cell cultures using primary duck embryo fibroblasts, preferably primary Muscovy duck embryo fibroblasts and Muscovy duck embryo liver cells, are reported as the methods for isolation of DEV (Wolf et al. 1976; Gough and Alexander 1990, OIE 2018). The cytopathic effects (CPE) may appear at 2–4 days post inoculation. Primary virus isolation can be carried out by inoculation onto the chorioallantoic membrane of 9–14-day-old embryonated duck eggs. The infected embryos may die and show characterized hemorrhages 4–10 days after inoculation (OIE 2018). Molecular detection such as conventional PCR (Hansen et al. 2000), real-time PCR (Yang et al. 2005; Zou et al. 2010), and loop-mediated isothermal amplification (LAMP) (Ji et al. 2009) has been reported for detecting DEV DNA in tissue samples and in cell culture or tissues of virus isolation. Recently, Xie et al.

(2017) reported a PCR assay to differentiate virulent from attenuated DEV strains. Neutralization assays performed in either embryonated eggs or cell cultures can be used to confirm the identity of the newly isolated virus. Serological tests have little value in the diagnosis of acute infections but serum neutralization has been used to monitor antibodies following exposure to the virus in wild waterfowl (OIE 2018).

The differential diagnoses in consideration of other diseases producing hemorrhagic and necrotic lesions include duck viral hepatitis, fowl cholera, necrotic enteritis, coccidiosis, and specific intoxications. Newcastle disease, avian influenza, and fowl pox may cause similar lesions but are infrequently reported in ducks (Metwally 2013).

There is no specific treatment of DEV infection. Prevention can be achieved by maintenance of susceptible birds in a disease-free environment or immunization. The susceptible birds should be avoided to contact with wild, free-flying waterfowl and direct or indirect contact with contaminated birds or materials (free-flowing water). Depopulation, removal of birds from the infected environment, sanitation, disinfection, and vaccination are the effective control measures in DEV infected flocks.

In countries where the disease is not enzootic and is truly exotic, quarantine of imported or clinically suspected Anseriforms should be strictly carried out. Surveillance of DVE in ornamental bird collections, zoos, and domestic growers of Anseriformes should be performed regularly using most advanced detection tools.

Currently, both live attenuated and killed vaccines are being given in broiler and breeder ducks older than 2 weeks of age (Dhama et al. 2017). Live attenuated vaccines are considered most effective against DEV. Since live vaccines are sensitive to environmental factors, during storage and transportation it is important to maintain the vaccine under optimal physical and physiological conditions of temperature, salt, and pH value to prevent the loss of vaccine molecule activity (Makhija and Kumar 2017). The attenuated vaccine is administered by subcutaneous or intramuscular routes and normally, the breeding flocks have to be revaccinated annually (Metwally 2013). The attenuated vaccine can be used in an outbreak, because it can provide protection rapidly after vaccination (Metwally 2013; OIE 2018). On the contrary, the killed DEV vaccine is effective in protecting ducks from the virulent strain infection (Shawky and Sandhu 1997) without the risk of introducing a live virus (Metwally 2013).

Recent studies demonstrated that attenuated DEV strain was an efficient vector for developing bivalent and/ or polyvalent live attenuated vaccine against avian influenza virus strain H5N1 (Liu et al. 2013a; Zou et al. 2017) and H9N2 (Sun et al. 2017), duck hepatitis A virus 1 and 3 (Zou et al. 2016), duck Tembusu virus (Zou et al. 2014; Zou et al. 2017; Chen et al. 2019), Newcastle virus (Ding et al. 2019), infectious bronchitis virus (Li et al. 2016), and fowl cholera (Apinda et al. 2020). The vaccine rapidly induced effective protection against delivering pathogens and DEV infections through a single-dose inoculation. It indicates that DEV has the potential to be utilized as a promising viral vector candidate for developing vaccines for poultry and waterfowl (Dhama et al. 2017).

DNA vaccine is a promising strategy for the prevention of DEV infection in ducks. Vaccination with plasmids encoding for DEV glycoprotein B, glycoprotein D, glycoprotein C, protein UL24, and protein tgB (Zhao et al. 2014a; Aravind et al. 2015; Liu et al. 2016) elicits both humoral and cell-mediated immunity against DEV in ducks. However, the relatively poor immunogenicity of DNA vaccines has hindered their use (Liu et al. 2016).

14.3.3 Avian Influenza

Avian influenza (AI) is referred to any infection or disease in birds caused by Type A Avian influenza virus (AIV). AIVs belong to the genus *Alphainfluenzavirus* of the family *Orthomyxoviridae* and are enveloped and negative-sense, single-stranded RNA viruses. The viral genome is composed of eight segments that encode for 10 or 11 proteins, depending on strains. AIVs are divided in to 16 hemagglutinin (H1–16) and 9 neuraminidase (N1–9) subtypes based on hemagglutinin inhibition (HI) tests and neuraminidase inhibition (NI) tests, respectively. AIVs are further classified as low pathogenic avian influenza (LPAI) or highly pathogenic avian influenza (HPAI) viruses based on their pathogenicity in chickens and/or the amino acid sequences of the connecting peptide of the hemagglutinin (HA) protein. Most AIVs are low pathogenic and all naturally occurring HPAI viruses have been H5 and H7 subtypes (Swayne et al. 2013; OIE 2015).

Wild waterfowl of the orders *Anseriformes* (such as ducks, geese, and swans) and *Charadriiformes* (gulls, terns, and shorebirds) are the natural reservoirs of AIVs (Kida et al. 1980; França and Brown 2014). In addition, viruses can be excreted in high concentrations in the feces of waterfowl. Waterfowl may transmit directly or indirectly AIs to susceptible hosts. Transmission between individual birds occurs by ingestion or inhalation. Vertical transmission has not been proved but the viruses can be recovered from the eggshell surface and the internal contents of the eggs from HPAI virus-infected hens (Kilany et al. 2010). Experimentally, AIV can replicate and be excreted from individual ducks for up to 30 days after infection. Moreover, AIV can be kept for much longer in duck farms or can re-emerge after significant stress on a population basis (Swayne et al. 2013).

AIVs can naturally infect a wide variety of wild and domestic birds, especially free-living birds in the aquatic habitats and the infection may result in extremely variable clinical outcomes which depend on virus strain, host species, age, sex, immunity, concurrent infections, and environmental factors (Swayne et al. 2013; França and Brown 2014).

With the exception of some lineages of Eurasian H5N1 HPAI viruses, both LPAI and HPAI viruses cause no or mild clinical diseases and lesions in ducks because of poor adaption to non-gallinaceous species or innate immunity and expression of RIG-I (Retinoic acid-induced gene I), an antiviral mechanism (Barber et al. 2010; França and Brown 2014). However, growing evidence suggests that infection with AIV of the host is significant and may cause subclinical effects, including changes in weight gain, behavior, immune function, and reproduction. When domestic ducks and geese exhibit diseases, there are often confounding factors, including bacterial co-infections, stress, unnatural routes of inoculation, or poor husbandry, making it

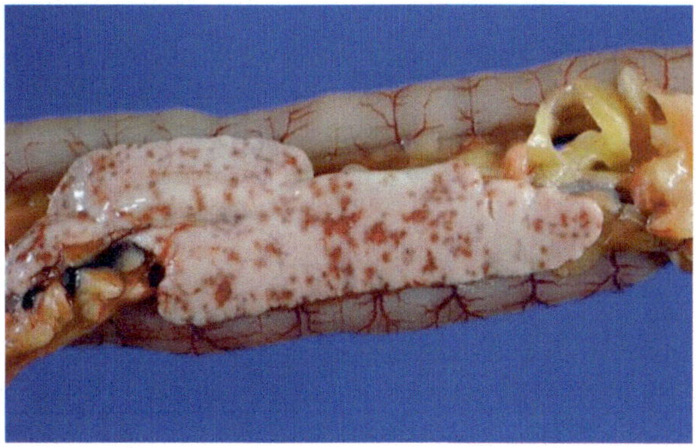

Fig. 14.2 Hemorrhage and necrosis in pancreas in Muscovy duck infected with highly pathogenic avian influenza virus

difficult to determine the association between clinical symptoms and LPAI virus infection (França and Brown 2014). Moreover, LPAI viruses may produce tracheitis, laryngitis, sinusitis, conjunctivitis, and pneumonia in domestic ducks and geese (Swayne et al. 2013; França and Brown 2014).

Before the emergence of the Asian HPAI H5N1 virus, only one case of AIVs had been reported to cause mortality in ducks. Muscovy ducks and domestic geese were reported dead or exhibiting severe neurologic clinical signs with viral replication and associated lesions in the central nervous system and pancreas during the outbreak of H7N1 HPAI virus in Italy during 1999–2000 (Pantin-Jackwood and Swayne 2007; França and Brown 2014).

Since 2002, some lineages of Eurasian H5N1 HPAI viruses cause morbidity and mortality in domestic and wild waterfowls under both natural and experimental conditions (França and Brown 2014; Lee et al. 2017; Mishra et al. 2017). The Eurasian H5N1 lineage HPAI viruses have evolved into 10 genetically distinct virus clades (0–9) and multiple subclades (e.g., 2.1.1, 2.1.2, 2.3.4.4, etc.). The pathogenicity of these viruses varies with strain, waterfowl species and age. The typical clinical signs are neurologic, including ataxia, depression, listlessness, seizures, and death (Swayne et al. 2013; França and Brown 2014). Other clinical signs include a decreasing egg production, decreased feed consumption, conjunctivitis, bloody ocular and nasal discharges, hyperemic shanks and feet, greenish diarrhea, and dyspnea to respiratory distress (Lee et al. 2016). In Taiwan, Muscovy ducks infected with the H5 clade 2.3.4.4 HPAI viruses show obvious clinical symptoms but Pekin ducks and Tsaiya ducks (layer ducks) show no or few clinical response in the field.

In waterfowl, viral replication and lesions are similar to gallinaceous poultry and the virus are most frequently detected in the pancreas, adrenal gland, brain, and heart (França and Brown 2014). At necropsy, the infected ducks may show cyanosis and hemorrhage in the face and shanks, and bleeding from the eyes, nostrils, and mouth. Necrotic foci are common in pancreas (Fig. 14.2) and heart. A variety of hemorrhagic lesions in several visceral organs, such as heart, pancreas, proventriculus, brain stem, larynx, and ovarian follicles are also reported (Pantin-Jackwood and Swayne 2007; Lee et al. 2016; Sun et al. 2016).

The histopathological lesions are most consistent in tissues having necropsy lesions. Severe, multifocal coagulative necrosis of pancreatic acinar cells can be seen. Myocarditis and acute necrosis of cardiac muscle cell are also observed. Hepatic lesions with congestion or hemorrhage, multifocal hepatic necrosis accompanied by infiltration of lymphocytes and macrophages, and chronic cholangitis have been reported. Neurological lesions include non-suppurative meningoencephalitis, perivascular cuffing with gliosis of the brain, and neuronophagia. Other histopathological lesions include parabronchitis, diffuse congestion in the lungs, splenic necrosis, renal tubular necrosis, and focal necrotizing enteritis (Sun et al. 2016; Haider et al. 2017; Kang et al. 2017).

Since infections in birds can cause a wide variety of clinical signs, the definitive diagnosis of AI is established by direct detection of AI viral proteins or genes in specimens or isolation and identification of AIVs (Swayne et al. 2013). The suitable samples from live ducks should include oropharyngeal and cloacal swabs. The samples from dead birds should include oropharyngeal and cloacal swabs or feces and various internal organs (OIE 2015). Several commercial antigen detection kits based on antigen capture immunoassays and various RT-PCR or real-time RT-PCR methods are developed to detect AI viral proteins or genes (Swayne et al. 2013). AIVs can be isolated by inoculating samples into the allantoic sac of 9–11-day-old embryonating chicken eggs and then the allantoic fluid is screened by hemagglutination test. The isolated viruses are further identified by influenza A type-specific test (such as agar gel immunodiffusion test), influenza A subtype-specific test (such as HI and NI tests), or a molecular test to detect viral RNA (such as RT-PCR or real-time RT-PCR). The pathogenicity of the AIV isolate is assessed by the mortality rate tests (intravenous inoculation of a minimum of eight susceptible 4–8-week-old chickens with infectious virus; AIV isolates are considered to be highly pathogenic if they cause more than 75% mortality within 10 days) or intravenous pathogenicity index (IVPI) test (inoculation of 10 susceptible 6-week-old chickens resulting in an IVPI of greater than 1.2). OIE also recommends the genetic analysis to assess pathogenicity: H5 or H7 viruses with an HA0 cleavage site amino acid sequence similar to any of those that have been observed in highly pathogenic viruses are considered as HPAI viruses (OIE 2015).

The prevention and control of AI are based on combination of five strategies: education, biosecurity, rapid diagnostics and surveillance, elimination of infected poultry, and decreasing host susceptibility. A key aspect in control is to educate all poultry and allied industry personnel about the knowledge of AI to minimize their risky behaviors. Biosecurity practices are the first line of defense and are discussed extensively in textbooks on poultry diseases and related publications. Biosecurity means prevention of the introducing pathogens into susceptible birds or the spread of pathogens from infected birds. This includes prohibiting any potential source of infection, such as live ducks, other poultry or animals. If it is necessary to bring live ducks to one's farm, the ducks must come from an established AI-free source, and should be quarantined for observation before being placed onto the farm. The potential carriers of infectious materials such as trucks, poultry cages and equipment must be appropriately disinfected before entering the farm. Duck caretakers should change clothes and boots and use disinfectant footbaths when entering duck houses or buildings. The biosecurity measures also apply to countries and areas around the epidemic sites, as well as practices such as cleaning and disinfection, and restrictions on movement of poultry and poultry-related items. Rapid detection and accurate identification of LPAI and HPAI is critical to early and successful control. After the identification of infected flocks, stamping-out followed by disposal of carcasses, eggs, and manure, and through disinfection is the method of choice for rapid elimination of the virus. Increase of host resistance to AIVs can be achieved by vaccination. In some situations, when immediate eradication is not feasible, vaccination can be added to as an additional control strategy, which will maintain livelihoods and food security, and control clinical disease (Swayne 2012; Swayne et al. 2013).

Various AI vaccines have been developed based on five technologies: 1) wild-type or reverse genetics whole AI virus grown in embryonated chicken eggs, then chemically inactivated and adjuvanted; 2) HA antigen or virus-like particles produced in insect cells by a genetically engineered baculovirus; 3) HA DNA vaccine adjuvanted; 4) recombinant technologies utilizing live virus vectors to express HA gene and NA gene in some cases (recombinant Herpes turkey virus (rHVT-AIV), recombinant Newcastle disease virus (rNDV-AIV), recombinant fowl pox virus (rFPV-AIV) and recombinant duck enteritis virus (rDEV-AI)) and, 5) defective replicating alphavirus (defective Venezuelan Equine Encephalitis virus) with H5 AI virus gene insert (Swayne and Kapczynski 2017; Guyonnet and Peters 2020). These vaccines have shown efficacy in experimental studies, mostly in gallinaceous poultry, to provide the protection from LPAI and HPAI viruses.

Various vaccines such as conventional inactivated whole virus vaccine, live recombinant virus-vectored vaccines with H5 influenza A virus hemagglutinin gene inserts vaccines and an RNA particle vaccine, based on Venezuelan equine encephalitis virus replicon particles, containing H5 Clade 2.3.4.4 Gs/GD HA insert (Schultz-Cherry et al. 2000) have been licensed and used in a few countries. Vaccines can prevent clinical signs and death, and reduce susceptible individuals in population and the virus shedding from infected birds. However, to use preventive vaccination against all possible types of AI viruses is not practical. If eradication is the desired outcome of the control of HPAI or H5/H7 LPAI, vaccination alone is not considered as a control strategy. Eradication can only be achieved through comprehensive strategies, such as the application of monitoring systems, strict biosecurity, and depopulation. Otherwise there is the possibility that HPAI and H5/H7 LPAI viruses could become endemic in vaccinated poultry populations (Swayne et al. 2013; OIE 2015).

14.3.4 Parvovirus Infection

The etiologic agents of waterfowl parvovirus infection are goose parvovirus (GPV) or Muscovy duck parvovirus (MDPV). According to the classification report of International Committee on Taxonomy of Viruses (ICTV), GPV and MDPV are defined as Anseriform *dependoparvovirus* 1, which is classified into the genus *Dependoparvovirus* of the family *Parvovirdae* (Zádori et al. 1995; Cotmore et al. 2014). Waterfowl parvovirus is a non-enveloped and single-stranded small DNA virus, and its genomic size is approximately 5.1 kb. Like most non-enveloped DNA virus, waterfowl parvovirus is resistant to chemical and physical treatments (Takehara et al. 1994). Therefore, it is not easy to eliminate waterfowl parvovirus from infected farms.

Waterfowl parvovirus is excreted in the feces of infected geese or ducks, resulting in rapid spread by direct or indirect contact. The viruses can be vertically transmitted from breeder geese or ducks to goslings or ducklings through transovarian transmission (Chen et al. 2016). Infected eggs could be the source of the virus. No other avian or biologic vectors have been identified.

As previous reports, both geese and Muscovy ducks are susceptible to GPV, but mule ducks and Pekin ducks could be infected with some distinct GPV strains (Palya et al. 2009; Chen et al. 2015; Ning et al. 2018). The MDPV was first isolated from Muscovy ducks in France and later in Europe, the USA, Japan, China, Taiwan, and Indonesia (Jestin et al. 1991; Lu et al. 1993; Chang et al. 2000; Woolcock et al. 2000; Zhao et al. 2014b; Mahardika et al. 2015). MDPV causes diseases in Muscovy ducks, several hybrid duck breeds, and Tsaiya ducks, but not in geese.

The clinical signs of waterfowl parvovirus infection depend on the age and immune status of the geese and ducks. The clinical signs can be acute, subacute or chronic. The course of the disease in susceptible avian aged less than 1 week old is usually rapid and the mortality may reach 100% (Glávits et al. 2005). The infected goslings or ducklings appear with anorexia, weakness and lead to death within a few days. Infection of the susceptible avian between 1 and 3 weeks of age may result in watery diarrhea, nasal or ocular discharge, and locomotor dysfunction; the diseased birds then die or the disease become chronic form. Those recovered flocks show growth retardation and prone to secondary infection (Poonia et al. 2006). The chronic clinical signs with a specific characterization in parvovirus affected Muscovy ducks is called beak atrophy and dwarfism syndrome (BADS). BADS was first reported in mule ducks in southern France and the diseased ducks showed atrophic beak with protruding tongues, podgy legs, feathering abnormality, and stunted growth with 10–30% morbidity (Palya et al. 2009). In 1989–1990 in Taiwan, a strain of MDPV could infect all breeds of ducklings, and the infected ducklings showed the similar syndrome as BADS but with high morbidity and mortality (86–100%) (Lu et al. 1993). In 2015, GPV caused BADS in Cherry Valley ducks has been reported in China (Chen et al. 2015). According to these reports, waterfowl parvovirus may cause BADS in ducklings and leads to huge economic loss in waterfowl farms.

Gross lesions observed in the diseased geese or ducks include myocarditis, cardiomegaly, myocardial pallor, hydropericardium, ascites, hepatitis, thickening of the intestine mucous membrane and fibrous casts inside, and polioencephalomyelitis (Glávits et al. 2005). The main microscopic lesions were enteritis with necrosis of epithelial cells and crypt of Lieberkühn. Fibrinous pericarditis, degeneration of cardiac muscle cells, periportal hepatitis, skeletal muscle fiber degeneration, and non-suppurative encephalitis could be seen in some cases. Intranuclear viral inclusion bodies could be seen within myocardial tissues or hepatocytes (Glávits et al. 2005; Poonia et al. 2006).

Diagnosis is based on the clinical symptoms and gross and histopathologic lesions. The confirmation of the diseases requires laboratory diagnosis. Methods for laboratory diagnosis include isolation of the virus by cell culture from embryonating eggs of geese and Muscovy ducks, molecular methods, and serologic tests. Differential diagnosis include duck viral enteritis (DVE) which is caused by a herpesvirus and duck viral hepatitis (DVH) which is caused by a picornavirus. DVE can infect geese and ducks of all ages with high mortality rates; on the contrary, waterfowl parvovirus infections are age-dependent. DVH has characteristic hemorrhagic hepatitis in liver. *Riemerella anatipestifer* and *Pasteurella multocida* may cause high mortality in goslings and ducklings, or even co-infection with parvovirus but can be confirmed by isolation and identification (Richard and Paul 2010).

There is no treatment for viral infection, but antibiotics are sometimes given to prevent secondary bacterial infections. Infected eggs may be the source of the virus, so surviving flocks cannot be used as breeding flocks. GPV and MDPV are very resistant to physical and chemical treatment and are difficult to eliminate. Vaccination of breeding flocks and their progeny is the best choice to prevent waterfowl parvovirus infection (Maurin-Bernaud et al. 2014). Both live vaccines and inactivated vaccines could be used to immunize breeding flocks to supply maternal antibodies to goslings or ducklings and to immunize ducklings that may become vulnerable after the drop in maternal antibodies (Kisary 1986; Gall-Reculé et al. 1996). A bivalent inactivated vaccine consisted of both GPV and MDPV antigens is used to ensure protection against the two waterfowl parvoviruses causing diseases in Muscovy ducks (Palya 2013). Recently, a live attenuated parvovirus vaccine, consisting of a live MDPV strain attenuated by passages on duck embryo cells, has been demonstrated the protection efficacy against a MDPV and a GPV challenge (Maurin-Bernaud et al. 2014). Maternal derived antibodies even at a very low level can neutralize live vaccines so the expected immune responses cannot be elicited. In maternally derived antibody-free ducks, live vaccines can stimulate rapid immune response and afford protection (Palya 2013). Inactivated vaccines induce a relatively slow immune response, but they are less sensitive to the interference with maternal antibodies than live vaccines (Takehara et al. 1995). Moreover, a recombinant subunit vaccine has been developed using baculo virus expression system and its antigenicity and immunogenicity has been proven (Gall-Reculé et al. 1996). Recently, the recombinant expression system SIP-409-VP2/NC8 is successfully constructed using *Lactobacillus plantarum* NC8 as a vector to express the VP2 gene of GPV. The expressed VP2 protein has favorable immunogenicity. However, its immune responses are not evaluated in birds (Liu et al. 2017). In addition, anti-waterfowl parvovirus immunoglobulin Y can be injected into one-day-old goslings or ducklings having a low level of maternal antibodies.

14.3.5 Reovirus Infection

A disease of Muscovy ducks caused by reovirus was first described in South Africa in 1950. However, the virus was not isolated until 1972 in France. The earliest onset of reovirus infection may be 7–10 days of age and may persist in the affected flock until 7–10 weeks of age. Morbidity can be 10–60% and mortality can reach 10%. Generally, mortality is higher in younger ducklings. The clinical signs include general malaise, diarrhea, and difficulties in moving. The surviving ducks may be stunted in growth and become lame. At necropsy, dead ducks show fibrinous pericarditis, marbled spleen, and enlarged friable livers. Arthritis and tenosynovitis are also seen. Microscopically, miliary foci of necrotic hepatocytes or granuloma-like foci with necrotic centers and proliferating macrophages can be found in the liver and spleen. It has been demonstrated that the duck reovirus is antigenically distinct from the chicken reovirus by cross neutralization. Specific prevention of reovirus infection in ducks has not been developed (The Poultry Site 2011; Woolcock 2013).

14.3.6 Duck Circovirus Infection

Duck circovirus mainly attack lymphoid tissues and results in immunosuppression, growth retardation, feathering disorders, and increased mortality due to secondary infections, such as *Riemerella anatipestifer* and *Aspergillus* sp. Microscopically, the main lesions in ducks are those of the primary and secondary lymphoid tissues. The bursa of Fabricius shows lymphofollicular hyperplasia, lymphoid necrosis, lymphocyte depletion, and cystic atrophy. Basophilic intracytoplasmic inclusions are frequently found in the medullar and cortical bursa follicular cells and bursal epithelial cells (Woolcock 2013; Hong et al. 2018).

14.3.7 Duck Egg Drop Syndrome

Duck egg drop syndrome is a newly emerging disease recognized in the People's Republic of China in 2010 and caused by Tembusu virus which is a member of genus *Flavivirus*, family *Flaviviridae*. *Culex* mosquitoes transmit the Tembusu viruses and evidences for airborne transmission and vertical transmission have also been reported. The virus is also reported in Thailand, Malaysia, and Taiwan. Duck egg drop syndrome is characterized by a sudden decline in feed uptake, dramatic decrease in egg production (lower than 10% in 10 days), and neurological signs in infected egg laying and breeder ducks. Morbidity is as high as 90% and mortality ranges from 5 to 30%. At necropsy, severe ovarian hemorrhage, oophoritis, and regression are consistently observed in the affected ducks. Other gross lesions observed include ruptured ovarian follicles, peritonitis, and enlarged spleen. Microscopically, the main lesions are ovarian hemorrhage and follicle atresia and rupture. In addition, the encephalitis with focal gliosis, lymphocyte infiltration, and necrosis is also observed. Isolation of Tembusu viruses from young meat-type ducks, geese, house sparrows, and a dead racing pigeon has been reported. The infected ducks show various degrees of neurological disorders, ranging from recumbence to leg and wing paralysis (Liu et al. 2013b; Zhang et al. 2017).

14.4 Bacterial Diseases

14.4.1 *Riemerella Anatipestifer* Infection

Riemerella anatipestifer infection, also known as new duck syndrome, duck septicaemia, and duck infectious serositis, is one of the important diseases, which impacting global waterfowl breeding industry. The infectious pathogen is *R. anatipestifer*, which was previously classified in genus *Moraxella* and *Pasteurella* (Sandhu 2008). *R. anatipestifer* is Gram negative, non-motile, non-spore-forming rod, appearing singly or in pairs under microscopy (Fig. 14.3). The cells vary from 0.2 to 0.4 μm in width and 1–5 μm in length. Primary isolation of the organism is usually accomplished using blood agar or chocolate agar at 37 °C with 5–10% carbon dioxide for 24–48 h incubation. The colonies of *R. anatipestifer* are usually convex, entire, transparent, glistening, and butyrous, and some strains give hemolysis on blood agar (Sandhu 2008). In biochemical characteristics, carbohydrates are not fermented by most strains. Gelatin is usually liquefied. There is no growth on

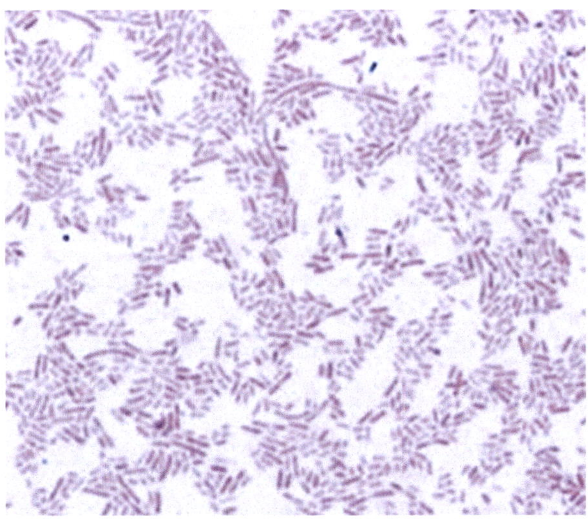

Fig. 14.3 Morphology and staining of *Riemerella anatipestifer* under microscopic examination (Gram stain)

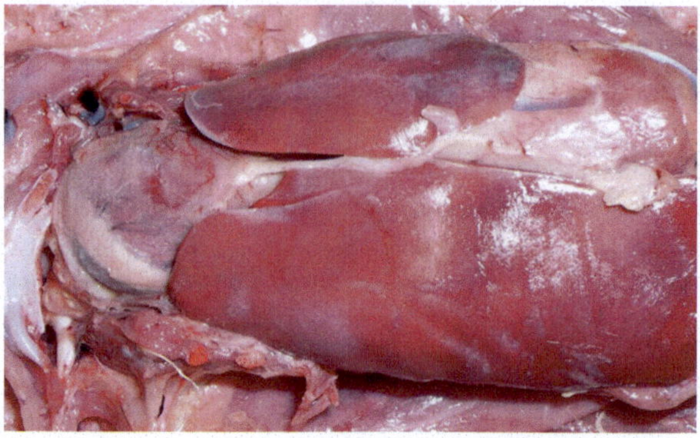

Fig. 14.4 Fibrinous polyserositis in ducks infected with *Riemerella anatipestifer*

MacConkey agar. It produces oxidase and catalase (Bangun et al. 1980; Hinz et al. 1998; Liu 2002).

R. anatipestifer infection is primarily a disease of domestic ducks, geese, and turkeys. Duck, goose, and turkey of 1–8 weeks of age are highly susceptible (Sandhu 2008). It has also been reported in chickens, pheasants, swans, quails, partridges, pigeons, and other terrestrial and waterfowl birds (Chen and Yu 2010). The disease can be transmitted in several ways. Birds are believed to be infected mainly by the respiratory route, through skin wounds (particular feet) or close contact (Hung 2002). The birds may be infected indirectly through contaminated drinking water, feed, and environment. Wild birds are important carriers. Egg transmission may occur (Sarver et al. 2005; Chen & Yu 2010).

The incubation period of the disease is approximately 2–5 days, and the course may last 2–10 days. Both morbidity and mortality are closely related to the climate changes, farm management, stress, and other environmental factors (Hung 2002). Mortality may vary from 5 to 75%. Clinically, the signs most often observed are ocular and nasal discharge, coughing, mouth breathing, ataxia, torticollis, lying on back with legs paddling, and swollen knee joint. In severe cases, birds show opisthotonus, coma, or even death. Even if recovered from infection, surviving birds may lead to stunted growth, and then poor carcass quality, which result in significant economic losses in poultry meat production (Sandhu 2008). The birds older than 8 weeks of age are less susceptible to the disease, but may carry infection without clinical sings and constantly shed the bacteria. The birds could contaminate drinking water, feed, and environment and act as long-term sources of infection. The breeder ducks occasionally exhibit reduced egg production (Chen and Yu 2010).

Lesions including beak cyanosis, congestion, and enlargement of liver are observed in acute stage of the disease. The most characteristic gross lesions are the deposit of fibrinous exudate on pericardium, liver capsule, or air sacs (Fig. 14.4). Infections of central nervous systems can produce purulent meningitis. Caseous salpingitis is also observed with chronic infection (Tsai 2002a).

Presumptive diagnosis may be made based on clinical symptoms and necropsy findings of typical serous fibrinous polyserositis. A definite diagnosis should be based on bacteria isolation and identification (Sandhu 2008). The bacteria can be isolated from heart blood, liver, spleen, kidney, lung and air sacs, and exudates in peritoneal cavity and knee joints. Further testing in laboratory includes serology, antibiotic sensitivity test, and PCR assay. The serotypes of bacteria are various and at least 20 serotypes have been reported to date (Pathanasophon et al. 2002). Serotypes are usually determined by plate agglutination test, tube agglutination test or agar gel immunodiffusion test.

Due to serotypes complexity, serotype-specific immunity protection and several serotypes co-infections toward individual bird or farm as well as different region or country with its own different predominant serotypes, the difficulty in disease prevention is thus increased (Chen and Yu 2010).

The disease should be differentiated from other septicemic diseases caused by *E. coli*, *Streptococcus*, *Pasteurella*, and *Salmonella* (Sandhu 2008). Since these diseases produce gross lesions indistinguishable from those caused by *R. anatipestifer* infection, diagnosis must include isolation and identification of the causal organism. *R anatipestifer* fails to grow in MacConkey agar while *E. coli* appears as pink colony and the colony of *Salmonella* is transparent. *R. anatipestifer* is unable to ferment glucose, sucrose and is not pathogenic on mice; on the contrary, *Pasteurella multocida* ferments some carbohydrates and is highly pathogenic to mice (Sandhu 2008).

Good husbandry is important to prevent *R. anatipestifer* infection. Careful management practices plus implementation of high level biosecurity measures are crucial for disease prevention. Good ventilation and dryness should be provided at brooder houses. Inappropriate temperature and overcrowding should always be avoided. The ponds in the farm should be kept clean. The skin or feet of birds should be always protected from getting punctured by sharp objects of straw bedding. The tools and utensils of the poultry farms should be disinfected completely. Thorough cleanout management between flocks should be practiced. Rigid sanitation and depopulation are necessary for elimination of the disease. The farmers and workers should be also well trained regularly. With good husbandry and careful management, the disease would only happen sporadically with low morbidity and mortality (Chen and Yu 2010).

In addition, a thorough vaccination program is an effective method for disease prevention. Inactivated and live vaccines are available. Since immunity induced by vaccine is serotype-specific, an ideal vaccine should contain bacterial cells of predominant serotypes to provide better effective protection (Sandhu 1979, 1991; Liu et al. 2013a).

Recently, various studies have attempted to develop subunit vaccines to provide the cross-serotype protection of *R. anatipestifer*. The *OmpA* gene encoding outer membrane protein A was mainly used for producing recombinant protein as the antigen of subunit vaccine (Huang et al. 2002; Chu et al. 2015; Phonvisay et al. 2019; Yang et al. 2019; Wu et al. 2020b; Xu et al. 2020). Moreover, a recent study demonstrated that specific anti-RA IgY has the potential to improve the degree of protection and responsiveness of ducklings to *R. anatipestifer* infections and provide them with passive immunity to *R. anatipestifer* although the IgY can provide short-term protection for only approximately 2 weeks (Yang et al. 2020).

If birds become infected or an outbreak of the disease occurs, farmers are advised to contact veterinarian immediately. Post-mortem examination should be performed, and specimens should be sent to laboratory for a definite diagnosis. Infected birds should be removed, segregated, or euthanized. Broad-spectrum antibiotics could be used for 3–5 days for early treatment. The contaminated carcasses should be cremated or buried deeply.

14.4.2 Fowl Cholera

Fowl cholera (avian pasteurellosis) is a contagious bacterial disease that affects domestic and wild birds worldwide. It usually occurs as an acute septicaemia with high morbidity and mortality, causing serious economic losses in poultry industry. The causative agent is *Pasteurella multocida*, a Gram negative, short rod-shaped, non-motile, non-spore-forming bacterium showing bipolar staining in tissue smears under microscopic examination. They are usually 0.2–0.4 × 0.6–2.5 μm in size, appearing singly or in pairs. Primary isolation is usually accomplished using dextrose starch agar, blood agar, or trypticase soy agar to incubate at 37 °C for 18–24 h (Glisson et al. 2013). Colonies on blood agar are discrete, convex, butyraceous, no haemolysis, and accompanied by a characteristic odor. It is oxidase-positive and catalase-positive. The bacteria can ferment some carbohydrates such as glucose and sucrose, but usually cannot ferment maltose and lactose. Indole and hydrogen sulfide are produced. The organism does not grow on MacConkey agar (OIE 2000; Wu 2004).

Dissemination of *P. multocida* within a flock and between houses is primarily by excretions from mouth, nose, and conjunctiva of diseased birds. The microorganism is also spread by contaminated feed, drinking water, forage, and tools, by water transmission in particular (Glisson et al. 2013). Egg transmission has not been confirmed. The chronic infected or recovered birds may shed the bacteria in their droppings and are considered major sources of infection (Su 2006).

All avian species, such as chickens, turkeys, ducks, geese, and wild birds, are susceptible to *P. multocida*, although turkeys may be particularly severely affected, followed by ducks and geese. Fowl cholera occurs primarily in ducks over 4 weeks of age, and adult birds are more sensitive than younger ones (Tsai 2002b; Wakenell 2016).

The incubation period of fowl cholera is approximately 4–9 days, and the course may last from 2 weeks to several months. Clinical signs vary greatly depending on the course of the disease. The mortality rate varies from 1 to 50%. Generally, the mortality in turkeys and ducks is higher than in chickens. Fowl cholera occurs throughout the year but often in high-temperature and high-humidity seasons. In peracute phase, the sudden and unexpected death could be the first sign of the disease, along with cyanotic comb and wattle (The poultry site 2014). Signs observed include fever, anorexia, ocular and nostril discharge, cyanosis on hairless skin, tachypnea, and diarrhea in acute cases. The dead birds are often found near water area. Chronic fowl cholera is characterized by localized infections, involving

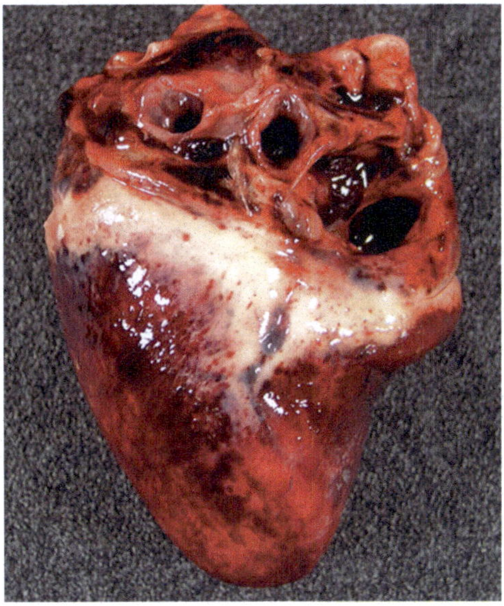

Fig. 14.5 Diffuse petechial and ecchymosed hemorrhage in heart in ducks infected with *Pasteurella multocida*

swollen joints, footpads, and infraorbital sinuses. Lesions result from accumulated fibrino-suppurative exudates. When invading to the brain and inner ear, torticollis may appear (Liu 1999; OIE 2000).

Lesions in acute form are predominantly related to vascular injuries. Gross lesions often observed include diffuse petechial and ecchymosed hemorrhages in heart (Fig. 14.5), gizzard, abdominal adipose tissue, and duodenum. Multiple focal necrosis throughout liver and spleen are also seen (Routh and Sanderson 2009). Yolk peritonitis may occur. The birds survived the acute stage may develop chronic infection. In chronic form of the disease, focal suppurative lesions may be widely distributed, including joints, tendon sheaths, footpads, conjunctiva, ears, and cranial bone (Tsai 2002b).

Presumptive diagnosis may be made based on the occurrence of a large number of sudden death, typical gross pathological findings such as multifocal petechial hemorrhages in heart (Fig. 14.5), gizzard, peritoneal fat tissue, and diffuse focal necrotic areas in liver and spleen, plus bipolar microorganism shown in Wright's or Giemsa stains of impression smears from heart blood or liver (OIE 2000; Tsai 2002b).

Further testing in laboratory includes bacterial isolation, inoculation into mice, serology, antibiotic sensitivity test, and PCR assay. *P. multocida* should be isolated and identified for confirmation. Isolation of the organisms is accomplished generally easily from visceral organs, such as liver, spleen, bone marrow, and exudates (OIE 2000). In addition, the pathogen can be obtained from heart blood and liver of dead mice after inoculation of infected bird's visceral emulsion for 24–48 h. The virulence of isolate varies greatly (Wu 2004; Glisson et al. 2013).

P. multocida can be sub-grouped by capsule serogrouping (A, B, D, E, and F) and somatic serotyping (1–16). Serotypes are usually determined by indirect hemagglutination test and agar gel immunodiffusion test, but either of them is highly sensitive (Rimler and Rhoades 1987). Serology tests are used for detection of specific humoral response from immunized animals, not for diagnosis of the disease (OIE 2000).

Differential diagnoses should include duck viral enteritis, infection due to *Escherichia coli* and *Erysipelas spp*., and in the chronic form, duck septicemia due to *Riemerella anatipestifer* infection. The most accurate method for confirming the causative organism is isolation and identification (Su 2006; Routh and Sanderson 2009).

The outbreak of the disease is mainly due to poor management of poultry housing. This opportunistic pathogen generally becomes affected under several circumstances including latent parasitic infection, malnutrition, and co-infection of other microorganisms, which result in immunosuppression of the birds. Careful management practices should be broadly applied to prevent the introduction of infectious agents, such as reducing crowding, the cleaning and disinfection of tools, equipment, and houses, as well as thorough cleanout between flocks (Glisson et al. 2008). If an outbreak of the disease occurs, farmers are advised to contact local veterinarian immediately. Broad-spectrum antibiotics could be administered to lower mortality in an acute phase, but have variable success depending on the sensitivity of the strain and the duration of the disease before treatment (Wakenell 2016). The disease is controlled effectively by eradication (Poultry Hub 2018).

Vaccination should be considered in the control of the disease. Inactivated bacterin, live vaccine, and autogenous bacterin are commercially available. Adjuvant bacterin is generally effective if prepared from the most prevalent serotypes of the locality (OIE 2000), but inactivated bacterin elicits only homologous immunity (Glisson et al. 2013). The use of oil-adjuvant vaccine may significantly reduce egg production in poultry breeders. However, if an aluminum hydroxide bacterin is used, revaccination may be required to afford immunity to a flock throughout laying cycle (Aucouturier et al. 2001). Live attenuated vaccines can induct heterologous and longer lasting protection, but they may cause chronic fowl cholera and mortality (Bierer and Derieux 1972; Mostaan et al. 2020).

Recent studies in genetic, biochemical, and virulence factors of *P. multocida* remarkably broaden our understanding of the disease mechanisms of the organism and led us to find the new vaccines and vaccine candidates against fowl cholera. Several *Pasteurella* outer membrane proteins have potential targets for vaccine development. The outer membrane protein H (OmpH) has displayed potential as a protective antigen with strong immunogenicity and may provide efficient cross-protectivity against heterologous strains of *P. multocida* (Luo et al. 1997; Sthitmatee et al. 2008; Thanasarasakulpong et al. 2015; Varinrak et al. 2017). Furthermore, Apinda et al. (2020) demonstrated that a vaccine containing live attenuated duck enteritis virus and recombinant OmpH of *P. multocida* elicit potent immune responses against fowl cholera in ducks. Moreover, live attenuated vaccines prepared by treating the bacteria in iron deficient environment (Gunawardana et al. 2004) and virulence gene deletion (Xiao et al. 2016; Zhao et al. 2016; Liu et al.

2019), subunit vaccines (Luo et al. 2019), recombinant vaccines (Wu et al. 2007; Thanasarasakulpong et al. 2015; Apinda et al. 2020), and DNA vaccines (Gong et al. 2018) are demonstrated to have potential efficiency to afford the protective immunity against avian *P. multocida* challenge.

14.5 Parasitic Diseases

In many countries, ducks are kept as free-range and may have more opportunities, comparing to housing system, to come into contact with parasites and/or the intermediate hosts of the parasites. Ducks can infect various nematodes, including *Ascaridia galli*, *Heterakis gallinarum*, *Heterakis isolonche*, *Amidostomum anseris*, *Cyathostoma bronchialis*, *Echinura uncinata,* and *Tetrameres crami* (McDougald 2013). These nematodes may be low pathogenic or cause mortality in ducks. Among these nematodes, *Heterakis gallinarum*, a cecal worm, is the most important because it is a reservoir and carrier of *Histomonas meleagridis* which is the causative agent of blackhead disease. However, *H. meleagridis* does not cause a substantial danger for duck production (Callait-Cardinal et al. 2006). A large worm, *Ascaridia galli,* is common in poultry and may cause decreasing of weight gain (Adang et al. 2012; McDougald 2013). In addition, ducks infected with *Cyathostoma bronchialis* may die or become stunted when they recover (McDougald 2013).

Ducks can also get infected by numerous species of cestodes, such as a cloacal tapeworm *Cloacotaenia megalops* and trematodes such as an oviduct fluke *Prosthogonimus* sp. introduced by wild birds (McDougald 2013; Guo 2016).

The etiologic agent of coccidiosis in ducks is protozoa which are intracellular parasites and may be *Eimeria, Wenyonella,* or *Tyzzeria.* Although coccidiosis is sporadic in ducks, moderate to heavy mortality caused by coccidiosis has been reported on domestic ducks in some countries; which calls for more attention of researchers (McDougald 2013; Rahman et al. 2019). *Eimeria mulardi* is an intestinal coccidia that mainly occur in the posterior jejunoileum and the caecum. It can cause congestive-hemorrhages with obstructive lesions and may lead to death (Pascalon-Pekelniczky 1998). *Wenyonella philiplevinei* can cause lesions in the ileum and rectum and result in decreased body weight gain and diarrhea, and can even lead to death (Leibovitz 1968; Wu et al. 2013). *Tyzzeria perniciosa* infection usually causes anorexia, weight loss, weakness, distress, and up to 70% of mortality. The lesion is hemorrhage in the anterior portion or entire area of the intestine (McDougald 2013).

Control of internal parasites may depend on change of production methods of the ducks, sanitation, interruption of the life cycle of the parasites and/or its intermediate hosts, and the use of chemotherapeutic agents. Moreover, accurate identification of parasite species is important to eliminate the intermediate hosts and then breaking the life cycle (McDougald 2013).

14.6 Conclusion

The main goals of the duck industry at present include disease control, high production, product quality and safety, and reasonable production costs. The emergence and re-emergence of diseases pose serious challenge to the duck industry. Therefore, in addition to biosecurity and development of biologicals for disease control, the duck industry should consider changes in production methods and management strategies to improve production efficiency.

References

Adang LK, Abdu PA, Ajanusi JO, Oniye SJ, Ezealor AU (2012) Effects of *Ascaridia galli* infection on body weight and blood parameters of experimentally infected domestic pigeons (*Columba livia domestica*) in Zaria, Nigeria. Revista Científica UDO 960 Agrícola 12:960–964

Apinda N, Nambooppha B, Rittipornlertrak A, Tankaew P, Punyapornwithaya V, Nair V, Sawada T, Sthitmatee N (2020) Protection against fowl cholera in ducks immunized with a combination vaccine containing live attenuated duck enteritis virus and recombinant outer membrane protein H of *Pasteurella multocida*. Avian Pathol 49:221–229

Aravind S, Kamble NM, Gaikwad SS, Shukla SK, Saravanan R, Dey S, Mohan CM (2015) Protective effects of recombinant glycoprotein D based prime boost approach against duck enteritis virus in mice model. Microb Pathogen 88:78–86

Asplin FD (1961) Notes on epidemiology and vaccination for virus hepatitis of ducks. Off Int Epizoot Bull 56:793–800

Aucouturier J, Dupuis L, Ganne V (2001) Adjuvants designed for veterinary and human vaccines. Vaccine 19:2666–2672

Bangun A, Tripathy DN, Hanson LE (1980) Studies of *Pasteurella anatipestifer*: an approach to its classification. Avian Dis 25:326–337

Barber MR, Aldridge JR Jr, Webster RG, Magor KE (2010) Association of RIG-I with innate immunity of ducks to influenza. Proc Natl Acad Sci U S A 107:5913–5918

Bierer BW, Derieux WT (1972) Immunologic response of turkeys to an avirulent *Pasteurella multocida* vaccine in the drinking water. Poult Sci 51:408–416

Burgess EC, Yuill TM (1981) Vertical transmission of duck plague virus (DPV) by apparently healthy DPV carrier waterfowl. Avian Dis 25:795–800

Callait-Cardinal MP, Chauve C, Reynaud MC, Alogninouwa T, Zenner L (2006) Infectivity of *Histomonas meleagridis* in ducks. Avian Pathol 35:109–116

Campagnolo ER, Banerjee M, Panigrahy B, Jones RL (2001) An outbreak of duck viral enteritis (duck plague) in domestic Muscovy ducks (*Cairina moschata domesticus*) in Illinois. Avian Dis 45:522–528

Chang PC, Shien JH, Wang MS, Shieh HK (2000) Phylogenetic analysis of parvoviruses isolated in Taiwan from ducks and geese. Avian Pathol 29:45–49

Chen YP, Yu CF (2010) Disease and control against *Riemerella anatipestifer* infection, handout for seminar of disease prevention and control against *Riemerella anatipestifer* infection, Animal Health Research Institute, COA, Taiwan

Chen LL, Xu Q, Zhang RH, Yang L, Li JX, Xie ZJ, Zhu YL, Jiang SJ, Si XK (2013) Improved duplex RT-PCR assay for differential diagnosis of mixed infection of duck hepatitis A, virus type 1 and type 3 in ducklings. J Virol Methods 192:12–17

Chen H, Dou Y, Tang Y, Zhang Z, Zheng X, Niu X, Yang J, Yu X, Diao Y (2015) Isolation and genomic characterization of a duck-origin GPV-related parvovirus from Cherry Valley ducklings in China. PLOS One 10:e0140284

Chen H, Tang Y, Dou Y, Zheng X, Diao Y (2016) Evidence for vertical transmission of novel duck-origin goose parvovirus-related parvovirus. Transboundary Emerging Dis 63:243–247

Chen L, Yu B, Hua J, Ni Z, Ye W, Yun T, Zhang C (2019) Optimized expression of duck tembusu virus E gene delivered by a vectored duck enteritis virus *in vitro*. Mol Biotechnol 61:783–790

Chu CY, Liu CH, Liou JJ, Lee JW, Cheng LT (2015) Development of a subunit vaccine containing recombinant *Riemerella anatipestifer* outer membrane protein A, and CpG ODN adjuvant. Vaccine 33:92–99

Cotmore SF, Agbandje-McKenna M, Chiorini JA, Mukha DV, Pintel DJ, Qiu J, Soderlund-Venermo M, Tattersall P, Tijssen P, Gatherer D, Davison AJ (2014) The family Parvoviridae. Arch Virol 159:1239–1247

Dhama K, Kumar N, Saminathan M, Tiwari R, Karthik K, Kumar MA, Palanivelu M, Shabbir MZ, Malik YS, Singh RK (2017) Duck virus enteritis (duck plague) – a comprehensive update. Vet Q 37:57–80

Ding L, Chen P, Bao X, Li A, Jiang Y, Hu Y, Ge J, Zhao Y, Wang B, Liu J, Chen H (2019) Recombinant duck enteritis viruses expressing the Newcastle disease virus (NDV) F gene protects chickens from lethal NDV challenge. Vet Microbiol 232:146–150

França MS, Brown JD (2014) Influenza pathobiology and pathogenesis in avian species. Curr Top Microbiol Immunol 385:221–242

Fu Y, Pan M, Wang X, Xu Y, Yang H, Zhang D (2008) Molecular detection and typing of duck hepatitis A, virus directly from clinical specimens. Vet Microbiol 131:247–257

Fu Y, Chen Z, Li C, Liu G (2012) Protective immune responses in ducklings induced by a suicidal DNA, vaccine of the VP1 gene of duck hepatitis virus type 1. Vet Microbiol 160:314–318

Gall-Reculé GL, Jestin V, Chagnaud P, Blanchard P, Jestin A (1996) Expression of Muscovy duck parvovirus capsid proteins (VP2 and VP3) in a baculovirus expression system and demonstration of immunity induced by the recombinant proteins. J Gen Virol 77:2159–2163

Glávits R, Zolnai A, Szabó É, Ivanics É, Zarka P, Mató T, Palya V (2005) Comparative pathological studies on domestic geese (*Anser anser domestica*) and Muscovy ducks (*Cairina moschata*) experimentally infected with parvovirus strains of goose and Muscovy duck origin. Acta Vet Hung 53:73–89

Glisson JR, Hofacre CL, Christensen JP (2008) Fowl cholera. In: Fadly AM, Glisson JR, McDougald LR, Nolan LK, Swayne DE (eds) Diseases of poultry, 12th edn. Blackwell, pp 739–757

Glisson JR, Hofacre CL, Christensen JP (2013) Fowl cholera. In: Swayne DE, Glisson JR, McDougald LR, Nolan LK, Suarez DL, Nair V (eds) Diseases of poultry, 13th edn. Wiley, pp 807–823

Gong Q, Kong LY, Niu MF, Qin CL, Yang Y, Li X, Ruan MD, Tian Y, Li ZL (2018) Construction of a *ptfA,* chitosan nanoparticle DNA, vaccine against *Pasteurella multocida* and the immune response in chickens. Vet J 231:1–7

Gough R (2002) Duck virus enteritis In: Jordan F Pattison M Alexander D Faragher T (ed) Poultry diseases 5th. WB Saunders. pp 235–239

Gough RE, Alexander DJ (1990) Duck virus enteritis in Great Britain 1980 to 1989. Vet Rec 126: 595–597

Gunawardana GA, Frost AJ, Wilkie I (2004) Evaluation of the immune response and efficacy in chickens of vaccines prepared from *P. multocida* capsular type A, grown in iron-depleted conditions. Br Poult Sci 45(Suppl 1):S18–S19

Guo A (2016) Characterization of the complete mitochondrial genome of the cloacal tapeworm *Cloacotaenia megalops* (Cestoda: Hymenolepididae). Parasit Vectors 9:490

Guyonnet V, Peters AR (2020) Are current avian influenza vaccines a solution for smallholder poultry farmers? Gates Open Res 4:122

Haider N, Sturm-Ramirez K, Khan SU, Rahman MZ, Sarkar S, Poh MK, Shivaprasad HL, Kalam MA, Paul SK, Karmakar PC, Balish A, Chakraborty A, Mamun AA, Mikolon AB, Davis CT, Rahman M, Donis RO, Heffelfinger JD, Luby SP, Zeidner N (2017) Unusually high mortality in waterfowl caused by highly pathogenic avian Influenza A, (H5N1) in Bangladesh. Transbound Emerg Dis 64:144–156

Hansen WR, Nashold SW, Docherty DE, Brown SE, Knudson DL (2000) Diagnosis of duck plague in waterfowl by polymerase chain reaction. Avian Dis 44:266–274

Hanson LE, Tripathy DN (1976) Oral immunization of ducklings with attenuated duck hepatitis virus. Dev Biol Stand 33:357–363

Hinz KH, Ryll M, Kohler B, Glunder G (1998) Phenotypic characteristics of *Riemerella anatipestifer* and similar micro-organisms from various hosts. Avian Pathol 27:33–42

Hong YT, Kang M, Jang HK (2018) Pathogenesis of duck circovirus genotype 1 in experimentally infected Pekin ducks. Poult Sci 97:3050–3057

Huang B, Subramaniam S, Frey J, Loh H, Tan HM, Fernandez CJ, Kwang J, Chua KL (2002) Vaccination of ducks with recombinant outer membrane protein (OmpA) and a 41 kDa partial protein (P45N') of *Riemerella anatipestifer*. Vet Microbiol 84:219–230

Hung SH (2002) *Riemerella anatipestifer* infection annex of animal infectious diseases, Bureau of Animal and Plant Health Inspection and Quarantine COA, Taiwan. pp. 199–204

ICTV (2017) International Committee on Taxonomy of Viruses) https://talkictvonlineorg/taxonomy/. Accessed 20 Mar 2021

Jestin V, Le Bras M, Cherbonnel M, Le Gall G, Bennejean G (1991) Demonstration of very pathogenic parvoviruses (Derzsy disease virus) in Muscovy duck farms. Recl Med Vet 167: 849–857

Ji J, Du LQ, Xie QM, Cao YC, Zuo KJ, Xue CY, Ma JY, Chen F, Bee YZ (2009) Rapid diagnosis of duck plagues virus infection by loop-mediated isothermal amplification. Res Vet Sci 87:53–58

Kang HM, Lee EK, Song BM, Heo GB, Jung J, Jang I, Bae YC, Jung SC, Lee YJ (2017) Experimental infection of mandarin duck with highly pathogenic avian influenza A, (H5N8 and H5N1) viruses. Vet Microbiol 198:59–63

Kang M, Roh JH, Jang HK (2018) Protective efficacy of a bivalent live attenuated vaccine against duck hepatitis A, virus types 1 and 3 in ducklings. Vet Microbiol 214:108–112

Kida H, Yanagawa R, Matsuoka Y (1980) Duck influenza lacking evidence of disease signs and immune response. Infect Immun 30:547–553

Kilany WH, Arafa A, Erfan AM, Ahmed MS, Nawar AA, Selim AA, Khoulosy SG, Hassan MK, Aly MM, Hafez HM, Abdelwhab EM (2010) Isolation of highly pathogenic avian influenza H5N1 from table eggs after vaccinal break in commercial layer flock. Avian Dis 54:1115–1119

Kim MC, Kwon YK, Joh SJ, Lindberg AM, Kwon JH, Kim JH, Kim SJ (2006) Molecular analysis of duck hepatitis virus type 1 reveals a novel lineage close to the genus Parechovirus in the family *Picornaviridae*. J Gen Virol 87:3307–3316

Kim MC, Kwon YK, Joh SJ, Kim SJ, Tolf C, Kim JH, Sung HW, Lindberg AM, Kwon JH (2007) Recent Korean isolates of duck hepatitis virus reveal the presence of a new geno- and serotype when compared to duck hepatitis virus type 1 type strains. Arch Virol 152:2059–2072

Kim MC, Kwon YK, Joh SJ, Kwon JH, Michael Lindberg A (2008) Differential diagnosis between type-specific duck hepatitis virus type 1 (DHV-1) and recent Korean DHV-1-like isolates using a multiplex polymerase chain reaction. Avian Pathol 37:171–177

Kim MC, Kim MJ, Kwon YK, Lindberg AM, Joh SJ, Kwon HM, Lee YJ, Kwon JH (2009) Development of duck hepatitis A virus type 3 vaccine and its use to protect ducklings against infections. Vaccine 27:6688–6694

Kisary J (1986) Diagnosis and control of parvovirus infection of geese (Derzsy's Disease) In: McFerran JB, McNulty MS (ed) Acute virus infections of poultry: a seminar in the CEC agricultural research Programme held in Brussels, 13–14 June 1985. pp 239-242

Lee MS, Chen LH, Chen YP, Liu YP, Li WC, Lin YL, Lee F (2016) Highly pathogenic avian influenza viruses H5N2, H5N3 and H5N8 in Taiwan in 2015. Vet Microbiol 187:50–57

Lee DH, Bertran K, Kwon JH, Swayne DE (2017) Evolution global spread and pathogenicity of highly pathogenic avian influenza H5Nx clade 2.3.4.4. J Vet Sci 18:269–280

Leibovitz L (1968) Wenyonella philiplevinei, n. sp., a coccidial organism of the white Pekin duck. Avian Dis 12:670–681

Li Y, Huang B, Ma X, Wu J, Li F, Ai W, Song M, Yang H (2009) Molecular characterization of the genome of duck enteritis virus. Virol 391:151–161

Li H, Wang Y, Han Z, Wang Y, Liang S, Jiang L, Hu Y, Kong X, Liu S (2016) Recombinant duck enteritis viruses expressing major structural proteins of the infectious bronchitis virus provide protection against infectious bronchitis in chickens. Antivir Res 130:19–26

Liu CS (1999) Fowl cholera. An introduction of animal infectious diseases-B, Agricultural Technology Research Institute, Taiwan. pp. 405–411

Liu YT (2002) Studies on the persistent and shedding of *Riemerella anatipestifer* in infected ducks developing polymerase chain reaction diagnostic technique and analysis of outer membrane protein A, gene sequences National Taiwan University

Liu M, Fanyi M, Li X-j, Zhang Z, Liu S, Zhang Y (2011) Goose haemorrhagic hepatitis caused by a new subtype duck hepatitis type 1 virus. Vet Microbiol 152:280–283

Liu J, Chen P, Jiang Y, Deng G, Shi J, Wu L, Lin Y, Bu Z, Chen H (2013a) Recombinant duck enteritis virus works as a single-dose vaccine in broilers providing rapid protection against H5N1 influenza infection. Antivir Res 97:329–333

Liu P, Lu H, Li S, Wu Y, Gao GF, Su J (2013b) Duck egg drop syndrome virus: an emerging Tembusu-related flavivirus in China. Sci China Life Sci 56:701–710

Liu H, Wang X, Ding C, Han X, Cheng A, Wang S, Yu S (2013c) Development and evaluation of a trivalent *Riemerella anatipestifer*-inactivated vaccine. Clin Vaccine Immunol 20:691–297

Liu X, Liu Q, Xiao K, Li P, Liu Q, Zhao X, Kong Q (2016) Attenuated *salmonella typhimurium* delivery of a novel DNA vaccine induces immune responses and provides protection against duck enteritis virus. Vet Microbiol 186:189–198

Liu YY, Yang WT, Shi SH, Li YJ, Zhao L, Shi CW, Zhou FY, Jiang YL, Hu JT, Gu W, Yang GL, Wang CF (2017) Immunogenicity of recombinant *Lactobacillus plantarum* NC8 expressing goose parvovirus VP2 gene in BALB/c mice. J Vet Sci 18:159–167

Liu Q, Hu Y, Li P, Kong Q (2019) Identification of Fur in *Pasteurella multocida* and the potential of its mutant as an attenuated live vaccine. Front Vet Sci 6:5

Lu YS, Lin DF, Lee YL, Liao YK, Tsai HJ (1993) Infectious bill atrophy syndrome caused by parvovirus in a co-outbreak with duck viral hepatitis in ducklings in Taiwan. Avian Dis 37:591–596

Luo Y, Glisson JR, Jackwood MW, Hancock RE, Bains M, Cheng IH, Wang C (1997) Cloning and characterization of the major outer membrane protein gene (ompH) of *Pasteurella multocida* X-73. J Bacteriol 179:7856–7864

Luo Q, Kong L, Dong J, Zhang T, Wang H, Zhang R, Lu Q, Chen H, Shao H, Jin M (2019) Protection of chickens against fowl cholera by supernatant proteins of *Pasteurella multocida* cultured in an iron-restricted medium. Avian Pathol 48:221–229

Mahardika GN, Putra MBAPA, Dewi NPS, Dewi NMRK, Winaya IBO (2015) Muscovy duck parvovirus infection with epicarditis in Bali Indonesia. J Veterinar Sci Techno 7:1000328

Makhija A, Kumar S (2017) Characterization of duck plague virus stability at extreme conditions of temperature pH and salt concentration. Biologicals 45:102–105

Maurin-Bernaud L, Goutebroze S, Merdy O, Chanay A, Cozette V, Le Gros FX (2014) Efficacy of a new attenuated duck parvovirosis vaccine in Muscovy ducks. Vet Rec 175:281

McDougald LR (2013) Internal parasites. In: Swayne DE, Glisson JR, McDougald LR, Nolan LK, Suarez DL, Nair V (eds) Diseases of poultry, 13th edn. Wiley, pp 1117–1146

Metwally SA (2013) Duck virus enteritis (duck plague). In: Swayne DE, Glisson JR, McDougald LR, Nolan LK, Suarez DL, Nair V (eds) Diseases of poultry, 13th edn. Wiley, pp 431–440

Mishra A, Vijayakumar P, Raut AA (2017) Emerging avian influenza infections: current understanding of innate immune response and molecular pathogenesis. Int Rev Immunol 36:89–107

Mostaan S, Ghasemzadeh A, Sardari S, Shokrgozar MA, NikbakhtBrujeni G, Abolhassani M, Ehsani P, Asadi Karam MR (2020) Pasteurella multocida vaccine candidates: a systematic review. Avicenna J Med Biotechnol 12:140–147

Ning K, Liang T, Wang M, Dong Y, Qu S, Zhang D (2018) Pathogenicity of a variant goose parvovirus from short beak and dwarfism syndrome of Pekin ducks in goose embryos and goslings. Avian Pathol 47:391–399

Niu Y, Liu B, Sun C, Zhao L, Chen H (2020) Construction of the recombinant duck enteritis virus delivering capsid protein VP0 of the duck hepatitis A virus. Vet Microbiol 249:108837

OIE (2000) Manual of diagnostic tests and vaccines for terrestrial animals Chapter 2711 Fowl Cholera (avian pasteurellosis)

OIE (2015) Manual of diagnostic tests and vaccines for terrestrial animals. Chapter 2.3.4. Avian Influenza (Infectious with Avian Influenza Viruses)

OIE (2017) Manual of diagnostic tests and vaccines for terrestrial animals Chapter 2.3.8. Duck viral hepatitis

OIE (2018) Manual of diagnostic tests and vaccines for terrestrial animals Chapter 2.3.7. Duck virus enteritis

Palya V (2013) Parvovirus infections in waterfowl. In: Swayne DE, Glisson JR, McDougald LR, Nolan LK, Suarez DL, Nair V (eds) Diseases of poultry, 13th edn. Wiley, pp 444–451

Palya V, Zolnai A, Benyeda Z, Kovacs E, Kardi V, Mato T (2009) Short beak and dwarfism syndrome of mule duck is caused by a distinct lineage of goose parvovirus. Avian Pathol 38: 175–180

Pantin-Jackwood MJ, Swayne DE (2007) Pathobiology of Asian highly pathogenic avian influenza H5N1 virus infections in ducks. Avian Dis 51:250–259

Pascalon-Pekelniczky A (1998) Plasma lipid changes in the female mule duck (*Cairina moschata* x *Anas platyrhynchos*) experimentally infected with *Eimeria mulardi*. Vet Res 29:197–206

Pathanasophon P, Phuektes P, Tanticharoenyos T, Narongsak W, Sawada T (2002) A potential new serotype of *Riemerella anatipestifer* isolated from ducks in Thailand. Avian Pathol 31:267–270

Phonvisay M, Lee JW, Liou JJ, Wang HY, Chu CY (2019) Evaluation of long-term antibody response and cross-serotype reaction in ducks immunised with recombinant *Riemerella anatipestifer* outer membrane protein A and CpGODN. J Vet Res 63:543–548

Poonia B, Dunn PA, Lu H, Jarosinski KW, Schat KA (2006) Isolation and molecular characterization of a new Muscovy duck parvovirus from Muscovy ducks in the USA. Avian Pathol 35:435–441

Poultry Hub (2018) Fowl Cholera (or Pasteurellosis), Poultry Hub. https://www.poultryhub.org/all-about-poultry/health-management/disease/fowl-cholera-or-pasteurellosis. Accessed 20 Mar 2021

Priz NN (1973) Comparative study of the characteristics of the manifestation of viral hepatitis in animals infected by different routes. Vopr Virusol 18:696–700

Rahman MA, Rahman MM, Abdullah MS, Sayeed MA, Rashid MH, Mahmud R, Belgrad JP, Hoque MA (2019) Epidemiological assessment of clinical poultry cases through the government veterinary hospital-based passive surveillance system in Bangladesh: a case study. Trop Anim Health Prod 51:967–975

Richard I, Paul H (2010) Diagnosis and control of goose parvovirus. In Pract 32:382–386

Rimler RB, Rhoades KR (1987) Serogroup F, a new capsule serogroup of *Pasteurella multocida*. J Clin Micro Apr 1987:615–618

Routh A, Sanderson S (2009) Chapter 12. Waterfowl. In: Tully TN Jr, Dorrestein GM, Jones AK, Cooper JE (eds) Handbook of avian medicine, 2nd edn. Elsevier, pp 275–308

Sandhu TS (1979) Immunization of white Pekin ducklings against *Pasteurella anatipestifer* infection. Avian Dis 23:662–669

Sandhu TS (1991) Immunogenicity and safety of a live *Pasteurella anatipestifer* vaccine in white Pekin ducklings: laboratory and field trials. Avian Pathol 20:423–432

Sandhu TS (2008) *Riemerella anatipestifer* infection. In: Fadly AM, Glisson JR, McDougald LR, Nolan LK, Swayne DE (eds) Diseases of poultry, 12th edn. Blackwell, pp 758–764

Sandhu TS, Calnek BW, Zeman L (1992) Pathologic and serologic characterization of a variant of duck hepatitis type I virus. Avian Dis 36:923–926

Sarver CF, Morishita TY, Nersessian B (2005) The effect of route of inoculation and challenge dosage on *Riemerella anatipestifer* infection in Pekin ducks (*Anas platyrrhynchos*). Avian Dis 49:104–107

Schultz-Cherry S, Dybing JK, Davis NL, Williamson C, Suarez DL, Johnston R, Perdue ML (2000) Influenza virus (A/HK/156/97) hemagglutinin expressed by an alphavirus replicon system protects chickens against lethal infection with Hong Kong-origin H5N1 viruses. Virol 278: 55–59

Shawky SA, Sandhu TS (1997) Inactivated vaccine for protection against duck virus enteritis. Avian Dis 41:461–468

Shawky S, Schat KA (2002) Latency sites and reactivation of duck enteritis virus. Avian Dis 46: 308–313

Song S, Li P, Zhang R, Chen J, Lan J, Lin S, Guo G, Xie Z, Jiang S (2019) Oral vaccine of recombinant *Lactococcus lactis* expressing the VP1 protein of duck hepatitis A virus type 3 induces mucosal and systemic immune responses. Vaccine 37:4364–4369

Spieker JO, Yuill TM, Burgess EC (1996) Virulence of six strains of duck plague virus in eight waterfowl species. J Wildl Dis 32:453–460

Sthitmatee N, Numee S, Kawamoto E, Sasaki H, Yamashita K, Takahashi N, Kataoka Y, Sawada T (2008) Protection of chickens from fowl cholera by vaccination with recombinant adhesive protein of *Pasteurella multocida*. Vaccine 26:2398–2407

Su CF (2006) Fowl cholera information center of veterinary technology and science, Animal Health Research Institute, COA, Taiwan

Sun H, Pu J, Hu J, Liu L, Xu G, Gao GF, Liu X, Liu J (2016) Characterization of clade 2.3.4.4. Highly pathogenic H5 avian influenza viruses in ducks and chickens. Vet Microbiol 182:116–122

Sun Y, Yang C, Li J, Li L, Cao M, Li Q, Li H (2017) Construction of a recombinant duck enteritis virus vaccine expressing hemagglutinin of H9N2 avian influenza virus and evaluation of its efficacy in ducks. Arch Virol 162:171–179

Swayne DE (2012) Impact of vaccines and vaccination on global control of avian influenza. Avian Dis 56:818–828

Swayne DE, Kapczynski DR (2017) Vaccines and vaccination for avian influenza in poultry. In: Swayne DE (ed) Animal Influenza, Wiley-Blackwell, pp 378–434

Swayne DE, Suarez DL, Sims LD (2013) Influenza. In: Swayne DE, Glisson JR, McDougald LR, Nolan LK, Suarez DL, Nair V (eds) Diseases of poultry, 13th edn. Wiley, pp 181–218

Takehara K, Hyakutake K, Imamura T, Mutoh K, Yoshimura M (1994) Isolation identification and plaque titration of parvovirus from Muscovy ducks in Japan. Avian Dis 38:810–815

Takehara K, Ohshiro T, Matsuda E, Nishio T, Yamada T, Yoshimura M (1995) Effectiveness of an inactivated goose parvovirus vaccine in Muscovy ducks. J Vet Med Sci 57:1093–1095

Thanasarasakulpong A, Poolperm P, Tankaew P, Sawada T, Sthitmatee N (2015) Protectivity conferred by immunization with intranasal recombinant outer membrane protein H from *Pasteurella multocida* serovar A:1 in chickens. J Vet Med Sci 77:321–326

The Poultry Site (2011) Major diseases of waterfowl and their control, The Poultry Site http://www.thepoultrysite.com/articles/2051/major-viral-diseases-of-waterfowl-and-their-control/. Accessed 20 Mar 2021

The Poultry Site (2014) Fowl Cholera, The Poultry Site http://www.thepoultrysite.com/publications/6/diseases-of-poultry/181/fowl-cholera/. Accessed 01 Feb 2021

Tsai SH (2002a) *Riemerella anatipestifer* infection handout for training workshop on pathological diagnosis of poultry disease. Chinese Soc Veterinary Pathol:75–76

Tsai SH (2002b) Fowl cholera handout for training workshop on pathological diagnosis of poultry disease. Chinese Soc Veterinary Pathol:66–68

Tseng CH, Tsai HJ (2007) Molecular characterization of a new serotype of duck hepatitis virus. Virus Res 126:19–31

Varinrak T, Poolperm P, Sawada T, Sthitmatee N (2017) Cross-protection conferred by immunization with an rOmpH-based intranasal fowl cholera vaccine. Avian Pathol 46:515–525

Wakenell P (2016) Chapter 15. Management and medicine of backyard poultry. In: Speer BL (ed) Current therapy in avian medicine and surgery, Elsevier, pp 550–565

Wang AP, Gu LL, Wu S, Zhu SY (2018) Duck hepatitis A virus structural proteins expressed in insect cells self-assemble into virus-like particles with strong immunogenicity in ducklings. Vet Microbiol 215:23–28

Wang AP, Liu L, Gu LL, Guo CM, Wu S, Feng Q, Xia WL, Wu Z, Zhu SY (2019a) Protection against duck hepatitis a virus type 1 conferred by a recombinant avian adeno-associated virus. Poult Sci 98:112–118

Wang AP, Liu L, Gu LL, Wu S, Guo CM, Feng Q, Xia WL, Yuan C, Zhu SY (2019b) Expression of duck hepatitis A virus type 1 VP3 protein mediated by avian adeno-associated virus and its immunogenicity in ducklings. Acta Virol 63:53–59

Wen H, Han L, Zhang X, Lian C, Zhao L, Si C, Chen H (2014) Duck hepatitis A virus (DHAV) genotype definition: comment on the article by Cha et al. Vet Microbiol 170:462–464

Wolf K, Burke CN, Quimby MC (1976) Duck viral enteritis: a comparison of replication by CCL-141 and primary cultures of duck embryo fibroblasts. Avian Dis 20:447–454

Woolcock PR (2013) Viral infections of waterfowl. In: Swayne DE, Glisson JR, McDougald LR, Nolan LK, Suarez DL, Nair V (eds) Diseases of poultry, 13th edn. Wiley, pp 417–421

Woolcock PR, Tsai HJ (2013) Duck hepatitis. In: Swayne DE, Glisson JR, McDougald LR, Nolan LK, Suarez DL, Nair V (eds) Diseases of poultry, 13th edn. Wiley, pp 422–431

Woolcock PR, Jestin V, Shivaprasad HL, Zwingelstein F, Arnauld C, McFarland MD, Pedersen JC, Senne DA (2000) Evidence of Muscovy duck parvovirus in Muscovy ducklings in California. Vet Rec 146:68–72

Wu YS (2004) Fowl cholera standard operating procedures for laboratory diagnosis of poultry and livestock diseases. Animal Health Research Institute, COA, Taiwan

Wu JR, Shien JH, Shieh HK, Chen CF, Chang PC (2007) Protective immunity conferred by recombinant *Pasteurella multocida* lipoprotein E (PlpE). Vaccine 25:4140–4148

Wu HL, Hu PH, Zhang J (2013) Molecular characterization of *Wenyonella philiplevinei* from ducks in Hunan province, China. Thai Vet Med 43:605–608

Wu F, Lu F, Fan X, Pan Q, Zhao S, Sun H, Zhang J, Liu C, Chao J, Zhang X (2020a) Development of a live attenuated duck hepatitis A virus type 3 vaccine (strain SD70). Vaccine 38:4695–4703

Wu HC, Chang WC, Wu MC, Wang HY, Chu CY (2020b) Assessment of immunization regimensof duck *Riemerella anatipestifer* vaccines. J Appl Microbiol 129:1185–1192

Xiao K, Liu Q, Liu X, Hu Y, Zhao X, Kong Q (2016) Identification of the avian *Pasteurella multocida phoP* gene and evaluation of the effects of *phoP* deletion on virulence and immunogenicity. Int J Mol Sci 17:12

Xie L, Xie Z, Huang L, Wang S, Huang J, Zhang Y, Zeng T, Luo S (2017) A polymerase chain reaction assay for detection of virulent and attenuated strains of duck plague virus. J Virol Methods 249:66–68

Xu X, Xu Y, Miao S, Jiang P, Cui J, Gong Y, Tan P, Du X, Islam N, Hu Q (2020) Evaluation of the protective immunity of *Riemerella anatipestifer* OmpA. Appl Microbiol Biotechnol 104:1273–1281

Yang FL, Jia WX, Yue H, Luo W, Chen X, Xie Y, Zen W, Yang WQ (2005) Development of quantitative real-time polymerase chain reaction for duck enteritis virus DNA. Avian Dis 49: 397–403

Yang S, Dong W, Li G, Zhao Z, Song M, Huang Z, Fu J, Jia F, Lin S (2019) A recombinant vaccine of *Riemerella anatipestifer* OmpA, fused with duck IgY Fc and *Schisandra chinensis* polysaccharide adjuvant enhance protective immune response. Microb Pathog 136:103707

Yang D, Mai K, Zhou Q, Zhu Y, Xing J, Luo C, Liu S, Zhou Q, Huang W, Luo J, Liu J (2020) The protective efficacy of specific egg yolk immunoglobulin Y (IgY) against *Riemerella anatipestifer* infections. Vet Microbiol 243:108642

Yuan GP, Cheng AC, Wang MS, Liu F, Han XY, Liao YH Xu C (2005) Electron microscopic studies of the morphogenesis of duck enteritis virus. Avian Dis 49:50–55

Zádori Z, Stefancsik R, Rauch T, Kisary J (1995) Analysis of the complete nucleotide sequences of goose and Muscovy duck pervoviruses indicates common ancestral origin with adeno-associated virus 2. Virol 212:562–573
Zhang H, Pi J, Tang C, Yue H, Yang F (2012) An experimental study of the pathogenicity of a duck hepatitis A virus genotype C isolate in specific pathogen free ducklings. Avian Pathol 41:613–620
Zhang W, Chen S, Mahalingam S, Wang M, Cheng A (2017) An updated review of avian-origin Tembusu virus: a newly emerging avian Flavivirus. J Gen Virol 98:2413–2420
Zhang R, Chen J, Zhang J, Yang Y, Li P, Lan J, Xie Z, Jiang S (2018) Novel duck hepatitis A virus type 1 isolates from adult ducks showing egg drop syndrome. Vet Microbiol 221:33–37
Zhang R, Yang Y, Lan J, Lin S, Xie Z, Zhang X, Jiang X (2020) A novel peptide isolated from a phage display peptide library modeling antigenic epitope of DHAV-1 and DHAV-3. Vaccine 8:121
Zhao Y, Cao Y, Cui L, Ma B, Mu X, Li Y, Zhang Z, Li D, Wei W, Gao M, Wang J (2014a) Duck enteritis virus glycoprotein D and B DNA vaccines induce immune responses and immunoprotection in Pekin ducks. PLoS One 9:e95093
Zhao H, Xie Z, Xie L, Deng X, Xie Z, Luo S, Huang L, Huang J, Zeng T (2014b) Molecular characterization of the full muscovy duck parvovirus isolated in Guangxi China. Genome Announc 2(6):e01249–e01214
Zhao X, Liu Q, Xiao K, Hu Y, Liu X, Li Y, Kong Q (2016) Identification of the crp gene in avian Pasteurella multocida and evaluation of the effects of crp deletion on its phenotype virulence and immunogenicity. BMC Microbiol 16:125
Zou Q, Sun K, Cheng A, Wang M, Xu C, Zhu D, Jia R, Luo Q, Zhou Y, Chen Z, Chen X (2010) Detection of anatid herpesvirus 1 gC gene by TaqMan™ fluorescent quantitative real-time PCR with specific primers and probe. Virol J 7:37–46
Zou Z, Liu Z, Jin M (2014) Efficient strategy to generate a vectored duck enteritis virus delivering envelope of duck Tembusu virus. Viruses 6:2428–2443
Zou Z, Ma J, Huang K, Chen H, Liu Z, Jin M (2016) Live attenuated vaccine based on duck enteritis virus against duck hepatitis A virus types 1 and 3. Front Microbiol 7:1613
Zou Z, Huang K, Wei Y, Chen H, Liu Z, Jin M (2017) Construction of a highly efficient CRISPR/Cas9-mediated duck enteritis virus-based vaccine against H5N1 avian influenza virus and duck Tembusu virus infection. Sci Rep 7:1478

Dr. Yen-Ping Chen is a Government Veterinarian. She holds a position as an Associate Researcher at the Animal Health Research Institute, Taiwan. She obtained her doctoral degree in the School of Veterinary Medicine, National Taiwan University. The subject of her Ph.D. program was antimicrobial resistant mechanisms of *Riemerella anatipestifer*. Dr. Chen has worked at the Animal Health Research Institute since 2001 and been one of the team members for diagnosis of avian diseases, served in the Epidemiology Division and focused on research and diagnosis of waterfowl diseases.

Dr. Chao-Fang Yu is one of the authors of Duck Diseases and Disease Management. She received her bachelor of Veterinary Medicine from the National Pingtung University of Technology and Science. She has been working as an Assistant Researcher of Animal Health Research Institute, Council of Agriculture in Taiwan for more than 17 years. Her research is focused on production and development of bacterial vaccine and diagnostics for *Riemerella anatipestifer*, *Salmonella pullorum*, and *Brucella abortus*.

Dr. Yu-Hua Shih is one of the authors of Duck Diseases and Disease Management. She received her bachelor and master of Veterinary Medicine from the National Chung Hsing University. She is an Assistant Researcher of Animal Health Research Institute, Council of Agriculture in Taiwan. Her research is focused on vaccine development of waterfowl diseases, such as waterfowl parvovirus live vaccine and duck hepatitis live vaccine.

Duck Genomics and Biotechnology 15

Surya Kanta Mishra and Adnan Naim

Contents

S. K. Mishra (✉)
ICAR- Directorate of Poultry Research, Regional Centre, Bhubaneswar, Odisha, India
e-mail: surya.mishra@icar.gov.in

A. Naim
KIIT-TBI, KIIT University, Bhubaneswar, Odisha, India
e-mail: adnan.naim@iiti.ac.in

A. Jalaludeen et al. (eds.), *Duck Production and Management Strategies*,
https://doi.org/10.1007/978-981-16-6100-6_15

Abstract

The ducks as a waterfowl and a commercially important domesticated poultry next to chicken, have been targets of Genomics and Biotechnology research, in evolving mode. The duck genome has already been sequenced, though it could take a dozen more studies to be sequenced in 3 to 6X scale, so that individual sequence data and polymorphism of the common breeds/strains of ducks are illustrated. These in turn, shall be the starting points to fuel more genomics and biotechnological applications, to the advantages of scientists and breeders. The available platforms like ex-vivo embryo culture, PGC bio-banking, embryonic stem cells and across-the species gene transfer, industrial application of eggs for recombinant protein production, from ducks are likely to be upcoming research hotspots in biotechnological research. The pace of developing newer DNA markers and their utility in discovery of QTLs are nearly in same footing, like chickens and other poultry species. The genomic selection in ducks is yet another area to harness the advantages of latest genomics knowhow, that covers: genome-sequence database; high-density SNP-array profiling (high-throughput marker-genotyping) and EBV (Expected Breeding Value) estimation, for facilitating MAS and molecular-breeding of ducks, in near future.

Keywords

Biotechnology · Ducks · Genomics · Molecular biology · Sequencing · Bio-banking

List of Abbreviations

AFLP	Amplified Fragment Length Polymorphism
BDC	Blastodermal Cells
Cas9	CRISPR associated protein 9
CNV	Copy Number Variation
CRISPR	Clustered Regularly Interspaced Short Palindromic Repeats
EBV	Expected Breeding Value
EDS	Egg Drop Syndrome
EGC	Embryonic Germ Cells
ESC	Embryonic Stem Cells
GWMAS	Genome-Wide Marker Assisted Selection
LD	Linkage Disequilibrium
MAS	Marker Assisted Selection
MHC	Multiple Histocompatibility Complex

NGS	Next Gen Sequencing
PCR	Polymerase Chain Reaction
PGC	Primordial Germ Cells
QTL	Quantitative Trait Loci
RAPD	Random Amplified Polymorphic DNA
RFLP	Restriction Fragment Length Polymorphism
SNP	Single Nucleotide Polymorphism
SPF	Specific Pathogen Free
SSCP	Single Stranded Conformation Polymorphism
STR	Short Tandem Repeats
TALEN	Transcription Activator-Like Effector Nucleases
TLR	Toll-Like Receptors
VNTR	Variations in Number of Tandem Repeats
WGA	Whole Genome Amplification
ZFN	Zinc Finger Nuclease

15.1 Introduction

Like all other commercial poultry species, the ducks as an alternate poultry, have attracted a fair share of Research and Development, into their Genetics and Genomics, while most of the developing world is content with its breeding and improvement. While chicken industry worldwide, have imbibed reasonable amount of biotechnological tools, for infusing faster rates of genetic improvement, by quantity and quality, lately much of this is catching up with ducks too. Accordingly, it is time to delve into few domains of Duck Genomics, especially the structural, functional and comparative aspects, besides transgenesis prospects, exclusively for *Anas platyrhynchos*, as a distinct poultry species. Therefore, few significant genomics and biotechnological advancements mainly of the preceding two decades, which merit some description in details, are laid out in the following pages.

15.2 Evolution of Duck's Genetic and Cytogenetic Maps

The first comparative and cytogenetic map of the ducks was published by Takagi and Makkino in the year 1966; where they postulated that diploid chromosome number of ducks was 80 unlike 78 chromosomes for chickens (Takagi and Makkino 1966). The authors reported upto 30 microchromosomes in the duck genome, besides 10 macrochromosomes, which included the chromosome 1 (the submetacentric and largest of duck chromosomes) and the conspicuous sex chromosome. Subsequent to this, two more studies on defining duck's genome further as Linkage maps, using current-day DNA markers have been documented (Huang et al. 2006; Huang et al. 2009). In as much as conserved DNA sequences between chicken and ducks remained the key to defining maps of ducks, using the already-published chicken genome data, Groenen et al. (2000) assumed the size of the genetic maps of chicken and mallard ducks could be comparable, and so speculated that DNA markers of chickens might measure ~36% of the total predicted length of 3800 cM.

Accordingly, one of the above marker-based linkage map as reported by Huang et al. (2006), which concerned the Pekin ducks, was synthesized using chicken-origin BAC sequence data as references, where 155 polymorphic microsatellites, including 84 novel ones were used for linkage analyses. The authors successfully delineated 19 linkage groups wherein, 115 microsatellite markers were placed for defining the duck genome. Thus a sex-corrected map spanning 1353.3 centiMorgans (cM) length, with a mean interval distance of 15.04 cM per marker was established, in which coverages for male and female maps were 1415 cM and 1387.6 cM, respectively. This study showed that, cross-species homology between duck and chicken DNA sequences, could be used as a background for integrating genetic and cytogenetic maps of duck genome. Here the authors made use of 28 BAC clones screened out of existing chicken BAC libraries, using specific PCR primers, which were successfully localized to duck chromosomes employing FISH (Fluorescent in situ Hybridization) techniques. In this map, 11 of these 19 linkage groups were seen distributed over 10 major duck chromosomes, with rest of 8 linked to other chromosomes.

The credit of reporting next other Linkage map in ducks should go to Huang et al. (2009); which was constructed using AFLP markers, based on typing of F2 segregants (N = 766) ducks drawn from six F1 families, which were crosses raised from two distinct and selected duck lines, i.e., brown Tsaiya and Pekin. This meticulous study contributed to raising 32 linkage groups defined by 260 codominant markers, covering a total length of 1766 cM, thus translating to an average interval distance of 7.75 cM between two adjacent markers (range: 0.0 cM to 33.3 cM). The marker-density analysis showed that, linkage group, LG-1 topped the marker-density chart, with 63 markers defining a length of 171 cM; whereas, the linkage group-11 (LG11) turned out to be the lowest in marker density, with 61.4 cM accounted by just three markers.

However, if one were to compare the results generated from both the marker-based maps (microsatellite versus vs AFLP maps), it was no doubt evident that the AFLP-system produced higher number of linkage groups (32 vs 19) yielding an obviously-higher marker density (average interval distance 7.75 cM vs 15.04 cM) favouring the latter (AFLP) map. This difference could mainly be attributed to the choice of AFLP markers; the resource population's design and analysis methods. However, for purpose of scientific analyses, the microsatellite map could be directly associated to a cytogenetic map, which was not feasible from using AFLP markers.

Skinner et al. (2009) published a detailed molecular cytogenetic map of the duck genome through FISH assignment employing 155 chicken clones. The authors discovered one inter- and six intra-chromosomal rearrangements between chicken and duck macrochromosomes and delineated conserved synteny among all microchromosomes, which were subject to analyses (Fig. 15.1). The authors also worked on comparative genomic hybridization, and fished out 32 Copy Number Variations (CNVs), out of which five overlapped the previously identified 'hotspot' regions between chickens and turkeys. The authors surmised extensive conservation

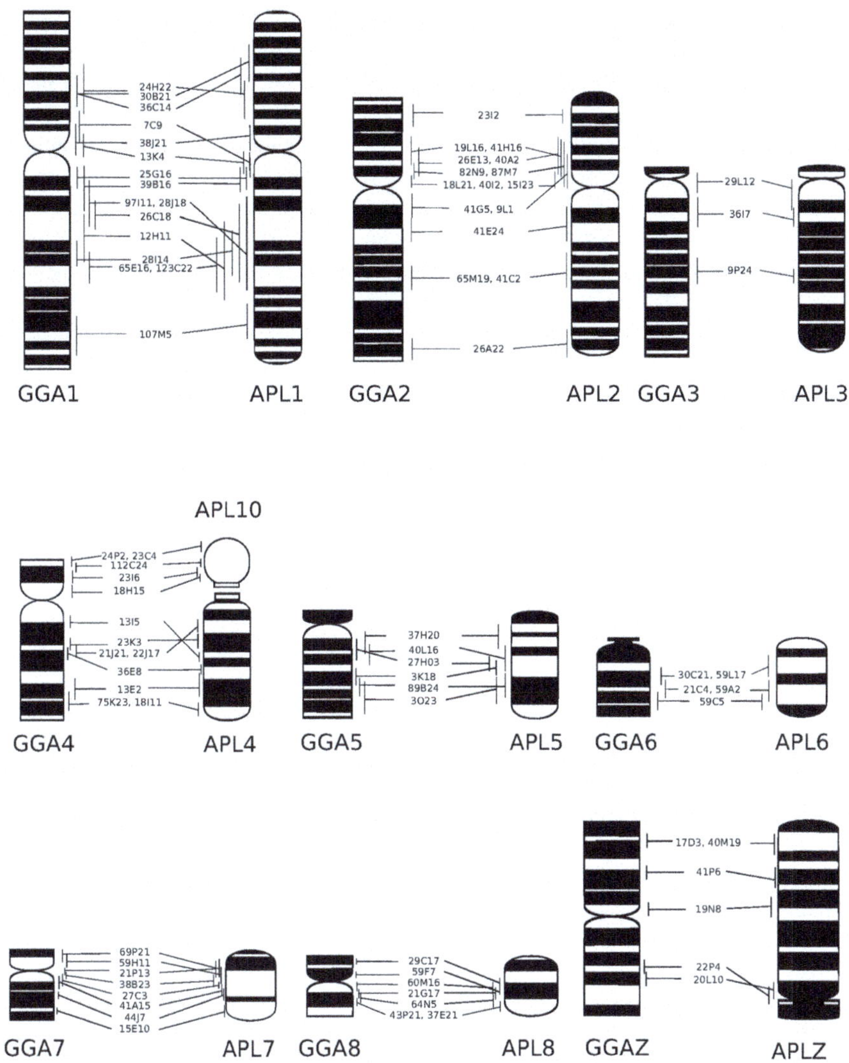

Fig. 15.1 Comparative map of chicken and duck chromosomes (Macrochromosomes 1–8 and Z are compared between chickens versus Ducks. Note how G-banded ideograms of chicken (*Gallus gallus*; GGA) match and differ from that of duck (*Anas platyrhynchos;* APL). These were raised using FISH based mapping. Intra-chromosomal rearrangements are visible on GGA and APL1, 2, 4, 7, 8 and Z. Interestingly, the GGA4p corresponds to APL10. (N.B: Figure: Courtesy, B.M. Skinner et al. (2009) for representation purpose only)

of avian genomes across an evolution process of 90 million years, across both macro and microchromosomes. Summarizing the comparison of CNVs between chicken and duck extended to that between chicken and turkey the conclusions were: that avian genomes in general, contained fewer CNVs than mammals, where genomes of evolutionarily distant species shared many regions of copy number variation.

Rao et al. (2012) reported a duck whole genome Radiation Hybrid (R.H) panel obtained by fusing female duck embryonic fibroblasts irradiated using a dose of 6000 rads, with HPRT-deficient Wg3hCl2 hamster cells. The authors examined ninety best hybrids, having average retention of 23.6% of the duck genome each, which were selected for final test-panel. To enable genotyping of large numbers of markers, necessary for whole-genome mapping, three different methods involving Whole Genome Amplification (WGA) using the latest ones for genotyping i.e., FluidigmBioMark™ Integrated Fluidic Circuits (IFC) Dynamic Array™, were tested. RH maps of APL12 (abbreviated for *Anas platyrhynchos*) and APL22 were generated, which clearly detected many intra-chromosomal rearrangements compared to chicken.

The authors established that, the FluidigmBioMark™ IFC Dynamic Array™ genotyping provided a rapid and cost-effective method for building RH linkage groups. Therefore, it was opined that RH mapping could form a complementary approach to NGS by permitting assignments of sequence-scaffolds to duck chromosomes including mapping of ESTs located on smallest of chromosomes, with inference that detailed RH maps will allow precise estimation of intra-chromosomal rearrangements existent between chickens and ducks.

A consortium led by Huang et al. (2013) took up detailed sequencing of the duck genome. This group sequenced the genome of a female Beijing duck using Illumina platform. In total, they generated some 77 Gb of paired-end reads amounting to 64-fold coverage of whole genome. Using SOAPdenovo (Supple Note), the short reads (50 bp ones) could be combined to generate a draft assembly, that consisted of 78,487 scaffolds and covered 1.1 Gb of the genome. This mega sequencing study yielded three major conclusions apart from delineating a high-quality draft duck genome assembly. Firstly, the immune-related gene repertoires in bird lineages (covering 3116–3355 genes) appeared to be contractive as compared to equivalent repertories of mammals (5715–6044 genes). The second one was the identification of many gene sequences which were responsive to avian Influenza viruses, namely, *AvIFT*, *AvDB7–AvDB10*, *IFITM5* and *IFITM10*. Lastly, it was concluded that LSDs of avian defensin and BTNL genes might be involved in duck's immune response to 2 H5N1 Influenza viruses, as analysed in this study.

15.3 Genetic Diversity in Ducks and Use of Molecular Markers

Similar to other poultry species, viz. chickens, turkeys, Japanese quails, etc., the ducks have not stayed behind, when it comes to characterization of genetic diversity and population genetic parameters using molecular markers. The ducks always, aroused great biological interest to students, entrepreneurs and researchers alike, in delineating their molecular evolution; their genetic diversity and immunological constitution. However, considering the quantum of work registered in other animals like cattle, sheep and goats and poultry species, namely chickens and turkeys, the molecular diversity studies in ducks, have remained relatively limited. However, with respect to molecular handles used to define genetic diversity in ducks, these correlated to the same class of molecular markers, i.e., single locus (RFLP, SNPs and similar ones) and multi-locus (Microsatellites, AFLPs and similar VNTRs) markers, which were amply reported in other poultry species.

As per literature, the work reported by Fields and Scribner (1997) was probably the first one which characterized many novel microsatellite loci from one waterfowl species (*Somateria fischeri*) after which, many more microsatellite markers for ducks were evolved and got continuously reported. Many of the isolated microsatellite loci were used to develop genetic maps of duck, which in turn, contributed to biodiversity studies and deduce genetic relationships between and among duck breeds. Considering that molecular markers have slowly became the gold standards in diversity analyses, let us list out some major molecular markers tried in ducks.

Despite having many types of genetic markers, i.e., biochemical marker, blood group polymorphism, single-gene-based phenotype, etc., the present day marker is synonymous with a DNA marker. The DNA markers currently available to scientists are plenty, say in millions, which provide the required flexibility to choose amongst, as per the need of selection. Further, the DNA markers remain the prerequisite for defining the quantitative trait loci (QTL) influencing various economic traits in poultry, otherwise required, to operate marker assisted Selection (MAS). The DNA markers, in practice, are used to increase the accuracy of selection for the traits, which are difficult to improve through traditional selection in poultry breeding programmes.

As MAS is synonymous with DNA markers, the next consideration is what type of marker should be opted for a particular breeding program? For this, a thorough understanding of available DNA marker systems is necessary. From technological points of view, for analysing DNA-level variance/polymorphism across breeds and lines, many types of variations have been investigated till date, i.e. (i) single nucleotide changes in form of SNPs; (ii) insertions or deletions (Indels) of variable lengths ranging from one to several hundred bases and (iii) VNTR, for variations in number of tandem repeats.

Complying with above principles, there could be three main categories of marker systems: the bi-allelic dominant, such as RAPDs (Random Amplified Polymorphic DNA), AFLPs (Amplified Fragment Length Polymorphism); the bi-allelic codominant, e.g., RFLPs (Restriction Fragment Length Polymorphism), SSCPs (Single Stranded Conformation Polymorphism) and the multi-allelic codominant, such as microsatellites or STRs (Short Tandem Repeats). Depending on the level of polymorphisms of these markers segregating in flock in question and the technical competence of the lab, the choice of marker is to be accomplished. While RFLP markers, proved to be the popular markers in eighties; with emergence of PCR, the microsatellites (multi-allelic) gradually became popular. Similarly, the other popular fingerprint marker types, i.e., RAPDs and AFLPs provided great advantages to workers, mostly in developing countries. However, drawback wise, such markers were generally dominant and generated at random. Nevertheless, these markers continued to be practiced in few labs for ducks (Dolmatova et al. 2000; Xiao et al.

2004; Gholizadeh et al. 2007) for their ease of handling during the first decade of twenty-first century, after which the same nearly disappeared from practice.

15.3.1 Amplified Fragment Length Polymorphism (AFLP)

The Amplified Fragment Length Polymorphism (AFLP, illustrated by Vos et al. 1995) remains the next other popular fingerprint marker that possesses the right quantity of polymorphism besides PCR specificity on the genome. The AFLP classically uses the Restriction Endonucleases (RE) to digest genomic DNA of animal, followed by ligation of adaptors to resultant sticky ends of various restriction fragments. The technique then selectively amplifies a subset of restriction fragments by smart choice of the adaptor sequences, where the primers are chosen based on complementarity to adaptor sequence; sequence at restriction site; besides some of the nucleotides within restriction site fragments. The fragments so amplified from PCR are usually resolved on denaturing agarose gels, and are visualized after an autoradiography or from fluorescent detection methods or capillary-sequencing instruments. Only one study of Huang et al. (2009) based on AFLP aimed to create Linkage map of ducks using F2 segregants from a cross between two selected duck lines, i.e., Brown Tsaiya and Pekin ended up in creating of 32 linkage groups and described 260 codominant markers.

15.3.2 Microsatellites

Compared to the abundance of microsatellites published from chicken DNA sequences, the reports of microsatellite loci in waterfowls have remained steady but limited (Buchholz et al. 1998; Huang et al. 2005, 2006; Maak et al. 2000; Maak et al. 2003; Paulus and Tiedemann, 2003; Stai and Hughes, 2003). Huang et al. (2005) have documented up to 261 microsatellite sequences and their use as genetic markers in ducks. The markers of those days being based on microsatellites, there was a big necessity to fish out as many microsatellites in ducks, through use of heterologous sequence information used for microsatellites of chicken and turkey origins. Since the year 2000 bearing the first report of seven microsatellites getting reported in ducks until 2020, the international literature witnessed some 34 studies on microsatellites reported in ducks over the NCBI-PubMed database. The perusal of the same showed that bulk of the microsatellite's results dealt with genetic diversity; whereas few were used for linkage mapping of duck genome.

The microsatellites, also known as Simple Sequence Repeats (SSRs), are representing a class of repetitive sequences widely distributed across most vertebrate genomes, which are largely used for studying biodiversity and polymorphism within and across closed populations. Microsatellite's hallmarks being, 'abundant, ubiquitous, easy to automate, codominant, robust, universal, reliable and reproducible' markers in chickens, also proved equally good for the ducks. Interestingly, the patterns of polymorphism shown by microsatellites proved higher compared to other DNA marker systems. In a recent study, Lai et al. (2020) demonstrated that, upon using just 24 polymorphic microsatellites, across two different egger duck populations (Tsaiya breed: brown vs white), they could accurately deduce their relation and genetic diversities, through population parameters, viz. F-statistics, coefficients of heterozygosity, polymorphic information content (PIC) and homozygosity indices accurately. The authors concluded that microsatellites could be a reliable tool for monitoring quality-breeding and conservation analysis of ducks in Taiwan.

Another recent study (Hariyono et al. 2019) involving application of 22 polymorphic microsatellites, established a more diverse potentials of microsatellites in judging genetic constitution of purebred duck populations. The authors studied eight local Indonesian duck breeds, documented as national germplasm of the country for conservation. The study tried to estimate the genetic diversity and phylogenetic relationship of these eight populations. The results revealed a moderate level of genetic diversity among populations, with detection of some 153 alleles over all loci and populations. The microsatellites based analyses could estimate various population parameters, i.e., Observed (Ho) and Expected (He) Heterozygosity as 0.440 and 0.566, respectively; PIC (0.513 across populations); F-statistics (FIT = 0.237, FIS = 0.114 and FST = 0.137); besides heterozygosity indices (He = 0.545 vs 0.368; detected in the most-diverse versus least-diverse of the populations), with indication that majority of the populations tested were relatively heterozygote-deficient, for reasons of inbreeding. Using the phylogenetic trees (based on molecular data) and principal coordinates' analysis, these populations could be grouped into two major clusters with obvious interpretations. The conclusions of this study was about detection of considerable genetic diversity in the studied populations and therefore, microsatellite used analyses aided in devising proper management strategies to retain genetic diversity and prevent loss of alleles.

In line of such diversity reports, few more microsatellite studies have also been reported in ducks (Seo et al. 2016; Li et al. 2007; Su et al. 2007; Huang et al. 2005; Stai and Hughes 2003), where the breeds and of species of ducks could efficiently been differentiated and defined for genetic purity, constitution and haplotype delineation, etc. These studies thus proved the versatility of microsatellites in analysing and interpreting diversity analyses of ducks.

The use of microsatellites has also been documented in constructing genetic and linkage maps of ducks, besides serving as markers for population specificity and QTL mapping for most economic traits (Mucha et al. 2014; Kileh-Wais et al. 2013; Huang et al. 2006, 2007) like carcass and meat traits; body conformation traits etc.. Microsatellite's usages also have been made for paternity determination (Ren et al. 2009) and for serving as tools for species differentiation (Tiedemann et al. 2011), thus proving their worthiness for ducks, more reliably than other marker systems.

15.3.3 Single Nucleotide Polymorphism (SNP)

In backdrop of above, in the previous two decades, the Single Nucleotide Polymorphism (SNP) has emerged as the latest and most dependent marker system, which is gaining popularity worldwide. By definition, an SNP marker emerges, when there is just a single base change in a DNA sequence, with usual alternative of two possible nucleotides at a given locus (position). For qualifying to be an efficient SNP, it is considered that the least frequent allele should have at least a frequency of 1% or greater. The latest report indicates the frequency of SNP markers in chicken genome was ~1 in every 200 base pairs of DNA. While the latest genome maps in chickens report more than six million SNPs, the same for ducks is in millions too (Lee et al. 2020).

15.3.4 Copy Number Variation (CNV)

Of late (in last two decades) a newer category of DNA variation namely, Copy Number Variation (CNV) has become popular, which is frequently used in DNA profiling and comparative genomics of animals and poultry. First reported by Redon et al. (2006) from Human DNA variation studies, the CNV as a phenomenon got noticed across mammalian and poultry genomes, where DNA sequences ranging from one Kb to several million base pairs (Mb), are found repeated with its number of repeats remaining variable between individuals within a given population. It is now used, as an important DNA marker in an individual, to profile variation, rather than SNP, because of its coverage across more genomic regions. Research documented till date, depicts many known phenotypes of chickens, which were associated with CNVs. Many classical major gene characters of chickens were studied. The pea-comb (Wright et al. 2009), the late-feathering gene localized on Z chromosome (Elferink et al. 2008), the dark-brown feather colour (Gunnarsson et al. 2011) and fibromelanosis or dermal hyperpigmentation character as found in Kadaknath and Silkie chickens (Dorshorst et al. 2010, 2011), were all reported to be linked to CNV markers in chickens. As such, the functional consequences of majority of Chicken CNVs still remain to be deciphered, where in individual chickens possessing most of CNVs, appear phenotypically normal, similar to reports in many other species.

However, despite lots of CNVs investigated in chickens, the lone CNV based study reported in ducks indicate that, birds in general, host a reduced number of CNVs compared to mammals and that genomic CNV hotspots previously described in chicken and turkey are also found across ducks vs chickens comparison, indicating conservation over evolutionary scale. Skinner et al. (2009) registered 32 CNVs in ducks (versus chickens), which worked out to be twice the variation, compared to CNV of turkey vs chickens (16 only).

15.3.5 QTL Mapping

The emergence of QTLs significantly influencing various economic traits of chickens has already been saturated. As is conceptualized, a chromosomal region that contains one or more genes, which can influence a complex trait, is defined as a QTL (Andersson 2001). As on date, there are numerous QTLs reported in ducks too, involving many economically important traits. As per NCBI's database, the cumulative reports until mid-2020 cited some 10 significant QTL studies involving mostly growth traits of ducks including body weight, carcass quality, meat quality and body confirmation, apart from general production and liver quality traits. With adoption of many strategies for construction of genetic maps and statistical tools for segregation analysis, several QTLs influencing growth, body composition and fertility have already been mapped with high accuracy in many livestock species (Andersson 2001).

For detecting and mapping QTL for any livestock species, the construction of a genetic linkage map remains a prerequisite; however, data is very limited in case of duck genetic map. Huang et al. (2006) reported a preliminary genetic linkage map using an inbred Pekin duck resource population employing microsatellite markers. In contrast, the advantage of AFLP is that, a large number of markers could be generated with a smaller number of primer pairs compared to that of microsatellites. This is especially true, while working with ducks, for which only few hundreds microsatellite markers have been validated till date, as generating large batteries of duck microsatellites are funds-intensive, by nature. But, once a dense-linkage map is developed, as a chain of events, it can facilitate mapping additional QTLs for many uncharacterized commercial traits, for future uses in duck breeding.

QTL Discovery and Resource Populations

As the prerequisite for a QTL experiment is the resource population, QTL theories have revolved around various kinds of genetic resources for research. Historically, most QTL mapping studies have been conducted using crosses of genetically and phenotypically divergent chicken lines. However, in ducks, the resource populations used for QTL detection, range from crosses of meat type to egger-type ducks; to crosses between two extreme meat type duck populations. Hocking's review dealing with chickens highlighted that bulk of experimental designs to identify QTL for production traits were, typically an F2 cross between two breeds with minimum size of 250–700 birds (Hocking 2005). Parallel to these experiments, few of the studies (Mucha et al. 2014; Huang et al. 2009) reported in ducks for QTL have used F2 populations too, mainly involving the Pekin ducks for meat and growth characters. A few studies also used direct genotyping-by-sequencing methods, wherein the SNPs explaining QTL associations for the trait of interest have been reported in literature. A couple of studies involving the mule ducks as backcrosses (Vitezica et al. 2010; François et al. 2017) have been analysed to fish out pQTLs for few important economic traits. Two studies emphasizing a typical half-sib analysis (Huang et al. 2007) have been reported for QTL studies, which primarily made use of microsatellite markers to fish out QTLs for mostly body weight and few meat characters. A genome-wide association analysis (Zhu et al. 2019) was conducted to identify potential QTLs coding for meat quality in Pekin ducks. Adopting an interspecies comparison, through transcriptomics and proteomic approaches, a series of significant QTLs have been reported in ducks (Zhang et al. 2020), which can be cited as innovative approach to detect QTLs in ducks.

Approaches based on above designs were successfully attempted in identifying QTL that explain differences between these populations. However, it cannot ensure if any of these identified QTLs are segregating within any acclaimed population which underwent selection for 30 generations. It would not be then improper to state that QTLs identified for any population of ducks, would be deemed as population specific and cannot just be applicable to other populations across the region or a country, in a generalized scale. For extending applications of the already identified QTLs with large effects to other lines or strains of a breed, one has to do several rounds of validation and confirmation for the QTL in question. Both for its significance and for a successful marker for selection, segregation of QTL needs to be established within the flock requiring the selection.

15.4 Marker Assisted Selection (MAS)

Application of quantitative techniques have been traditionally relied upon in selection and breeding of ducks. The advent of molecular markers in genetic selection has revolutionized the modes of genetic selection, which are now practised by many groups of scientists worldwide, as Marker Assisted Selection (MAS). As the name suggests, MAS in avian species, is a selection method carried out on a breeding population wherein the birds are selected based on the marker information alone, obviating the actual phenotyping of the bird for respective target trait. Accordingly, the simplest definition is MAS is a selection process, in which a trait is selected indirectly on a marker linked to that trait and not based on the measurements of the trait (Ribaut and Ragot 2007). Therefore, by employing MAS, the scientists can integrate the two selection methods, namely, conventional selection based on quantitative genetics with that of a selection based on molecular genetics to improve the selection response in both magnitude and quality.

Advantages of MAS

Among the major advantages of MAS for poultry, it can enable a selection based on such traits where the phenotyping a bird is not easy, as can be the case for disease resistance genes. Further, the selection can be done for recessive and mutant genes in the populations. It can also hasten the selection process, since an individual's phenotype can be predicted at a very early stage, even at the time of birth or hatch. The MAS is also more profitable than conventional selection especially for sex-limited traits (egg production, semen parameters), low-heritability traits or traits which are a poor predictors of breeding value (fertility, hatchability etc.), carcass traits (meat quality), difficult or expensive-to-measure traits (disease resistance, morbidity etc.), product quality traits (serum proteins, flavour of eggs and meat) and also for traits which are expressed late in life. As such, the MAS is used as a tool to reduce generation intervals by effective early selection, say at the hatch, and usually used as the first choice for improving lowly-heritable traits. Obviously, the efficiency of Marker Assisted Selection (MAS) becomes very high, if the markers are tightly linked to the trait genes or quantitative trait loci (QTL) of interest. Employing this concept, during the last three decades, hundreds of such studies were conducted to identify DNA markers, bearing tight linkage with the QTLs of interest, meaning strong Linkage Disequilibrium (LD). In general, the MAS can increase reliability of breeding value of the birds and so enable, its easy selection for propagation.

However, on the flip side, there remain many limiting considerations in implementation of the MAS. As multiple genes usually influence most economic traits, tracking a small number of them through DNA markers can only explain a small proportion of the genetic variance. Further, individual genes are likely to cast small-average effects and accordingly, a large volume of data is usually needed to accurately estimate their effects. Further, in situations where the marker is in linkage equilibrium (LE) with the QTL, all QTL alleles in source animals are likely to be different and hence the number of QTL alleles (whose effects are to be evaluated) is further increased. Though Boichard et al. (2006) have showed how gains can be made and LE-marker's independent-segregation can be actually tackled, when a large amount of genotyping would have to be involved.

15.5 GWMAS or Genomic Selection

For overcoming most of such relevant issues, the latest available alternative is genomic selection, otherwise known as: genome-wide marker assisted selection (GWMAS). This was first proposed by Meuwissen et al. (2001) as a variant of MAS. In this, evenly-spaced markers spanning the entire genome are genotyped on individuals to estimate the breeding value, which by assumption, could substantially increase the rate of genetic response compared to conventional methods of selection. Here, one has to assume the markers, to be in linkage disequilibrium (LD) with the QTL so that, the number of effects per QTL to be estimated remains small. Using simulation, the authors demonstrated that the breeding value could be predicted with an accuracy of 0.85 from the marker data alone. Usually, the power of GWMAS increases with the density of markers and also with extent of population-wide LD. Coincidentally, costs of genotyping many-thousands of SNPs in one go, has become affordable and may further go down in near future, with advent of newer and innovative technologies. Thus, with passage of time, the GWMAS would turn to be the most promising solution covering both theoretical and technical prospects of MAS. The thrust of a GWMAS programme being the estimation of expected breeding value (EBV) from that of phenotyped value, the successful statistical approaches used currently can achieve the same in three important steps, viz.,

(i) Use markers to deduce the genotype of each animal across each QTL.
(ii) Estimate the effects of each QTL genotype on the trait of interest.
(iii) Sum all the QTL effects for selection candidates to arrive at their genomic EBV (GEBV).

Although the accuracy of estimating GEBV remains as the most important parameter determining response of selection, it is important to understand the changes in the genome structure through implementation of multiple generations of GWMAS. Accordingly, it is important to understand the basic inputs and concerns in a GWMAS system, which can be discussed as follows.

The inputs to implement a GWMAS in a breeding programme are systematic. Sequentially, there are three distinct requirements, viz., discovery, validation and selection:

(a) A discovery dataset bearing a large number of SNPs which are already typed on a moderate number of birds having phenotypic records for all relevant traits [referred as Training Generation (TG)].
(b) A prediction equation taking the markers as input to predict the breeding value.
(c) Thereafter, remaining validation samples where a larger number of individuals are recorded for the traits and genotyped at least for the markers which are to be used for commercial purpose.

The prediction equation is usually tested first, to assess its accuracy, followed by the selection candidates getting genotyped for the markers and the prediction equation estimated for the discovery data used, finally to calculate the GEBV. The essence of this exercise remains that, the selection candidates, per se, do not need to be phenotyped at all. GWMAS can be a dynamic program in twenty-first century's breeding that can transform the poultry breeding programmes.

Advantages of GWMAS

As on date, above genomics information have found limited applications in duck breeding industry, for which the main reason is that detected-QTL normally explain only a small fraction of total breeding goal. GWMAS has the potential to radically change the structure of genetic selection. Many selection steps can be included which were not possible earlier, and as the result, the selection process could be dynamic. Hence, the implementation of GWMAS can help to efficiently select for a broader breeding goal, reduce generation interval and increases genetic gains in long-term breeding program.

Using the Genome wide Association analyses (GWAS), a major recent study has been reported by Zhu et al. (2019), where growth and feeding based phenotypes for 639 Pekin ducks were recorded, with each individual duck's genotype being deduced by following a genotyping-by-sequencing (GBS) protocol. After the estimation of genetic parameters for traits like growth and feeding efficiency, the GWAS was performed, from which a total of 15 QTLs for growth and feeding efficiency and 12 significant SNPs for feed efficiency could be detected, from using a mixed linear model analysis. The authors reported some-significant loci of feed intake (FI), which could explain 2.3% of total phenotypic variation. This locus was found significantly associated with residual feed intake (RFI), explaining 3% of phenotypic variation. This was acclaimed as the first GWAS study targeting the traits like growth and feeding efficiency in ducks, highlighting worthiness of population-wide association analysis weighted across millions of marker points.

15.6 Candidate Gene Polymorphism Affecting Key Production and Growth Traits of Ducks

The term candidate gene gets grossly defined as a known-gene sequence located in host's genome, that is responsible either for coding a protein product or determinant of a disease in question (being a biological candidate itself), or a chromosomal region which is linked to the expression of the gene in question. So, in concept, candidate genes are genomic segments of the host, which are studied based on prior knowledge of the gene's influence on a trait of interest or disease process of the host, under analyses. Accordingly, the candidate-gene-based approach in genomics is aimed at identifying polymorphism, in the candidate gene segments, per se, queries if a given allele of a candidate gene is more frequently discovered in poultry or animal for conferring an advantage to the trait in question, or the disease process. Accordingly, the candidate gene's polymorphism itself hypothesizes their determinants by association with complex traits, which are now revealed over GWAS.

A review of the candidate-gene-based studies in ducks, when updated from NCBI's database, as on date (July 2020), documents only some 21 candidate gene studies, being of direct relevance to duck research and improvement. Out of the same, 10 meat and growth characters were reportedly probed through candidate genes; while a 6 polymorphism of candidate genes related to duck's production and reproduction characteristics were registered thereof. Similarly, five such polymorphisms in candidate genes coding for qualitative characters (plumage colour, etc.) of ducks were also reported and lodged in NCBI databases.

Among the genes, which were of relevance for egg production and reproductive functions in ducks, Amponsah et al. (2019) have carried out a detailed review. The authors summarized the status of various works taken in candidate genes having roles in ovarian function or follicular developments in ducks. The prominent genes researched in ducks, having or bearing on functions in differentiation, growth, build-up of tissues and secretions, egg production are namely, prolactin (symbolically: PRL, Bai et al. 2019), growth hormone (GH; Wu et al. 2014), insulin-like growth factor-2 (IGF-2; Ye et al. 2017), melatonin receptor (MTNR; Feng et al. 2018), follicle stimulating hormone receptor (otherwise known as FSHR; Xu et al. 2017) and ovoinhibitor (OIH; Wu et al. 2018) genes.

As a sequelae to most of the candidate gene studies, the efforts have always been to identify the SNPs intrinsically involved within each polymorphisms leading to alterations in constituent DNA sequences: where one base in the gene (per a certain stretch of DNA) gets changed, with this difference of one nucleotide changing the constitution of normal sequence, for example a change of nucleotide: guanine (G) being substituted by nucleotide thymine (T) at some specific location (Brookes 2007). Thus, when such an SNP occurs within the coding region of the concerned gene, it reflects a distinct variant of the allele, which can lead to functional differences in the resultant protein, by virtual of a possible amino acid change (Collins et al. 1998). In most cases, many of these SNPs are found located in noncoding areas, which can result an effect on the gene splicing; noncoding RNAs and even transcription factor binding (Barreiro et al. 2008). Interestingly, most or all of these genes (PRL, GH, IGF-2, MTNR, FSH, OIH, etc.) have been reported with significant SNPs segregating in the analysed-hosts (ducks) with prospects of their uses, as potential markers for duck egg production and reproductive performance (Bai et al. 2019; Wu et al. 2014; Ye et al. 2017; Feng et al. 2018; Xu et al. 2017 and Wu et al. 2018).

Similar SNP effects linked to polymorphisms discovered in many candidate genes of importance to broiler characteristics, i.e., meat yield, carcass fatness, meat quality etc. are reported from studies on meat type ducks too. Zhang et al. (2013) analysed the perilipin (PLIN) gene, which is a candidate gene for carcass and fat traits in ducks, wherein they discovered many SNPs in PLIN, resulting in significant difference between SNP genotypes for net carcass weights and prove their utility at potential DNA markers. Wu et al. (2012) had their studies on GH gene involving three types of ducks (including Muscovy duck, Cherry Valley duck and Jingjiang duck) where distinct genotypes: entailing one SNP (variation in intron 2 of GH) were detected. The authors reported significant marker effect for one SNP associated with superior growth and carcass traits (carcass weight, dressing % and eviscerated-weight %) and recommended the same for its use in MAS for improving growth and carcass traits in Pekin duck populations. Another candidate gene: fat mass and obesity-associated gene (FTO) having known effect on energy metabolism (in cattle, pig and rabbits) was studied by Gan et al. (2015) for its polymorphism in crossbred meat ducks to identify its use as potential marker. The authors discovered significant differences between genotypes arisen for an SNP location in FTO gene, which impacted many meat quality traits i.e., intramuscular fat content (IMF), cooking yield (CY), pH values (45 min post-slaughter: pH45m) and drip losses from breast and leg muscles, thereby qualifying this SNP to be used as a DNA marker for above meat traits. Gong et al. (2014) investigated the Polymorphisms of duck Myostatin (MSTN) gene in meat type ducks ($N = 413$) sequence analysis (PCR-RFLP) where impact of 4 SNPs discovered across the coding regions were analysed. Upon analysis of five main haplotypes arisen from these SNP loci, it was concluded that SNPs across duck-MSTN gene possessed significant marker association with juvenile body weight and therefore these SNPs provided effective markers for MAS for juvenile duck growth. With advent of whole genome resequencing techniques, the research programs become more dynamic, of late, where from a single resequencing reaction of a bird's genome, lakhs of SNP locations can be detected in just one go. Zhang et al. (2020) conducted just a resequencing project for a Crested duck breed and reported generation of 785,202 unique SNPs and 105,596 unique InDels in this crested breed. The authors employed two specific analyses: PPI and KEGG analysis of these sequences, upon which they realized that formation of crest in ducks is because of a complex process. It was surmised that, for formation of crest, feather development; fatty acid metabolism block and skull hypoplasia together might be the factors, for which roles of four candidate genes were linked, and recommended as markers to define molecular mechanism of the crest formation in ducks.

15.7 Genomics of Disease Resistance

15.7.1 Candidate Gene Studies in Ducks for Duck's Immunity and Disease Resistance

Under any discussion involving disease resistance/ tolerance in poultry, the ducks have always captivated the scientist's association for duck's differential susceptibilities to avian influenza viruses compared to chickens. Historically, the ducks have been described as near-asymptomatic carriers of influenza virus (H5N1 etc.), though it can spell a big lethal for chickens (Webster et al. 2002). Therefore, a comparative genomics of duck versus chicken can provide the readers a detailed logic behind molecular basis of this differential susceptibility. The duck genome sequencing has already been completed and released online in form of a sequence map alone by Huang and coworkers (Huang et al. 2013), which throw many lights, on such aspects. Currently, there are more than 500 genes or candidate genes, having relevance to innate immunity, and immune functions in general which were already characterized in chickens. Using these clue, and many other cytokine genes such as TLR7, TLR3, RIG-I, MDA5, IFN-a, IFN-b, IL-6, IL-8 besides the immune-expression related genes like Mx, MHC I, the B cell activating factor (BAFF), interleukin-2 (IL-2) and interferon gamma etc. have been identified and sequenced out (Guan et al. 2007; Schultz and Chisari 1999).

15.7.2 Molecular Studies on Disease Resistance and MHC Genes

Molecular genetics involved in the avian immune system for developing disease resistance and other immune-modulations have been widely studied through the characterization of the genes and their translated products using different genomics tools and techniques. With the assembling of the first genome sequence of the chicken Major Histocompatibility Complex (MHC) (Kaufman et al. 1999), little progress has been made in the isolation of cytokines and important signalling molecules in avian model.

With the advancement in genomics studies the available information becomes simplified thus assigning functional role to each part of the sequence components, which makes it meaningful. A recent publication by Wu et al. (2019) has elucidated the duck specific candidate genes, in details that included interleukins, transcription factors, chemokines, differentiation antigens, receptors, genes involved in the Toll pathway and MHC-associated genes.

15.7.3 The MHC Loci and Their Roles in Duck Genomics

The genomic region 'Major Histocompatibility Complex (MHC)' is a chromosomal segment of animals or birds (a region of DNA) that is composed of closely linked polymorphic genes coding for multiple cell surface proteins having roles in adaptive immune system of the host. The nomenclature of this genome-segment is ostensibly drawn from studies on tissue compatibility upon transplantation in recipient mammalian systems. The MHC is already known to encode for three classes of cell membrane antigens, which are distinct and defined by serological, histogenetic, biochemical and biotechnological methods. Out of three, the two are homologous to classes I and II of mammals (B-F and B-L), whereas the third class: B-G antigen, is specific of birds and absent in the mammals. The MHC genes play vital roles in the immune-regulation, disease resistance and regression of Rous sarcomas (Plachy et al. 1992).

MHC class- I molecules are known to play critical roles in adaptive immune responses in host, by presenting pathogen-derived epitope peptides to specific T-cell receptors, which leads to the proliferation of cytotoxic T lymphocytes (CTLs) entailing removal of pathogens from the host (Jondal et al. 1996). As such, the chicken and duck, predominantly express a single MHC I gene (Kaufman 1999; Moon et al. 2005; Chan et al. 2016). In chicken, BF2, adjacent to transporter associated with antigen processing (TAP) gene, is the only dominant MHC I locus expressed (Kaufman et al. 1999; Shaw et al. 2007). In ducks, there remain five different loci (UAA, UBA, UCA, UDA and UEA) in the MHC I genome (*Anpl-MHC1*) out of which, the UAA, lying adjacent to polymorphic TAP2 gene, gets highly expressed (Mesa et al. 2004; Chan et al. 2016; Fleming-Canepa et al. 2016).

Among the chicken's highly variable polymorphic molecules BF2, associated with polymorphic TAP genes, determine their peptide-binding characteristics and influence adaptive immune response and so play an important role in disease resistance and susceptibility (Kaufman et al. 1995; Wallny et al. 2006; Zhang et al. 2012). Similarly, *Anpl-MHC* I also plays a role in susceptibility to viral infections. The high expression of MHC I and the high resistance of some duck lines to avian influenza infections, Newcastle disease virus or duck Tembusu virus indicate that MHC I alleles of such lines bind more virus peptide epitopes (Bingham et al. 2009; Liang et al. 2011; Li et al. 2015).

However, only few Anpl-MHC I alleles have been characterized in the duck lines; therefore, the associations between Anpl-MHC I alleles and infectious disease is poorly understood at present (Fleming-Canepa et al. 2016; Zhang et al. 2017), and only one Anpl-MHC I crystal structure (Anpl-UAA*01) has been resolved (Wu et al. 2017). The polymorphism and peptide-binding specificity of Anpl-MHC I in other duck breeds remain unknown. Limited studies on Anpl-MHC I polymorphism and its functional classification remain the major bottlenecks hampering duck CTL immunity research and the development of duck lines resistant to viruses.

To address some of above, Zhang et al. (2019) analysed MHC I polymorphism in 2 duck breeds of China, i.e., Shaoxing (SX) and Jinding (JD). From these two breeds, 27 unique UAA alleles were targeted from MHC I genes which later were interpreted for their amino acid composition, homology, and phylogenetic relationships. The sequence homology in the amino acid, allelic groups of duck-MHC I (Anpl-MHC I) were established and their distribution was analysed. The outcome of the full-length sequence homology suggests that the newly identified UAA alleles exhibit high polymorphism. Upon adding the alleles discovered by the authors onto known Anpl-MHC I genes from domestic ducks, 17 groups could be formed in total, from which four novel groups were assigned to the SX and JD ducks breeds analysed. The authors thus contributed to enrich the existing Anpl-MHC I polymorphism data and delineate the distribution of newer alleles with different peptide-binding specificities.

Wu and Han (2012) summarized their studies on DNA sequence analyses of MHCs conserved region covering MHC I, MHC II and few functional genes, across five species of birds (chicken, turkey, duck, goose and Japanese quails), which they inferred as simple and contracted. The lengths of introns of MHC I gene were found shorter than the mammals; with most of the species having eight exons and seven introns each, in MHC I region (expect Japanese quails). All the species have six exons and five introns in MHC II genes. Further, the structural patterns of BG genes in chicken, turkey, and quail are reported identical, with all the species containing microsatellite repetitive motifs. However, differences among species were also noticed with MHC I and MHC II genes duplicated, with several copies reported in duck, goose and quail and with difference in the numbers of BG genes across species. Thus, the MHC structures across major poultry species are found variable, in term of ‘gene sequences and repeated motifs inside’.

Wu et al. (2017) examined the structural role of duck MHC Class I molecules for seeking clues, which could describe potential cytotoxic T-Lymphocyte resistance against Influenza-A virus. They examined whether a single dominantly expressed allele of MHC class I in ducks was responsible for high immunity to the virulent influenza A virus (HP-IAV) compared to lowly tolerant chicken. In their study, the crystal structures were determined for duck MHC I (Anpl-UAA*01) and duck β2-microglobulin (β2m) with two peptides from the H5N1 strains. Two remarkable features of Anpl-UAA*01 complex distinguished it from other known MHC I structures. A disulphide bond formed by Cys95 and Cys112 and connecting the β5 and β6 sheets at the bottom of peptide-binding groove (PBG) were identified in Anpl-UAA*01 complex, which can enhance IAV peptide binding. Further, the interface area between MHC I and β2m in ducks was found to be larger than in other poultry species. The outcome of the study confirmed the similarity of the peptide-binding domain with the HLA-A*0201, which further set a basis for screening approximately 600 peptides from the IAV strains as the candidate epitope peptides for Anpl-UAA*01, which worked out as a greater number, than those for chicken BF2*2101 and BF2*0401 molecules. The authors concluded that enhanced

IAV peptide interaction would permit ducks with this Anpl-UAA*01 haplotype to exhibit immunity for the IAV infections.

15.8 Bioinformatics Studies in Duck Research

Bioinformatics has played a significant role in duck genomics and transcriptomic research. Bioinformatics tools and other publicly available softwares at different platforms have enabled the researchers from different parts of the world to communicate, store data in cloud-based database system, interpret and further exchanges of the data. Ensembl (https://www.ensembl.org), a gene annotation and distribution platform for vertebrate species, including duck has enhanced worldwide research in sequencing of whole genome, further resequencing of duck genome and its comparative analysis with different breeds and species has enabled the scientists to develop research and commercial activities based on selective breeding, propagation and commercialization.

Whole genome re-sequencing of Jiaji duck, a highly valuable indigenous duck breed of China was performed to study its genetic characteristic and relationship with other existing duck species (Mallard, French Muscovy duck and Beijing duck) using mutation identification, selective screening and Gene Ontology (GO) as parameters. GO and Kyoto Encyclopaedia of Genes and Genomes (KEGG) analyses showed that the positive selection genes for JJ vs. BD ducks mainly involved in pigmentation, muscle contraction and stretch, gland secretion and immunology, while the positive selection genes obtained from JJ vs. YD ducks mainly involved in embryo development, muscle contraction and stretch and gland secretion (Gu et al. 2020). Bioinformatics has also helped in the characterization of certain disease related genes in ducks as well as genes involved in innate immune mechanism, for example, Duck Enteritis Virus Glycoprotein D (gD) gene, Duck Plague Virus UL22 gene and Toll-Like Receptor (TLRs) (Jia et al. 2012).

The latest development in the field of Avian related Bioinformatics is the availability of Avian Immunome DB platform (https://avimm.ab.mpg.de) to provide a user-friendly information related to genes involved in regulating avian immune system. Avian Immunome DB acts like a single-window resource for avian immuno-genomics by aggregation information regarding genes involved in avian immune biology from other publicly available portals like Ensembl, UniProt and B10K and narrowing down the gene name ambiguity by connecting multiple names of a gene to a single authentic gene identification number in the database (Mueller et al. 2020).

15.9 High-Throughput Sequencing Including Next Gen Sequencing (NGS) and RNA-seq in Ducks

With the advancement in DNA sequencing methodologies from Sanger sequencing, a low-throughput method to high-throughput method, the Next Gen Sequencing, it has become easier and efficient to study an individual organism cellular genomics and transcriptomic and to correlate them with the phenotypic and genotypic characteristic features involved in the evolutionary, developmental biology, egg production related and disease related parameters in avian species including ducks.

In ducks, High-Throughput Sequencing using Illumina HiSeq2500 platform has been used to study the 16S rRNA genes expression in a variety of enteric microorganism causing pathogenicity and comparing it with the microbiome of the Specific Pathogen Free (SPF) ducks (Zhao et al. 2018).

High-throughput sequencing has helped the poultry scientists in exploring the role of expression of key MicroRNAs (miRNAs) involved in the egg laying of ducks (Qiu et al. 2020). In another latest study, High-throughput sequencing along with Meta-genomic analysis has been conducted to study the viral diversity existing in Peking and Muscovy ducks (Liais et al. 2012).

Next Gen Sequencing has also enabled to study the superior and distinctive characteristics i.e., morphological, physiological and behavioural features among the indigenous native duck breeds. For example, Jianchang ducks in China are superior in terms of growth rate than the other domestic breeds. Jianchang ducks are mostly liked for their improved ability of lipid deposition in liver. High-throughput sequencing helped us in identifying the key genes involved in growth development of organs and tissues. Also, re-sequencing of whole genome has enabled the researchers to study the crested trait in ducks and also the genetic signatures involved in the artificial selection of ducks (Wang et al. 2020). High throughput RNA-sequencing and further analysis of whole transcriptome has enabled the researchers to study the role of different genes regulating the development of hepatic steatosis among different species of duck (i.e., Muscovy, Pekin, mule and hinny) as a model organism (Hérault et al. 2019). RNA-sequencing (RNA-seq) has also been used to explore the potential role of differentially expressed key genes involved in the follicle development in Peking ducks (*Anas platyrhynchos*) (Ren et al. 2019). Based on these results, it was inferred that out of the six upregulated genes analysed, among the 500 odd candidate genes, the EPOX (a member of cytochrome P450 family) emerged as an important factor regulating duck-follicle development. In duck's life cycle, EPOX gene exhibited 100 times higher expression level during the laying period compared to the youth and 98% decreased expression than the ones in stopped-laying period.

15.10 Embryology of Ducks

As a variation from other avian species, the ducks (*Anas platyrhynchos*) have a relatively longer incubation cycle, i.e., 28 days, enabling the duck embryo to grow little slower when compared to the chicken embryo. However, the complete cycle of development of duck embryo, from fertilization to hatch of a duckling, still remains to be efficiently elucidated at par with chickens (*Gallus gallus),* which is already redefined, through classification of developmental stages as per the popular Hamburger-Hamilton stages (Hamburger and Hamilton, 1951). Perusal of available literature (comparative study) of different avian species, i.e., chicken, duck, goose embryos suggests that all the early stage phenotypic developments of the embryos were similar. The prominent or significant changes occur post 27th stage of embryo development where the morphological development of duck beak shape differs as from other avian species. Duck, being an aquatic bird also differs from chicken embryonic development in maintaining the second and third interdigital webs of hind limbs post stage 31.

In the year 1932, Chen reported development of duck embryo from pre-streak to the somite stage (Chen 1932). Dupuy et al. (2002) further studied the complete somite stage of duck. In 2018, Lumsangkul and coworkers studied the pre-72 h embryonic development in the Brown Tsaiya Duck (Lumsangkul et al. 2018) (Fig. 15.2).

According to Li et al. (2019), the development of a duck embryo from stage 1–20 is summarized in the following lines. The Stage 1 is also called 'pre-streak' which is followed with a 'primitive streak' stage 2. In stage 3, the streak gets broader and by stage 4, it attains its maximum length. In stage 5, the notochord starts appearing and in stage 6, the 'head fold' is visible. Stage 7 remarks the appearance of single somite which reaches to 22 somites by stage 14. The optic cup is formed when the embryo reaches the stage 15 and the wing buds are visible at stage 16. The Stage 17 characterizes the formation of wings and leg buds and the tail bud attains

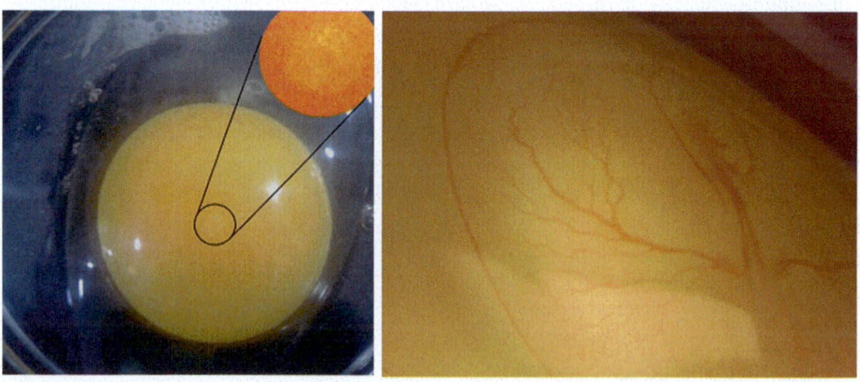

(a) (b)

Fig. 15.2 Embryo development in duck eggs

sharpness at Stage 18. Allantois formation can be observed in the embryo at stage 19. At stage 20, the eye pigmentation has been seen as pale grey in colour.

Distinct features designated to ducks starts to appear post stage 27, where the specific blunt beak development in duck embryos can be observed on 7th day post incubation as compared to a pointed and sharp beak in chicken embryo. Webbing between the first and second toes in duck embryos can be noticed on 8.5 days of embryonic development which continues between the second and third toe on 9th day and between third toe and fourth toe on 9.5–10th day. There is no comb formation in duck embryos as compared to the chicken embryos.

15.11 Embryo Biotechnology in Ducks

The duck's embryonic cells, as usual, present many advantageous features for their manoeuvrability in being used for biotechnological research, including pharmaceutical purposes; the germplasm's bio-banking and transformation potentials. Though in chickens, using the newly laid egg for experimental production of many recombinant proteins have been reported already, similar cases for ducks have also been developed. Many duck embryonated eggs were used as a bioreactor for the commercial manufacturing of Egg drop syndrome (EDS) virus vaccines (Solyom et al. 1982).

15.11.1 Bio-banking

Avian species including ducks, which bear certain important genotypic and phenotypic characteristics, needs to be conserved for economic as well as research purpose. However, due to less attention paid towards the conservation of indigenous and commercially important duck breeds, they are generally maintained as live birds at designated research institutes or with the small-scale poultry farmers. Maintaining a parent stock of live birds adds huge economic burden to the stakeholders. Semen cryopreservation has been viewed as one of common methods for conservation of avian germplasm including ducks. However, due to complexity of the chromosome in avian species (ZZ—male and ZW—female), the semen cryopreservation only leads to the preservation of male genetic information and the counter-sex partner's (i.e., female) genetic information is lost. Therefore, in order to overcome abovementioned challenges, an alternative and fool proof strategy shall be to conserve the important duck germplasm, in the form of embryonic cells, which should be pluripotent, by nature.

Currently, bio-banking of avian species including chickens and ducks has been successfully attempted by different research groups through harvesting the potential early embryonic cells, i.e., Blastodermal cells (BDCs), Primordial Germ Cells (PGCs) and Embryonic Stem Cells (ESCs), in different parts of the world (Fig. 15.3). Researchers around the world are developing advanced technologies to

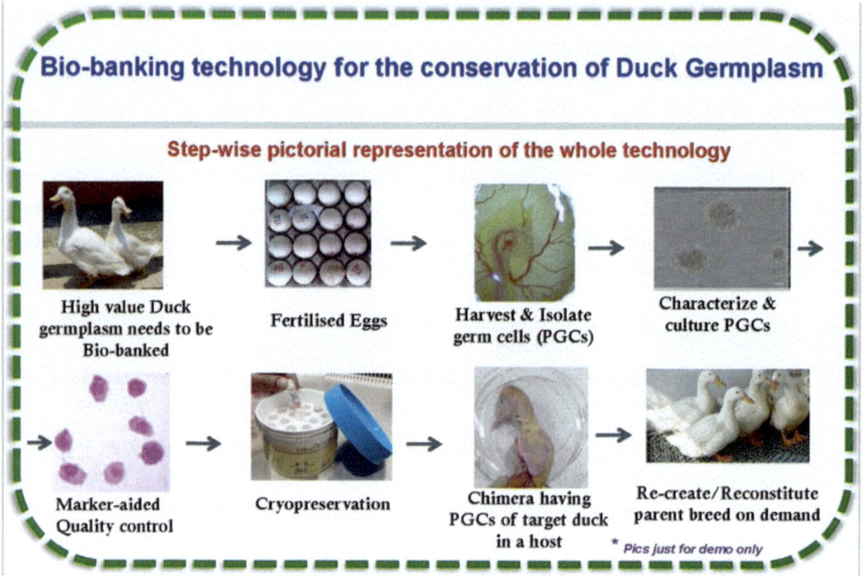

Fig. 15.3 Schematic diagram of Bio-banking of duck germplasm using Primordial Germ Cells. (Image Courtesy: A. Naim, KIIT-TBI and S.K. Mishra, ICAR-DPR, India)

use chicken as a surrogate to host exogenous germplasm and help in rescuing an endangered bird (Mishra and Naim 2017).

Recent reports documented from Indian scientists working with bio-banking of avian germplasm suggest that they have been able to harvest and propagate chicken Primordial Germ Cells (PGCs) from Native Indian chicken breeds, the Hypermelanotic Kadaknath and game bird: Aseels. They have also transferred chicken PGCs to duck embryos post 2.5 days of incubation successfully to raise duck-chicken chimeras (Naim et al. 2020a, b, c).

In some of the recent ongoing studies, the potentials of bio-banking of waterfowls to a terrestrial avian model system and vice versa were evaluated, by transferring indigenous chicken's Primordial Germ Cells (PGCs) into the chickens and ducks respectively. The results suggest that both waterfowl and non-waterfowl embryonic developmental system can act as suitable hosts for conservation of donor germplasm (Naim et al. 2021).

As such, PGCs-mediated germline chimera technology has successfully been reported as novel systems for conserving endangered avian species, using chickens and ducks as the donor and recipients for cross-species germline transfer, way back in 2012 (Liu et al. 2012). The author reported the presence of chicken oogonia, primitive and primary follicles and chicken-derived oocytes in the ovary of chimeric ducks, suggesting donor chicken-derived PGCs to colonize, multiply and differentiate in the recipient duck ovaries and thus involved in the process of duck ovarian folliculogenesis. The research findings thus demonstrate that donor chicken PGCs might interact with the recipient duck germinal epithelium and complete spermatogenesis leading to the production of fertile sperm, using male duck chimeras, however, parallel demonstration could not be made for the generation of donor progeny from chimeric female ducks, thereby leaving a challenge to be overcome in this approach.

Guan et al. (2010) reported the successful in-vitro culture and propagation of Beijing duck (*Anas platyrhynchos domesticus*) Embryonic Germ (EG) cells from primordial germ cells isolated from genital ridges of stage 28 duck embryo following the protocol of development of chicken EG cells.

Ma et al. (2018) explored the potential of Beijing duck Amniotic Mesenchymal Stem Cells (AMSC) for their known characteristic of having a role in cell-based therapies targeting regenerative medicine and bio-medical applications. Beijing duck AMSCs were found to exhibit multi-lineage potential when induced by a suitable cue.

Gao et al. (2011) have worked on generating quail-duck chimera chicks by transplanting blastodermal cells from stage X quail donor embryos. Later, the migration of donor cells was studied by PCR method in heterogenous chimeras.

15.11.2 Preservation of Gametes, as a Bio-banking Option

The techniques like avian gonadal tissue cryo-conservation and their transplantation into a suitable recipient host leads to the establishment of an innovative technology for rescuing any indigenous, commercially important or an endangered avian species. However, limited research work has been documented in the area using waterfowl as a model organism. Most of the research scientists have used chicken gonadal tissues for the purpose of germplasm conservation.

Song et al. (2012) have carried out outstanding work using duck as a model. They explored the potential of interspecific ovarian tissue transplantation from Muscovy ducks as donors using Pekin ducks as recipients, just after hatch, by administering immune-suppressant to recipients to avoid any tissue rejection.

Under one such currently undertaken research, at an ICAR institute, the potentials of gonadal grafts of chicken and duck breeds were evaluated through surgical approach, into chicken and duck systems respectively, through orthotopic transplantations. The available results suggest that duck and chickens can act as surrogates for bio-banking purposes, through hosting inter- or intra-species gonads without any significant risk of tissue rejection. However, the fertility potential of transplanted donor gonads remains to be delineated (Naim et al. unpublished data).

Other than gonadal tissue isolation and transplantation into same species host, several other research groups have also worked on generating interspecies chimera using duck—chicken or quail-duck models, via early blastodermal cells and primordial germ cells.

Li et al. (2002) have reportedly produced duck-chicken chimeras using stage X blastodermal cells from Maya duck embryos, much before the authors claimed produced chicken – duck chimeras by using Primordial Germ Cells (PGCs) from chicken donors using duck embryo as recipients.

Such technology of harvesting early embryonic cells, which have germ cells properties and are pluripotent, can be used for conservation of avian germplasm of domestic and wild birds including waterfowls like ducks.

As such being authors compiling embryo Biotechnology findings in ducks, it is our cherished vision that, such technologies can be revolutionary if applied to field for the conservation of prized and endangered avian species. One needs to be expert in veterinary and molecular biology for the isolation of gonads and for their long-term preservation in liquid nitrogen. Due to sudden outbreaks of avian influenza virus and other natural calamities, bio-banking of indigenous avian breeds via early embryonic cells preservation or gonads preservation and reconstitution of target breed will be in huge demand, but one needs to be perfect at such skills besides the preparedness to take up such challenges.

15.11.3 Transgenesis in Ducks

The term 'Transgenesis' is defined as introduction of a foreign gene (transgene) into the genome of a living organism; so that, the recipient organism not only exhibit a new character but also transfer the specific characteristics to its offspring. Therefore, a transgenic bird, be it a chicken or duck, hosts a foreign DNA sequence in its own genome which transcribes and further translates leading to an expression of the 'trans' gene introduced. The development of transgenic mice in 1980s laid the path for scientist to explore the possibility of developing transgenic animals other than mice. The first transgenic chicken was produced in 1986 by using retrovirus mode of introducing a transgene into the host genome (Salter et al. 1986). For generating transgenic chickens, different methods like viral, nonviral, genome editing tools and transposon-mediated methods have been developed and used with the time. In viral mediated protocol, retroviruses, Lenti-viruses are used to introduce a foreign gene into primordial germ cells, which further differentiate into sperm or ovum and will transfer the foreign gene to the next generation. However, due to some drawbacks, i.e. off target (dividing cells or non-dividing cells), pay load limit of foreign DNA, nonviral mediated transgenic protocols like, microinjection, lipofection and electroporation methods came into practice. Currently, more precise gene editing tools like ZFN, TALEN, CRISPR/Cas9 are in use to generating transgenic birds.

Konoval et al. (2018) have been able to produce transgenic chimeric duck through blastoderm cell transfer CRISPR/Cas9-mediated gene integration into the duck genome (Konoval et al. 2018). Shaoxig ducks embryos at blastodermal stage were used as donors whereas Shan partridge breed of duck embryos were the surrogates. UV-irradiation was used to sterilize the endogenous blastodermal cells of the recipients. The donor BDCs were transfected with the foreign genetic material using CRISPR/Cas9 as gene integration into the duck genome by using lipid based transfection reagent underneath the recipient's embryos. Hearteningly, the survival of recipient embryos post transfection was 6.98%, with a final rate of 44.4% for generating transgenic ducks using this method.

Researchers have exploited chickens more than ducks for the purpose of biotechnological advancements and research purposes. Genetically modified chickens have been developed which produce eggs containing anti-arthritis and anti-cancer drugs. Producing drugs in chicken eggs have significantly reduced the manufacturing cost of such medicines in pharmaceutical industries. Dr. Lisa Herron from Roslin Technologies, Edinburgh foresees the use of chicken eggs than other farm animals for the commercial production of important drugs to reduce the production cost. Current focus for Dr. Herron and her group was for the two proteins, IFNalpha2 and macrophage-CSF. IFNalpha2 is known for anti-viral and anti-cancer properties whereas macrophage-CSF has potential to activate self-repair mechanism against damaged tissues. According to her, three chicken eggs can produce a single dose of the drug. In contrast to chickens, ducks have advantage of having bigger egg size and hence higher albumen content, may help in scaling up the production of drugs to meet the market demand at a much-reduced cost of production.

In an attempt to use chicken as a bioreactor, transgenic chicken producing human erythropoietin in egg whites have been produced. This shows that similar work can be replicated in ducks, which again will yield such high economic value recombinant proteins in large quantities as compared to chickens (Kwon et al. 2018).

15.12 Gene Editing

Genome or gene editing refers to the technologies involved in adding, deleting or manipulating a genetic sequence of an organism's genome in a precise and efficient manner, in order to modify its genotype, phenotype or certain characteristic traits. Gene editing in poultry especially, chickens have gained momentum since the first genetically modified mice was produced (Costantini and Lacy 1981; Gordon and Ruddle 1981). However, negligible or very less work has been done in gene editing in ducks as a model organism.

Since the last decade, many more dynamic options for gene editing have emerged, before the scientists including the Crispr® based systems, which have revolutionized site specific nucleotide manipulation and corrections onto the genome. Obviously, if the same are applied on pluripotent cells, a net editing of the genome could be achieved and be handed out to the scientists, to overcome inherent defects of the genomes.

Different methods used for gene editing of an avian genome includes Meganucleases, Zinc Finger nucleases (ZFN), TALEN mediated, CRISPR/Cas9 mediated. CRISPR/Cas9 is one of the best methods considered for efficient and precise editing of an organism genome.

Kostenko et al. (2018) have used three different methods to generate transgenic ducks (Kostenko et al. 2018). By using CRISPR/Cas9 mediated-homology directed repair for inserting a foreign DNA into the duck genome, they successfully generated a genetically modified duck. They used sperms transfection prior to artificial fertilization; implantation of transfected blastoderm cells into duck blastoderm-stage recipient embryos; implantation of the foreign gene into the space below the germinal disc of recipient duck embryos. They found that in sperm-mediated gene transfer, 20.6% of the founder ducks were positive for the transgene. In the other two approaches, 65% and 77.8% chimeric founders transmitted the transgene to the next generation. All the three routes demonstrate transgene transmission to the progeny, thus confirming suitable genome manipulation.

Currently, gene editing is mainly used for the purpose of disease resistance by targeting the avian influenza virus receptor gene, conservation of avian germplasm by targeting the DDX4 gene and developing a universal recipient host using chickens as an avian model organism (Taylor et al. 2017). However, in a near future, gene targeting technology will be widely used under the aim of 'Gene-Drive', which is basically used to modify a target gene loci in both the chromosomes so that the offspring will exhibit the targeted character and which will be followed in a complete population. For example, Gene-Drive may be applied for producing only male or female chicks from a particular parent population, which will benefit the poultry (layer or broilers industry) stakeholders in a significant manner.

15.13 Future Directions

The twenty-first century being the era of high-end research, in all spheres of science, a leading alternate poultry species like ducks deserve to be researched for all its frontier aspects, at par with chickens. A thorough review indicates the existence of large gaps in research on ducks, with respect to genomics and biotechnology. The primary reason for these voids is for popularity of duck-farming being largely concentrated in Asian and South-East Asian countries. The ducks being storehouse of many rare genes and characters, these genome sequences need to be deciphered with respect to their uniqueness and constitution. All the genomics and biotechnological tools being fully developed for chickens and turkeys now, it remains just as a matter of time to seek similar applications in ducks, purposefully. Further, the developments in fields of transgenesis and gene editing are promising and can be replicated in ducks, raising hopes of duck eggs being used as bioreactor, which for its bigger egg size can be more economical and innovative for expressing pharmaceutical products. Similarly, potential of embryo-biotechnologies like bio-banking of duck germplasm through embryonic precursors (e.g., stem cells, PGCs, germ cells), gonads, gametes cryopreservation or transplantation into a suitable host, through chimeric intermediates (intra-or interspecies) can be used of for conservation as well as rescuing any endangered avian species Further, latest genomics and molecular biology tools like, Next Generation Sequencing; resequencing of rare and diverse duck breeds; designing high-density SNP-array (Affy-chips, Illumina bead-array etc.); transcriptomics and proteomics of ducks just being in catching up stage. It is likely that the coming decade (Third decade of twenty-first century) will see a saturation of Genomics and Biotechnology applications in ducks, at par with Human and Mouse genomes, which remain as the highest researched mammalian species in history of Genomics.

References

Amponsah CA, Zou K, Lu LL, Zhang SW, Xue Y, Su Y, Zhao Z (2019) Genetic effects of polymorphisms of candidate genes associated with ovary development and egg production traits in ducks. Anim Reprod Sci 211:106219

Andersson L (2001) Genetic dissection of phenotypic diversity in farm animals. Nat Rev Genet 2(2):130–138

Bai DP, Hu YQ, Li YB, Huang ZB, Li A (2019) Polymorphisms of the prolactin gene and their association with egg production traits in two Chinese domestic ducks. Br Poult Sci 60(2):125–129

Barreiro LB, Laval G, Quach H, Patin E, Quintana-Murci L (2008) Natural selection has driven population differentiation in modern humans. Nat Genet 40(3):340–345

Bingham J, Green DJ, Lowther S, Klippel J, Burggraaf S, Anderson DE, Wibawa H, Hoa DM, Long NT, Vu PP, Middleton DJ, Daniels PW (2009) Infection studies with two highly pathogenic avian influenza strains (Vietnamese and Indonesian) in Pekin ducks (*Anas platyrhynchos*) with particular reference to clinical disease tissue tropism and viral shedding. Avian Pathol 38(4):267–278

Boichard D, Fritz S, Rossignol M, Guillaume F, Colleau JJ, Druet T (2006) Implementation of marker assisted selection: practical lessons from dairy cattle. In: Proceedings of the 8th World Congress on genetics applied to livestock production, 13–16 August 2006, Belo Horizonte, MG, Brazil

Brookes AJ (2007) Single Nucleotide Polymorphism (SNP) eLS:3–6. https://onlinelibrary.wiley.com/doi/abs/10.1038/npg.els.0005006. Accessed 21 Mar 2021

Buchholz WG, Pearce JM, Pierson BJ, Scribner KT (1998) Dinucleotide repeat polymorphisms in waterfowl (family Anatidae): characterization of sex-linked (Z-specific) and 14 autosomal loci. Anim Genet 29:323–325

Chan WF, Parks-Dely JA, Magor BG, Magor KE (2016) The minor MHC class I gene UDA of ducks is regulated by let-7 microRNA. J Immunol 197(4):1212–1220

Chen BK (1932) The early development of the duck's egg with special reference to the origin of the primitive streak. J Morphol 53(1):133–187

Collins FS, Brooks LD, Chakravarti A (1998) A DNA polymorphism discovery resource for research on human genetic variation. Genome Res 8:1229–1231

Costantini F, Lacy E (1981) Introduction of a rabbit β-globin gene into the mouse germ line. Nature 294:92–94

Dolmatova I, Saiłbatałov TF, Gareev FT (2000) RAPD-analysis of genetic polymorphism of ducks: differences in breeds. Genetics 36:682–687

Dorshorst B, Molin AM, Rubin CJ, Johansson AM, Stromstedt L, Pham MH, Chen CF, Hallböök F, Ashwell C, Andersson A (2010) A complex genomic rearrangement involving the endothelin 3 locus causes dermal hyperpigmentation in the chicken. PLoS Genet 7(12):e1002412101371/journalpgen1002412

Dorshorst B, Okimoto R, Ashwell C (2011) Genomic regions associated with dermal hyperpigmentation polydactyly and other morphological traits in the Silkie chicken. J Hered 101:339–350

Dupuy V, Nersessian B, Bakst MR (2002) Embryonic development from first cleavage through seventy-two hours incubation in two strains of Pekin duck (*Anas platyrhynchos*). Poult Sci 81(6):860–868

Elferink MG, Vallée AA, Jungerius AP, Crooijmans RP, Groenen MA (2008) Partial duplication of the PRLR and SPEF2 genes at the late feathering locus in chicken. BMC Genomics 9:391. https://doi.org/10.1186/1471-2164-9-391

Feng P, Zhao W, Xie Q, Zeng T, Lu L, Yang L (2018) Polymorphisms of melatonin receptor genes and their associations with egg production traits in Shaoxing duck. Asian Australas J Anim Sci 31:1535–1541

Fields RL, Scribner KT (1997) Isolation and characterization of novel waterfowl microsatellite loci: cross-species comparison and research application. Mol Ecol 6:199–202

Fleming-Canepa X, Jensen SM, Mesa CM, Diaz-Satizabal L, Roth AJ, Parks-Dely JA, Moon DA, Wong JP, Evseev D, Gossen DA, Tetrault DG, Magor KE (2016) Extensive allelic diversity of MHC class I in wild mallard ducks. J Immunol 197(3):783–794

François Y, Vignal A, Molette C, Marty-Gasset N, Liaubet L, Marie-Etancelin C (2017) Deciphering mechanisms underlying the genetic variation of general production and liver quality traits in the overfed mule duck by pQTL, analyses. Genet Sel Evol 49:38

Gan W, Song Q, Zhang NN, Xiong XP, Wang DM, Li L (2015) Association between FTO polymorphism in exon 3 with carcass and meat quality traits in crossbred ducks. Genet Mol Res 14(2):6699–6714

Gao J, Yuan F, Tang X, Han H, Sha J, Yuan J, Shao Y, Jin X, Liu H, Rui L, Li Z (2011) Contribution of blastoderm cells to japanese quail (*Coturnix coturnix japonica*) – Peking Duck (*Anas platyrhynchos*) chimeras. Anim Sci J 82(6):729–734. https://doi.org/10.1111/J1740-0929201100905X

Gholizadeh M, Mianji GR, Ghobadi A, Zadeh HS, Montazami S (2007) Measurement of within and between genetic variability in duck breeds by RAPD markers. Pak J Biol Sci 10(6):982–985

Gong P, Yang YP, Yang Y, Ye SQ, Deng B, Wang LX, Yu T, Qian YG, Gong YZ (2014) Effects of polymorphisms and haplotypes within the MSTN gene on duck growth trait. Br Poult Sci 55(1):37–43

Gordon JW, Ruddle FH (1981) Integration and stable germ line transmission of genes injected into mouse pronuclei. Science 214:1244–1246

Groenen MA, Cheng HH, Bumstead N, Benkel BF, Briles WE, Burke T, Burt DW, Crittenden LB, Dodgson J, Hillel J, Lamont S, de Leon AP, Soller M, Takahashi H, Vignal A (2000) A consensus linkage map of the chicken genome. Genome Res 10(1):137–147

Gu L, Wang F, Lin Z, Xu T, Lin D, Xing M, Yang S, Chao Z, Ye B, Lin P, Hui C, Lu L, Hou S (2020) Genetic characteristics of Jiaji Duck by whole genome re-sequencing. PLoS One 15(2): e0228964

Guan ZB, Ye JL, Dan WB, Yao WJ, Zhang SQ (2007) Cloning expression and bioactivity of duck BAFF. Mol Immunol 44:1471–1476

Guan W, Wang Y, Hou L, Chen L, Li X, Yuw W, Ma Y (2010) Derivation and characteristics of pluripotent embryonic germ cells in duck. Poult Sci 89(2):312–317

Gunnarsson U, Kerje S, Bed'hom B, Sahlqvist AS, Ekwall O, Tixier-Boichard M, Kämpe O, Andersson L (2011) The dark brown plumage color in chickens is caused by an 83-kb deletion upstream of SOX10. Pigment Cell Melanoma Res 24(2):268–274

Hamburger V, Hamilton HL (1951) A series of normal stages in the development of the chick embryo. J Morphol 88(1):49–92

Hariyono DNH, Maharani D, Cho S, Manjula P, Seo D, Choi N, Hasoloan J, Sidadolog JHP, Lee JH (2019) Genetic diversity and phylogenetic relationship analyzed by microsatellite markers in eight Indonesian local duck populations. Asian Australas J Anim Sci 32:31–37

Hérault F, Houée-Bigot M, Baéza E et al (2019) RNA-seq analysis of hepatic gene expression of common Pekin, Muscovy, mule and hinny ducks fed ad libitum or overfed. BMC Genomics 20: 13

Hocking PM (2005) Review of QTL, mapping results in chickens. Worlds Poult Sci J 61:215–226

Huang Y, Tu J, Cheng X, Tang B, Hu X, Liu Z, Feng J, Lou Y, Lin L, Xu K, Zhao Y, Li N (2005) Characterization of 35 novel microsatellite DNA markers from the duck (Anas platyrhynchos) genome and cross-amplification in other birds. Genet Sel Evol 37:455472

Huang Y, Zhao Y, Haley CS, Hu S, Hao J, Wu C, Li N (2006) A genetic and cytogenetic map for the duck (*Anas platyrhynchos*). Genetics 173:287–296

Huang Y, Haley CS, Wu F, Hu S, Hao J, Wu C, Li N (2007) Genetic mapping of quantitative trait loci affecting carcass and meat quality traits in Beijing ducks (*Anas platyrhynchos*). Anim Genet 38(2):114–119

Huang CW, Cheng YS, Rouvier R, Yang KT, Wu CP, Huang HL, Huang MC (2009) Duck (*Anas platyrhynchos*) linkage mapping by AFLP fingerprinting. Genet Sel Evol 41:28

Huang Y, Li Y, Burt DW, Chen H, Zhang Y, Qian W, Kim H, Gan S, Zhao Y, Li J, Yi K, Feng H, Zhu P, Li B, Liu Q, Fairley S, Magor KE, Du Z, Hu X, Goodman L, Tafer H, Vignal A, Lee T, Kim KW, Sheng Z, An Y, Searle S, Herrero J, Groenen MAM, Crooijmans RPMA, Faraut T, Cai Q, Webster RG, Aldridge JR, Warren WC, Bartschat S, Kehr S, Marz M, Stadler PF, Smith J, Kraus RHS, Zhao Y, Ren L, Fei J, Morisson M, Kaiser P, Griffin DK, Rao M, Pitel F, Wang J, Li N (2013) The duck genome and transcriptome provide insight into an avian influenza virus reservoir species. Nat Genet 45(7):776–783

Jia H, Li G, Li J, Tian Y, Wang D, Shen J, Tao Z, Xu J, Lu L (2012) Cloning expression and bioinformatics analysis of the duck TLR 4 gene. Br Poult Sci 53(2):190–197

Jondal M, Schirmbeck R, Reimann J (1996) MHC class I-restricted CTL responses to exogenous antigens. Immunity 5:295–302

Kaufman J, Volk H, Wallny HJ (1995) A "minimal essential MHC" and an "unrecognized MHC": two extremes in selection for polymorphism. Immunol Rev 143(1):63–88

Kaufman J, Milne S, Gobel TW, Walker BA, Jacob JP, Auffray C, Zoorob R, Beck S (1999) The chicken B locus is a minimal essential major histocompatibility complex. Nature 401:923–925

Kileh-Wais M, Elsen JM, Vignal A, Feves K, Vignoles F, Fernandez X, Manse H, Davail S, André JM, Bastianelli D, Bonnal L, Filangi O, Baéza E, Guéméné D, Genêt C, Bernadet MD, Dubos F, Marie-Etancelin C (2013) Detection of QTL, controlling metabolism meat quality and liver quality traits of the overfed interspecific hybrid mule duck. J Anim Sci 91(2):588–604

Konoval O, Doroshenko M, Kostenko S, Lu L, Tabaka P, Korol P, Chepiha A, Bu X, Huang L, Huang X, Li L (2018) Generation of transgenic chimeric ducks using blastoderm cell transfer crispr/ CAS9-mediated gene insertion into the duck genome. J Biotechnol Biomater 8:53

Kostenko S, Konoval O, Tabaka P, Lu L, Korol P, Chepiha A, Doroshenko M, Bu X, Huang L, Huang X, Li L (2018) Generation of genetically modified ducks by CRISPR/Cas9-mediated gene insertion into the duck genome. J Biotechnol Biomater 8:48

Kwon MS, Koo BC, Kim D, Nam YH, Cui XS, Kim NH, Kim T (2018) Generation of transgenic chickens expressing the human erythropoietin (hEPO) gene in an oviduct-specific manner: production of transgenic chicken eggs containing human erythropoietin in egg whites. PLoS One 13(5):e0194721

Lai FY, Chang YY, Chen YC, Lin EC, Liu HC, Huang JF, Wang PH (2020) Monitoring of genetically close Tsaiya duck populations using novel microsatellite markers with high polymorphism. Asian Australas J Anim Sci 33:888–901

Lee D, Lee J, Heo KN, Kwon K, Moon Y, Lim D, Lee KT, Kim J (2020) Population analysis of the Korean native duck using whole-genome sequencing data. BMC Genomics 21(1):554–563

Li ZD, Deng H, Liu CH, Song YH, Sha J, Wang N, Wei H (2002) Production of duck-chicken chimeras by transferring early blastodermal cells. Poult Sci 81(9):1360–1364

Li HF, Li BC, Ma YH, Tang QP, Chen KW, Tu YJ (2007) Molecular genetic diversity of Fujian domestic duck breeds. Chin J Appl Ecol 18:463–466

Li N, Wang Y, Li R, Liu JY, Zhang JZ, Cai YM, Liu SD, Chai TJ, Wei LM (2015) Immune responses of ducks infected with duck Tembusu virus. Front Microbiol 6:425

Li S, Bai S, Qin X, Zhang J, Irwin DM, Zhang S, Wang Z (2019) Comparison of whole embryonic development in the duck (Anas platyrhynchos) and goose (*Ansercygnoides*) with the chicken (*Gallus gallus*). Poult Sci 98(8):3278–3291

Liais E, Croville G, Mariette J, Klopp C, Lagoutte F (2012) Molecular detection and characterization of duck respiratory viruses using unbiased high-throughput sequencing. ESVV 2012 European Society for Veterinary Virology (ESVV) Madrid, Spain -hal-01191333

Liang QL, Luo J, Zhou K, Dong JX, He HX (2011) Immune-related gene expression in response to H5N1 avian influenza virus infection in chicken and duck embryonic fibroblasts. Mol Immunol 48:924–930

Liu C, Khazanehdari KA, Baskar V, Saleem S, Kinne J, Wernery U, Chang IK (2012) Production of chicken progeny (Gallus gallusdomesticus) from interspecies germline chimeric duck (Anas domesticus) by primordial germ cell transfer. Biol Reprod 86(4):101

Lumsangkul C, Fan YK, Chang SC, Ju JC, Chiang HI (2018) Characterizing early embryonic development of Brown Tsaiya Ducks (*Anas platyrhynchos*) in comparison with Taiwan Country Chicken (*Gallus gallus domestics*). PLoS One 13(5):e0196973

Ma C, Wang K, Ji H, Wang H, Guo L, Wang Z, Ren H, Wang X, Guan W (2018) Multilineage potential research of Beijing duck amniotic mesenchymal stem cells. Cell Tissue Bank 19:519–529

Maak S, Neuman K, Lengerken G, Gattermann R (2000) First seven microsatellites developed for the Peking duck. Anim Genet 31:233–235

Maak S, Wimmers K, Weigend S, Neumann K (2003) Characterization of 18 microsatellites in the Peking duck (*Anas platyrhynchos*) and their application in other waterfowl species. Mol Ecol Notes 3:224–227

Mesa CM, Thulien KJ, Moon DA, Veniamin SM, Magor KE (2004) The dominant MHC class I gene is adjacent to the polymorphic TAP2 gene in the duck *Anas platyrhynchos*. Immunogenetics 56(3):192–203

Meuwissen THE, Hayes BJ, Goddard ME (2001) Prediction of total genetic value using genome-wide dense marker maps. Genetics 157:1819–1829

Mishra SK, Naim A (2017) Surrogate chickens: a molecular approach to safeguard endangered avian germplasm. In: Proceedings of national seminar on opportunities and challenges of translational research in frontier areas of animal biotechnology and V convention of SVSVT, 22–23 September 2017, Bhubaneswar, India, pp 112–115

Moon DA, Veniamin SM, Parks-Dely JA, Magor KE (2005) The MHC of the duck (*Anas platyrhynchos*) contains five differentially expressed MHC genes. J Immunol 175:6702–6712

Mucha S, Grajewski B, Gomowicz E, Lisowski M, Radziszewska J, Szwaczkowski T (2014) Mapping quantitative trait loci affecting some carcass and meat traits in duck (*Anas platyrhynchos*). J Appl Genet 55:497–503

Mueller RC, Mallig N, Smith J, Eöery L, Kuo RI, Kraus RHS (2020) Avian Immunome DB: an example of a user-friendly interface for extracting genetic information. BMC Bioinf 21:502

Naim A, Mishra SK, Nayak D, Jena C (2020a) Avian Germplasm Bio-Banking: Primordial Germ Cells serving the purpose. In: Proceedings of emerging areas in Biosciences and Biomedical Technologies (e-BBT2), International conference 7–9 February 2020 at IIT, Indore India

Naim A, Mishra SK, Nayak D, Jena CR, Dash S (2020b) Chicken-chimera: primordial germ cells mediated poultry germplasm conservation. In: Proceedings of World Veterinary Poultry Association (WVPA) conference, 28 February 2020, NIANP, Bengaluru India

Naim A, Mishra SK, Nayak D (2020c) Bio banking of indigenous chicken germplasm by cryopreservation of germline potent early embryonic cells and gonadal tissues. Indian J Poult Sci 55:1–9

Naim A, Mishra SK, Nayak D (2021) Primary assessment of bankability of rare native chicken breed as cryo-preserved PGCs: chicken–chicken *vs* chicken–duck transfer system. In Proceedings 26th World Poultry Congress, Paris, France (accepted)

Paulus KB, Tiedemann R (2003) Ten polymorphic autosomal microsatellite loci for the Eider duck *Somateria mollissima* and their cross-species applicability among waterfowl species (Anatidae). Mol Ecol Notes 3:250–252

Plachy J, Pink JR, Hala K (1992) Biology of the chicken MHC (B complex). Crit Rev Immunol 12 (1–2):47–79

Qiu M, Zhang Z, Xiong X, Du H, Li Q, Yu C, Gan W, Liu H, Peng H, Xia B, Chen J, Hu C, Song X, Yang L, Jiang X, Yang C (2020) High-throughput sequencing analysis identified microRNAs associated with egg production in ducks ovaries. PeerJ 8:e8440

Rao M, Morisson M, Faraut T, Bardes S, Fève K, Labarthe E, Fillon V, Huang Y, Li N, Vignal A (2012) A duck RH panel and its potential for assisting NGS genome assembly. BMC Genomics 13:513

Redon R, Ishikawa S, Fitch KR, Feuk L, Perry GH, Andrews TD, Fiegler H, Shapero MH, Carson AR, Chen W, Cho EK, Dallaire S, Freeman JL, González JR, Gratacòs M, Huang J, Kalaitzopoulos D, Komura D, MacDonald JR, Marshall CR, Mei R, Montgomery L, Nishimura K, Okamura K, Shen F, Somerville MJ, Tchinda J, Valsesia A, Woodwark C, Yang F, Zhang J, Zerjal T, Zhang J, Armengol L, Conrad DF, Estivill X, Tyler-Smith C, Carter NP, Aburatani H, Lee C, Jones KW, Scherer SW, Hurles ME (2006) Global variation in copy number in the human genome. Nature 444:444–454

Ren J, Lu L, Liu X, Tao Z, Zhang C, Wang D, Shen J, Liu W, Tian Y, Zhu Z (2009) Paternity assessment: application on estimation of breeding value in body-weight at first egg trait of egg-laying duck (*Anas platyrhynchos*). Mol Biol Rep 36:2175–2181

Ren J, Sun C, Chen L, Hu J, Huang X, Liu X et al (2019) Exploring differentially expressed key genes related to development of follicle by RNA-seq in Peking ducks (*Anas Platyrhynchos*). PLoS One 14(6):e0209061

Ribaut JM, Ragot M (2007) Marker-assisted selection to improve drought adaptation in maize: the backcross approach perspectives limitations and alternatives. J Exp Bot 58:351–360

Salter DW, Smith EJ, Hughes SH, Wright SE, Fadly AM, Witter RL, Crittenden LB (1986) Gene insertion into the chicken germ line by retroviruses. Poult Sci 65:1445–1458

Schultz U, Chisari FV (1999) Recombinant duck interferon gamma inhibits duck hepatitis B, virus replication in primary hepatocytes. J Virol 73(4):3162–3168

Seo D, Bhuiyan M, Sultana H, Heo J, Lee J (2016) Genetic diversity analysis of South and East Asian duck populations using highly polymorphic microsatellite markers. Asian-Australas J Anim Sci 29(4):471–478

Shaw I, Powell TJ, Marston DA, Baker K, Andrew van Hateren AV, Riegert P, Michael V, Wiles MV, Milne S, Beck S, Kaufman J (2007) Different evolutionary histories of the two classical class I, genes BF1 and BF2 illustrate drift and selection within the stable MHC, Haplotypes of Chickens. J Immunol 178(9):5744–5752

Skinner BM, Robertson LB, Tempest HG, Langley EJ, Ioannou D, Fowler KE, Crooijmans RP, Hall AD, Griffin DK, Völker M (2009) Comparative genomics in chicken and Pekin duck using FISH, mapping and microarray analysis. BMC Genomics 10:357

Solyom F, Nemesi M, Forgacs A, Balla E, Perenyi T (1982) Studies on EDS vaccine. Dev Biol Stand 51:105–121

Song Y, Cheng KM, Robertson MC, Silversides FG (2012) Production of donor-derived offspring after ovarian transplantation between Muscovy (*Cairinamoschata*) and Pekin (*Anas platyrhynchos*) ducks. Poult Sci 91(1):197–200

Stai SM, Hughes CR (2003) Characterization of microsatellite loci in wild and domestic Muscovy ducks (*Cairinamoschata*). Anim Genet 34:387–389

Su Y, Long R, Chen G, Wu X, Xie K, Wan J (2007) Genetic analysis of six endangered local duck populations in China based on microsatellite markers. J Genet Genomics 34:1010–1018

Takagi N, Makkino S (1966) A revised study on the chromosomes of three species of birds. Caryologia 19:443–455

Taylor L, Carlson DF, Nandi S, Sherman A, Fahrenkrug SC, McGrew MJ (2017) Efficient TALEN-mediated gene targeting of chicken primordial germ cells. Development 144:928–934

Tiedemann R, Paulus KB, Havenstein K, Thorstensen S, Petersen A, Lyngs P, Milinkovitch MC (2011) Alien eggs in duck nests: brood parasitism or a help from Grandma? Mol Ecol 20:3237–3250

Vitezica ZG, Marie-Etancelin C, Bernadet MD, Fernandez X, Robert-Granie C (2010) Comparison of nonlinear and spline regression models for describing mule duck growth curves. Poult Sci 89: 1778–1784

Vos P, Hogers R, Bleeker M, Reijans M, van de Lee T, Hornes M, Frijters A, Pot J, Peleman J, Kuiper M (1995) AFLP: a new technique for DNA, fingerprinting. Nucleic Acids Res 23:4407–4414

Wallny HJ, Avila D, Hunt LG, Powell TJ, Riegert P, Salomonsen J, Skjodt K, Vainio O, Vilbois F, Wiles MV, Kaufman J (2006) Peptide motifs of the single dominantly expressed class I

molecule explain the striking MHC-determined response to Rous sarcoma virus in chickens. Proc Natl Acad Sci 103:1434–1439
Wang L, Guo J, Xi Y, Ma Yanying S, Li He H, Wang J, Han C, Bai L, Mustafa A, Liu H, Li L (2020) Understanding the genetic domestication history of the Jianchang duck by genotyping and sequencing of genomic genes under selection. G3 (Bethesda) 10:1469–1476
Webster RG, Guan Y, Peiris M, Walker D, Krauss S, Zhou NN, Govorkova EA, Ellis TM, Dyrting KC, Sit T, Perez DR, Shortridge KF (2002) Characterization of H5N1 influenza viruses that continue to circulate in geese in southeastern China. J Virol 76:118–126
Wright D, Boije H, Meadows JR, Bed'hom B, Gourichon D, Vieaud A, Tixier-Boichard M, Rubin CJ, Imsland F, Hallböök F, Andersson L (2009) Copy number variation in intron 1 of SOX5 causes the pea-comb phenotype in chickens. PLoS Genet 5:e1000512
Wu YS, Han LX (2012) Advance in research of poultry major histocompatibility complex-structure. Yi Chuan 34(6):673–678
Wu Y, Pan AL, Pi JS, Pu YJ, Du JP, Liang ZH, Shen J (2012) One novel SNP of growth hormone gene and its associations with growth and carcass traits in ducks. Mol Biol Reprod 39:8027–8033
Wu X, Yan MJ, Lian SY, Liu XT, Li A (2014) GH, gene polymorphisms and expression associated with egg laying in Muscovy ducks (*Cairinamoschata*). Hereditas 151:14–19
Wu Y, Wang J, Fan S, Chen R, Liu Y, Zhang J, Yuan H, Liang R, Zhang N, Xiaa C (2017) Structural definition of duck major histocompatibility complex class I molecules that might explain efficient cytotoxic T lymphocyte immunity to influenza A virus. J Virol 91:e02511–e02516
Wu Y, Liang H, Zhang H, Pi JS, Pan AL, Shen J, Pu YJ, Du JP (2018) The differential expression and SNP analysis of the ovoinhibitor gene in the ovaries of laying duck breeds (*Anas Platyrhynchos*). Braz J Poult Sci 20:281–286
Wu S, Zhang J, Huang J, Li W, Liu Z, He Z, Chen Z, He W, Zhao B, Qin Z, Jiao P, Liao M (2019) Immune-related gene expression in ducks infected with waterfowl-origin H5N6 highly pathogenic avian influenza viruses. Front Microbiol 10:1782
Xiao T, Zheng X, Wu Q (2004) Genetic diversity of Fujian local duck populations in different ecological type. Ying Yong Sheng Tai Xue Bao 15(5):879–882
Xu J, Gao X, Li X, Ye Q, Jebessa E, Abdalla BA, Nie Q (2017) Molecular characterization expression profile of the FSHR gene and its association with egg production traits in Muscovy duck. J Genet 96(2):341–351
Ye Q, Xu J, Gao X, Ouyang H, Luo W, Nie Q (2017) Associations of IGF2 and DRD2 polymorphisms with laying traits in Muscovy duck. PeerJ 5:e4083
Zhang JH, Chen Y, Qi JX, Gao F, Liu Y, Liu J, Zhou X, Kaufman J, Xia C, Gao GF (2012) Narrow groove and restricted anchors of MHC class I molecule BF2 *0401 plus peptide transporter restriction can explain disease susceptibility of B4 chickens. J Immunol 189(9):4478–4487
Zhang HL, Fan HJ, Liu XL, Wu Y, Hou SS (2013) Molecular cloning of the perilipin gene and its association with carcass and fat traits in Chinese ducks. Genet Mol Res 12:1582–1592
Zhang L, Liu W-J, Wu J-Q, Xu M-L, Kong Z-J, Huang Y-Y, Yang S-H (2017) Characterization of duck (*Anas platyrhynchos*) MHC, class I gene in two duck lines. J Genet 96(2):371–375
Zhang L, Lin D, Yu S, Bai J, Jiang W, Su W, Huang Y, Yang S, Wu J (2019) Polymorphism of duck MHC class molecules. Immunogenetics 71(1):49–59
Zhang F, Yin ZT, Zhang JF, Zhu F, Hincke M, Yang N, Hou ZC (2020) Integrating transcriptome proteome and QTL data to discover functionally important genes for duck eggshell and albumen formation. Genomics 112:3687–3695
Zhao LL, Yin HC, Lu TF, Niu YJ, Zhang YY, Li SQ, Wang YP, Chen HY (2018) Application of high-throughput sequencing for microbial diversity detection in feces of specific-pathogen-free ducks. Poult Sci 97(7):2278–2286
Zhu F, Cheng SR, Yang YZ, Hao JP, Yang FX, Hou ZC (2019) Genome-wide association study of growth and feeding traits in Pekin ducks. Front Genet 10:702

Dr. Surya Kanta Mishra, a Doctor of Veterinary Medicine, hails from RajaRanpur, Odisha (India). He earned his master's degree (poultry science) from IVRI, Izatnagar and Ph.D(Animal Biotechnology) from TANUVAS, Chennai. He joined ICAR-ARS in 1991 as a poultry scientist and is currently at DPR-RC, Bhubaneswar as a principal scientist, working in Duck Genomics for more than a decade. Dr. Mishra has wide international experience, through postdoctoral research in biotechnology and participation in many international conferences. He has published more than 70 peer-reviewed research publications in national and international journals. He has earned many National awards and International recognitions.

Dr. Adnan Naim, is a native of Bareilly, Uttar Pradesh (India). He holds a Master of Science degree in Biotechnology from Barkatullah University, Bhopal and further he gained his PhD degree from Griffith University, Queensland, Australia after receiving the prestigious Endeavour Postgraduate Award Fellowship from the Australian Government. He did his postdoctoral research at the Indian Institute of Technology, Indore and currently leading a research project in the area of Primordial Germ Cells mediated avian germplasm bio-banking and conservation at the KIIT-TBI, Bhubaneswar. Dr. Naim has supervised some national and international students for their graduate and postgraduate dissertation work.

Duck Farming: Opportunities, Constraints and Policy Recommendations 16

R. Richard Churchil and A. Jalaludeen

Contents

R. R. Churchil (✉)
Faculty of Veterinary and Animal Sciences, Tamil Nadu Veterinary and Animal Sciences University, Chennai, Tamil Nadu, India
e-mail: richardchurchil.r@tanuvas.ac.in

A. Jalaludeen
Faculty of Poultry Science, Kerala Veterinary and Animal Sciences University, Wayanad, India

A. Jalaludeen et al. (eds.), *Duck Production and Management Strategies*,
https://doi.org/10.1007/978-981-16-6100-6_16

Abstract

During early civilization, the localized expansion of human populations took place mainly along the river banks. Frequent encounter of human with waterfowl, especially ducks, led to domestication of this poultry species early, during first century BC in China. Duck farming is a major component of agriculture in paddy cultivating Asian countries. Ducks possess a number of advantages like disease tolerance, excellent wetland foraging ability and flocking behaviour. The synergy of duck is utilized as predator of insects and producer of natural fertilizer in paddy cultivation. The ducks reduce weed population and improve the physical property of the soil; altogether contribute to higher paddy yield and profit. On the other hand, shrinking water bodies, pollution of grazing fields, difficulty in sourcing of inputs like birds, feed and medicines, marketing hardships and disease outbreaks like epizootics of avian influenza are the major constraints. Comprehensive action plans and policies from the local governments like conservation of duck genetic resources, ensuring the supply of critical inputs like birds, feed, biologicals, medicines, veterinary services, finance and training, formation of Farmer Producer Groups and marketing supports like establishment of quality standards and grading are necessary to safeguard the interests of duck farming community.

Keywords

Constraints · Duck farming · Opportunities · Policy recommendations

Abbreviation/Symbol: Full Form/Explanation

°C	Degree Celsius
AI	Avian Influenza
HPAI	Highly Pathogenic Avian Influenza
INR	Indian Rupee
K–Pg	Cretaceous–Paleogene
kg	Kilogram
m	Metre
PGC	Primordial germ cell
Tk	Taka (Bangladeshi currency)

16.1 Introduction

Duck rearing is a source of cash income in short time with limited investment and an employment opportunity for the rural landless farmers and womenfolk. The evolution of ducks took place 66 million years ago in Cretaceous–Paleogene (K–Pg) boundary and domestication took place in China (500 BC) and in Western Europe (800 AD) (Liu and Churchil 2021). The egg and meat produced by ducks in the world is second next only to chicken among all poultry species. The total quantity of eggs produced in the world by poultry other than chicken, in which ducks contribute a major share in 2019 was 6.04 million tons and the duck meat produced was 4.86 million tons (FAOSTAT 2019). This is 6.75% of total eggs produced by all poultry and 3.87% of poultry meat produced. Ducks are reared mainly for egg production in Asian countries and the meat is only a byproduct. The geographical distribution and density of ducks, duck egg and meat production, per capita duck egg and meat availability and overseas trade of duck products are discussed in detail by Jalaludeen and Churchil (2021).

Ducks had prominent traditional role in the history, cultures and livelihood of rice producing countries in Asia. A substantial portion of ducks are maintained under free grazing, otherwise called scavenging or nomadic system of management in countries like Philippines (Chang and Dagaas 2004), Bangladesh (Islam et al. 2016), India (BAHS 2019), Vietnam (FAO 2010), Indonesia (CIVAS 2006) and many other Asian countries. This directly benefits the socially downtrodden sector of the population in developing countries, particularly in the rural areas and thereby acts as one of the sources of low-cost animal protein and income (Chang et al. 2003). Duck production is distinct in its character, by virtue of ducks being adaptive to variable environments, less demanding in terms of housing requirements, productive for long period, able to grow faster than chicken, easy for management and hardy to diseases.

On the front of sophistication, duck farming on the other hand, is at disadvantage in the developing countries compared to commercially grown broiler and layer chicken. The chicken farms are large-structured, integrated and much advanced. The marketing chain of commercial chickens is short, well organized with efficiently integrated production-processing-marketing operations. Conversely, duck farming is generally a small-scale backyard operation in developing countries and depends only on much less efficient traditional marketing methods. As the economy develops, a traditional farming sector like duck rearing is increasingly facing competition from commercial chicken and other meats in the domestic and overseas markets. Statistics have shown that the production of duck egg and meat in the traditional duck producing countries is not keeping pace with the fast growing chicken farming.

The factors, which can be attributed to the declining trend in the share of smallholder backyard duck production are: shrinking scavenging land area and water bodies, poor sourcing of inputs like chicks, feed, medicine, veterinary services and biologicals, lack of access to credit, emerging diseases and limited access to latest technical know-how. These factors trigger low productivity and variability in performance (Chang et al. 2003). The constraints faced by duck farming across different systems of management are variable, although, some factors are common to all systems. The Governments at developing nations should come to the rescue of poor duck farmers by implementing need based and country specific action plans to sustain this sector of agriculture, which is providing livelihood to a large section of people in the rural areas.

16.2 Opportunities

The duck farming has multiple benefits and opportunities. Uniqueness of this species and its habitat, and flexibility in terms of management and feeding requirements open up tremendous possibilities. Taking advantage of their ability to integrate with agriculture and fisheries, ducks can be used in environment friendly agriculture with reduced usage of fertilizers and pesticides. The opportunities of duck farming are discussed below.

16.2.1 Advantage of Topography

Marshy land, backwater bodies, lagoons, riverine and coastal lines, ponds and lakes form attractive habitat to waterfowls like ducks and geese; although this environment is not well suited to other species of poultry like chicken or livestock. Ducks also play an important role in environment protection in the areas along coastal line interspersed with inland backwater bodies by their coexistence. The plants, water and soils of wetlands provide homes to the ducks; whereas, a rich foundation for food chains of aquatic plants, fish, crustaceans, molluscs, insects, invertebrates, bacteria, algae and decaying plants provides incredible variety and abundance of food to ducks. Lowland in countries like Bangladesh (Islam et al. 2016), Kuttanad backwater region of Kerala, India (Jalaludeen et al. 2004), delta regions in Vietnam (FAO 2010) and similar topographies all over the world are very much suitable for duck rearing.

16.2.2 Role in Women Empowerment and Livelihood Opportunities for Weaker Sections

Duck farming offers hard-cash income and employment at a reasonably less time and cost to the rural landless women farmers (Huque et al. 2011). Duck farming creates a source of employment for unemployed or under qualified or unskilled people. Ghosh et al. (2012) found that most of the primary village duck raisers (80%) are women in Bangladesh. Duck rearing can be considered as a tool for food production and poverty alleviation for the rural womenfolk (Khanum et al. 2005). In Bangladesh, rural women are more comfortable to participate in duck farming (Hossain and Bayes 2009) that provides them food security and employment opportunities (Islam et al. 2016).

16.2.3 Low Cost of Production

The ducks need comparatively less space under shelter, only during night, as they go for scavenging during daytime. Little to no monetary investment on housing and farm equipment is a major advantage in low-budget duck farming. The duck houses protect the birds from bad weather and predator animals. The ducks need only a less expensive, simple and non-elaborate housing facilities which can be made of locally available low-cost materials. In Rajshahi region of Bangladesh, the duck houses are made using tin-shed (74%) or bamboo-straw (15%) or soil (11%) (Alam et al. 2012). Similar duck houses are located in coastal region of Noakhali and Lakshimpur districts of Bangladesh (Rahman et al. 2009). The tin-shed houses are permanent and long lasting with low-cost involvement, more secured and hygienic for ducks. As a result, housing cost for setting up of a small-scale commercial duck farming business is much less than commercial chicken farming.

Duck rearing is economical as compared to chicken by virtue of their foraging feeding habits, which cuts the cost on feeding. The ducks can be fed with a wide variety of feed. A duck's regular feed includes nonconventional raw materials like cassava, copra, corn, rice, fruits and any other low cost and easily available local feedstuffs. They also have the natural tendency of foraging on aquatic weeds, algae, green legumes, fungi, earthworms, maggots, snails and various types of insects; which directly reduce feeding cost. Ducks also utilize the aquatic feed resources and wastage feed materials better (Islam et al. 2016).

16.2.4 Ease of Brooding

Ducklings grow very fast and need artificial heat only up to 5–7 days of age in tropics; although they require a little longer heating period during cold months. The brooding arrangements for ducklings are not elaborate unlike other species of poultry like chicken and Japanese quail. In village flocks of tropical conditions, ducks are reared without artificial heating. The practice of brooding of ducklings adopted by the farmers is entirely different from that of chicks and natural. In such conditions, temperature of duckling nursery shed (brooder house) is maintained by constructing low roof, spreading new layer of sand on the floor and by increasing stocking density; so that the body heat of ducklings is conserved to keep the microenvironment of the house warm. The roof of this type of brooding set up is made of thatch or plastic sheets, and the sides are made of plaited coconut leaves or bamboo mat. A brooder house of 35 sq. metre area with 2.25 m height at centre and just 1.25 m at eaves can hold 5000–6000 ducklings (Jalaludeen et al. 2004).

16.2.5 Ease of Maintenance

The ducks can adapt easily to all types of environmental conditions. They start laying eggs in the early morning and complete before the dawn. This enables the farmer to collect and save the eggs before sending them to foraging. Cannibalism and other agnostic behaviours, which are very common in chicken are not usually encountered with ducks, especially in grazing ducks. It is simpler to sex ducklings than chicks even when they are very young. They also do not require continuous physical attention in maintaining them. Once trained, the ducks can go to pond or grazing fields in the dawn and return to the shelter in the dusk on their own. The requirement of only little skill is another reason for wide duck farming (Saleque and Mustafa 1996).

16.2.6 Fast Juvenile Growth

The pace of juvenile growth is much higher in waterfowl compared to any other poultry species. The average daily weight gain of meat type White Pekin ducks is more than 70 g from 0 to 14 days of life (Kaewtapee et al. 2018); which is not possible in broiler chicken. Owing to this early pickup, broiler ducks can attain 2 kg body weight in less than 30 days of age compared to 35 days in broiler chicken. This growth rate is possible, because increase in the weight of gastrointestinal tract is much faster in ducklings that support the secretory activities of enzymes in the pancreas to achieve maximal growth during the early growth stage (Lu 1999).

16.2.7 Hardiness to Cold Climates

The waterfowl are not at risk of death even in extreme winter weather due to adaptations like dense layers of insulating feathers, counter-current blood flow that reduce heat loss through their feet and legs, behavioural modifications to reduce exposure to the elements and the ability to carry large fat reserves (Brasher 2021). During severe cold snaps, waterfowl often simply hunker down to conserve energy and burn fat reserves to keep warm, which ultimately causes loss of body mass. The birds can rapidly replenish fat reserves when warmer weather returns. Duck in small village flocks in Bangladesh are highly adaptable to stressful environmental conditions than chicken (BER 2012).

16.2.8 Hardiness to Diseases

Ducks are highly resistant to the common avian diseases and they have less mortality rate and usually live longer than chickens. Nevertheless, waterfowls are affected badly by some devastating diseases like duck virus enteritis and duck viral hepatitis, these diseases can be managed with vaccination. The influenza viruses that are highly pathogenic for chicken rarely harm ducks (Hulse-Post et al. 2005). It causes only mild clinical signs and the ducks recover from infection quickly (Perkins and Swayne 2003). However, during the first months of 2021, the last influenza crisis occurring in the southwest part of France was highly pathogenic for ducks (ANSES

2021). Ducks in small village flocks in Bangladesh are comparatively more resistant to common diseases (BER 2012) and disease occurrence was ranked as comparatively less serious problem (fifth important one) by the duck farmers unlike chicken farmers (Roy 2000). If proper management and vaccination are followed, duck rearing will be more economic and sustainable (Maijer 1987). Despite frequent exposures to potential pathogens due to their scavenging character, ducks rarely contract diseases due to the strong innate immune response they elicit. Stimulation of certain genes triggers gene expression involved in the innate immune response and further instructs development of antigen-specific adaptive immunity (Kolluri et al. 2014).

16.2.9 More Egg Production

Ducks are highly productive (BER 2012) and in small holder farming systems, they lay on an average 40–50 more eggs than the *desi* hen layers. Ducks have an economic productive life during second year also; that reduces the cost of replacement to half. Duck eggs are heavier by about 15–20 g and fetches higher price in many places.

16.2.10 Suitability in Integrated and Mixed Farming Models

Ducks, unlike other poultry species can fit well in the integrated farming with fish, rice, rice with fish and rice with fish and azolla. In integrated farming, duck sheds are constructed over the fishpond which is a contiguous part of a rice field. The duck droppings and spilled feed fall directly into the fishpond through the gaps between slats and welded mesh. The plankton blooms by utilizing droppings of duck shed and forms the feed for fish. The nutrient enriched water from the fishpond can be irrigated to the rice field. Azolla growing in rice fields helps to check weed growth besides forming feed for fish and ducks and natural fertilizer for rice crop (Cagauan et al. 2000; Furuno 2001). The continuous stirring of the water and plodding of soil increase aeration of water and stimulate rooting and tillering of paddy plant. The strengths of duck integrated rice farming (with or without fish and azolla integration) include natural incompatibility, soil enrichment, control of weed and insects and ultimately more income to the farmers (Suh 2014). The global warming potential is reduced and net ecosystem increased in duck-rice integrated farming (Sheng et al. 2018). In mixed farming, ducks and other waterfowl like geese can coexist in harmony with other poultry (Kumar et al. 2013). The mixed farming consisting cattle, goats, chicken and ducks was found beneficial system, which can augment the income and socioeconomic status of tribal population (Ramrao et al. 2008).

16.2.10.1 Biological Control of Pests

When duck rearing is integrated with paddy cultivation, the ducks feed on the larvae and insect pests of rice like brown hopper and case worm. This kind of control of biological pest maintains the ecosystem intact without the need for application of chemical or pesticides. Hossain et al. (2005) conducted the efficiency of duck-rice farming in Sylhet and Barisal areas of Bangladesh and observed green leafhopper, brown plant hopper, zigzag leaf hopper, rice bug, short-horned grasshopper and long-horned grasshopper were significantly lower in rice-duck plots compared to farmers' plots without ducks. However, no significant difference in the populations of the white-backed plant hopper and stem borer were observed. Ducklings catch the insects efficiently in the rice-duck plots; thereby, reducing the insect population. On the other hand, beneficial insects like carabid beetle, ladybird beetle, damselfly, spider and dragonfly were found in similar number both in the rice-duck and sole-rice plots. The ducks kill apple snails, slugs, harmful insects and other crop pests from the gardens. The duck when used for pest control has double benefit of getting good scavengeable feed resource thereby improving feed efficiency and reducing insect problems. It is quite interesting to note that in China, ducks were specially trained to ingest harmful insects like gross hopper in agricultural lands.

16.2.10.2 Weed Control

The rice yield is reduced considerably due to the weed population in paddy fields. Ducks destroy the weeds by feeding on them as well as by disturbing its growth when they wade through the rice plants. In an experiment in Sylhet and Barisal areas of Bangladesh, the weed counts per square meter of land was found to be significantly lower in the rice-duck plots when compared to sole-rice plots (Hossain et al. 2005). The predominant weed species like *Echinocloa crus-galli*, *Scirpus mucronatus*, *Monochoria vaginalis* and *Fimbristylis miliaceae* were in the paddy fields of experimental areas. Ducks were seen consuming weed plants and weed seeds. Moreover, the trampling activity of ducks also kept the weeds under control by as much as 90%. The total weed biomass was controlled better in rice plots integrated with ducks in Japan when compared to the agrochemical treatment (Manda 1992; Furuno 1996). Cagauan (1997) reported a reduction of 52–58% of total weed biomass in the rice field by ducks in Philippines; whereas, Wei et al. (2005) reported control of weeds of 70–80% by introducing ducks to the rice fields of China. Compared to the rice monoculture control, Zhang et al. (2009) found that the weed biomass was reduced by 98% in the treatment with duck grazing; of which, 84% was due to disturbance and 14% due to grazing. Van Groenigen et al. (2003) reported that the grassy weed biomass more than doubled from 89 to 204 kg/ha in the absence of waterfowl.

16.2.10.3 Effect on Soil Health

As the ducks wade through paddy plants, they add natural manure by way of their droppings. With dabbling of their beaks and feet, the ducks stir up the soil to create an aerated environment that decreases methane production and methane emissions from paddy fields. By preventing the accumulation of harmful gases in the rhizosphere, the growth of the rice plants is stimulated. The N, P, K, Ca and S levels in the paddy fields increase after duck-rice integrated farming (Zheng et al. 1997; Hossain et al. 2005). The physical property of the soil is improved due to ducks' movement and feeding activity resulting in enhanced root system of rice plant (Furuno 1996).

16.2.10.4 Economic Gains

The integrated rice-duck system provided 50–60% higher net returns than rice monoculture in Bangladesh (Hossain et al. 2005). The increase of net returns ranged from 8455 to 16,103 Tk. (1US $ ≈ 66 Bangladeshi Tk) per hectare depending upon season and region. The income was higher in rice-duck system, because of (a) higher rice yield with reduced cost of production, (b) additional revenue from the ducks, (c) reduced requirements for insecticides and chemical fertilizer and (d) premium price for organic produces.

16.2.10.5 Role in Organic Paddy Cultivation

The organic farming system employs management practices to nurture ecosystems to achieve sustainable productivity and provide weed, pest and disease control through a diverse mix of mutually dependent life forms and recycling plant and animal residues (FAO 2007). It emphasizes the use of natural inputs like mineral and products derived from plants and the renunciation of synthetic fertilizers and pesticides. In general, the average yield of cereal crops in organic farming is less than that obtained from conventional production practices (Bhattacharyya et al. 2003). This is primarily due to difficulties in plant nutrient management and lack of effective pest management options. The ducks when integrated with organic paddy cultivation perform the functions of pest management, weed management, tillers stimulation, cultivation and fertilization by eating the harmful insects and weeds, thus eliminating the need for pesticides and herbicides. The rice–duck mutualism in organic farming has many economic, environmental and ecological benefits compared to conventional rice cultivation (Pirdashti et al. 2015).

The dabbling and plodding activities also oxygenate the water and as a result the roots of the rice plants to grow vigorously. Rice-duck system improves the number of tillers per hill as well as grains per panicle and average grain weight; altogether resulting in, on an average, 20% higher yield than those of the sole-rice sub-plots (Hossain et al. 2005). In a simulation study, Huang et al. (2012) showed average length of rice-ear increase by 8.88%, empty grain number per rice-ear decrease by about 36.12% and rice seed setting rate increase by 3.79% as impacts of duck integration in rice cultivation. The superiority of the duck integrated rice cultivation is consistent across locations and seasons.

The organically grown rice fetches premium price in Malaysia, South Korea and many other countries taking advantage of organic food certification system in those countries. Farmers can sell certified organic rice at about 60–70% higher price compared to rice grown in conventional agriculture in South Korea (Suh 2014).

16.2.11 Specialty Duck Products

The duck products prepared out of their eggs and meat are in great demand in the local and international markets. Duck meat is also a delicacy and relished by many people. The down and fluffy body feathers of the ducks are valuable and used for different industrial purposes. The duck eggs have many special attributes that are liked for baking and other products making industries. The specialty products like thousand-year-old-eggs, salted duck eggs and balut (Jalaludeen and Churchil 2006; Jalaludeen et al. 2009) are unique to this species of poultry. Similarly, some special duck meat products can only be made from ducks; which include roasted garlic breast, confit, rotisserie breasts, 'à l'orange', 'rillettes' and *foie gras* (Jalaludeen et al. 2009).

16.3 Constraints

The duck production in recent time follows a reverse trend due to various difficulties confronting village duck farming. They include difficulty in sourcing of inputs like ducklings, feed and medicines, marketing hardships, disease outbreaks, etc. across different systems of management. However, there are some constraints, which are specific to certain system of management. For example, problems associated with shrinking scavengeable area and pollution of grazing fields, and epizootics of avian influenza are pointed towards nomadic flocks; while behavioural and welfare problems are restricted to intensive farming system. Therefore the constraints are discussed separately for different systems of management like a) free-grazing stationary, b) free-grazing nomadic, c) rice-duck, d) fattening and e) intensive duck farming.

16.3.1 Stationary Free-Grazing Duck Flocks

The stationary flock is the one that spends its night in the same shelter day after day. These flocks include

(a) Small village flocks reared as backyard units

These flocks normally contain around a dozen of ducks. The women households of the family maintain these flocks mainly for earning subsidiary income. The birds are reared under low-input/no-input system with feeding of only kitchen waste and scavenging in the backyards and nearby water bodies during daytime. The night shelter is usually a simple, modest and mostly made of locally available materials.

(b) Short distance free range ducks

The flock size of ducks reared under this system ranges from few tens to few hundreds. The birds are kept overnight in a duck house of a farm and taken for feeding on rice fields within the village boundaries. These flocks are herded by a family member, towards scavenging fields and driven back to the duck house every night.

The difficulties faced by the stationary free ranging duck flocks are as follow:

16.3.1.1 Limited or Zero Land Holdings

The livestock owners are not always land owners in this system. Basuno (1983) found that in a village in West Java up to 40% of livestock owners were landless. The size of land holdings of duck farmers influences the flock size simply because the famers with high land ownership can rear more ducks as they have more area for scavenging. As described by Veeramani et al. (2017) neighbouring field owners sometimes deny the entry of ducks for foraging in their field. In Bangladesh, 28, 46, 22 and 4% of the total duck farmers are landless (<0.049 acre), marginal/small (0.05–2.49 acres), medium (2.50–7.49 acres) and large (more than 7.50 acres) land holders respectively (Khanum et al. 2005). The landless farmers had in general kept small flocks (less than 200 ducks per farm); whereas, most of the marginal/small and medium farmers (68%) kept up to 536 ducks. The flock size increased with the increase of land holding in Bangladesh (Huque and Sultana 2002). In Kerala state of India, Jalaludeen et al. (2003) reported that 62% of duck farmers had a land holding of only 0.1–0.25 acre.

16.3.1.2 Supply of Ducklings

The nonsteady supply of ducklings is considered as one of the most important constraints in duck farming in small village flocks (Alam et al. 2012). The ducks are preferred and commonly raised in smallholder farming system in most of the rural areas of Southeast Asian countries. In Philippines, majority of the duck eggs are used for producing balut, leaving a little (2%) for the production of replacement stocks. Poor availability of quality breeding stock is the major constraint in layer-type duck production in the Philippines in the 1980s (Arboleda et al. 1985) and the problem persists (Lambio 2001). Penoy (infertile egg separated from other incubated eggs by candling) is the byproduct of balut production. The eggs not fit for hatching purpose were used for producing salted eggs and century eggs (Lambio 2001). Therefore, the duck raisers in general depend only on day-old ducklings or ready-to-lay ducks available from the balutan operators or duck pullet growers. Another aspect is that the duck raisers give less importance to the quality of stocks used as replacement. They do not follow appropriate selection procedures and mating systems, which in turn leads to loss of vigour and fertility. Ultimately, the performance in terms of egg production, fertility, hatchability and egg quality of a majority of duck farms suffered a major setback (Chang et al. 2003).

The mechanized hatching facilities are not available in many least developed and poor countries due to the nonavailability of electricity in village level. In most of the hatcheries of Cambodia due to lack of electricity, traditional rice husk system of heating of hatching eggs involving daily intensive work to warm the rice husks and turn the eggs was used (FAO 2009).

In small village flocks, the production of ducklings by natural incubation is also a major challenge. The ducks are not so broody and are poor mothers. This is overcome by using broody hens for incubation and brooding purposes. Most of the duck farmers adopt this practice. Fifty percent of the farmers feel that the hens are good in broodiness and/or mothering ability compared to ducks (Islam et al. 2016). The other reasons are that broody hens engaged in natural incubation scavenge near the house (12%), stop egg production (16%), more hatchability (10%) and more sitting time (2%).

16.3.1.3 Availability of Scavengeable Feed Resources

Scavenging ducks forage on different types of faunas such as snails, fish, earthworms and flora such as duckweed and algae. The average egg production increases from 30% in the dry period to 62% in the scavenging period (Huque et al. 1993). Women dominate the duck rearing and a good amount of marshy land is available in the coastal area of Bangladesh (Ghosh et al. 2012). Ducks are scavengers on the natural water sources. Of late, many stationary free-grazing duck farmers are forced to give up duck farming due to drying up of water bodies and conversion of paddy fields for residential and industrial uses.

The nutrient availability of the scavenging ducks varies on daily basis as they graze through different fields with varying levels of scavengeable feed resources. Poor nutrition has much greater effect on production parameters than the genetic factors in ducks under scavenging system of rearing. Improved feeding system of scavenging ducks was suggested by Huque (1991) to achieve optimum production. Traditionally, the ducks are reared on scavenging system with some feed supplementation but results have been disappointing because of imbalanced use of available supplements/nutrients. The field trials indicated that even though medium nutrient density diet is appropriate and cost-effective to explore growth potentials of growing indigenous (*desi*) ducklings under scavenging system, low nutrient density diets could be recommended for the resource poor farmers because of low cost that will lead to profitable production (Pervin et al. 2013).

16.3.1.4 Limited Knowledge on Technical Know-How

The backyard duck farmers practice the rearing methods that they inherited from their ancestors. Most of the farmers follow unscientific practices in feeding of their ducks. This might be due to lack of proper knowledge about scientific advances on feeding management in duck farming (Alam et al. 2012). The production technologies in duck farming continuously improved over the years with the advent of new sophisticated technological breakthroughs. However, the backyard duck farmers have not benefited from the advanced technologies resulting in less efficient production. This is mainly due to limited access to information and partly due to the unavailability of the extension services. Low-level education also limits their capacity to adopt these new techniques (Chang et al. 2003). A survey study indicated that the lack of training as the third most important constraint in Bangladesh among the small-scale duck farmers (Alam et al. 2012). Therefore, the intensive training on farm management, duck diseases and its control need to be provided regularly by

Department of Livestock Services and nongovernment agencies to ensure good management for profitable duck farming.

16.3.1.5 Financial Supports

The financing from government is required to set up a small-scale household duckling production unit by resource poor farmers in developing countries. The resource poor duck farmers in Bangladesh go for low-cost housing. Nearly 74% farmers make tin-shed houses, because they are not only cheap but also permanent and long lasting. Cost involvement is not so high. Another reason is that the tin-shed houses are more secured and hygienic for ducks (Alam et al. 2012).

The local governments are the major source of cash loan as a part of welfare programs. NGOs and nationalized banks are the main sources of microcredit. Duck farming require only short-term loans and the same can be recovered in a single season. However, the farmers are reluctant to take microcredit fearing the weekly repayments (Orr et al. 2004) and natural calamities. High interest rates, requirement for collateral and lengthy paper work deter smallholder duck farmers from availing agricultural loans of private banks. On the other hand, credit provided by informal lenders is more accessible; even though, the interest rates are much higher than that of commercial banks (Chang et al. 2003).

16.3.1.6 Diseases and Mortality

The success in duck smallholder farming is greatly determined by mortality. Duck plague is the major disease threat among the duck farmers. In South India, duck plague caused up to 90–100% mortality until 1980s (Gajendran and Karthickeyan 2009). Later, the farmers adopted vaccination against the disease using a live attenuated vaccine supplied by public institutes. Other important disease conditions prevalent in small household village flocks are fowl pox, coryza, hock joint swelling in ducklings, Newcastle disease, duck cholera and duck viral hepatitis. The helminths such as *Echinostoma, Capillaria, Notocotylus, Coccidia, Railletiena, Choanataenia, Hymenolepis* and *Cotugnia* species are common in field ducks. In semi-scavenging duck flocks of Bangladesh, duck plague (32%) and concurrent occurrence of both duck plague and duck cholera (32%) are the major disease problems followed by duck cholera alone (21%), concurrent colibacillosis and salmonellosis (5.2%) and duck viral hepatitis (1.6%).

In Cambodia (FAO 2009), as per farmers' perception, access to services for disease prevention and control is the major step to improve smallholder duck farming. The farmers also felt that the access to reliable medicines and vaccines for their ducks is the next important factor. The farmers reported mortality is high in early stages of their growth and adult stage as well; the vaccination against duck plague helps to reduce this risk. The diseases were more common during winter (63%) than other seasons of the year in Southern area of Bangladesh (Ghosh et al. 2012). Inadequate veterinary services are one of the major constraints in preventing disease occurrence in Bangladesh (Alam et al. 2012) and other developing countries. From a survey, Rahman et al. (2009) reported that outbreak of diseases, insufficient preventive health care as well as lack of knowledge, unavailability of veterinary medicines and services, and high price of veterinary drugs and medicines are considered as constraints, in the order of decreasing importance by the farmers of small village duck flocks.

Among all diseases, avian influenza (AI) outbreaks impacts the village duck flocks heavily. The serious impacts of AI outbreaks as perceived by the farmers in Cambodia are (a) difficulty in selling eggs (47.6%), (b) difficulty in selling ducks (31.9%) and (c) restrictions on the movement of ducks to markets (17.4%). Above all, the consumption of duck and duck products is heavily affected in countries where human casualties happened. For example, Takeo and Kampong Cham provinces of Cambodia were among the first places of outbreaks of H5N1. As a result, the consumption of poultry meat and products in large towns and cities of Cambodia reduced drastically between 2004 and 2006, because of concerns about human infection and death to H5N1 (FAO 2009). The outbreaks of AI create busts of market dullness even in countries where human death has not been reported so far. Many consumers are still concerned about the risk of contracting HPAI from poultry products. Therefore, the demand for duck products in the urban areas is still limited compared to pre-HPAI era (FAO 2010). Many local governments make TV and radio announcements to educate the consumers about the safety of consumption of poultry meat and products during outbreaks.

16.3.1.7 Predation

The predation is also a serious threat to the duck safety causing mortality in village flocks if the housing is not predator proof. Nocturnal predators such as civet cats and foxes can potentially gain entry to duck houses by digging under the walls. The types and population size of predators, as well as the frequency of attacks vary according to the geographical area. The set of predators of hilly areas is different from that of plains. Predation is the more serious threat causing 8.3% out of total 15.0% total mortality in ducks in Chatkhil of the Noakhali district in Bangladesh. Importantly, predation caused a significantly higher level of duck mortality than diseases (5.9%) (Hoque et al. 2011).

16.3.1.8 Natural Calamities

Flood is one of the calamities that affect poultry. The ducks are less prone for mortality in flooded conditions compared to chicks, as they can swim. However, incessant rains make their waterproof plumage to lose its oily coat and become wet, finally drown. Ahmed et al. (2019) estimated that on an average, the households in flooded areas of Bangladesh lost 44% of their total birds (chicken and ducks).

16.3.1.9 Waste Management

The proper disposal of dead ducks is highly essential to break the disease transmission cycle in small village flocks. The domestic and wild waterfowl exposed to natural infections of duck viral enteritis periodically shed the virus for 1–4 years after infection depending on the virus strain and the species of waterfowl (Burgress et al. 1979). The presence of latently infected ducks is common for many duck diseases therefore has epidemiologic consequences; because, the carrier ducks provide a source of infection for other birds and enable the infectious agent to persist in wild and domestic populations even at low infection rates (Shawky and Schat 2002). Not giving proper attention for the disposal of dead ducks, nonexistence of proper burial/ incineration facilities for the dead ducks and throwing the birds into the nearest water bodies are encountered in some countries (Khan et al. 2018). In Indonesia, only 32.14% dead birds are disposed by burial or burning; whereas, 53.57% dead birds are thrown in the water bodies (Yuyun et al. 2020). In a study, In Rajshahi region of Bangladesh, 11% farmers threw the dead ducks outside the farm as they lack proper awareness on the impact of doing so (Alam et al. 2012).

16.3.1.10 Marketing

Although the duck farming is an old agribusiness, it has no organized marketing system; therefore, the duck rearing farmers are facing difficulties in selling the duck meat and duck eggs. The duck industry is small and produced mainly by geographically diverse small-scale backyard production units. This means basic infrastructure such as processing, storage and transport facilities must be put in place in an efficient marketing system before value can be created (Chang et al. 2003). The low price of duck egg and meat is considered as the topmost constraint in duck farming in Bangladesh (Alam et al. 2012). They also found price fluctuation of duck egg is also a major constraint. The government intervention on the market players is necessary to control the price fall of eggs (FAO 2009).

16.3.2 Free-Grazing Nomadic System

The nomadic free-grazing duck flocks, also called moving flocks or long-distance free-grazing flocks, which are herded to long distances to feed on harvested rice fields and are confined in temporary enclosures en route on the rice fields during nights (Henning et al. 2013). The nomadic adult free-grazing ducks are normally above 3 months of age, scavenge in flooded rice fields after harvest and feed on leftover grains, insects and molluscs (Minh et al. 2010). Nomadic free-grazing duck flocks are transported from one place to another scavenging field at long distance by foot, truck and boat. The journey of nomadic free-grazing duck flocks involves long-distance movements of more than 100 km, most of the times. For example, 68 and 33% of duck grazing sites are located outside of the commune and province of residence of the farmer, respectively, in southern Vietnam (Meyer et al. 2017).

The scavenging system of duck rearing revolves around the villages where the duck producers live as long as the natural feed resources and leftover fallen paddy are available in the post-harvest rice fields. When the scavengeable feed resources are exhausted in the local proximity, the flocks are moved to neighbouring districts, or even to nearby provinces (FAO 2010). The duck producers plan the movements of their flocks based on their experience and network of contacts for optimal exploitation of the feed resources.

The problems encountered in free-grazing duck farming are as follow:

16.3.2.1 Shrinking Foraging Lands and Water Bodies

The nomadic system of duck farming benefit out of scavengeable feed resources in the harvested paddy field and flora and fauna in the water bodies. To get optimum nutrition for 1000 ducks, four to five acres of post-harvested paddy fields are required every day (Tamizhkumaran et al. 2013). Shrinkage of grazing land for ducks mainly due to urbanization of rural areas and narrowing of water channels and encroachment of lakes and ponds is considered as a serious threat in India (Gopinathan et al. 2015; Vignesh et al. 2020). The situation is same in many other developing countries. The ducks are therefore driven to farthest areas in search of scavenging land, and this long-distance migration leads to considerable loss in profit margin (Gajendran and Karthickeyan 2009). Shrinkage of land for foraging is considered by the free range duck farmers as fourth major constraint and denial of entry by the land owners to the ducks for foraging is fifth major constraint and shrinking water bodies as seventh major constraint among the ten listed constraints confronting free range duck farming in Tamil Nadu, India (Veeramani et al. 2017).

16.3.2.2 Nonavailability of Capital Funding

In developing countries like India, a nomadic farmer should possess around INR 0.15 million ($\approx$ 2000 US$) to start nomadic duck farming with one thousand ducks of around 12 weeks of age. This is beyond the scope of imagination for the poor nomadic duck farmers (Tamizhkumaran et al. 2013). The farmers are at disadvantage, as the public sector banks require immovable properties like land or movable property like gold to pledge as security to grant them a bank loan. This forces the duck farmers to depend solely on egg vendors/private money lenders for all the operational expenses of duck farming in India (Gajendran and Karthickeyan 2009) and many other developing countries (Sankaralingam and Mahanta 2021). About 76.9% of the duck farmers depend on the duck contractors and agents for initial investments in and around the Puducherry region of India (Tamizhkumaran et al. 2013). The farmers have to pay the debt back to the moneylenders along with huge interest. The intermediaries who finance the farmers also act as egg dealers to fix price for egg and collect them (Ramachandran and Ramakrishnan 1982). In this way, the vendors exploit the duck breeders to such an extent that the farmers are indebted to the vendors and can never come out of the clutches of the vendors. The financial position to start or scale-up duck farming is considered as the topmost constraint; whereas, the nonavailability of loan from banks is the third most important constraint among top ten constraints of nomadic duck farmers in Tamil Nadu state of India (Veeramani et al. 2017).

16.3.2.3 Nonavailability of Quality Ducklings

Free-grazing system of duck rearing consist flocks of 500–5000 ducks. In this system of management, to start with, the ducklings of 12 weeks old are needed. As the paddy harvesting is a seasonal activity, many such flocks start travel at the same time from a starting point. This requires a large number of grownup ducklings available at a particular point in time. The duck contractors/agents who mobilize the grownup ducks from neighbouring provinces and exploit this precarious situation to their favour (Tamizhkumaran et al. 2013). Private hatchery operators also seasonally grow the ducklings up to 12 weeks of age to exploit the seasonal demand and make the farmers bonded to them. Nonavailability of good quality chicks is considered as one of the important constraint, next only to the nonavailability of finance by the free range duck farmers in Tamil Nadu state of India (Veeramani et al. 2017). Gopinathan et al. (2015) also identified this as a major constraint in the same geographical area.

16.3.2.4 Contract Farming System

In many instances, the duck breeders involved in nomadic farming do not own the ducks they graze. This system of rearing is seen in countries like India (Tamizhkumaran et al. 2013). The farmers are bonded to the contractors/vendors from the start of the rearing cycle itself as they accept the grown-up ducklings of around 12 weeks of age from the contractors/vendors as loan. The flock size ranges from 500 to 3000 depending upon the family size. These nomads migrate in groups as per the information provided by the contractors. The contractors collect the revenues, in terms of eggs and spent ducks, on weekly basis. The contractors give only a meagre rearing charge to the breeders. The nomadic duck farmers are indebted to, and exploited by the contractors and made entangled in their clutches. The nomad duck breeders are mostly illiterates who do not know anything other than rearing the ducks. They are kept in dark on entire process of duck marketing like the prices of the ducklings supplied to them initially, price of eggs collected from them and transportation charges. The socioeconomic status of these farmers remains poor with no marked changes since many generations; which is the testimony for the magnitude of exploitation they suffer at the hands of vendors (Tamizhkumaran et al. 2013).

Although the duck rearers are aware of the exploitation by the vendors, they are contented with this arrangement because: (i) no capital investment and less risk (ii) getting periodical foraging charges for their labour from the vendors (iii) permitted to sell as well as consume some of the eggs and duck meat (iv) able to educate their children who are settled in the village(s) and (v) facility to avail loan at any time from the vendors to meet any exigency (Tamizhkumaran et al. 2013).

16.3.2.5 Limited Open Time in Paddy Fields

The genetic improvement in rice plant resulted in not only increased production, but also decreased life cycle. The early maturing, high-yielding rice varieties are commonly cultivated nowadays; consequently, the land has to be prepared quickly for the next crop. Therefore, the time available for the ducks to scavenge in the fields between two crops has decreased. As the fallen grains in the harvested paddy fields is the major source of scavengeable feed resource for nomadic ducks, the short open period for grazing in paddy fields is also a major constraint.

16.3.2.6 Risks and Issues Associated with Transport Between Grazing Fields

In free-grazing system, most of the herders who moves to nearby villages or sub-districts herd the flock by foot. The ducks are migrated to neighbouring districts, and sometimes inter-state/inter-province movement is needed. The farmers use waterways also wherever available. For example, in Northeastern Bangladesh, about 16% of the duck flocks are transported by boat and 5% through roads by the motor vehicle to nearby districts (Sarkar et al. 2017). From Puducherry region of India, transport of duck to neighbouring states (provinces) like Kerala, Karnataka and Andhra Pradesh has been reported (Gajendran and Karthickeyan 2009; Tamizhkumaran et al. 2013). The ducks are transported in specialized vehicles if the grazing fields are distant. About 150–200 dozens of ducks placed in cages are loaded one above other and transported. Sometimes, few farmers join together to hire a bigger vehicle to transport their ducks in a single trip to save money. Sankaralingam and Mahanta (2021) reviewed the methods of transport and cost involved in transport of ducks in different places. In Indian conditions, the transport cost varies from INR 40 to 55 (1 US$ ≈ INR 75) per kilometre depending on the capacity of the truck. Indian farmers consider transport risk as one of the major constraints in free-grazing system. The transporting cost of scavenging ducks from one province to the other is on constant increase because of hike in petrol and diesel prices; which cuts down the earnings of farmers.

16.3.2.7 Lean Season Feeding

In general, the cost of feeding of ducks in the rice fields with strewn grains and other feed resources after harvest is almost zero. The natural food becomes unavailable during dry and flood seasons. If proper nutrition management is not taken, ducks lose weight and become prone to diseases. In a Bangladesh grazing model of duck farming, there are 2 months of dry and 2 months of flood seasons. The better-off farmers with 300–500 ducks generally feed their birds with broiler chicken feed when natural feed is unavailable; whereas, poor farmers with less than 100 ducks feed rice husks with weeds (duckweed *Lemna perpusilla*), frogs and snails collected from the flooded plains (Khan et al. 2005). Although readily available, duckweed can also be cultivated to have a more steady supply (Journey et al. 1993).

In Puducherry region of India, hand feeding of ducks with paddy is being practiced to meet only the maintenance requirements of ducks during the lean season; although, it does not support egg production (Tamizhkumaran et al. 2013). In Kuttanad area of Kerala state of India during lean season birds are hand fed. Lean season in Kerala include both monsoon and summer months and this constitutes about 4–5 months in a year. Availability of water to a depth of one to two inches in the paddy field is essential for effective scavenging. When the depth of water goes below one inch, that occurs 1–2 months every year, ducks cannot collect the feed resources including grains. Similarly, in monsoon, incessant rain and flooded fields deprive ducks from scavenging. This condition lasts for around 2–3 months. Lean season period vary from country to country and among provinces/states within the country. Exploring locally available low-cost feed resources can help the farmers to keep the cost of hand feeding to a minimum. The feed items like palm pith, sorghum, wheat, whole and broken rice, paddy chaff, unsalted dried fish, peanut cake and rice polish are used for lean season feeding of nomadic ducks (Annual Report 2002). Chopped pith of umbrella palm (*Corypha umbraculifera*) of the family Palmae is another major cheap alternative feed item. This palm is seen abundantly in hilly areas of Central Kerala that grows up to a height of 21 m with a massive trunk. It blooms only once in lifetime and after blooming and fruiting, the plant dies. The palm trees intended for feeding ducks are to be cut just before the onset of blooming. On dry matter basis, the palm pith contains 10.90% crude protein and 65.80% nitrogen free extracts (Annual Report 2002). Nonavailability of cheap feed ingredients for lean season feeding is considered as sixth major constraint in scavenging duck farming in Tamil Nadu state of India (Veeramani et al. 2017).

16.3.2.8 Poor Performance

Poor performance of free scavenging ducks compared to intensively reared ducks is mainly due to genetic makeup and low plane of nutrition. The indigenous *desi* ducks like Jinding (Uddin et al. 2020) Tau ducks (Burgos et al. 2008) and Arni (Gajendran and Karthickeyan 2009) are reared by scavenging or free range farmers in different countries. Local breeds, with smaller body size and a more active foraging capacity, are reputed scavengers than 'improved' varieties or exotic breeds. The productive performance under scavenging conditions is generally low. Nevertheless, utilization of no-cost/low-cost supplementary feeds can make this system highly profitable. Although the indigenous ducks can attain good body weight in intensive rearing systems, they perform much poorly in extensive system (Kolluri et al. 2015). In some parts of world, Kakhi Campbell breed is also used widely in scavenging system of rearing as a purebred or in crosses. In Kerala state of India, the farmers prefer indigenous ducks, because of their hardiness, adaptability to local conditions, ease of management and attractive egg size (Jalaludeen et al. 2004). Under nomadic system of duck rearing, the farmers of Kerala use indigenous Kuttanad breed extensively (88%; Ravindran 1983) in scavenging system.

The egg production from indigenous ducks is 100 eggs per duck per year in Tamil Nadu (Tamizhkumaran et al. 2013) and 96.2 eggs in West Bengal (Halder et al. 2007) states of India, 117.5 eggs in Mymensingh district of Bangladesh (Islam et al. 2016) and 51–70 eggs in coastal areas of Bangladesh (Pervin et al. 2013). The weight of eggs produced by the ducks in scavenging system also low ranging from 55 to 62 grams only (Vignesh et al. 2020; Gajendran and Karthickeyan 2009).

Insufficient nutrition from grazing and excessive movements also results in poor production efficiency in this system. Although, the high-yielding breed like Khaki Campbell which has the genetic potential to yield more than 300 eggs in a year in intensive system produce only around 100 eggs in free-grazing system. The grazing area of less than four acres of post-harvested paddy field for every 1000 ducks per day supply inadequate scavengeable feed resources and the production performance suffers (Tamizhkumaran et al. 2013). Marked influence on the egg production also resulted from the pesticide application in agricultural fields, which in turn cause damage to the proteinaceous scavengeable feed resources like snails, tadpole and earth worm population (Gajendran and Karthickeyan 2009).

16.3.2.9 Health Problems

The most important constraint in free-grazing system is mortality, which was estimated to be more than 30% per annum. Various diseases recurrently affect ducks, mainly duck plague and other health problems like poisoning by polluted water and pesticides, mechanical injuries caused during foraging and debility due to other disease conditions. The nomadic duck farmers do not get their sick ducks treated normally, rather slaughter and consume.

(a) ***Infectious diseases***

The major diseases of free-grazing ducks are duck plague, pasteurellosis, diarrhoea and coccidiosis. The scavenging ducks have weakened immune system because of fluctuating feed supply and exposure to pesticide residues in the fields. Therefore, the nomadic flocks are vulnerable to severe infection and high mortality (FAO 2010). The diseases in descending order of importance as perceived by farmers in Mymensingh district of Bangladesh are duck cholera (52%), duck plague (26%), limber neck poisoning (12%) and avian influenza (2%) (Islam et al. 2016). The duck farmers in India are aware of the importance of vaccination and protect their birds by using live attenuated vaccine from Government vaccine production institutes (Tamizhkumaran et al. 2013). In Bangladesh, majority of the duck flock owners vaccinate their flocks against either duck plague and duck cholera; although all of them are not vaccinating their ducks against avian influenza A (H5N1) (Sarkar et al. 2017). The farmers in different parts of the world also consider the diseases and mortality as one of the major constraints (Hoque et al. 2011; Tamizhkumaran et al. 2013; Veeramani et al. 2017). The nonavailability of vaccines in remote villages, where scavenging duck farming is practiced is also a constraint in many developing countries.

(b) ***Poisoning***

The indiscriminate use of pesticides, fertilizers and other industrial pollutants had made the environment increasingly unclean. The scavenging ducks are recurrently affected by various health problems due to poisoning by polluted water and contaminated feed. Nitrate toxicosis can result from accidental ingestion of fertilizer or other chemicals, which are the largest sources of nitrates in water. Premalatha and Ramya (2016) reported nitrate toxicity in ducks exposed to ponds that received extensive feedlot or fertilizer runoff during a flood in Tamil Nadu state of India.

(c) ***Injuries***

The mortality due to mechanical injuries to foot web during foraging is also one of the major factors of mortality (Gajendran and Karthickeyan 2009).

(d) ***Debility***

As the ducks are moved through the grazing lands, they may be exposed to high pathogen load arising from other flocks or infected wild birds. Moreover, the movement of ducks usually coincides with the winter–spring rice season when the daily weather fluctuates from high in the day (32–35 °C) to low at night (19–21 °C). During this period, the rice fields have low water levels both vertically and horizontally. This leads to the duck flocks to concentrate in high densities around the shallow water bodies of ditches and canals for drinking and swimming purposes (FAO 2010). The transport and movement stress coupled with exposure to high pathogenic load during movement causes many nonspecific ailments that make a good number of ducks weak and unthrifty.

16.3.2.10 Avian Influenza (AI)

The long-distance free-grazing ducks are likely to play an important role in the maintenance and spread of the AI viruses (Henning et al. 2016). First, the free-grazing ducks act as local reservoirs and amplification hosts of AI virus transmitted by migratory birds, followed by secondary spread to other domestic birds (Gilbert et al. 2006). Second, while grazing or when transported long-distance from one grazing place to another, free-grazing ducks may come to direct or indirect contact with other free-grazing duck flocks, potentially leading to virus transmission across larger distances. Third, transport of free-grazing duck flocks across province and even national borders (Bui 2010) may lead to the trans-boundary spread of AI viruses. Finally, transport of ducks in high density is a risk factor for influenza outbreaks, due to increased stress levels due to herding which is likely to increase virus shedding (Beaudoin et al. 2014).

The spread of HPAI through free-grazing ducks prompted movement restrictions in many countries to prevent the birds from becoming infected and spreading the disease. It is evident from the pattern of avian influenza outbreaks in the past, limiting the movement of flocks is the best management practice to reduce the risk of HPAI infection. The government imposed movement restrictions on free-grazing duck farming adds pressure on the producers to adopt alternate intensive systems with supplemental feeding to meet their nutrient requirements. The movement restrictions imposed by local governments in some countries prevents duck flock movements across provinces (FAO 2010). The duck production based on scavenging, has traditionally provides livelihoods of smallholder duck producers in many parts of the developing countries; this movement restriction is an emerging constraint which need to be addressed.

The avian influenza epizootics occurred in 2004 has affected the domestic and export marketing of duck products. For example, in Vietnam, the low-income people consume low priced duck products than other poultry products. Although, the social strata that consume the duck products have not changed much, the volume of consumption has reduced drastically. The exports of salted duck eggs and feathers to China have ceased after the first HPAI epidemics and most of the duck eggs are consumed domestically resulting in price slumps.

16.3.2.11 Dead Bird Disposal

The lack of facility for proper disposal of birds in nomadic duck farming is also a potent factor of recurrent disease outbreaks. The farmers on the move leave the dead birds in the paddy fields or throw them in the water bodies en route. The virus can survive in the soil and water that spread to the moving flocks across.

16.3.2.12 Predation

The predators in the open area including mongooses, black backed jackal and otters are considered as one of the constraint in free range duck farming. Measures like electrified movable fencing can restrict the predation and limits major losses (McCullough 2019). Considerable loss of scavenging ducks is due to predators; mainly dogs and jackals in Tamil Nadu, India (Tamizhkumaran et al. 2013).

16.3.2.13 Natural Calamities

Duck rearing in extreme cold climate is a challenge. Although waterfowl like ducks have exceedingly good ability to thrive in cold conditions (Brasher 2021), the temperature during the winter months below zero degrees centigrade makes management of the ducks difficult. This may require usage of heaters to ensure adequate warmth that will increase the cost of production during winter months (McCullough 2019).

16.3.2.14 Lack of Inheritance by Young Generation

The scavenging nomadic duck farming is physically demanding, as the herder needs to travel long distance daily and spend night hours in tents braving inclement weather. This farming system mainly earns livelihood of illiterate and lowly educated people in developing countries. The youth of new generation is educated and is reluctant to practice the nomadic duck farming. This is a serious threat in sustaining this farming system (Gopinathan et al. 2015).

16.3.2.15 Marketing of Duck Products

The acceptance of duck and its products vary from country to country and among provinces of a country. In India, duck egg and meat are preferred in states like Kerala, West Bengal and Assam. The eggs and ducks produced in Tamil Nadu state of India therefore need to be transported to neighbouring Kerala state to get good price and demand which leads to additional cost on transport (Gajendran and Karthickeyan 2009). The problems related to marketing of duck and its products is considered as one of the major constraints in free range faming system in Tamil Nadu, India (Gopinathan et al. 2015; Veeramani et al. 2017). Higher cost (43.85%), nonavailability (32.98%) and lack of cleanliness (21.75%) are the constraints limiting duck meat consumption (Kanagaraju et al. 2013).

16.3.2.16 Poor Government Supports

The Governments can sustain the nomadic duck farming mainly by way of: (i) capital funding, (ii) input supply (mainly ducklings and feed), (iii) disease diagnosis and other veterinary services, (iv) marketing support and (v) technical advice. In survey studies, these needs are felt by the farmers (Gopinathan et al. 2015; Veeramani et al. 2017). The government intervention is also needed to stabilize production price of inputs like feed and day-old-chicks. Government run feed mills and hatcheries and nongovernment establishments are required for the farmers to make available the feed and chicks at reasonable cost (Alam et al. 2012).

16.3.3 Rice-Duck System

In the integrated rice–duck farming (IRDF) system, rice paddies provide food (weeds and pests) for ducks, and ducks in turn play a role in fertilizing rice paddies. The system is practiced in Asian countries like Japan, Korea, Vietnam, China and Indonesia. Rice-duck culture is a form of organic farming; where weeds and insects are controlled effectively using ducks without the usage of pesticides. This method is popularly known as 'Aigamo rice cultivation' in Japan. The integration of rice cultivation with duck farming provides subsidiary products like duck meat and eggs along with paddy from main crop. The droppings of ducks provide almost all the essential nutrients to the rice crops and fertilize the land. A systematic review and SWOT analysis on integrated rice-duck farming in Asian countries by Suh (2014) gives critical insights on systems associating rice culture and ducks.

16.3.3.1 Fragmented Paddy Fields

The optimum stocking density in rice-duck farming is 150 ducks on one acre of land. The fragmented land ownership is the major constraints in duck-rice coculture. This can be overcome by implementing rice-duck at the community level with groups of farmers having adjacent plots, rather than by individual farmers. This reduces the management costs hugely and improves the efficiency of duck herding and protection from predators while reducing fencing costs (Khan et al. 2005).

16.3.3.2 Labour-Intensiveness

The most challengeable weakness of duck–rice system is its labour-intensiveness. Firstly, land preparation before transplanting for duck–rice farming requires extra care and is time consuming. Secondly, labour is required to take the ducklings to barns at night to protect from predators like foxes, dogs and eagles. The ducks can be kept in the rice paddies at night if there is no predator problem, because ducks are active at night. Thirdly, feeding of ducks regularly and monitoring the scavengeable food source in rice paddies for the ducks need labour. Fourthly, fencing is indispensable to keep predators out even at the daytime and to entice active paddling of ducks within confined rice fields. Fencing around every rice plot requires additional labour input (Furuno 2001).

16.3.3.3 Special Care for Rice Culture

The rice-duck coculture requires some special care compared to rice monoculture system. Firstly, line or row transplanting rather than random transplanting is to be followed, which enables easy paddling of ducks between rows and less damage to the rice crop. The paddy fields need to be puddled and levelled in such a way that the surface is evenly flat and has no islands or shallow spots. The water level also needs to be maintained as deep as the feet or bills of ducks can barely touch the paddy bed (Furuno 2001). Further, voracious paddling of ducks in a confined paddy field area can cause the paddy water muddy, which is helpful to not only the control of the germination and growth of weeds but also rice cultivation (Furuno 2009). The age of ducklings is also critical for successful rice-duck farming because mature ducks may overgraze rice paddies by eating the tips of rice leaves (Furuno 2001).

16.3.3.4 Fencing and Housing Infrastructure

Fencing around every unit of rice-duck involves additional labour input and cost. Due to additional cost, a growing number of integrated rice farmers in South Korea started using freshwater snails, *Cipangopaludina chinensis malleata* instead of ducks; because, unlike ducks freshwater snails do not necessitate fences, sheds and feeding regularly. The rice-snail farming method is known to be as effective for weeding as ducks (Rutz 2008).

16.3.3.5 Timing of Ducklings Production

The supply of ducklings at the right time is the key to success of duck-paddy integrated farming system. The availability of good quality ducklings in right quantity at the right age and on time in duck-rice farming is one of the major constraints. As the paddy cultivation is a seasonal activity, a huge number of grownup ducklings are required in a particular point in time. To tackle this problem, Khan et al. (2005) stressed the need for setting up village hatcheries or duckling production units for timely supply of ducklings to the rice-duck farms and these village hatcheries should have the capacity of supplying around 600 ducklings every week. In rice-duck farming system, the ducklings are reared in intensive system until they are 20–25 days of age when they can graze in the paddy fields. In Bangladesh, women generally conducted this household activity (Khan et al. 2005). The families involved in duck-rice co-culture purchase ducklings from outside or hatch out them themselves. The homemakers become experienced in this activity and the mortality rate significantly decreases as they are doing this activity year after year.

16.3.3.6 Climatic Hazards

The climate for ducklings exposed in paddies may be too cold in some countries. For example, in South Korea after seedling transplantation in late May or early June, the climate is still cold for newly introduced ducklings. The extreme weather changes vulnerably affect the ducklings in the rice fields causing mortality. Hossain et al. (2005) reported duckling mortality of only 8% and 12% in Barisal and Sylhet regions respectively and attained an average weight of 1050 g in Barisal and 990 grams in Sylhet at the time of crop flowering during T. Aman season. On the other hand, during Boro season, high mortality rates (18 and 17% respectively) and low mean weights (940 and 950 g respectively) which could be attributed as the effect of prevailing severe cold weather.

16.3.3.7 Overloaded Weeds

In some strange conditions in tropical Malaysia and Vietnam, wherein during initial days of rice-duck culture the weed populations build up is fast that is too much for the young ducklings to clear (Suh 2014). The ducklings should be allowed into the paddy field within a couple of weeks after transplanting for effective control of weeds from the start itself (Furuno 2001).

16.3.3.8 Availability of Biologicals and Disease Treatment Support

The timely availability and provision of duck vaccines is the essential element next to supply of ducklings in duck-rice integrated farming in developing countries (Khan et al. 2005). The availability of duck vaccine is a bottleneck in village conditions. If vaccines are available at village level, the farmers readily go for vaccination. The vaccine supply is the weakest link of the system in duck-rice coculture in countries like Bangladesh as described by Khan et al. (2005). It is of utmost importance to establish good links with the vaccine supplier in the region when selecting partners to implement rice-duck farming. The distribution system of duck vaccines could become more readily available to farmers at crucial times if private companies also produce duck vaccines in addition to government vaccine producing centres. Professional vaccinators can be trained or in some cases, village women can be trained to become vaccinators for all rice-duck farms in their community.

16.3.3.9 Capacity Building

There is a prevailing perception amongst non-rice-duck farmers that rice-duck farming is not suitable for large-scale rice farming due to the high labour-intensiveness. They prefer rice monoculture to rice-duck farming because of being misinformed or uninformed. Some farmers also believe that rice plant and ducks are mutually conflicting, and therefore should be separated from each other. Many people do not have technical knowledge on duck farming leading to improper management of the ducks also.

To optimize the rice-duck coculture, specific training is essential for paddy and duck farmers to integrate the two farming methods. In developing countries where, duck-rice farming is followed, the farmers need to be made aware of this. The misconception of ducks harming the rice and building confidence can be done by village group discussions, training to women in Front Line Demonstrations (FLD) and farmers' workshops. To educate farmers, communication materials such as videos, posters, leaflets and monthly newsletters are useful. In a successful Bangladesh model, there are many groups of men and women, village federations, farmer extension agents, community-learning centres and community libraries involved in implementation and management of field activities to ensure long-term sustainability. This enabled the farmers capable of managing rice-duck systems by themselves within few years (Khan et al. 2005). The local governments can also come up with training centres for the duck farmers.

16.3.3.10 Financing

The financing and government support is required to set up rice-duck farming by resource poor farmers in developing countries. The major investments in rice-duck are expenditures on shelter, feeding equipment for lean season feeding, fencing, feed and vaccines.

16.3.4 Fattening Duck Industry

The duck production system is different in France from South East Asia as the system of production and environmental features differ. During the 2016–2017 epizootic, most of the outbreaks were located in southwest France, a region known for its '*foie gras*' production, accounting for more than 70% of the world's *foie gras* production. The fattening duck production system in France is characterized by three stages, namely, rearing (indoor rearing of day-old to 3 weeks) and breeding (outdoor rearing with access to shelter from 4 to 12 weeks) followed by force-feeding (indoor overfeeding for a period of up to 12 days). The fattening ducks are reared in first two stages (rearing and breeding) in large flocks of several thousand during first 12 weeks of age. They are then divided into small batches of hundreds of ducks, moved to overfeeding units, and reared for next 12 days before slaughter. This small size of overfeeding units coupled with short production cycle facilitates direct and indirect contacts between initial rearing and fattening duck flocks and transport networks. The entire France was exposed to HPAI H5N8 despite the severely affected neighbouring countries such as Germany and Switzerland are located in eastern side, the clustered distribution of outbreaks is noted in southwest France. These findings of Guinat et al. (2019) shed light on the need of strengthening surveillance of duck production systems and on making transportation of ducks more secure in France. The biosecurity on farms and transport system need to be enhanced with additional disease surveillance strategies to support fattening duck industry. The fattening duck transport trucks should comply with transport guidelines of cleaning and disinfection of cages during the high-risk period between December and March. Diagnostic tests need to be performed on duck flocks before being moved to improve effectiveness in preventing the spread of HPAI virus.

16.3.5 Intensive System

Ducks are waterfowls, spends their time happily on water or around water. This makes the extensive rearing system much suited for this poultry species. Confinement rearing is another alternative to scavenging systems of rearing, through which vertical expansion of duck production is possible. The intensive production system was developed over the last 60 years. The breeding and broiler ducks are reared in closed buildings at relatively high densities. Rearing in total confinement involves keeping the birds on slatted floors made of wood, metal or plastic instead of even straw or wood-shaving litter, use of special lighting programs and the absence of ponds. The construction of buildings with impoverished environment for ducks and improved working conditions for workers creates some welfare problems.

16.3.5.1 Behavioural Anomalies

The behavioural anomalies in intensive units was reviewed by Raud and Faure (1994). Various forms of behavioural disorders observed among intensively reared ducks are mainly pecking feathers, fear to humans and leg problems.

(a) ***Pecking***

Pecking, which may degenerate into cannibalism in severe cases has a direct effect on the well-being of the birds. Feather pecking is mainly a problem in Muscovy ducks, especially when the ducks are kept on slatted floors without litter. Pulling of feathers of other ducks with skin injuries occurs frequently during the growth of the adult plumage in young ducks, or after moulting in breeders, which may necessitates a beak trimming, which is traumatic to the bird as the beak is innervated up to the tip (Rauch et al. 1993). Dim lighting (Heil and Torges 1990), and balanced nutrition and sufficient space allocation (Klemm et al. 1992) can avert this behaviour in intensive housing. Pecking of the cloacal region (vent pecking) is directed to attractive red regions, namely the cloaca of male and female breeders leading to cloacitis in the female ducks, which interrupts egg laying and females 'harassing' males by nipping their penis, sometimes results in irreversible mutilation. The impact of such behaviour has very deleterious effects on health and reproductive performance of the flock (Raud and Faure 1994).

(b) ***Fear and stress***

The fear reactions can lead to injuries and sometimes even death by suffocation as the birds pile up on top of each other. The injuries induced by sharp claws during transport can largely be prevented by claw trimming in Muscovy ducks (Dayen and Fiedler 1990). Henderson et al. (2001) showed that ducks appear to perceive an approaching human as a greater threat than an approaching object. Pekin ducks are more fearful than Muscovy ducks and had higher corticosterone levels following exposure to humans (Faure et al. 2003).

(c) ***Leg problems and body conditions***

The quality of the litter is important in deep litter system, because a bad litter quality (wet and dirty) has undesirable effects (Raud and Faure 1994). In male Muscovy ducks, splay leg or spraddle leg is a considerable problem, presumably associated with the high growth rate, low light intensities and restricted locomotion (Dayen and Fiedler 1990). The slatted floors of wood or metal or plastic in breeder rearing units cause problems like difficulty in balancing, particularly in duck lines selected for rapid growth, and skin irritation, particularly at the palmar junction of digits resulting in the development of calluses as the bird grows, which are often fissured (Raud and Faure 1994). Severe problems are seen in wire floors rather than plastic or wooden slats. The skeletal body is changed in spraddle leg ducks and that there are degenerative alterations of the muscles and bones and footpad dermatitis and injury in Muscovy ducks (Dayen and Fiedler 1990). Jones and Dawkins (2010) reported the deterioration of body condition of Pekin ducks in intensive commercial farms at 41 days of age with only 84% of ducks having clean eyes, 67% clean feathers and 79% no gait abnormalities. Gait worsened with increasing temperature, litter moisture and atmospheric ammonia concentrations. The incidence of footpad lesions is positively correlated with increasing humidity and ammonia.

16.3.5.2 High Feed Cost and Low Product Pricing

The high feed cost also hit the intensive production system more than scavenging system as the former relies only on stall-feeding. Begum et al. (2020) identified that higher price of feed was found to be the topmost serious problem in duck farming in the study area in small-scale intensive duck farmers of Bangladesh. The authors also identified lower price of duck egg and meat is the second major problem next to feed cost.

16.3.5.3 Marketing

The marketing system for duck described for Philippines by Chang and Dagaas (2004) is common for many developing countries. The duck production is fragmented and inefficient due to contributing factors such as: multilayered chain involving a large number of small market players; the absence of formal product standards and grading systems; and inadequate marketing infrastructure. This resulted in higher cost and lower product quality.

The identified issues related to duck production and marketing enterprises were: (1) declining demand for duck products due to intense competition with commercial chicken sector; (2) inefficient marketing system due to absence of product standards; (3) impact of increasing commercialization of smallholder enterprises and increasing dependence on formulated feeds and imported inputs and (4) threats from imports.

(a) ***Seasonality of demand***

The demand for duck eggs and other products is very seasonal and highly influenced by the availability of substitutes like chicken products, fish and other meats. The climatic factors with low egg and meat consumption in summer, religious fasting, festival seasons, outbreaks of epizootics, etc. also has influence on the duck products marketing. For example, the demand for balut, penoy and salted eggs in Philippines is seasonal in different months of a year and the egg production also fluctuates in an annual rhythm. This leads to price variability which in turn tends to distort the price signal leading to bust and boom production cycles (Chang et al. 2003). The intensive production houses are highly affected by the bust and boom price cycles due to high scale of production concentrated in a close circle.

(b) ***Absence of grading system***

There exists no grading system for duck products in many countries. Although there are some grading schemes existing in countries like Philippines only for fresh eggs, no such criteria is available for balut, penoy and salted eggs. The products supplied are usually not consistent in quality if no grading system is available; which tend to have a negative impact on demand (Maravilla 1994). In some developing countries, despite grades and standards exist they are not enforced. Chang et al. (2003) described the consequences of absence of quality standards and market information for duck products.

(c) ***Absence of market information system***

The market information system in developing countries is very much undeveloped. Prices of a product between two markets may differ significantly even if they are just a few kilometers apart. For example, in Philippines, 'viajeros' or middlemen who are better informed exploit the producers. The lack of market information leads to acute regional variations even between two closely located markets. The asymmetric information tends to place producers in a disadvantageous position when it comes to price negotiation leading to 'boom and bust cycles'. In such cases, production decisions are based on personal intuition or past experience rather than market conditions. As a result, the producers tend to miss the peak or hit the bottom of the market. The results are a less stable market and a less profitable operation (Chang et al. 2003).

(d) ***Inefficient distribution and marketing***

The marketing chain of ducks relies mainly on traditional and much less efficient marketing channels unlike relatively short, efficient and integrated marketing channel of commercial chickens. The duck eggs tend to pass through several intermediaries before reaching the final consumer. In an imperfect market where market information and quality standards are lacking, a few players, who overcharge for their services, dominate markets. The efficient marketing system depends on accurate price signals; whereas, the inefficient marketing creates underpaid producers who tend to produce less while an overcharged consumer who tend to consume less (Chang et al. 2003).

(e) ***Local taxes***

In some countries, the registered duck farms need to pay more tax. The duck farms in Cambodia, for example, evade paying of tax or any other unofficial payments by skipping the registration of their businesses (FAO 2009). The farmers should be educated and encouraged to register their firms with government to receive the supports; however, agribusiness like duck farming need to be exempted from taxation or to be kept at low slab of taxation.

16.3.5.4 Capacity Building

Most of the duck farmers have low technical knowledge about duck farming; therefore, intensive training should be provided to make duck farming more profitable. Short training programs can be arranged with the involvement of government and nongovernment authorities (Alam et al. 2012). Lack of extension services was considered as the third serious problem in Bangladesh among small-scale intensive duck farmers next to feed cost and low product price (Begum et al. 2020).

16.4 Policy Recommendations

The duck farming in developing nations is in deep crisis due to high pricing and scarcity of inputs, high production cost of eggs and meat and emerging, reemerging and trans-boundary diseases, which resulted in negative trend in growth during recent past. Hence, it has become necessary to safeguard the interest of duck farming through comprehensive action plans and policies for economic and nutritional security.

16.4.1 Conservation of Duck Genetic Resources

A detailed description on wealth of duck genetic resources, their distribution, production potential and risk status is given by Liu and Churchil (2021). The Domestic Animal Diversity Information System of FAO enlists 400 common duck and 63 Muscovy duck genetic groups of breeds, strains, lines, etc. (DAD-IS 2020). Out of total 400 duck genetic groups of common ducks, 130 are endangered, four are endangered maintained, 20 are critical, one is critical maintained, three are vulnerable and 16 are extinct. Among the 63 genetic groups of Muscovy duck, one is endangered. The conservation efforts are necessary to check the erosion of duck genetic diversity occurring due to substitution of local breeds by exotic breeds or uncontrolled crossbreeding or farmers'/consumers' preference. This is the high time that the local governments need to involve in conservation of duck genetic resources by in situ orgamete cryobanking methods. Although, in situ live bird conservation carries the risk of loss due to pathogen outbreaks, genetic problems, breeding cessation and natural disasters, this system is robust that preserves the birds in natural state in the habitat where they naturally occur, maintains the genetic diversity extant in the population and makes the samples of the preserved material readily available. The structure of avian eggs makes the cryopreservation of female gamete (ova) and fertilized embryos impossible and that prevents conservation of the W chromosome and mitochondrial DNA. PGC transplantation combined with recent advanced PGC manipulation techniques (Nakamura 2016) enable ex situ conservation of poultry genetic resources in poultry PGC-banks. The conservation of domestic duck diversity must encompass identifying, characterizing and monitoring for their best short-term use and ensuring their long-term ready availability. The local governments have paramount responsibility in this endeavour.

16.4.2 Supply of Critical Inputs

The rural duck farming is jeopardized mainly by the lack of availability of critical inputs like birds, feed, capital investment, advancements in technical know-how, and veterinary services in developing countries.

16.4.2.1 Birds

Good quality duckling is one of the critical inputs for duck farming; its nonavailability at the required point in time greatly hampers this sector. The government and nongovernment interventions are required to stabilize the production and supply of ducklings/grown-up ducks by setting up of hatcheries and nurseries in places where duck farmers have easy access. The government can incentivize private players to ramp up production and supply of ducklings throughout the country to bridge the gap between demand and supply at the time of need (Alam et al. 2012). Region specific germplasm, *desi* breeds or hybrid crosses depending upon the adaptability of birds to the local conditions and acceptability of farmers can be propagated for supply. Breeding research is necessary to develop genetically improved duck varieties suitable to the rural areas to get higher production.

16.4.2.2 Feed

Feed constitute more than two-third of production cost in intensive duck production and also plays an important role in extensive duck farming in the form of supplemental or lean season feeding. The nonavailability of duck specific feeds at village level can be addressed by setting up of feed mills in different locations to supply quality feed at reasonable cost (Alam et al. 2012). Research is also needed to assess the nutrient requirements of local germplasm and the utility of locally available nonconventional feedstuffs.

16.4.2.3 Biologicals and Medicines

Duck farmers do not have sufficient knowledge of disease management. They normally do not know the symptoms of duck diseases and remedial measures. Any short supply of vaccines, especially of duck plague may incur huge losses to the farmers. Vaccines and medicines for ducks should be made available in the local markets or from the government veterinary services round the year to ensure routine vaccination against common diseases (Islam et al. 2016).

16.4.2.4 Financial Support

Nomadic duck farming is a seasonal operation that requires high investment during the start in every season. Providing low interest microfinance will break the shackle of local moneylenders and can bring revolution in the duck farming. Loans free from collateral, short processing time, flexible repayment, minimum terms and conditions, door-to-door service and availability just in time of need are necessary to solve the insufficiency of working capital. In nomadic duck farming, lending money from public sector banks to the resource poor nomadic farmers can save them from being exploited by the vendors. The financial support to duck farmers will help them to expand their business as well. The farmers will be able to buy modern equipment and make better housing facilities for their birds (Alam et al. 2012). The loans provided to the farmers in Hoar area of Bangladesh by a nongovernmental development organization transformed the village duck farming. The loans were made available to the farmers without any precondition on selling produce to, or purchasing inputs from any selective person, so that farmers were able to sell his/her produce in local market at competitive prices. (Islam et al. 2012).

16.4.2.5 Veterinary Services

The veterinary services are either not available or not given due priority in the remote duck growing areas in developing countries. The disease outbreaks causing mortality of ducks is one of the major problems; therefore, preventive measures are necessary thorough vaccination program. In order to provide necessary veterinary services to the duck farmers, the government should establish veterinary care centres with adequate machineries, vaccines, medicines and technical staff in duck farming pockets (Alam et al. 2012). All types of duck vaccines should be made available at the doorsteps of the farmers at minimum price (Alam et al. 2012).

16.4.3 Farmer Producer Groups for Collective Bargaining Power

The collective bargaining power of the farmers is very low, when they operate individually, no matter how large in numbers. The formation of Village Organizations (VO) consisting of a group of farmers promptly guided with counselling and training is necessary to develop cooperative sense among the farmers. Using the power of being united, farmers can sell their produce jointly in the market in large numbers with high power of bargaining. Buying medicines and vaccines jointly by the group members will be cost-effective and in turn benefit them economically. As a backward linkage, supports like supply of ducklings or pullet ducks, feed, vaccine and technical or duck healthcare services can be extended to the farmers. The marketing of farm products like eggs and live birds for meat purpose can be provided by forward linkage support. Passing information on commercial hatcheries, feed companies, sources of pharmaceutical products, live bird sellers and large traders in their locality can be easily done if groups are formed to develop backward and forward market linkages among duck farmers.

16.4.4 Establishment of Quality Standards and Grading

The grading of duck egg, meat and their products and the tolerance levels are need to be fixed by every country for promoting good manufacturing practices among the farmers. The grading standards for external and internal characteristics of eggs, tolerance levels of drug and pesticide residues and other contaminants, standards for microbial contamination, codes for different grades and labelling procedures are to be established and enforced in each country.

Developing and promoting appropriate product standards by the government and grading and labelling schemes for individual duck products is one possible solution to market failure as a result of imperfect market information. Not only standardization and grading can facilitate trade and reduce marketing costs, but it also helps to reduce the market powers enjoyed by bigger players who may be able to take advantage of more arbitrary or less rigorous grading and pricing systems (Chang and Dagaas 2004). Establishment of official product standards will improve market efficiency for duck eggs, meat and their products (Chang et al. 2003; Chang and Dagaas 2004; Alam et al. 2012). The countries need to set specifically standards where USDA grading system may not be available or suitable for locally important specialty products; for example, a system that is specific to balut is needed in Philippines (Chang and Dagaas 2004).

16.4.5 Marketing Support

Developing a market system where the duck farmers can sell their products at a comparatively good price is also important. Low price for meat and egg of ducks and its frequent fluctuations discourage the duck farmers. Price stabilization should be ensured for the producers and other stakeholders. The Governments should strictly prohibit illegal trading of duck produces from neighbouring countries (Alam et al. 2012). The government intervention, whether direct or indirect, is minimal in countries in which the marketing system is in the hands of private sector. This is why marketing services such as research and development, market information, product standards and marketing infrastructure are not sufficiently provided by the Government and the food marketing system in those countries are under-regulated (Chang and Dagaas 2004). Market research to be taken up by the local Governments in technical details which affect the quality of the products, for example, links between the quality of the breeder stock, the quality of fresh eggs and the quality of balut. These linkages are important in determining the market value of each of these key products and making sure their prices are consistent with their values to the supply chain (Chang and Dagaas 2004). Support should be given for market research on demand trends and market competition and development of new technologies to improve product quality and shelf life of duck (Chang et al. 2003; Chang and Dagaas 2004). Measures like curb on imports may be necessary to protect domestic market. About 19% of respondents in Cambodia are concerned about importing ducks and their products from other countries, as this makes their own businesses difficult (FAO 2009).

16.4.6 Training

Most of the duck farmers are from lower strata of the society and have low technical knowledge about duck farming. Therefore, intensive training on farm management and duck diseases and their control should be provided by Department of Veterinary

Services (Alam et al. 2012). Trainings are also necessary on scientific feeding of duck to get better production (Islam et al. 2016). Short training programs arranged by government or nongovernment authorities can then increase the profit from duck farming. Long-term focused hands-on trainings on techniques like feed formulation, vaccination and hatchery management are also necessary to develop skilled workers that can help the farmers around. For this purpose, governments should come up with training centres exclusively for the duck farmers (Alam et al. 2012). Experts can be hired to conduct on- and off-farm trainings using multimedia presentations on modern farming practices and emerging scientific developments.

16.4.7 Other Government Supports

Apart from technical, financial, input supply, market and veterinary supports already described in this chapter, the local governments should play proactive role in promoting duck farming through policy initiatives.

1. Establishing National Research Centre on Ducks at country level as the nodal agency for conceiving, monitoring and evaluating duck research (Ramakrishnan 1996).
2. The duck farmers should be assured of support prices for their products (eggs and meat) so as to tide over difficult situations of price slumps.
3. The consumption of duck products should be promoted by making wholesome duck egg, meat and their products among nutritionally weaker sections to alleviate malnutrition. Duck eggs can be introduced in midday meal schemes in schools to counter protein deficiency in children.
4. Promotion of establishment of small-scale processing and value addition units with government subsidies in duck farming hubs.
5. The government can establish cold storage warehouses in different parts of the country and the facility has to be extended to the farmers at minimum rent to stock their products during surplus season.

16.5 Conclusion

Duck production traditionally plays a vital role in the history and culture of rice producing countries in Asia. It provides women empowerment and livelihood opportunities for weaker sections in developing countries. Although the traditional duck farming have many advantages like easy to manage, hardy to diseases and cold climate, ability to flourish in marshy area and produce more eggs, it faces stiff competition from commercial poultry as the economy develops and lifestyle of people changes. The crucial issues confronting duck farming are: the lack of supply of quality breeder stocks and an inefficient marketing system. Many of the problems, on close examination, are in fact a manifestation of market imperfection and a reflection of an under-regulated marketing system. Since an efficient marketing system depends on market information and marketing infrastructure, the absence of pertinent marketing services has prevented the market from functioning properly and resulted in misallocation of resources and market inefficiency. However, to improve efficiency, more research is needed in several areas. Areas for duck research should focus on identifying issues and solutions facing the commercial sector and the smallholders, better understanding of consumer demand, developing product standards, and improving generation and dissemination of market intelligence. Research should also focus on the selection of breeds, disease control, feeds and feeding and waste management that suit smallholders and different scale of duck production, as well as utilization of agricultural by-products in duck farming and new product development from duck egg and meat. Support measures in any form to duck farming will result in improvement of livelihoods, employment generation, food security and women empowerment.

References

Ahmed S, Haque TS, Rahman S, Sultana MR, Rahman MM, Islam MS, Kabir A, Moni MIZ, Haque MH, Khaton R (2019) Loss of domestic poultry due to flood and the state of veterinary Care Services in Flood-Prone Areas of Bangladesh. Int J Poult Sci 18:231–237

Alam MB, Uddin ABMS, Bablu MAZH, Kamaly MHK, Rahaman MM (2012) Socio-economic profile of duck farmers and duck management practices in Rajshahi region, Bangladesh. J Anim Sci 41:96–105

Annual Report (2002) National Agricultural Technology Project on 'Productivity Enhancement of Ducks', Centre for Advanced Studies in Poultry Science (Lead Centre), Kerala Agricultural University, Mannuthy, Thrissur, Kerala

ANSES (2021) Avian influenza in six questions. https://www.anses.fr/en/content/avian-influenza-six-questions. Accessed on May 7 2021

Arboleda CR, Coligado EC, Alejar AS, Luis ES, Novero RP, Capitan SS (1985) Breeding and management of ducks for meat and egg production. Terminal Report, NSTA Assisted, Project No. 7901Ag

BAHS (2019) Basic animal husbandry statistics. Department of animal Husbandry and Dairying, Ministry of Fisheries, Animal Husbandry and Dairying, Government of India

Basuno E (1983) Small-scale animal husbandry at the village level: a case study in Desa Pandansari, Kecamatan Ciawi, Bogor, West Java. M. Phil. Thesis, Griffith University, Brisbane, Australia

Beaudoin AL, Kitikoon P, Schreiner PJ, Singer RS, Sasipreeyajan J, Amonsin A, Gramer MR, Pakinsee S, Bender JB (2014) Risk factors for exposure to influenza a viruses, including subtype H5 viruses, in Thai free-grazing ducks. Transbound Emerg Dis 61(4):362–374

Begum M, Farid MS, Rahman MM, Barua S, Begum M, Sharkar MS (2020) Competitive advantage and constraints associated with duck farming in North Eastern region of Bangladesh. IOSR J Huma Soc Sci (IOSR-JHSS)

BER (2012) Bangladesh economic review, ministry of finance. Government of the People's Republic of Bangladesh, Dhaka

Bhattacharyya P, Chakraborty A, Bhattacharya B, Chakrabarti K (2003) Evaluation of MSW compost as a component of integrated nutrient management in wetland rice. Compost Sci Utiliz 11(4):343–350

Brasher M (2021) Understanding Waterfowl: Winter Weather and Waterfowl, Ducks unlimited. https://www.ducks.org/conservation/waterfowl-research-science/understanding-waterfowl-winter-weather-and-waterfowl. Accessed on 16 Jan 2021

Bui XM (2010) Duck farming systems and avian influenza in the Mekong Delta of Vietnam. Food and Agriculture Organization of the United Nations, Rome

Burgos S Hinrichs J, Otte J, Pfeiffer D, Roland-Holst D (2008) Poultry, HPAI and Livelihoods in Vietnam – a review Mekong Team Working Paper No. 2 of A Collaborative Research Project Funded by Department of International Development, Rome https://assets.publishing.service.gov.uk/media/57a08bbbed915d622c000e6d/wp02_2008.pdf. Accessed on 14 January 2021

Burgress EC, Ossa J, Yuill TM (1979) Duck plague: a carrier state in waterfowl. Avian Dis 24:940–949

Cagauan AG (1997) Final report of research project on 'Integrated rice-fish azolla-duck farming system' funded by the Food and Agriculture Organization (FAO), Catholic University of Louvain, Belgium and Freshwater Aquaculture Center, Central Luzon State University Philippines, p 265

Cagauan AG Branckaert RDS van Hove C (2000) Integrating fish and azolla into rice–duck farming in Asia Naga the ICLARM Quart 20(1):4–10

Chang, HS, Dagaas CT (2004) The Philippine Duck Industry: Issues and research needs Working Paper Series in Agricultural and Resource Economics ISBN: 1863898794. University of New England Graduate School of Agricultural and Resource Economics and School of Economics. https://ageconsearch.umn.edu/record/12904?ln=en. Accessed on 9 July 2020

Chang, HS, Dagaas C, Castro N, Ranola R, Lambio A, Malabayuabas ML (2003) An overview of the Philippine duck industry. Paper presented to the 47th Annual Conference of the Australian Agricultural and Resource Economics Society at Fremantle, February 12–14, 2003

CIVAS (2006) A report from the Centre for Indonesian Veterinary Analytical Studies (CIVAS) for the Food and Agricultural Organization (FAO) of the United Nations, http://www.fao.org/docs/eims/upload/213701/agal_duckfarmingindonesia_210906.pdf. Accessed on 9 Jan 2021

DAD-IS (2020) Domestic animal diversity information system, Food and Agriculture Organization of the United Nations. http://www.fao.org/dad-is/dataexport/en/. Accessed 25 May 2020

Dayen M, Fiedler HH (1990) Intensive raising of Muscovy ducks. Dtsch Tierarztl Wochenschr 106: 55–59

FAO (2007) Organically produced foods. 3rd ed, Food and Agriculture Organization of the United Nations, Vialedelle Terme di Caracalla, 00153, Rome

FAO (2009) Investigation of duck production and hatcheries, and duckling supply in Cambodia. Prepared by Khieu Borin, Pok Samkol and Olaf Thieme. AHBL – Promoting strategies for prevention and control of HPAI, Rome

FAO (2010) Duck farming systems and avian influenza in the Mekong Delta of Vietnam, by Bui Xuan Men. FAO Smallholder Poultry Production Paper No. 1, Rome

FAOSTAT (2019) FAOSTAT-Data-Production-Live Animals-Ducks, Food and Agriculture Organization of the United Nations, Rome, Italy Available: http://www.fao.org/faostat/en/#data/QA. Accessed 27 Apr 2021

Faure JM, Val-Laillet D, Guy G, Bernadet M-D, Guémené D (2003) Fear and stress reactions in two species of duck and their hybrid. Horm Behav 43:568–572

Furuno T (1996) Significance and practice of integrated rice cultivation and duck farming-sustainable agriculture. Kyushu international center, Japan International Cooperation Agency and Kitakyushu Forum on Asian Women, p 12

Furuno T (2001) The power of duck: integrated rice and duck farming. Tagari, Sisters Creek, Tasmania

Furuno T (2009) Traditional and modern matters on rice cultivation associated with duck. Paper presented at the 6th International Rice–Duck Conference held at Cebu, the Philippines, 23–27 February, 2009

Gajendran K, Karthickeyan SMK (2009) Status of indigenous ducks in Tamil Nadu, India: a traditional perspective. Livest Res Rural Dev 21:Article #175. http://www.lrrd.org/lrrd21/10/gaje21175.htm. Accessed 5 July 2020

Ghosh S, Haider N, Khan MKI (2012) Status of Household's ducks and their associated factors under scavenging system in a Southern Area of Bangladesh. Int J Nat Sci 2(4):108–112

Gilbert M, Chaitaweesub P, Parakamawongsa T, Premashthira S, Tiensin T, Kalpravidh W, Wagner H, Slingenbergh J (2006) Free-grazing ducks and highly pathogenic avian influenza, Thailand. Emerg Infect Dis 12(2):227–234

Gopinathan A, Murugan M, Sivakumar T (2015) A farming system perspective of rural duck farming of Tamil Nadu. Shanlax Int J Vet Sci 2:10–14

Guinat C, Artois J, Bronner A, Guérin JL, Gilbert M, Paul MC (2019) Duck production systems and highly pathogenic avian infuenza H5N8 in France, 2016–2017. Sci Rep 9:6177. https://doi.org/10.1038/s41598-019-42607-x

Halder G, Ghoshal TK, Samanta G (2007) Socio-economic background of duck owners and status of duck rearing in West Bengal. Indian Res J Ext Educ 7(2):55–59

Heil G, Torges HG (1990) Effect of light intensity on growth, activity and feather pecking of Muscovy ducks from two commercial strains. In: 8th European Poultry Conference, Barcelona, 26–28 June. Feria de Barcelona, pp 634–636

Henderson JV, Nicol CJ, Lines JA, White RP, Wathes CM (2001) Behaviour of domestic ducks exposed to mobile predator stimuli. 1. Flock responses. Br Poult Sci 42:433–438

Henning J, Henning KA, Long NT, Ha NT, Vu le T, Meers J (2013) Characteristics of two duck farming systems in the Mekong Delta of Vietnam: stationary flocks and moving flocks, and their potential relevance to the spread of highly pathogenic avian influenza. Trop Anim Health Prod 45(3):837–848

Henning J, Pfeiffer DU, Stevenson M, Yulianto D, Priyono W, Meers J (2016) Who is spreading avian influenza in the moving duck flock farming network of Indonesia? PLoS One 11(3): e0152123

Hoque MA, Skerratt LF, Cook AJC, Khan SA, Grace D, Alam MR, Vidal-Diez A, Debnath NC (2011) Factors limiting the health of semi-scavenging ducks in Bangladesh. Trop Anim Health Prod 43:441–450

Hossain M, Bayes A (2009) Rural economy and livelihoods: insights from Bangladesh AH Development Publishing House, Bangladesh

Hossain ST, Sugimoto H, Ahmed GJU, Islam MR (2005) Effect of integrated Rice-duck farming on Rice yield, farm productivity, and Rice-provisioning ability of farmers. Asian J Agric Dev 2:79–86

Huang Z, Zhang J, Liang K, Quan G, Zhao B (2012) Mechanical stimulation of duck on rice phytomorphology in rice-duck farming system. Chin J Eco Agric 20(6):717–722

Hulse-Post DJ, Sturm-Ramirez KMS, Humberd J, Seiler P, Govorkova EA, Krauss S, Scholtissek C, Puthavathana P, Buranathai C, Nguyen TD, Long HT, Naipospos TSP, Chen H, Ellis TM, Guan Y, Peiris JSM, Webster RG (2005) Role of domestic ducks in the propagation and biological evolution of highly pathogenic H5N1 influenza viruses in Asia. Proc Natl Acad Sci U S A 102:10682–10687

Huque QME (1991) Duck production system in Bangladesh. Asian Livest 16(2):18–23

Huque KS, Sultana N (2002) Study on the existing duck production systems in Bangladesh. A report of Bangladesh Livestock Research Institute, Bangladesh

Huque QME, Ukil MA, Hossain MJ (1993) Supplementary feeding of laying ducks, under scavenging condition. Bangladesh J Livest Res 1:57–62

Huque KS, Saleque Md A, Khatun R (2011) Commercial poultry production in Bangladesh. In: Proceedings of 7th international show and seminar organized by World's Poultry Science Association (WPSA), Bangladesh branch

Islam MM, Bhuiyan MNK, Harun MY (2012) Development of value chain: an effective way of profitable duck farming in Haor areas of Bangladesh. INFPD good practices of family poultry production Note No 04. http://www.fao.org/3/aq631e/aq631e.pdf. Accessed 10 Apr 2021

Islam MA, Howlider MAR, Alam MA, Heyamet MA, Debnath M (2016) Present status, problem and prospect of duck farming in rural areas of Mymensingh district, Bangladesh. Asian J Med Biol Res 2:202–212

Jalaludeen A, Churchil RR (2006) Duck eggs and their nutritive value, Poult Line, October 2006, pp 36–39

Jalaludeen A, Churchil RR (2021) Duck production: an overview. In: Jalaludeen A, Richard Churchil R, Baéza E (eds) Duck production and management strategies. Springer Nature. (in press)
Jalaludeen A, Peethambaran PA, Leo J, Lal JJ (2003) Socio-economic status and indigenous knowledge of duck farmers of Kerala, India. Paper presented at 2nd World Waterfowl Conference, 7–9 October, Alexandria, Egypt
Jalaludeen A, Peethambaran PA, Joseph L, Manomohan CB (2004) Duck production in Kerala NATP on productivity enhancement of ducks, Centre for advanced studies in poultry science, College of veterinary and animal sciences, Kerala Agricultural University, Mannuthy, Thrissur, Kerala, India, pp 44
Jalaludeen A, Churchil RR, Joseph L, Anitha P (2009) Duck meat egg and their products. In: Souvenir of IV world waterfowl conference, Thrissur, pp 57–64
Jones TA, Dawkins MS (2010) Environment and management factors affecting Pekin duck production. Br Poult Sci 51:12–21
Journey WK, Skillicorn P, Spira W (1993) Duckweed aquaculture: a new aquatic farming system for developing countries. The World Bank, Washington. Available from http://documents1.worldbank.org/curated/en/499791468739284085/pdf/multi-page.pdf. Accessed on 7 May 2021
Kaewtapee C, Prahkarnkaeo K, Bunchasak C (2018) Effect of sex on growth curve production performance and carcass quality of Cherry valley ducks. J Appl Anim Sci 11(2):9–18
Kanagaraju P, Churchil RR, Rajini A, Rathnapraba S, Veeramani P (2013) Factors influencing consumption pattern of duck and duck products among people of Kerala. Indian Vet J 90(3):137–138
Khan MA, Ahmed GJU, Magor NP, Salahuddin A (2005) Integrated rice-duck: a new farming system for Bangladesh. In: Van Mele P, Salahuddin A, Magor NP (eds) Innovations in rural extension: case studies from Bangladesh. CABI Publishing, Oxfordshire, pp 143–155
Khan KA, Saha S, Hossain MT, Haque ME, Haq MM, Islam MA (2018) Epidemiological investigation of recurrent outbreaks of duck plague in selected Haor (wetland) areas of Bangladesh. J Adv Vet Anim Res 5(2):131–139
Khanum J, Chwalibog A, Huque KS (2005) Study on rural duck production systems in selected areas of Bangladesh. Livest Res Rural Dev 17(10):Article#113. http://www.lrrd.org/lrrd17/10/khan17113.htm. Accessed 10 Jan 2021
Klemm R, Pingel H, Reiter K, Bierschenk F, Rauch W (1992) Investigations on feather pecking in Muscovy ducks. In: Ponsen, Looijen (eds) Proceedings of 19th World Poultry Congress, Amsterdam, vol 2, 19–24 September, Wageningen, 390–393
Kolluri G, Ramamurthy N, Churchil RR, Raj GD, Kannaki TR (2014) Influence of age sex and rearing systems on toll-like receptor 7 (TLR7) expression pattern in gut lung and lymphoid tissues of indigenous ducks. Br Poult Sci 55(1):59–67
Kolluri G, Ramamurthy N, Churchil RR, Sundaresan A, Gawdaman G (2015) Carcass studies in native ducks reared under different housing systems. Indian J Vet Anim Sci Res 44(1):1–11
Kumar P, Churchil RR, Jalaludeen A, Narayanankutty K, Joseph L, Kannan A, Anitha P (2013) A survey on village chicken production in Kerala state of India. Worlds Poult Sci J 69(4):917–930
Lambio AL (2001) Status, problems and prospect of the duck industry in the Philippines Unpublished manuscript, Institute of Animal Science, UPLB-CA, College, Laguna
Liu MHC, Churchil RR (2021) Duck genetics and breeding. In: Jalaludeen A, Richard Churchil R, Baéza E (eds) Duck production and management strategies. Springer Nature. (in press)
Lu JJ (1999) Development of protein digestive enzymes and effects of nutrient treatment on their activities in broilers and mule ducks. PhD thesis, National Taiwan University, Taipei, pp 28–36
Maijer AM (1987) Backyard poultry in the coastal area of rural Bangladesh. Department of Tropical Animal Production, National Agricultural University, Wageningen
Manda M (1992) Paddy rice cultivation using crossbred ducks. Farming Jpn 26(4):35–42
Maravilla JS (1994) A micro enterprise study: the case of 'Balut' marketing in Calamba, Laguna. Special problem report, Department of Agribusiness and Management, CEM, UPLB, College, Laguna

McCullough C (2019) South Africa: Challenges of free-range duck farming, Poultry World, June 5, 2019. Retrieved from https://www.poultryworld.net/Meat/Articles/2019/6/South-Africa-Challenges-of-free-range-duck-farm-farming-435078E/. Accessed 5 July 2020

Meyer A, Dinh TX, Nhu TV, Pham LT, Newman S, Nguyen TTT, Pfeiffer DU, Timothée V (2017) Movement and contact patterns of long-distance free-grazing ducks and avian influenza persistence in Vietnam. PLoS One 12(6):e0178241. https://doi.org/10.1371/journal.pone.0178241

Minh PQ, Stevenson MA, Schauer B, Morris RS, Quy TDA (2010) Description of the management of itinerant grazing ducks in the Mekong river delta of Vietnam. Prev Vet Med 94:101–107

Nakamura Y (2016) Poultry genetic resource conservation using primordial germ cells. J Reprod Dev 62:431–437

Orr A, Seema F, Nabi SA, Peter JI (2004) From technology to livelihoods: case studies of four PETRRA sub-projects, Consultancy report. IRRI, Dhaka

Perkins LE, Swayne DE (2003) Comparative susceptibility of selected avian and mammalian species to a Hong Kong-origin H5N1high-pathogenicity avian influenza virus. Avian Dis 47: 956–967

Pervin W, Chowdhury SD, Hasnath MR, Khan MJ, Ali MA, Raha SK (2013) Duck production strategy and profile of duck farmers in the coastal areas of Bangladesh. Livest Res Rural Dev 25: Article #129. http://www.lrrd.org/lrrd25/7/perv25129.htm. Accessed 16 Jan 2021

Pirdashti H, Pirdashti M, Mohammadi M, Baigi MG, Movagharnejad K (2015) Efficient use of energy through organic rice–duck mutualism system. Agro Sustain Dev 35:1489–1497

Premalatha N, Ramya R (2016) Nitrate poisoning in ducks. Haryana. Veterinarian 55:237–238

Rahman MM, Khan MJ, Chowdhury SD, Akbar MA (2009) Duck rearing system in southern coastal districts of Bangladesh. Bangladesh J Anim Sci 38(1&2):132–141

Ramachandran CK, Ramakrishnan A (1982) A survey on the status of duck farming in suburbs of Trissur town. Poult Guide 19(9):65–68

Ramakrishnan R (1996) Duck production scenario in India – scope for improvement, In: Proceedings: XX World's poultry congress Vol I, World's Poultry Science Association & India Branch, New Delhi, India, 2–5 September, 1996. https://naldc.nal.usda.gov/download/CAT10862694/PDF. Accessed 7 May 2021

Ramrao WY, Tiwari SP, Saraswat SS, Pathak R, Gupta R (2008) Integration of crop-livestock-poultry-duck farming system by large farmers in the plain tribal areas of Chhattisgarh in Central India. Livest Res Rural Dev 20:Article #42. http://www.lrrd.org/lrrd20/3/ramr20042.htm. Accessed 24 Apr 2021

Rauch HW, Pingel H, Bilsing A (1993) Welfare of waterfowl. In: Savory CJ (ed) Proceedings of 4th European Symposium on Poultry Welfare, Edinburgh, 18–21 September. Animal and Food Research Council, Roslin Research Station

Raud H, Faure JM (1994) Welfare of ducks in intensive units. Rev Sci Tech Int Epiz 13(1):125–129

Ravindran TK (1983) A survey on the status of duck farming in Kerala state. MVSc Thesis submitted to Kerala Agricultural University, Mannuthy, Thrissur, Kerala, India

Roy BC (2000) Identification of constraints of raising broiler of small flock size and determination of their remadial measures. MS thesis, Department of Poultry Science, Bangladesh Agricultural University, Mymensingh, Bangladesh, pp 34–50

Rutz D (2008) Adaptation and evaluation of integrated rice and duck farming in the Hongdong Community of South Korea and the Poolmoo Schools. Masters thesis, Department of Geography, The University of Zurich, Zurich, Switzerland

Saleque MA, Mustafa S (1996) Landless women and poultry: the BRAC model in Bangladesh, In: Dolberg F, Petersen PH (eds), Integrated farming in human development, Proceeding of the workshop, 25–29 March 1996, Tune, Landboskole, Denmark, pp 38–55

Sankaralingam S, Mahanta JD (2021) Nomadic (transhumant) duck farming practices. In: Jalaludeen A, Richard Churchil R, Baéza E (eds) Duck production and management strategies. Springer Nature. (in press)

Sarkar S, Khan SU, Mikolon A, Rahman MZ, Abedin J, Zeidner N, Sturm-Ramirez K, Luby SP (2017) An epidemiological study of avian influenza A (H5) virus in nomadic ducks and their raising practices in northeastern Bangladesh. Influenza Other Respir Viruses 11:275–282

Shawky S, Schat KA (2002) Latency sites and reactivation of duck enteritis virus. Avian Dis 46: 308–313

Sheng F, Cao CG, Li CF (2018) Integrated rice-duck farming decreases global warming potential and increases net ecosystem economic budget in Central China. Environ Sci Pollut Res 25: 22744–22753

Suh J (2014) Theory and reality of integrated rice–duck farming in Asian developing countries: a systematic review and SWOT analysis. Agric Syst 125:74–81

Tamizhkumaran J, Rao SVN, Natchimuthu K (2013) Nomadic duck rearing in and around Puducherry region - an explorative study. Indian J Poult Sci 48:377–382

Uddin MJ, Tarin TS, Zaman MH, Uddin AHMM (2020) Scavenging duck at Haor Basin: plane of nutrition and response to production. Int J Adv Res Innov Ideas Educ 6:935–942

van Groenigen JW, Burns EG, Eadie JM, Horwath WR, van Kessel C (2003) Effects of foraging waterfowl in winter flooded rice fields on weed stress and residue decomposition. Agric Ecosyst Environ 95:289–296

Veeramani P, Prabakaran R, Sivaselvam SN, Sivakumar T, Selvan ST, Senthilkumar K (2017) Constraints in duck farming practices in northern districts of Tamil Nadu. Indian Vet J 94(04):12–14

Vignesh KS, Saravana Pandian A, Veeramani P, Prabu M (2020) Feeding practices of ducks and ducklings and their productivity in north-eastern zone of Tamil Nadu. J Entomol Zool Stud 8(4):2415–2417

Wei SH, Qian S, Ma B, Wei JG, Chen JW, Wu JQ (2005) Influence of rice–duck and other weed methods on weed community. Chin J Appl Ecol 16:1067–1071

Yuyun I, Wibawa H, Setiaji G, Kusumastuti TA, Nugroho WS (2020) Determining highly pathogenic H5 avian influenza clade 2.3.3.1c seroprevalence in ducks, Purbalingga, Central Java, Indonesia. Vet World 13:1138–1144

Zhang J, Xu R, Chen X, Quan G (2009) Effects of duck activities on a weed community under a transplanted rice–duck farming system in Southern China. Weed Biol Manag 9:250–257

Zheng YH, Deng GB, Lu GM (1997) A study on economic benefits of rice-fish-duck complex ecosystem. Chin J Appl Ecol 8(4):431–434

Dr. R. Richard Churchil, PhD (Poultry Science)

Dr. R. Richard Churchil hails from Kanyakumari district, Tamil Nadu, India. He earned his graduation from Madras Veterinary College, doctoral degree from Indian Veterinary Research Institute, many accolades including gold medals in postgraduation and Jawaharlal Nehru Award of ICAR for his doctoral research on development of transgenic spermatozoa. He worked as poultry breeding scientist in Kerala state for a decade before joining TANUVAS. He was trained on 'Transgenic animals' at University of Nantes, France. He has published 89 peer-reviewed research papers and seven book chapters. He teaches Poultry Science to undergraduate and Poultry Breeding and Biotechnology to postgraduate/doctoral students in TANUVAS.

Dr. A. Jalaludeen, PhD (Poultry Science)

Dr. A. Jalaludeen hails from Thiruvananthapuram District, Kerala, India. He earned his graduation from College of Veterinary and Animal Sciences (COVAS), Kerala Agricultural University (KAU) and postgraduation in the discipline of Poultry Science from the same institute. He pursued Doctoral degree in Poultry Science with specialization in Poultry Nutrition. Dr. Jalaludeen, served as teacher, researcher and extension worker in Kerala Agricultural University and Kerala Veterinary and Animal Sciences University (KVASU) for more than 36 years. He functioned as Professor and Head of Dept. of Poultry Science and Director of Centre for Advanced Studies in Poultry Science, Kerala Agricultural University for more than 10 years. He also functioned as Dean, College of

Veterinary and Animal Sciences, Pookode and later as Director of Academics and Research of Kerala Veterinary and Animal Sciences University, Wayanad, Kerala, India. He was made special officer in KVASU and was instrumental in establishing a separate faculty of poultry science which is the first of its kind in India. He guided many research students both for postgraduation and doctoral degree programmes. He has published many peer-reviewed research articles in journals of high repute.

The manufacturer's authorised representative in the EU is Springer Nature Customer Service Centre GmbH, Europaplatz 3, 69115 Heidelberg, Germany. If you have any concerns regarding our products, please contact ProductSafety@springernature.com

Printed and bound by CPI Group (UK) Ltd, Croydon, CR0 4YY
30/09/2026
02225253-0020